Unauthorized Guide to

SMURF

Around the W🌎rld

JOYCE AND TERRY LOSONSKY

4880 Lower Valley Road, Atglen, PA 19310 USA

Every effort has been made to identify copyright holders of materials referenced in this book. Should there be omissions, the authors apologize and shall be pleased to make appropriate acknowledgements in future editions.

This guide is neither authorized nor approved by the I.M.P.S., International Merchandising, Promotion & Services s.a. Company of Rue Du Cerf, 85, B-1332 Genval, Belgium.

The following trademarks, service marks and registered marks are owned by the following corporations:

I.M.P.S. International Merchandising, Promotion & Services— all licensed products. IMPS in Brussels is the official licensing agent of Smurf figures and all Smurf related items around the world. They have really created the history and spread the story of Smurfs around the world.
Schleich—licensed products, figurines
Peyo—licensed products, figurines
Hanna-Barbera Productions, Inc.—licensed products
Sepp International, s.a.—licensed products
Wallace Berrie & Company, Inc.—licensed products
Editions du Lombard—Smurf Comic Books
Ralph Co.—golf accessories
EMI—music
Segura—perfumes
JPI—watches
Time & Diamonds—jewelry
Waltonia—watches, jewelry
New Tec—licensed products
Flanders—licensed products
Anagrams—party supplies

Designed by Bonnie M. Hensley
Type set in Impress BT/Humanist 521 BT

ISBN: 0-7643-0959-5
Printed in China

Published by Schiffer Publishing Ltd.
4880 Lower Valley Road
Atglen, PA 19310
Phone: (610) 593-1777; Fax: (610) 593-2002
E-mail: Schifferbk@aol.com
Please visit our web site catalog at
www.schifferbooks.com

This book may be purchased from the publisher.
Include $3.95 for shipping. Please try your bookstore first.
We are interested in hearing from authors
with book ideas on related subjects.
You may write for a free printed catalog.

In Europe, Schiffer books are distributed by
Bushwood Books
6 Marksbury Avenue
Kew Gardens
Surrey TW9 4JF England
Phone: 44 (0)181 392-8585; Fax: 44 (0)181 392-9876
E-mail: Bushwd@aol.com

Dedication

We would like to expressly acknowledge the owners of the wonderful collection that is portrayed and described within these pages, Tina and Randy Tank of Florida. Without a doubt, Tina is the ultimate Smurf collector! She has only been collecting Smurfs for ten years but as you will see, her collection already numbers over four thousand pieces. As a young child, Tina was given some Smurfs by her parents. One day, Tina saw some Smurfs at a flea market and the memories started to flow back of the many hours of enjoyment she had playing with her Smurfs as a child. Well, that's all it took! Tina was off to collect Smurfs in all varieties, relive old precious memories and start new ones. Tina's wonderful parents, Andrew and Mary Karls, and her husband, Randy, have all contributed immensely to Tina's collection of Smurfs. They can't help but see the enjoyment and fascination on Tina's face when a new item is found—Tina's enthusiasm is contagious! She is the smurfiest! Tina also enjoys surfing the net for new items to add to her collection plus going to toy shows to obtain the latest smurfy item just waiting to be taken home to a loving collection. Our heartfelt thanks are sent to Tina and her family for sharing the memories! All of this collecting and sharing is in the name of fun and in search of adventure. You can send Tina a "Smurf Gram" telling her what a fantastic collection she has or share knowledge with her by e-mailing Tina at <rtank@bigfoot.com>.

Tina and Randy Tank.

Acknowledgments to the Smurfiest People

The authors would like to thank the following smurfy individuals who have contributed to our research and assisted us in organizing the data for this book:

Tina & Randy Tank
Andrew & Mary Karls
Elaine Blaylock
Charlie Kashian
Rick Atkins
Alan Rennard
Michele Moore
Colleen Lewis
Christopher Behrends
Frans Seuren
Frederic Pierard
Dimitri Elve
Major Mike Davis, USA
Ken Clee
Bill & Pat Poe
E.J. Ritter

David Stanley
Mary Kirby
Eileen & Ron Corbett
Mari-Faith & Rudy Storz
Pat Sentell
Manuella & Lari Poli
Jim & Sally Christoffel
Sabine Ertl
Natalie & Jean Claude Royer
Jimmy and Pat Futch
Donna Baker
Bruce Waters
Blair Loughrey
The Ultimate Web Smurf Club

SMURF Collectors Club International
24 Cabot Road W.
Massapequa, NY 11758

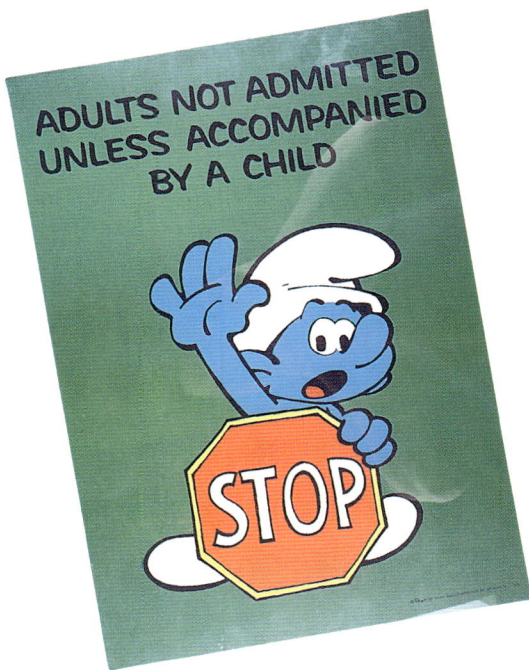

Contents

About this Smurfy Book

This comprehensive book encompasses over 4,000 Smurf collectibles divided into 279 sets. The Photo Gallery and the Set Listings are both arranged alphabetically, starting with "A"—Arts & Crafts Activities, proceeding to "B"—Baby Items, Baby Rattles, Badges, Balls, Banks, Banners, Bathroom Items, and Battery Operated Toys, continuing to "P"—PVC Figures, and ending with "Z"—Zipper Pulls. Some sets, like Set #73 listed under "P", describe 473 PVC Figures alone! Readers should note that set numbers were collector assigned and are therefore not sequential. For a Smurfbreak, check out Set #1, listed under "M" Mugs—Ceramic, where 63 different mugs are listed, or view the 93 different Plush Smurfs, listed under "P"—Plush Smurfs. You will find a wonderful variety of items: 22 different Smurf PEZ collectibles are listed under "P" for Pez, 5 telephones are found under "T" for Telephones, and 16 different ceramic banks plus 8 different plastic and tin banks are listed under "B" for Banks. These listings will surely keep you smurfing along!

The sheer number of sets listed—a grand total of 279—is an exhaustive number, plus the number of items listed within the sets is the smurfiest! Items listed are truly from around the world; the number of different countries selling Smurf collectibles continues to expand each year. Of course, Holland, Belgium, and Germany lead the parade of countries distributing Smurf items, followed closely by the United States and the rest of Europe. But, the driver of the "Smurf Express" is certainly the consumer, the Smurf collector who lives and collects around the world. The Internet has accelerated the distribution of Smurf collectibles, so that items produced in Belgium are available in the United States in a matter of weeks or even days, in some cases. The global Smurf collecting village is a closely knit group of collectors who share items and information leading them to continual fun and adventure, which in the end is the real purpose of collecting. So, it is with a smurfy smile and hug that we, the authors, welcome you to Bullyland and the world of Smurf collectibles! Enjoy the adventure!

Sign for Bullyland.

The museum at Bullyland.

In the beginning . . .

Welcome to the world of Smurf collectibles from around the world! The world certainly became a Smurfier place some forty-one years ago in 1958 when these little blue characters were designed by an illustrator named Peyo. It was only twenty years ago, in 1979, that Smurfs migrated to the United States and American collectors began to turn blue with delight. The beginning of this Smurf book is a good place to try and understand the collectible nature of these little blue friends.

Simply stated, Smurfs have been successful around the world because of their basic appeal: they are cute and they have personalities! The creator of Smurfs, Pierre Culliford (better known as "Peyo") created these little blue characters on October 23, 1958 in a comic strip in the journal, *Le Journal de Spirou*. Johan and Peewit were among his most popular characters until that time. Then he discovered the Smurfs and his career as an illustrator soared to unimaginable new height! Actually, as his comic strip story goes, the early characters Johan and Peewit could not believe their ears when these little blue people, who bounce around like kids and who are one hundred years old, invited them to come play. The little blue figures, like Papa Smurf with his white beard who was the oldest of them all, had personalities of their own. Papa had just turned 542 in October 1958 when he met Johan and Peewit. Once present, Smurfs never grow older. So, Papa Smurf will always be 542, Baby Smurf will always be just Baby Smurf, and Smurfette—a sassy lady who is very flirtatious at times—will always be young, because no one knows how old she is. After all, a lady never tells and it is not polite to ask one's age; Smurfs are always polite! A Smurf is a Smurf and that is all there is to it!

Far, far away in a village of mushroom shaped houses live the Smurfs. The little characters have blue skin and are dressed, for the most part, in white trousers and a cap. They are young, happy—an easy going bunch except, naturally, for a few of them. Johan and Peewit, Peyo's original characters, actually met the Smurfs because they were invited to meet them—one can only enter the land of the Smurfs if a Smurf invites you. As the story goes, with the help of an enchanter called Homnibus, Johan and Peewit were invited to meet the village members. Johan lived in the magical Middle Ages and was the King's young page. Peewit was Johan's faithful companion. Together they were portrayed as youngsters who faced thousands of dangerous situations, came to the aid of victims of injustice, and could fend off the plans of heartless villains. Naturally, a wicked Smurf named Gargamel is ever present. What would a story be without a villain? So Gargamel is still trying to meet the Smurfs while he adds levity to the character development!

At first, the Smurfs were just secondary characters in Peyo's comic strip development. Their smurfy tale began to unfold in the United States in 1979 when a few mini-albums appeared. Then a small film called *The Smurfs and the Magic Flute* appeared on television and the tiny blue figurines were no longer just smurfing along, they were on a supersonic path. Soon, figurines, toys, plush figures, and records were being made in their image in mind boggling numbers.

By 1981, the cartoon duo of Hanna & Barbera, who created Tom & Jerry, the Flinstones and Yogi Bear, started work on a TV series for the NBC network. Ultimately, 250 plus episodes were produced as the Smurf's popularity continued to soar. Over ten million copies of CDs have been sold, along with zillions of advertising books, comic books, figures, and related merchandise exceeding the ten million number.

To answer the questions, "What are Smurfs? Why are they collectible?" one needs to understand the history of these unique little blue figures. First, the Smurfs speak a different language, i.e. some words in a sentence are replaced by the word "Smurf" or a derivative of it. Additionally, the verb in the sentence can be replaced by the "Smurf" word. Instead of saying, "Have a wonderful adventure reading this book!" one might actually hear, "Have a smurfy adventure!" or "Isn't Losonsky's book the smurfiest!" At latest count, Smurfs know some twenty-five different languages and can converse smurfly in all of them! So, your invitation to visit them might sound like, "Let's have a smurfy time together!"

The Smurf story unfolds in a land where little blue people enjoy fun and adventure, starting over forty-one years ago. In fact, the last forty-one years are so full of so many great Smurf collectibles that the story of smurfy smiles being sent around the world is just beginning to unfold. Now it is time to meet some of the Smurf characters and view the world of Smurf collectibles!

Meet Some of the
Smurfs—the Good, the Bad, and the Ugly!

Smurfs were created in 1958 by Peyo and came to the United States in 1979. Smurfly, they are:

Apprentice—is a Smurf who wants to create magic just like Papa Smurf. However, to keep the story line interesting, apprentices take a recipe from Gargamel's book of spells. After drinking a concoction, the little blue Smurf turns into a green Smurf with a large nose and body covered with green skin. They are all covered in scales with a very long fat tail. Not a pretty looking sight! Seeking a return to being a blue Smurf and not green, the Smurf Apprentice must sneak into Gargamel's territory and find a potion that works in reverse! This leads to some very exciting story lines and merchandise galore.

Astro Smurf—is a Smurf who has transformed himself into an imaginary character on another planet by way of a dream. Actually, it is when dreams come true. That is, when a Smurf dreams he is on another planet, he can enter a spaceship, don a helmet and spacesuit, and begin an adventure with the help of all the other Smurfs in the village. It is an escape to fun and adventure that Smurfs rarely miss! Just dream you are the smurfiest and you are there!

Azrael—is a scraggly cat owned somewhat by Gargamel. He is known to be the "underdog" in the story line. He follows Gargamel around in hopes he will catch a Smurf and dinner will be served. Being somewhat stupid and always evil like Gargamel, Azreal's efforts are in vain, to date.

Baby Smurf—was naturally delivered by a stork. Everyone in the village adores the Baby Smurf who tends to be a Baby all the time and wrecks the routine of the Smurf village. A baby does what a baby does!

Bigmouth—is a ferocious eater. He is a one-minded type of person who thinks only of eating. The evil Gargamel convinced Bigmouth that the best food was Smurf soup. Consequently, Bigmouth is always on the lookout for Smurfs, which creates havoc within the best of situations.

Brainy Smurf—is a bright, moralistic Smurf who takes everything seriously. He truly believes everything Papa Smurf has to say and tends to be boring at times. He has been called a pain in the Smurf at times because of his personality!

Clockwork—is sure and steady. Invented by Handy Smurf to assist him with his domestic work, Clockwork appears and disappears from the village. He forages up terrific sarsaparilla soup for the village when he comes back and has his mechanical parts overhauled by Handy.

Cook—wears a tall chef's hat and white apron. His mushroom house is full of the aroma of freshly cooked bread, muffins, pies, and a wide assortment of sweets, which the Smurfs so dearly love. Others simply call him "Cordon Bleu."

Enamored—is the Smurf who dreams about Smurfette all the time. All the Smurfs love Smurfette, but none more than The Enamored Smurf. He daydreams about her all the time, carves her name in trees, peels off petals from daisies hoping she will love him the best! When she appears to love him not, he cries all day and night. He is basically "in love" !!!!

Farmer—works the earth. However, when he needs rain the sun typically shines and vice versa. Using rough speech, he sorts out the best remedies for his earth, producing the tastiest of vegetables and fruits for the Smurf village.

Flying—wants to fly endlessly like a bird, over the village. In the beginning, he tried everything to fly . . .sticking feathers on his back, building fabric wings, attaching himself to a kite, using a magic broomstick, riding on the wings of a windmill, and even building a rocket. Nothing worked until the magic potion appeared. He drank it and was lighter than air. Only eating bricks would bring him down to ground again! This combination of adventure makes the Flying Smurf a village favorite!

Gargamel—is the dangerous sorcerer who is basically evil. He simply hates the Smurfs and is in constant pursuit of them. Fortunately, the Smurfs are a smart lot and always outsmart Gargamel and escape his best of plans. Naturally, after each episode, Gargamel seeks additional revenge which leads to continual chaos.

Grandpa Smurf—is the oldest of all Smurfs. He has been absent from the Smurf village for the last 500 years but when he appears, his enthusiasm is fantastic! He is known to wear a bright yellow cap and pants along with glasses. Being an avid storyteller, he loves to weave his tales around souvenirs he has collected over the years.

Greedy—is greedy when it comes to food! He loves to steal food from Cook Smurf and basically his motto is, "Live to Smurf and not Smurf to live!" Everything is tasty to him.

Grouchy—is just that grouchy about everything! His mind is set on grouchy speed control level, which leads to negative thoughts all the time. Know someone like that? But underneath it all, he is truly a loving and tender Smurf who adores Sassette and Baby. He tends to hate it if and when his loving nature shows.

Handy—is handy at all mechanical work in the village. Whatever needs fixing, Handy is ready to tackle the job. He wears a pencil behind his ear to use to create and sketch things.

Harmony—is everyone who is tone deaf but has the soul of a musician, just not the talent. He plays a cornet while other Smurfs wear earplugs.

Hefty—is the strongest Smurf in the village. He wears a heart shaped tatoo on his arm and works out daily. He encourages all the other Smurfs to join in his exercise programs, even organizing Olympic Smurf events from time to time.

Johan and Peewit—are two human friends the Smurfs invited to come play. They live in the Middle Ages time frame, working for the King as Page and Assistant. Actually, while living in the King's castle, they met the Smurfs for the first time on their quest for the magic flute. Johan rides a horse and strives to protect the poor and the oppressed. Peewit, who is a midget and cheats at most everything, tends to be a boastful fellow as he rides along on his goat, Biguette. Peewit can always be heard before he is seen, as he loves music and sings out of tune.

Jokey—constantly seeks to play a prank on the others. He appears to have an endless repertoire of jokes, which sometimes gets on the Smurfs' nerves. At times his jokes backfire and the fireworks begins!

King Smurf—When Papa Smurf is away, the Smurfs select a new leader called "King Smurf." Each "King Smurf" relishes his time of power and tends to act like a dictator. Naturally, the Smurfs all rise up against dictatorial powers and erupt into a civil war. It is not until Papa Smurf returns that the village settles down into smurfy adventures and fun again. Being "King Smurf" for a day is the smurfiest of times for each Smurf!

Lazy—only wants to sleep. Actually, he should have been called "Sleepy" but the name was already taken. He is either sleeping or waiting for evening to go back to sleep. Brainy is always after him to do something and not doze off, but to no avail.

Nat—just naturally loves nature. He is plumb crazy about animals, especially the caterpillars that turn into beautiful Monarch butterflies. Running barefoot with a straw hat chasing butterflies is Nat. He is an ecologysmurf!

Painter—lives only to paint and create arts and crafts. He is dressed in an apron and bow tie when he is ready to paint his favorite model, Smurfette.

Papa Smurf—is the village leader with his white beard and red cap and trousers. He is 542 years old today, tomorrow, and yesterday. He is known to use his wisdom to advise other Smurfs when trying situations occur, but he advises only and does not tell others what to do. At times he is called upon to rescue the younger Smurfs from danger when situations get out of the Smurf code of justice. He accomplishes these rescues with the use of "magic" sometimes, because he is a pretty good alchemist.

Poet—spends his days and night writing poetry. With a quill and parchment paper in hand, he daydreams the days away. This lack of attention to his surroundings tends to get him into plenty of adventures.

Puppy—is Baby Smurf's companion, since Baby Smurf was the only one who opened the locket around his neck. Puppy protects his Smurf friends from Gargamel and Azrael's attacks.

Sassette—keeps Smurfette company. She was created by the Smurflings with red pigtails and pink overalls. She is actually a tomboy who constantly asks the questions "Who?" or "Why?" She and Smurfette are the best of friends, like Frick and Frat, Peanut Butter and Jelly, or Amos and Andy. They just complement each other and add to the levity of the story line. They can tease and please.

Schlips—are what the Smurfs turn into when they have transported the spaceship into the crater of an inactive volcano. To create Astrosmurf fun, this is a necessary transformation. For this role playing, they change their skin color from blue to orange, have long black hair, and wear loincloth wraps. Schlips are all the unusual orange colored Smurfs you will see in the pictures. Want to be a Schlip? All it takes is imagination.

Slouchy—is one of the Smurflings who tends to take everything in stride. That is, if he is asked to do something, he will do it . . .it's just that it takes quite a while. He is s l o w! Actually he plays a sax in the Smurf band and is relaxed all the time.

Smurfette—is a flirty lady who was originally created by Gargamel to tempt the little Smurfs. But, Papa Smurf's magic powers changed Smurfette into a loving little Smurf that everyone adores. This tenderness and adoring behavior teases everyone endlessly.

Smurflings—are three Smurfs—Slouchy, Nat, and Snappy—who got caught up in Father Time's clock. They are the only Smurfs who are not at least 100 years old. They actually get younger as times flies.

Snappy—is a '90s type of Smurf, busy all the time! The only drawback is that he goes wacky when he doesn't get his way. He snaps in an instant and only Papa Smurf can manage his temper and calm him down.

Vanity—adores himself. He's so vain! He loves talking fashion with Smurfette and thinks of himself as the Smurfiest Smurf in the village. He usually has a mirror in his hand and a flower in his bonnet.

Smurfs have been successful around the world because of their basic appeal: they are very cute and they have amazing personalities! If you want to visit Smurf Park, don't miss the exciting adventure at Walibi Schtroumpf park near Metz, France. The Smurf Park is located close to the borders of Belgium, Luxembourg, and Germany and features a number of unique attractions including Europe's biggest wooden roller-coaster ride. You have to be Smurf enough to ride it!

Smurf Names

Smurfs are called different names around the world:

English countries:	Smurfs	Portugal:	Estrumpfe
Spanish countries:	Pitufo	Hungary:	Torpikek
French countries:	Schtroumpf	Greece:	Et Poymp
Netherlands:	De Smurfen	Iceland:	Strumpar
Belgium:	De Smurfen	Finland:	Smurffi
Germany:	Schlumpf	Czech Republic:	Smoulove
Italy:	Puffi	Japan:	Sumafu
Denmark & Norway:	Smolf	Israel:	Ha - Dardasim
Japan:	Sumafu	Turkey:	Pirinler
China:	Nam Ching Ling	Catalonia:	Barrufet
Poland:	Smerf	Sweden:	Smurferna

Pricing, Name, and Numbering Information

MINT price range for all items is listed.

Set names are selected to represent the items within the category. Some sets will naturally overlap and others will stand alone.

Items are numbered sequentially for each listing. They are not organized by year of distribution, although if known the year of distribution is provided.

Now, from the land where little blue people appear . . . you're invited to **COME SMURF WITH US!**

Enamored Smurf and Smurfette

Gargamel and Azrael

King Smurf

Hefty Smurf

Nat Smurf

Papa Smurf

Poet Smurf

Sign for Walibi Schtroumpf Park, near Metz, France.

Alphabetical Photo Gallery

Set #57, 1-2, & 7 Arts & Crafts. Crafty activities begin with the Smurf Peg Desk, Smurf Magic Rub-It-Ups, or Papa Smurf's Crayon Keeper.

Set #57 Arts & Crafts. Crafty activities continue with the Doodle Bag, Smurf Paint-By-Number, Smurf Sewing Cards, and Smurf's Acrylic Paint-By-Number kits.

Set #57 Arts & Crafts. Children's entertainment becomes fun with the Lite Catcher, Painting Kit, Slates, and the Take Alongs.

Set #57 Arts and Craft Activities. Magnetic chalkboards and Rub-ons are the perfect toy for the little Smurf or Smurfette at home.

Set #57 Arts and Craft Activities. Drawing, stenciling, or creating Smurf scenes is all part of the fun in this category.

Set #94 Baby Items. Smurfy dressed babies are the cutest!

Set #94 Baby Items. Baby bottles and pacifiers help to keep the family smurfing along!

Set #94 Baby Items. From gum soothers to teether toys, this category encompasses a myriad of baby supplies.

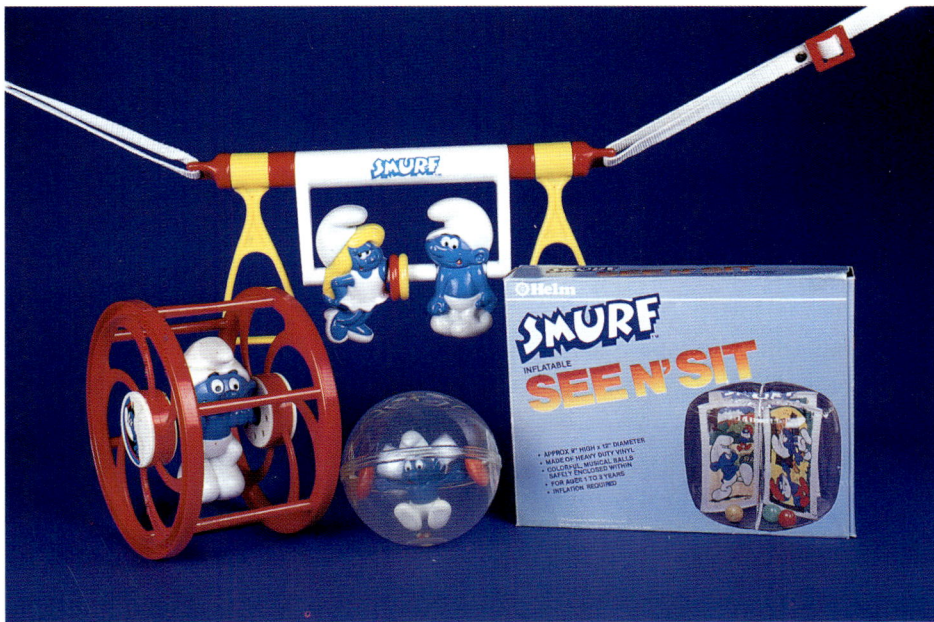

Set #94 Baby Items. A hanging crib pull toy, a See N' Sit, a red rolling ball and a See Thru tub toy provide babies with hours of fun.

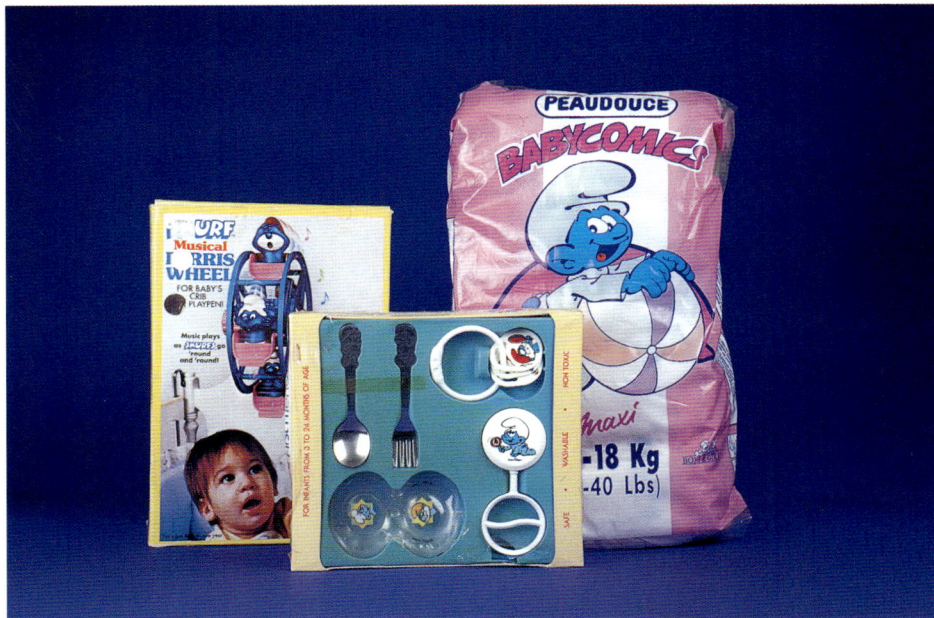

Set #94 Baby Items.

Set #90 Baby Rattles. A high chair rattle and numerous flat rattles sound the Smurf beat in this category.

Set #230 Balls. Sporting a Smurf ball keeps the blues away! This assortment ranges from a hard sponge ball to several soft rubber balls and a baseball.

Set #133 Banks—Ceramic. "A penny saved" is a real treat in one of these ceramic banks.

Set #109 Banks—Plastic & Tin. Five plastic banks help little Smurfs save for the future. They include: Smurf Bank, Papa Bank, Smurf with a Gift Bank, Brainy Sitting on a Tree Stump Bank, and Traveler Smurf Bank.

Set #160 Banners—Ice Capades. The follies of the Ice Capades yielded many Smurf collectibles, including this Smurfs Alive felt pennant and assorted felt mini banners.

Set #60, 1 & 25 Bathroom Items. Pictured are a Talking Scale and a Smurf Toothbrush Dispenser.

Set #60 Bathroom Items. Three fun toothbrush and cup sets are pictured along with the Smurfy Cup Dispenser.

Set #60 Bathroom Items. Using a Vanity Mirror and Toothbrush and Toothpaste, children come out smurfin' clean with this category.

Set #60 Bathroom Items. Smurf Submarine and assorted bathroom toys provide the incentive for cleanup time.

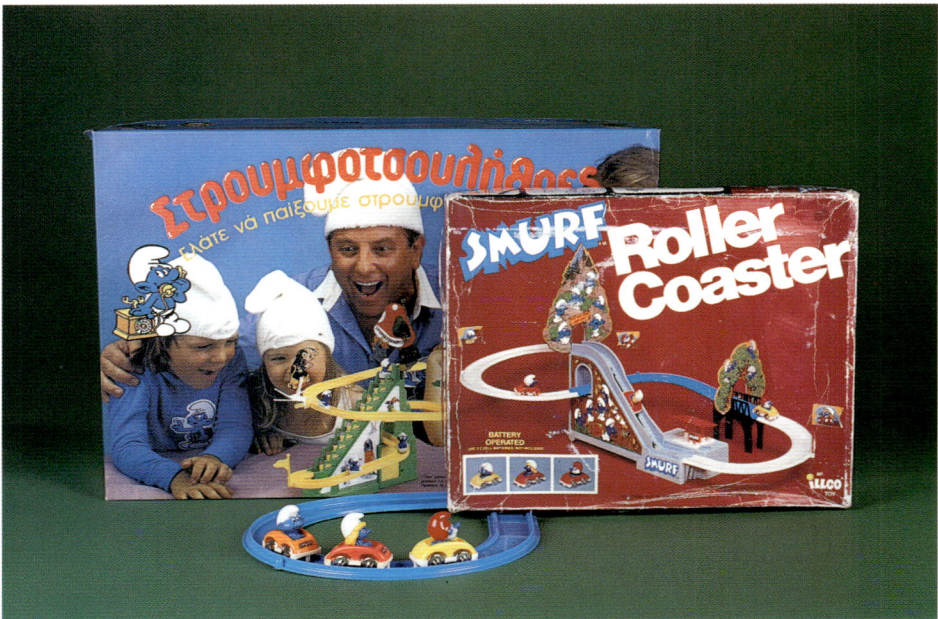

Set #54, 1 & 8 Battery Operated Toys. Two different mint in the box roller coasters are pictured.

Set #54, 2-5 Battery Operated Toys. Battery operated toys provide imaginative hours of play with toys like Bumper Cars, Speed Boat, Red Smurf Fun Buggy, and Smurf Helicopter.

Set #54, 6-7 Battery Operated Toys. The slot car set, Smurf Road Race Village Set by Talbot Toys, and Crawling Baby by Puffi are fantastic examples of smurfing along fun!

Set #80 Beach Items. Sailing away to the beach with these collectibles is like heading into the sunset, pure enjoyment!

Set #80 Beach Items. Filling up a Smurf Inner Tube with all sorts of smurfy beach collectibles is the best choice on a rainy day.

Set #80 Beach Items. Life's a beach with these Smurf collectibles!

Set #144 Bean Bag Minis. Pictured is an assortment of mini bean bag Smurfs: Happy Smurf with his tongue out, Crying Smurf, Smiling Smurf, and King Smurf.

Set #268 Bedding and Blankets. Snooze time is perfect on these sheets and blankets.

Set #268 Bedding and Blankets. This assortment of Sleeping Bags provides the perfect cozy spot for the littlest Smurf or Smurfette.

Set #268 Bedding and Blankets. The selection of pillowcases and sheets is almost Smurf-endless!

Set #268 Bedding and Blankets. Making a bed with Smurf sheets or decorating a room with these Smurf curtains makes Smurfland that much closer to home.

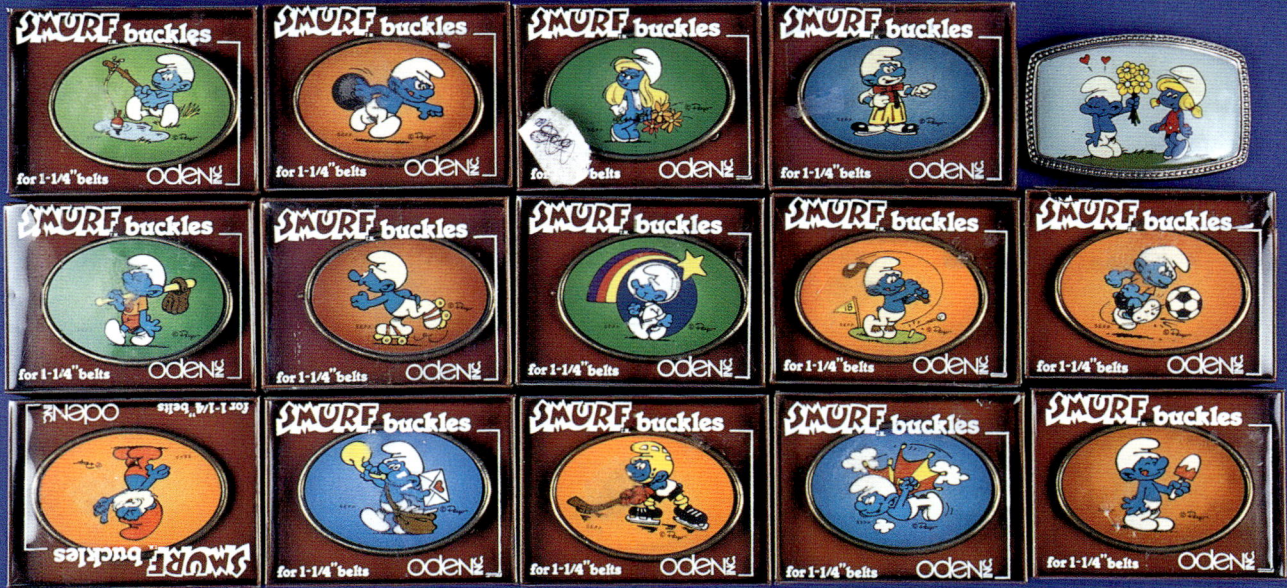

Set #51 Belt Buckles. Oden Manufacturing provides an assortment of belt buckles. From Smurfette with Flowers, Clown Smurf, Roller Skating Smurf, Golfer, Baseball Smurf, Papa with Finger in the Air, Popcycle Smurf, Bowling Smurf, Flying Smurf, Mailman, Fishing Smurf, Astro Smurf, Hockey Smurf, Soccer Smurf to Smurf giving Smurfette Flowers, Oden has produced them all.

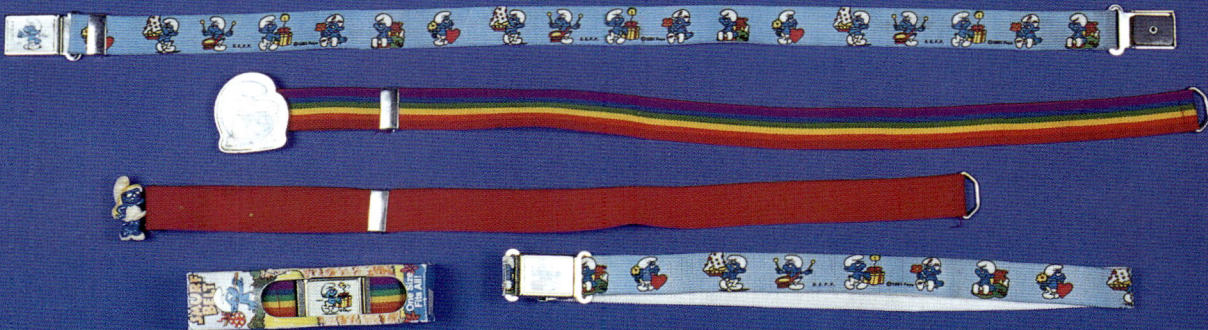

Set #52 Belts. Elastic belts and belt buckles add just the right zip to a properly attired Smurf collection.

Set #225 Bendies. This picture shows a combination of Bendies available to Smurf collectors. Included are five Bendies by Ceji (mint on the card): Smurf with Ax, Papa with Test Tube, Brainy with Book, Smurf with Spoon, and Smurf with Horn. Also included is the Large Bendy Smurf listed under Set #209 and the small Papa Smurf Bendy listed under Dolls.

Set #186, 1- 6 Bicycle Bells. Peyo produced a series of six children's bells to be attached to bikes *(Top Row)*: Smurf Hitting Another Smurf, Smurf Sleeping in a Hammock, Smurfette Coloring a Picture; *(Bottom Row)*: Smurfette and a Smurf on a Bicycle Built for Two, Papa, Smurfette and Baby, and Smurf Racing.

Set #120 Binders—3 Ring. To smurf or not to smurf, that is the question when choosing three ring binders for supplies! When dreaming of projects, one can select a blue or black binder with a Smurf laying on the grass and a dreaming design.

Set #273 Bobbing Heads + #115 Store Display. Two different bobbing heads are pictured: Smurf with an ax, which is the oldest, and Smurf with a trumpet.

23

Set #113 Book Bags & Back Packs. This photo shows an assortment of purses and bags classy enough to carry special Smurf collectibles.

Set #113 Book Bags & Back Packs. Going to school is a Smurf adventure with this selection of backpacks and book bags.

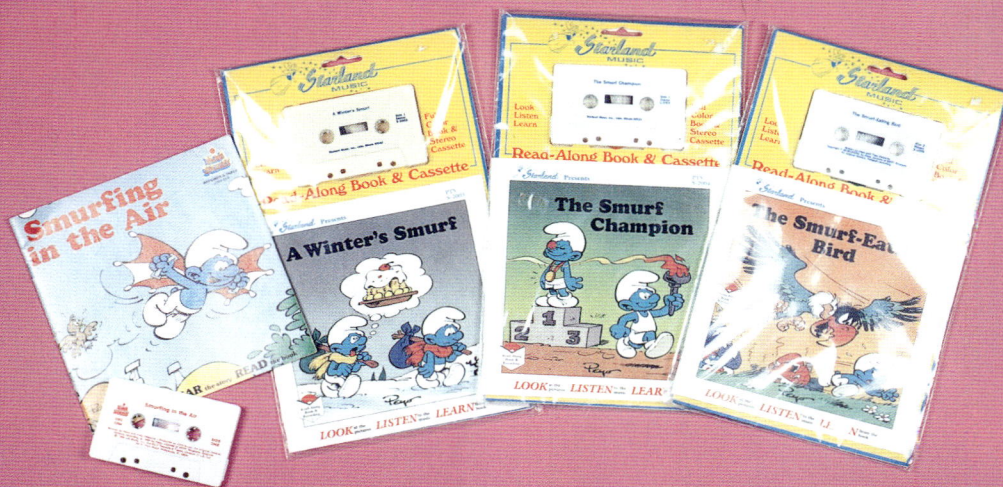

Set #17 Books and Cassettes—Read-Along. Children continue to smurf along with these cassettes and tapes: *A Winter's Smurf, The Smurf Champion, The Smurf Eating Bird,* and *Smurfing in the Air.*

Set #18 Books and Records—Read-Along. Read-Along Books include: *A Winter's Smurf, The Smurf Champion, There's a Smurf in my Soup, The Smurf Eating Bird, Smurf's Daydream, Smurfing in the Air,* and *La Mouche Bzz.*

Set #10 Books—Activity and Miscellaneous. Smurf activities begin with this assortment of Activity Books.

Set #10 Books—Activity and Miscellaneous. Punch-out Activity Books and German Activity Books continue the allure that smurfing along is the best route!

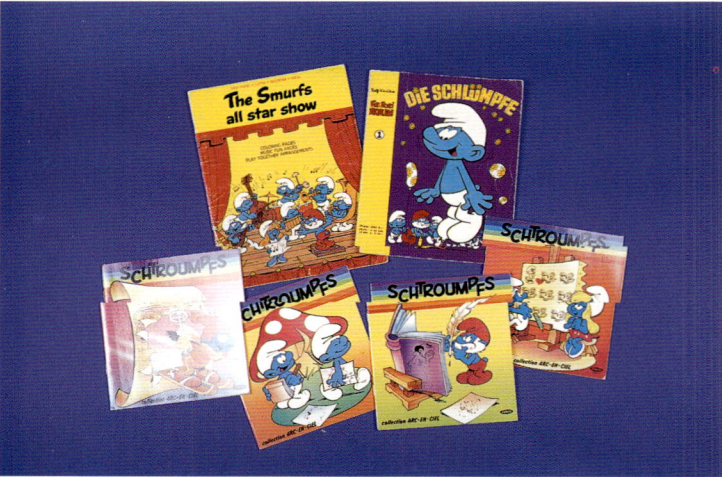

Set #10 Books—Activity and Miscellaneous. The All Star Show is only the beginning with these activity books!

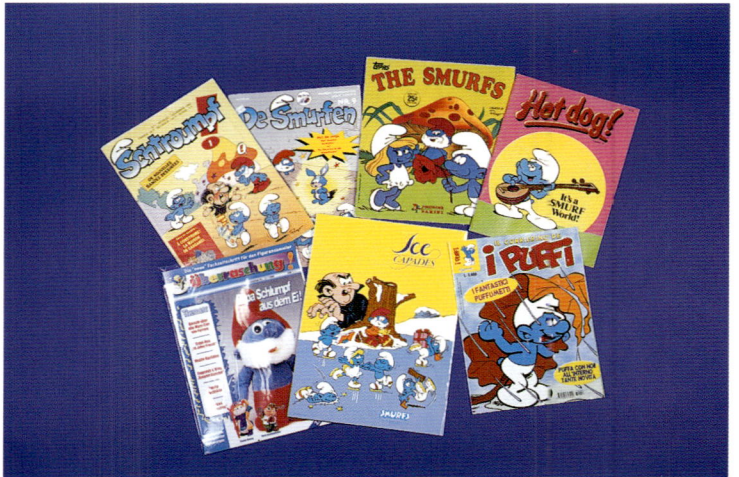

Set #10 Books—Activity and Miscellaneous. I Puffi's Activity Books prove it is a Smurf world!

Set #188 Books—Dutch. Holland has contributed to the Smurfland library through a variety of book titles, such as: *De Zwarte Smurfen, De Smurfen En De Krwakakrwa, De Ruimtesmurf, Baby Smurf Will Een Beer, De Supersmurf, Gargamel En De Fonkelstenen, De Fluit Met Zes Smurfen, De Smurfuhrer, De Smurfin* and *Smurfekoppen En Koppige Smurfen.*

Set #227 Books—France. France has provided many wonderful smurfy adventures in books. A variety is pictured.

Set #175 Books—German Large Die Schlumpfe. Carlsen Comics have created five soft cover books that keep Smurfs smurfin' along.

Sets #7 and #9 Books—Hard Cover and Soft Cover. Viking Studios, Random House and Langens have provided a variety of hard cover and soft cover books.

Sets #7 and #9 Books—Hard Cover and Soft Cover.

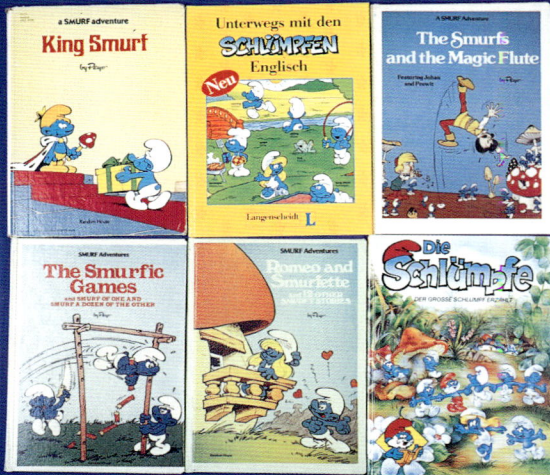

Set #7 Books—Hard Cover.

Set #9 Books—Soft Cover. Random House, Semic, and Hodder have provided a variety of soft cover Smurf books.

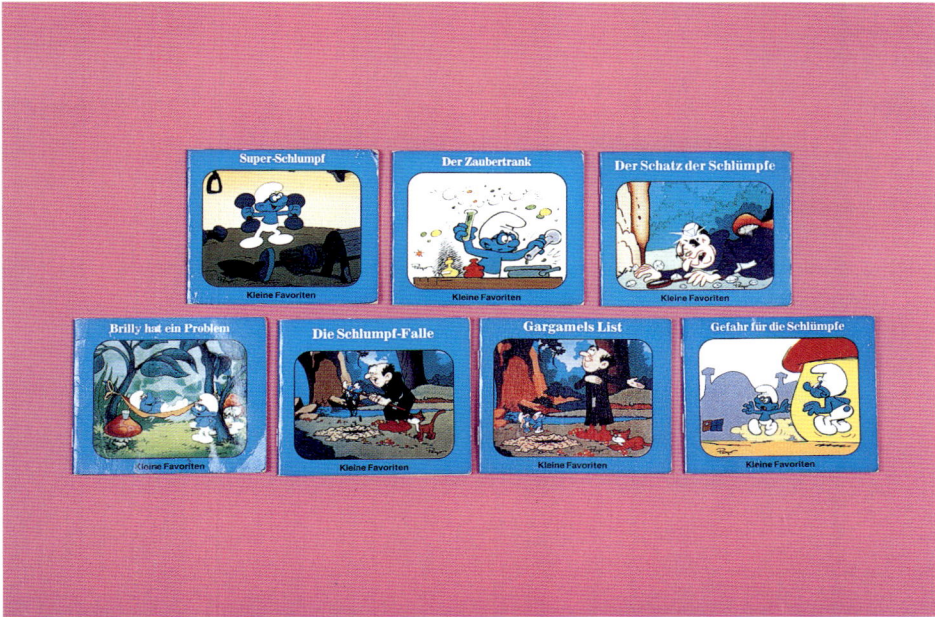

Set #155 Books—Mini Mini German. This assortment of mini German books includes the classic Smurf tales.

Set #14 Books—Pop-Up. Random House and Hemma have printed wonderful examples of hard cover pop-up Smurf books, such as *The Smurf Catching Trap, A Little Smurf Bedtime Story, Smurf On the Grow, A Smurf Picnic, Der Geburtstag..., Ein Fest bei..* or *De Baby Smurf.*

Set #8 Books—
Puppet. This
category includes
books with
puppets to extend
the role playing.

Set #11 Books—Small. Random House, Bastei, Ravensburger, and Zuidnederland have produced an assortment of mini books to keep the young reader smurfing along.

Set #6 Books—Coloring. Creating the perfect Smurf picture is easy with this assortment of coloring books.

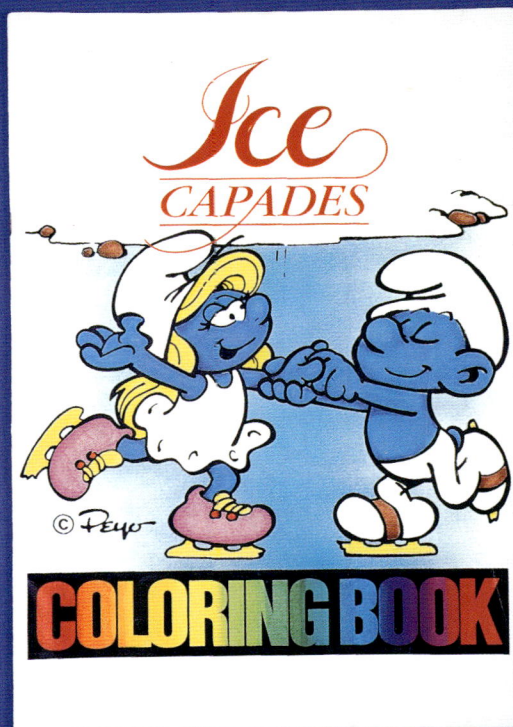

Set #6 Books—Coloring. Smurfing on ice with this Ice Capades coloring book is a breeze!

Set #271 Bracelets. Jewelry and dresser items decorate this category. What little Smurfette wouldn't want to wear the bracelet sitting on the mirror?

Set #195 Bubble Gum Wrappers. Smurfy delicious are these bubble gum wrappers, for the serious collector.

Set #92 Bubble Pipes. Creating the perfect array of bubbles with this assortment of bubble pipes is easy!

Set #82, 1-2 Bucket of Bolts—Smurf in a Bucket. Plastic bolts and nuts form the basics of these autos with Smurf N' Papa Smurf riding; made by Toysville.

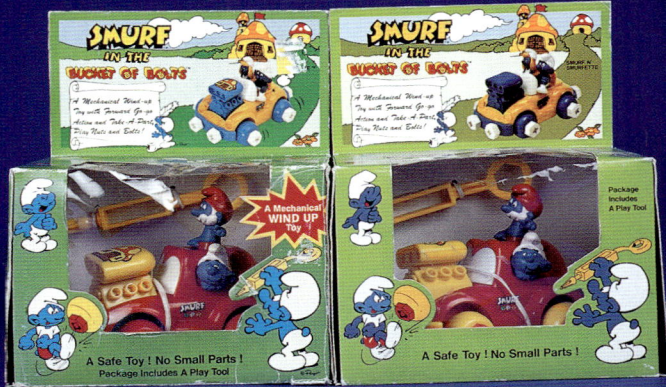

Set #248 Bundesliga Soccer Smurfs.

Set #223, 1-2 Burago Diecast Jeeps. Burago produced two distinctive diecast jeeps: a pink jeep and a blue Lamborghini Cheetah.

Sets #27 and #28 Buttons. Pictured is an assortment of promotional buttons, metal pin backs, and stick-ons.

Set #27 Buttons—McDonald's Fast Food. The two pictured buttons promote "Free Smurf Forest Cups" from McDonald's of Canada. On each occasion, McDonald's of Canada gave away a series of Smurf plastic cups.

Set #28 Buttons—Stick-ons. This set of four stick-on buttons was designed as an award presentation for: Most Daring, Most Friendly, Most Funny, and Most Mischievous. Any true blue smurfer was eligible!

Set #96 Buttons—BP Promotional. Round metal buttons come from Wavery Productions and include: Smurf and Smurfette Riding a Motorcycle, Smurf in an Airplane, Smurf Driving a Go Cart, Smurf Driving a Car, Smurf Driving a Gas Truck, 3 Smurfs in a House Boat, Smurf with a Space Ship and Gas Attendant Smurf!

Set #96 Buttons—Assorted. Pictured is a tray of eighteen assorted buttons.

Set #96 Buttons—Assorted. Pictured is a tray of twenty assorted buttons.

Set #47, 1-3 Cake Pans. Wilton has manufactured three different Smurf cake pans: Smurf Holding a Sign, Smurfette Holding a Balloon and Mini Smurf Cake Pan.

Set #183, 1-25 Cakes—Porcelain Figures. French bakeries distributed a spectacular set of ten Smurf Baby figurines in children's cakes during 1996: Baby Laying Down, Baby Laying in the Moon, Baby with a Sailboat, Baby Sucking His Thumb, Baby Holding a Popsicle, Baby Playing with a Rattle, Baby with a Teddy Bear, Baby with blocks, Baby Making a Funnyface and the Smurfs Baby Sign! Additionally, fifteen Smurf figures were distributed within the children's birthday cakes: Schtroumpfette, Schtroumpfissime (King), Schtroumpf Musician, Gargamel, Grand Schtroumpf (Papa), Schtroumpf Farceur (Jokey), Bebe Schtroumpf (Baby), Schtroumpf Gourmand (Cake), Schtroumpf Cocuet (Shy), Schtroumpf Amoureus (Cupid), Pirouit (Peewit), Lux-Vieux Schtroumpf (Grandpa), Schtroumpt Paysan (Gardner), Puppy and Johan.

Set #123 Calendars. The early 1980s and 1990s yielded a proliferation of Smurf calendars and even almanacs detailing the not so blue life of a Smurf. In fact, adventure and fun abound on these colorful event planners. Noteworthy is a 1998 calendar marking the 40th Anniversary of Smurfs!

Set #108, 1-3 Cameras. One's smurfy moment can be captured with a Clown Smurf flash camera, a blue Musical Camera (play camera), or a Smurf N' See Camera.

Set #221 and #153 Candle Holders and Candles—Figures. Additional candles include Watchman, Chimney Sweep, Sleepwalker, and Mailman along with two different candle holders for birthday cakes: Mailman mini candle holder and Soccer Smurf mini candle holder.

Set #153 Candles—Figures. Candles by Geis and Bougies burn for: Clown Smurf, Tyrolese in dark blue, Tyrolese in light blue, Lute, Smurfette, Jolly, and Smurfette.

Set #48 Candles — Numerals. Birthday celebrations are smurfy events with numeral wax candles.

Set #30, 1-4 Cars—Diecast Getaways. Irwin Toys produced a series of four diecast Getaway vehicles: Cowboy Smurf and his diecast Rocker, Boating Smurf and his diecast Speed Boat, Pilot Smurf and his diecast Airplane, and Smurfette and her diecast Roadster.

Set #2, 1-11 Cars—ERTL Diecast. ERTL manufactured a series of eleven diecast cars (Top Row): Ol' McSmurf, Handy in a tan car, Handy in a light green car. and Handy in a red VW auto; (Bottom Row): Smurfette car, Aero Smurf car with Smurf wearing a mask, Aero Smurf car with Smurf not wearing a mask, Smurf About in a Log, Papa Smurf Car, Gargamobile car, The Smurf & Ladder truck.

Set #3, 1-9 Cars—Italian Puffi Diecast. Puffi manufactured a series of nine diecast cars: Black Smurf in a Homemade car, Papa Smurf driving a brown shoe car, Smurfette's light green frog car, Smurfette's dark green frog car, Love-struck Smurf in a Roadster, Smurf in a red Roadster, Super Smurf's Can Am Racer, Smurfette in a Corncob car, and Gargamel in a purple Roadster.

Set #45 Cassette Tapes. Dreaming of Smurfland is possible with this selection of Smurf cassettes: Smurfing Sing Song, The Smurfs All Star Show, Best of Friends, Father Abraham in Smurfland, Cool New Toons, Smurfin!, Handy's Liebling..., Smurfin 10th Anniversary Commemorative Album, or The Smurfs Featuring...Together Forever.

Set #237 Catalogs—Price Guides. Keeping track of your latest Smurf collectibles is made easier with these catalogues!

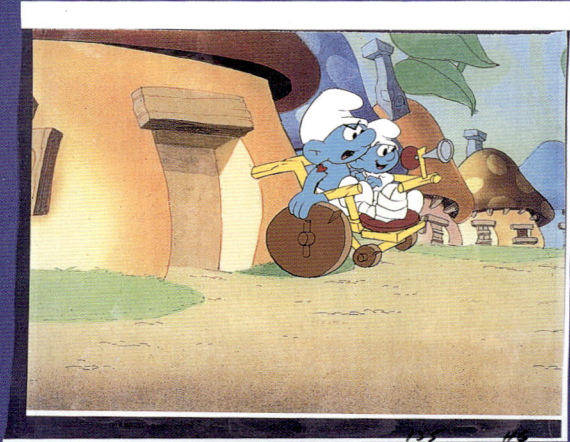

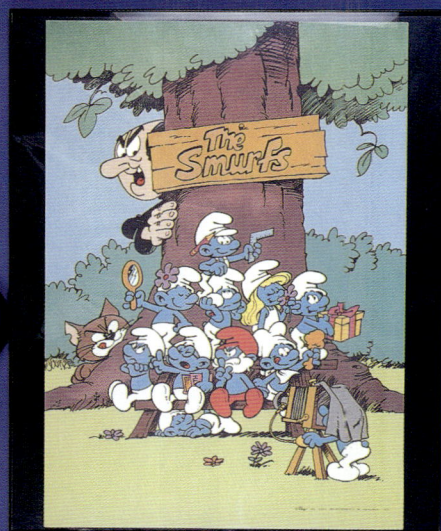

Set #241 Cels—Animation. This black and white line art Cel shows the Smurfs at play during the holiday season.

Set #22 Ceramics—Birthday Smurfette. This series includes twelve different ceramic birthday candles decorated with Smurfette figurines.

Set #23 Ceramics— Christmas Figures. Four ceramic collectibles are pictured: Smurf with a snowman, Smurf playing a drum with Papa blowing a horn, Smurfette and Smurf singing Christmas carols, and Smurfette with a Christmas tree.

Set #24 Ceramics —Figures. Eleven ceramic figures are pictured: Brainy, Gargamel and Azreal, Greedy, Hefty pounding a Smurf Village sign into the ground, Jokey, Papa holding a potion book, Smurf and Smurfette sitting on a log, Smurf blowing a horn while Papa has his ears plugged, Smurfette Rollerskating, Valentine Smurf, and Valentine Smurfette.

Set #262 Ceramics—Juice Cups. Six white ceramic juice cups are pictured: Smurf with Mallet, Jokey Smurf, Smurfette Flirting, Soccer Smurf, Smurf Singing, and Smurf Covering His Ears.

Set #106 Chalk Boards. Drawing is easy with one of these smurf decorated chalk boards.

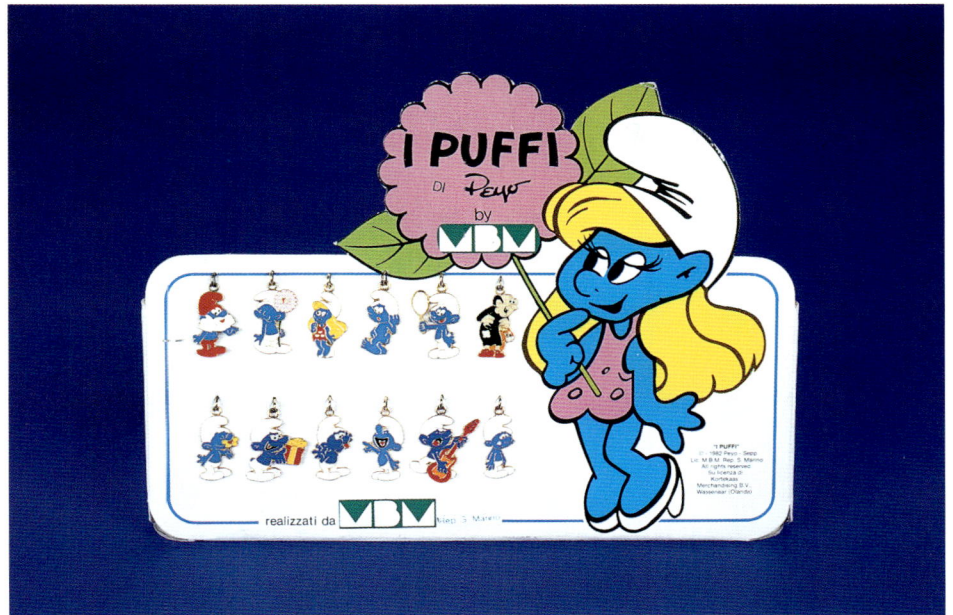

Set #215 Charms + #115, 23 Store Display. Twelve charms are pictured: Papa Smurf pointing, Gargamel and Azrael, Nude Smurf, Smurf crying, Smurf holding pink daisy, Smurf laughing, Smurf playing guitar, Smurf smelling a daisy, Smurf sticking out his tongue, Smurf with gift, Smurfette, and Tennis Smurf.

Set #110, 1-2 Chatter Chums. Smurf and Smurfette Chatter Chums by Wallace Berrie have a pull string for movement.

Set #87 Christmas Items. Lighting up the holidays is not hard to do with these lights, musical toys, cloth stockings, and gift tags.

Set #64 1-3 Clay—Models. Adica Pongo, DAS, and Fila have produced modeling clay for children's imagination. Das Smurf Deluxe Modeling Clay Playset, Das Modeling Clay, and Smurf Het Gezellige Smurfendorp all leave their impression.

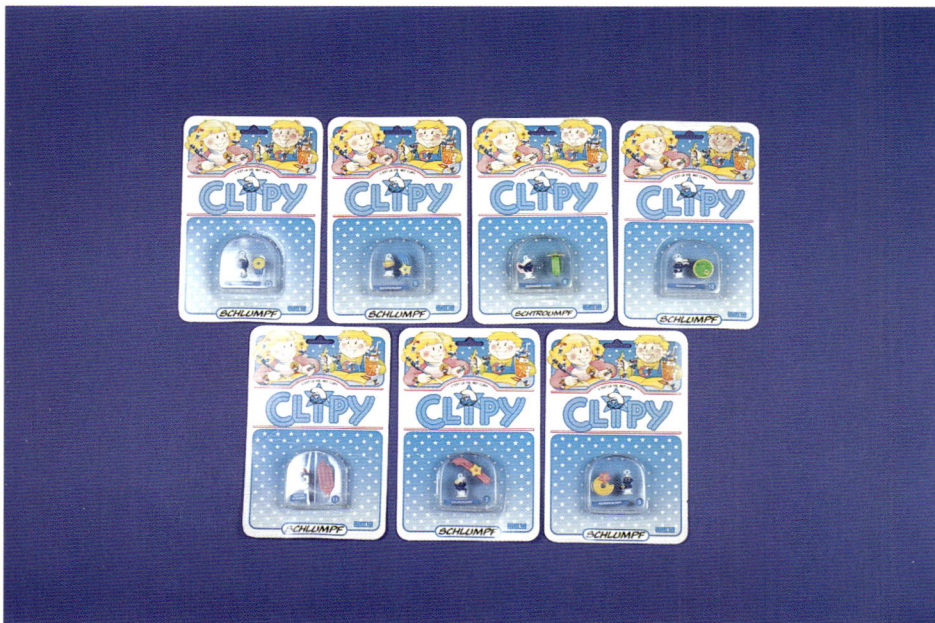

Set #246 Clipys. These plastic clips attach themselves anywhere!

Set #58 Clocks. Since a Smurf or Smurfette's work is never done, clocks only appear to tell time. This assortment is timeless.

Set #242 Club Figures—SCCI. Smurf with elephant (Club figures from 1978 mold that was issued later), Papa and his prehistoric pets, Smurf holding a gold pig, Smurf holding a pink pig, Aerobic Smurfette with ladybug, and Smurf with a dove.

Set #89 Coasters. Assorted drink coasters made of felt and hard plastic come six to a set.

Set #43 1-3 Colorforms. Three Colorform sets dress-up this category: Smurf Land Colorform, Smurf Mini Playset, and Smurfette Dress-up Set.

Set #210 Comic Albums—Die Schlumpfe. Pictured is an assortment of these Comic Albums.

Set #124 Comic Books. Marvel Comic produced some wonderfully smurfy adventures in this series of comic books.

Set #126 Comic Books—Mini. Marvel comics created the perfect collectibles for the littlest of fingers and pockets with these mini comic books. Packaged together for added enjoyment, this assortment ranges in size and description.

Set #46, 1-10 Compact
Discs. Ten assorted
compact discs are shown:
De Smurfen Houseparty,
Irene Moors & De
Smurfen, De Smurfen Party
House Hits, Die Schlumpfe
Mesaparty, Die Shlumpfe
Tekkno ist Cool, Irene
Moors & De Smurfen
Smurf the House, Het
Smurfenfeest, Telesmurfen,
De Smurfen Surprise, and
De Smurfen Feest.

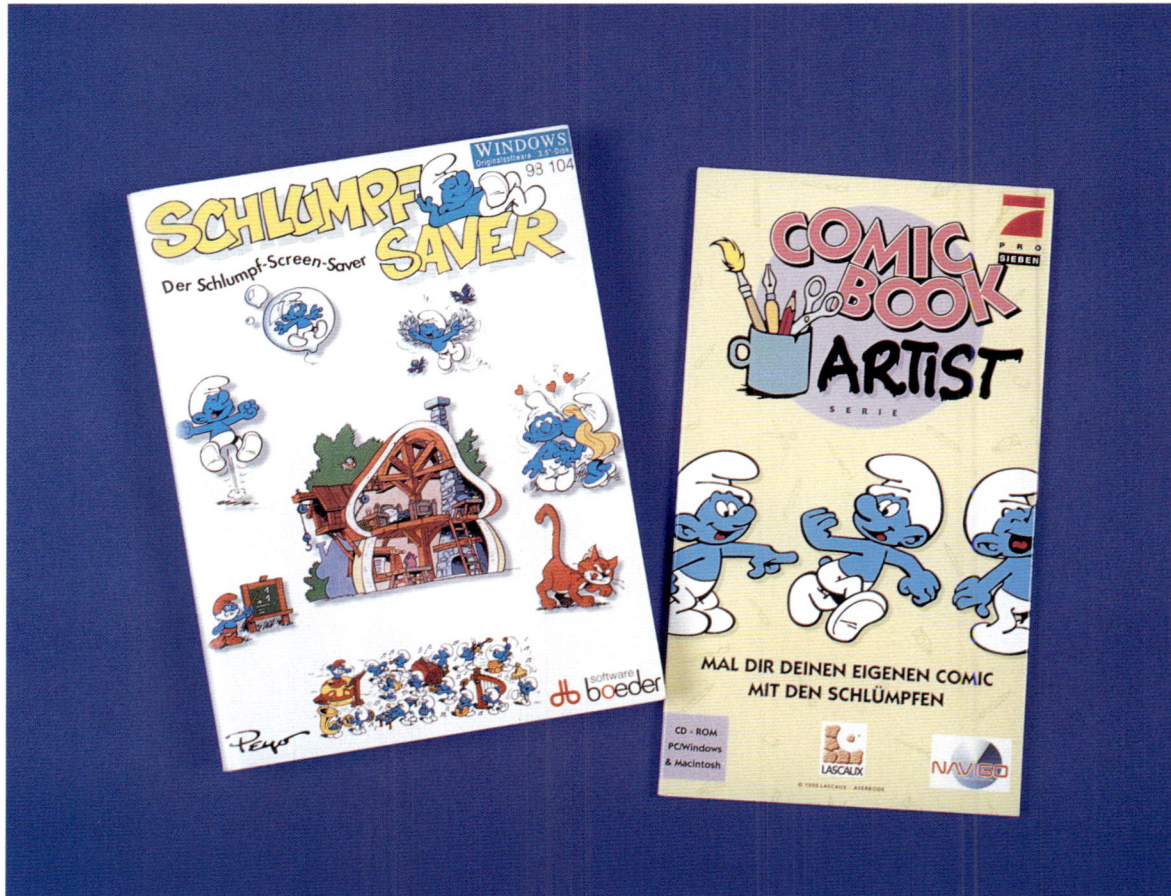

Set #256 Computer
Programs. The perfect
screen saver or computer
program is shown in this
category.

Set #102 Cups—Plastic Tumblers. In the 1980s Canadian McDonald's provided an assortment of plastic cups featuring Canada's Wonderland Adventure along with the Smurf family of characters.

Set #102 Cups—Plastic Tumblers. King's Dominion and Ponderosa Steak House have provided an assortment of plastic cups with Smurf adventures portrayed.

Set #180, 1-3 Cutting Boards. Fackelmann has produced three designs in cutting boards: Picnic on Blanket scene, Papa and Smurfette at a table, and Smurfs in the Kitchen.

Set #148 Dart Supplies. This Velcro dart game by Synergistic Research is a "10", even though a top score of "5" is the magic number.

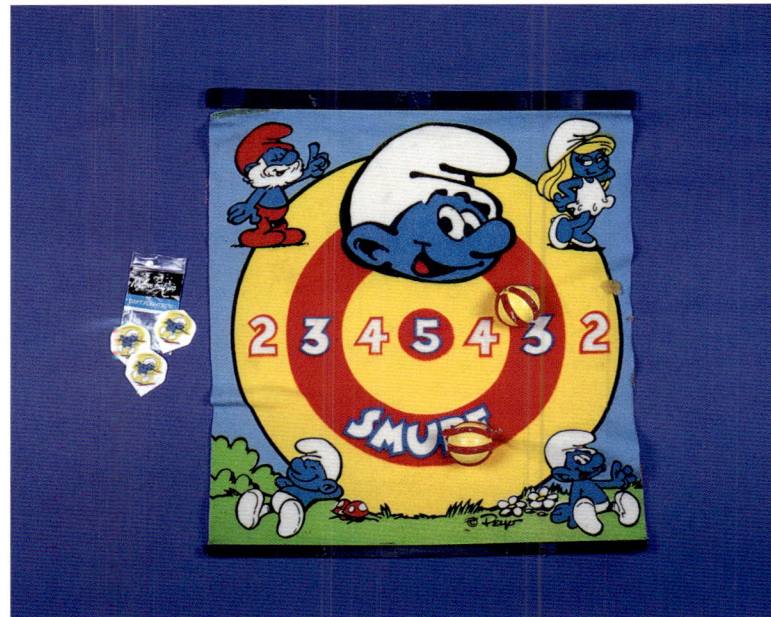

Set #206, 1-4 Desk Sets. From a Smurfy Memo Dispenser with pencil to a Paper Clip Tray or Smurfy Pencil Pot, this Smurfy collection has it all for the perfect Smurf desk.

Set #13, 1-5 Dishes. Numerous manufacturers produced plastic dishes for children to play with, from Deka Inc, Fackelmann, Helm, and Decor to Peyo and Irwin Toys. Plastic dishes include coffee cups, water cups, juice cups, cereal bowls, flat plates, shallow bowls, three-section baby's plate, soup bowls, Melmac plates (not pictured), water squeeze bottles, Smurf plastic straws, Smurfette plastic straws, Smurf cookie containers, Smurf tin pails, and Smurf-A-Getti from Irwin toys.

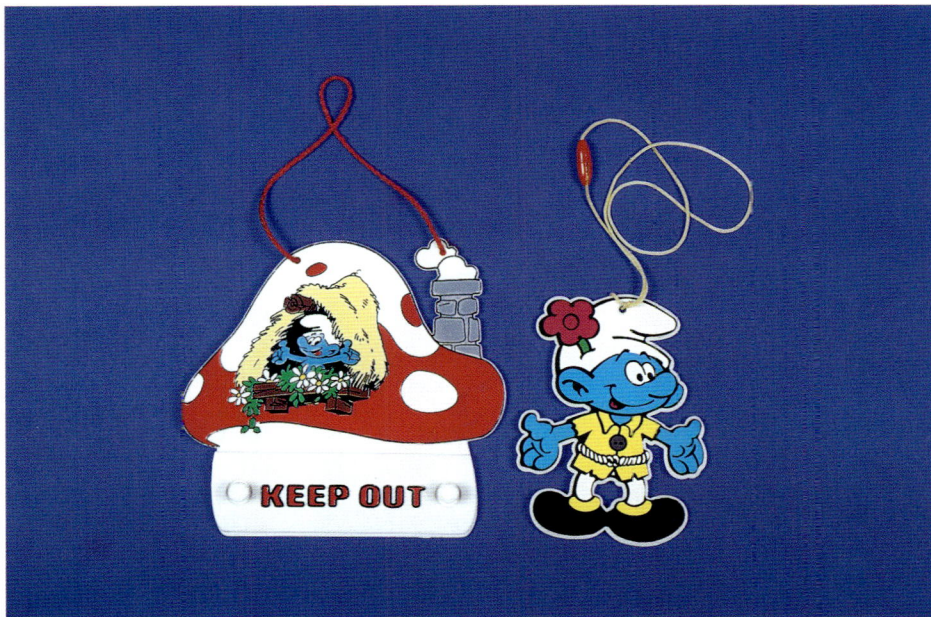

Set #157 Door Hangings. Door signs lead the way to Smurfland rooms.

Set #116 Dress N' Style. Two large spectacular dolls highlight this category: Brainy Judge Doll and Clown Doll are shown along with the Smurfette Dress 'n Style Doll.

Set #214 Earrings. Earrings made from silver, gold and metal decorate this category, along with two charms.

Set #156 Easter Items. One's Easter season is the brightest with the addition of these papier mâché egg decorations.

Set #159 Fabric. Yards and yards of homemade Smurf collectibles could originate with this assortment of Smurf fabric. The smurfiest of attire leads to SMURFUN!

Set #239, 1- 14
Figures - 1" Blue.
Over the years, Peyo has produced a series of blue 1" tall rubber figurines which include (Top Row): Gymnast, Rock N Roll, Jungle, Flying, Trumpet Player, Sledgehammer and Policeman; (Bottom Row) Telephone and Captain (Light Blue) and Papa, Brainy, Drummer, Mechanic, and Laughing in dark blue.

Set #197
Figures—Brass. Six brass Smurfs weigh in on this category: Normal, Laughing, Sleepwalker, Naughty, Jolly, and Shy.

Set #50, 1-6
Figures—Ferrero. Six different Kinder Ferrero figures are illustrated: Smurf with a Flower, Blindfold Smurf, Smurf on a Red Jumping Ball, Smurf in a Potato Sack, Papa Referee, and Smurf with an Ice Cream Cone.

Set #274 Figures—France. Twenty-four figures include: Angler, Artist, Blackboard, Bowler, Bricklayer, Captain, Clown, Conjuror, Cricket, Emperor, Graduation, Hairdresser, Hammer, Judo, Lover, Mallet, Oboist, Pirate, Policeman, Rock N Roll, Rollerskater Smurfette, Scot, Telephone, and Tennis Smurfette.

Set #202 Figures—Jubilee. Telephone Smurf, Papa, and Sailor Smurf are known to Smurf collectors as Jubilee Smurfs; they were issued to commemorate the 1965-1985 era.

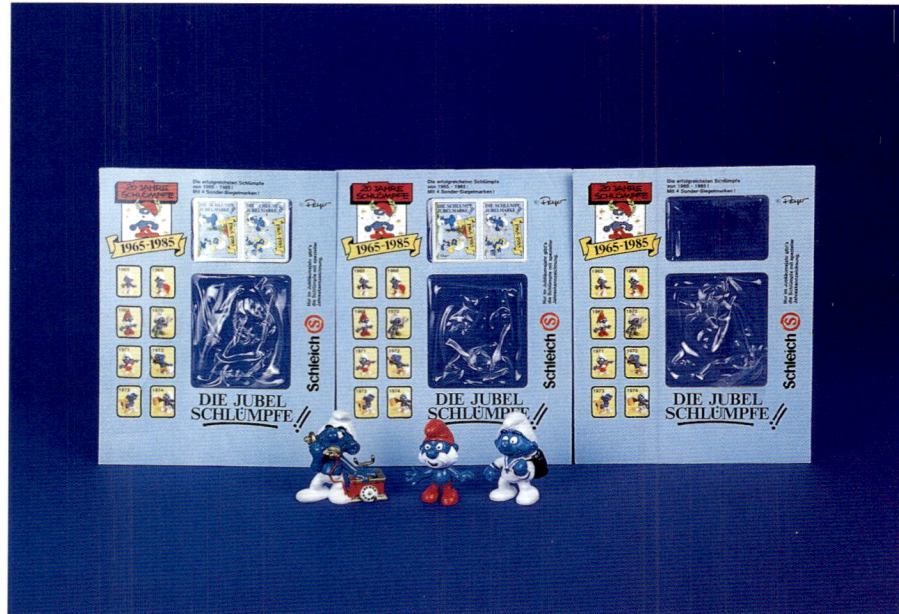

Set #31, 1-5 Figures—Poseable. Smurf plastic poseable figures by Irwin include Baby Smurf, Handy Smurf, Artist Smurf, Papa Smurf, and Smurfette. See Set #216 for additional poseable figures.

Set #163, 1-2 Figures—Rubber 6", and Set #76, 1-12 Squeeze Toys. An assortment of rubber squeeze toys is pictured. The sizes vary along with the selection of colors. Assortment includes: Shy Smurf (6") and Normal Smurf (6") as well as Papa Standing, Smurf Carrying a Candle, Flirting Smurfette, Cowboy Smurf, Baby Smurf Sleeping, Baby with a Rattle, Baby Sitting Sucking his Thumb, Schlroumpfs Regular Smurf, Smurf Holding a Cake, Baby with Bottle, Smurfette Standing, and Papa Smurf (mint in package).

Set #264, 1-2 Finger Puppets. A black Smurf on a black base and a blue Smurf on a purple base represent this category. Six different Finger Puppets are known to exist.

Set #71 Folders. When it's time to get smurfin' organized, folders such as Smurfy Friends Stay in Touch, Gargamel and Azrael, I Love Work..., Smurf Sports, Smurf 'em Cowboy! or Dreamy Smurf provide the perfect reason.

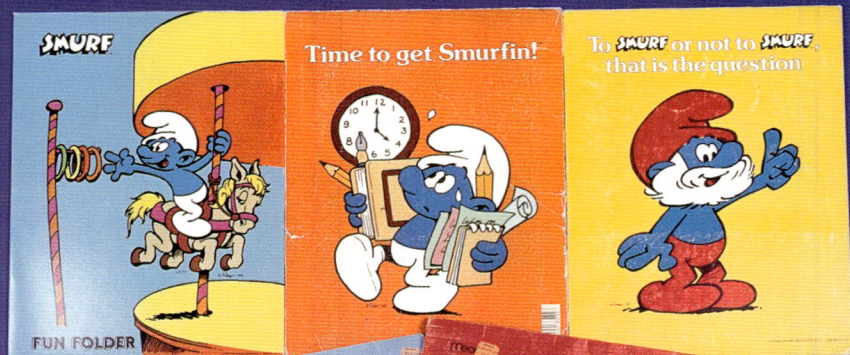

Set #71 Folders. Wallace Berrie provides the perfect folders, starting with Time To Get Smurfin!, To Smurf Or Not To Smurf, That is The Question, Smurf on a Carousel Horse, Smurf and Smurfette Rollerskating, and Jokey Giving a Smurf a Present.

Set #19 Food Items. An assortment of food items have Smurf characterizations on them, including Smurf cookies, vitamins, Gummie candies, juice boxes, bubble gum, chocolates, assorted candies, and cereal boxes.

Set #19 Food Items. Peyo and Schleich try to make the Smurf name stick, with advertisements on cookie packages, vitamins, pasta, and Gummi candies as well as on a company designed gum ball machine!

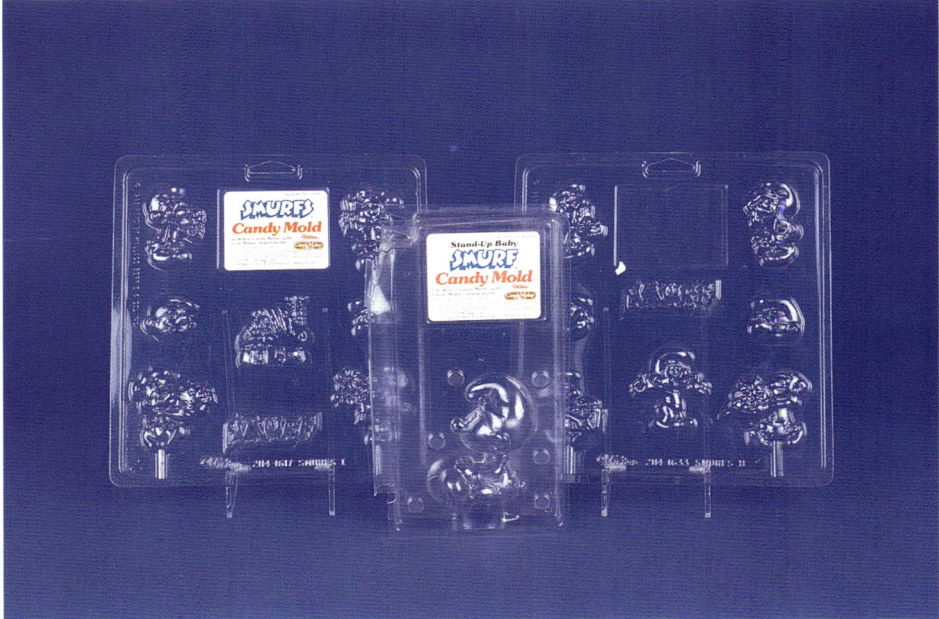

Set #19 Food Items. Chocolate and ice cream turn into delicious Smurf edibles with these two candy molds and one ice cream mold.

Set #42 Games. This category deals in a variety of table top games. From cards to dominos, the games stack up to FUN!

Set #42 Games. The Smurf Game (green box) comes with Smurf figures for game pieces and the mazes come in red or yellow for "amazing" fun!

Set #42 Games. An additional assortment of games provides collectible figures and hours of smurfy fun.

Set #42 and #111 Games. A puzzle cube, a Rubix Cube and a slide puzzle illustrate a trio of smurfy games.

Set #233 Games—Atari and Coleco. Atari didn't miss a click when they produced these game cartridges: Rescue in Gargamel's Castle, Smurf Paint 'n' Play Workshop, and Smurfs Save the Day. Combining these with the available talking video game cartridges, Cassettes #1 and #2, provides for an entertaining time.

Set #199 Games—Hand Held. This category features the hand held games by Tomy and Tiger Electronics. These exciting electronic games— Smurf Ball, Smurf Look Alike, Hat Trick, and Schtroumpf—yield hours of entertaining fun.

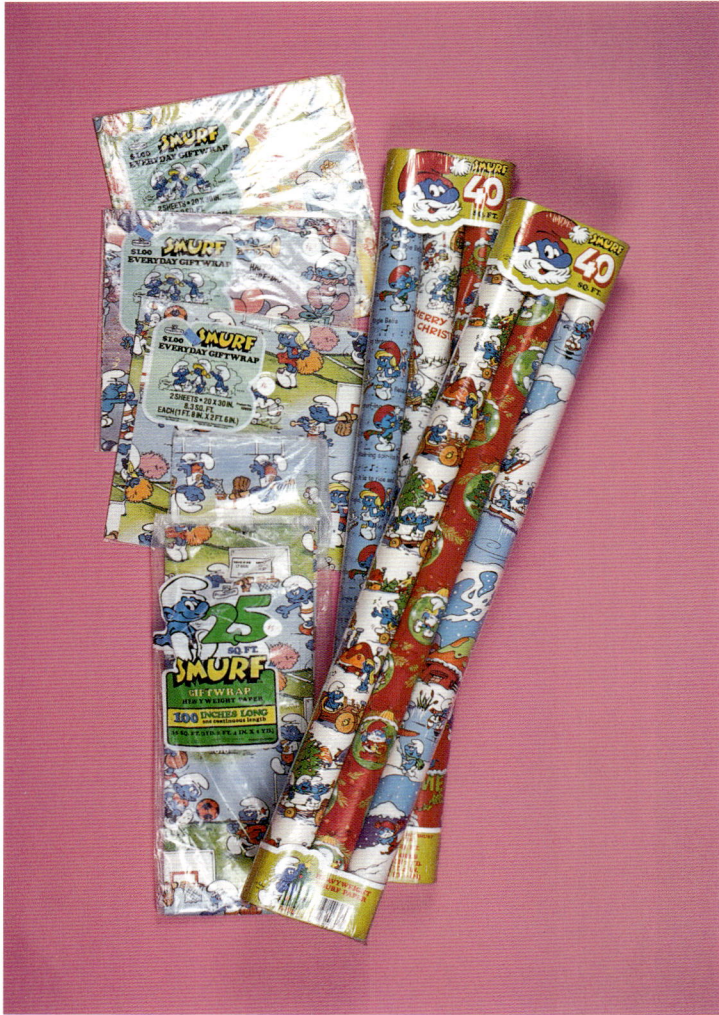

Set #125 Gift Wrap. An assortment of gift wrap is pictured.

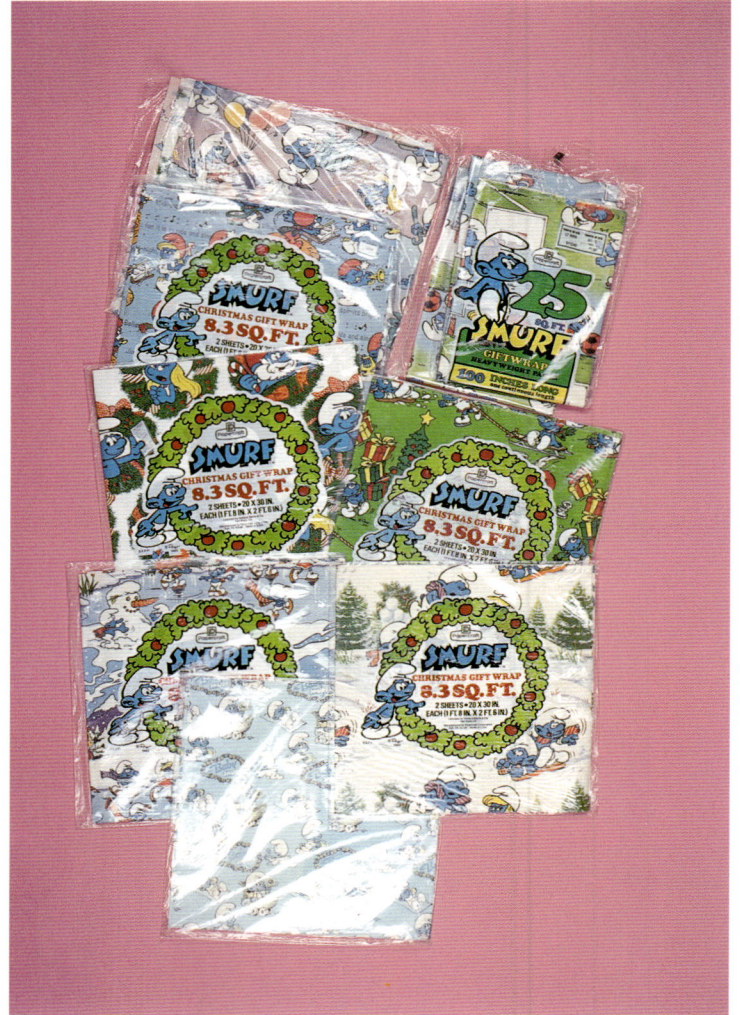

Set #125 Gift Wrap. An assortment of eight different gift wrap designs is pictured.

Set #257 Glasses—Australian. Shown are four High Ball glasses that originated from Australia: Drummer, Runner, Smurf Hitch Hiking, and Smurf on a Tricycle.

Set #173. 1-11 Glasses—Canadian Pedestal. Peyo provides a wonderful collection of Canadian pedestal glasses with this series: Sport Smurf, King Smurf, Fishing Smurf, Drummer, Lute, Flying Smurf, Trumpet, Guard with a Lance, Prisoner Smurf, Papa Bandleader, Hockey Smurf, and Spy Smurf (missing from the picture). See Set #213 for Canadian Water Glasses.

Set #213, 1-12 Glasses—Canadian Water. Canada provides a smurfy collection of water glasses with this set: Sport Smurf, King Smurf, Fishing Smurf, Drummer, Lute, Flying Smurf, Trumpet, Guard with a Lance, Prisoner Smurf, Papa Bandleader, Hockey Smurf and Spy Smurf. See Set #173 for Canadian Pedestal Glasses.

Above: Set #235 Glasses—France. This wonderful assortment of small glasses containing mustard originated in France. Over the years, they have included: 1976 - Smurf Giving Smurfette Flowers and Smurf Village; 1983 - Smurf looking at Self in Water, Smurfette Ballerina, Smurfette Mermaid, Smurfette Roller Skating, Tennis Smurfette, Smurf Running a Marathon, Smurf Shaking a Referee, Smurfling in a Ship; 1985 - Smurf Blowing Fire From His Mouth; 1986 - Five Smurfs Dancing Around a Christmas Tree, Papa and Smurfette with another Smurf Ice Skating, and Smurf and Smurfette Sitting on a Log.

Below: Set #235 Glasses—France, and Texaco glasses. In the 1980s mustard glasses continued to be distributed in France: 1987 - Jokey Carrying a Present to Brainy, Smurf Pulling a Wood Sheep, Papa giving Smurfette a Present, Smurfs Carrying a Cob of Corn; 1988 - One Smurf Downhill Skiing (Texaco glass), Two Smurf Carolers and Papa, Gargamel and Azrael (Texaco glass), Papa Standing on a Mushroom (30th Anniversary of Smurfs), Smurf with Bucket of Fish (Texaco glass), Smurfette Ringing School Bell (Texaco glass), Smurfette Standing on Top of a Birthday Cake (30th Anniversary of Smurfs); 1989 - Two Smurfs Downhill Skiing, and Smurfs on a Roller Coaster.

Set #235 Glasses—France. In the 1990s mustard glasses continued to be distributed in France: 1990 - Two Smurfs Rolling a Snowball off a Snowdrift, Three Smurfs at the Beach, Slouchy Laying Against Puppy, Smurf Flying a Kite, Smurf Kicking a Goal (Soccer); 1991- Three Smurflings Make a Sandcastle, Papa is Putting Glasses on a Mole, Smurf and Dragon Eating a Book, Smurf Raking Leaves, Smurflings Splashing in a Puddle, and Smurfs are Carrying Apples.

Set #235 Glasses—France. More mustard glasses from the 1990s. 1992 - Smurfs Shaking Acorns From a Tree; 1993 - Puppy Chasing a Stick, Smurf Fishing; 1994 - Smurf Lighting Firecrackers in a Mole Hole, and Smurfs Walking Towards a Door (Toc Toc); 1996 - Le Grande Schtroumpt (Papa), Le Schtroumpf Gourmand (Greedy), Le Schtroumpf A Lanettes (Brainey), Le Schtroumpf Coquest (Vanity), Le Schtroumpf Farceur (Jokey), Le Schtroumpf Musicien (Trumpet Smurf), and Le Schtroumpfette (Smurfette).

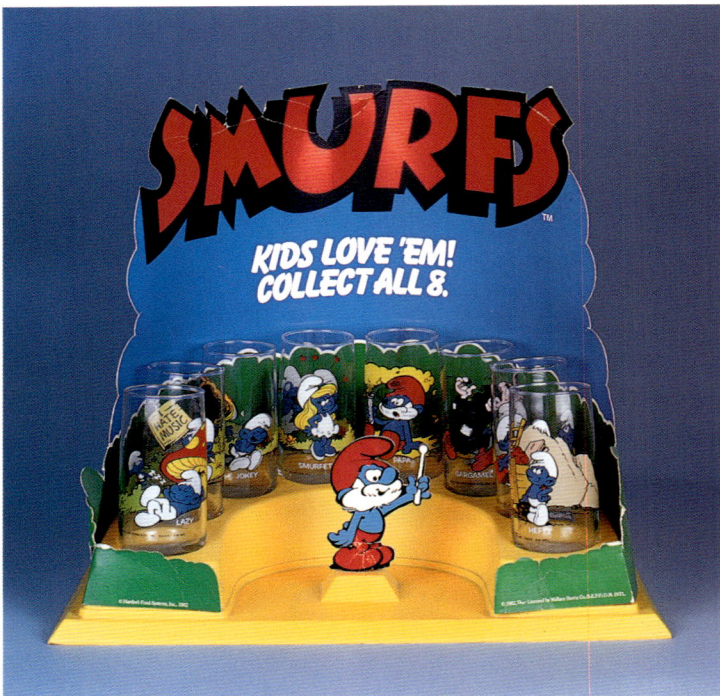

Set #258, 8-15 Glasses—Hardee's + #115 Display - 1982 Hardee's. In 1982 Hardee's Fast Food Restaurant issued a series of eight Smurf glasses: Lazy, Grouchy Smurf, Jokey Smurf, Smurfette, Papa, Gargamel and Azrael, Brainy, and Hefty.

Set #258, 2 -7 Glasses—Hardee's. In 1983 Hardee's followed up its successful 1982 Smurf glass promotion with another set of six glasses: Papa Smurf Party glass, Harmony Smurf Party glass, Baker Smurf Party glass, Smurfette Party glass, Handy Smurf Party glass, and Clumsy Smurf Party glass.

Set #258, 1 Glasses—Hardee's. 1982 Hardee's Manager Glass Stein with Hardee's and Anchor Hocking logos.

Set #263
Glasses—
Miscellaneous.
Four miscella-
neous drinking
glasses are
pictured: Albino
Smurf,
Schtroumpfissime,
Smurf Jumping
(large pedestal
glass), and
Smurfette and
Smurf (large
pedestal glass).

Set #224 Greeting Cards. Sending that special message in one of these perfect greeting cards is a great way to keep in touch with Smurf friends.

Set #265 1-3 Greeting Stands–German. German Greeting Stands Schleich are pictured: Unser Goldstück! Smurfette; Bleib Frisch Und Gesund! Apple Smurf; Verzeihur Devil Smurf; and Kopf Hoch! Baby with bowl and spoon.

Set #182 1-8 Greeting Stands—German Soccer Team. The German soccer team is represented on these triangle stands. Holding the poseable rubber PVC figurines are: Ha-Ess-Vau!, Vau Eff Ell!, Heja, Heja, Ole-Ole Vfb, I bin a Bayern Fan and EINTRACHT...!

Set #238 1-3
Greeting Stands—
Promotional Figures.
Schleich produced this
assortment of
promotional greeting
stands.

Set #219 Greeting Stands—Triangle. Fourteen different Smurf figurines attached to triangle shaped greeting stands are shown in each picture. These represent a sampling of the category listed.

Set #84 Hair Accessories. Shown is an assortment of hair care accessories, from a Deluxe Dresser set to a Travel Case including a Brush and Shampoo Holder.

Set #84 Hair Accessories. Mirrors, combs, brushes, clips, and barrettes hold this category together for a stunning appearance.

Set #61 Halloween Costumes and Masks. Halloween Trick-or-Treat would not be complete without Smurf or Papa Smurf ringing the doorbell! Made by Ben Cooper, these Smurf Halloween costumes come complete with costume and mask.

Set #129 Happy Meal Boxes—Hardee's. Fast Food restaurants, such as Hardee's, have not missed a beat by producing Funmeal boxes such as Papa with Mushroom House, Baby Smurf Sitting Under a Tree, Smurfs Singing, 2 Smurfs Swimming, or Smurf on an Orange Skateboard.

Set #129 Happy Meal Boxes—Quick Restaurant, France. Quick Fast Food restaurant of France gave out their Smurf collectibles in this Magic Box.

Set #129 Happy Meal Boxes—McDonald's. McDonald's of Germany gave out their sets of Smurf figurines in Junior Tute / Happy Meal bags. McDonald's of Europe, where boxes are permitted, gave out their collectibles in the red mushroom shaped box.

Set #174 Hardee's Figures. Seven figurines from Hardee's Fast Food restaurant are shown: Puppy on blue skateboard with pink tongue, Puppy on blue skateboard with white tongue, Smurfette on purple skateboard, Papa Smurf on red skateboard, Smurf on orange skateboard, Smurf on yellow skateboard, and Sassette on green skateboard.

Set #161 Hats & Visors. Smurf collectors could easily top off any occasion with one of these hats!

Set #161 Hats and Visors. More hats and visors are shown in this assortment.

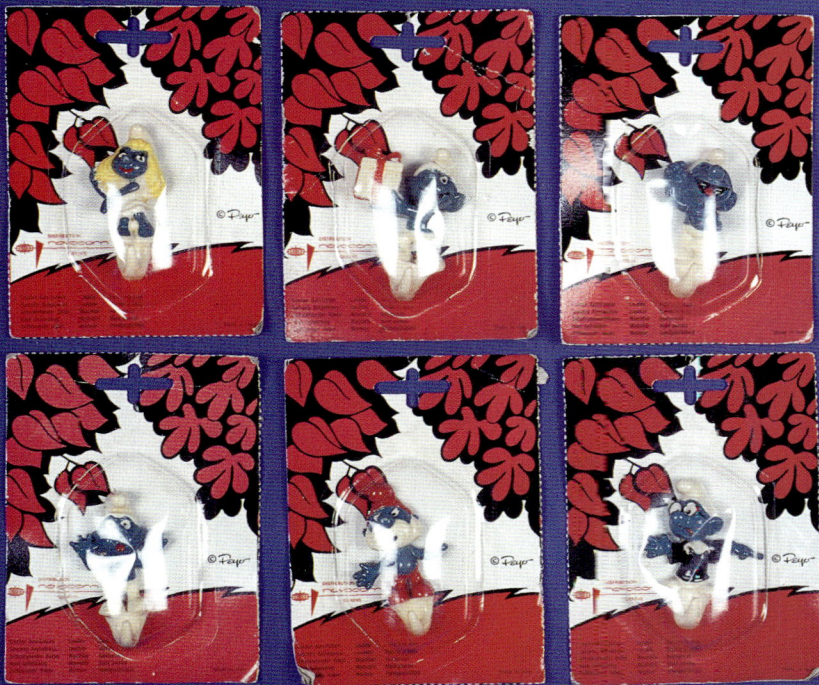

Set #211 Hooks. Hang up your Smurf collectibles on these specially designed hooks. Papa, Judge, Smurfette Flirting, Naughty, Present, or Jolly can hold any smurfy item.

Set #158 Household Items. Cleaning house with the likes of these Smurf collectibles makes playing house the rule.

Set #69, 1, 9, 18 Houses—Smurf. The three plastic houses shown include a large red roof mushroom house, a farmhouse, and a large Bully house.

Set #69 Houses—Smurf. Eleven small Smurf mushroom cottages include two versions of the red cottage (dark red and light red) and three versions of the blue cottage (blue with tan door, blue with gray door and blue/red with brown door), as well as items 4, 6, 7, 12, 13 and 14 from this set.

Set #69, 1 and 11 Houses—Smurf. Pictured are Gargamel's Castle and The Windmill plastic play house.

Set #69 15- 17 Houses—Smurf. Three plastic cottage playlets are illustrated, mint in the box: Orange Roof cottage with Snappy Smurf by Irwin, Purple Cottage with Baby Smurf, and All Blue Cottage with Smurf Playing a Flute.

Set #254 Iron-ons. Pressing on these little decorations adds just the right touch to attire.

Set #251 Jam Caps—Die Smurfen. Jam Caps were the rage in 1995. Referred to as POGS in some countries and Jam Caps in others, they caught the imagination of children around the world. Pictured is an assortment of Jam Caps.

Set #104 Keychains—Acrylic. Thirty-three acrylic key chains are pictured.

Set #79 Keychains—Pencil Sharpeners. Taking note of the latest in Smurf collectibles is easy with these key chain pencil sharpeners. They are always handy to get to the point!

Set #253 Keychains—Figure Clips.

Set #14 Keychains—Miscellaneous. Twenty-one assorted key chains are pictured, including slide key chains and figure key chains.

Set #249 and #269 Keychains—PVC Figures. Smurf figures make great friends and hook onto any occasion. Made by Schleich / Peyo, these rubber figures are the basic character line of figures produced over the years.

Set #249 and #269 Keychains—PVC Figures. Smurf figures make carrying a key chain a sporting affair to turn anyone blue.

Set #78 Keychains—Soft Rubber + #115, 2 Store Display. Six soft rubber key chains are displayed: Papa Waving, Smurf on a Skateboard, Baby with a Teddy Bear, Baby Sitting on the Moon, Mermaid Smurfette, and Sexy Smurfette.

Set #139 and #140 Keychains—Slide Puzzle & Slide Card. Key chains not only hold the key to Smurfland, they provide entertaining times with Smurf slide puzzles, like Jokey Smurf key chain, Sassette's first key chain or Cake Smurf key chain. Slide cards attached to the key chain keep traveling smurfy with scenes like Smurf Pond, Smurf with a Horn, and Papa and Smurf in an Airplane.

Set #34, 1-7 Kinder Egg Playsets. Kinder Egg, Ferrero of Germany, issued a series of two Kinder Egg Playsets in 1991: Smurf in a Log Car and Smurf Tower. In 1995, five additional sets were issued: Bricklayer Smurf with a Lighthouse, Cook Smurf with a Mushroom House, Farmer Smurf with a Tractor, Lazy Smurf with a Hammock (Top Left) and Miller Smurf with a Windmill (Top Right).

Set #276 Kinder Egg Puzzles. Pictured is an assortment of Kinder Egg puzzles, four puzzles make up one scene with two different scenes photographed.

Set #33 Kinder Egg Smurfs - To Assemble. Five musical and five garden Kinder egg figurines are pictured (Garden): Smurf Carrying a Sprinkling Can, Smurf Leaning on a Shovel, Smurf with a Backpack, Smurf with a Rake, Smurf with a Wheelbarrow; (Musical) Papa Bandleader, Smurf with a Trumpet, Smurf Playing a Violin, Smurf Playing an Accordion, and Smurfette Dancing.

Set #33 Kinder Egg Smurfs - To Assemble. In 1991, nine additional Kinder Egg figurines were issued: Papa Referee, Smurf Offering a Box of Kinder Eggs, Smurf with a Jack-in-the Box, Smurfette, Soccer Referee with Flag, Soccer Smurf Squatting, Standing Soccer Smurf, Vanity Smurf, and Winter Smurf.

Set #33 Kinder Egg Smurfs - To Assemble. Two figures originated in Italy: Black Smurf with Whip and Smurf with Torch.

Set #40 Latch Hook Rugs. Hook a memory into a rug is the heart of this crafty category.

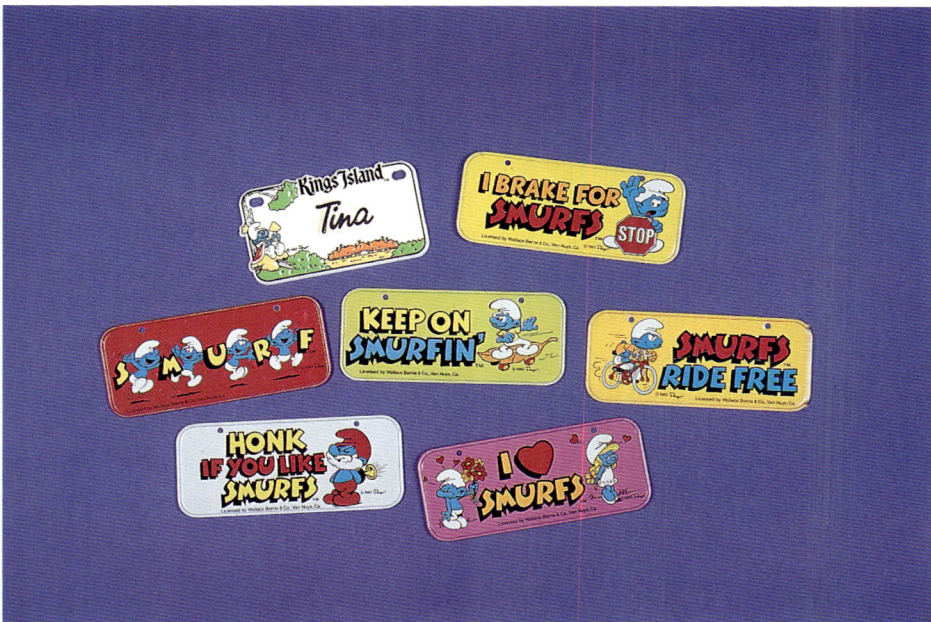

Set #53 License Plates. Smurfing along with your bicycle is not complete without one of these. Simply call your Smurf mobile a SMJRF, Smurfs Ride Free, I Brake For Smurfs, Keep on Smurfin, Honk If You Like Smurfs, I Love Smurfs, and Kings Island Tina.

Set #220. Light Fixtures. Ceramic smurf lamp with mushroom shade.

Set #220 Light Fixtures. In addition to the Smurf Mushroom Lamp, there exists a Smurf on the Moon Globe, Smurfs on Carousel Horses Overhead Globe, and a Smurf Lamp Shade.

Set #164 Lite-Brite. Children light up with the fun of the Smurf Lite-Brite set and the Smurf Picture Play Lite set by Janex Corporation and Hasbro.

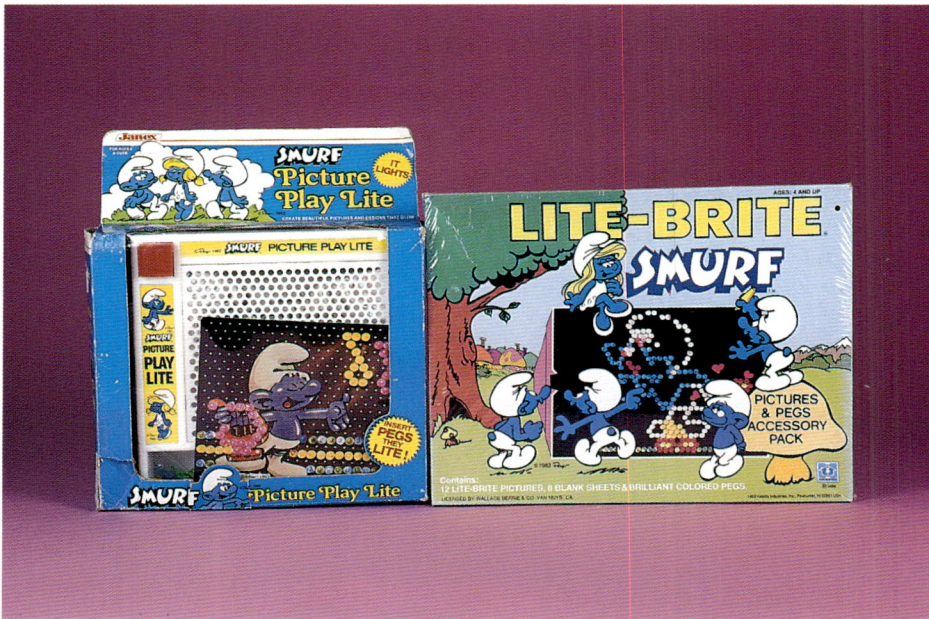

Set #65 Loc Blocs. Dr. Smurf's Office or Papa Smurf's Laboratory provide the perfect connection between child and play.

Set #166, 1, 4, 6 and 7 Luggage. Four luggage carriers are pictured, made from four different manufacturers: Rollerskating Smurfetter Suitcase by Wallace Berrie, Orange Suitcase by Dupuis, Yellow Hard Suitcase by BP, and Smurfs Standing in Line at Puffi Aeroporto by I Puffi.

Set #25, 1-6 and 8-10 Lunch Boxes & Thermos. Plastic and metal lunch boxes came in assorted designs, including Queen Smurfette being Honored (Metal), Queen Smurfette being Honored (Plastic) and all plastic ones: Smurfling Surfing in light blue and in dark blue, Smurfette, Smurfs and Smurfling Eating, Smurf Pond Scene, Drummer Smurf in green from Australia, and Papa Smurf in blue from Australia.

Set #25 Lunch Boxes & Thermos. These specially designed brown bags are just not the ordinary lunchtime experience. Any Smurfette would eat the vegetables if she were adorned with this special Smurfette lunch purse.

Set #130 Magazines—German Die Schlumpfe Large. Displayed is an assortment of large comics from Germany.

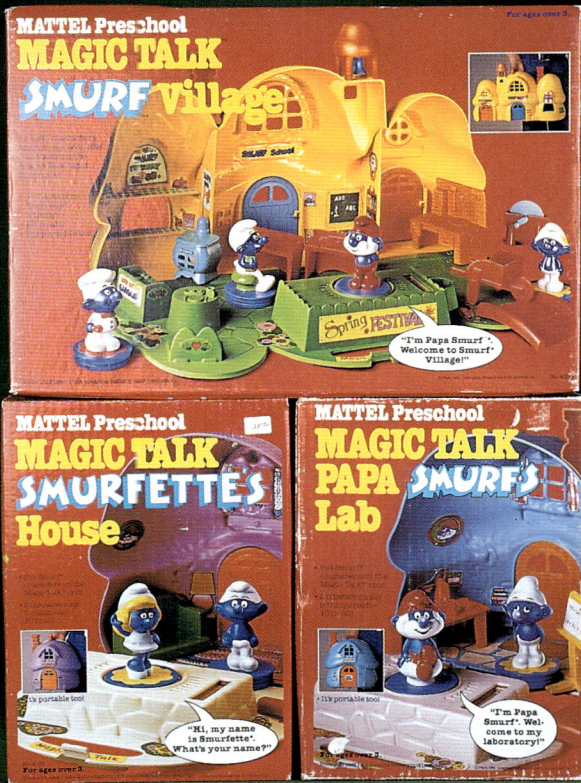

Set #41, 1-3 Magic Talk Houses. Three Magic Talk houses are shown: Magic Talk Smurf Village, Magic Talk Smurfette's House, and Magic Talk Papa's Lab.

Set #250 Magic Tricks. One can perform amazing feats with these Magic Sets, using the wizard with blue guillotine or the wizard with magic disc and beads routine. Either one could provide hours of mystical fun.

Set #100 Magnets—Figures, and Set #171 Magnets—Square Ceramic. Magnets hold the experience in your mind long after the event. It is for this reason and others that Wallace Berrie, the Robinson Design Group, and Peyo have distributed a myriad of magnets, from the ceramic square design to the figural design.

Set #190 Magnets—Flat. Any refrigerator would be properly attired with this assortment of flat magnets adorning the door.

Set #179 McDonald's —Promotional PVC's. The 1997 McDonald's set distributed internationally is shown: Smurf on Skateboard, Smurf on Rollerblades, Basketball Smurf, Basketball Smurfette, Smurfette on Rollerblades, Baseball Smurf with Bat, and Baseball Smurf with Ball and Glove. An additional Smurf with Happy Meal Box was given out to McDonald's customers only in the UK.

Set #101 Memo Sets—Mini. Children love to play grown-up by carrying along their mini memo sets to record the adventure. This selection includes: Measure Set (yellow), Memo with Phone Book, Memo with a Corkboard and Mini Sketch (green).

Set #216, 1-8 Minimates. Smurf Minimate poseable figures by Toy Island include: Baby with Rocking Chair, Fireman with Ladder, Sassette on a Surfboard, Clown Smurf with a Tuba, Smurf Downhill Skier, Smurfette with Removeable Dress, Smurf with Video Camera and Smurfette with Stove. See Set #31 (Figures—Poseable) for additional poseable figures.

Set #1, 1-6 Mugs—Ceramic. Lazy, Jokey, Smurfette, Papa, Greedy, and Grouchy start off the ceramic mug category.

Set #1, 7-13, and 44 Mugs—Ceramic. Travel America Mugs highlight this category with I Love NY, I'm a Rocky Mountain Smurf, California Smurfin'. I'm a Florida Sunshine Smurf, I Smurf America, Smurfin' Thru the USA, Smurf Loves Dixie, and We Smurf Things Big in Texas!

Set #1, 14-21 Mugs—Ceramic. Smurfing along, one can find ceramic mugs by Wallace Berrie with a variety of phrases: Guess Who Loves Ya?, Have a Happy Day, Lover Smurf, A Smurfing We Will Go!, #1 GRAD, SMILE SMILE SMILE, Jogger Smurfs, and Sporty Smurf.

Set #1, 22-29 Mugs—Ceramic. Wallace Berrie continues their line of ceramic Happy Birthday mugs with Super Smurf, OOPS Happy Birthday, Wow! Another Birthday!, Have a Bang-Up Birthday, Happy Birthday from 1981, For Your Birthday, Hooray For Birthdays, and Happy Birthday with Papa displayed from 1982.

Set #1 30 -36, and 42 Mugs—Ceramic. Wallace Berrie continues their line of ceramic mugs with Is it Break Time Yet?, Bless This Mess!, I Love You, You're My Sweetheart!, Gotcha! I Love You, Smurfs & NBC, and Smurf Chasing Hearts.

Set #1, 38, 39, 41, 43, 45-46, and 48-51 Mugs—Ceramic. Wallace Berrie, Peyo, Grindley, and IMA produced an assortment of ceramic mugs, including Smurf Pole Vaulter Beach Life, Smurfette Posing. Merry Christmas 1981, SMURF pink cup, SMURF blue cup, and SMURF white cup.

Set # , 47, 52-54 Mugs—Ceramic. From New Zealand, these ceramic mugs include 4 Different Smurfette Poses, 4 Sports Smurfs Swinging Smurfs, and Smurfs are Fun! varieties.

Set #67, 1-6 Musical Instruments. Six miscellaneous musical instruments from different manufacturers are pictured: Horn from Ohio Art, Metal Drum, Strummin' Smurf Guitar from Ohio Art, Orange Musical Ge-tar from Wallace Berrie, Smurf Mini Organ from Galoob, and Smurf Small Plastic Drum from Ohio Art.

Set #67 Musical Instruments.
Musical Drum set.

Set #56 Musical Toys. Musical toys lead the parade to Smurfland! Starting with Smurf in the Box popping out and serenaded by the Musical Smurf Radio, the parade continues with a Carousel, Roll Along Smurf, Smurf on a Mushroom Music Box, Smurf Tin Can, and Musical Cottage with Animated Picture, concluding with Schlumpf Musikhaus with Bully Figures. Eight musical toys comprise the Smurfland Parade!

Set #56 Musical Toys. Continuing the Parade to Smurfland, this set of six musical toys includes Musical Smurfette, Spinning Tops in green and red, Tree Go Around, Papa Smurf Player, and the Musical Cottage. Each provides entertainment for the audiences—family and friends from around the world.

Set #56 Musical Toys. See N Say is a classic toy with children around the world, as well as the Belgium Read and Press musical book. Children are thrilled with the sights and sounds of Smurfland.

Set #192 Necklaces—Assorted.

Set #192 Necklaces. One's jewelry box would not be complete without an assortment of necklaces to hold the vision close to one's heart.

Set #192 Necklaces. Mini Smurf figures add just the right touch to a jewelry collection of necklaces.

Set #132 Needlepoint Stitchery. Needlepointing or cross stitching the perfect Smurf collectible weaves together memories for a lifetime.

Set #37 Night Lights. Papa Smurf stained glass night light illuminates the scene!

Set #37 Night Lights. At least twenty-one different Night Lights brighten the view!

Set #49 Notebooks— Spiral. Keeping track is easy with the Smurf spiral notebooks.

Set #222, 1-9 Ornaments— Alderbrook Christmas Plastic Figures. Alderbrook produced a series of nine plastic Christmas Ornaments: Skier, Smurf on Sleigh, Smurfette Wearing Santa Suit (mint in package), Papa Wearing Santa Suit (mint in package), Skier (not pictured), Bell Ringer, Ice Skater, Angel, and Papa Holding a Candy Cane.

Set #170 Ornaments—Satin Christmas. Six satin ball ornaments, mint in the boxes, bring in the Christmas season, Smurf style.

Set #170 Ornaments—Satin Christmas. Fifteen holiday balls and bells bring Smurfland home for the holidays.

Set #252 Ornaments—Christmas. Thirteen different Christmas ornaments from around the world add to the holiday cheer.

Set #165 Outdoor Riding Toys. Winter fun comes with a Smurf-A-Boggin Sled, a Smurf Snow Disc, or Smurf Skis.

Set #165 Outdoor Riding Toys— Lawn Ornaments. Assorted Smurf lawn ornaments can adorn a smurfy yard.

Set #165 Outdoor Riding Toys. Children can have hours of fun on the Smurf Sit N' Spin and the Poppin Train Engine.

Set #165 Outdoor Riding Toys. A Smurf Doll Carriage and a Smurf Piggyback Stroller roll this category right along.

Set #212 Paper Clips. Figures attached to large clips hold the smurfiest of messages.

Set #193 Paper Items & Advertisements—Miscellaneous. Smurf collectibles span the globe when it comes to paper advertising.

Set #193 Paper Items & Advertisements—Miscellaneous. Paper Smurf hats from McDonald's fast food restaurants in Europe added to the 40th Anniversary celebration in 1998.

Set #193 Paper Items & Advertisements—Miscellaneous. From mini catalogues to assorted Smurf paper collectibles, the supply appears endless around the globe.

Set #193 Paper Items & Advertisements—Miscellaneous. The early 1980s were heydays for Smurf collectors around the world. Skippy Peanut Butter and Chef Boyardee, along with cereal manufacturers and retail catalogues like Montgomery Wards, cashed in on the advertising appeal of Smurfs in Smurfland.

Set #15 Party Supplies. Party, party, party time begins with these supplies!

Set #15 Party Supplies. "You're Invited" is the theme of this assortment of party supplies!

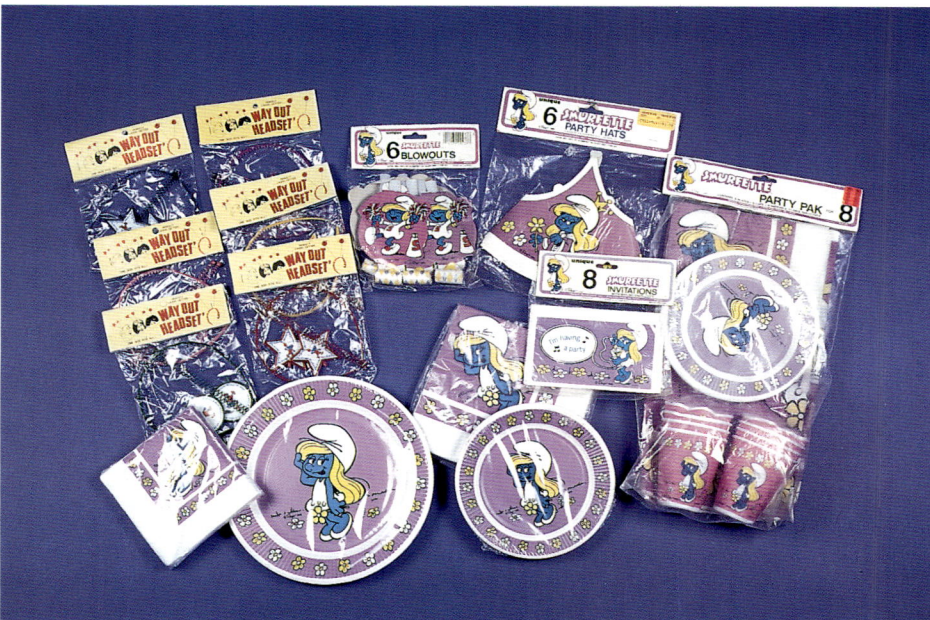

Set #15 Party Supplies. It is Smurfette party time with this assortment.

Set #15 Party Supplies. Launching a smurfy party is not difficult with these balloons and party supplies.

Set #15 Party Supplies. Having a Happy Smurf Day is the order of business with these party supplies! Makes holidays the smurfiest!

Set #15 Party Supplies. This collection of holiday party supplies could turn the serious collector green with envy.

Set #15 Party Supplies. Join the New Year's celebration with this assortment of party supplies.

Set #142 and #143 Patches. Pictured is an assortment of Smurf patches.

Set #218 Patterns—Cloth Pillow Patterns Unfinished. Cutting out and stitching up a Smurf pillow provides just the right incentive to enter Smurf Dreamland.

Set #146 Pen Sets. A smurfy looking desk begins with these pens: Baseball Smurf, Cowboy Smurf, Cupid, or Ballerina Smurf.

Set #266 Pencil Sharpeners—PVC Figures on Tree Stump. Schleich has produced an assortment of PVC figures on tree stumps which double as pencil sharpeners along with a variety of pencil toppers and pencils.

Set #95 Pens and Pencils. Writing is easy with this vast assortment of pens and pencils to guide one through the lesson.

Set #236 Pewter Smurfs. Eighteen pewter Smurfs from the 1980s include: Gold, Drummer, Mechanic, Mirrow, Crying, Clown, Smurfette, Singer, Gift, Trumpeter, Baseball Batter, Kayak, Cyclist, Sunbather, Painter, Normal, Baby with blocks and Smurf carrying lantern.

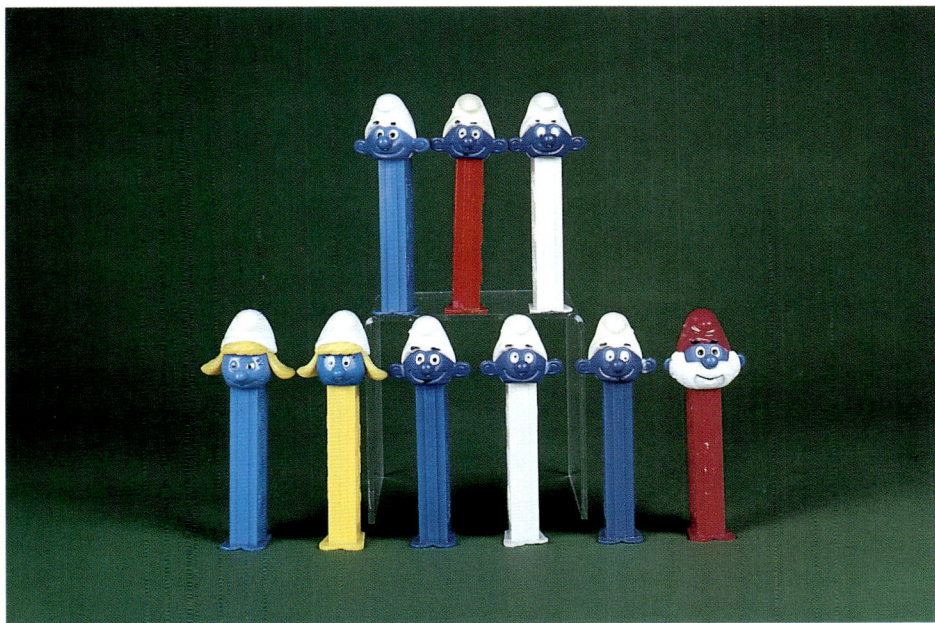

Set #32 Pez. Over the years and around the world, an assortment of Pez candy holders were made with Smurf designs, Differences exist between stem colors, eyes and no eyes, hair colors, and characters. Older Pez have no feet. An assortment of Pez candy holders is pictured, with 1980s Pez on the top row: and 1990s Pez on the bottom row:

Set #32 Pez. An assortment of carded Pez is pictured.

Set #117 Pez—Ball Games. Pez ball games come in four varieties of case colors: black, blue, lime green, or pink with the game being essentially the same.

Set #66, 1-2 Phonograph. Two Smurf phonographs are pictured: A Smurf and Smurfette Dancing on a Phonograph and a Smurf Band phonograph.

Set #208 Pillows—Red Satin Heart. Love is not far in Smurfland, including red satin pillows to dream away one's thoughts.

Set #88 Pillows—Shaped. Smurf stuffed pillows give just the right amount of comfort when needed.

Set #167 Pins and Set #192 Necklaces. Pictured are nineteen assorted Smurf necklaces and a variety of thirty-nine Smurf related pins.

Set #167 Pins. Pins add just that touch of class to a Smurf collector.

Set #167 Pins. An assortment of fifty-two pins gives even the serious Smurf collector a selection to turn blue for.

Set #275, 1-3 Pixi—Minis. Photographed along with the Chess Set Pixi-Mini Jeu D' Echecs is the seven piece Smurf Pixi Set along with the Black Smurf and Angry Smurf.

Set #105 Place Mats. Four place mats came from Texaco.

Set #105 Place Mats. Assorted place mats by Wallace Berrie, Peyo, and Partytime.

Set #112 Plaques—German Wood. Staying smurf positive is easy with this assortment of plaques! There is always one for the right occasion.

Set #147 Plastic and Rubber Smurfs. Peyo has produced numerous size plastic figures over the years. Specifically, the 4 1/2" size includes: Smurfette Holding a Book, Smurf with a Guitar, Papa Bandleader.

Set #172 Plates—Ceramic & Accessories Sets. The 1982 Merry Christmas cup, plate, and music box from The Smurf Carolers is pictured on the left. The 1983 series, The Night Before Christmas, is on the right.

Set #172 Plates—Ceramic & Accessories Sets. Ceramic plate and accessories include Hot Air Balloon Plate, Hot Air Balloon Mug, Hot Air Balloon Bell, Hot Air Balloon Trinket Box, Smurf with Ice Cream Cone Trinket Box, and the matching cup from the Ceramic Cup series, Set #1.

Set #172, 6-10 Plates—Ceramic & Accessories Sets. Ceramic plates and accessories continue with Have a Smurfy Day! Smurfette Trinket Box, Im Land Der Schlumpfe Plate, Smurf Plate with various Smurfs, Smurf Doing Various Activities Bowl, and Smurfs Doing Various Activities Cup.

Set #44, 1-3 Playsets—Flocked Figures Jointed. Four flocked figures come in each playset. Three playsets pictured are: The Workers, Cleaning U Crew, and The Circus.

Set #44, 4-6 Playsets—Flocked Figures Jointed. Three additional sets continue the category: Seaside, the Choir and Baby, Smurfette and Papa.

Set # 87, 1-4 Playsets—Transforming. Four mint in the boxes transforming playsets by Irwin are pictured: Daisy Wheel & Snappy, Dinin Out & Greedy, Picnic Wish & Sassette, and Saturn Ride & Astrosmurf.

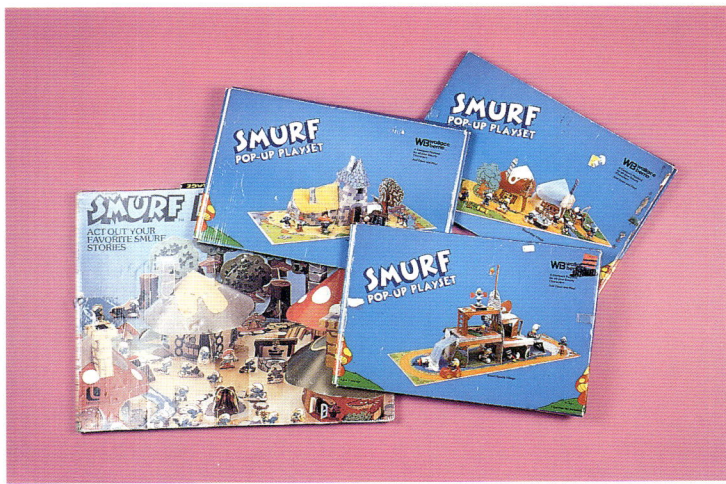

Set #44, 7-9 Playsets—Flocked Figures Jointed. The Smurf, 2 Smurfs and Brainy, and Musicians complete the playset category.

Set #12 Playsets—Pop-Up. Children's imagination soars with these pop-up playsets, such as: Gargamel's Castle, Smurf Sports Village, Smurf Deluxe Play Village, and Smurf Play Village.

Set #70, 1-5 Playsets—Regular. The Garden, Fences, Gate, Mushroom with Flower, and Sailboat make up into regular plastic playlets by Schleich.

Set #70, 6-10 Playsets—Regular. The Gas Station, Wishing Well, Snail Carriage, Forklift, and Conveyor Belt continue the line of regular Playsets by Schleich.

Set #70, 11-13 Playsets—Regular. The Gate with the Fence and figures, the Wishing Well and figure, and the Snailcart and figure conclude the regular plastic playlets by Schleich.

Set #83 Plush—Kinder Eggs. Kinder plush from the 1990s are pictured: Brainy, Papa, Baby Smurf, Farmer Smurf, Normal Smurf, Papa, Sassette, Papa Valentine, and Smurf Valentine.

Set #29 Plush Smurfs. Numerous plush Smurf figures have been made and distributed around the world over the last forty years. Starting with Baby Smurf with a rubber head and hands and smurfing along to Berry Lovin Smurf, these two plushes are just the start of an immense category of Smurfy Plushes!

Set #35 Plush Smurf & Smurfette Wardrobe. Pictured is an assortment of Smurf and Smurfette outfits ranging from jogging suits for the perfect Smurf workout to pajamas for the perfect night's rest. All in all, sixteen different outfits are pictured for the smurfiest of occasions!

Set #29 Plush Smurfs. Taking a plush home for the holidays is the theme for this photo.

Set #29 Plush Smurfs. Papa Smurf and Baby Smurf make ideal Valentine greetings and bring out the widest of smiles. Smurfs love to be loved.

Set #29 Plush Smurfs. This 3' tall Papa Smurf in his rolling wheels Toy Chest is ready for adventure.

Set #247 Pogs—Les Schtroumpfs The 1990s saw the rage of the POG collectibles come and go, wi[th] Smurf collectible[s] not missing a flip. Pictured is an assortment of PCGS. See Categories #251 or #277 for additional listings

Set #277 Pogs—Schtroumpf. Pictured is a collection of thirty-one Schtroumpf Pogs from Germany, excluding #11 and #24 (Total #33). They are numbered consecutively.

Set #184 Postcards. Sending a smurfy greeting home could easily be accomplished with this assortment of postcards from around the world.

Set #184 Postcards. Postcard greetings during the holidays or messages to one's smurfiest of friends bring smiles to the face of Smurf memorabilia collectors.

Set #149 Posters. This sign says it all!

Set #149 Posters. This sign opens the door to imagination.

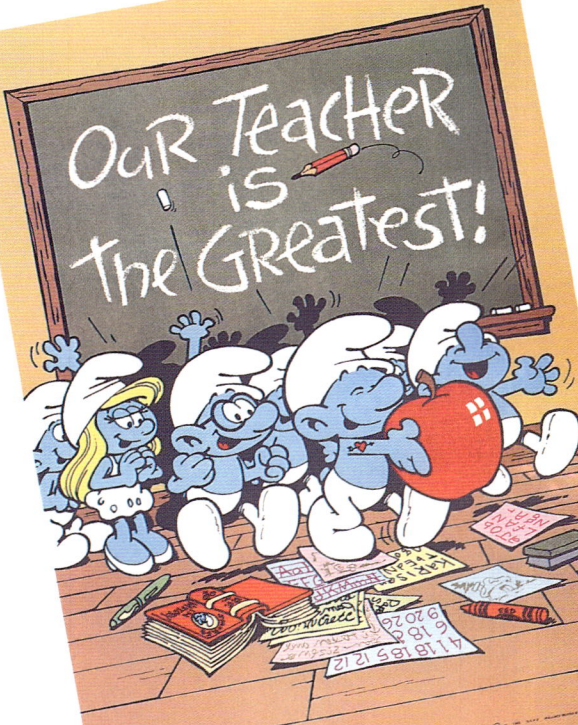

Set #149 Posters. All Smurfs and Smurfettes love learning.

IF YOU NEED A FRIEND...
I'M HERE

DIE SCHLÜMPFE
Hier erhältlich!
Spiel und Spaß mit tollen Stempeln
MARBURGER KINDERSTEMPEL
© Peyo 1995 I.M.F.S. (Brüssel) - Lizenz durch AT & TV Merchandising Concepts, München

Set #77 Pull String Flapping Toys. Five pull string toys are pictured, including Smurf as Superman, as an Ice Cream Smurf, as a Boxer, as a Smurf with Silverware, and as a Tennis Player.

Set #134 Puppets—Hand. Using one of these hand puppets adds life to a Children's Theater.

Set #201 Purses. The properly attired Smurfette has her choice from this assortment of mini purses.

116

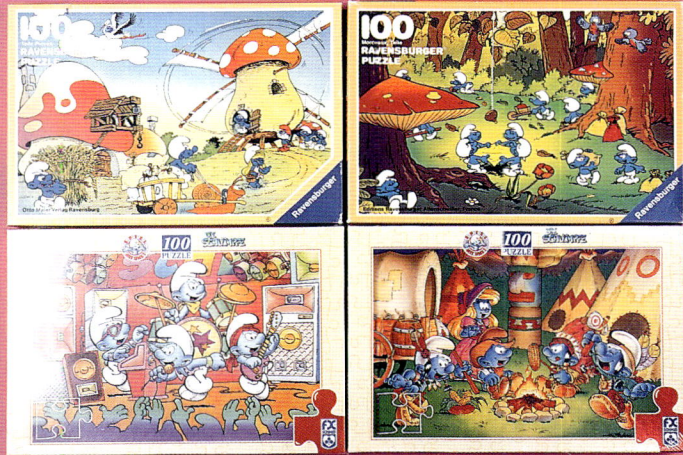

Set #229, 1-2 Puzzle Blocks. One hard and one soft puzzle block are illustrated.

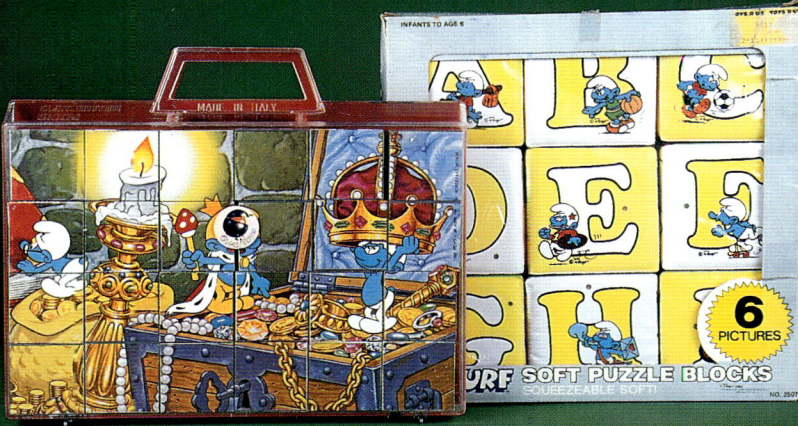

Set #62 Puzzles—100 Pieces. Pictured are four of the 100 piece puzzles from the broad category of puzzles.

Set #16 Puzzles—Miscellaneous. This collection represents three 2 x 20 and three 200 piece puzzles.

Set #16 Puzzles—Miscellaneous. An assortment of miscellaneous puzzles are pictured in this group.

Set #16 Puzzles—Miscellaneous. One 140 piece puzzle, one 250 piece puzzle, and two 320 piece puzzles piece together this puzzle assortment.

Set #16 Puzzles—Miscellaneous. One 280 piece puzzle and two 450 piece puzzles show little Smurfs in action.

Set #16 Puzzles—Miscellaneous. Cardboard tray puzzles with Baby Smurf can bring smiles to the littlest Smurfs or Smurfettes of the future.

Set #131 Puzzles—Plastic. Smurfette and Papa Smurf puzzles are plastic within a plastic tray by Peyo.

Set #20 Puzzles—Wood. Milton Bradley's puzzles include a Space Scene, Smurf Village Scene, Smurf Daydreaming Scene. and a Smurf Eating Scene.

Set #20 Puzzles—Wood. The Smurf and Smurfette Astronaut could be just the puzzle to launch one into becoming a Smurf collector.

Set #20 Puzzles—Wood. Three wood puzzles display the Smurf characters in hard copy.

Set #4 Puzzles—Wood Tray. In the early 1980s Smurfette and Smurf appeared on several 8 piece wood puzzles, from Adventure Bound to Soccer Star.

Set #4 Puzzles—Wood Tray. Smurf, Papa Smurf, and Smurfette continue their sporting fun on this assortment of 8 piece wood puzzles.

Set #176, 1-5 and 6-12 PVCs—Christmas Cord Smurfs. *(Top Row)*: Smurf with wreath, Smurf riding on a candycane, Smurf with drum, Smurfette with music sheet and candle, Smurf praying, Smurfette praying, and Smurfette with candycane. *(Bottom Row)*: Smurf with Christmas tree, Smurf with a present with light color body, Smurf with a present with dark color body, Pappa Smurf with a sack of gifts, Smurf with a sack of gifts, and Smurf with sheet music and a candle.

Set #245 PVCs—Dupuis Figures. Oldest 1957 Hand Painted Smurfs: Mirror Smurfette, Naughty, Papa, Drummer, and Judge.

Above: Set #181, 1-7, 8-12, 15-17 PVCs—Easter Figures. *(Top Row):* Smurf in white bunny costume with light pink ears, Smurf in white bunny costume with dark pink ears, Smurfette in pink bunny costume with "Made in China," Smurfette in pink bunny costume with "Made in Portugal," Smurfette in pink bunny costume with "Made in Hong Kong" inscribed on the bottom, Smurf carrying an egg on his back, and Smurf eating an egg. *(Bottom Row):* Smurfette with dark purple Easter egg, Smurfette with light purple Easter egg, Smurf with Easter egg and orange bow, Smurf with Easter egg and red bow, Smurf painting an Easter egg with a green tip brush, Smurf painting an Easter egg with a yellow tip brush, Smurf in white egg, and Smurf in yellow egg.

Left: Set #177, 1-6 PVCs—History Smurfs. *(Top Row):* Thomas Edison, George Washington, and Abraham Lincoln. *(Bottom Row):* Paul Revere, Benjamin Franklin, and Christopher Columbus.

Below: Set #243 PVCs—Novelty Figures.

Set #74 PVCs—
Promotional Figures.
Assorted figures,
1960s-1979: Smurf
holding a piggy bank,
Brainey, Tennis player 2
OMO, ASB First Aid,
ASS Card Player, Rode
Kruis Vlandereen First
Aid, Thirsty, Torch-
bearer, 75 Year Smurfs
for BP, Sign Bearer "Bon
Nadal," Sign Bearer
"Feliz Navidad," Sign
Bearer "Feliz
Aniversanrio," BP
Cleaner, Fluocaril
Toothbrush Smurf, Ice
Skater Holiday on Ice,
Paracodin Smurf with
red scarf and white hat,
Paracodin Smurf with
yellow scarf and yellow
hat, Paracodin Smurf
with yellow scarf and
white hat, Scot Silan,
Coca Cola Smurf with
brown coke, Coco Cola
Smurf with green coke,
Rugby OMO, Smurf
with OMO Soapbox
and Colgate Smurf.

Set #74 PVCs—Promotional Figures. Assorted figures, 1980-1990s: GB Smurf, Heart De Tout Mon Coeur, Phillips White Light Bulb
with white bulb, Baseball Batter, Jogger Silan, Papa with book Silan, Philips Light Bulb Smurf with clear bulb, Scot Silan, Schimmel
Pianos, Soccer Smurfette Silan, Bodybuilder Sports & Fitness, Smurf in Wheelchair, Smurf with Crutch, Ice Lolly with
Schlumpfhausen Flag, Ice Lolly with Scholler Flag, Miller Bag Smurf Volks Raiffeisen Bank, and Money Smurf Volks Raiffeisen Bank.

Top: Set # 73, 2.0001-0008 PVCs—Regular Figures. (*Bottom Row*): 2.0001 Papa with dark blue body; 2.002 Normal with dark blue body; 2.002 Normal with light blue body; 2.0003 Astro Bully with red tie; 2.0003 Astro Schleich with white tie; 2.0004 Shiver with red scarf; 2.0004 Shiver with yellow scarf. (*Top Row*): 2.0005 Gold with gold hat; 2.0006 Brainy with yellow glasses; 2.0006 Brainy with black glasses; 2.0006 Brainy with red glasses; 2.0007 Angry with red eyes, black teeth lines, foot raised; 2.0007 Angry with red eyes, black teeth lines; 2.0007 Angry with red eyes, red teeth lines; 2.0007 Angry with black eyes, black teeth lines; 2.0008 Spy with light body; 2.0008 Spy with dark body.

Center: Set # 73, 2.0009-00013 PVCs—Regular Figures. (*Bottom Row*): 2.0009 Drummer with yellow drum sticks; 2.0009 Drummer with red drum sticks and light blue body; 2.0009 Drummer with red drum sticks and dark blue body; 2.0010 Prisoner with dark blue body; 2.0010 Prisoner with medium blue body; 2.0011 Laughing with hand over mouth and pointing. (*Top Row*): 2.0012 Mechanic with white pants and black wrench; 2.0012 Mechanic with dark green pants and dark gray wrench; 2.0012 Mechanic with bright green pants and light gray wrench; 2.0013 Lute with light body and reddish orange lute; 2.0013 Lute with dark body and dark red lute; 2.0013 Lute with yellow lute;

Bottom: Set # 73, 2.0014 - 0020 PVCs—Regular Figures. (*Bottom Row*): 2.0014 Sunbather with yellow trunks and black stripes and light mold; 2.0014 Sunbather with yellow trunks and black stripes and dark mold; 2.0014 Sunbather with white trunks and red stripes; 2.0014 Sunbather with red trunks and black stripes; 2.0014 Sunbather with green trunks and dark mold; 2.0014 Sunbather with green trunks and light mold; 2.0015 Earache with medium blue body; 2.0016 Judge with red gown; 2.0016 Judge with red/orange gown. (*Top Row*): 2.0017 Mirror with red mirror; 2.0017 Mirror with plastic orange mirror; 2.0018 Crying with dark yellow handkerchief; 2.0019 Flower with red flower; 2.0020 Gymnast with light blue body and red tank top; 2.0020 Gymnast with dark blue body and red tank top; 2.0020 Gymnast with light blue body and no tank top; 2.0020 Gymnast with light blue body and yellow tank top.

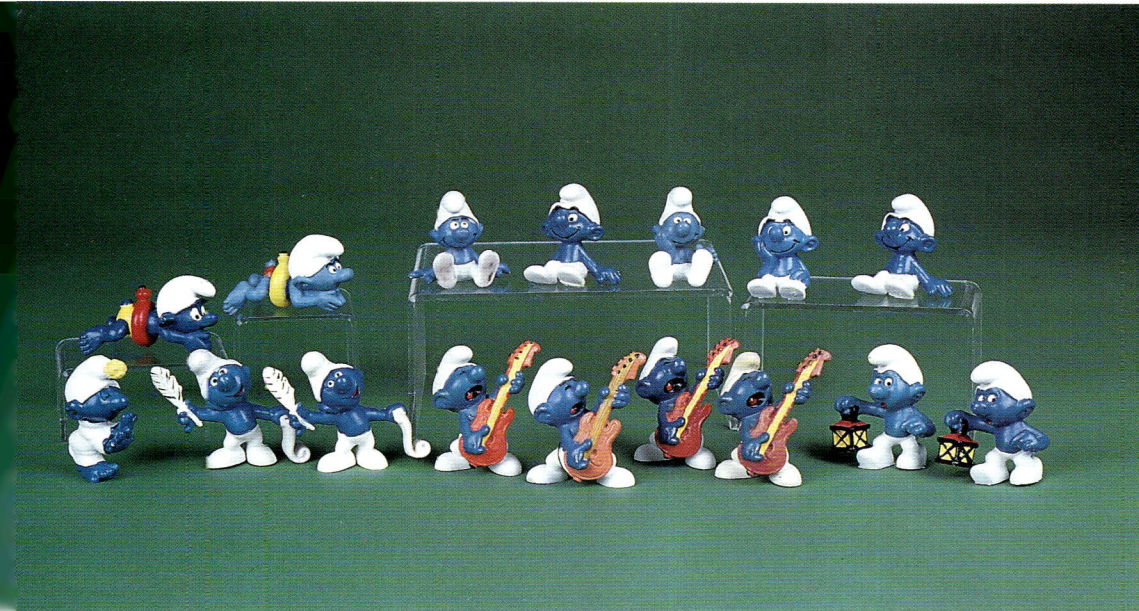

Top: Set # 73, 2.0021 - 0027 PVCs—Regular Figures. (Bottom Row): 2.0021 Sleepwalker with dark blue body; 2.0022 Author with dark blue body; 2.0022 Author with medium blue body; 2.0023 Rock N Roll with brownish/orange guitar; 2.0023 Rock N Roll with medium shiny blue body and light orange guitar; 2.0023 Rock N Roll with dark orange guitar and dark blue body; 2.0023 Rock N Roll with medium orange guitar; 2.0024 Watchman with light blue body; 2.0024 Watchman with dark blue body. (Top Row): 2.0025 Swimmer with red inner tube; 2.0025 Swimmer with yellow inner tube; 2.0026 Sitting with hands flat on both sides and medium shiny body; 2.0026 Sitting with hands flat on both sides and dark dull blue body; 2.0027 Thinker with light blue body and small mold; 2.0027 Thinker with medium blue body and big mold and red dot on bottom; 2.0026 Sitting with hands flat on both sides and large mold, bright blue body.

Center: Set # 73, 2.0028 - 0034 PVCs—Regular Figures. (Bottom Row): 2.0028 Gardner with green apron; 2.0029 Money with dark coin; 2.0029 Money with light coin; 2.0030 Torchbearer with red shorts; 2.0030 Torchbearer with white shorts; 2.0031 Postman with light mold and heart on envelope; 2.0031 Postman with dark mold and heart on envelope; 2.0031 Postman with holly on envelope. (Top Row): 2.0032 Ice Hockey with dark yellow helmet; 2.0032 Ice Hockey with light yellow helmet; 2.0033 Clown with medium blue body; 2.0033 Clown with dark blue body; 2.0034 Smurfette with bright yellow hair; 2.0034 Flirting Smurfette with dark yellow hair; 2.0034 Flirting Smurfette with bright yellow hair.

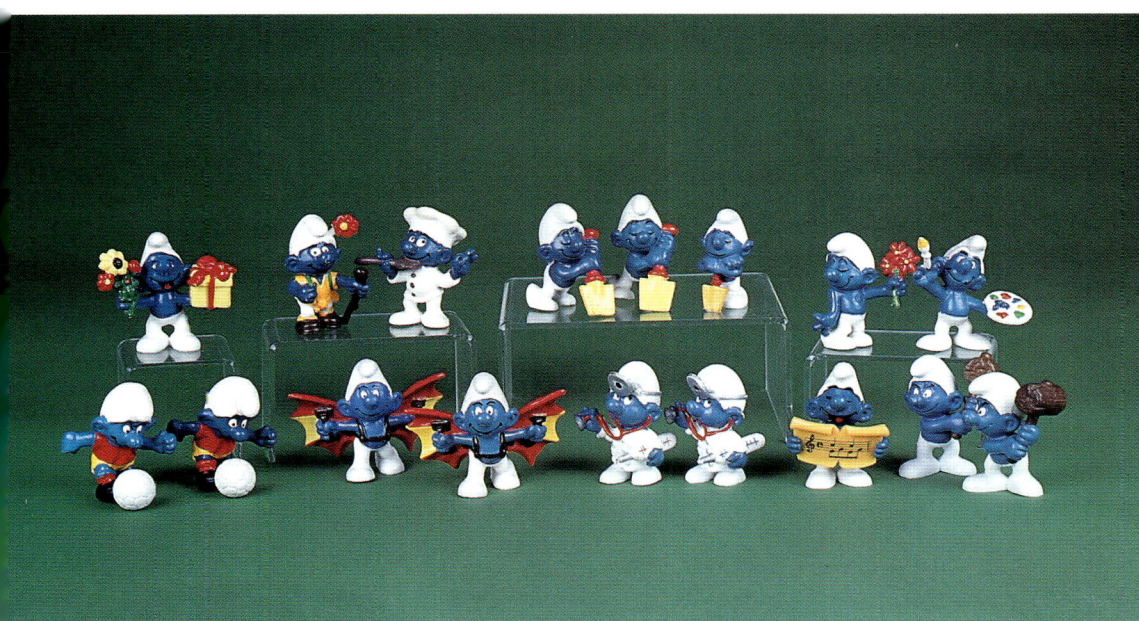

Bottom: Set # 73, 2.0035 - 0045 PVCs—Regular Figures. (Bottom Row): 2.0035 Soccer Footballer light mold; 2.0035 Soccer Footballer dark mold; 2.0036 Hang Glider with light mold; 2.0036 Hang Glider with dark mold; 2.0037 Doctor with red lines on thermometer; 2.0037 Doctor with blue lines on thermometer; 2.0038 Singer; 2.0039 Mallet with light brown mallet; 2.0039 Mallet with dark brown mallet. (Top Row): 2.0040 Gift; 2.0041 Hiker; 2.0042 Chef; 2.0043 Digger with dark mold; 2.0043 Digger with light yellow shovel; 2.0043 Digger with small mold; 2.0044 Lover; 2.0045 Artist.

Top: Set # 73, 2.0046 - 0051 PVCs—Regular Figures. *(Bottom Row):* 2.0046 Emperor with light blue mold; 2.0046 Emperor with dark blue mold; 2.0046 Emperor with yellowish gown; 2.0046 Emperor with white hat and gold crown; 2.0047 Trumpeter with light body and yellow trumpet; 2.0047 Trumpeter with light body and tan trumpet; 2.0047 Trumpeter with light body and mustard color trumpet; 2.0047 Trumpeter with dark body and dark mustard color trumpet. *(Top Row):* 2.0048 Flautist with maroon shirt; 2.0048 Flautist with red/orange shirt; 2.0048 Flautist with red shirt and dark blue body; 2.0049 Tennis Star with light blue body; 2.0049 Tennis Star with dark blue body; 2.0050 Pointing with light blue body; 2.0050 Pointing with dark blue body; 2.0051 Bowler with light orange/reddish ball; 2.0051 Bowler with red ball.

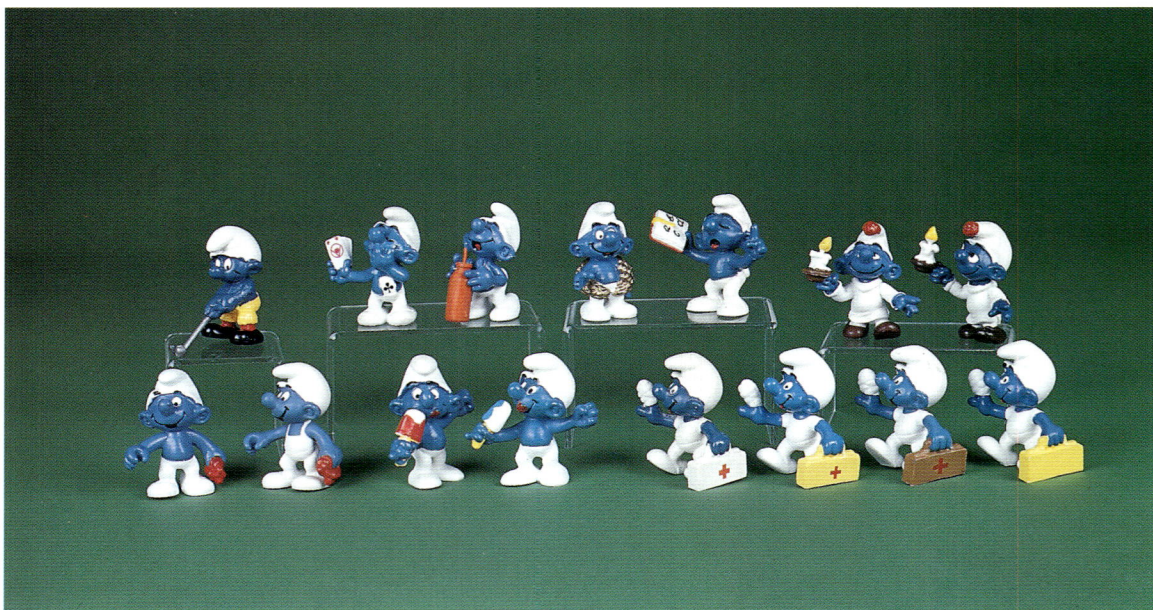

Center: Set # 73, 2.0052 - 0060 PVCs—Regular Figures. *(Bottom Row):* 2.0052 Cleaner with no shirt; 2.0052 Cleaner with white shirt; 2.0053 Ice Lolly with red popsicle; 2.0053 Ice Lolly with blue and white popsicle; 2.0054 First Aid with white case with red cross; 2.0054 First Aid with yellow case with red cross; 2.0054 First Aid with brown case with red cross; 2.0054 First Aid with yellow case with no cross. *(Top Row):* 2.0055 Golfer; 2.0056 Card Player; 2.0057 Thirsty; 2.0058 Champion; 2.0059 Teacher; 2.0060 Candle with light mold; 2.0060 Candle with dark mold.

Bottom: Set # 73, 2.0061 - 0072 PVCs—Regular Figures. *(Bottom Row):* 2.0061 Bandleader; 2.0062 Telephone with dark body; 2.0062 Telephone with light body; 2.0063 Tailor; 2.0064 Toothbrush; 2.0065 Rugby; 2.0066 Cricket. *(Top Row):* 2.0067 Congratulations; 2.0068 Football Player (Soccer player); 2.0069 Jungle with light brown body; 2.0069 Jungle with dark brown body; 2.0070 Harp; 2.0071 Flying; 2.0072 Trumpet Player.

Top: Set # 73, 2.0073 - 0081 PVCs—Regular Figures. *(Bottom Row)*: 2.0073 Cook with light body; 2.0073 Cook with dark body; 2.0074 King with light body; 2.0074 King with dark body; 2.0075 Quack by Schleich; 2.0076 Courting with light body; 2.0076 Courting with dark body. *(Top Row)*: 2.0077 Naughty with light body; 2.0077 Naughty with dark body; 2.0077 Naughty with bright blue body; 2.0078 Beer; 2.0079 Jolly; 2.0080 Biscuit; 2.0081 Tyrolese with light body and orange pipe; 2.0081 Tyrolese with dark body and red pipe.

Center: Set # 73, 2.0082 - 0092 PVCs—Regular Figures. *(Bottom Row)*: 2.0082 Shy; 2.0083 Hammer; 2.0084 Handstand with yellow shorts; 2.0084 Handstand with red shorts; 2.0085 Cushion with light orange cushion; 2.0085 Cushion with dark orange cushion; 2.0086 Present with orange bow; 2.0086 Present with red bow. *(Top Row)*: 2.0087 Woodcutter; 2.0088 Traveler; 2.0089 Painter; 2.0090 Jester with olive green outfit; 2.0090 Jester with bright green outfit; 2.0091 Skier; 2.0092 Conductor.

Bottom: Set # 73, 2.0093 - 0100 PVCs—Regular Figures. *(Bottom Row)*: 2.0093 Tennis Player 2 with a light body; 2.0093 Tennis Player 2 with a dark body; 2.0094 Bookworm; 2.0095 Oboist; 2.0096 Sledgehammer with light brown hammer; 2.0096 Sledgehammer with dark brown hammer; 2.0097 Injured with a light brown walking stick; 2.0097 Injured with a dark brown walking stick. *(Top Row)*: 2.0098 Ballerina with light body; 2.0098 Ballerina with dark body; 2.0099 Head Cook with light body; 2.0099 Head Cook with dark body; 2.0100 Cake with an orange cake; 2.0100 Cake with a red cake; 2.0100 Cake with a dark red cake.

Top: Set # 73, 2.0101 - 0107 PVCs—Regular Figures. *(Bottom Row):* 2.0101 Angler with dark mold and light brown pole; 2.0101 Angler with light mold and dark brown pole; 2.0101 Angler with separate yellow and orange pole; 2.0101 Angler with separate dark brown pole; 2.0102 Archer with bow and arrow attached; 2.0102 Archer with bow and arrow separate; 2.0103 Scholar. *(Top Row):* 2.0104 Pirate with light mold; 2.0104 Pirate with dark mold; 2.0105 Scot with light mold; 2.0105 Scot with dark mold; 2.0106 Hunter; 2.0107 Carnival with dark mold; 2.0107 Carnival with light mold.

Center: Set # 73, 2.0108 - 0114 PVCs—Regular Figures. *(Bottom Row):* 2.0108 Sauna with light mold; 2.0108 Sauna with dark mold; 2.0108 Sauna with light mold and bright yellow soap; 2.0109 Knight with light mold; 2.0109 Knight with dark mold; 2.0110 Hairdresser with light mold; 2.0110 Hairdresser with dark mold. *(Top Row):* 2.0111 Cupid; 2.0112 Carpenter with light mold; 2.0112 Carpenter with dark mold; 2.0113 Baker; 2.0114 Conjuror with light mold; 2.0114 Conjuror with dark mold.

Bottom: Set # 73, 2.0115 - 0122 PVCs—Regular Figures. *(Bottom Row):* 2.0115 Lion Tamer; 2.0116 Alchemist with light blue mold; 2.0116 Alchemist with dark blue mold; 2.0117 Cornucopia with light blue mold; 2.0117 Cornucopia with dark blue mold; 2.0118 Umbrella with light mold; 2.0118 Umbrella with dark mold. *(Top Row):* 2.0119 Smurferman on yellow base; 2.0119 Smurferman on white base and light blue mold; 2.0119 Smurferman on white base and dark blue mold; 2.0120 Frogman with light mold and bright orange scuba; 2.0120 Frogman with dark mold and dark orange scuba; 2.0121 Ice Skater with light blue mold; 2.0121 Ice Skater with dark blue mold; 2.0122 Cowboy with dark blue mold; 2.0122 Cowboy with light blue mold.

Top: Set # 73, 2.0123 - 0129 PVCs—Regular Figures. *(Bottom Row):* 2.0123 Policeman with black uniform; 2.0123 Policeman with white uniform; 2.0124 Santa; 2.0125 Heart with EIN HERZ FUR KINDER; 2.0125 Heart with DE TOUT MOR COEUR; 2.0126 Roller Skating with dark blue mold; 2.0126 Roller Skating with light blue mold. *(Top Row):* 2.0127 Superman; 2.0128 Amour with a yellow bow; 2.0128 Amour with a tan bow; 2.0128 Amour with an orange bow; 2.0129 Baseball Batter with light cream color bat; 2.0129 Baseball Batter with next light peach color bat; 2.0129 Baseball Batter with light tan color bat; 2.0129 Baseball Batter with dark tan color bat; 2.0129 Baseball Batter with dark brown color bat.

Center: Set # 73, 2.0130 - 0137 PVCs—Regular Figures. *(Bottom Row):* 2.0130 Graduation with blue cap; 2.0130 Graduation with light maroon cap; 2.0130 Graduation with dark maroon cap; 2.0131 French Fries with dark blue mold; 2.0131 French Fries with light blue mold; 2.0132 American Football; 2.0133 Field Hockey with red and white shirt; 2.0133 Field Hockey with green shirt. *(Top Row):* 2.0134 Judo with light blue mold; 2.0134 Judo with dark blue mold; 2.0135 Tennis Smurfette with light blue mold; 2.0135 Tennis Smurfette with dark blue mold; 2.0136 Halloween with Pumpkin with light body and pumpkin face; 2.0136 Halloween with Pumpkin with dark body and pumpkin face; 2.0136 Halloween with Pumpkin with no face on pumpkin; 2.0137 Surfer.

Bottom: Set # 73, 2.0138 - 0144 PVCs—Regular Figures. *(Bottom Row):* 2.0138 Gardener with rake and orange and tan hat; 2.0138 Gardener with rake and tan hat; 2.0138 Nurse with white uniform and blue apron; 2.0138 Nurse with blue uniform and white apron; 2.0139 Secretary with white dress; 2.0139 Secretary with pink dress; 2.0140 Captain with white jacket and red pants; 2.0140 Captain with blue jacket and white pants. *(Top Row):* 2.0142 Mermaid with olive green tail; 2.0142 Mermaid with dark green tail 2.0142 Mermaid with blue and silver tail; 2.0142 Mermaid with blue and silver tail and no necklace; 2.0143 CB Operator; 2.0144 Indian with black and white feathers; 2.0144 Indian with light blue mold and multicolored feathers; 2.0144 Indian with dark blue mold and multicolored feathers.

Top: Set # 73, 2.0145 - 0157 PVCs—Regular Figures. *(Bottom Row)*: 2.0151 Graduate Smurfette; 2.0152 Miller; 2.0153 Santa Smurfette; 2.0154 Patrol Crossing; 2.0155 Traffic; 2.0156 Valentine Smurfette; 2.0157 Grouchy; 2.0150 Australian Football. *(Top Row)*: 2.0149 Cheerleader with bright white uniform and medium green base; 2.0149 Cheerleader with dark white uniform and medium green base; 2.0149 Cheerleader with white uniform on dark green base; 2.0145 Farmer with sycle; 2.0146 Baseball Catcher; 2.0147 Cowgirl; 2.0148 Bricklayer with no shirt; 2.0148 Bricklayer with a white shirt.

Center: Set # 73, 2.0158 - 0167 PVCs—Regular Figures. *(Bottom Row)*: 2.0158 Hamburger; 2.0159 Violin; 2.0160 Apple #1 Teacher with red apple; 2.0160 Apple #1 Teacher with green apple; 2.0161 Clumsy; 2.0162 Waiter with brown tray; 2.0163 Waiter with silver tray; 2.0164 Soccer Smurfette with light orange shirt; 2.0164 Soccer Smurfette with dark orange shirt. *(Top Row)*: 2.0164 Papa with orange Lab Glass; 2.0164 Papa with tan Lab Glass; 2.0165 Greedy with light blue mold; 2.0165 Greedy with dark blue mold; 2.0166 Baseball Pitcher; 2.0167 Indian Smurfette with light color outfit; 2.0167 Indian Smurfette with dark color outfit.

Bottom: Set # 73, 2.0168 - 0180 PVCs—Regular Figures. *(Bottom Row)*: 2.0168 Smurfette Jumping Rope with thin rope; 2.0168 Smurfette Jumping Rope with thick rope; 2.0169 Hot Dog; 2.0170 Football Quarterback with #3; 2.0171 Handy; 2.0172 Jogger wearing orange jacket with white pants; 2.0172 Jogger wearing a red jacket with white pants; 2.0173 School Girl. *(Top Row)*: 2.0174 Papa with book; 2.0175 Clockwork; 2.0176 St. Patrick's Day with green outfit; 2.0176 St. Patrick's day with red and black outfit; 2.0177 Thanksgiving Day with turkey on platter; 2.0178 Tracker; 2.0179 Baby with Rattle with light white outfit; 2.0179 Baby with Rattle with dark white outfit; 2.0180 Papa with pizza.

Top: Set # 73, 2.0181 - 0193 PVCs—Regular Figures. *(Bottom Row)*: 2.0181 Gargamel with flesh colored skin; 2.0181 Gargamel with peach colored skin; 2.0181 Smurfette with comb and mirror; 2.0183 Aerobic Smurfette; 2.0184 Bullfighter; 2.0185 Sailor; 2.0186 Baseball Smurfette; 2.0187 Handy Plumber. *(Top Row)*: 2.0188 Majorette with bright pink boots; 2.0188 Majorette with purple boots; 2.0189 Smurf with Dust Pan in light blue mold; 2.0189 Smurf with Dust Pan in dark blue mold; 2.0190 Smurfette with Ice Cream; 2.0191 Brainy Referee; 2.0192 Smurfette with Baby; 2.0193 Smurf with Mop and Pail.

Center: Set # 73, 2.0194 - 0208 PVCs—Regular Figures. *(Bottom Row)*: 2.0194 Smurfette Jogger; 2.0195 Graduation with Diploma; 2.0196 Smurfette with Thanksgiving Pie; 2.0197 Thanksgiving with ear of Corn; 2.0198 Smurfette Witch; 2.0199 Smurf with Gargamel mask with light blue body; 2.0199 Smurf with Gargamel mask with dark blue body; 2.0200 Christmas Smurfette; 2.0201 Christmas Smurf with lantern. *(Top Row)*: 2.0202 Baby with Rattle wearing pink outfit and light blue mold; 2.0202 Baby with Rattle wearing pink outfit and dark blue mold; 2.0203 Baby with Rattle wearing blue outfit; 2.0204 Smurfette with Flute; 2.0205 Baby with Teddy Bear with light blue mold; 2.0205 Baby with Teddy Bear with dark blue mold; 2.0206 Baby with Ice Cream; 2.0207 Christmas Smurf with a Gift; 2.0208 Christmas Smurfette.

Bottom: Set # 73, 2.0209 - 0221 PVCs—Regular Figures. *(Bottom Row)*: 2.0209 Dentist; 2.0210 Smurfette Golfer; 2.0211 Smurf basketball; 2.0212 Angel with light blue body; 2.0212 Angel with dark blue body; 2.0213 Devil with light red body; 2.0213 Devil with dark red body. *(Top Row)*: 2.0214 Baby with pastel blocks in pink, yellow and blue; 2.0214 Baby with yellow, blue and red blocks; 2.0214 Baby with yellow, green and red blocks and light blue mold; 2.0214 baby with yellow, green and red blocks and dark blue mold; 2.0215 Baby with car; 2.0216 Smurf in Tuxedo; 2.0218 Baby with Butterfly Rattle; 2.0217 Smurfette in Gown; 2.0219 Surprise Bag in green; 2.0220 Surprise Bag in pink; 2.0221 Surprise Bag in orange.

Top: Set # 73, 2.0222 - 0232 and 2.0401-2.0404 PVCs—Regular Figures. *(Bottom Row):* 2.0222 Smurfette Stewardess; 2.0223 Papa Pilot; 2.0224 Baby with a Bowl and Spoon; 2.0225 Smurf with Accordion; 2.0226 Grandpa; 2.0227 Table Tennis Player; 2.0228 Fitness with light blue mold; 2.0228 Fitness with dark blue mold; 2.0229 Architect Smurf. *(Top Row):* 2.0230 Wild; 2.0231 Hula Smurfette; 2.0232 Gargamel holding pink laboratory flask in right hand; 2.0232 Bully Gargamel pointing finger with right hand; 2.0401 Snappy Smurfling; 2.0402 Slouchy Smurfling; 2.0403 Nat with caterpillar; 2.0404 Sassette Smurfling with orange hair color; 2.0404 Sassette Smurfling with reddish orange hair color.

Center: Set # 73, 2.0405 - 0420 PVCs—Regular Figures. *(Bottom Row):* 2.0405 Puppy in gray color; 2.0405 Puppy in brown color; 2.0406 Smoogle pink rabbit; 2.0407 Chitter gray squirrel; 2.0408 Nanny; 2.0409 Patriot Smurf; 2.0410 Smurfette with Mouse; 2.0411 Azreal. *(Top Row):* 2.0412 Bride; 2.0413 Bridegroom; 2.0414 Video Smurf; 2.0415 Handball; 2.0416 New Football Smurf; 2.0417 Bodybuilder; 2.0418 Gargamel with Raised Hands; 2.0419 Boxer; 2.0420 Scruple.

Bottom: Set # 73, 2.0421 - 0441 PVCs—Regular Figures. *(Bottom Row):* 2.0421 Smurfette with Flower; 2.0422 Tramp; 2.0423 Sitter; 2.0424 Papa with hands behind his back; 2.0425 Gargamel with hands on his hips; 2.0426 Schoolboy; 2.0427 Stoneage Smurf; 2.0428 Stoneage Smurfette; 2.0429 Tattoo Smurf; 2.0430 Viking Smurf; 2.0431 Monk. *(Top Row):* 2.0432 Jockey Smurf; 2.0433 Jockey Smurfette; 2.0434 Slouchy with Cone; 2.0435 Sassette with Cone; 2.0436 Saxophone; 2.0437 Techno Smurf; 2.0438 Mobile Smurf; 2.0439 Azreal Sitting; 2.0440 Sport Swimmer; 2.0441 Sprinter.

Set # 73, 2.0442 - 0499
PVCs—Regular Figures.
(Bottom Row): 2.0442 Inline
Skater; 2.0443 Inline Skater
Smurfette; 2.0444 Disco
Smurf; 2.0445 Disco
Smurfette; 2.0446 Girl with
Baby; 2.0448 Boy with Truck;
2.0447 Girl Bathing; 2.0449
Lead Guitar Player; 2.0450
Bass Guitar Player; 2.0451
Singer. (Top Row): 2.0452
Snowboarder; 2.0453
Snowboarder Smurfette;
2.0454 Soccer Player; 2.0498
Johan; 2.0498 Johan Bully;
2.0499 Pewit with light blue
mold, blue shirt and orange
shoes; 2.0499 Pewit with dark
blue mold, blue shirt and
orange shoes; 2.0499 Pewit
with light blue mold, green
shirt and red shoes.

Set #204 PVCs—Solid Color Molds. Brainy, Judge, Rock N Roll, Courting, Beer, and Present.

Set #178 Quick Promotional
PVCs. Six promotional Smurfs
were given out by the Quick
Fast Food Restaurants in France:
Rocking chair, Smurf in a log car,
Vanity table, Smurfette in bath
tub, Smurf in bed, and Smurfette
in plug car.

Set #59, 1-5 Radios AM only radios provided by Power Tronic highlight this category: Square Smurf Radio with Headset, Smurf Face Radio, and Smurfette Square Radio & Headphones. In addition, there is a Smurf plastic radio s called Smurf Radio and Speakers and a Square Schlumpf Radio with Headset.

Category #232 Rain Gear. Galoshes, rain coat, and the rare umbrella shield the little Smurfs and Smurfettes from the weather.

Set #5 Records—33 & 45. Music provides the vehicle for the All Star Show in this category. It is Smurf party time with this assortment of albums.

Set #5 Records—33 & 45. Smurfing sing-along continues its tune with this second assortment of albums.

Set #169 Rings. Love is blue when wearing a Smurf ring! This assortment is half metal and half silver, made by Peyo.

NECKLACE & RING

HANDCRAFTED IN 14K GOLD ELECTROPLATE, SUITABLE FOR CHILDREN 4 YRS. OF AGE AND OLDER.

Howard Eldon, Ltd.
13901 Saticoy St.
Van Nuys, CA 91402
(213) 873-7711 989-4444
Telex: 651379
85-001

SMURF

© Peyo   SEPP
LICENSED BY WALLACE BERRE & CO
VAN NUYS, CA

Set #194 Roller Skates. Tying these on made for a smurfin time!

Set #103 School and Office Supplies. Don't rule out finding the perfect eraser or office supplies in this assortment!

Set #114, 1, 2 See N Say. Two See N Says are pictured: Sport Scene and Talking Telephone by Mattel.

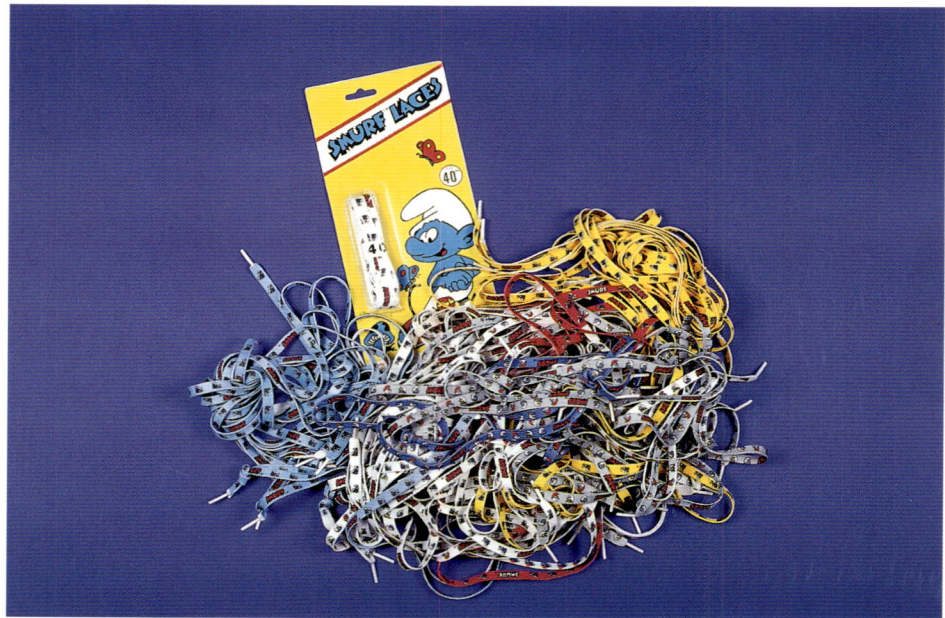

Set #128 Shoe Laces. Selecting the perfect set of laces to match one's outfit is easy with this broad assortment of shoe laces.

Set #137 Shoes. If "the shoe fits" adage were real, many a collector would be Smurf-arella.

Set #217 Shopping Bags. Carrying home one's latest "find" is highlighted by the addition of these shopping bags.

Set #21, 1-6 Shrinky Dinks. Six sets of Shrinky Dinks by Milton Bradley and Colorforms shrink this category down to size. Included are: Smurf Carrying Knapsack Collectibles Figure Kit, Smurf Playhouse Delux Shrinky Dink Set, Superman Collector Set, Papa Collector Set, and Jumping Smurf Collector Set.

Set #26 Silverware. Made by Danara, I.M.P.S., and Peyo, these individual pieces of children's silverware add the perfect touch to the dinner hour.

Set #228 1-2 Squeeze Bottles. Peyo and Applause have manufactured two different colored plastic bottles, called Walk Pot, a yellow and a blue. Photograph shows mint in package examples.

Set #267 1-2 PVC Stamp & Ink Pads—PVC Figures. Stampers like an Indian and a Mailman add fun to the writing venture!

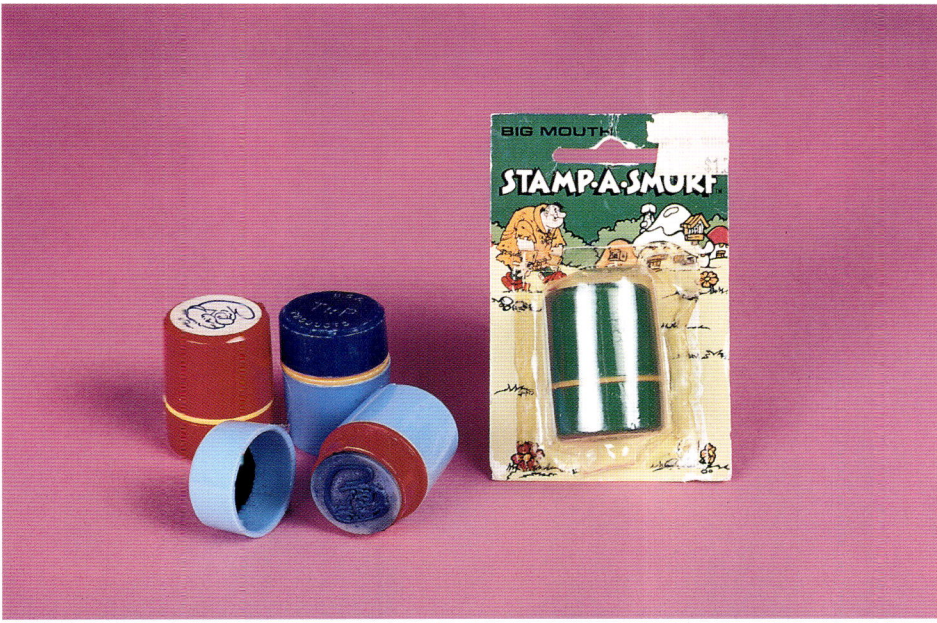

Set #98 Stampers. Ink stampers make the Naughty Smurf appear more often and the other Smurfs appear everywhere. Made by Marburger, these ink stampers come in an enclosed container for repeated usage.

Set #98 Stampers + #115, 6 Store Display. Number on the stamper box coincides with the listing number within the category.

Set #240 Stickers—Miscellaneous. Whether applying scratch n' sniff stickers, puffy, scented or just plain sticker fun, this category provides enough glue to hold one's imagination.

Set #185 Stickers— National Gas Station. Stickers galore is the name of this category!

Set #240 Stickers—Miscellaneous. Larger Smurf stickers provide twice the fun.

Set #240 Stickers—Miscellaneous. The fast food restaurants in France, Quick Restaurants, provided this assortment of Smurf stickers.

Set #118 Stickers—Puffy. The rage of the 1980s didn't miss a beat with Smurf collectibles. Puffy stickers came in an assortment of shapes, colors, and sizes.

Set #119 Stickers—Scratch 'n Sniff. Shown is an assortment of scratch 'n sniff and scented stickers.

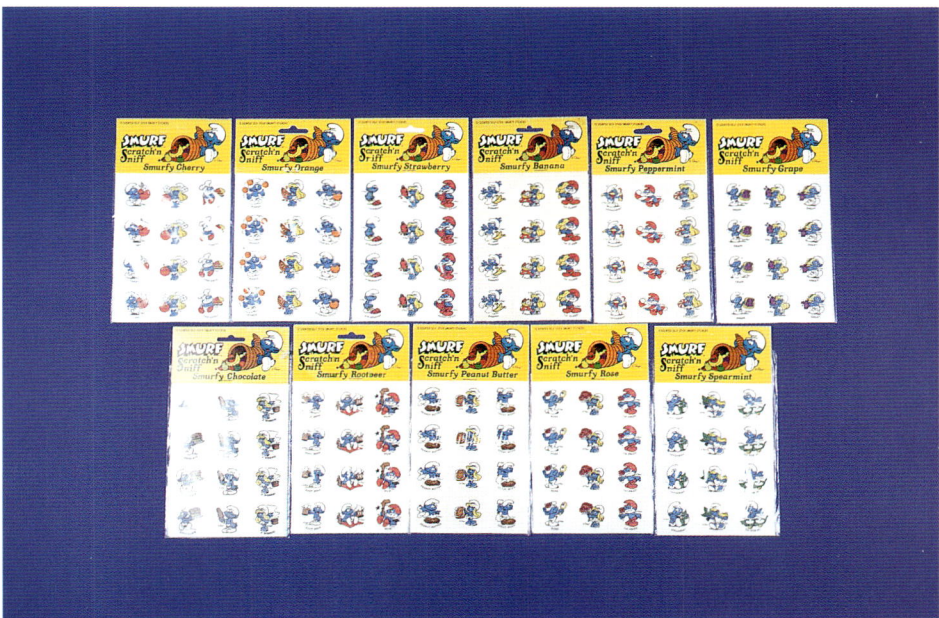

Set #189 Stickers—Vitamins. A Smurf vitamin a day keeps one smurfin' along, plus the added incentive of Smurf flicker stickers in each package brings smiles to the littlest of Smurf or Smurfette.

Set #115 Store Displays and Display Cases. This display for mini plushes is music to the ears of a serious Smurf collector.

Set #115 Store Displays and Display Cases. From Germany, this Kinder Egg Surprise Display and advertising card illustrate the colorful appeal of Smurfs.

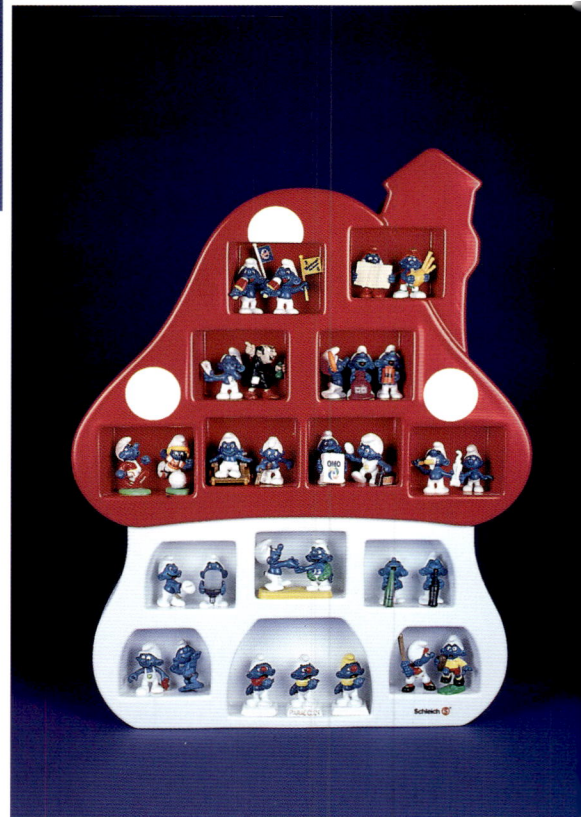

Set #115 Store Displays and Display Cases. Mushroom Display with assorted figurines.

Set #115, 13 Store Displays and Display Cases.
Super Smurf Display with assorted figurines.

Set #115 Store Displays and Display Cases.
Smurf Collectors Center with assorted
figurines.

Set #115 Store Displays and Display Cases.
Super Smurf Collectors Center with
assorted figurines.

Set #115 Store Displays and Display Cases.
Store Display and McDonald's figurines:
McDonald's 25th Anniversary Junior Tute Display
from Germany and Smurfs 40th Anniversary set
of ten figurines given out by McDonald's.

Set #115 Store Displays and Display Cases + Set #86.
Fourteen toppers are displayed along with the store display
from 1995. (*Top Row*): Smurf Toppers holding Jelly Beans.
(*Bottom Row*): Toppers holding M & M candies. Toppers
include: Baby Smurf riding on Fapa's back, Gargamel &
Azreal, Jack N A Box Smurf, Magician, Papa laying under
mushroom, Papa Santa, Smurf and Smurfette on a turtle,
Smurf in a package, Smurf playing a drum, Smurf stringing
Christmas lights, Smurfette rollerblading, Smurf dancing
around a tree, and Vacation Smurf.

Set #115 Store Displays and Display
Cases. Olympic Stadium Display with
Kinder Smurfs.

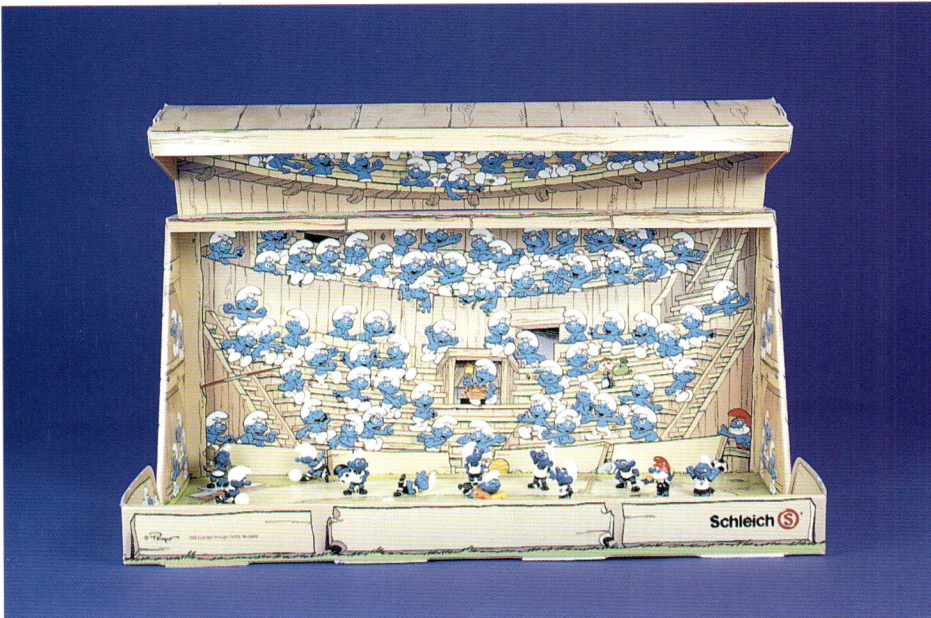

Set #115, 5 Store Displays and Display Cases. Papa Smurf Clip-On. Two different clip-ons are pictured: Papa Smurf with a yellow scarf and Papa Smurf with a red scarf.

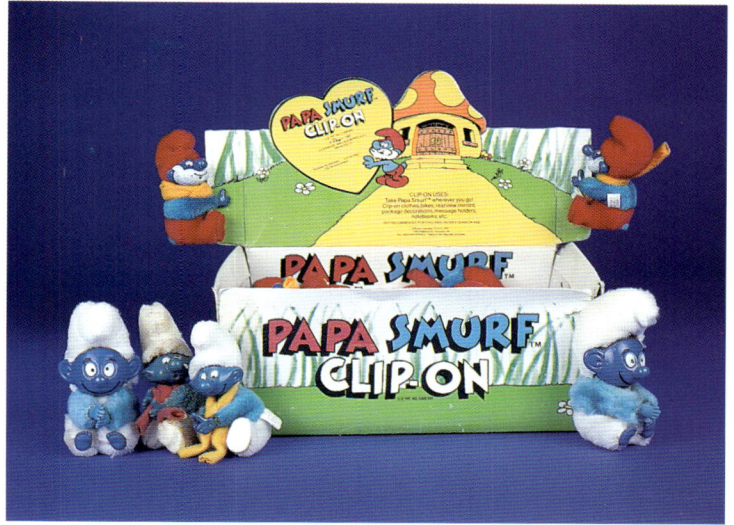

Set #115, 5 Store Displays and Display Cases. Papa Smurf Clip-On. Photo illustrates the assorted Smurfs used in the clip-on promotion: Papa Smurf and Baby Smurf with different scarf colors.

Set #115, 3 Store Displays and Display Cases. A Smurf Christmas display with assorted Christmas Smurfs.

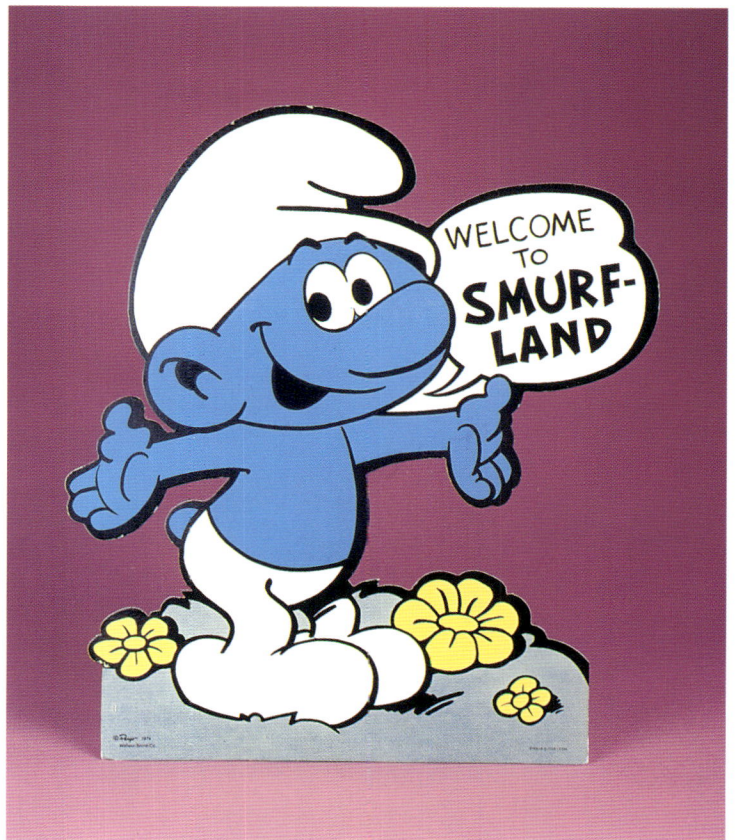

Set #115, 12 Store Displays and Display Cases. Welcome to SMURFLAND.

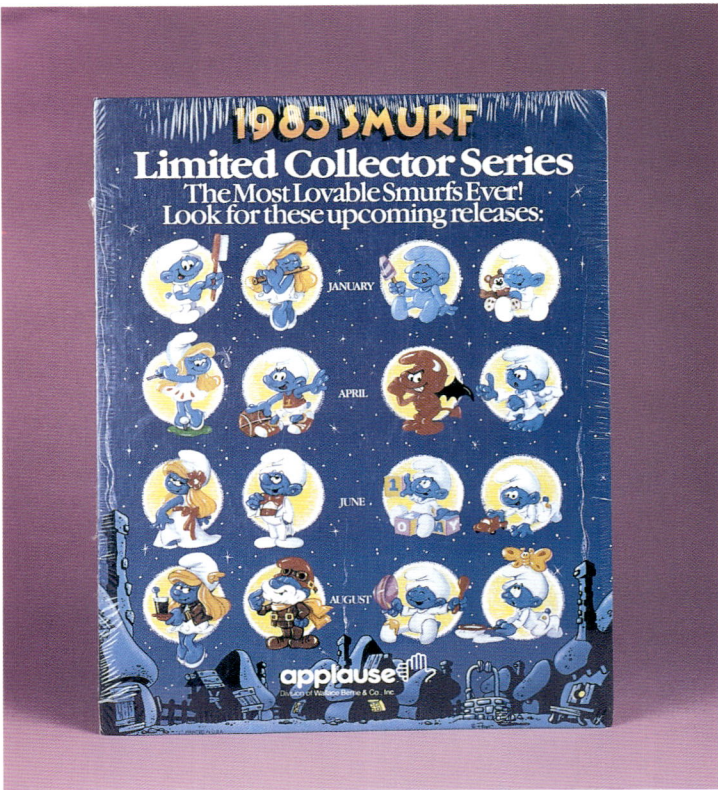

Set #115, 16 Store Displays and Display Cases. 1985 Smurf Limited Collector Series Display.

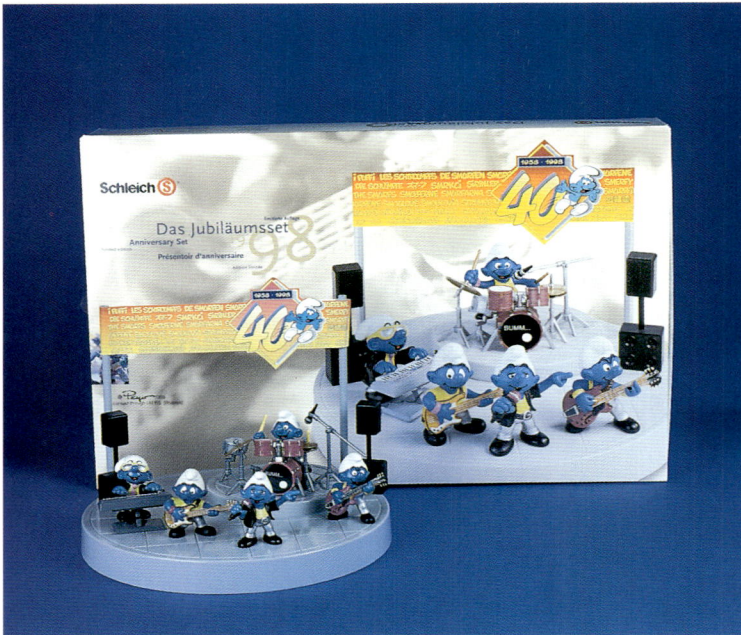

Set #115 Store Displays and Display Cases. Smurf's 40th Anniversary is celebrated with this 40th Anniversary Band Display.

Set #152 Sun Catchers. Falling off a log is not in store for this Smurf. He's more inclined to catch a few rays of sunshine with this sun catcher.

Set #36, 6 Super-Super Playsets. Latex Tree Stump Playset by Schleich.

Set #36, 2, 3, 4, 7, 8 Super-Super Playsets. Smurfs Super-Super Playsets include the Moon Landing set, Gargamel's Lab, Smurfette's Bedroom, Reissued Western Chuck Wagon, and the Drummer. See Set #70 for Regular Playsets.

Set #36 Super-Super Playsets. A Smurf latex Castle leads this category!

Set #127 Super Smurfs. Sixteen Super Smurfs pictured are: Sledder, Chimney Sweep, Tricycle, Skateboard, Skier, Gardener, Angler with light body from West Germany, Angler with dark body from Hong Kong, Signbearer, Butterfly Catcher with yellow net and red decoration, Butterfly Catcher with yellow net and no decoration, Car driver in red car, Car driver in orange car, Car driver in yellow car, Gargamel and Azreal, and Smurf in a Cage pictured in box.

Set #127 Super Smurfs. Fourteen additional Super Smurfs pictured include: Chain gang, Hobby Horse with light brown horse, Hobby Horse with dark brown horse, Wind surfer, Fireman with yellow jacket from Hong Kong, Fireman with black jacket from Germany, Photographer, Go-Kart, Row boat with light brown boat from Hong Kong, Row boat with dark brown boat from Germany, School desk, Rocking Horse, Volleyball Smurf, and Airplane Smurf in box.

Set #127 Super Smurfs. Sixteen additional Super Smurfs pictured include: Blackboard Smurf with numbers on the blackboard, blackboard Smurf with drawings on the blackboard, Lawn mower Smurf, Smurfette with shopping cart, Papa in rocking chair, Piano Smurf, Scooter, Motor cycle with green shirt, Motorcycle with yellow bike suit, Smurf in a log car, Helicopter, Vanity table Smurfette, Bath tub with light blue Smurf, Bath tub with dark blue Smurf, and Bicycle Smurfette.

Set #127 Super Smurfs. Fourteen additional Super Smurfs pictured include: Trapeze, Smurfette in kitchen, Artist with easel, Smurf in bed, Smurfette in plug car, Lifeguard with yellow flag, Lifeguard with red flag, National gas station, High diver, Tea set, River raft, Jokey with trick box with dark color face, Jokey with trick box with flush face color, and Stork delivering baby.

Set #127 Super Smurfs. Sixteen additional Super Smurfs pictured include: Computer Smurf, Cyclist, Kayak, Discus Thrower, Fencer, Ice Hockey with net, Pole Vaulter, Weight lifter with light brown outfit, Weight lifter with dark brown outfit, Weight lifter with pink outfit, Boxer, Bars Gymnast, Rings gymnast pictured in box, Hurdler, Basketball player, and Keyboarder.

Set #136 Tattoos. The selection of Smurf seems to stick around with this assortment of 24 different tattoos.

Set #226 Tea Sets. Party time begins with a proper tea set: Smurf Tea and Serve Set or Los Pitufos set.

Set #121, 1-5 Telephones. Five different telephones from H-G Industries, Eurostil, Irwin, and Enterprise highlight this category, starting with a 6 ½" Smurf standing on a yellow base telephone, 6 ½" Smurf standing on a red base telephone, Schtroumpf phone, talking Flip phone, and Smurf & Mushroom House telephone.

Set #226 Tea Sets. Smurfette Cook N Serve Tea Set (yellow and green box) and Smurf Cook N Serve Tea Set (blue box) both make role playing a memorable experience for children of all ages around the world.

Set #138 Trading Cards. You could either choose one super trading card and one stick of bubble gum or select the larger packages of 36 different trading cards and forgo the gum. Either selection makes Smurf collecting stick in your mind!

Set #55 Trains. Beginning with a Choo-Choo Train with headlights by Durham and continuing with a Crib Train, Mini Train Set, Smurf Land Wind-up Train, Smurf Express (not pictured), Mini Yellow Engine, and ending with a small rubber Smurf train by Peyo, the Smurfland Express continues to roll.

Set #38 Trays—Lap. Wallace Berrie and Willow provide the perfect metal tray for Smurf fun, starting with Three Smurfs Eating, Smurf Giving Smurfette Flowers, or Sport Smurf. They all provide the perfect spot for a Smurf break.

Set #162 Trinket Boxes. Saving one's treasure in a trinket box is a smurfy idea, selecting from Let's Be Friends, Love is Something You Share...Not Own, Gargamel and Azreal, Papa Looking Troubled, or Schoolgirl.

Set #200 Video Movies. Watching the perfect Smurf adventure is not far off with this selection of videos: *The Smurfs and the Magic Flute*, *L'Isola Del Tesoro*, *Gargamel's Giant*, *Baby's First Christmas*, *The Secret of Shadow Swamp*, *Village Tales*, *Pussywillow Pixies*, *Never Smurf Off...*, *The Whole Smurf..*, *Cartoon All-Star*, and *The Smurfs*.

Set #97, 1-6 View Master and Reels. View Master International has produced at least six examples of Smurf viewing: Flying Smurf Gift Set, Smurfette, Traveling Smurf, Show Beam Smurphony in C, the Smurf's Apprentice Show Beam Cartridge (not pictured), and the Smurf Theater.

Set #150 Wall Hanging Pictures & Plaques and Set #151 Wall Mirrors.
An assortment of framed wall mirrors and plaques is pictured.

Set #150 Wall Hanging Pictures & Plaques. Porcelain wall hangings add
the perfect touch.

Set #150 Wall Hanging Pictures & Plaques. Smurfette and Smurf add
just the right touch for a perfect spot on one's walls.

Set #150 Wall Hanging Pictures & Plaques. All
Smurfs are bashful when it comes to love! This
wall hanging is the loveliest of the assortment.

Set #85 Wallets and Coin Purses. Assorted varieties of wallets and coin purses hold this category together!

Set #93 Watches. Time smurfs along with these watches. Made by Bradley Time, Royal, Durham, Frontier and Peyo, Smurf collectors watch for the latest digital product.

Set #93 Watches. This assortment of digital and "old timey" windup
watches makes one say it is, "Smurf Time" when asked.

Set #205, 1-3
Weebles. Weebles
wobble, but they don't
fall down! Smurfette,
Papa, and Brainy
Smurf prove the rule.

Set #81 Wind-up Toys. This fast moving category includes: Fun House wind-up, the Smurf Chase, Smurf with a Wheelbarrow, Smurf on a Train Cart, Smurf Runabout Red Car, Airplane, 3 Smurfs Jumping Rope, Smurf on a Swing, Smurfs on a Teeter Totter, Smurf Swimmer and Smurf with a Pull Cart. These wind-ups, some of which are white knob style, add hours of fun to the role playing games.

Set #107 Wind-up Walkers. White knob wind-ups add Smurf appeal to the parade of toys with Papa, Smurfette, Smurf, Smurf with a Present, Gargamel, Flying Smurf, Drummer Smurf, Guitar Smurf and Smurf Play a Trumpet.

Set #255 Window Transfers. Adding a window transfer to a pane of glass provides just the right class.

Set #75 Zipper Pulls. These zipper pulls add just the right tug to the sweater, jacket, or heart to make them a zippy collectible!

Alphabetical Listing of Sets

Set 57: Arts & Craft Activities

ID #	ITEM NAME	MANUFACTURER	ISSUE DATE	COLOR	TYPE	SIZE	COUNTRY	VALUE	DESCRIPTION
057/001	Magic Rub-It-Ups	Hasbro	1/1/83	Blue	Rub-It-Ups		U.S.A.	$18.00-25.00	A blue mushroom frame comes with paper, 12 double-sided picture plates, 8 jumbo crayons, Smurfette crayon sharpener, and Papa Smurf crayon pusher with tracing crayon.
057/002	Papa Smurf's Crayon Keeper	Hasbro	1/1/83	Brown	Plastic Keeper		U.S.A.	$15.00-20.00	The box shows Papa standing next to a tree stump that holds crayons. Included are 24 jumbo crayons, Papa Smurf figure that's a pencil sharpener, and a drawing stencil.
057/003	Zeichen-Schablonen Drawing Stencil	Apollo	1/1/87	Yellow	Plastic Stencil		Germany	$3.00-4.00	The stencil sheet is clear yellow plastic. It has whole body Smurfs.
057/004	Smurf Doodle Bag	Campus Craft			Doodle Bag		Canada	$11.00-15.00	There are 6 posters and 6 color pens in the bag. The posters are a beach scene, poetry scene, flirting scene, space scene, superman scene and cowboy scene.
057/005	Smurfette Figure Painting Kit	Avalon	1/1/82	White	Vinyl Figure	5 ½" high	U.S.A.	$7.50-10.00	Smurfette is made of vinyl. She comes with 5 non-toxic acrylic paints, ceramic glaze and a brush. Smurfette is standing in a flirting pose.
057/006	Smurf, Papa And Smurfette Light Catchers	Avalon	1/1/82	Clear	Light Catchers		U.S.A.	$6.00-8.00	There are 3 lie catchers. Smurfette is standing in flowers smelling one. Papa is playing a drum. A Smurf is standing pointing with 3 mushrooms by his feet. Included are 4 transparent paints and a brush.
057/007	Peg Desk	Avalon	1/1/83		Peg Desk	13"x 13"	U.S.A.	$25.00-35.00	The Smurf peg desk is blue and white. The desk comes with a sturdy wipe off board and locking storage compartment, colorful mushroom shaped pegs, a peg hammer, crayons and a wipe-off cloth. Painter Smurf and handy Smurf stand on the top edge of the board. Painter holds a crayon sharpener and Handy is a hammer holder.
057/008	Let's Be Friends Take-a-long Color By Number Set	Paradise		Blue	Travel Set	5"L X 4"W	Hong Kong	$3.50-5.00	The set includes a mini wallet made of plastic. The cover is blue and has a Smurf sitting looking happy, holding a balloon that says "Smurf" in red letters. In the corner of the cover it say' "Let's Be Friends". Inside are 8 mini color pencils, 6 color- in postcards and a wallet picture frame.
057/009	Papa And Band Paint By Number Set	Avalon	1/1/82		Paint By Number	9" X 12"	U.S.A.	$3.00-4.00	The paint by number set has one 9" X 12" picture. The picture is of Papa standing on a mushroom leading a band. There is a Smurf playing a cello, a Smurf playing a X-Phone, and the third Smurf is leading music with his hand. The picture is slightly painted. Included are a brush and paint.
057/010	Grouchy Paint By Number	Avalon	1/1/83		Paint By Number	8" X 10"	U.S.A.	$3.00-4.00	The set comes with one 8" X 10" picture. The picture has Grouchy Smurf standing with his hands crossed over his chest. There is a round border and at the bottom it says "Grouchy".
057/011	Smurf Set Of Six Sewing Cards	Colorforms			Sewing Card Set		U.S.A.	$3.00-4.00	There are 6 colorful yarns and 6 sewing cards. Two 12" x 7 5/8" sewing cards: The pictures are of a Smurf waving and Smurfette posing. Four 6" x 7 5/8" sewing cards: Clown Smurf, Papa pointing, a Smurf carrying a cake and a Smurf playing a horn.
057/012	Ministeck Set	Otto Simon	1/1/95		Plastic Peg Set	26,6 x33,3 cm	Brussels	$25.00-35.00	The board is a plastic canvas and there are pegs to fit inside. You can make 4 different Smurfs: Brainy, soccer Smurf, dumbbell lifter, and cello player. The box is blue and has a Smurf playing a cello.
057/013	Magnetic Chalkboard	Avalon	1/1/82		Magnetic Chalkboard		U.S.A.	$30.00-40.00	The chalkboard is black with blue plastic around the board. Included are 6 colorful magnets, 1 - 10 magnetic numbers, built in stand and handle, storage compartment, chalk and eraser. The outside of the board has embossed alphabet.
057/014	Smurf Fun Screens	Ohio Art	1/1/83		Fun Screens		U.S.A.	$6.00-8.00	There are 6 clear screens that fit over the Etch a Sketch™ screen. Included are: school days maze, draw the letters, draw the shapes, find Smurfette, connect-the-dots, and Smurf football
057/015	Let's Be Friends Tak-a-long Color By Number Set	Paradise		Red	Travel Set	5"L X 4"W	Hong Kong	$3.50-5.00	The set includes a mini wallet made of plastic. The cover is red and has a Smurf walking carrying the word Smurf. On the side of the cover it says "Let's Be Friends". Inside are 8 mini color pencils, 6 color-in postcards and a wallet picture frame.
057/016	Color Me Blue Tak-a-long Color By Number Set	Paradise		Blue	Travel Set	5"L X 4"W	Hong Kong	$3.50-5.00	The set includes a mini wallet made of plastic. The cover is blue and has a Smurf sitting looking happy, holding a balloon that says "Smurf" in red letters. In the corner of the cover it says "Color Me Blue". Inside are 8 mini color pencils, 6 color- in postcards and a wallet picture frame.
057/017	Smurfs At Play Tak-a-long Color By Number Set	Paradise		Green	Travel Set	5"L X 4"W	Hong Kong	$3.50-5.00	The set includes a mini wallet made of plastic. The cover is green and has a Smurf standing, holding a red pencil and yellow notebook. In the corner of the cover it says "Smurfs At Play". Inside are 8 mini color pencils, 6 color- in postcards and a wallet picture frame.
057/018	Astro Smurf Magic Slate	Golden	1/1/88		Magic Slate	13 ½" x 8 ½"	U.S.A.	$2.00-3.00	The slate has a plastic piece of paper you draw on and pull the paper up to erase. The top has a picture of a Smurf in a brown rocket. The Smurf is wearing a clear helmet and red suit.
057/019	Smurf And Smurfette Dancing Magic Slate	Golden	1/1/88		Magic Slate	13 ½" x 8 ½"	U.S.A.	$2.00-3.00	The slate has a plastic piece of paper you draw on and pull the paper up to erase. The top has a picture of a Smurf and Smurfette dancing.
057/020	Pressers #1	Spec. Products	1/1/84		Transfers		Holland	$4.50-6.00	There are 16 transfers in the package. The top corner one is Cupid. There are 6 Smurfettes.
057/021	Smurf And Smurfette Paint by Number Set	Avalon	[illegible]		Paint by Number		U.S.A.	[illegible]	The set comes with a #261 Smurf carrying a flower towards Smurfette. Smurfette is cooking. #262 a Smurf riding on a horse, #263 Papa leading a band.
057/022	Smurf Peel 'N Stick	Peyo-SEPP	1/1/82		Peel 'n Stick	13 ½" x 18"	Denmark	$7.50-10.00	The set comes with plastic reusable Smurfs, props and Smurf village play board. The Smurf plastic stickers include several different Smurfs doing different things. A Smurf carrying an ax, poet, drummer, traveler, fisherman, jockey, etc.
057/023	Pressers #4	Spec. Products	1/1/84		Transfers		Holland	$4.50-6.00	There are 34 transfers in the package. The top corner has a Smurf carrying a cake. There are no Smurfettes or Papa's in the pack.
057/024	Schtroumpf Pressers # 2	Spec. Products	1/1/92		Transfers		Holland	$4.50-6.00	There are 20 transfers in the package. There are 4 Smurfette's and 1 Papa. The top corner has a Smurf playing a trumpet.
057/025	Pitufolandia Transfers	Recorta			Transfers		Spain	$4.50-6.00	The cover of the package is blue. It has a Smurf band on the front. Papa is conducting. The transfers comes with an open paper scene.
057/026	El Bosque De Los Pitufos	Recorta			Transfers		Spain	$4.50-6.00	The cover is red. The front of the package has Smurfs working on a bridge. The transfers comes with an open up paper scene.
057/027	El Colegio Los Pitufos	Recorta			Transfers		Spain	$4.50-6.00	The cover is orange. It has a picture of a Smurf lying on the floor writing and another Smurf holding a stick. They are in a classroom. The transfers come with a fold open paper scene.
057/028	Smurf Rub-On	W. Berrie/Peyo	1/1/84		Rub-On		U.S.A.	$3.00-4.00	There are 2 halves. The sheet is from a cereal box. One sheet has: A Smurf laughing, one walking looking sleepy, one jumping, one holding a pitch fork. one pointing, Papa walking looking troubled, a mushroom house. The other half has a Smurf holding a music sheet. one walking carrying a chicken leg, one standing with his arms crossed looking angry, one running carrying a level, one sitting, holding a finger in the air, and Smurfette holding a flower.
057/029	Pitufa Tus Con Avebturas Calca Pitufos	Recorta			Transfers		Spain	$4.50-6.00	The cover is green. It has a picture of a Smurf laying in the grass on the bank by a river. There is a Smurf fishing and one swimming. The transfers come with a play sheet.

Set 94: Baby Items

ID #	ITEM NAME	MANUFACTURER	ISSUE DATE	COLOR	TYPE	SIZE	COUNTRY	VALUE	DESCRIPTION
094/001	Smurf & Smurfette Gum Soother	Danara	1/1/82		Gum Soother		U.S.A.	$3.00-4.00	The gum soother is clear plastic, double ring. It has Smurfette's face in 1 ring and the other ring has a Smurf face with a pencil behind his ear. The pictures have a green and orange square background.
094/002	Esslernschale Mit Ansaugring Baby Schlumpfe	Helly	1/1/94	Clear	Baby Bowl	7" Dia	Brussels	$7.50-10.00	The bowl is clear. There is a picture of Baby Smurf crawling on the bottom of the bowl. Baby is holding a red and yellow rattle. He is wearing a pink sleeper and a pink hat. The bowl has a lip. The lip has 4 Baby Smurf's and pink and blue dots. The bottom has white rubber to hold the bowl on the table.
094/003	Teether Toy	Danara	[illegible]	White	Teether Toy		U.S.A.	[illegible]	There are three hard plastic pieces attached to a white ring. Each plastic piece has a picture. One has a Smurf holding balloons, one has Papa holding a sucker and the third has Baby playing with a rattle.
094/004	Baby Pacifier Clip	Helly	1/1/94		Baby Clip		Brussels	$3.50-5.00	The clip is a yellow circle with Baby Smurf . Baby has his hands out. He is wearing a white sleeper and a white hat. There is a red plastic chain with a blue clip on the end attached to the yellow clip.
094/005	Baby Schlumpfe Baby-Flasche	Helly	1/1/94	Clear	Baby Bottle	8" High	Brussels	$5.00-7.00	The Baby bottle is clear and has three pictures of Baby Smurf. Baby is wearing a pink outfit in all three pictures. Baby is crawling holding a yellow and red rattle. Baby is sitting, holding a bottle in one hand and a red and orange block in the other. Baby is standing, leaning on a big red and white ball. The bottle holds 9 oz. and has a blue cap.
094/006	Baby Schlumpfe Teeflasche	Helly	1/1/94	Clear	Mini Baby Bottle	5" High	Brussels	$5.00-7.00	The Baby bottle is clear and has three pictures of Baby Smurf. Baby is wearing a pink outfit in all three pictures. Baby is crawling holding a yellow and red rattle. Baby is sitting, holding a bottle in one hand and a red and orange block in the other. Baby is standing, leaning on a big red and white ball. The bottle holds 4 oz. and has a blue cap.
094/007	Baby Schlumpfe Greifflasche	Helly	1/1/94	Clear	Grip Baby Bottle	8 ½" High	Brussels	$5.00-7.00	The Baby bottle is clear and has two pictures of Baby Smurf. Baby is wearing a pink outfit in both pictures. Baby is sitting, holding a yellow rattle. Baby is sitting, holding a bottle in one hand and a yellow and white block in the other. On the bottom of the bottle is Smurfette and another Smurf dancing around Baby Smurf. The bottle holds 8 oz. and has a blue cap. There is a slot in the middle of the bottle for the Baby's hand to go through to hold the bottle.
094/008	Baby Schlumpfe Mini Greifflasche	Helly	1/1/94	Clear	Mini Grip Bottle	6" High	Brussels	$5.00-7.00	The Baby bottle is clear and has four pictures of Baby Smurf. Baby is wearing a pink outfit in all four pictures. Baby is crawling holding a white and red rattle. Baby is sitting, holding a red rattle in the air. Baby is standing, leaning on a big red and white ball. Baby is sitting, holding a red rattle on the ground. The bottle holds 4 oz. and has a blue cap. There is a slot in the middle of the bottle for the Baby's hand to go through to hold the bottle.
094/009	Smurf & Smurfette Gum Soother	Danara	1/1/82	Clear	Gum Soother		U.S.A.	$3.00-4.00	The gum soother is a clear plastic, double ring. It has Smurfette's face in 1 ring and the other ring has a Smurf face. The Smurf looks angry. The pictures have a green and orange square background.
094/010	Cook & Papa Gum Soother	Danara	1/1/82	Clear	Gum Soother		U.S.A.	$3.00-4.00	The gum soother is a clear plastic, double ring. It has a Smurf with a chef's hat in 1 ring and the other ring has Papa's face. The pictures have a green and orange square background.
094/011	Smurfette Trainer Cup	T. Tippee	1/1/84	Blue	Sippee Cup		U.S.A.	$3.00-4.00	The cup is light blue with white. The cup has handles on each side. The cup has Smurfette holding a brown teddy bear. The cup only has Smurfette's upper body. Smurfette is wearing a white dress.
094/012	Papa Smurf Kuschel-Warmflasche	Fashy	1/1/96		Hot Water Bottle		Brussels	$22.50-30.00	The water bottle is inside a cloth Papa Smurf. Papa has a red hat and red pants on. Papa has a white fuzzy beard.

094/013	See Through Roll Ball	Illco		Clear	Roll Ball	4" Diam	Hong Kong	$6.00-8.00	The ball is clear. Inside is a Smurf that turns when the ball rolls. There are tiny balls also rolling around inside.
094/014	Baby Comics Diapers	Peaudouce	1/1/81	White	Diapers	20-40 lbs.	Brussels	$18.00-25.00	The diapers are white with little pictures of Smurfs on them. They come in a pink package with Baby Smurf on the front. Baby is wearing a blue sleeper. Baby is playing with a huge white, orange and purple ball.
094/015	See N' Sit	Helm	1/1/82		Inflatable Toy	9" x 12"	U.S.A.	$9.00-12.00	The see n' sit inflates. Inside are different pictures and different colored balls. The top is log shape.
094/016	Yellow Baby Bunting			Yellow	Baby Wrap		Hand made	$11.00-15.00	The bunting is yellow and has a zipper all the way down the front. There are various Smurf scenes. The Smurfs are all either singing or playing instruments.
094/017	No Smurfing! Bib	T. Tippee	1/1/81	White	Plastic & Terry Bib		U.S.A.	$3.00-4.00	The bib has a Smurf diving into a pond. There are frogs swimming with him. There is an orange sign on the side that says "No Smurfing!" The bib is white with green border.
094/018	Smurf Carrying A Cake Bib "Hungry Smurfs"	Peyo		White	Bib Pattern		U.S.A.	$3.00-4.00	The bib is white with an embroidery pattern. The pattern is of a Smurf carrying a cake. There is another Smurf pointing. There are 3 mushroom houses in the back. The pattern is incomplete.
094/019	2 Smurfs Playing Outside "Happy Smurfs"	Peyo		White	Bib Pattern		U.S.A.	$3.00-4.00	The bib is white with an embroidery pattern. The pattern is of a Smurf doing a handstand in the grass. Another Smurf is jumping in the air. There is a mushroom house in the back. The pattern is incomplete.
094/020	Baby's First Friends Set	Danara	1/1/83		Baby Items		U.S.A.	$9.00-12.00	The set has a spoon with Smurfette on the end, a fork with an outline of a Smurf on the end, a gum soother with Papa's face and Cook Smurf's face in the center, a rattle with a picture of Baby Smurf in a blue sleeper crawling holding a rattle, and a tether toy. It has 3 plastic pieces attached. One is a picture of Papa, another is Baby crawling, and the third is a Smurf holding balloons.
094/021	Papa Pacifier	Peyo	1/1/86	Blue	Baby Pacifier		Spain	$9.00-13.00	The pacifier has Papa's head on the outside. The bottom plastic piece between Papa's head and the nipple is light blue.
094/022	Smurf Pacifier	Peyo	1/1/86	White	Baby Pacifier		Spain	$9.00-13.00	The pacifier has a Smurf head on the outside. The bottom plastic piece between the Smurf's head and the nipple is light white.
094/023	Smurf That Tune Bib	T. Tippee	1/1/82		Bib		U.S.A.	$3.00-4.00	The bib is white with blue trim. It has a picture of a Smurf holding a music sheet singing. The top says "Smurf That Tune" in white and blue letters.
094/024	Smurf Face Diaper Pin	Peyo	1/1/83		Diaper Pin	2"	Hong Kong	$1.50-2.00	The diaper pin has a Smurf face on the end. The Smurf has a smile on his face.
094/025	Graduate Smurfette Bib	Peyo/SEPP	1/1/82		Bib	6 Months		$3.00-4.00	The bib is white. It has Smurfette wearing a pink graduation gown. She is holding a diploma and a red book. The bottom says "Graduate Smurfette.
094/026	Gone Smurfing Bib	Peyo	1/1/81		Bib			$3.00-4.00	The bib is white with yellow trim. It has a Smurf holding a fishing pole on the front. There is a plastic pocket on the bottom in front to catch the food. Under the pocket is a water scene with a red fish.
094/027	Christmas Joy Bib	Peyo/SEPP	1/1/81		Bib			$3.00-4.00	The bib is white with red trim. It has Papa Smurf dressed in a Santa outfit. Papa is holding a green sack with presents in it. He is also holding a candy cane. The bib says "Christmas Joy".
094/028	Smurf Bib	Handmade			Bib			$3.00-4.00	The bib is a white terry- cloth. It has a Smurf sewn on the front. The Smurf has his hands out in front of him.
094/029	Smurf Kushel-Waarmflasche	Fashy	1/1/96		Hot Water Bottle		Germany	$22.50-30.00	The water bottle is inside a cloth Smurf. The Smurf is fuzzy like a stuffed animal.
094/030	Lovesick Smurf And Smurfette Bottle	I.M.P.S./Peyo	1/1/92		Baby Bottle	7 ½"	Brussels	$6.00-8.00	The bottle is clear. It has a red cover. There is a picture of a Smurf pulling petals from a flower. Smurfette is posing and a cupid is flying above her head.

Set 90: Baby Rattles

090/001	Yellow Rattle With Smurf With A Butterfly On Finger	Peyo	1/1/83	Yellow	Flat Rattle	4 3/4" Long		$2.00-3.00	The rattle is white with a Smurf holding a butterfly on his finger. The back of the rattle is yellow. The picture only shows the Smurfs upper body. The handle is round and open in the center.
090/002	Blue Rattle With Smurf With A Butterfly On Finger	Peyo	1/1/83	Blue	Flat Rattle	4 3/4" Long		$2.00-3.00	The rattle is white with a Smurf holding a butterfly on his finger on the front. The back of the rattle is blue. The picture only shows the Smurfs upper body. The handle is round and open in the center.
090/003	Orange Rattle With Smurf/Butterfly On Finger	Peyo	1/1/83	Orange	Flat Rattle	4 3/4" Long		$2.00-3.00	The rattle is white with a Smurf holding a butterfly on his finger on the front. The back of the rattle is orange. The picture only shows the Smurfs upper body. The handle is round and open in the center.
090/004	Blue Rattle With A Smurf Eating A Popsicle	Peyo	1/1/83	Blue	Flat Rattle	4 3/4" Long		$2.00-3.00	The rattle is white with a Smurf holding a purple and pink popsicle on the front. The back of the rattle is blue. The picture only shows the Smurfs upper body. The handle is round and open in the center.
090/005	Blue Rattle With Baby Crawling	Peyo	1/1/83		Flat Rattle	4 3/4" Long		$2.00-3.00	The rattle is white with Baby wearing a blue sleeper holding a red rattle in his hand on the front. Baby is crawling. The back of the rattle is blue. The handle is round and open in the center.
090/006	Smurf Face High Chair Rattle	Peyo	1/1/83		High Chair Rattle	8 ½" Tall	Hong Kong	$11.00-15.00	The top of the rattle is a big Smurf face. The handle is red and there is a blue suction cup on the bottom to attach on to a high chair. The rattle jingles.
090/007	Orange Rattle With A Smurf Eating A Popsicle	Peyo	1/1/83	Orange	Flat Rattle	4 3/4" Long		$2.00-3.00	The rattle is white with a Smurf holding a purple and pink popsicle on the front. The back of the rattle is orange. The picture only shows the Smurfs upper body. The handle is round and open in the center.
090/008	Pink Baby With Block	Peyo	1/1/83	Pink	Flat Rattle	4 3/4" Long		$2.00-3.00	The rattle is white with Baby Smurf crawling on the front. Baby is wearing a blue sleeper and a white hat. There are marbles in front of Baby and a pink block behind him. The back of the rattle is pink. The handle is round and open in the center.
090/009	Heart-Shaped Mirror Rattle	Danara	1/1/83		Flat Rattle	4 3/4" Long	Hong Kong	$4.50-6.00	The mirror is heart-shaped with a red plastic border. A Smurf is standing off to one side waving. The Smurf has a yellow daisy in his hat. The back of the mirror is white. The mirror is a rattle.

Set 28: Badges

028/001	Brainy For President Smurf In Prosperity	S.E.P.P.	1/1/84		Stick On Buttons	3"	U.S.A.	$3.50-5.00	The badge is red, white, and blue. In the center is the face of Brainy.
028/002	Vanity For President Smurf For You & Vanity Too!	S.E.P.P.	1/1/84		Stick On Buttons	3"	U.S.A.	$3.50-5.00	The badge is red, white, and blue. In the center is the face of vanity.
028/003	Handy For President For A Smurfy Deal	S.E.P.P.	1/1/84		Stick On Buttons	3"	U.S.A.	$3.50-5.00	The badge is red, white, and blue. In the center is the face of Handy.
028/004	Smurfette For President	S.E.P.P.	1/1/84	Red	Stick On Buttons	3"	U.S.A.	$3.50-5.00	The background of the button is red and in the center is Smurfette's face.
028/005	Most Daring True Blue Smurf Award	W. Berrie	1/1/83	Yellow	Stick On Buttons	2"	U.S.A.	$2.00-3.00	The background is yellow. The picture is of a Smurf scaring Gargamel.
028/006	Most Friendly True Blue Smurf Award	W. Berrie	1/1/83	Yellow	Stick On Buttons	2"	U.S.A.	$2.00-3.00	The background is yellow. The picture is of a Smurf talking on a telephone.
028/007	Most Mischievous True Blue Smurf Award	W. Berrie	1/1/83	Red	Stick On Buttons	2"	U.S.A.	$2.00-3.00	The background is red. The picture is of a Smurf holding a burnt up test tube.
028/008	Most Lovable True Blue Smurf Award	W. Berrie	1/1/83	Pink	Stick On Buttons	2"	U.S.A.	$2.00-3.00	The background is pink. The picture is of a love struck Smurf.
028/009	Most Funny True Blue Smurf Award	S.E.P.P.	1/1/83	Yellow	Stick On Buttons	3"	U.S.A.	$2.00-3.00	The background is yellow. The picture is of a Smurf jumping in the air laughing.

Set 230: Balls

230/001	Smurf Pink Rubber Mini Ball	Peyo		Pink	Hard Sponge Ball	2 ½" Dia	U.S.A.	$2.00-3.00	The ball is a hard rubber. On one side there is a picture of a Smurf jumping in the air. There are blue lines around the Smurf. The bottom says Smurf in white letters, and in black letters, super sponge ball. The ball is pink.
230/002	Smurf Light Blue Rubber Mini Ball	Peyo		Light Blue	Soft Rubber Ball	3"	U.S.A.	$2.00-3.00	The ball is a soft rubber. On one side there is a picture of a Smurf jumping in the air. There are blue lines around the Smurf. The bottom says Smurf in white letters. The ball is light blue.
230/003	Smurf Dark Orange Rubber Mini Ball	Peyo		Dark Orange	Soft Rubber Ball	2 ½" Diam.	U.S.A.	$2.00-3.00	The ball is a soft rubber. On one side there is a picture of a Smurf jumping in the air. There are blue lines around the Smurf. The bottom says Smurf in white letters. The ball is dark orange.
230/004	Smurf White Rubber Mini Ball	Peyo		White	Soft Rubber Ball	4" Diam.	U.S.A.	$2.00-3.00	The ball is a soft rubber. On one side there is a picture of a Smurf jumping in the air. There are blue lines around the Smurf. The bottom says Smurf in white letters. The ball is white.
230/005	Yellow, Blue And Orange Ball	Peyo			Soft Rubber Ball	7" Diam.	U.S.A.	$2.00-3.00	The ball is yellow, blue and orange. On one side is a picture of a Smurf playing a drum. The other side has a Smurf carrying a white package.
230/006	Smurf Baseball	W. Berrie/Peyo		White	Baseball	2 ½"	U.S.A.	$3.50-5.00	The baseball is white. On the front is a Smurf in a baseball uniform. The Smurf is swinging a brown bat.

Set 133: Banks- Ceramic

133/001	Smurf Mushroom House Bank	W. Berrie	1/1/82		Ceramic Collectable	4"W X 4"H	Taiwan	$18.00-25.00	The bank is in a shape of a mushroom house. The house is yellow and has an orange and red roof. There is a yellow and brown chimney, brown windows and a brown door. A Smurf is sticking out one of the windows. There is green grass painted around the base.
133/002	Smurf Head Bank				Ceramic Bank	6"H X 6"W	Hand made	$11.00-15.00	The bank is the shape of a Smurf face. The Smurf is has a blue face, yellow eyebrows, a black mouth and white and black eyes. He is wearing a white hat. The Smurf has a glazed finish.
133/003	Smurf Face On A Base Bank - Yellow				Ceramic Bank	6 ½"H X 5"W	Hand made	$7.50-10.00	The bank is a yellow circle with a black base. In the middle of the circle is a raised Smurf face. The Smurf face is dark blue. He has white and black eyes, black eyebrows and a white and red mouth. The face is on both sides.
133/004	Smurf In A Red Car	W. Berrie	1/1/83		Ceramic Collectable	5" X 5 ½"	Korea	$18.00-25.00	A Smurf is driving a red ceramic car. The car has yellow bumpers, a yellow grill, yellow headlights and a yellow steering wheel. The tires are white with yellow wheel covers. The Smurf is a medium bright blue. He is wearing a white hat with red goggles that have yellow lenses. There is a slot in the trunk of the car for coins.
133/005	Pirate Smurf Sitting On A Treasure Chest	W. Berrie	1/1/83		Ceramic Collectable	6" x 4"	Korea	$18.00-25.00	The Smurf is a medium bright blue. The Smurf is sitting on a brown and yellow treasure chest. The Smurf has a black patch over his left eye. The Smurf is wearing a red belt with a yellow and brown knife stuck in the side. The Smurf has a yellow earring in his right ear. There is a slot in the front of the treasure chest for coins.
133/006	Pilot Smurf In A Red Airplane	Ganz Bros.	1/1/82		Ceramic Collectable	5 ½" X 6"	Korea	$18.00-25.00	The Smurf is a medium bright blue. He is weaning a brown coat with white trim and brown goggles with yellow lenses. The goggles are over his hat. The Smurf is driving a red plane with yellow stripes on the wings. The propeller is brown, with blue and yellow behind it. The plane has black wheels. There is a slot in the back of the Smurf's hat for coins.

133/007	Jockey Smurf With A Yellow Box				Ceramic Bank	9 ½" High	Mexico	$7.50-10.00	The Smurf has a light blue body. He is standing, wearing white pants and a white hat. The Smurf has a grin on his face. The Smurf is holding a yellow box with red ribbon around it in both his hands. The present is resting on a brown rock. The slot for the money is in the back of the Smurf's hat.
133/008	Smurf Face On A Base Bank - White			White	Ceramic Bank	6 ½" High	Hand Painted	$5.00-7.00	The bank is a white circle with a white base. In the middle of the circle is a raised Smurf face. The Smurf face is light blue. He has white and black eyes, black eyebrows and a white and red mouth. The face is on both sides.
133/009	Dark Blue Smurf With Hands Out				Ceramic Bank	9" High	Hand Painted	$7.50-10.00	The Smurf has a dark blue body. He looks short and fat. The Smurf has both his hands in the air at his sides. The Smurf has a big smile on his face. The Smurf is wearing white pants and a white hat. The slot for the money is in the back of his hat. The paint is dull.
133/010	Dark Blue Glossy Smurf With Hands Out				Ceramic Bank	9" High	Hand Painted	$7.50-10.00	The Smurf has a dark blue body. He looks short and fat. The Smurf has both his hands in the air at his sides. The Smurf has a big smile on his face. The Smurf is wearing white pants and a white hat. The slot for the money is in the back of his hat. The paint has a glossy finish.
133/011	Smurf Face On A Base Bank - Light Blue			Light Blue	Ceramic Bank	6 ½"High	Hand Painted	$9.00-13.00	The bank is a light blue circle with a light blue base. In the middle of the circle is a raised Smurf face. The Smurf face is dark blue. He has white and black eyes, black eyebrows and a white and pink mouth. The face is on both sides.
133/012	Smurfette And A Smurf Standing Together				Ceramic Bank	8 ½"L X 6 ½"W	Hand Painted	$15.00-20.00	A Smurf and Smurfette are standing next to each other on a patch of grass. Smurfette has on a white dress. Her one hand is by her mouth. The Smurf looks like he is standing there serenading her. The Smurf has on yellow pants. The paint has a glossy finish. The slot for money is in the back of Smurfette's hat.
133/013	Smurfette Holding An Envelope				Ceramic Bank	6 ½" High	Hand Painted	$6.00-8.00	Smurfette has a dark blue body and yellow hair. She has on a white dress, shoes and hat. Smurfette is holding a blue envelope to her chest. The envelope has a red heart on the back. Smurfette is standing against some dark green grass. The slot for money is in the back of her hat.
133/014	Smurf With A Present	W. Berrie	1/1/82		Ceramic Collectable	10"L X 4"W		$18.00-25.00	The Smurf is standing with one hand behind his back and one hand on his tummy. The Smurf is painted a bright blue. He has white pants and a white hat on. There is a white package with a red ribbon sitting by his feet.
133/015	Smurf With Hand On Chest				Ceramic Collectable	7 12/" High	Mexico	$18.00-25.00	The Smurf is light blue with a shiny finish. He is standing against a tree stump. The tree stump is brown. The Smurf has a white hat with a black tip on and white pants. The Smurf has pink ears and a pink tongue sticking out. The Smurf has his left hand on his chest. He has black fingernails. The Smurf has slanted eyes.
133/016	Smurf With A Big Pot	Peyo	1/1/96		Ceramic Collectable	6" High		$15.00-20.00	A Smurf is standing and trying to lift a big yellowish orange pot. There is a slot in the pot for money.

Set 109: Banks- Plastic & Tin

109/001	Smurf Bank	Peyo	1/1/82	White	Plastic Bank	11" High		$7.50-10.00	A Smurf is standing, wearing a white hat, shirt and pants. The Smurf's right hand is holding a white heart against his side. There is a slot on top for coins. The bank is made of soft plastic.
109/002	Papa Bank	Juwa	1/1/82		Plastic Bank	6" High	West Germany	$11.00-15.00	Papa is standing with his hands out at his side. His arms move. Papa is wearing a light red hat and pants. There is a slot on top for coins. Papa comes with a key to unlock the bank. (Rare)
109/003	Smurf With A Gift	Peyo	1/1/84		Plastic Bank	7" High	U.S.A.	$7.50-10.00	The Smurf is standing on a green base painted like grass. He is holding a big white box with yellow ribbon in front of him. The bank is made of hard plastic and the Smurfs' hat has a slot for coins.
109/004	Brainy Sitting On A Tree Stump	Peyo	1/1/84		Plastic Bank	7" High	U.S.A.	$5.00-7.00	Brainy is sitting on a brown tree stump. He has a red book open in his lap. He is holding one hand in the air and looks like he's declaring something. Brainy is wearing a white hat, black glasses and white pants. Brainy has a dark blue body.
109/005	Traveler Smurf	Danara	1/1/84		Plastic Bank	7" High	U.S.A.	$7.50-10.00	The Smurf is standing by a green bush. He is holding a brown stick over his shoulder. There is a yellow knapsack tied on the end of the stick. The Smurf has a medium blue body. The slot for money is in the back of his hat.
109/006	Basketball Smurf Yellow Tin Bank	IMA	1/1/96	Yellow	Tin Bank	4 ½"	Brussels	$7.50-10.00	The bank is round. It is yellow. It has a picture of a Smurf dunking a basketball. Another Smurf is standing behind trying to stop him. One Smurf is in green and the other is wearing red and yellow. The same picture is on 2 sides.
109/007	Rollerblading Smurfs Tin Bank	IMA	1/1/96	Green	Tin Bank	4 ½"	Brussels	$7.50-10.00	The bank is round. It is green. It has a picture of 2 Smurfs rollerblading. One Smurf is wearing a pink shirt and blue shorts. The other Smurf is wearing an orange shirt and yellow shorts. The same picture is on 2 sides.
109/008	Soccer Smurf Tin Bank	IMA	1/1/96	Blue	Tin Bank	4 ½"	Brussels	$7.50-10.00	The bank is round. It is blue. It has a picture of 2 Smurfs playing soccer. The ball is in the air. The 2 Smurfs are running after the ball. The picture is the same on 2 sides.

Set 160: Banners

160/001	Smurfs N' Ice Capades	Peyo	1/1/83		Felt Banner	18"L X 9"W	U.S.A.	$7.50-10.00	The banner has Smurfette posing on ice skates. Smurfette is wearing a pink dress and pink ice skates. Below her in a blue box it says "Smurfs 'n' Ice Capades" in red and white letters.
160/002	Smurfs Alive! Ice Capades	Peyo	1/1/83		Felt Banner	19 ½"L X 9"W	U.S.A.	$7.50-10.00	The banner has Smurfette posing on ice skates. Smurfette is wearing a red dress and white ice skates. Below her in a white box it says "Smurfs Alive!" in dark blue letters and "Ice Capades" in multicolor letters.
160/003	Smurf's N' Ice Capades	Peyo	1/1/83	Red	Felt Banner	19 ½"L X 9"W	U.S.A.	$7.50-10.00	The banner has Papa skating on ice skates. Papa is holding one hand in the air and pointing with the other. Below him in a blue box it says "Smurfs 'n' Ice Capades". "Smurfs" is in red and "Ice Capades" in white letters.
160/004	Smurfs Alive! in Ice Capades	W. Berrie/Peyo		Red	Felt Banner	12"w x 28" Long		$11.00-15.00	The banner is red. It has Smurfette, Papa and 4 other Smurfs ice skating on a pond. On the side of the pond are different colored mushrooms. At the opposite end of the banner is Gargamel and Azreal sneaking towards the Smurfs. There is a lab table behind Gargamel. The Banner says "Smurfs Alive!" in red letters and "Ice Capades" in multi colored letters.

Set 60: Bathroom Items

060/001	Talking Scale With Tape Measure	Helm	1/1/83	White	Talking Scale		Hong Kong	$63.75-85.00	A white scale that has a Smurf face up were you read the pounds. The Smurfs nose pulls out to a tape measure.
060/002	Smurf Battery Operated Toothbrush	Helm		Blue/White	Toothbrush	5 ½" High	Hong Kong	$18.00-25.00	A Smurf is standing and is 5 ½" high . The Smurf is holding a tube of toothpaste. There is a red and yellow toothbrush to go in the top of the Smurf.
060/003	Smurf Toothbrush & Cup Set	Talbot Toys	1/1/82		Toothbrush Set		Hong Kong	$22.50-30.00	A Smurf is holding a tube of toothpaste in one hand and it's flowing all over. In his other hand he's holding a toothbrush. The Smurf is standing on a grass base with a white cup that has a red mushroom cover.
060/004	Smurfette Toothbrush & Cup Set	Talbot Toys	1/1/83		Toothbrush Set		Hong Kong	$22.50-30.00	Smurfette is standing and holding a tube of toothpaste in her left hand. Her right hand is extended to hold a yellow toothbrush. She is standing on a green grass base. The cup is white with a purple and yellow mushroom top.
060/005	White Mini Cup			White	Plastic Cup	3" High	Hong Kong	$1.50-2.00	The cup has a decal of a Smurf sitting on a stool brushing his teeth.
060/006	Baby Smurf Toothbrush & Cup Set	Talbot Toys	1/1/84		Toothbrush Set		Hong Kong	$22.50-30.00	Baby Smurf is sitting on a base holding red toothpaste in his right hand and a brown teddy bear in the other. The base is light blue. There is a red mushroom cup with a yellow cover. The cover has a dark blue spots on it. Baby is wearing a white outfit.
060/007	Smurf On A Raft	Talbot Toys	1/1/83		Bath-Tub Toy		Hong Kong	$15.00-20.00	A Smurf sitting on yellow raft. The Smurf is sitting on a dark brown stool. He is holding a dark brown captains wheel. It floats!
060/008	Smurfette Soap			White	Bath Soap	3"L X 2"W		$1.50-2.00	The bar of soap is white and square. The front of the soap has a decal of Smurfette standing and wearing a white dress with pink dots.
060/009	Submarine	Galoob	1/1/83		Floating Submarine	6 ½" Long	Hong Kong	$25.00-35.00	The submarine is brown and yellow. It has a white periscope on top. In front there is a clear plastic window with a Smurf face inside. The submarine has an air pump that makes the submarine dive and float in water. It also has an automatic rising periscope.
060/010	Smurf Dixie Cup Holder	Helm	1/1/83		Dixie Cup Holder	8"Tall	Hong Kong	$11.00-15.00	A plastic Smurf 5 ½" tall is standing on a green grass base with flowers. The Smurf has his left hand up in the air and a log hooks on. The part that holds the dixie cups looks like part of a tree, it's brown and on the top there is a little gray squirrel. The top of the log comes off to insert the dixie cups. The Smurf has a suction cup on the bottom to hold the dixie cup dispenser in place
060/011	Dixie Cups	Canada Cup			Paper Dixie Cups	3 OZ	Canada	$6.00-8.00	100 Dixie cups in the box. The cups have different designs with different sayings.
060/012	Dixie Cups	James River-Dixie			Paper Dixie Cups	3 OZ	U.S.A.	$7.50-10.00	200 Dixie cups in the box. The cups have different designs with different sayings. Completely different from the 100 cup set.
060/013	Kleenex Box Cover			Blue	Kleenex Box Cover	8" X 4 ½"	Hand made	$3.50-5.00	The Kleenex box cover is in the shape of a house. The sides are blue and have windows with different Smurfs in them. The top is made of Smurf fabric with different Smurfs and is white with blue lines. There is a chimney and that's were the Kleenex stick out. The whole cover is made of cloth.
060/014	Smurf Bubble Bath Cap				Shampoo Cap	4"L X 2"W	Korea	$1.50-2.00	The Smurf head is made of rubber and has a wide cap bottom.
060/015	Smurf Soap Container	Sugar	1/1/95		Soap Container	7"L X 4"W	Spain	$22.50-30.00	The container is in the shape of a mushroom. The bottom is white, the top of the mushroom is red with white dots and has a green caterpillar crawling on it. There is a Smurf sitting on top of the mushroom holding a red paint brush with a white tip. The top of the mushroom pops off to an unscrew top.
060/016	Papa Bubble Bath Cap	SEPP	1/1/83		Shampoo Cap	[illegible]	U.S.A.	$1.50-2.00	Papa Smurf's head is made of rubber and has a wide cap bottom. He has a red hat on and a white beard.
060/017	Smurfette Battery Operated Toothbrush	Helm	1/1/83		Toothbrush	5 ½" High	Hong Kong	$11.00-15.00	Smurfette is standing and is 5 ½" high . Smurfette looks like she's sitting on a red and pink mushroom. Her hands are at her sides. A white toothbrush is stuck in her hat.
060/018	Bubble Bath Smurf In Tub	Suga	1/1/96		Bubble Bath		Spain	$35.00-45.00	The bubble bath is in a blue bottle. There are pictures of Smurf's faces in bubbles on the bottle. The cap is a rubber Smurf sitting in a bathtub. The bathtub is brown with white bubbles in. The Smurf is scrubbing his back. Included is a bottle of eau de toilette water. The bottle is small and has a Smurf on the front. The Smurf is holding a hand out. The cover is a mushroom house. The 2 things are in a clear plastic bag.
060/019	Poet Smurf Towel	AT & TV Mrch..	1/1/95	White	Towel & Washcloth		Brussels	$9.00-12.00	The towel and washcloth are white. On the bottom is Poet Smurf laying down with a tan piece of paper in front of him. In his other hand he is writing with a feather pen. He is looking up. Above him is Smurfette posing. There is a flower cloud around her.
060/020	Smurf Bath mat	Peyo/SEPP		White	Bath mat	17" x 24"	U.S.A.	$18.00-25.00	The bath mat is white. It has Papa, Smurfette, Brainy, a Smurf with a daisy, a Smurf jumping in the air, and one walking that looks half awake on it. The rug says "SMURF" in red letters on the bottom.
060/021	Smurf Toothbrush Shelf	Helm	1/1/82	White	Toothbrush Self	10" x 6"	Hong Kong	$15.00-20.00	The Smurf shelf is white. The shelf has a mirror. There is a plastic Smurf on the side of the mirror. There is a picture of a Smurf carrying a toothbrush walking towards Smurfette. The shelf is missing a toothbrush and cup.

060/022	Smurfette Bathtub Toy	Peyo	1/1/83	Yellow	Bathtub Toy	3 ½" x 5"	U.S.A.	$11.00-15.00	Smurfette is sitting in a yellow boat. Smurfette is holding a black steering wheel. The boat has a blue steering wheel. It floats!
060/023	Smurfette Body Bath Paint Blue	W. Berrie/Peyo	1/1/83	Blue	Body Paint	2 oz.	U.S.A.	$1.50-2.00	The paint is blue and is grape scented. The paint is in a pink squeeze tube. Smurrfette in a bathtub is on the front of the tube.
060/024	Smurfette Body Bath Paint	W. Berrie/Peyo	1/1/83	Red	Body Paint	2 oz.	U.S.A.	$1.50-2.00	The paint is red and is strawberry scented. The paint is in a pink squeeze tube. Smurrfette in a bathtub is on the front of the tube.
060/025	Smurf Toothpaste Dispenser	Helm	1/1/83		Toothpaste Dispenser	9" High	Hong Kong	$11.00-15.00	The toothpaste dispenser is a tree. The toothpaste comes out of the bark of the tree then you push down on the mushroom at the tree base. There is a Smurf standing on a red and white mushroom next to the tree. He is brushing his teeth.
060/026	Bathroom Vanity	Helm		White	Shower Vanity	14"L x 8 ½" W	Hong Kong	$11.00-15.00	The vanity hangs from the shower hook. It has 2 shelves and a rack on the bottom to hang a wash cloth on. The vanity has a plastic Smurf on top holding a finger in the air. The backboard of the vanity has a picture of Smurfette sliding down a waterfall, 2 Smurf's floating in a washtub of bubbles and another Smurf jumping in the water.
060/027	Smurf Strawberry Toothpaste	Suga Perfumes	1/1/96		Toothpaste	1.7 Fl oz		$3.00-4.00	The toothpaste is in a white tube. It has a picture of Smurfette sitting on a toothbrush on one end and a Smurf standing on a tube of paste squirting at Smurfette's toothbrush.
060/028	Smurf Non Skid Mats			Pink	Skid Mats	6 ½"		$15.00-20.00	The mats are pink. They are of a Smurf standing with his hands at his side. There are 6 of them. They are all the same design.
060/029	Papa Smurf Eau De Toilette	Suga Perfumes	1/1/95		Toilette Water	5 ½"	Spain	$7.50-10.00	The bottle has Papa Smurf's body painted on it. The cap is a rubber head. The head is of Papa Smurf.

Set 54: Battery Operated Toys

054/001	Smurf Roller Coaster	Illco	1/1/83		Plastic		Hong Kong	$30.00-40.00	The ramp for the cars to go up is light blue and has a picture of Smurf's playing in the beams. There is Papa in a yellow car, Smurfette in a red car, and a Smurf in a orange car. There is a black bridge with a cardboard Smurf berry tree over it.
054/002	Bumper Cars	Galoob	1/1/82	Blue Box	Battery Operated		Hong Kong	$30.00-40.00	There is a play mat with cardboard fencing. There is a red bumper car and a yellow bumper car, both have blinking headlights. There is a cardboard track.
054/003	Speed Boat	Galoob	1/1/82		Battery Operated		Hong Kong	$18.00-25.00	The speed boat has a green bottom and a blue top. There is a Smurf in the middle(only his upper body). There is a silver engine on the front of the boat. The boat has an orange propeller on the back. There is a yellow flag on the back that says "Smurf Speed Boat".
054/004	Red Smurf Fun Buggy	Galoob	1/1/82	Red	Fun Buggy	5"L X 4"H	Hong Kong	$18.00-25.00	The jeep is red and has a plastic Smurf sitting in the drivers seat. The roll bar, bumpers and wheels are black.
054/005	Helicopter	Galoob	1/1/82		Battery Operated		Hong Kong	$25.00-35.00	The helicopter is white and green. The landing gear is yellow and there are black tires on both sides. The top of the helicopter has a blue light and a red rotor. There are a stickers on each side that look like windows with 2 Smurf's and Smurfette inside. The front of the helicopter has a clear window with a Smurf sitting inside holding the control. The helicopter lights up, has sound, a spinning rotor, and adjustable steering.
054/006	Smurf Road Race Village Set	Talbot Toys	1/1/83		Slot Cars		Hong Kong	$55.00-75.00	Include: 1 Smurf brown log car with a blue mushroom top, 1 Smurfette brown log car with a red mushroom top, 1 yellow and 1 orange hand held controller, 3 black straight tracks, 1 black power in-take track, 8 black curved track and 4 sheets die-cut cardboard Smurf village.
054/007	Crawling Baby Smurf	Furga/Puffi	1/1/83		Crawling Baby	8" x 6"	Italy	$37.50-50.00	The Baby crawls across the floor. His head turns while he is crawling. Baby has a hard rubber face and a plastic body. He is wearing a white and pink knit outfit.
054/008	Smurf Roller Coaster From Greece	Lyra	1/1/85		Plastic		Greece	$40.00-55.00	The ramp for the cars to go up is light blue. The track is orange. The side of the ramp has a picture of a clown Smurf, a Smurf playing a harp and a Smurf holding flowers. There are 4 Smurf's in blue carts. Each Smurf is wearing a different white hat. At the top of the ramp is a cardboard mushroom house,

Set 80: Beach Items

080/001	Beach Boat Sand Toy	H-G Industries	1/1/81		Plastic Beach Toys		U.S.A.	$11.00-15.00	There is a plastic Smurf body sand mold, a red oar shovel, a red mini sieve and a yellow sand mold boat.
080/002	Sand Toys With A Yellow Sieve	H-G Industries	1/1/81		Plastic Beach Toys		Hong Kong	$15.00-20.00	There is a large yellow sieve, a plastic Smurf face sand mold, a blue shovel and a 4 ½" white sand pail with a blue handle. The sand pail has a picture of five Smurfs marching with an orange background. The same picture is on two sides.
080/003	Sand Toys With A Red Sieve	H-G Industries	1/1/81		Plastic Beach Toys		Hong Kong	$15.00-20.00	There is a large red sieve, a plastic Smurf face sand mold, a blue shovel and a 4 ½" white sand pail with a blue handle. The sand pail has a picture of five Smurfs marching with an orange background. The same picture is on two sides.
080/004	Blue Kids Sunglasses	Foster Grant	1/1/83	Blue	Plastic Sunglasses		Mexico	$6.00-8.00	The sunglasses have blue rims with shaded lenses. In the corner of the right lens is an outline of a Smurf. The Smurf has his hand in the air. The glasses are in a white and clear package with Brainy on the front of the package.
080/005	Smurf Sunbathing Orange Mini Frisbee	Wham-O	1/1/80	Bright Orange	Mini Frisbee	4 ½"	U.S.A.	$3.00-4.00	The frisbee is small and bright orange. The picture on the front is of a Smurf laying on a multi color beach towel sunbathing. The Smurf is wearing orange sunglasses and white swim trunks He has a radio on one side of the towel and his beach bag on the other side.
080/006	Cheerleader Smurfette Blue Mini Frisbee	Wham-O	1/1/80	Blue	Mini Frisbee	4 ½"	U.S.A.	$3.00-4.00	The frisbee is small and blue. The picture on the front is of Smurfette wearing a pink and white cheerleader outfit. There is a mushroom house in the background.
080/007	Smurf's Up! Pink And Clear Beach Bag	Peyo	1/1/81		Plastic Beach Bag	35cm x 22cm	Taiwan	$7.50-10.00	The beach bag is clear plastic with a pink border on top and a pink bottom. The bag has a picture around the whole thing. The top says Smurf's Up! in red letters. There is a picture of Smurfette in a red swimsuit sitting on a blue towel with a yellow duffel bag next to her. A Smurf is laying on a red and white towel with a yellow radio on his right side, the sun is over head. Another Smurf is surfing on a yellow surfboard, he is wearing a white tank shirt and white shorts.
080/008	Beach Scene Beach Bag	Granmark	1/1/82		Plastic Beach Bag			$7.50-10.00	The beach bag is clear plastic with a yellow border on the top and bottom and a red bottom. The bag has pictures around the whole thing. The picture has a Smurf laying on a white beach towel with the sun behind him. Another Smurf is sitting on a white towel with a red, white and yellow umbrella overhead. There is a Smurf carrying a basket and an umbrella and another Smurf is standing, holding a shovel. There is a red draw string.
080/011	Large Beach Scene Sand Pail	W. Berrie	1/1/81	White	Sand Pail	7" High	Italy	$9.00-13.00	The pail is white with a blue handle and yellow shovel. There is a picture of a Smurf sitting on an orange beach towel with an orange umbrella over top. There is a Smurf carrying an orange umbrella and picnic basket and another Smurf is standing, holding a shovel pointing. The sand is yellow and the sky is blue. The same picture is on both sides.
080/012	24" Blue And White Beach Ball	W. Berrie	1/1/82	Blue/White	Vinyl Beach Ball	24"	Taiwan	$4.50-6.00	The ball has 3 blue sections and 3 white sections. The white sections have pictures. There is a picture of a Smurf on a yellow surf board, Smurfette holding a rope water-skiing on green skis, and another Smurf snorkeling with green fins on his feet. All three white sections say "Smurfs" in blue and white letters.
080/013	Small Multi Colored Beach Ball	Puppy	1/1/93	Multi	Vinyl Beach Ball	12"	Vietnam	$3.50-5.00	The beach ball has multi colored sections. In the white section it has a picture of a Smurf being chased by a red crab. The Smurf has a red and white swimsuit on and yellow slippers. The ball says Daniel Cooperman on the back.
080/014	Clown Smurf Kite	H-G Industries	1/1/83	Blue	Plastic Kite		Taiwan	$7.50-10.00	The kite is huge. The kite is blue and has a clown Smurf on the front holding several colored balloons in one hand and a candy cane in the other hand. The clown has an orange and yellow outfit on. The kite is missing the back plastic poles and string.
080/015	Baseball Smurf Yellow Mini Frisbee	Wham-O	1/1/80	Yellow	Mini Frisbee	4 ½"	U.S.A.	$3.00-4.00	The frisbee is small and yellow. The picture on the front is of a Smurf dressed in a baseball uniform. The Smurf is wearing a red shirt, white shorts, white shoes, white hat and red socks. The Smurf is carrying a baseball bat over his shoulder. The bat has a dark brown mitt hanging off of it. The Smurf is throwing a white baseball in the air with his other hand.
080/016	Smurf Water Pal	Coleco	1/1/83		Water Pal	36" x 36 "	Taiwan	$9.00-12.00	The Smurf is inflatable. The Smurf has a head, arms and feet. He is a round tube to sit on. Has a tow rope.
080/017	Beach Boat Sand Toy	H-G Industries	1/1/81	Yellow	Plastic Beach Toys		U.S.A.	$11.00-15.00	There is a plastic Smurf body sand mold, a yellow oar shovel, a yellow mini sieve and a yellow sand mold boat.
080/018	Medium Sand Pail 5 Smurfs Marching	H-G Industries	1/1/82		Sand Pail	5 1/4"	U.S.A.	$4.50-6.00	The sand pail has a picture of five Smurfs marching with an orange background. The same picture is on two sides. The pail is white with a blue handle.
080/019	Small Sand Pail With 5 Smurfs Marching	H-G Industries	1/1/81		Sand Pail	4 ½"		$3.00-4.00	The sand pail has a picture of five Smurf's marching with an orange background. The same picture is on two sides. The pail is white with a blue handle.
080/020	Blue Smurf Sand Pail	BP Australia	1/1/79	Blue	Sand Pail	6 ½" x 6 ½"	Australia	$9.00-12.00	The pail is blue. It has a white handle. The top of the pail is wide and the bottom is narrow. There is a picture on the front of a Smurf wearing a red inner tube and flippers. Smurfette is wearing a white shirt and has a blue mermaid tail.
080/021	Zwemband Bouee	Puppy	1/1/90		Swim Ring		China	$6.75-9.00	The swim ring has the Smurflings on it. They are in the water and on the sand. Sassette is laying on a red towel. Nat is diving in the water. Slouchy is waving at a fish. There are 5 Smurfs in the picture.
080/022	20" Blue And White Beach Ball	Coleco	1/1/82	Blue/White	Beach Ball	20"	Taiwan	$5.00-7.00	The ball has 3 blue sections and 3 white sections. The white sections have pictures. There is a picture of a Smurf on a yellow surf board, Smurfette holding a rope water-skiing on green skis, and another Smurf snorkeling with green fins on his feet. All three white sections say "Smurfs" in blue and white letters.
080/023	Fun Float Raft	Coleco	1/1/82	White	Raft	34" x 22"	Taiwan	$7.50-10.00	The raft is white and has a picture of a Smurf on the top. The Smurf has see through eyes. The Smurf is wearing green flippers, white shorts and an orange snorkeling mask. The raft comes with a tow rope.
080/024	Smurf Swim Ring	Coleco	1/1/82	Blue/White	Swim Ring		Taiwan	$4.50-6.00	The top of the swim ring is white. It has a picture of Smurfette walking, carrying a pail of sand and a shovel. A Smurf sailing a dark brown raft. A Smurf laying on a red and white towel. Another Smurf sitting under a green and red umbrella. The bottom side of the raft is blue.
080/025	Smurf Sit N' Float Pool Toy	Coleco	1/1/82	Yellow	Sit N' Float		Taiwan	$4.50-6.00	The sit 'n float is a large tube with holes for legs and a big Smurf head to hold on to. The top has a Smurf body. The Smurf has a blow up head and tail. The center is red where the legs go. The bottom is yellow.
080/026	Smurf's Up! Keep Dry Holder	Applause		White	Keep Dry Holder	4 ½"		$3.00-4.00	The holder is tube- shape. It has a white screw on top with a white rope for hanging. The front says "Smurf's Up!" in blue letters. There are shells under the words. Used at the beach to keep money dry.
080/027	48" Beach Ball	Coleco	1/1/81	Blue & White	Beach Ball	48"	Taiwan	$7.50-10.00	The ball is blue and white. On one side is a picture of Smurfette walking carrying a pail of sand and a Smurf sitting on a beach blanket.
080/028	Smurf Riding Bike Mini Orange Frisbee	Wham-O	1/1/80	Orange	Mini Frisbee	4 ½"	U.S.A.	$3.00-4.00	The frisbee is small and orange. The picture on the front is of a Smurf riding a bike. The Smurf is wearing a red shirt, white shorts, white shoes, white hat, brown gloves and red socks. The Smurf is riding a white bike with purple wheels.
080/029	Smurfs Playing In The Ocean Pail	Sucf	1/1/81	White	Sand Pail	5 1/4"	Italy	$7.50-10.00	The pail has a picture of Smurfette water skiing. Smurfette is being pulled by a red sailboat. Another Smurf is surfing in a wave. The Smurf is on a yellow and red surf board. The picture goes all the way around the pail.

080/030	Smurfette Beach Towel	Latex Industries	1/1/81	White	Beach Towel			$9.00-12.00	The towel is white. On the front is Smurfette standing and posing. Smurfette is wearing a white dress with yellow dots. Smurfette is wearing a red sash that says "SMURFETTE" in yellow letters. There are white and red daisies all over the front of the towel.
080/031	Wind Smurfer	HG	1/1/81		Floating Toy		U.S.A.	$22.50-30.00	The toy looks like a surf board. It has a red board with a yellow sail and fin. A Smurf is holding on to the sail. The toy floats and sails.
080/032	Smurf Surfing Mini Frisbee	Wham-O	1/1/80	Blue	Mini Frisbee	4 ½"		$3.00-4.00	The frisbee is small and blue. The picture on the front is of a Smurf riding a red surfboard. The Smurf is wearing a white hat and white swim trunks.
080/033	White Smurf Sunglasses Case	BerDel	1/1/84	White	Sunglass Case	6 1/4"L		$3.50-5.00	The case is white and blue. It is a soft material. It has a picture of Brainy Smurf's face on the front. The re is a flap on one end to put the glasses in.
080/034	Smurf Beach Scene Beach Towel	AT & TV Mrch..	1/1/95		Beach Towel	4' 11" x 2' 5"		$22.50-30.00	The beach towel has 3 Smurfs playing on the beach. One Smurf is standing on his hands in the sand. Another Smurf is laying on a beach towel sunbathing. The third Smurf is diving into the water. The towel has a red border.

Set 144: Bean Bags- Mini

144/001	Happy Smurf With Tongue Out	Kortekaas Merch	1/1/83		Cloth Bean Bag	4"	Italy	$22.50-30.00	The Smurf has a bean bag body and a hard rubber head. The Smurf is wearing an orange scarf, white pants and a white hat. The Smurf has a happy expression on his face and his tongue is hanging out on one side of his mouth.
144/002	Crying Smurf	Kortekaas Merch	1/1/83		Cloth Bean Bag	4"	Italy	$22.50-30.00	The Smurf has a bean bag body and a hard rubber head. The Smurf is wearing an orange scarf, white pants and a white hat. The Smurf has his mouth open in the form of an "O" and a tear is running down the left cheek
144/003	Smiling Smurf	Kortekaas Merch	1/1/83		Cloth Bean Bag	4"	Italy	$22.50-30.00	The Smurf has a bean bag body and a hard rubber head. The Smurf is wearing an orange scarf, white pants and a white hat. The Smurf has a big smile on his face.
144/004	King Smurf	Kortekaas Merch	1/1/83		Cloth Bean Bag	3"	Italy	$25.00-35.00	The Smurf has a bean bag body and a hard rubber head. The Smurf is wearing a white scarf, white pants and a yellow hat with a red crown on top. The Smurf has a smile on his face.

Set 268: Bedding And Blankets

268/001	Smurf Village Scene Curtains	Latex Industries			Curtains	60 x 24	U.S.A.	$7.50-10.00	The curtain has various village pictures. It has some Smurfs in a boat. It has Smurfette looking out a window with 3 Smurfs outside the door. It has a Smurf running through a finish line with a bunch of other Smurfs hanging around. It has a Smurf orchestra.
268/002	Fitted Sheet Village Scene	Latex Industries			Fitted Sheet	Twin	U.S.A.	$7.50-10.00	The fitted sheet has various village pictures. It has some Smurfs in a boat. It has Smurfette looking out a window with 3 Smurfs outside the door. It has a Smurf running through a finish line with a bunch of other Smurfs hanging around. It has a Smurf orchestra.
268/003	Flat Sheet Village Scene	Latex Industries			Flat Sheet	Twin	U.S.A.	$7.50-10.00	The flat sheet has various village pictures. It has some Smurfs in a boat. It has Smurfette looking out a window with 3 Smurfs outside the door. It has a Smurf running through a finish line with a bunch of other Smurfs hanging around. It has a Smurf orchestra.
268/004	Village Scene Yellow Comforter				Comforter			$22.50-30.00	The comforter has various village pictures. It has a picture of a castle with a King Smurf and a bunch of other Smurfs in front. It has a picture of a couple Smurfs looking out a window of a house at Papa. It has another scene of Smurfs with wheelbarrows and tools working around by the houses. The comforter has a yellow border. The back is white
268/005	Red Quilt With Smurf Holding A Daisy			Red	Quilted Blanket		Hand made	$11.00-15.00	The front of the quilt is red and has a Smurf holding a red daisy in the center. The quilt has a light blue border. The back is yellow with small red and blue flowers.
268/006	Yellow Home Sweet Home Blanket			Yellow	Blanket		Hand made	$7.50-10.00	The blanket is white with different colored dots and a yellow border. In the center is Papa sitting in a rocking chair reading a book to 3 Smurfs and Smurfette. The Smurfs have milk and cookies sitting next to them. The back of the blanket is yellow.
268/007	Bedspread With Castle Scene	Latex Industries			Bedspread	Twin	U.S.A.	$37.50-50.00	The bedspread has various village pictures. It has a picture of a castle with a King Smurf and a bunch of other Smurfs in front. It has a picture of a couple Smurfs looking out a window of a house at Papa. It has another scene of Smurfs with wheelbarrows and tools working around by the houses.
268/008	Party Scene Curtains	Latex Industries			Curtains		U.S.A.	$11.00-15.00	The curtains have a village scene. It has a scene where Papa and a bunch of Smurfs are sitting at the table eating. There are lanterns hanging around. There is another scene where Smurfette is in a balcony and a Smurf is down below serenading her.
268/009	Smurf Village River Scene				Comforter			$37.50-50.00	The bedspread has woods and a river on it. There are a bunch of Smurfs playing in the pond at the end of the river. There are houses and Smurfs all over. Papa is standing on a mushroom.
268/010	Smurfette Pink Fitted Sheet	Latex Industries	1/1/83	Pink	Fitted Sheet	Twin	U.S.A.	$11.00-15.00	The sheet is pink. It has several different pictures of Smurfette. In the center Smurfette is wearing a long purple and green robe. Another scene is Smurfette standing by a table holding 2 daisies. Smurfette is dressed as a ballerina. Smurfette is talking on a phone.
268/011	Smurfette Pink Pillowcase	Latex Industries	1/1/83	Pink	Pillow Case		U.S.A.	$3.50-5.00	The pillow case is pink. It has Smurfette on the front dressed as a queen. Smurfette is wearing a yellow crown with a purple and white robe.
268/012	6 Smurf Pillowcases	Latex Industries		White	Pillowcase		U.S.A.	$3.50-5.00	The pillowcase is white. It has Papa, Smurfette and 5 other Smurfs on it. The picture is the same on both sides.

Set 51: Belt Buckles

051/001	Smurfette With Flowers	Oden		Green	Metal Belt Buckle	For 1-1/4" Belts	U.S.A.	$9.00-12.00	Smurfette is standing and holding 3 orange flowers. The background is green. The belt buckle is metal.
051/002	Clown Smurf	Oden		Blue	Metal Belt Buckle	For 1-1/4" Belts	U.S.A.	$9.00-12.00	Clown Smurf is wearing yellow pants with red stripes, a white shirt, red bow tie and a white hat. He is standing and pointing his left hand at something. The background is blue. The belt buckle is metal.
051/003	Roller- skating Smurf	Oden		Brown	Metal Belt Buckle	For 1-1/4" Belts	U.S.A.	$9.00-12.00	A Smurf is wearing red straps and yellow wheeled rollerskates. The background is brown. The belt buckle is metal.
051/004	Golfer	Oden		Orange	Metal Belt Buckle	For 1-1/4" Belts	U.S.A.	$9.00-12.00	The Smurf is swinging a brown golf club. The Smurf is standing on a green patch of grass with a yellow flag for the 18th hole behind him. The background is orange. The belt buckle is metal.
051/005	Baseball Smurf	Oden		Green	Metal Belt Buckle	For 1-1/4" Belts	U.S.A.	$9.00-12.00	The Smurf is wearing red shorts, an orange shirt with a red "B" on the front and a white hat. He is carrying a yellow bat over his shoulder. The bat has a brown glove hanging off the end. The background is green. The belt buckle is metal.
051/006	Papa With Finger In The Air	Oden		Orange	Metal Belt Buckle	For 1 1/4" Belts	U.S.A.	$9.00-12.00	Papa is standing and holding a finger in the air. He has his eyes closed and a smile on his face. The background is orange. The belt buckle is metal.
051/007	Popsicle Smurf	Oden		Brown	Metal Belt Buckle	For 1 1/4" Belts	U.S.A.	$9.00-12.00	A Smurf is standing and eating a popsicle. The top of the popsicle is white and the bottom is pink. The Smurf is licking his lips. The background is brown. The belt buckle is metal.
051/008	Bowling Smurf	Oden		Brown	Metal Belt Buckle	For 1 1/4" Belts	U.S.A.	$9.00-12.00	A Smurf is running like he's going to throw the bowling ball. The Smurf is holding a black ball in his right hand. The background is brown. The belt buckle is metal.
051/009	Flying Smurf	Oden		Blue	Metal Belt Buckle	For 1 1/4" Belts	U.S.A.	$9.00-12.00	A Smurf has red and yellow wings attached to his arms. The Smurf is flying in the air with white clouds around him. The background is blue. The belt buckle is metal.
051/010	Mailman	Oden		Blue	Metal Belt Buckle	For 1 1/4" Belts	U.S.A.	$9.00-12.00	A Smurf has a brown sack with white envelopes inside over his shoulder. He is carrying a white envelope with a red heart in his left hand. He is holding a yellow horn to his mouth with his other hand. The background is blue. The belt buckle is metal.
051/011	Fishing Smurf	Oden		Green	Metal Belt Buckle	For 1 1/4" Belts	U.S.A.	$9.00-12.00	A Smurf is sitting next to a pond of water. The Smurf's holding a brown fishing pole with a yellow string and a red and white bobber on the end. The background is green. The belt buckle is metal.
051/012	Astro Smurf	Oden		Green	Metal Belt Buckle	For 1 1/4" Belts	U.S.A.	$9.00-12.00	A Smurf is wearing a white spacesuit and helmet. The Smurf is walking in a blue bubble with a yellow star in one corner and a rainbow behind. The background is green. The belt buckle is metal.
051/013	Hockey Smurf	Oden		Orange	Metal Belt Buckle	For 1 1/4" Belts	U.S.A.	$9.00-12.00	The Smurf is wearing black shorts, a white tank shirt, a yellow helmet and socks, black skates and red gloves. He is holding a brown hockey stick. The background is orange. The belt buckle is metal.
051/014	Soccer Smurf	Oden		Orange	Metal Belt Buckle	For 1 1/4" Belts	U.S.A.	$9.00-12.00	The Smurf is wearing black shorts, an orange shirt, white shoes and a white hat. He is kicking a black and white soccer ball. The background is orange. The belt buckle is metal.
051/015	Smurf Giving Smurfette Flowers				Metal Belt Buckle	For 1 1/4" Belts		$15.00-20.00	The belt buckle is silver metal. The front has a Smurf handing Smurfette yellow daisies. Smurfette is wearing a red dress. The background is light blue.

Set 52: Belts

052/001	Jockey Smurf	Pyramid Belt Co.		Multi	Expandable Belts	Fits 18" - 28"	U.S.A.	$6.00-8.00	The belt is rainbow colored and it has a magnetic buckle that has a Smurf carrying a yellow box with a red bow. The background for the buckle is white.
052/002	Brainy	Pyramid Belt Co.		Light Blue	Expandable Belts	Fits 18" - 28"	U.S.A.	$3.50-5.00	The belt is light blue colored and has different Smurfs all the way around the belt. It has a magnetic buckle that has Brainy Smurf with his hand in the air. The background for the buckle is white.
052/003	Smurfette's Face	Pyramid Belt Co.		Multi	Expandable Belts	Fits 18" - 28"	U.S.A.	$3.00-4.00	The belt is rainbow colored and it has Smurfette's face on the buckle.
052/004	Drummer	Peyo/SEPP	1/1/81		Expandable Belts	Fits 18" - 28"		$3.00-4.00	The belt is light blue colored and has different Smurfs all the way around the belt. It has a magnetic buckle that has a Smurf playing a red drum. The background for the buckle is white.
052/005	Red Smurfette Belt			Red	Expandable Belts	Fits 18" - 28"		$2.00-3.00	The belt is red. The buckle clip has a ceramic Smurfette on it. She is in a flirty pose.

Set 209: Bendies

Item #	Name	Manufacturer	Year	Color	Type	Size	Country	Price	Description
209/001	Regular Smurf	Bendy Toys			Soft Bendy	7 1/2" High	England	$45.00-60.00	The Smurf is made of a soft foam. The Smurf bends. The Smurf is a light blue. He has a white hat and white pants on.

Set 225: Bendies- Les Schtroumpfs

Item #	Name	Manufacturer	Year	Color	Type	Size	Country	Price	Description
225/001	Smurf With Ax	Ceji	1/1/83		Soft Bendy	4"	Macau	$11.00-15.00	The Smurf is made of a flexible plastic. The Smurf has a blue body, white hat and white pants. He comes with a yellow plastic ax.
225/002	Papa With Test Tube	Ceji	1/1/83		Soft Bendy	4"	Macau	$11.00-15.00	Papa is made of a flexible plastic. Papa has a blue body, red hat and red pants. He comes with a yellow plastic test tube.
225/003	Brainy With Book	Ceji	1/1/83		Soft Bendy	4"	Macau	$11.00-15.00	The Smurf is made of a flexible plastic. The Smurf has a blue body, white hat and white pants. He comes with a yellow plastic book.
225/004	Smurf With Spoon	Ceji	1/1/83		Soft Bendy	4"	Macau	$11.00-15.00	The Smurf is made of a flexible plastic. The Smurf has a blue body, white hat and white pants. The Smurf comes with a yellow plastic spoon.
225/005	Smurf With Horn	Ceji	1/1/83		Soft Bendy	4"	Macau	$11.00-15.00	The Smurf is made of a flexible plastic. The Smurf has a blue body, white hat and white pants. He comes with a yellow plastic horn.

Set 186: Bicycle Bells

Item #	Name	Manufacturer	Year	Color	Type	Size	Country	Price	Description
186/001	Smurfette And A Smurf on A Bicycle For 2	Peyo	1/1/95	Blue	Bike Bell	2" Diam	Brussels	$9.00-13.00	Smurfette and a Smurf are riding a bicycle for 2. There is a 3 mile marker in the background. The bell is blue metal with a yellow ringer switch.
186/002	Papa, Smurfette And Baby	Peyo	1/1/95	Yellow	Bike Bell	2" Diam	Brussels	$9.00-13.00	The bell is yellow metal with a red switch. It has a picture of Papa waving and Smurfette standing, holding Baby Smurf. Baby is wearing a pink outfit. They are standing in the grass.
186/003	Smurf Racing	Peyo	1/1/95	Yellow	Bike Bell	2" Diam	Brussels	$9.00-13.00	The bell is yellow metal with a blue switch. It has a picture of a Smurf riding a three wheel bike to a checkered flag. Papa is standing at the finish line waving the flag.
186/004	Smurf Hitting Another Smurf	Peyo	1/1/95	Red	Bike Bell	2" Diam	Brussels	$9.00-13.00	The bell is red metal with a yellow switch. It has a picture of a Smurf hitting Brainy Smurf over the head with a mallet. There are stars floating around Brainy's head.
186/005	Smurf Sleeping In A Hammock	Peyo	1/1/95	Red	Bike Bell	2" Diam	Brussels	$9.00-13.00	The bell is red metal with a blue switch. It has a picture of a Smurf laying in a hammock sleeping. There is a little table next to the hammock. The table has a glass and bowl on it. The picture is outside.
186/006	Smurfette Coloring A Picture	Peyo	1/1/95	Blue	Bike Bell	2" Diam	Brussels	$9.00-13.00	The bell is blue metal with a red switch. It has a picture of Smurfette and a Smurf standing by an easel. Smurfette is coloring a picture. The Smurf is standing around the corner of the easel offering Smurfette a piece of bread.

Set 120: Binders - 2 & 3 Rings

Item #	Name	Manufacturer	Year	Color	Type	Size	Country	Price	Description
120/001	To Smurf or Not To Smurf, That Is The Question	Mead	1/1/87	Yellow	3 Ring Binder	12" X 10"	U.S.A.	$7.50-10.00	Papa is holding 1 finger in the air on his left hand. Inside is a pad clip and pencil holder, 6 filing pockets and a 36 sheet note pad.
120/002	Smurf Laying In Grass Dreaming (Blue)	Mead	1/1/82	Blue	Organizer	12" X 10"	U.S.A.	$7.50-10.00	The binder has a picture of a Smurf laying in the grass sleeping and dreaming. There is a cloud above him. In the cloud is a picture of a bunch of Smurfs playing by a water hole. The picture is on 3 flaps.
120/003	Smurf Laying In Grass Dreaming (Black)	Mead	1/1/82	Black	Organizer	12" X 10"	U.S.A.	$7.50-10.00	The binder has a picture of a Smurf laying in the grass sleeping and dreaming. There is a cloud above him. In the cloud is a picture of a bunch of Smurfs playing by a water hole. The picture is on 3 flaps.
120/004	Smurf Band	Clairefontaine	1/1/98	Red	2 Ring Binder	9" x 8 1/2"	France	$6.00-8.00	On the cover a Smurf is singing, one is playing the drums and the third is playing guitar. On the back
120/005	Smurfs Snowboarding	Clairefontaine	1/1/98		2 Ring Binder	9" x 8 1/2"	France	$6.00-8.00	Smurfette is singing and a Smurf is playing the keyboard. They are on a stage. The background is red.
120/006	Smurf Rollerblading	Clairefontaine	1/1/98		2 Ring Binder	9" x 8 1/2"	Brussels	$6.00-8.00	The cover has 2 Smurfs and Smurfette rollerblading. Smurfette is racing a Smurf. The background is of mushroom houses. in the way of a Smurf. The background is of mushroom houses.

Set 273: Bobbing Heads

Item #	Name	Manufacturer	Year	Color	Type	Size	Country	Price	Description
273/001	Smurf With Ax	Simex			Bobbing Head Smurf	3"	West Germany	$15.00-20.00	The Smurf is made of plastic. His head is on a spring so it moves. The Smurf's arm also moves. The Smurf is standing and holding a red ax.
273/002	Smurf With Trumpet	Simex			Bobbing Head Smurf	3"	West Germany	$15.00-20.00	The Smurf is made of plastic. His head is on a spring so it moves. The Smurf's arm also moves. The Smurf is standing and holding a silver trumpet.

Set 113: Book Bags & Back Packs

Item #	Name	Manufacturer	Year	Color	Type	Size	Country	Price	Description
113/001	Techno Smurf Book Bag	Schleich	1/1/96	Cream	Cloth Book Bag	10"W x 12"L	Brussels	$3.50-5.00	The bag is cream color and has 2 cloth handles. There is a Smurf wearing purple pants, white tennis shoes, a white tank top and a white hat with red and yellow on it. The Smurf has his hat on backwards and looks like he's brake dancing.
113/002	Smurflings Book Bag	Schleich	1/1/96	Cream	Cloth Book Bag	18"L x 15"W	Brussels	$5.00-7.00	The bag is cream color and has 2 cloth handles. The bag has Puppy on the front. Puppy is gray with brown spots... Slouchy is sitting on Puppy's head. Nat and Sassette are riding on Puppy's back. Snappy is being pulled behind by the leash.
113/003	Jokey Smurf Red Book Bag	BBC Imports		Red	Cloth Bag	11"L x 12"W	Taiwan	$7.50-10.00	The bag is red with 2 red handles and a snap to hold it closed. The front has a Smurf carrying a yellow package with red ribbon. It says "Smurf" on the top in yellow.
113/004	Smurf Carrying A Knapsack Over Shoulder Book Bag	Schleich		Light Blue	Cloth Book Bag	12 1/2" X 12 1/2"		$7.50-10.00	The bag is light blue with 2 light blue handles and a snap to hold it closed. The front has a Smurf carrying a stick over his shoulder. The stick has 2 red books hanging from it. The Smurf is whistling. The bag says "Smurf" on the top in red.
113/005	Smurf Sitting On Books	Schleich		Tan	Cloth Bag	13"L x 12"W		$4.50-6.00	The bag is tan with 2 tan handles and a snap to hold it closed. The front has a Smurf sitting on three books. There are 2 red books and a yellow book. The Smurf has his hands out at his side.
113/006	Papa Writing In A Book Briefcase	BBC Imports		Blue	Cloth Briefcase	10"L x 12"W		$7.50-10.00	The briefcase is dark blue with a close flap and a big yellow book. The front has Papa standing with his hand on his chin. He has a feather pen in his other hand. In front of Papa is a big yellow book. A candle is behind him.
113/007	Soccer Smurf Backpack	Schleich	1/1/83	Red	Nylon Backpack			$11.00-15.00	The backpack is red. It has foam padded shoulder straps and a large zipper opening. The front has a Smurf kicking a white and black soccer ball. The Smurf has a white hat, white shirt, black shorts and white shoes on. The top of the backpack says "Smurf" in white letters with blue trim.
113/008	Smurfette Cheerleader Book bag	Schleich		Tan	Cloth Book Bag	12" x 12"		$4.50-6.00	Smurfette is wearing a red cheerleader outfit. She is holding red pompoms. The book bag says "Smurfette" in red letters. The bag is tan with 2 tan handles. The bag snaps together in the middle.
113/009	Smurfette Cheerleader Book bag	Schleich		Red	Cloth Book Bag	12" x 12"		$4.50-6.00	Smurfette is wearing a red cheerleader outfit. She is holding red pompoms. The book bag says "Smurfette" in red letters. The bag is red with 2 red handles. The bag snaps together in the middle.
113/010	Smurf Carrying A Knapsack Over Shoulder Backpack	Schleich		Light Blue	Cloth Backpack Bag	13" x 11"		$4.50-6.00	The bag is light blue with 2 dark blue shoulder straps and flip over cover. The front has a Smurf carrying a stick over his shoulder. The stick has a red handkerchief sack hanging from it. The Smurf is whistling.
113/011	Smurf Holding Notebook Book bag	Schleich		Dark Blue	Cloth Book bag Bag	11" x 12 1/2"		$4.50 6.00	The bag is dark blue with 2 dark blue handles. It snaps shut in the center. The front has a Smurf holding a red notebook in one hand and a pencil in the other.
113/012	Smurf Band	Schleich	1/1/98	Cream	Cloth Bag	20" x 16 1/2"		$3.50-5.00	The front of the bag has 2 Smurfs playing guitars, a Smurf on a keyboard, a Smurf on drums and one Smurf singing into a microphone. The Smurfs are on a blue stage.
113/013	Smurfs Gardening Backpack	Schleich	1/1/97		Backpack	14" x 12"	Brussels	$18.00-25.00	The front has a big yellow mushroom house with an orange roof. There is a Smurf in the window of the house. 6 Smurfs are doing gardening stuff.

Set 17: Books & Cassettes- Read Along

Item #	Name	Manufacturer	Year	Color	Type	Size	Country	Price	Description
017/001	A Winter's Smurf	Starland Music			Cassette & Book		Brussels	$9.00-13.00	The cover of the book has 2 sad Smurfs carrying knapsacks, walking through the snow.
017/002	The Smurf Champion	Starland Music			Cassette & Book		Brussels	$9.00-13.00	The cover of the book has a Smurf carrying a torch and another Smurf standing on steps with a metal around his neck.
017/003	The Smurf-Eating Bird	Starland Music			Cassette & Book		Brussels	$9.00-13.00	The cover of the book has a Smurf running from a large bird.
017/004	Smurfing In The Air	Kid Stuff	1/1/84		Cassette & Book		Brussels	$9.00-13.00	The cover has a Smurf with red and white wings flying.

Set 18: Books & Records- Read Along

018/001	A Winter's Smurf	Starland Music			45 Record & Book		Brussels	$9.00-13.00	The cover of the book has 2 sad Smurfs carrying knapsacks, walking through the snow.
018/002	The Smurf Champion	Starland Music			45 Record & Book		Brussels	$9.00-13.00	The cover of the book has a Smurf carrying a torch and another Smurf standing on steps with a metal around his neck.
018/003	There's A Smurf In My Soup	Starland Music			45 Record & Book		Brussels	$9.00-13.00	The cover has Bigmouth watching the Smurf's make soup.
018/004	The Smurf-Eating Bird	Starland Music			45 Record & Book		Brussels	$9.00-13.00	The cover of the book has a Smurf running from a large bird.
018/005	Smurf's Daydream	Kid Stuff	1/1/84		45 Record & Book		Brussels	$9.00-13.00	A Smurf is walking down a path into the sunset carrying a red and white knapsack.
018/006	Smurfing In The Air	Starland Music	1/1/84	Light Blue	45 Record & Book		Brussels	$9.00-13.00	The cover has a Smurf with red and white wings flying.
018/007	La Mouche Bzz	Annihanbar Music	1/1/81	Yellow	45 Record & Book		Brussels	$11.00-15.00	The cover is yellow and has an angry purple Smurf chasing a blue Smurf. A black bee is chasing the blue Smurf also.

Set 10: Books- Activity And Misc.

010/001	The Smurf Stamp Fun	Happy House	1/1/82		Soft Cover Book	8"x 5"	Brussels	$3.00-4.00	The book has 32 Smurf stamps, dot-to-dots, and mazes
010/002	The Smurfs Sticker Book	Topps	1/1/82		Sticker Book	9 1/2" x 10 1/2"	Italy	$7.50-10.00	The cover has a Smurf, Smurfette and Papa standing in front of a mushroom.
010/003	The Smurf All Star Show	Hal Leonard Corp.	1/1/75		Soft Cover Book	9" x 12"	U.S.A.	$3.00-4.00	The cover has a band playing music.
010/004	Baker Smurf's Sniffy Book	Random House	1/1/82		Sticker Book	6 1/2" x 6"	Brussels	$3.00-4.00	The cover has a Smurf carrying a cake. A story with 6 fragrances to scratch and sniff.
010/005	Smurf Water Fun	Random House	1/1/83	Yellow	Soft Plastic Book		U.S.A.	$3.50-5.00	The book is soft plastic and is water proof. The cover has a Smurf taking a bath.
010/006	Ice Capades Book	Ice Capades	1/1/87		Soft Cover Book	13"L X 9 1/2"W	U.S.A.	$7.50-10.00	The book is for the 1987 ice capades. The Smurfs were in the program. The back cover has a picture of the Smurfs ice skating and Gargamel hiding behind a tree watching.
010/007	Bananas Visits The Smurfs	Scholastic	1/1/83		Magazine	11" L X 8"W	U.S.A.	$3.00-4.00	The cover has a foot stepping on 5 Smurfs. The Smurfs on the cover are different. Smurfette and Papa have red hair. Inside the book is a cartoon on Smurfs.
010/008	Kein Schlumpf Wie Die Anderen	AT & TV Mrch..	1/1/95		Mini, Mini Book	2"L X 1 1/2"W	Brussels	$1.50-2.00	The book is very tiny. The cover is yellow and has a picture of a Smurf carrying a red knapsack over his shoulder. There are 3 Smurfs and Papa looking at the Smurf leaving.
010/009	Baby Schlumpfe Book	Helly/Peyo	1/1/94	White	Soft Cover Book	5" x 5"	Germany	$3.50-5.00	The book is soft plastic. It is water proof. On the cover is Baby Smurf standing and leaning on a red and white ball. Baby has a pink sleeper and hat on. The book squeaks.
010/010	Jouons Schtroumpf Volume 1	Les Editions Heritage	1/1/80	Red	Activity Book	10 1/2"L x 8"W	Canada	$6.00-8.00	The cover is red and has a Smurf doing a cross word puzzle on it. The Smurf is laughing and holding a big green pencil. The activities are in French.
010/011	Jouons Schtroumpf Volume 2	Les Editions Heritage	1/1/80	Yellow	Activity Book	10 1/2"L x 8"W	Canada	$6.00-8.00	The cover is yellow and has Papa Smurf standing on a letter block. Papa is wearing a white hat and white pants. Papa is holding a stick with a rope that is attached to a Smurf's hat, like a puppet. The Smurf is Brainy.
010/012	Hot Dog Magazine	Scholastic	1/1/82	Pink	Soft Cover Book		U.S.A.	$4.50-6.00	The cover is pink with a yellow sun and green grass on it. A Smurf is standing and playing a brown guitar. The article in the book is called "It's a SMURF World!"
010/013	Smurfing Sing Song And Fun Book	Hal Leonard	1/1/81	Blue	Soft Cover Book	12"L X 9" W	U.S.A.	$7.50-10.00	The cover is blue and has mushroom houses on the front. There is a Smurf playing a tuba, a Smurf playing a drum, a Smurf holding sheet music and a Smurf carrying a pail in the background.
010/014	Doll World	Hs. Of Wht. Britches	2/1/91		Magazine	10 1/2"L X 8"W	U.S.A.	$6.00-8.00	Doll World has an article called Smurfin' U.S.A.!
010/015	Le Village Des Schtroumpfs	Peyo/SEPP/PAF	1/1/80	Orange	Activity Book	11"L X 8"W	France	$6.00-8.00	The book is a punch out book. It makes a village scene. The cover is orange and has a picture of 3 mushroom houses and Gargamel's castle. There are 7 Smurfs in front of the houses.
010/016	Schtroumpf De Nouvelees Bandes Dessinees	Cartoon Crtn /Peyo	11/1/89	Yellow	Activity Book	12"L X 8 1/4"W	Brussels	$9.00-12.00	The cover is yellow and has a Smurf opening a book and a Gargamel puppet popping out. Papa and another Smurf is standing and laughing. Brainy is standing and preaching. There are mushroom houses in the background. Inside the book is a punch out Gargamel castle.
010/017	Schtroumpfs Vol. #1	Jesco	1/1/84	Pink	Activity Book	8 1/2"L x 7 1/2"W	France	$6.00-8.00	The cover is pink with rainbow colors on the top. On the center of the cover is a script of paper and a Smurf is busting through it. The Smurf is in a red ski outfit. The Smurf is wearing skis and holding ski poles.
010/018	Schtroumpfs Vol. #2	Jesco	1/1/84	Orange	Soft Cover Book	8 1/2"L x 7 1/2"W	France	$6.00-8.00	The cover is orange with rainbow colors on the top. On the center of the cover is a mushroom with 2 Smurfs. The 2 Smurfs are doing activity pages.
010/019	Schtroumpfs Vol. #3	Jesco	1/1/84	Yellow	Soft Cover Book	8 1/2"L x 7 1/2"W	France	$6.00-8.00	The cover is yellow with rainbow colors on the top. On the center of the cover is Papa standing in front of a big book. Papa is holding a white feather pen. The book that is in front of Papa has Gargamel on the front.
010/020	Schtroumpfs Vol. #4	Jesco	1/1/84	Dark Orange	Soft Cover Book	8 1/2"L x 7 1/2"W	France	$6.00-8.00	The cover is dark orange with rainbow colors on the top. On the center of the cover is Smurfette and a Smurf sitting in front of a canvas picture. The canvas picture has a Smurf in the corner and birds.
010/021	The Smurf Activity Book	Random House	1/1/83	Yellow	Activity Book	9" X 8"	Brussels	$3.50-5.00	The book has a yellow cover and it has 3 Smurfs standing by a table. One Smurf is holding a plate of cookies, another Smurf is looking at a book and the other Smurf is pulling ribbon from a box. The book has mazes, board games, word puzzles, riddles, and other things to make.
010/022	Toy Collector And Price Guide	Krause	6/1/94		Soft Cover Magazine	10 1/2"L X 8"W	U.S.A.	$3.00-4.00	There is an article about Smurfs.
010/023	The Smurfs Sticker Activity Book	Introduct Holland	1/1/97		Sticker Book	6 1/4" x 9"	Brussels	$3.50-5.00	The cover has a Smurf laying in the grass poking a ladybug. Inside are color pages and 35 stickers.
010/024	The Smurfs And Their Woodland Friends	Random House	1/1/83		Chunky Book	3 1/2"L x 3"W	Brussels	$3.00-4.00	The cover has 3 Smurfs walking with wild animals.
010/025	Uberraschung Papa Schlumpf Aus Dem Ei!	Alle Abbildungen	1/1/96	Blue	Soft Cover Book	11 1/2" x 8 1/4"	Germany	$6.00-8.00	The magazine has Papa Smurf on the cover. The magazine has kinder Smurfs inside.
010/026	de 14-Daagse 1996	Rode Kruis	1/1/96		Soft Cover Book	8" x 5 1/2"	Brussels	$15.00-20.00	The cover has a girl wearing a gray sweatshirt with the Rode Kruis emblem on. The girl is holding a fold- out of Smurf pictures. Papa Smurf is on the bottom. There is a white and red van behind the girl. There are Smurfs in the pamphlet. The pamphlet came with a promotional pin. The pin is black and white. The pin has an outline of a Smurf banging a drum. The bottom of the pin has a red cross and say's Fiesta 96.
010/027	TV Guide April 21-27, 1990	TV Guide	4/21/90		TV Guide	7 1/2" x 5"	U.S.A.	$7.50-10.00	The cover has Arnold Schwarzenegger. Inside is an article about the "Cartoon All-Star to the Rescue". There is a Smurf shown in the picture.
010/028	Smurfing Safari	Budget Books	1/1/83		Soft Plastic Book	5 1/2" x 5 1/2"	Australia	$3.50-5.00	The book has a Smurf carrying a surfboard and another Smurf carrying an umbrella and lawn chair. The Smurfs are going to the beach. The book is a water book.
010/029	WB Licensing Smurf Book	SCCI			Soft Cover Book	11 x 8 1/2		$18.00-25.00	The book is from the Smurf collectors club. It is a book on licensing material. The pages are all black and white.
010/030	Smurf Collector Club Book	SCCI		Blue	Soft Cover Book	11 x 8 1/2"		$37.50-50.00	The book has a blue cover with a Smurf sticker on the front. Inside are old newsletters from the club.
010/031	Smurf Club Figure Book	SCCI		Blue	Soft Cover Book	11 x 8 1/2		$18.00-25.00	The cover has different figures on the front. The book is in black and white. It has pictures of the figures with a list of names and numbers.
010/032	A Smurf Village Punch-Out Book	Random House	1/1/82	Blue	Soft Cover Book	8 1/2" x 12"		$9.00-12.00	The book is blue. Inside is a punch out village and characters. The cover has a picture of a village.
010/033	Die Schlumpfe Fix Und Foxi Album	Von Peyo & Delporte		Purple	Soft Cover Book	11" x 8 3/4"	Italy	$6.00 - 7.50	The cover is purple . It has Papa Smurf pumping up a big Smurf balloon. 5 other Smurfs are dancing around.
010/034	Puffa Con Noi! All'mInterno Tante Novita	Edigamma	11/1/97	Blue	Activity Book	11" x 8"	Brussels	$3.00 - 3.50	The cover has a Smurf holding leaves over his head to cover himself from the rain. The inside has activities and mazes.

Set 188: Books- Dutch Books

188/001	De Zwarte Smurfen	Dupuis	1/1/75	Yellow	Soft Cover Book	11 1/2"L x 8 1/2"W	Netherlands	$5.00-7.00	The cover is yellow and has a Smurf standing and holding a red flower. Behind him a black Smurf is coming out of the bushes.
188/002	De Smurfen En De Krwakakrwa	Dupuis	1/1/76	Light Blue	Soft Cover Book	11 1/2"L x 8 1/2"W	Netherlands	$5.00-7.00	The cover is light blue and has a green and black hawk swooping down on 4 Smurfs and Papa.
188/003	De Ruimtesmurf	Dupuis	1/1/76	Dark Blue	Soft Cover Book	11 1/2"L x 8 1/2"W	Netherlands	$5.00-7.00	The cover is dark blue and has Astro Smurf walking in space. The red space ship is in back. Wild Smurf is on the cliff watching Astro Smurf.
188/004	De Leerling Smurf	Dupuis	1/1/75		Soft Cover Book	11 1/2"L x 8 1/2"W	Netherlands	$5.00-7.00	The cover has a Smurf sitting in the middle holding a test-tube that exploded. The Smurf looks dazzled and there is a mess around him.
188/005	Baby-Smurf Wil Een Beer	Cartoon Crtn	1/1/91	Pink	Soft Cover Book	11 1/2"L x 8 1/2"W	Netherlands	$5.00-7.00	The cover is pink and has Baby Smurf hugging a teddy bear. There is a shadow of a big bear on the side. There are toys laying around.
188/006	De SuperSmurf	Dupuis	1/1/83	Blue	Soft Cover Book	11 1/2"L x 8 1/2"W	Netherlands	$5.00-7.00	The cover is blue. It has a Smurf flying through the air with a red cape on and black glasses. The Smurf is flying towards a house. Coming out the door of the house is Bigmouth.
188/007	Gargamel En De Fonkelstenen	Dupuis	1/1/83	Green	Soft Cover Book	11 1/2"L x 8 1/2"W	Netherlands	$5.00-7.00	The cover is green. It has Gargamel crawling in the grass patting a Smurf on the head. 6 other Smurfs are watching.
188/008	De Fluit Met Zes Smurfen	Dupuis	1/1/75	Blue	Soft Cover Book	11 1/2"L x 8 1/2"W	Netherlands	$5.00-7.00	The cover has Peewit playing a flute. Johan is flying upside down in the air. A bunch of Smurfs are watching the activity.
188/009	De Smurfuhrer	Dupuis	1/1/78	Tan	Soft Cover Book	11 1/2"L x 8 1/2"W	Netherlands	$5.00-7.00	The cover has King Smurf standing on the top of a step. Another Smurf is walking up to him carrying a present. The cover is tan
188/010	De Smurfin	Dupuis	1/1/76	Green	Soft Cover Book	11 1/2"L x 8 1/2"W	Netherlands	$5.00-7.00	The cover is green. It has a Smurf handing Smurfette a bouquet of flowers.
188/011	Smurfekoppen En Koppige Smurfen	Dupuis	1/1/75	Blue	Soft Cover Book	11 1/2"L x 8 1/2"W	Netherlands	$5.00-7.00	The cover has 2 Smurfs fighting. One is holding a mallet and the other has a stick. Papa is walking towards them looking worried.
188/012	De GeldSmurf	Lombard	11/25/92		Soft Cover Book	11 1/2"L x 8 1/2"W	Brussels	$5.00 - 7.00	The cover is brown and purple. The cover has a safe. One Smurf is walking with a gold penny. Another Smurf is holding the key to the safe and a script. The safe has money and papers inside.
188/013	De Juwelen-Smurfer	Le Lombard	11/1/94	Blue	Soft Cover Book	11 1/2"L x 8 1/2"W	Brussels	$5.00 - 7.00	The cover is blue. It has a masked Smurf sneaking out of a cave with jewels.
188/014	Dokter Smurf	Le Lombard	10/1/96	Light Blue	Soft Cover Book	11 1/2"L x 8 1/2"W	Brussels	$5.00 - 7.00	The cover has Papa laying in bed sick. Doctor Smurf is taking Papa's pulse. Smurfette is dressed as a nurse. She is carrying a tray with a thermometer and water on it.
188/015	Smurfensoep	Dupuis	1/1/76	Yellow	Soft Cover Book	11 1/2"L x 8 1/2"W	Brussels	$3.50-5.00	The cover is yellow. It has Smurfs swimming in a kettle that is over a fire. A Smurf is standing on the side holding a spoon.
188/016	De Wilde Smurf	Le Lombard	1/1/98	Green	Soft Cover Book	11 1/2"L x 8 1/2"W	Brussels	$6.00-8.00	The cover has trees and grass. A Smurf wearing leaf clothing is hiding in the trees watching the Smurfs work below.

Set 227: Books- French

227/001	La Schtroumpfette	Dupuis	1/1/67	Light Green	Hard cover Book	8 1/2" x 11 3/4"	France	$11.00-15.00	The cover is light green and has a Smurf giving Smurfette flowers.
227/002	L'oeuf Et Les Schtroumpfs	Dupuis	1/1/78	Red	Hard cover Book	8 1/2" x 11 3/4"	France	$11.00-15.00	There are 2 Smurfs playing with an egg and sausage on the cover. The sausage has a Smurf hat. One Smurf is holding a yellow spoon.
227/003	Le Cosmoschtroumpf	Dupuis	1/1/70	Dark Blue	Hard cover Book	8 1/2" x 11 3/4"	France	$11.00-15.00	The cover has a space Smurf walking with a spaceship behind him. Jungle Smurf is standing on a rock watching Astro Smurf.
227/004	Le Bebe Schtroumpf	Dupuis	1/1/84	Pink	Hard cover Book	8 1/2" x 11 3/4"	France	$11.00-15.00	The cover has Smurfette, Papa and 4 other Smurfs standing around in a circle looking at Baby Smurf. The Baby is laying in a basket holding a red rattle.
227/005	Les P'tits Schtroumpfs	Dupuis	1/1/88	Light Blue	Hard cover Book	8 1/2" x 11 3/4"	France	$11.00-15.00	The cover has Nat, Sassette, Slouchy and Snappy walking in the grass. There are mushroom houses in the back.
227/006	Le Schtroumpfissime	Dupuis	1/1/65	Tan	Hard cover Book	8 1/2" x 11 3/4"	Belgium	$11.00-15.00	The cover is tan. It has a Smurf bringing a present to King Smurf.
227/007	Les Schtroumpfs Et Le Cracoucass	Dupuis	1/1/69	Light Blue	Hard cover Book	8 1/2" x 11 3/4"	Belgium	$11.00-15.00	The cover is light blue. It has a big hawk swooping down after 3 Smurfs and Papa. The Smurfs look scared.
227/008	L' Apprenti Schtroumpf	Dupuis	1/1/71		Hard cover Book	8 1/2" x 11 3/4"	Belgium	$11.00-15.00	The cover has a Smurf sitting and looking dazzled. He is holding an exploded test-tube. Things are laying all over.
227/009	Histoires De Schtroumpfs	Dupuis	1/1/72	Yellow	Hard cover Book	8 1/2" x 11 3/4"	Belgium	$11.00-15.00	The cover is yellow. It has a Smurf painting flowers in a cloud. Another Smurf is dreaming of the flowers.
227/010	Les Schtroumpfs Olympiques	Dupuis	1/1/83	Orange	Hard cover Book	8 1/2" x 11 3/4"	Belgium	$11.00-15.00	The cover is orange. It has 3 Smurfs running a marathon.
227/011	L' Etrange Reveil Du Schtroumpf Paresseux	Le Lombard	1/1/96		Hard cover Book	8 1/2" x 11 3/4"	Brussels	$11.00-15.00	The cover has a Smurf village on it. It has Smurfette, Papa and another Smurf all looking old. A young Smurf is looking on worried.
227/012	Le Schtroumpf Sauvage	Le Lombard	1/1/98	Green	Hard cover Book	8 1/2" x 11 3/4"	Brussels	$11.00-15.00	The cover has trees and grass on it. A Smurf wearing leaf clothing is hiding in the trees watching the Smurfs work below.
227/013	Les Schtroumpfs Noirs	Hemma	1/1/95		Hard cover sound book	12" x 8 1/2"	China	$22.50-30.00	The cover has a Smurf holding a red flower on it. A black angry Smurf is walking out of the bushes. The book has buttons on the side that make noise.

Set 175: Books- German

175/001	Blauschlumpfe Und Schwarzschlumpfe	Carlsen Comics	1/1/96		Soft Cover Books	11 1/2"L X 8 3/4"W	Germany	$4.50-6.00	The cover is yellow and has a black Smurf coming out of some bushes. There is another Smurf standing and holding a red flower with his back to the black Smurf.
175/002	Schlumpfissimus, Konig Der Schlumpfe	Carlsen Comics	1/1/96		Soft Cover Books	11 1/2"L X 8 3/4"W	Germany	$4.50-6.00	The cover is tan and has Jokey taking a present to King Smurf.
175/003	Schlumpfine	Carlsen Comics	1/1/96		Soft Cover Books	11 1/2"L X 8 3/4"W		$4.50-6.00	The cover is light green and has a Smurf giving Smurfette flowers.
175/004	Johann Und Pfiffikus Der Zauber Von Schwarzenfels	Carlsen Comics	1/1/96		Soft Cover Books	11 1/2"L X 8 3/4"W	Germany	$4.50-6.00	The cover has Johan standing in a room with his hand over his mouth. 6 Smurfs and Papa are coming through the door and they are laughing. There is a bulldog wearing Peewit's clothes.
175/005	Das Zauberei Und Die Schlumpfe	Carlsen Comics	1/1/96		Soft Cover Books	11 1/2"L X 8 3/4"W	Germany	$4.50-6.00	The cover is red and has a Smurf holding a spoon with an egg in front of him. There is another Smurf standing and looking at a sausage that is wearing a Smurf hat and running away.

Set 7: Books- Hard Cover

007/001	What Do Smurfs Do All Day?	Random House	1/1/83	Blue	Dr. Seuss Book	6 1/4" x 8 3/4"	U.S.A.	$3.50-5.00	The cover has various Smurfs doing different activities in the village. Regular hard cover blue.
007/002	The Smurfs And The Toy Shop	Random House	1/1/83	Yellow	Hard Cover Book	10" x 7 1/4"	U.S.A.	$3.50-5.00	The cover has Smurfs making toys.
007/003	The Smurfs And The Miller	Random House	1/1/84	Blue/Green	Hard Cover Book	10" x 7 1/4"	U.S.A.	$3.50-5.00	The cover has Smurfs carrying sacks of grain to a windmill.
007/004	The Art Of Hanna-Barbera	Viking Studio	1/1/89		Hard Cover Book	11 1/4" x 11 1/4"	Japan	$37.50 50.00	The cover has different cartoon characters on it. The book is Fifty Years Of Creativity.
007/007	King Smurf/The Astrosmurf	Random House	1/1/78		2 Story Book	8 1/4" x 11"		$3.50-5.00	The book has one story on one side, flip the book over for a second story.
007/008	Unterwegs mit den Schlumpfen English	Langenscheidt	1/1/96	Yellow	Hard Cover Book	11" x 8 1/2"	Brussels	$3.50-5.00	The cover has a picnic scene on it. A Smurf is cooking on a grill, one is jumping rope, Smurfette's sitting on a rock, a Smurf is laying on a blanket, another Smurf is sitting by the water and one is running. The book has English and German words in it.
007/009	What Do Smurfs Do All Day?	Random House	1/24/83		Hard Cover Book	9 1/4"L X 6 3/4"W	U.S.A.	$3.50-5.00	The cover has various Smurfs doing different activities in the village. White laminated cover.
007/010	Baby Smurf's First Words	Random House	1/1/84	Pink	Hard Cover Book	8"L X 8"W	U.S.A.	$3.50-5.00	The cover has Baby Smurf on his hand and knees looking at a book. Baby is wearing a light blue sleeper. There are 2 picture blocks on one side of Baby and another block on the other.
007/011	Romeo And Smurfette	Random House	1/1/78	Blue	Hard Cover Book	11"L X 8"W	U.S.A.	$3.50-5.00	The cover has a Smurf declaring his love to Smurfette. Smurfette is on a balcony. There are 12 other Smurfy stories in the book.
007/012	The Smurfs And The Magic Flute	Random House	1/1/75	Blue	Hard Cover Book	11" L X 8"W	U.S.A.	$3.50-5.00	The cover has Peewit playing the flute with several Smurfs listening. Johan is upside down in the air.
007/013	The Smurfic Games	Random House	1/1/75		Hard Cover Book	11"L X 8"W	U.S.A.	$3.50-5.00	The cover has a Smurf pole vaulting while a different Smurf sawed the pole in half and is standing there laughing.
007/014	Die Schlumpfe Der Grosse Schlumpf Erzahlt	Random House	1/1/97		Hard Cover Book	11 1/2" x 8 1/2"	Brussels	$7.50-10.00	The cover has 4 Smurfs, Papa and Smurfette dancing in the woods. 2 Smurfs are hiding behind a mushroom. There are mushroom houses in the background.
007/015	Schlumpfonie In C	AT & TV Mrch..	1/1/94		Sound Book	12" x 8 1/2"		$8.25-12.00	The cover has a Smurf band on it. Papa is standing on a mushroom with a conductors stick. The side of the book has buttons that when pushed make different sounds.

Set 198: Books- Mini Die Schlumpfe German

198/001	Die Schlaftorte	Bastei	1/1/92	Yellow	Small book	5"L X 5 1/4"W	Brussels	$2.00-3.00	The cover is yellow and has a Smurf at a table. The table has a small Smurf cake sitting on top. The Smurf is reaching for the desert and licking his mouth.
198/002	Das Zauberwasser	Bastei	1/1/92	Orange	Small book	5"L X 5 1/4"W	Brussels	$2.00-3.00	The cover has 3 Papa Smurfs and a Smurfette sitting at a table eating. They all look old and are wearing gray hats.
198/003	Die Verwandlung	Bastei	1/1/93	Light Blue	Small book	5"L X 5 1/4"W	Brussels	$2.00-3.00	The cover is light blue. A Smurf is swimming in water with a green monster. The Smurf is holding a wand.
198/004	Viel Gebrull Um Den Mull	Bastei	1/1/93	Yellow	Small book	5"L X 5 1/4"W	Brussels	$2.00-3.00	The cover has a Smurf sitting on a bank of grass while fishing. The Smurf looks mad. There is a big pile of junk sitting on the bank behind him.
198/005	Wie Ein Schlumpf Zum Griesgram Wurde	Bastei	1/1/93	Light Blue	Small book	5"L X 5 1/4"W	Brussels	$2.00-3.00	The cover has 3 Smurfs playing ball in front of a mushroom house. There is another Smurf standing off by himself behind a bush looking sad.
198/006	Ein Hauustier Fur Schlumpfinchen	Bastei	1/1/93	Pink	Small book	5"L X 5 1/4"W	Brussels	$2.00-3.00	The cover has a Smurf looking at Smurfette with a silly grin on his face. Smurfette is standing in front of a mirror. Smurfette is wearing a green live caterpillar fur. Smurfette has a shocked expression on her face.
198/007	Ewige Nacht im Schlumpf-Dorf	Bastei	1/1/93	Dark Blue	Small book	5"L X 5 1/4"W	Brussels	$2.00-3.00	The cover has 4 Smurfs that look troubled with Papa. The background is dark blue for nighttime.

Set 155: Books- Mini- Mini German

155/001	Brilly Hat Ein Problem	Favorit-V-R	1/1/83	Blue	Mini-Mini Book	3 3/4" X 4 1/2"	Germany	$3.00-4.00	The book is a mini-mini. The cover has a Smurf laying in a hammock between 2 vines. Brainy is standing at the foot of the hammock.
155/002	Gefahr Fur Die Schlumpfe	Favorit-V-R	1/1/83	Blue	Mini-Mini Book	3 3/4" X 4 1/2"	Germany	$3.00-4.00	The book is a mini-mini. The cover has Brainy running from a mushroom house with smoke behind him. There is another Smurf standing on the side of the house looking startled.
155/003	Der Zaubertrank	Favorit-V-R	1/1/83	Blue	Mini-Mini Book	3 3/4" X 4 1/2"	Germany	$3.00-4.00	The book is a mini-mini. The cover has Brainy Smurf standing at a table holding a test tube.
155/004	Gargamel's List	Favorit-V-R	1/1/83	Blue	Mini-Mini Book	3 3/4" X 4 1/2"	Germany	$3.00-4.00	The book is a mini-mini. The cover has Gargamel and Azreal in the woods standing in front of a hole. A Smurf and Smurfette are standing on the other side of the hole looking scared.
155/005	Die Schlumpf-Falle	Favorit-V-R	1/1/83	Blue	Mini-Mini Book	3 3/4" X 4 1/2"	Germany	$3.00-4.00	The book is a mini-mini. The cover has Gargamel and Azreal in the woods. Gargamel is holding a net with Smurfette and another Smurf in it and there is a hole below the Smurfs.
155/006	Der Schatz Der Schlumpfe	Favorit-V-R	1/1/83	Blue	Mini-Mini Book	3 3/4" X 4 1/2"	Germany	$3.00-4.00	The book is a mini-mini. The cover has Gargamel laying on the ground with a diamond falling on his head.
155/007	Super - Schlumpf	Favorit-V-R	1/1/83	Blue	Mini-Mini Book	3 3/4" X 4 1/2"	Germany	$3.00-4.00	The book is a mini-mini. The cover has Brainy Smurf lifting dumbbells. There are weights sitting around him.

Set 14: Books- Pop-Up

014/001	The Smurf-Catching Trap	Random House	1/1/82	Green	Hard Cover Book	5 1/3" x 6"	Brussels	$2.00-3.00	The cover has 2 Smurfs walking by a tree with Gargamel hiding behind the tree.
014/002	A Little Smurf Bedtime Story	Random House	1/1/82	Yellow	Hard Cover Book	5 1/3" x 6"	Brussels	$2.00-3.00	The cover has Papa reading a Smurf a story.
014/003	Smurf On The Grow	Random House	1/1/82	Purple	Hard Cover Book	5 1/3" x 6"	Brussels	$2.00-3.00	The cover has Gargamel looking at a Smurf in a cage.
014/004	A Smurf Picnic	Random House	1/1/82	Yellow	Hard Cover Book	5 1/3" x 6"	Brussels	$2.00-3.00	The cover has Papa and 2 Smurfs walking out of the mushroom house with picnic things.
014/005	Der Geburtstag Des Grossen Schlumpfs	Hemma	1/1/93		Hard Cover Book	8"L X 9 3/4"W	Brussels	$11.00-15.00	The cover has a Smurf whispering to another Smurf. Papa is walking up a hill away from them. Papa is grinning. One of the Smurf's is pointing at a mushroom.
014/006	Ein Festbei Den Schlumpfen	Hemma	1/1/93		Hard Cover Book	8"L X 9 3/4"W	Brussels	$11.00-15.00	The cover is of the Smurfs having a party. Papa is standing on stage leading a band. Smurfette and 3 other Smurfs are dancing. A Smurf is eating cake. Lanterns are hung on wires.
014/007	De Baby Smurf	Hemma	1/1/93		Hard Cover Book	5 1/3" x 6"	Brussels	$11.00-15.00	The cover has Baby Smurf crawling in the grass towards Smurfette. Papa is standing and watching Baby. There are 2 mushroom houses in the background. There is a stork flying in the air.

Set 8: Books- Puppet

008/001	High And Dry Smurf	Intervisual Comm.	1/27/83	Multi	Puppet Book	7 1/3" x 9 1/3"	U.S.A.	$3.00-4.00	The cover has 2 Smurfs fishing in a dried up river.
008/002	Smurf King For A Day	Intervisual Comm.	1/27/83	Multi	Puppet Book	7 1/3" x 9 1/3"		$3.00-4.00	The cover has 2 Smurfs looking at a list on the wall.

008/003	Smurfette's Choice	Intervisual Comm.	1/1/83		Puppet Book	7 1/3" x 9 1/3"	U.S.A.	$4.50-6.00	The cover is pink and orange. The cover has a Smurf playing a horn and another Smurf playing a guitar under a window.

Set 278: Books- Schtroumpferies

278/001	1 Smurf Walking With Fish Bowl On Head	Le Lombard	1/1/94		Hard cover Book	11 1/2" x 9"	Brussels	$11.00-15.00	The cover has a Smurf walking with a fish bowl over his head. One Smurf is looking on worried and another is laughing.
278/002	2 Smurf Playing Ball Toss	Le Lombard	1/1/96		Hard cover Book	11 1/2" x 9"	Brussels	$11.00-15.00	The cover has a Smurf tossing a ball through the hole of a board. The board has Brainy painted on it. Brainy is on the other side and got hit by the ball. A third Smurf is standing in front laughing.

Set 11: Books- Small

011/001	The Smurf's Apprentice	Random House	1/1/82	Red	Soft Cover Book	5 1/2" x 5"	Brussels	$1.50-2.00	The cover has a Smurf mixing potion on a work bench.
011/002	The Fake Smurf	Random House	1/1/80	Dark Green	Soft Cover Book	5 1/2" x 5"	Brussels	$1.50-2.00	The cover has Papa yelling at a Smurf with an ax.
011/003	The Wandering Smurf	Random House	1/1/80	Orange	Soft Cover Book	5 1/2" x 5"	Brussels	$1.50-2.00	The cover has a Smurf sitting with his head in his hand dreaming about paradise.
011/004	The Weather-Smurfing Machine	Random House	1/1/82	Blue	Soft Cover Book	5 1/2" x 5"	Brussels	$1.50-2.00	The cover has a Smurf skiing in snow and a Smurf sitting on grass in a bathing suit.
011/005	Smurf Cake	Random House	1/1/00	Yellow	Soft Cover Book	5 1/2" x 5"	Brussels	$1.50-2.00	The cover has a Smurf stealing a piece of cake and the other Smurf has a rolling pin and knife yelling at the Smurf that stole the cake.
011/006	Smurphony In C	Random House	1/1/82	Yellow	Soft Cover Book	5 1/2" x 5"	Brussels	$1.50-2.00	The cover has a Smurf playing a shazalakazoo and one Smurf looks like he's getting blown away.
011/007	The Hundredth Smurf	Random House	1/1/82	Light Green	Soft Cover Book	5 1/2" x 5"	Brussels	$1.50-2.00	The cover has a Smurf looking at his reflection through a wood frame.
011/008	A Smurf In The Air	Random House	1/1/80	Light Blue	Soft Cover Book	5 1/2" x 5"	Brussels	$1.50-2.00	The cover has a Smurf with white wings flying.
011/010	Der Spiegelschlumpf	Bastei	1/1/81		Soft Cover Book	5 1/2" x 5"	Germany	$3.50-5.00	The cover has a mad Smurf fishing with a pile of junk sitting on the grass behind him.
011/011	Der Winterkonig	Ravensburger	1/1/83	Yellow	Soft Cover Book	5 1/2" x 5"	Italy	$3.50-5.00	The cover has a picture of a snow monster. Smurfette is standing on a chunk of ice by the monsters face. The monster has sad eye's.
011/012	Die Geheimnisvolle Hohle	Ravensburger	1/1/83	Yellow	Soft Cover Book	7"L X 4 3/4"W	Italy	$3.50-5.00	The Smurfs are following Papa through a tunnel. There is food and diamonds in the cliff.
011/013	De Winterkoning	Zuidnederlandse		Red	Soft Cover Book	7" x 4 3/4"	Netherlands	$3.50-5.00	The cover has a picture of a snow monster. Smurfette is standing on a chunk of ice by the monsters face. The monster has a weird expression on his face.
011/014	De RidderSmurf	Zuidnederlandse		Blue	Soft Cover Book	7" x 4 3/4"	Netherlands	$3.50-5.00	The cover has a Smurf wearing a suit of armor. Smurfette and 2 Smurfs are looking at the Smurf dressed in the suit of armor with a puzzled expression on their face.
011/015	De Geheimzinnige	Zuidnederlandse		Yellow	Soft Cover Book	5 1/2" x 5	Netherlands	$3.50-5.00	The cover has a Smurf in a cave holding a big diamond on it.

Set 9: Books- Soft Cover

009/001	King Smurf	Random House	1/1/77	Yellow	Soft Cover Book	8" x 11"	Brussels	$2.00-3.00	The cover has Jokey taking a present to King Smurf.
009/002	The Smurfs And The Magic Flute	Random House	1/1/75	Light Blue	Soft Cover Book	8" x 11"	U.S.A.	$2.00-3.00	The cover has Peewit playing the flute with several Smurfs listening. Johan is upside down in the air.
009/003	Romeo And Smurfette	Random House	1/1/77	Light Blue	Soft Cover Book	8" x 11"	U.S.A.	$2.00-3.00	The cover has a Smurf declaring his love to Smurfette. Smurfette is on a balcony. There are 12 other Smurfy stories in the book.
009/004	The astrosmurf	Random House	1/1/78	Dark Blue	Soft Cover Book	8" x 11"	U.S.A.	$2.00-3.00	The cover has astrosmurf walking in front of a spaceship.
009/005	Papa Smurf And The Show-Offs	Random House	1/1/83		Soft Cover Book	6 1/2" x 11"	Brussels	$2.00-3.00	Papa Smurf is on the cover.
009/006	The Smurf ABC Book	Random House	1/1/83	Light Blue	Soft Cover Book	8" x 8"	U.S.A.	$2.00-3.00	The cover has ABC on it with Smurfs playing on the letters.
009/007	Coloring Magic With Painter Smurf	Random House	1/1/83	Multi	Soft Cover Book	7 1/2" x 5 1/4"	Brussels	$2.00-3.00	The cover has Painter Smurf on it, and only half the book opens.
009/008	Through The Seasons With Smurfette	Random House	1/1/77		Soft Cover Book		Brussels	$2.00-3.00	The book is a half book and has Smurfette on the top.
009/009	Rotschlumpfchen Und Schlumpfkappchen	Semic	1/1/78	Yellow	Soft Cover Book	8 1/2" x 11"	Brussels	$3.50-5.00	The cover of the book is yellow and there are 2 Smurfs hitting each other over the head with signs.
009/011	The Smurfic Games	Random House	1/1/75	Light Blue	Soft Cover Book	8" x 11"	U.S.A.	$2.00-3.00	The cover has a Smurf pole vaulting and a different Smurf sawing the pole in half and standing there laughing.
009/012	The Smurfette	Hodder & Stoughton	1/1/76	Green	Soft Cover Book	8" x 11"	Great Britain	$2.00-3.00	The cover has a Smurf giving Smurfette flowers.
009/013	Smurphony In C & Flying Smurf	Hodder & Stoughton	1/1/78	Yellow	Soft Cover Book	8" x 11"	Great Britain	$2.00-3.00	The cover has a Smurf blowing a shazalakazoo that knocked out 2 Smurfs. The book has 2 stories.
009/014	Baby Smurf's First Words	Random House	1/1/84	Pink	Soft Cover Book	8" x 8"	U.S.A.	$2.00-3.00	The cover has Baby Smurf on his hands and knees looking at a book. Baby is wearing a light blue sleeper. There are 2 picture blocks on one side of Baby and another block on the other.
009/015	Rainy Day A Smurf Book Of Feelings	Random House	1/1/02		Soft Cover Book	6 1/2"H X 6"W	Brussels	$2.00-3.00	The cover is the shape of a mushroom. Papa is standing outside in the rain. There is a Smurf looking out the window and another standing in the door holding an umbrella.
009/016	My First Dictionary With The Smurfs	PAF	1/1/80	Yellow	Soft Cover Book	6 1/2"L X 4 1/4"W	Canada	$2.00-3.00	The cover is yellow and has a Smurf holding a red letter A, with a green letter Z laying on the ground. Another Smurf is holding a pointing stick pointing at the A.
009/017	Schlumpfe Figure Book	Schleich Gmbh		White	Soft Cover Book	8 1/4" x 8 1/4"	Germany	$18.00-25.00	The book is white. The cover has Smurf figures pictured. The book pictures figures from 1966 to 1986. The book has the history of Peyo. It also has little comics. The book is done in German.

Set 6: Books- Coloring

006/001	The Smurfs Learn-To-Read Coloring Book	Happy House	1/1/82	Light Blue	Coloring Book	8" x 11"	Brussels	$2.00-3.00	The cover has Papa Smurf leaning against a tree stump writing in a book.
006/002	Gargamel Strikes Again!	Happy House	1/1/82	Blue/Green	Coloring Book	8" x 11"	Brussels	$2.00-3.00	The cover has two Smurfs walking outside of Gargamel's castle on it.
006/003	A Smurf For All Seasons	Happy House	1/1/82	Red	Coloring Book	8" x 11"	Brussels	$2.00-3.00	The cover has a Smurf walking in the rain, a Smurf in a meadow, a Smurf diving off a diving board, and Papa dressed as Santa.
006/004	Smurf's Up! The Smurf Coloring Book Of Sports & Games	Happy House	1/1/82	Yellow	Coloring Book	8" x 11"	Brussels	$2.00-3.00	The cover has a Smurf carrying a bunch of sports equipment.
006/005	Great Moments In Smurf History	Happy House	1/1/82	Light Blue	Coloring Book	8" x 11"	Brussels	$2.00-3.00	The cover has Papa and 2 Smurfs in a civil war scene.
006/006	The Smurf Year-round Coloring Book	Happy House	1/1/83	Light Blue	Coloring Book	8" x 11"	Brussels	$2.00-3.00	The cover has a Smurf and Smurfette in a meadow of flowers on it.
006/007	The Smurf Puzzle Book	Happy House	1/23/83		Coloring/Puzzle Book	8" x 11"	Brussels	$2.00-3.00	The cover has a Smurf doing a word search that's pinned on to a tree.
006/008	De Smurfen Kleurboek	Hema B.V.	1/1/96	Red	Coloring Book	8' x 11'	Brussels	$2.00-3.00	The coloring book is red. On the front is a picture of a Smurf holding cut out numbers and a paint brush. Smurfette sitting on a stool cutting pink paper. Another Smurf is standing and cutting a moon out of a large piece of paper.
006/009	Smurfette Schtroumpf Coloring Book	Hemma	1/1/93		Coloring Book	8' x 11'	Belgium	$3.50-5.00	The cover of the coloring book has Smurfette holding a palette and a paintbrush.
006/010	Smurf Schtroumpf Schlumpf Color	Hemma	1/1/93		Coloring Book	8' x 11'	Belgium	$3.50-5.00	The cover of the book has a Smurf walking while carrying books and pencils. The Smurf has a bead of sweat running down his face.
006/011	[illegible] Coloring Book	Peyo	1/1/88	white	Coloring Book	16" x 11"	U.S.A.	$9.00-12.00	The front cover has a Smurf and Smurfette on ice-skates. They are holding hands.
006/012	Numbers With The Smurfs	PAF	1/1/80	White	Workbook/ Coloring	8' x 11'		$2.00-3.00	The cover has a Smurf pushing a wheelbarrow with numbers in it. Another Smurf is walking next to him carrying a number 5.
006/013	Here Comes The Smurfs	Happy House	1/1/83	Yellow	Coloring Book	8' x 11'	U.S.A.	$3.50-5.00	The coloring book is yellow and shows 7 different Smurfs on the front. A Smurf is fishing. A Smurf is playing in the rain. A Smurf is sunbathing. A Smurf is skiing. A Smurf is holding flowers. A Smurf is surfing. A Smurf is raking hay.
006/014	Paint Schlumpf Color Book	Hemma	1/1/93		Coloring Book	11 1/2" x 8 1/4"	Belgium	$3.50-5.00	The cover has a Smurf standing on a ladder painting pink and yellow. He has a bucket hanging on a hook on the side of the ladder.
006/015	Schtroumpf On A Stool Painting	Hemma	1/1/93		Coloring Book	11 1/2" x 8 1/4"	Belgium	$3.50-5.00	A Smurf is sitting on a swivel stool. He is painting a picture of a Smurf on a purple background..
006/016	Clown Smurf Coloring Book	Hemma	1/1/96	Pink	Coloring Book	12" x 8 1/2'	Brussels	$3.50-5.00	The coloring book is pink. It has a clown Smurf painting shapes on it. In the corner it has Smurfette painting a picture.

Set 271: Bracelets

271/001	Smurfette Ballerina Heart Bracelet	Applause		Pink/Purple	Plastic Bracelet		Taiwan	$4.50-6.00	The bracelet is pink and purple hearts attached together. One pink heart has Smurfette dressed as a ballerina on it. A pink heart on the other side has Smurfette written in it.

Set 195: Bubble Gum Wrappers

195/001	Brainy	T.M./Peyo		Red	Bubble Gum Wrapper	2 1/2"L X 2"W	France	$1.50-2.00	The wrapper is red with white trim. The wrapper has Brainy's face on the front. Brainy is wearing black glasses. The wrapper is for cherry gum.
195/002	Spy	T.M./Peyo		Red	Bubble Gum Wrapper	2 1/2"L X 2"W	France	$1.50-2.00	The wrapper is red with white trim. The wrapper has a masked Smurf face on the front. The Smurf is wearing a black mask over his eye's. The wrapper is for cherry gum.
195/003	Smurfette	T.M./Peyo		Red	Bubble Gum Wrapper	2 1/2"L X 2"W	France	$1.50-2.00	The wrapper is red with white trim. The wrapper has Smurfette's face on the front. The wrapper is for cherry gum.
195/004	Football Smurf	T.M./Peyo		Red	Bubble Gum Wrapper	2 1/2"L X 2"W	France	$1.50-2.00	The wrapper is red with white trim. The wrapper has Football Smurf's face on the front. The Smurf is wearing a football helmet. The wrapper is for cherry gum.
195/005	Clown	T.M./Peyo		Red	Bubble Gum Wrapper	2 1/2"L X 2"W	France	$1.50-2.00	The wrapper is red with white trim. The wrapper has a clown face on the front. The Smurf is wearing a hat with red stars on it and has a big red nose. The wrapper is for cherry gum.
195/006	Papa	T.M./Peyo		Red	Bubble Gum Wrapper	2 1/2"L X 2"W	France	$1.50-2.00	The wrapper is red with white trim. The wrapper has Papa's face on the front. The wrapper is for cherry gum.

195/007	Smurf	T.M./Peyo		Red	Bubble Gum Wrapper	2 1/2"L X 2"W	France	$1.50-2.00	The wrapper is red with white trim. The wrapper has a Smurf face on the front. The wrapper is for cherry gum.
195/008	Laughing Smurf	T.M./Peyo		Red	Bubble Gum Wrapper	2 1/2"L X 2"W	France	$1.50-2.00	The wrapper is red with white trim. The wrapper has a Smurf face on the front. The Smurf is laughing. The wrapper is for cherry gum.
195/009	Flower Smurf	T.M./Peyo		Red	Bubble Gum Wrapper	2 1/2"L X 2"W	France	$1.50-2.00	The wrapper is red with white trim. The wrapper has a Smurf face on the front. The Smurf has a pink daisy in his hat. The wrapper is for cherry gum.
195/010	Hungry Smurf	T.M./Peyo		Red	Bubble Gum Wrapper	2 1/2"L X 2"W	France	$1.50-2.00	The wrapper is red with white trim. The wrapper has a Smurf face on the front. The Smurf is licking his lips. The wrapper is for cherry gum.

Set 92: Bubble Pipes

092/001	Papa	Peyo	1/1/82		Blow Bubble Pipe	4 1/2" Long	Hong Kong	$3.50-5.00	The pipe is white and on the end it has Papa's face. You put bubbles in Papa's face. Papa's face is blue and he is wearing a red hat.
092/002	Smurf	Peyo	1/1/82		Blow Bubble Pipe	4 1/2" Long	Hong Kong	$3.50-5.00	The pipe is white and on the end it has a Smurf face. You put bubbles in the Smurf's face. The Smurfs face is blue and he is wearing a white hat.
092/003	Bubble Set	Imperial	1/1/90		Blow Bubble Pipe	8" X 10"	Hong Kong	$11.00-15.00	6" Blue wand is a Smurf figure, red tray is the shape of a Smurf. There is a green bottle of bubbles. The package is yellow and has a Smurf face in the right corner.

Set 82: Bucket Of Bolts- Smurf In A Bucket

082/001	Smurf N' Papa Smurf	Toysville	1/1/81	Red/Yellow	Plastic Nuts & Bolts		Hong Kong	$15.00-20.00	The car is red with a yellow engine and blower sticking out the hood. The wheels and bumpers are red and have yellow bolts holding them on. The floor boards are light blue. Papa is sticking out the top of the car and a Smurf with a flower in his hat is out the side window. There is a yellow key to take the car apart with. The car is a mechanical wind-up that goes forward. The car is plastic.
082/002	Smurf N' Papa Smurf	Toysville	1/1/81	Red/Yellow	Plastic Nuts & Bolts		Hong Kong	$15.00-20.00	The car is red with a yellow engine and blower sticking out the hood. The wheels and bumpers are yellow and have red bolts holding them on. The floor boards are red. Papa is sticking out the top of the car and a Smurf with a flower in his hat is out the side window. There is a yellow key to take the car apart with. The car is a mechanical wind-up that goes forward. The car is plastic.

Set 248: Bundesliga Soccer Smurfs

248/001	B. Monchengladbach	Schleich/Peyo			Promotional Smurf	2" High	West Germany	$37.50-50.00	The Smurf has a medium blue body. He has his arms out at his sides. His right foot has a white soccer ball on the tip. He is wearing a white shirt, white socks and white shorts. The shirt and pants have a green stripe on them. He has on black shoes. The Smurf represents the German soccer team B. Monchengladbach.
248/002	Eintracht Frankfurt	Schleich/Peyo			Promotional Smurf	2" High	West Germany	$37.50-50.00	The Smurf has a medium blue body. He has his arms out at his sides. His right foot has a white soccer ball on the tip. He is wearing a red shirt with black stripes, black socks and black shorts. He has on black shoes. The Smurf represents the German soccer team Eintracht Frankfurt.
248/003	MSV Duisburg	Schleich/Peyo			Promotional Smurf	2" High	West Germany	$37.50-50.00	The Smurf has a medium blue body. He has his arms out at his sides. His right foot has a white soccer ball on the tip. He is wearing a white shirt with blue stripes and sleeves, white and blue socks and white shorts. He has on black shoes. The Smurf represents the German soccer team MSV Duisburg
248/004	Arminia Bielefeld	Schleich/Peyo			Promotional Smurf	2" High	West Germany	$37.50-50.00	The Smurf has a medium blue body. He has his arms out at his sides. His right foot has a white soccer ball on the tip. He is wearing a dark blue shirt, blue socks and blue shorts with a white stripe. He has on black shoes. The Smurf represents the German soccer team Armina Bielefeld.
248/005	Herta BSC Berlin	Schleich/Peyo			Promotional Smurf	2" High	West Germany	$37.50-50.00	The Smurf has a medium blue body. He has his arms out at his sides. His right foot has a white soccer ball on the tip. He is wearing a light blue shirt with a white collar, light blue socks and white shorts with a dark blue stripe. He has on black shoes. The Smurf represents the German soccer team Herta BSC Berlin.
248/006	1860 Munchen	Schleich/Peyo			Promotional Smurf	2" High	West Germany	$37.50-50.00	The Smurf has a medium blue body. He has his arms out at his sides. His right foot has a white soccer ball on the tip. He is wearing a dark blue shirt, white socks and white shorts with a dark blue stripe. He has on black shoes. The Smurf represents the German soccer team 1860 Munchen.

Set 223: Burago Diecast Jeeps

223/001	Pink Jeep	Burago	1/1/84	Pink	Diecast Jeep	1:24 Scale	Italy	$75.00-100.00	The jeep is pink with a yellow plastic interior and blue tires. There are 2 Smurfs in the front seat. There are 2 green jerry cans in the back. On one side in back is a gray shovel and there are 2 red gas cans on the other side. There is a spare blue tire attached to the back. The hood has a Smurf standing by a mushroom house decal. The windshield folds down.
223/002	Blue Lamborghini Cheetah	Burago	1/1/84	Light Blue	Diecast Jeep	1:24 Scale	Italy	$75.00-100.00	The jeep is blue with a pink plastic interior and red tires. Features are steer-able front tires, and an opening rear deck with a pink air cleaner. There is a yellow surfboard in the back seat. There is a red tire attached to the trunk. On the hood is a decal of a Smurf on a surfboard. On one front fender there is a gray shovel. There is one Smurf in the front seat.

Set 27: Buttons

027/001	Smurf Super Fan	W. Berrie	1/1/80	Light Blue	Round Metal Buttons	2 1/4"	U.S.A.	$2.00-3.00	The button has a blue background. The picture is of a Smurf holding different sports equipment.
027/002	Soccer Smurf	W. Berrie	1/1/80	Light Green	Round Metal Buttons	2 1/4"	U.S.A.	$2.00-3.00	The background of the button is light green. The picture is of a Smurf kicking a soccer ball.
027/003	Hockey Is My Schtick	W. Berrie	1/1/80	Light Blue	Round Metal Buttons	2 1/4"	U.S.A.	$2.00-3.00	The background of the button is light blue. The picture is of a Smurf in a yellow hockey outfit with a hockey stick.
027/004	Love Too	W. Berrie	1/1/80	Yellow	Round Metal Buttons	2 1/4"	U.S.A.	$2.00-3.00	The background of the button is yellow. The picture is of a Smurf playing tennis.
027/005	Downhill Smurfer	W. Berrie	1/1/80	Light Blue	Round Metal Buttons	2 1/4"	U.S.A.	$2.00-3.00	The background of the button is a light blue sky with white snow. The picture is a Smurf skiing.
027/006	Veux-tu Schtroumpfette	W. Berrie	1/1/80	Orange	Round Metal Buttons	2 1/4"	U.S.A.	$2.00-3.00	The background of the button is orange. The picture is of a love-struck Smurf sitting and holding a pink flower.
027/007	Want To Smurf Around?	W. Berrie	1/1/80	Orange	Round Metal Buttons	2 1/4"	U.S.A.	$2.00-3.00	The background of the button is orange. The picture is of a love-struck Smurf sitting and holding a pink flower.
027/008	I'm Too Busy To Think	W. Berrie	1/1/80	Blue/Green	Round Metal Buttons	2 1/4"	U.S.A.	$2.00-3.00	The background of the button has a blue sky, green grass, and a tree. The picture is of a Smurf leaning against the tree with his eyes shut resting.
027/009	I've Been Smurfed!	W. Berrie	1/1/80	yellow/green	Round Metal Buttons	2 1/4"	U.S.A.	$2.00-3.00	The button has a yellow background with grass on it. The picture is of a Smurf jumping with his hand on his head looking like he forgot something.,
027/010	I've Been Smurfed!	W. Berrie	1/1/80	Yellow	Round Metal Buttons	2 1/4"	U.S.A.	$2.00-3.00	The button has a yellow background. The picture is of a Smurf sitting and looking dazzled with stars around his head,
027/011	Happy Smurfday	W. Berrie	1/1/80	Pink	Round Metal Buttons	2 1/4"	U.S.A.	$2.00-3.00	The button has a pink background. The picture is of a Smurf carrying a cake with one candle on top.
027/012	Smurf Your True Blue Friend	W. Berrie	1/1/80	Yellow	Round Metal Buttons	2 1/4"	U.S.A.	$2.00-3.00	The background of the button is yellow. The picture is of a proud Smurf.
027/013	It's Not Easy Being A Smurfette	W. Berrie	1/1/80	Light Blue	Round Metal Buttons	2 1/4"	U.S.A.	$2.00-3.00	The button has a blue background. The picture is of Smurfette standing with her hands at her side.
027/014	T.G.I.F.	W. Berrie	1/1/80	Orange	Round Metal Buttons	2 1/4"	U.S.A.	$2.00-3.00	The button has an orange background. The picture is an exhausted Smurf leaning on his desk with papers all over it.
027/015	I Live For Saturdays	W. Berrie	1/1/80	Orange/Yellow	Round Metal Buttons	2 1/4"	U.S.A.	$2.00-3.00	The button has an orange border with a yellow center. The picture is of a Smurf carrying books and looking at a clock.
027/016	I Hate Homework	W. Berrie	1/1/80	Red/Yellow	Round Metal Buttons	2 1/4"	U.S.A.	$2.00-3.00	The button has a red border and the middle is yellow. The picture is a grumpy Smurf.
027/017	Smile! It Doesn't Hurt	W. Berrie	1/1/80	Yellow	Round Metal Buttons	2 1/4"	U.S.A.	$2.00-3.00	The button has a yellow background. The picture is of a Smurf's smiling face.
027/018	Go For It!	W. Berrie	1/1/80	Yellow	Round Metal Buttons	2 1/4"	U.S.A.	$2.00-3.00	The button has a yellow background. The picture is of a Smurf on a leaf skateboard.
027/019	Equal Rights For Smurfettes	W. Berrie	1/1/80	Light Blue/Green	Round Metal Buttons	2 1/4"	U.S.A.	$2.00-3.00	The button has a light blue border with a green center. The picture is of Smurfette carrying a briefcase.
027/020	Smurf Collectors Club International	W. Berrie	1/1/80	White	Round Metal Buttons	2 1/4"	U.S.A.	$6.00-8.00	The button has a white background. The picture is of a Smurf sitting by a mailbox reading a letter.
027/021	Canada's Wonderland	W. Berrie	1/1/80	White	Round Metal Buttons	2 1/4"	Canada	$6.00-8.00	The button has a white background. The picture is of Papa, Smurfette and a Smurf in the front of a roller coaster.
027/022	Smurf Holding An Orange Daisy Behind His Back	W. Berrie	1/1/80	Yellow	Round Metal Buttons	2 1/4"	U.S.A.	$2.00-3.00	The button has a yellow background. The picture is of a Smurf holding an orange daisy.
027/023	Smurf Holding Yellow Daisy In Front Of His Body	W. Berrie	1/1/80		Round Metal Buttons	2 1/4"	U.S.A.	$2.00-3.00	The button has a white and blue background. The picture is of a Smurf holding flowers in front of him.
027/024	Smurf Sticking Out His Tongue	W. Berrie	1/1/80	Orange	Round Metal Buttons	2 1/4"	U.S.A.	$2.00-3.00	The button has an orange background. The picture is of a Smurf with his fingers in his ears while sticking out his tongue.
027/025	Smurfs Alive In Ice Capades	W. Berrie	1/1/80	Yellow	Round Metal Buttons	3 1/2"	U.S.A.	$6.00-8.00	The button has a yellow background. The picture is Papa and Smurfette ice-skating.
027/026	It's Not Easy Being A Smurfette	W. Berrie	1/1/80	Pink	Heart Metal Buttons	2 1/4"	U.S.A.	$2.00-3.00	The button is a pink heart with purple hearts inside. The picture is of Smurfette holding a trophy.
027/027	No. 1 Fan! Smurfette	W. Berrie	1/1/80	Pink	Heart Metal Buttons	2 1/4"	U.S.A.	$2.00-3.00	The button is a pink heart with purple hearts inside. The picture is of Smurfette's face.
027/028	Hug A Smurfette Today!	W. Berrie	1/1/80	Purple	Heart Metal Buttons	2 1/4"	U.S.A.	$2.00-3.00	The button is a purple heart with pink hearts inside. The picture is of Smurfette standing and holding a daisy.
027/029	Keep On Smurfin!	W. Berrie	1/1/80	Purple	Heart Metal Buttons	2 1/4"	U.S.A.	$2.00-3.00	The button is a purple heart with pink hearts inside. The picture is of Smurfette roller-skating.
027/030	Give Me A S M U R F !	W. Berrie	1/1/80	Purple	Heart Metal Buttons	2 1/4"		$2.00-3.00	The button is a purple heart with pink hearts inside it. The picture is of Smurfette cheerleading.
027/031	I Love The Smurfs	W. Berrie	1/1/80	White	Round Metal Buttons	2 1/2"	U.S.A.	$2.00-3.00	The button has black lettering and a white background.
027/032	Chef Boyardee Is Going Smurfy	W. Berrie	1/1/85	Red	Round Metal Buttons	3"	U.S.A.	$6.00-8.00	The button has a red background. The picture is of a chef Smurf.
027/033	Smurf My Kind Of Friend	W. Berrie	1/1/80		Round Metal Buttons	2 1/4"	U.S.A.	$2.00-3.00	The official Smurf fun club. The button has 2 Smurfs shaking hands, with trees around them.
027/034	Be Happy Be Smurfy!	W. Berrie	1/1/80	Pink	Heart Metal Buttons	2 1/4"	U.S.A.	$2.00-3.00	The button is a pink heart with purple hearts inside it. The picture is of Smurfette in a white and pink ballerina outfit.
027/035	I Love Smurfs	W. Berrie	1/1/82	Yellow	Round Metal Buttons	2 1/4"	U.S.A.	$2.00-3.00	The background of the button is yellow. The button has a heart with a Smurf head in it. The button says "I (with the heart in the middle) Smurfs".
027/036	The Answer Is No!	W. Berrie	1/1/80	Pink	Round Metal Buttons	2 1/4"	U.S.A.	$2.00-3.00	The background is a rose pink. The badge is Smurfette walking with an angry expression on her face. The badge says "The Answer Is No!" in black letters.
027/037	Nuke The Smurfs		1/1/80	Blue	Round Metal Buttons	2 1/4"		$2.00-3.00	The button is blue. It says "Nuke The Smurfs" in black letters.
027/038	Super Biker	W. Berrie	1/1/80	Light Blue	Round Metal Buttons	2 1/4"	U.S.A.	$2.00-3.00	The background is a light blue. The badge is of a Smurf on a bicycle. The bike is brown. The Smurf is wearing a green shirt, yellow gloves, yellow and red shorts and shoes. The button says "Super Biker" in black letters.

027/039	Catch A Rising Star	Peyo			Round Metal Buttons	3"		$2.00-3.00	The button is white and has a Smurf standing with a shy expression on his face. He has his head turned to one side and his hands behind his back. There is a yellow star behind the Smurf. The top of the button says "Catch A Rising Star" in red letters.
027/040	Girls Can Do Anything!	W. Berrie	1/1/82		Round Metal Buttons	2 1/4"	U.S.A.	$2.00-3.00	The button has blue sky, a gray race track and green grass on it. Smurfette is running through a finish line. There are 2 Smurfs running behind her. The button says "Girls Can Do Anything".
027/041	Smurf On A Beach Blanket	SEPP			Round Metal Buttons	2 1/4"		$2.00-3.00	The background is brown. A Smurf is laying on a red and yellow beach towel. The Smurf is wearing yellow sunglasses, a white hat and white swim trunks. He has a green radio on one side of him and an orange duffel bag on the other side.
027/042	I Love Smurfs			White	Round Metal Buttons	1 3/4"		$2.00-3.00	The button has black lettering and a white background. There is a red heart representing love on it.
027/043	Smurfette			Gray	Round Metal Buttons	1 3/4"		$2.00-3.00	The button has a gray background. Smurfette is standing with her hands in front of her. Smurfette is wearing a white dress, shoes and hat. The button says "Smurfette" in multi colors on the bottom.
027/044	Smurf			Gray	Round Metal Buttons	1 3/4"		$2.00-3.00	The button has a gray background. A Smurf looks like he's running. The button says "Smurf" in multi colors on the bottom.
027/045	Smurf With Ice Cream				Round Metal Buttons	2 1/4"		$2.00-3.00	The button has a Smurf holding an ice cream on it. The Smurf is licking his lips. There is a corner of a mushroom behind him. The background is white and gray.
027/046	Hug Me I'm Lovable			White	Round Metal Buttons	2 1/2"		$2.00-3.00	The button is white with a King Smurf puffy sticker in the center. The Smurf has a red and white robe, yellow crown, yellow hat and yellow pants. He is carrying a red and white septor. The button says "Hug Me I'm Lovable".
027/047	I May Not Be Perfect But Parts Of Me Are Excellent			White	Round Metal Buttons	2 1/2"		$2.00-3.00	The button is white with a Smurf playing soccer puffy sticker in the center. The Smurf is wearing black pants, orange socks and an orange and yellow shirt. The button says "I May Not Be Perfect But Parts Of Me Are Excellent".
027/048	Nat & Slouchy	Peyo	1/1/88	White	Round Metal Buttons	2 1/4"	Brussels	$2.00-3.00	The button has Nat with a yellow butterfly on his wrist. Nat is standing and wearing a yellow straw hat, brown pants and a brown sash. Slouchy is walking towards Nat and yawning. There is a green caterpillar by Slouchy's foot. Slouchy is wearing a droopy white hat, red shirt and white pants. The background is white.
027/049	Snappy & Sassette	Peyo	1/1/88	White	Round Metal Buttons	2 1/4"	Brussels	$2.00-3.00	The button has Slouchy walking towards Sassette. Sassette is wearing pink bibs. Slouchy is wearing white pants, a white hat and a yellow shirt with a lightening bolt. The button has a white background.
027/050	Have A Nice Smurf!	W. Berrie	1/1/79	Yellow	Round Metal Buttons	2 1/4"		$2.00-3.00	The button has a yellow background. A Smurf is standing pointing with his left hand. His right hand is in the air behind him. The Smurf has a happy expression on his face.
027/051	I Love Smurf				Round Metal Buttons	2 1/4"		$2.00-3.00	The button is white and has an angry Smurf in the center. The Smurf is jumping. The buttons says "I Love (Heart symbol) Smurfs" in black letters.
027/052	Smurfs 'N' Ice Capades	W. Berrie	1/1/83	Blue	Round Metal Buttons	3 1/4" Circle	U.S.A.	$6.00-8.00	The button is light blue. It has Smurfette and Papa wearing ice skates. Papa is sitting and Smurfette is standing next to him looking down. Smurfette is wearing a pink dress. The top says "Smurf 'N' Ice Capades" in white letters.
027/053	Smurf Playing A Drum, Smurfette Waving			Blue	Round Metal Buttons	2 1/4"		$2.00-3.00	The button is light blue. It has a white square outlined background. In the center is a Smurf walking and playing a drum. Smurfette is standing in front of him and waving.
027/054	Smurfette With A Camera			Blue	Round Metal Buttons	2 1/4"		$2.00-3.00	The button is light blue. It has a white square outlined background. In the center is Smurfette taking a picture with a camera that's set on a tripod.
027/055	Smurfette Taking A Picture Of A Smurf			Light Blue	Round Metal Buttons	2 1/4"		$2.00-3.00	The button is light blue. It has a white square outlined background. In the center is Smurfette taking a picture with a camera that's set on a tripod. In front of the camera is a Smurf posing. He has his hands resting on his hips.
027/056	Smurf Laughing At A Smurf Posing			Light Blue	Round Metal Buttons	2 1/4"		$2.00-3.00	The button is light blue. It has a white square outlined background. In the center is a Smurf laughing at a Smurf that is posing. He has his hands resting on his hips and a sly grin on his face.
027/057	Smurf Holding Daisy's			Light Blue	Round Metal Buttons	2 1/4"		$2.00-3.00	The button is light blue. It has a white square outlined background. In the center is a Smurf holding a bouquet of red and yellow daisies. The Smurf has red hearts floating above his head. He has a shy grin on his face.
027/058	Smurf Giving Smurfette A Flower			Light Blue	Round Metal Buttons	2 1/4"		$2.00-3.00	The button is light blue. It has a white square outlined background. In the center is a Smurf giving Smurfette a red wilted daisy. The Smurf has red hearts floating above his head. He has a shy grin on his face. Smurfette has hearts floating above her head also.
027/059	Smurfette Holding A Tulip			Light Blue	Round Metal Buttons	2 1/4"		$2.00-3.00	The button is light blue. It has a white square outlined background. In the center is Smurfette holding a red tulip.
027/060	Smurfette Has Red Flowers Around Her Head			Light Blue	Round Metal Buttons	2 1/4"		$2.00-3.00	The button is light blue. It has a white square outlined background. In the center is Smurfette standing and looking shy. She has red hearts floating above her head.
027/061	Smurf Giving Another Smurf A Cake			Light Blue	Round Metal Buttons	2 1/4"		$2.00-3.00	The button is light blue. It has a white square outlined background. In the center is a Smurf handing another Smurf a red cake with a candle on top.
027/062	Have You Hugged Your Smurf Today? (White)	W. Berrie	1/1/82	White	Round Metal Buttons	2 1/4"	U.S.A.	$2.00-3.00	The button is white. In black and pink letters it says "Have You Hugged Your Smurf Today?" There is a Smurf standing with his hands behind his back and a smile on his face on one side of the button.
027/063	Have You Hugged Your Smurf Today? (Yellow)	W. Berrie	1/1/82	Yellow	Round Metal Buttons	2 1/4"	U.S.A.	$2.00-3.00	The button is yellow. In black and red letters it says "Have You Hugged Your Smurf Today?" There is a Smurf standing with his hands behind his back and a smile on his face on one side of the button.
027/064	Free Smurf Forest Cup Baby	W. Berrie	1/1/85		Round Metal Buttons	3 1/2"		$9.00-13.00	The button has a pinkish/orange background. The button has Baby Smurf waving. He has 3 balloons attached to his button on his butt. Baby is wearing a white sleeper. The button says "Free Smurf Forest Cup With Any Medium Soft Drink McDonald's".
027/065	Canada's Wonderland Cups (Smurfette)	Applause/Peyo	1/1/80		Round Metal Buttons	3 1/2"		$9.00-12.00	The button is yellow with orange and blue. It has Smurfette on the front sitting and waving. Smurfette is wearing a white hat, a white dress and white shoes. The button says "Canada's Wonderland Cups Free With Every Medium Soft Drink".
027/066	Have A Good Smurf?	Granger Mtl		Green	Round Metal Buttons	3" Diam	Canada	$3.50-5.00	A Smurf is walking with his eye's closed while playing a yellow trumpet. On the bottom it says "Have A Good Smurf?" in black letters.
027/067	Golf Pro	W. Berrie	1/1/80	Light Blue	Round Metal Buttons	2 1/4"	U.S.A.	$2.00-3.00	The Smurf is swinging at a golf ball. The Smurf has a mad expression on his face. The Smurf is wearing brown checkered pants.
027/068	Free Smurf Forest Cup Smurfette	Dm Int'l Ltd	1/1/84	Yellow	Round Metal Buttons	3 1/2"	Canada	$5.00-7.00	The button has Smurfette sitting and holding a cup with Papa Smurf on it. Smurfette is wearing a white dress, white shoes, and a white hat. The button says "Free Smurf Forest Cup with Any Medium Soft Drink".

Set 96: Buttons- BP Promotional

096/001	Smurf And Smurfette Riding A Motorcycle	Wavery Prod.	1/1/84	White	Round Metal Buttons	2 1/4"	Germany	$2.00-3.00	The button is white. The picture is of a Smurf driving a motorcycle and Smurfette sitting on the back. The motorcycle is yellow with gray tires. The BP symbol is on the front fender of the motorcycle and in the right upper corner of the button. The BP symbols are green and yellow.
096/002	Smurf In An Airplane	Wavery Prod.	1/1/84	White	Round Metal Buttons	2 1/4" Round	Germany	$2.00-3.00	The button is white. The picture is of a Smurf flying an airplane. The airplane is white on top and yellow on the bottom, the wings are green. The BP symbol is on the side of the airplane and in the right upper corner of the button. The BP symbols are green and yellow.
096/003	Smurf Driving A Go-Cart	Wavery Prod.	1/1/84	White	Round Metal Buttons	2 1/4" Round	Germany	$2.00-3.00	The button is white. The picture is of a Smurf driving a blue go-cart. The BP symbol is on the side of the airplane and in the lower left corner of the button. The BP symbols are green and yellow.
096/004	Smurf Driving A Car	Wavery Prod.	1/1/84	White	Round Metal Buttons	2 1/4" Round	Germany	$2.00-3.00	The button is white. The picture is of a Smurf driving a red convertible car with gray trim. The BP symbol is on the side of the car and in the lower right corner of the button. The BP symbols are green and yellow.
096/005	Smurf Driving A Gas Truck	Wavery Prod.	1/1/84	White	Round Metal Buttons	2 1/4" Round	Germany	$2.00-3.00	The button is white. The picture is of a Smurf driving a gas truck. The truck is green with a yellow hood and the gas tank on the back is white. The Smurf is hanging out the window waving. The BP symbol is on the side of the gas tank and in the lower left corner of the button. The BP symbols are green and yellow.
096/006	3 Smurfs In A House Boat	Wavery Prod.	1/1/84	White	Round Metal Buttons	2 1/4" Round	Germany	$2.00-3.00	The button is white. The picture is of 3 Smurfs in a brown house boat. One Smurf is in front looking through a periscope, one is sitting on the back of the boat, and the third is fishing out the cabin window. The BP symbol is on the smoke stack and the other is in the lower left corner of the button. The BP symbols are green and yellow.
096/007	Smurf With A Space Ship	Wavery Prod.	1/1/84	White	Round Metal Buttons	2 1/4" Round	Germany	$2.00-3.00	The button is white. The picture is of a Smurf floating out of a spaceship. The spaceship is red with a gray top. The Smurf is wearing a white spacesuit. The BP symbol is on the bottom of the spaceship and in the lower left corner of the button. The BP symbols are green and yellow.
096/008	Gas Attendant Smurf	Wavery Prod.	1/1/84	White	Round Metal Buttons	2 1/4" Round	Germany	$2.00-3.00	The button is white. The picture is of a Smurf standing next to a gas pump holding the gas hose. The Smurf is wearing white coveralls with a BP symbol on the front. The gas pump is white with a red super sign on top of the pump. The BP symbol is on the front of the pump and in the right lower corner. The BP symbols are green and yellow.

Set 47: Cake Pans

047/001	Smurf Holding A Sign	Wilton	1/1/83	Silver	Metal Cake Tins	10"W x 15"L	Korea	$9.00-13.00	Smurf cake tin. The Smurf is holding a sign.
047/002	Smurfette Holding A Balloon	Wilton	1/1/83	Silver	Metal Cake Tins	11"W x 16"L	Korea	$9.00-13.00	Smurfette is holding a balloon in her hands.
047/003	Mini Smurf Cake Pan	Wilton	1/1/83	Silver	Metal Cake Tins	9"W x 13 1/2"L	Korea	$11.00-15.00	The cake pan makes 6 individual Smurf face cakes.

Set 183: Cakes- Porcelain Figures

183/001	Schtroumpfette	Peyo	1/1/96		Porcelain Figure	1" High	Brussels	$3.50-5.00	Smurfette is standing with her left hand out at her side and her right hand is by her mouth. There are flowers behind her.
183/002	Schtroumpfissime (King)	Peyo	1/1/96		Porcelain Figure	1" High	Brussels	$3.50-5.00	The Smurf is wearing a red and white robe, yellow pants, yellow hat, and a yellow crown. He is holding a red and white septor. The Smurf is pointing with his left hand.
183/003	Schtroumpf Musician	Peyo	1/1/96		Porcelain Figure	1" High	Brussels	$3.50-5.00	The Smurf is playing a yellow trumpet.
183/004	Gargamel	Peyo	1/1/96		Porcelain Figure	1 1/2"High	Brussels	$3.50-5.00	Gargamel is standing and wearing a black robe. He has one hand under his chin. Azreal is standing by his feet.

183/005	Grand Schtroumpf (Papa)	Peyo	1/1/96		Porcelain Figure	1" High	Brussels	$3.50-5.00	Papa Smurf is standing with both hands behind his back. He is wearing a red hat and red pants.
183/006	Schtroumpf Farceur (Jokey)	Peyo	1/1/96		Porcelain Figure	1 1/2"High	Brussels	$3.50-5.00	The Smurf is walking and carrying a yellow package with a red ribbon.
183/007	Bebe Schtroumpf (Baby)	Peyo	1/1/96		Porcelain Figure	1" High	Brussels	$3.50-5.00	Baby Smurf is sitting in a highchair. He is wearing a pink outfit. Baby is holding a red rattle in the air with his right hand.
183/008	Schtroumpf Gourmand (Cake)	Peyo	1/1/96		Porcelain Figure	1" High	Brussels	$3.50-5.00	The Smurf is carrying a yellow plate with a brown cake on the plate. The cake has white frosting and red cherries on top.
183/009	Schtroumpf Cocuet (Shy)	Peyo	1/1/96		Porcelain Figure	1" High	Brussels	$3.50-5.00	The Smurf has a shy expression on his face. He has one hand in front of his mouth and the other is behind his back. The Smurf has a pink flower on his hat.
183/010	Schtroumpf Amoureux (Cupid)	Peyo	1/1/96		Porcelain Figure	1" High	Brussels	$3.50-5.00	The Smurf has white and black wings. He is wearing a white cloth around his waist. The Smurf is holding a red heart with a yellow arrow through it in his left hand. In his other hand he is holding a brown and yellow bow.
183/011	Pirlouit (Peewit)	Peyo	1/1/96		Porcelain Figure	1 1/2" High	Brussels	$3.50-5.00	Peewit is standing and wearing a green vest, white shirt, green shorts, red socks and red shoes. He is waving with his left hand. Peewit has dark yellow hair.
183/012	Lux-Vieux Schtroumpf (Grandpa)	Peyo	1/1/96		Porcelain Figure	1" High	Brussels	$3.50-5.00	Grandpa is standing slouched over and leaning on a brown stick. He is wearing a yellow hat, yellow pants and black glasses. He has one hand on his lower back.
183/013	Schtroumpf Paysan (Gardner)	Peyo	1/1/96		Porcelain Figure	1" High	Brussels	$3.50-5.00	The Smurf is standing and holding a green watering can in his right hand. The Smurf is wearing white pants, white suspenders and orange shoes.
183/014	Puppy	Peyo	1/1/96		Porcelain Figure	1" High	Brussels	$3.50-5.00	Puppy is tan with brown spots, a brown nose, a brown tail, and 1 brown ear. Puppy has his big red tongue hanging out. Puppy is wearing a yellow collar with a yellow flower on it. There is dark green grass by his feet.
183/015	Johan	Peyo	1/1/96		Porcelain Figure	1 1/2"High	Brussels	$3.50-5.00	Johan is wearing a tan shirt, red pants, army green boots, and a blue cape with a black lining. Johan is holding a blue sword in his left hand and his right hand is in the air waving. Johan has black hair.
183/016	Baby Laying Down	Peyo	1/1/96		Porcelain Figure	1 1/2"Long	Germany	$3.50-5.00	Baby Smurf is wearing a pink sleeper. He is laying on his tummy. Baby is on a white base.
183/017	Baby Laying In The Moon	Peyo	1/1/96		Porcelain Figure	1 1/2"High	Germany	$3.50-5.00	Baby is wearing a pink sleeper. He is laying in a yellow half moon. Baby has his hands out in the air. The moon is on a white base.
183/018	Baby With A Sailboat	Peyo	1/1/96		Porcelain Figure	1 1/2"High	Germany	$3.50-5.00	Baby is sitting and holding a sailboat. Baby is wearing a pink sleeper. The sailboat is yellow and red. Baby is on a white base.
183/019	Baby Sucking His Thumb	Peyo	1/1/96		Porcelain Figure	1 1/2"High	Germany	$3.50-5.00	Baby is sitting and sucking his thumb. Baby is wearing a pink sleeper. Baby is on a white base.
183/020	Baby Holding A Popsicle	Peyo	1/1/96		Porcelain Figure	1 1/2"High	Germany	$3.50-5.00	Baby is sitting while holding a yellow and red popsicle. Baby is wearing a pink sleeper. Baby has a big grin on his face. Baby is on a white base.
183/021	Baby Playing With A Rattle	Peyo	1/1/96		Porcelain Figure	1 1/2"High	Germany	$3.50-5.00	Baby is sitting while holding a red and yellow rattle. Baby is wearing a pink sleeper. Baby is leaning on his left hand. Baby is on a white base.
183/022	Baby With A Teddy Bear	Peyo	1/1/96		Porcelain Figure	1 1/2"High	Germany	$3.50-5.00	Baby is sitting and holding a teddy bear. The bear is brown with a white face and white tummy. Baby is wearing a pink sleeper. Baby is on a white base.
183/023	Baby With Blocks	Peyo	1/1/96		Porcelain Figure	1 1/2"High	Germany	$3.50-5.00	Baby is sitting and holding a block in the air. Baby has 2 blocks in front of him. The blocks are yellow/ red/ blue. Baby is wearing a yellow sleeper. Baby is on a white base.
183/024	Baby Making A Funny Face	Peyo	1/1/96		Porcelain Figure	1 1/2"High	Germany	$3.50-5.00	Baby is sitting and sticking out his tongue. He has his thumbs in his ears. Baby is wearing a yellow sleeper. Baby is on a white base.
183/025	The Smurfs Babies Sign	Peyo	1/1/96		Porcelain Figure	1 1/2"High	Germany	$3.50-5.00	The sign is flat. The sign is diamond shape. On top it says "The Smurfs" in red letters. The background is blue behind the word Smurfs. In the center of the diamond is Baby's face. Baby has a pink hat. Under Baby's face it says "BABIES" in blue/yellow/pink letters. The diamond is dark pink with a yellow border.

Set 123: Calendars

123/001	16 Month Character Calendar 1982-83	S.E.P.P.	1/1/82		16 Month	9 1/2"W X 12"L	U.S.A.	$15.00-20.00	The front of the calendar has Smurfette holding a baton, with Papa and 5 Smurfs following her. The Smurfs are walking down a path and Gargamel and Azreal are hiding behind a bush watching them. Each month has a certain character with a colorful picture.
123/002	Smurf 1983 Calendar	Campus Craft	1/1/81		12 Month Calendar	12" X 12"	Canada	$15.00-20.00	The cover is light blue and has Papa sitting in a rocking chair reading to 3 Smurfs and Smurfette. The Smurfs have milk and cookies sitting next to them. There is a sign on the wall that says home Smurf home. Each month has a different picture.
123/003	De Smurfen Scheurkalender 1997	Vrijbuiter	1/1/96		Calendar	7" x 5"	Holland	$11.00-15.00	The cover is yellow and has a mushroom house with Papa, Smurfette and a Smurf holding hands outside. The calendar is day by day and has activities and puzzles.
123/004	Valentine's Day Calendar Page	Peyo	1/1/82		Calendar Page	8 1/2" x 10 1/2"		$2.00-3.00	Smurfette is standing in the grass with 7 Smurf's bringing her valentines. Papa is standing on a hill in the background watching.
123/005	Verjaardags Kalender	Peyo	1/1/95	Yellow	Calendar	13"L x 5"W	Amsterdam	$9.00-13.00	The cover of the calendar has a Smurf carrying a huge birthday cake. He is wearing a chefs hat. The calendar has months and numbers but not days, so it can be used any year. There are different colored pictures on each page.
123/006	The Smurfs Are A Year Old	W. Berrie	1/1/80		Calendar	24" x 17 3/4"	U.S.A.	$18.00-25.00	The calendar is to celebrate the Smurfs 1st birthday. The top of the calendar is a lithograph. It is a village scene with a lot of Smurfs doing different things. The bottom is a rip off calendar. On one side of the calendar pages is the Smurf story. The other side says the Smurfs are a year old.
123/007	Schtroumpf 1992	Cartoon Crtn	1/1/91	Blue	Calendrier	12" x 10"	Germany	$15.00-20.00	The cover of the calendar is blue. It has Papa Smurf dressed as Santa. He is carrying a sack of gifts over one shoulder. In his hand he is carrying a gray lantern.
123/008	1998 40th Anniversary Desk Calendar	Puppy	1/1/82	Blue	Desk Calendar	7 1/2" x 8 1/4"	Brussels	$7.50-10.00	The calendar is blue. The front has Papa with Johan and Peewit in a comic strip. The calendar is for the Smurfs 40th anniversary. Each page of the calendar has a comic strip and shows a new comic book.
123/009	Die Schlumpfe Wand-Kalender 1997	Bastei	1/1/96	Blue	Wall Calendar	11" x 12 1/2"	Brussels	$11.00-15.00	The cover has Smurfs out playing in the grass. A Smurf is carrying a book were Gargamel's head jumped out. Another Smurf is hugging himself laughing. Smurfette is by a bush kissing a Smurf. A Smurf is by a mole hole. Papa is standing with his hand over his mouth laughing. Each page of the calendar has a colored picture.

Set 108: Cameras

108/001	Flash Camera Clown Smurf	Helm		White	Flash Camera	4 1/2"W X 5"H	Hong Kong	$25.00-35.00	The camera is white, square and uses flash cubes. On the front of the camera is an orange mushroom around the lens. There is a Smurf sitting on the mushroom with red and white ruffles around his neck. The camera uses 126 film and flash cubes. The camera is in a yellow box.
108/002	Blue Musical Camera	Illco	1/1/82	Blue	Play Camera	4 1/2"W X 5"H	Macau	$15.00-20.00	The camera is blue with a pink handle. The focus button is white and winds to play music. There is a white push button on top and a Smurf face. Don't work.
108/003	Smurf N' See Camera	Irwin Toys	1/1/96		Toon Viewer	7" x 5 1/2"	China	$18.00-25.00	The camera is in the shape of a yellow mushroom house. It has a red roof. The top has a Smurf that turns when you push the button. Inside are 20 different pictures. Includes a red carrying strap.

Set 221: Candle Holders

221/001	Mailman Mini Candle Holder			White	Candle Holder	1"	Germany	$6.75-9.00	The candle holder is 1 inch high and made of ceramic. It has a picture of a Smurf carrying a white envelope and playing a yellow trumpet. The Smurf has a brown mail bag over his shoulder. The background is part red. The picture is on 2 sides.
221/002	Soccer Smurf Candle Holder			White	Candle Holder	1"	Germany	$6.75-9.00	The candle holder is 1 inch high and made of ceramic. It has a picture of a Smurf kicking a white and black soccer ball. The Smurf is wearing green shorts, a red shirt, brown shoes, a white hat, and red socks. The Smurf is standing in grass. The background is brown. The candle holder is white. The picture is on 2 sides.

Set 153: Candles- Figures

153/001	Smurfette	Gies			Wax Candles	3 1/2" High	Germany	$6.75-9.00	Smurfette has a blue body and yellow hair. Smurfette is wearing a white dress, hat and shoes. Smurfette is standing in a flirting pose. The figure is wax.
153/002	Sleepwalker	Gies			Wax Candles	3 1/2" High	Germany	$6.75-9.00	The Smurf has a blue body. The Smurf is wearing a white nightshirt and a white hat with a red tassel on the end. The Smurf has his eyes closed and is walking with his hands out in front of him. The figure is wax.
153/003	Watchman	Gies			Wax Candles	3 1/2" High	Germany	$6.75-9.00	The Smurf has a blue body. He has white pants and a white hat on. The Smurf is carrying a red/yellow/black lantern in his right hand. The figure is wax.
153/004	Mailman	Gies			Wax Candles	3 1/2" High	Germany	$3.50-5.00	The Smurf has a blue body. He has a white hat and white pants on. The Smurf has a black bag with letters in it. The Smurf is carrying a white envelope with a red heart on it in his left hand. He is holding a horn to his mouth with his other hand.
153/005	Football Player	Gies			Wax Candles	3 1/2" High	Germany	$6.75-9.00	The Smurf has a blue body. He is wearing a white shirt, a white hat, black shorts and black shoes. The Smurf is holding a black and white soccer ball to his side. The figure is made of wax.
153/006	Jolly	Gies			Wax Candles	3 1/2" High	Germany	$6.75-9.00	The Smurf has a blue body. The Smurf has white pants and a white hat on. He is standing with one hand in the air and the other on his belly. The Smurf looks like he's laughing. The figure is made of wax.
153/007	Tyrolese	Gies			Wax Candles	3 1/2" High	Germany	$6.75-9.00	The Smurf has a blue body. He is wearing green shorts with green suspenders over white pants. The Smurf has a red pipe in his mouth. There is a red feather in his white hat. The figure is made of wax.
153/008	Chimney Sweep	Gies			Wax Candles	3 1/2" High	Germany	$6.75-9.00	The Smurf has a blue body. The Smurf is wearing a black hat, black jacket with a red scarf, gray pants, and black shoes. The Smurf is holding a black ladder. The figure is made of wax.
153/009	Lute	Bougies			Wax Candles	3 1/2" High	Korea	$6.75-9.00	The Smurf has a blue body. The Smurf is holding a yellow lute. The Smurf has his eyes closed. He has a white hat and white pants on.
153/010	Clown	Bougies			Wax Candles	3 1/2" High	Korea	$6.75-9.00	The Smurf has a blue body. The Smurf is wearing a clown outfit. He has on yellow pants with black suspenders, a white shirt, white shoes with a black tip, white gloves, a red bow tie and a white hat. The Smurf has white paint around his face.
153/011	Mushroom Cottage	Gies			Wax Candles	5"L x 4"W	Germany	$6.75-9.00	The candle is in the shape of a mushroom cottage. The walls are white and the roof is red. The cottage has gray trim windows and a gray door with yellow trim. Papa Smurf is peaking out the window of the door.

Set 48: Candles- Numerals

048/001	Smurf With A #0	Unique	1/1/82	Light Blue	Wax B-Day Candles	3 1/4" High	Hong Kong	$3.50-5.00	A Smurf is standing with his hand on the side of a zero. The candle is light blue. The number is white and has a yellow border.
048/002	Smurf With A #7	Unique	1/1/82	Light Blue	Wax B-Day Candles	3 1/4" High	Hong Kong	$3.50-5.00	A Smurf is blowing a yellow trumpet next to a #7. The candle is light blue. The number is white and has a yellow border.
048/003	Smurf With A #9	Unique	1/1/82	Light Blue	Wax B-Day Candles	3 1/4" High	Hong Kong	$3.50-5.00	A Smurf is sitting next to a #9. The candle is light blue. The number is white and has a yellow border.
048/004	Smurf With A #6	Unique	1/1/82		Wax B-Day Candles	3 1/4" High	Hong Kong	$3.50-5.00	A Smurf looks like he is jumping with his hands out next to the #6. The candle is light blue. The number is white and has a yellow border.
048/005	Smurf With A #8	Unique	1/1/02		Wax B-Day Candles	3 1/4" High	Hong Kong	$3.50-5.00	A Smurf is walking and holding a drum stick next to a #8. The candle is light blue. The number is white and has a yellow border.

Set 86: Candy Items

086/001	Stand-Up Baby Candy Mold	Wilton	1/1/83	Clear	Plastic Candy Mold		U.S.A.	$3.50-5.00	The mold is plastic and you pour the mix in the bottom. It makes Baby Smurf sitting and holding a ball in front of him. The mold is a full figure mold.
086/002	Smurfs I	Wilton	1/1/83	Clear	Plastic Candy Mold		U.S.A.	$3.50-5.00	The mold makes a mushroom house, a Smurf eating an ice cream cone, a chef Smurf holding a spoon, a smiling Smurf face, Brainy's face, a Smurf carrying a cake, Smurfette holding flowers, and the word Smurf. All the figures are 2" and the mold only makes half figures.
086/003	Smurf II	Wilton	1/1/83	Clear	Plastic Candy Mold		U.S.A.	$3.50-5.00	The mold makes Papa holding a cupcake, Smurfette holding flowers, a smiling Smurf face, Brainy's face, a Smurf carrying a cake, a Smurf with his hands out, Gargamel carrying Azreal, and the word Smurf. All the figures are 2" and the mold only makes half figures.
086/004	Papa Smurf Face Candy Container	Topps	1/1/82		Plastic Container	4 Oz	Hong Kong	$1.50-2.00	The container is 2" high and opens up to put candy inside. The container is Papa Smurf's face and he has a red hat on. There is a clear blue string through the top. The bottom comes off. There are little candies inside.
086/005	Smurf Face Candy Container	Topps	1/1/83		Plastic Container	4 Oz.	Hong Kong	$1.50-2.00	The container is 2" high and opens up to put candy inside. The container is a Smurf face and he has a white hat on. There is yellow string through the top. The bottom comes off. Inside are little candies.
086/006	Smurf In A Package - Smurf On A Stick Smolf	Belga	1/1/95		Jelly Beans	12" Long	Germany	$7.50-10.00	There is a long clear tube filled with jelly beans. The top has a rubber Smurf on it. The Smurf is in a yellow package with a red bow. The Smurf has only his head, feet and arms sticking out of the package..
086/007	Gargamel & Azreal - Smurf On A Stick Smolf	Rodako	1/1/95		Candy Tube	13" Long	Holland	$7.50-10.00	There is a long clear tube filled with M & M's. The top has a rubber Gargamel & Azreal on it. Gargamel is carrying a yellow pail with a Smurf inside. Azreal is walking by his feet. Azreal is brown.
086/008	Gargamel & Azreal - Smurf On A Stick Smolf	Belga	1/1/95		Candy Tube	13" Long	Holland	$7.50-10.00	There is a long clear tube filled with jelly beans. The top has a rubber Gargamel & Azreal on it. Gargamel is carrying a yellow pail with a Smurf inside. Azreal is walking by his feet. Azreal is orange.
086/009	Vacation Smurf - Smurf On A Stick Smolf	Belga	1/1/95		Candy Tube	13" Long	Holland	$7.50-10.00	There is a long clear tube filled with M&M's. The top has a rubber Smurf that is wearing red glasses and red shorts with white flowers. The Smurf has a yellow backpack on his back. He is carrying a black camera. The Smurf is kneeling in the grass.
086/010	Smurf And Smurfette On A Turtle - Smurf On A Stick Smolf	Belga	1/1/95		Candy Tube	13" Long	Holland	$7.50-10.00	There is a long clear tube filled with jelly beans. The top is made if rubber. It has a Smurf and Smurfette sitting on a green turtle on it. Smurfette is holding a bunch of bright pink flowers.
086/011	Jack-N-A-Box Smurf - Smurf On A Stick Smolf	Belga	1/1/95		Candy Tube	13" Long	Holland	$7.50-10.00	There is a long clear tube filled with jelly beans. The top has a rubber Smurf in a jack-n-a-box on it. The Smurf is in a yellow jack-n-a-box with red stars. The Smurf has a square red and green body. The Smurf is wearing a big red nose. The Smurf is sticking out his tongue and he has his thumbs in his ears.
086/012	Papa Laying Under Mushroom - Smurf On A Stick Smolf	Belga	1/1/95		Candy Tube	13" Long	Holland	$7.50-10.00	There is a long clear tube filled with jelly beans. The top has a rubber Papa Smurf laying under a mushroom sleeping on it. The mushroom is red and white with a yellow base. There is a brown squirrel on top of the mushroom.
086/013	Baby Smurf On Papa's Back - Smurf On A Stick Smolf	Belga	1/1/95		Candy Tube	13" Long	Holland	$7.50-10.00	There is a long clear tube filled with M&M's. The top is made of rubber. It has Baby Smurf riding on Papa's back. Papa is crawling in the grass. Baby is sitting on his back holding a red and yellow lollipop. Baby has on a pink sleeper.
086/014	Smurf Playing A Drum - Smurf On A Stick Smolf	Belga	1/1/95		Candy Tube	13" Long	Holland	$7.50-10.00	There is a long clear tube filled with M&M's. The top has a rubber Smurf playing a drum on it. The Smurf has his face painted like a clown and is wearing a big red nose. The drum is red, green and yellow. There is a brown bird sitting on top of the drum.
086/015	Magician - Smurf On A Stick Smolf	Belga	1/1/95		Candy Tube	13" Long	Holland	$7.50-10.00	There is a long clear tube filled with M&M's The top has a rubber Smurf dressed as a magician on it. The Smurf has a black hat on the ground in front of him. A brown bird is flying out of the hat. The Smurf is wearing a blue hat with yellow stars, blue pants with yellow stars, and a yellow cape with blue stars.
086/016	Smurfette Rollerblading - Smurf On A Stick Smolf	Belga	1/1/95		Candy Tube	13" Long	Holland	$7.50-10.00	There is a long clear tube filled with M&M's. The top has a rubber Smurfette on rollerblades on it. Smurfette is wearing a white shirt, a green skirt with purple flowers, and green rollerblades with purple laces and wheels. Smurfette has a brown walkman with purple headphone on.
086/017	Smurf Stringing Christmas Lights - Smurf On A Stick Smolf	Belga	1/1/95		Candy Tube	13" Long	Holland	$7.50-10.00	There is a long clear tube filled with M&M's. The top has a rubber house on it. There are 2 Smurfs on top of the mushroom house hanging Christmas lights. The house is yellow with a red roof and blue chimney. There is snow on top of the roof.
086/018	Smurfs Dancing Around Tree - Smurf On A Stick Smolf	Belga	1/1/95		Candy Tube	13" Long	Holland	$7.50-10.00	There is a long clear tube filled with M&M's. The top has a rubber Christmas tree on it. There are 3 Smurfs holding hands and dancing around the tree. The tree is green with a purple tree top and has green and orange candles, red garland, and purple and orange ornaments on it.
086/019	Papa Santa - Smurf On A Stick Smolf	Belga	1/1/95		Candy Tube	13" Long	Holland	$7.50-10.00	There is a long clear tube filled with M&M's. The top has a rubber Papa Smurf dressed as Santa on it. Papa is wearing a red and white Santa outfit and black boots. Papa is walking in snow and carrying an orange lantern. Over his shoulder he is carrying a green bag filled with presents.
086/020	Smurfette And Smurf Sledding - Smurf On A Stick Smolf	Belga	1/1/83		Candy Tube	13" Long	Holland	$7.50-10.00	There is a long clear tube filled with M&M's. The top has a rubber Smurfette sitting on a Smurfs back on it. The Smurf is laying on a sled. Smurfette is wearing hot pink winter clothes. The Smurf is wearing a yellow snowsuit with green trim and a red hat. The sled is maroon.
086/021	Smurf Jogging - Smurf On A Stick Smolf	Belga	1/1/98		Candy Tube	13" Long	Holland	$3.50-5.00	There is a long clear tube filled with M&M's. The top has a rubber figure on it. The figure is a Smurf jogging. He is wearing a white shirt, red shorts, and red and white shoes. The Smurf has a pink towel over his shoulder. There is a squirrel running next to him.
086/022	Golfer Smurf - Smurf On A Stick Smolf	Belga	1/1/98		Candy Tube	13" Long	Holland	$3.50-5.00	There is a long clear tube filled with M&M's. The top has a rubber figure on it. It is a Smurf swinging a golf club at a ball. The Smurf is wearing a white hat, yellow an orange plaid pants, and white shoes.
086/023	Aerobic Smurfette - Smurf On A Stick Smolf	Belga	1/1/98		Candy Tube	13" Long	Holland	$3.50-5.00	There is a long clear tube filled with M&M's. The top has a rubber figure on it. Smurfette is laying on her back lifting a leg. She is wearing a pink leotard, white leg warmers, and pink slippers. Smurfette has dumbbells laying beside her.
086/024	Soccer Champion - Smurf On A Stick Smolf	Belga	1/1/98		Candy Tube	13" Long	Holland	$3.50-5.00	There is a long clear tube filled with M&M's. The top has a rubber figure on it. A Smurf is holding a gold trophy cup. He has a soccer ball by his foot. The Smurf is wearing a red shirt and yellow shorts.
086/025	Magician - Smurf On A Stick Smolf	Belga	1/1/98		Candy Tube	13" Long	Holland	$3.50-5.00	There is a long clear tube filled with M&M's. The top has a rubber Smurf dressed as a magician on it. The Smurf has a black hat on the ground in front of him. A white bird is flying out of the hat. The Smurf is wearing a yellow hat with blue stars, blue pants with yellow stars, and a yellow cape with blue stars.
086/026	Bicycle Smurf - Smurf On A Stick Smolf	Belga	[illegible]		Candy Tube	13" Long	Holland	$3.50-5.00	There is a long clear tube filled with M&M's. The top has a rubber Smurf figure on it. The Smurf is riding a red bike with yellow tires. The Smurf is wearing a green jacket and dark green pants.

Set 261: Carded Super Smurf Figures

261/001	Papa In Rocking Chair	Irwin/Schleich/Peyo	1/1/96		Carded Super Smurfs	3"	Ger/China	$6.00-8.00	Papa has a dark blue body. Papa is smoking a tan and gray pipe. The rocking chair is brown and has a yellowish/orange cushion. The Smurf is on a blue card with a picture of 2 Smurfs and Smurfette on top.
261/002	Smurf In Log Car	Irwin/Schleich/Peyo	1/1/96		Carded Super Smurfs	3"	Ger/China	$6.00-8.00	The Smurf has a dark bright blue body. The Smurf is wearing black goggles. The car is brown with a brown cushion and has dark brown wheels, dark yellow candles, a dark yellow steering wheel, and a dark yellow stem on the mushroom. The top of the mushroom is dark red with 5 white spots. The Smurf is on a blue card with a picture of 2 Smurfs and Smurfette on top.
261/003	Vanity Table	Irwin/Schleich/Peyo	1/1/96		Carded Super Smurfs	3"	Ger/China	$6.00-8.00	Smurfette has a medium blue body. Smurfette is sitting and wearing a white dress while holding a pink face puff. The table has a pink doily top, dark brown legs, a dark brown stool, and a dark brown mirror frame with a silver mirror. Smurfette is on a blue card with a picture of 2 Smurfs and Smurfette on top.
261/004	Bath Tub	Irwin/Schleich/Peyo	1/1/96		Carded Super Smurfs	3"	Ger/China	$6.00-8.00	The Smurf has a medium bright blue body. He is wearing a red pair of [illegible] with green scrub brush with white soap and soap suds. The tub is dark brown and the sprinkling can is yellow. The Smurf is on a blue card with a picture of 2 Smurfs and Smurfette on top.
261/005	Smurf In Bed	Irwin/Schleich/Peyo	1/1/96		Carded Super Smurfs	3"	Ger/China	$6.00-8.00	The Smurf has a medium bright blue body. The Smurf is wearing a red night shirt and a white hat with a red tassel. The bed is dark brown with red and white mushrooms on the bed post. The mattress is yellow. The Smurf is on a blue card with a picture of 2 Smurfs and Smurfette on top.
261/006	Smurfette In Car	Irwin/Schleich/Peyo	1/1/96		Carded Super Smurfs	3"	Ger/China	$6.00-8.00	Smurfette has a medium bright blue body. Smurfette is wearing a white dress and a pink ribbon in her hair. The car is bright pink with white/yellow daisy decals, yellow wheels, and a red heart- shaped steering wheel. Smurfette is on a blue card with a picture of 2 Smurfs and Smurfette on top.

Set 231: Carrying Cases

231/001	Mallette De Schtroumpfs (Red)	Ganz Bros.	1/1/81	Red	Carrying case	12 1/2" x 10"	Canada	$37.50-50.00	The case is red. The cover has a village scene on it. There is a Smurf pushing a wheelbarrow down a path. Smurfette is standing in the grass pulling petals off of a flower. There are a couple of Smurfs fishing in a river. A Smurf is carrying a hoe. Papa just walked off a bridge. A Smurf is carrying a butterfly net. A Smurf is by a well. The case holds 48 Smurfs.

231/003	Mallette De Schtroumpfs (Blue)	Ganz Bros.	1/1/81	Blue	Carrying Case	12 1/2" x 10"	Canada	$37.50-50.00	The case is blue. The cover has a village scene on it. There is a Smurf pushing a wheelbarrow down a path. Smurfette is standing in the grass pulling petals off of a flower. There are a couple of Smurfs fishing in a river. A Smurf is carrying a hoe. Papa just walked off a bridge. A Smurf is carrying a butterfly net. A Smurf is by a well. The case holds 48 Smurfs.

Set 30: Cars- Diecast Getaways

030/001	Cowboy Smurf And His Die Cast Rocker	Irwin Toys	3/1/96	Brown	Diecast Vehicles	3"	China	$7.50-10.00	A cowboy Smurf is riding a brown rocking horse.
030/002	Boating Smurf And His Die Cast Speed Boat	Irwin Toys	3/16/96	Brown	Diecast Vehicles	4"	China	$7.50-10.00	A Smurf driving a brown speed boat.
030/003	Pilot Smurf And His Die Cast Airplane	Irwin Toys	3/16/96	Green	Diecast Vehicles	3"	China	$7.50-10.00	Pilot Smurf in a green airplane with brown wings.
030/004	Smurfette And Her Die Cast Roadster	Irwin Toys	3/16/96	Pink	Diecast Vehicles	4"	China	$7.50-10.00	Smurfette driving a pink roadster with gray wheels.

Set 2: Cars- ERTL Diecast

002/001	Smurfette Car	Ertl	1/1/82	Red	Diecast Car	3"	U.S.A.	$7.50-10.00	Smurfette driving a red VW bug convertible with a yellow butterfly painted on the hood.
002/002	Aero Smurf	Ertl	1/1/82	Yellow	Diecast Car	3"	U.S.A.	$7.50-10.00	A Smurf in a yellow BI-plane with red markings and red plastic propeller.
002/003	Smurf-About	Ertl	1/1/82	Brown	Diecast Car	3"	U.S.A.	$7.50-10.00	A brown hollowed-out log vehicle with a yellow plastic parasol protecting the Smurf driver.
002/004	Papa Smurf Car	Ertl	1/1/82	Dark Green	Diecast Car	3"	U.S.A.	$7.50-10.00	Papa's driving a green roadster oddly reminiscent of a hot rod's street machine, with a flat-head V-8 out in the open and a turtle deck rear end.
002/005	Gargamobile	Ertl	1/1/82	Blue	Diecast Car	3"	U.S.A.	$7.50-10.00	Gargamel is driving a blue roadster with outside exhaust pipes and side-mounted spare tire.
002/006	The Smurf & Ladder Truck	Ertl	1/1/82	Red	Diecast Car	3"	U.S.A.	$7.50-10.00	A Smurf with a fireman's hat is driving a red fire truck.
002/007	Ol' McSmurf	Ertl	1/1/82	Orange	Diecast Car	3"	U.S.A.	$7.50-10.00	A red neckerchief and yellow floppy-brimmed stocking cap Smurf is driving an orange, older model tractor.
002/008	Handy	Ertl	1/1/82	Beige	Diecast Car	3"	U.S.A.	$7.50-10.00	Handy Smurf driving a beige soapbox derby racer
002/009	Aero-Smurf	Ertl	1/1/82	Yellow	Diecast Car	3"	U.S.A.	$7.50-10.00	A Smurf in a yellow BI-plane with red markings and red plastic propeller. The Smurf is wearing a firemen's hat.
002/010	Handy	Ertl	1/1/82	Light green	Diecast Car	3"	U.S.A.	$7.50-10.00	Handy Smurf driving a light green soapbox derby racer. He has a red pencil behind his ear.
002/011	Handy In Red Car	Ertl	1/1/82	Red	Diecast Car	3"	U.S.A.	$7.50-10.00	Handy is driving Smurfette's red car. The car is all red with black bumpers. Handy has a red pencil behind his ear.

Set 3: Cars- Italian Puffi Diecast

003/001	Black Smurf In A Homemade Car	ESCI	1/1/83	Yellow	Car -Diecast Car	3"	Italy	$18.00-25.00	A yellow car constructed from pieces of a log which forms the rear deck, a wooden box for the middle, and a barrel for the bonnet with a black Smurf driving.
003/002	Papa Smurf Driving A Brown Shoe	ESCI	1/1/83	Brown	Diecast Car	3"	Italy	$18.00-25.00	Papa is driving a brown shoe car.
003/003	Smurfette's Light Green Frog Car	ESCI	1/1/83	Light Green	Diecast Car	3"	Italy	$18.00-25.00	Smurfette is driving a light green frog car. Smurfette has a white shirt on and a red flower in her hair.
003/004	Love-struck Smurf In A Roadster	ESCI	1/1/83	Pink	Diecast Car	3"	Italy	$18.00-25.00	Love-struck Smurf carrying flowers and driving a pink roadster with cycle fenders up front and a black plastic base- plate, which also forms injector tubes in the bonnet, exhaust pipes out the side, and a rear bumper on the back.
003/006	Smurf In A Red Roadster	ESCI	1/1/83	Red	Diecast Car	3"	Italy	$18.00-25.00	The red roadster looks like it's from the 1930's. The car has a black plastic base- plate extended out to form the front and rear bumpers, running boards and the radiator. The Smurf driving has a big smirk on his face.
003/007	Smurfette's Dark Green Frog Car	ESCI	1/1/83	Dark Green	Diecast Car	3"	Italy	$18.00-25.00	Smurfette is driving a dark green frog car. Smurfette has a white shirt on and a red flower in her hair.
003/008	Super Smurf's Can Am Racer	ESCI	1/1/83	Red	Diecast Car	3"	Italy	$18.00-25.00	Super Smurf disguised by a black mask, bearing a yellow "S" on his chest and wearing a flowing red cape is driving a red can-am racer that looks like a batmobile.
003/009	Smurfette In A Corncob Car	ESCI	1/1/83	Orange	Diecast Car	3"	Italy	$18.00-25.00	Smurfette is driving a car that looks like a cob of corn. The car is orange. Smurfette has black headphones over her ears. Smurfette is wearing a white hat and white dress.
003/010	Gargamel In A Purple Roadster	ESCI	1/1/83	Purple	Diecast Car	3"	Italy	$18.00-25.00	Gargamel is driving a purple roadster.

Set 45: Cassette Tapes

045/001	Smurfing Sing Song	Sessions	1/1/79		Cassette Tapes	Canada	$6.00-8.00	The cover has 4 mushroom houses and 4 Smurfs playing musical instruments on it.
045/002	The Smurfs All Star Show	Sessions	1/1/81		Cassette Tapes	Canada	$7.50-10.00	The Smurf band is playing on a stage with red curtains on the side.
045/003	Best Of Friends The Smurfs	Sessions	1/1/82		Cassette Tapes	Canada	$7.50-10.00	The cover is white and has a Smurf walking across it carrying a record.
045/004	Father Abraham In Smurfland	Sessions	1/1/82		Cassette Tapes	Canada	$7.50-10.00	Smurf band on the cover with Father Abraham's face, red border.
045/005	Cartoon Network's Cool New Toons	Turner	1/1/95		Cassette Tapes	U.S.A.	$7.50-10.00	The cover has a Smurf playing a trumpet and 2 other non Smurf characters on it. The cover is green and purple squares.
045/006	Smurfin! 10th Anniversary Commemorative Album	Sessions	1/1/89	White	Cassette Tapes	U.S.A.	$9.00-13.00	The cover has a beach scene with a white background on it. Three Smurfs are dressed like Arabs. A Smurf and Smurfette are walking on a pier. A Smurf is surfing. Nat, Sassette and Snappy are jumping in the sand. A Smurf is playing the drums, another is playing a horn, and Brainy is singing.
045/007	Handy's Liebling Schlumpfe/Willenskraft Clumsy's Gluck	Karussell	1/1/95		Cassette Tapes	Brussels	$10.50-14.00	The cover of the tape has a green mermaid Smurfette in a wash tub on it. Handy Smurf is holding the mermaid's hands. There are bushes in the background. There is a watch in the package. The watch has a white band with a Smurf in it. The face is digital. The face is blue and has a Smurf laying on top of the window box sleeping.
045/008	Smurfin! 10th Anniversary Commemorative Album	Sessions	1/1/89	Blue	Cassette Tapes	U.S.A.	$9.00-13.00	The cover has a beach scene with a blue background on it. Three Smurfs are dressed like Arabs. A Smurf and Smurfette are walking on a pier. A Smurf is surfing. Nat, Sassette and Snappy are jumping in the sand. A Smurf is playing the drums, another is playing a horn, and Brainy is singing.
045/009	The Smurfs Featuring Father Abraham Together Forever	Quality	1/1/90		Cassette Tapes	Canada	$7.50-10.00	The tape cover has a Smurf village on it with Father Abraham sitting in the grass under a tree. There are Smurfs gathered all around Father Abraham.
045/010	I've Got A Little Puppy	EMI	1/1/96	Multi	Cassette Tapes	Brussels	$7.50-10.00	The cover has a Smurf and Smurfette dancing with a brown puppy.
045/011	Die Schlumpfe Fette Fete! Vol. 7	EMI	1/1/98	Pink	Cassette Tapes	Brussels	$11.00-15.00	The cover is pink. It has a Smurf and Smurfette dancing on a birthday cake. It has a magician Smurf in the corner.
045/012	De Smurfen Smurfenparade	EMI	1/1/97	Red	Cassette Tapes	Holland	$11.00-15.00	The cover has a Smurf band on it. It has 3 Smurfs and Smurfette on a stage. There are Smurfs below the stage with their hands in the air.
045/013	De Smurfen Fiesta	EMI	1/1/98	Orange	Cassette Tapes	Holland	$11.00-15.00	The cover is all yellow and orange with different colored stars. It has Smurfette and a Smurf dancing on it. Both Smurfette and the Smurf are wearing headphones and walkmans. Smurfette is in a pink dress.
045/014	De Smurfen Holiday	EMI	1/1/97	Blue	Cassette Tapes	Holland	$11.00-15.00	The cover is a beach scene. It has a Smurf wind-surfing on it, Smurfette dancing on the beach, and a Smurf putting a record on a record player.
045/015	De Smurfen De Wilde Smurf	EMI	1/1/98		Cassette Tapes	Holland	$11.00-15.00	The cover has a Smurf swinging from a vine on it. The Smurf has a leaf hat on and a brown cloth wrapped around his waist.
045/016	Die Schlumpfe Alles Banane! Vol. 3	EMI	1/1/96		Cassette Tapes	Holland	$15.00-20.00	The cover has 2 Smurfs rollerblading in a banana. Another Smurf is on a skateboard.
045/017	Die Schlumpfe Voll Der Winter! Vol. 4	EMI	1/1/96		Cassette Tapes	Holland	$15.00-20.00	The cover has 2 Smurfs and Smurfette snowboarding on it. Smurfette is holding Baby Smurf.
045/018	Die Schlumpfe Balla Balla Vol. 5	EMI	1/1/97		Cassette Tapes	Holland	$15.00-20.00	The cover has 2 Smurfs and Smurfette playing soccer on it.
045/019	Die Schlumpfe Irre Galaktisch! Vol. 6	EMI	1/1/97	Purple	Cassette Tapes	Holland	$15.00-20.00	The cover is of outer space. It has a Smurf in a green space ship on it. Another Smurf and Smurfette are holding hands floating through space. A third Smurf is floating through the air holding a Smurf flag.
045/020	Die Schlumpfe Oh Du Schlumpfige! Vol. 8	EMI	1/1/98		Cassette Tapes	Holland	$15.00-20.00	The cover has Papa Smurf dressed as Santa. He is falling from the sky holding an umbrella. Papa is also holding a bag of toys.
045/021	Schtroumpf Party 3	Royal River	1/1/97		Cassette Tapes	France	$15.00-20.00	The scene on the cover is a desert. It has a Smurf playing a guitar on it. He is wearing a Mexican hat. Another Smurf is dressed as an Arab and is dancing in the background. Smurfette is dressed in a grass skirt and lei's.

Set 279: Cassettes- Die Schlumpfe Story

279/001	Handy's Liebling	Karussell	1/1/95	Story Cassette	Brussels	$6.00-8.00	The cover has a green mermaid in a wood tub on it. Handy Smurf is holding her hands.
279/002	Der Unglaublich Schrumpfende Hexenmeister	Karussell	1/1/95	Story Cassette	Brussels	$6.00-8.00	The cover has Papa looking at Gargamel. Gargamel looks shocked. They are outside Gargamel's castle.
279/003	Marco Schlumpfund Die Pfefferpiration	Karussell	1/1/95	Story Cassette	Brussels	$6.00-8.00	The cover has a Smurf wrapped to a pole on a boat on it. Another Smurf with a life jacket on is looking at the Smurf worriedly.
279/004	Roboter-Schlumpfe	Karussell	1/1/95	Story Cassette	Brussels	$6.00-8.00	2 Smurfs are taking a picture of Gargamel, Azreal and Scrupples in the woods. Gargamel, Azreal and Scrupples have a frazzled expression on their faces.
279/005	Denisas Puppe	Karussell	1/1/95	Story Cassette	Brussels	$6.00-8.00	The cover has a little girl yelling at Gargamel. Gargamel is standing in his boxer shorts holding a Smurf in one hand and a brown sack in the other hand. Papa Smurf and 2 Smurfs are watching.
279/006	Ein Schlumpf Fur Denisa	Karussell	1/1/95	Story Cassette	Brussels	$6.00-8.00	The cover has a little girl holding out her arm with a bird on it. Sassette is standing there looking at the bird. Gargamel is looking through the trees watching.
279/007	Das Fremde Wesen	Karussell	1/1/95	Story Cassette	Brussels	$6.00-8.00	Painter Smurf is looking troubled at a picture. The picture has a little man holding a painter palate on it.
279/008	Land Der Verlorenen	Karussell	1/1/98	Story Cassette	Brussels	$6.00-8.00	A wicked male witch is after the Smurfs. 4 Smurfs are trying to get away from him. The guy is wearing a pink robe and a black witch hat.
279/009	Der Astroschlumpf	Karussell	1/1/95	Story Cassette	Brussels	$6.00-8.00	The cover has a jungle Smurf standing on a cliff watching a Smurf. A Smurf is walking on a planet. The Smurf has an astronaut suit on. In the background is a red space shuttle.

279/010	Der Valentinstag	Karussell	1/1/95		Story Cassette		Brussels	$6.00-8.00	The cover has a little old man dressed as cupid on it. Smurfette is standing with him. Smurfette is holding her hands together and laughing. The background looks like purple spiky things and blue flowers.
279/011	Guter Nachbarschlumpf	Karussell	1/1/96		Story Cassette		Brussels	$6.00-8.00	The cover has a Smurf hanging on to a log on it. The log is attached to a red balloon and is hanging in the air. There are 2 Smurfs on the ground with troubled expressions on their faces.
279/012	Die Drei Schlumpfketiere	Karussell	1/1/97		Story Cassette		Brussels	$6.00-8.00	The cover has Smurfette inside a castle on it. She is dressed in a pink dress and veil. There are 3 Smurfs coming through the door wearing armor. Smurfette has a troubled expression on her face.
279/013	Die Schlumpfe Und Heulvogel	Karussell	1/1/97		Story Cassette		Brussels	$6.00-8.00	The cover has a big hawk swooping down after 3 Smurfs and Papa on it. The Smurfs look scared. One Smurf is trying to hit the bird with a sling shot.
279/014	Die Schlumpfsuppe	Karussell	1/1/98		Story Cassette		Brussels	$6.00-8.00	The cover has 3 Smurfs swimming in a kettle of green stuff on it. The kettle is cooking over a fire.
279/015	Die Blaue Plage	Karussell	1/1/98		Story Cassette		Brussels	$6.00-8.00	The cover has 3 Smurfs sitting outside a mushroom house looking sick. Brainy Smurf is wearing a stethoscope and carrying a medical bag. He is dictating something to the 3 sick Smurfs.
279/016	Die Schlumpfischen Spiele	Karussell	1/1/98		Story Cassette		Brussels	$6.00-8.00	The cover has 3 Smurfs running a marathon on it. The winning Smurf is carrying a torch. There are Smurfs in the stands watching the race.

Set 237: Catalogs- Price Guides

237/001	Der Schlumpf Katalog 1997/98	Oswald/Pfanzelter	1/1/96	Green	Price Guide	8 1/4" x 5 3/4"	Germany	$15.00-20.00	The catalog is green. The cover has a Smurf pointing on it. There are 4 Smurfs surrounding him. Majorette is in a red dress. Clown promotional for McDonald's. One soapbox Smurf and a Smurf in a log car. The guide has pictures of every Smurf and promotional. The catalog is German. There are 224 pages.
237/002	Comicfiguren -Preiskatalog- 1995/96	Koller/Maier/Sterz	1/1/95	Orange	Price Guide	8 1/4" x 5 3/4"	Germany	$15.00-20.00	The catalog has 4 cartoon characters on the cover. The number one graduation Smurf is on the front. The catalog has all different characters in it. There is a part dedicated to the Smurfs.
237/003	Smurf Collectibles Handbook	Lindenberger	1/1/96	Yellow	Price Guide	9" x 6"	U.S.A.	$15.00 - 20.00	The cover has a sporty Smurf mug on it. A Smurf standing on a yellow telephone. A Smurf sticking out his tongue figure. 2 Baby Smurf figures crawling and holding a rattle, one is in a pink sleeper and the other is in a blue sleeper. There are 159 pages.
237/004	Katalog Spielzeug Aus Dem Ei 1997/1998	SU Verlag	1/1/98	Blue	Kinder Egg Catalog	8 1/4" x 5 1/4"	Germany	$18.00-25.00	The catalog has all different kinder egg prices. The front of the catalog shows all different figures. The Smurfs are in the book. The prices for the figures are done in German Marks.
237/005	More Smurf Collectibles	Schiffer Books	1/1/98	Red	Price Guide	9" x 6"	China	$15.00 - 20.00	The book is red. The cover has a Smurf plush on it, a Happy Smurfday plate, and Smurfette in a pink car. The book has 160 colored pages.
237/006	W. Berrie Dealers Catalogs	W. Berrie			Dealer Catalog		U.S.A.	$24.00-32.00	This is a multi-piece SMURFS 1980 dealer's packet! 1. Original mailing envelope with SMURF logos. Some wear and soiling. Overall in very good condition. 2. Fold out full- size catalog with SMURF characters, play sets, houses, Olympian Super Smurfs, plush's, key rings, cloisonné jewelry, and displays for the retailer! Order/price sheet included. Excellent condition 3. Another Smurf catalog with Huggin Stuff and other items by W. Berrie included. Price sheet included. Excellent condition.

Set 241: Cels- Animation

241/001	Smurf Carrying Baby				Cell	10 1/2" x 12 1/4"		$100.00-130.00	The cell is of a Smurf carrying Baby Smurf. There is a laser color background. The background is of a pink Smurf house. The Smurf is standing on a yellow path outside the house.
241/002	Smurfs In A Boat		1/1/80		Cell	10 1/2" x 12 1/4"		$55.00-75.00	The cell is of a pink boat. Smurfette and 3 other Smurfs are in the boat. The cell has a laser background. The background is of a marsh. The boat is in the marsh.
241/003	Smurf Dressed As Robin Hood				Cell	10 1/2" x 12 1/4"		$75.00-100.00	The cell is of a Smurf laying with his hands in front of him.. The Smurf is wearing a hat with a red feather on the side, a red toga and white pants. The laser background is a wood scene. The Smurf is laying in the grass with a hole in front of him. The Smurf looks like he's peaking in the hole. There is a big tree behind the Smurf's head. The cell comes with a pencil drawing of the Smurf.
241/004	Smurf In Tuxedo				Cell	10 1/2" x 12 1/4"		$95.00-125.00	The cell is of a Smurf in a dark blue tuxedo. The Smurf is wearing a dark blue top hat with a pink flower on the side, blue coat with tails, a white shirt, white pants, and a red bow tie. The Smurf is holding a mirror that is attached to a stick. The background is hand painted. It is a balcony. The Smurf is standing on the balcony at night with stars around.
241/005	4 Smurfs Dancing On A Stage				Cell	10 1/2" x 12 1/4"		$75.00-100.00	The cell is of 4 Smurfs dancing on stage. 2 cell setup. The second cell is a stage. The stage is dark brown with brown curtains. There are 3 lights on the bottom of the stage. The Smurfs all have their arms linked. The Smurfs are wearing purple sashes with a yellow button by their waist. One Smurf has a bandage on his nose.
241/006	2 Smurfs On A Boat				Cell	10 1/2" x 12 1/4"		$75.00-100.00	The cell is of 2 Smurfs standing in a boat looking up at the sky. The picture is a 2 cell setup. The second cell is the side of a boat. The boat is brown. There are 2 ropes in front of the Smurfs. The boat is hand painted also. There is a laser copy of the sky.
241/007	Snow Beast, Gargamel And Azreal		1/1/80		Cell	10 1/2" x 12 1/4"		$75.00-100.00	The picture is a 2 cell setup. The first cell is a snow beast holding Gargamel in his hand. Azreal is falling out from the hand. The snow beast has blue feet, blue hands and blue around his eye's. The second cell is a snowdrift. There is a blue sky background with a white bottom.
241/008	Papa Dancing With Sassette				Cell	10 1/2" x 12 1/4"		$75.00-100.00	The cell is of Papa dancing with Sassette. Sassette is wearing a pink dress with white bows. Sassette has orange hair. The cell has a painted background. The background is a purple room.
241/009	Smurf Chopping Wood				Cell	10 1/2" x 12 1/4"		$75.00-100.00	The cell is of a Smurf chopping wood. Another Smurf is standing in front of the other Smurf watching. There is a pile of wood laying on the ground. The cell is a 2 cell setup. There is a laser copy background. The background is a wooded scene by a river. There are a lot of bushes in the background. The cell is framed and matted. The frame is copper colored.
241/010	Gargamel Reading A Book				Cell	10 1/2" x 12 1/4"		$82.50-110.00	The cell is of Gargamel standing and holding a pink book. Azreal is sitting on Gargamel's shoulder. The background is gray matting. The cell is framed and matted. The frame is black.
241/011	5 Smurfs running By A River				Cell	10 1/2" x 12 1/4"		$70.00-95.00	The picture is a 5 cell setup. It is of Papa, Sassette, Nat, Slouchy, and Snappy running. There is a yellow butterfly above Sassette's head. There is a green caterpillar behind Slouchy. There is a laser copy background. The background is a wooded scene by a river. There are a lot of bushes in the background. The cell is framed and matted. The frame is blue.
241/012	Smurf And Azreal Looking In A Mirror				Cell	10 1/2" 12 1/4"		$150.00-200.00	The picture is a 3 cell setup. The cells are of a Smurf standing and looking in a floor length mirror. The Smurf has a cloth ruler around his neck and a red cloth folded over one arm. Azreal is standing behind the Smurf. Azreal's face is reflected in the mirror. The picture has a laser copy background. The background is of a cabinet in a room. The cell is framed and matted. The frame is blue.
241/013	Gargamel With A Green Pig				Cell	10 1/2" x 12 1/4"		$18.00-25.00	The cell is of Gargamel standing next to a green pig. Gargamel is holding a finger up in the air. Gargamel has a tan apron around his waist.
241/014	Brainy Getting Hit In The Head With A Ball				Cell	10 1/2" x 12 1/4"		$18.00-25.00	The cell is of Brainy Smurf getting knocked over. Brainy is getting hit in the head with an orange ball. There is a brown table behind him with lab jars sitting on the table. Brainy is falling backwards into the table.
241/015	Papa Walking				Cell	10 1/2" x 12 1/4"		$18.00-25.00	The cell is of Papa walking and looking upset. Papa has his hands clenched into fists. Papa has an angry expression on his face.
241/016	Smurf Walking On A Tightrope				Cell	10 1/2" x 12 1/4"		$22.50-30.00	The cell is of a Smurf balancing on a tightrope. The Smurf is holding a pole in both hands. The Smurf has a red heart tattooed on his arm.
241/017	Smurf Looking At His Hand				Cell	10 1/2" x 12 1/4"		$25.00-35.00	The cell is of a Smurf looking at his hand. The Smurf has a goofy expression on his face. The cell is only of half of the Smurfs body.
241/018	Sassette As A Puppet				Cell	10 1/2" x 12 1/4"		$22.50-30.00	The cell is of Sassette. Sassette is wearing a pink gown, dark pink shoes, a white hat, and a yellow crown. Sassette has black stings attached to her and she is hanging in mid air. She is holding her hands up like she is trying to scare someone.
241/019	Smurf In A Trench Coat				Cell	10 1/2" x 12 1/4"		$22.50-30.00	The cell is of a Smurf wearing a long brown coat. The Smurf is standing with his hands down at his sides.
241/020	Gargamel Holding A Hammer				Cell	10 1/2" x 12 1/4"		$22.50-30.00	The cell is of Gargamel holding a hammer in the air. Gargamel has soot all over him. He is looking over his shoulder.
241/021	Smurf Carrying A Caterpillar				Cell	10 1/2" x 12 1/4"		$22.50-30.00	The cell is of a Smurf walking and carrying a caterpillar. The Smurf is holding a green caterpillar above his head. The Smurf has a droopy white hat and saggy white pants on.
241/022	Gargamel Sitting Pointing At His Head				Cell	10 1/2" x 12 1/4"		$22.50-30.00	The cell is of Gargamel sitting and pointing up at his head. He has one hand clenched in a fist.
241/023	Papa Yawning				Cell	10 1/2" x 12 1/4"		$37.50-50.00	The cell is of Papa yawning. Papa has on a red shirt and a red hat. He has his hand covering his mouth. Papa has sleepy eye's. The picture is of Papa's upper body.
241/024	Smurf With A Tattoo				Cell	10 1/2" x 12 1/4"		$25.00-35.00	The cell is of a Smurf with a tattoo on his arm. The Smurf has a red heart tattooed on his arm. The Smurf is looking sad. The picture is of the Smurfs upper body. The cell is a 2 cell setup. The cell comes with a pencil drawing of the Smurf.
241/025	Smurf Holding A Rock				Cell	10 1/2" x 12 1/4"		$22.50-30.00	The cell is of a Smurf holding a big gray rock. The Smurf has a shocked expression on his face. The Smurf has a yellow cloth yard stick around his neck. The Smurf has 2 pins stuck in his hat.
241/026	Smurfette Holding A Ball				Cell	10 1/2" x 12 1/4"		$22.50-30.00	The cell is of Smurfette holding a ball. The ball is orange.
241/027	Smurf Painting A Mask On His Face				Cell	10 1/2" x 12 1/4"		$22.50-30.00	The cell is of a Smurf wearing a funny costume. The Smurf has orange and yellow paint around his eye's. The Smurf has 2 white fang teeth. The Smurf is wearing a pink coat, white pants, a white hat, and a black bow tie. The Smurf is holding a paintbrush in one hand. In his other hand he is holding a mirror up. The cell comes with a pencil drawing of the Smurf.
241/028	Brainy With A Book And Wheels Above His Head				Cell	10 1/2" x 12 1/4"		$22.50-30.00	The cell is of Brainy Smurf standing and holding an orange book. There are 2 grayish blue machinery wheels above his head. The cell is a 2 cell setup. The cell comes with a pencil drawing of the wheels.
241/029	King Smurf And His Throne				Cell	10 1/2" x 12 1/4"		$95.00-125.00	The cell is of King Smurf. The Smurf is wearing a yellow crown and a purple robe with white trim. The background is of a red chair with a white rug underneath. There is armor hanging on the wall. The room is orange. The cell is in a purple matting and frame.

241/030	Nat With Butterfly And Caterpillar			Cell	13 1/2" x 16"		$90.00-120.00	The cell is of Nat, he is holding a purple flower up to a butterfly. The butterfly is yellow. There is a green caterpillar by the Smurf's foot. Nat is standing on a dirt path. There are mushroom houses in the background. Nat is wearing brown pants, brown sash, and a tan straw hat.
241/031	Smurf By River With Spears			Cell	13 1/2" x 16"		$60.00-80.00	The cell is of 2 Smurfs chasing Smurfette and 2 other Smurfs with spears. Smurfette is carrying a pick and the 2 other Smurfs are carrying a ladder. The background is of a river with rocks behind it. The Smurfs all have mad expressions on their face's.
241/032	Smurfette On Snowshoes			Cell	13 1/2" x 16"		$95.00-125.00	Smurfette is coming out of a mushroom house. Smurfette is wearing brown snowshoes. The scene is a snow scene.
241/033	Hefty Smurf In Wheelchair			Cell	13 1/2" x 16"		$37.50-50.00	Hefty Smurf is in a wheelchair. He is carrying Baby Smurf in his lap. Hefty has one leg bandaged.
241/034	Papa And Baby Smurf In A Rocking Chair			Cell	12 1/2" x 10 1/2"		$37.50-50.00	Papa is sitting in a brown rocking chair. He has Baby Smurf in his lap. Papa is reading a book to Baby Smurf.

Set 22: Ceramics- Birthday Smurfette

022/001	Smurfette Birthday Series #2	W. Berrie	1/1/83	Ceramic Collectibles	4" high	Taiwan	$22.50-30.00	Smurfette sitting on a #2 holding balloons. The balloons are pink, yellow and orange. The number 2 is blue. Smurfette has a pink bow in her hair.
022/002	Smurfette Birthday Series #7	W. Berrie	1/1/83	Ceramic Collectibles	4" high	Taiwan	$22.50-30.00	Smurfette is standing behind a number 7. The number is green. There is a yellow butterfly on top of the 7. Smurfette's wearing a white dress and has a pink flower in her hair.
022/003	Smurfette Birthday Series #8	W. Berrie	1/1/83	Ceramic Collectibles	4" high	Taiwan	$22.50-30.00	Smurfette is behind the number 8. The number is blue. Smurfette is dressed in a white dress with pink trim, wearing pink mittens and white ice skates.
022/004	Smurfette Birthday Series #9	W. Berrie	1/1/83	Ceramic Collectibles	4" high	Taiwan	$22.50-30.00	Smurfette is standing next to a number 9. The number is green. She is flying a kite. Smurfette is wearing a white dress.
022/005	Smurfette Birthday Series # 1	W. Berrie	1/1/83	Ceramic Collectibles	4" high	Taiwan	$22.50-30.00	Smurfette is standing next to a number 1. There is a blue teddy on top of the number and a brown teddy leaning against the number. Smurfette has flowers in her hat. The number one is pink.
022/006	Smurfette Birthday Series # 3	W. Berrie	1/1/83	Ceramic Collectibles	4" high	Taiwan	$22.50-30.00	Smurfette is leaning against a number 3. Smurfette is holding a bouquet of flowers. The flowers are pink, orange and yellow. The number 3 is pink.
022/007	Smurfette Birthday Series # 4	W. Berrie	1/1/83	Ceramic Collectibles	4" high	Taiwan	$22.50-30.00	Smurfette is standing next to a number 4. The number 4 is green. There is a pot of flowers sitting on the 4. The flowers are pink. Smurfette is holding a purple watering can.
022/008	Smurfette Birthday Series # 5	W. Berrie	1/1/83	Ceramic Collectibles	4" high	Taiwan	$22.50-30.00	Smurfette is leaning on a number 5. The number 5 is purple. Smurfette is holding an umbrella. The umbrella is purple with pink and yellow flower trim. There is a pink and green bow in the center. The umbrella handle is purple.
022/009	Smurfette Birthday Series # 6	W. Berrie	1/1/83	Ceramic Collectibles	4" high	Taiwan	$22.50-30.00	Smurfette is standing next to a number 6. The number 6 is green. Smurfette is wearing a ballerina outfit. Her dress is white with pink trim. Smurfette has purple tights, pink slippers, and a pink flower on her hat.
022/010	Smurfette Birthday Series # 10	W. Berrie	1/1/83	Ceramic Collectibles	4" high	Taiwan	$22.50-30.00	Smurfette is sitting in the 0. The number 10 is purple. Smurfette is wearing roller-skates. There is a blue bird on the tip of her skate and on top of the 1. Smurfette looks like she feel into the 0.
022/011	Smurfette Birthday Series # 11	W. Berrie	1/1/83	Ceramic Collectibles	4" high	Taiwan	$22.50-30.00	Smurfette is standing next to the number 11. The numbers are blue. Smurfette is wearing a pink and white cheerleader outfit. Smurfette is holding pink pompoms.
022/012	Smurfette Birthday Series # 12	W. Berrie	1/1/83	Ceramic Collectibles	4" high	Taiwan	$22.50-30.00	Smurfette is sitting on the back of the 2. The number 12 is green. Smurfette is holding a pink daisy with a purple stem.

Set 23: Ceramics- Christmas Figures

023/001	Smurf With a Snowman	W. Berrie	1/1/82	Ceramic Collectibles	3 1/2" high	Taiwan	$25.00-35.00	A Smurf is making a snowman. The snowman has a red hat, green mittens, and a scarf on. The Smurf is wearing red mittens and a red scarf.
023/002	Smurf Playing A Drum & Papa Blowing A Horn	W. Berrie	1/1/82	Ceramic Collectibles	3 1/2" high	Taiwan	$25.00-35.00	Papa is playing a yellow horn and a Smurf is playing a red drum.
023/003	Smurfette & Smurf Singing Christmas Carol	W. Berrie	1/1/82	Ceramic Collectibles	3 1/2" high	Taiwan	$25.00-35.00	Smurfette and Smurf are holding a caroling book that says Joy to the World.
023/004	Smurfette With A Christmas Tree.	W. Berrie	1/1/82	Ceramic Collectibles	3 1/2" high	Taiwan	$25.00-35.00	Smurfette is holding a pink and red candy cane and has her arms around the tree. The tree has a big yellow star on top.

Set 24: Ceramics- Figures

024/001	Greedy	W. Berrie	1/1/82	Ceramic Collectibles	3 1/2" high	Taiwan	$15.00-20.00	Greedy Smurf is holding a pink and yellow birthday cake.
024/002	Brainy	W. Berrie	1/1/82	Ceramic Collectibles	3 1/2" high	Taiwan	$15.00-20.00	Brainy's holding up a finger and in the other arm he's holding a red book.
024/003	Papa Holding A Potion Book	W. Berrie	1/1/82	Ceramic Collectibles	3 1/2" high	Taiwan	$15.00-20.00	Papa's holding a white and green lab glass in one hand and a purple Smurf magic book in the other hand.
024/004	Smurf & Smurfette Sitting On A Log	W. Berrie	1/1/82	Ceramic Collectibles	3 1/2" high	Taiwan	$22.50-30.00	A Smurf is sitting next to Smurfette and handing her some pink flowers.
024/005	Hefty Pounding A Smurf Village Sign	W. Berrie	1/1/82	Ceramic Collectibles	3 1/2" high	Taiwan	$15.00-20.00	Hefty is pounding a brown sign into the ground.
024/006	Gargamel And Azreal	W. Berrie	1/1/82	Ceramic Collectibles	3 1/2" high	Taiwan	$18.00-25.00	Gargamel is walking with his hands together and Azreal is behind his left foot.
024/007	Jokey	W. Berrie	1/1/82	Ceramic Collectibles	3 1/2" high	Taiwan	$15.00-20.00	Jokey's got one hand in front of his mouth and there's a yellow box with a red ribbon sitting by his feet.
024/008	Smurfette Roller-skating	W. Berrie	1/1/82	Ceramic Collectibles	3 1/2" high	Taiwan	$15.00-20.00	Smurfette's got her hands out at her sides and she is wearing pink and red roller-skates.
024/009	Smurf Blowing A Horn & Papa Has His Ears Plugged	W. Berrie	1/1/82	Ceramic Collectibles	3 1/2" high	Taiwan	$22.50-30.00	A Smurf is playing a yellow horn and Papa's making a weird face and has his fingers in his ears.
024/010	Valentine Smurfette	W. Berrie	1/1/82	Ceramic Collectibles	3 1/2" high	Taiwan	$15.00-20.00	Smurfette is standing on a white base that has red and pink hearts around the outside. Smurfette is holding a pink heart that says I Love You.
024/011	Valentine Smurf	W. Berrie	1/1/82	Ceramic Collectibles	3 1/2" high	Taiwan	$15.00-20.00	A Smurf is standing on a white base that has red and pink hearts around the outside. The Smurf is holding a pink heart that says I Love You.

Set 145: Ceramics- Hand Painted

145/001	Papa With Hands Out			Ceramic Smurf	6" High	Hand Painted		Papa is standing and he has both his arms straight out at his sides. Papa has a dark blue body, red hat, red pants and a white beard.
145/002	Smurfette With Her Left Hand On Her Chest			Ceramic Smurfette	6" High	Hand Painted		Smurfette is standing and wearing a white dress with red poke-a-dots, white shoes, and a white hat. Smurfette has a dark blue body. Smurfette has her left hand on her chest and her other hand is out at her side.
145/003	Smurf With Flowers			Ceramic Smurf	6" High	Hand Painted		The Smurf is standing and he is holding 2 red daisies in his left hand, his right arm is at his side. The Smurf has a dark blue body, white hat and white pants. The Smurf has a shy grin on his face and his head is turned away from the hand with the flowers.
145/004	Smurfette Standing In Flowers			Ceramic Smurfette	6" High	Hand Painted		Smurfette is standing on a green base with white flowers all around her. Smurfette is wearing black shoes and a white dress. Smurfette has one hand in her hair and the other behind her back. Smurfette has a flirting look on her face.
145/005	Smurfette Holding An Orange Flower			Ceramic Smurfette	9" High	Hand Painted		Smurfette is standing and wearing a white dress, white shoes, and a white hat. Smurfette has a medium blue body. Smurfette has her left hand on her chest and her other hand is holding an orange flower against her chest. The paint has a glossy coat over it.
145/006	Smurfette Holding A Yellow Flower			Ceramic Smurfette	9" High	Hand Painted		Smurfette is standing and wearing a dark pink dress, white shoes, and a white hat. Her hair is a dark yellow with black stripes in it. Smurfette has a light blue body. Smurfette has her left hand on her chest and her other hand is holding a yellow flower against her chest. The paint is dull.
145/007	Smurfette Holding A Pink Flower			Ceramic Smurfette	9" High	Hand Painted		Smurfette is standing and wearing a white dress with light blue poke-a-dots, white shoes, and a white hat. Her hair is dark yellow. Smurfette has a dark blue body. Smurfette has her left hand on her chest and her other hand is holding a light pink flower against her chest. The paint is dull.
145/008	Smurfette Holding An Orange Glittery Flower			Ceramic Smurfette	6" High	Hand Painted		Smurfette is standing and wearing a light pink dress with silver glitter, gray shoes and a pink hat with silver glitter. Her hair is yellow with gold glitter. Smurfette has a light blue body. Smurfette has her left hand on her chest and her other hand is holding an orange glittery flower against her chest. The paint is dull.
145/009	Smurf With Hands Out At Side			Ceramic	9" High	Hand Painted		The Smurf has a light blue body. He looks short and fat. The Smurf has both his hands in the air at his sides. The Smurf has a big smile on his face. The Smurf is wearing white pants and a white hat. The paint has a glossy finish. There is a light blue platform that spins and plays music.
145/010	Papa With His Head Turned To The Side			Ceramic	6"High	Hand Painted		Papa is standing with his hands at his side and his head turned to the right side. He has red pants and a red hat on. His body is painted bright blue.
145/011	Smurf Sitting Under A Flower			Ceramic	3"High	Hand Painted		A Smurf is sitting under an orange and a yellow flower. The Smurf has a dark blue body. The Smurf looks like he's sitting on a brown stump.
145/012	Papa Looking Angry			Ceramic	2 1/2"High	Hand Painted		Papa is standing with his hands on his hips. He has a red hat and white pants on. Papa is standing on a green patch of grass.
145/013	Baseball Smurf			Ceramic Smurf	2 1/2"High	Hand Painted		The Smurf is standing and holding a tan baseball bat. A white and black ball is on the end of the bat. The Smurf is wearing white pants and shoes. He is standing on a patch of light green grass.
145/014	Baseball Smurf			Ceramic Smurf	2 1/2"High	Hand Painted		The Smurf is standing and holding a brown baseball bat. A white ball is on the end of the bat. The Smurf is wearing orange pants and black shoes. He is standing on a patch of dark green grass.
145/015	Swimmer Smurf			Ceramic Smurf	1"L X 2 1/2"W	Hand Painted		The Smurf is on a light blue ripple of water. He has a black inner-tube around his waist. The Smurf is wearing yellow swim trunks. He is in a diving position.
145/016	Papa Waving			Ceramic	7 1/2"L X 8"W			Papa is standing and he has one arm straight out at his side and the other is turned up. Papa has a light blue body, red hat, red pants and a whitish silver beard. Papa has big black eyebrows. He has a knobby tail that sticks out 1 1/2".
145/017	Smurfette Holding A red Flower			Ceramic	8 1/2" High			Smurfette is standing and wearing a whitish silver dress, shoes and hat. Her hair is pale yellow. Smurfette has a medium blue body. Smurfette has her left hand on her chest and her other hand is holding a red flower with a green stem against her chest. The paint is shiny.
145/018	Smurf Waving			Ceramic	7 1/2" High	Hand made		A Smurf is standing and he has one arm straight out at his side and the other is turned up. The Smurf has a light blue body, white hat, white pants. He has a knobby tail that sticks out 1 1/2".

145/019	Smurf Eating				Ceramic	5 1/2" High			The Smurf is standing and licking his mouth. He is holding something yellow in his right hand. The Smurf is painted a bright medium blue. He has a white hat and pants on.
145/020	Mini King Smurf				Ceramic	2 1/4 "High	Hand Painted		The Smurf is holding a green and orange septor. The Smurf is wearing a blue crown, white hat, yellow gown, and a white robe with orange trim.
145/021	Smurf Sitting, holding A Pumpkin				Ceramic	6 1/2" x 6 1/2"	Hand Painted	$7.50-10.00	A Smurf is sitting on a log holding a pumpkin. There are flowers in front of the log.
145/022	Mini Smurfette				Ceramic	1 3/4"High	Hand Painted		Smurfette has a light glossy blue body. She is wearing a pink dress, white shoes, and a white hat. Smurfette has one hand on her chest and the other out at her side.
145/023	Mini Smurfette				Ceramic	1 3/4"High	Hand Painted		Smurfette has a dark blue body. She is wearing a white dress with red dots and a white hat. Smurfette has one hand on her chest and the other out at her side.
145/024	Mini Smurf Wearing An Apron				Ceramic	2"High			The Smurf has a dark blue body. He is wearing a brown gardeners apron. He has his hands out at his side like he's suppose to push something.
145/025	Smurf Holding A Present				Ceramic	5" High	Hand Painted		The Smurf has a light blue body. He is standing and holding an orange present. The Smurf has big ears.
145/026	Brainy Smurf With Big Ears				Ceramic Figure	5 High	Hand Painted		The Smurf is standing with one hand on his chest and the other behind his back. The Smurf has black glasses on. He has big ears. The Smurf has a light blue body.
145/027	Football Smurf				Flat Ceramic	2 1/4"	Hand Painted	$1.50-2.00	The Smurf is running with a brown football under his arm. He is wearing a green helmet, green pants, and a green shirt. The ceramic is flat like a magnet.
145/028	Traveler Smurf				Flat Ceramic	2 1/4"	Hand Painted	$1.50-2.00	The Smurf is walking and carrying a yellow knapsack over his shoulder. The knapsack is on the end of a brown stick. The ceramic is flat like a magnet.
145/029	Soccer Smurf				Flat Ceramic	2 1/2"	Hand Painted	$1.50-2.00	The Smurf is kicking a white and black soccer ball. The Smurf is wearing white pants, a green shirt, and a white hat. The ceramic is flat like a magnet.
145/030	Smurf Holding Flower				Flat Ceramic	2 3/4"	Hand Painted	$1.50-2.00	The Smurf looks like he is holding a red flower in his hand. He is looking over his shoulder. The ceramic is flat like a magnet.
145/031	Baseball Smurf				Flat Ceramic	2 1/4"High	Hand Painted	$1.50-2.00	The Smurf is walking and carrying a baseball bat over his shoulder. The bat has a brown glove hanging off the end of it. The Smurf is wearing a yellow shirt, red shorts, white shoes, and a white hat. The ceramic is flat like a magnet.

Set 262: Ceramics- Juice Cups

262/001	Smurf With Mallet	Peyo/Scharfen		White	Juice Cup	3 1/2"		$7.50-10.00	On the front is Papa pointing. He is larger than the rest of the Smurfs. Next to Papa there is a Smurf holding a mallet, a Smurf carrying a wrench, a doctor Smurf, and Painter Smurf. The picture goes all the way around the cup. The cup is white.
262/002	Jokey Smurf	Peyo/Scharfen		White	Juice Cup	3 1/2"		$7.50-10.00	On the front is Papa pointing. He is larger than the rest of the Smurfs. Next to Papa there is a Smurf holding a present and flowers, a Smurf chef, a Smurf carrying a lantern, and a Smurf holding a script and feather pen.
262/003	Smurfette Flirting	Peyo/Scharfen		White	Juice Cup	3 1/2"		$7.50-10.00	On the front is Papa pointing. He is larger than the rest of the Smurfs. Next to Papa there is Smurfette flirting, a Smurf handing Smurfette flowers, a Smurf delivering mail, and a Smurf holding a penny in the air.
262/004	Soccer Smurf	Peyo/Scharfen		White	Juice Cup	3 1/2"		$7.50-10.00	On the front is Papa pointing. He is larger than the rest of the Smurfs. Next to Papa there is a Smurf kicking a soccer ball, a Smurf holding a dumbbell, a Smurf carrying a torch, and a Smurf riding a tricycle.
262/005	Smurf Singing	Peyo/Scharfen		White	Juice Cup	3 1/2"		$7.50-10.00	On the front is Papa pointing. He is larger than the rest of the Smurfs. Next to Papa there is a Smurf singing, a Smurf playing the trumpet, a Smurf playing guitar, and a Smurf playing the drums. The picture wraps around the cup.
262/006	Smurf Covering His Ears	Peyo/Scharfen		White	Juice Cup	3 1/2"		$7.50-10.00	On the front is Papa pointing. He is larger than the rest of the Smurfs. Next to Papa there is a Smurf covering his ears, a Smurf smiling, another Smurf covering his mouth and laughing, and a Smurf looking at himself in the mirror.

Set 270: Ceramics- Unpainted

270/001	King			White	Ceramic Figure	2"		$1.50-2.00	The King Smurf is solid white and unpainted. He is holding a finger in the air. The Smurf is holding a septor in the other hand.
270/002	Smurf Crawling			White	Flat Ceramic	3"		$1.50-2.00	The Smurf looks like he's crawling. The ceramic is flat like a magnet. He is solid white.
270/003	Smurf Waving			White	Flat Ceramic	2 3/4"		$1.50-2.00	The Smurf is waving with one hand. The ceramic is flat like a magnet. He is solid white.
270/004	Smurf Sitting Thinking			White	Flat Ceramic	2"		$1.50-2.00	The Smurf is sitting and holding his head in his hands. The ceramic is flat like a magnet. He is solid white.
270/005	Smurfette			White	Flat Ceramic	2 1/2"		$1.50-2.00	Smurfette is standing and posing. The ceramic is flat like a magnet. She is solid white.
270/006	Traveler Smurf			White	Flat Ceramic	2"		$1.50-2.00	The Smurf is walking and carrying a knapsack over his shoulder. The ceramic is flat like a magnet. He is solid white.
270/007	Smurf Carrying A Book			White	Flat Ceramic	1 3/4"		$1.50-2.00	The Smurf is running and carrying a book. The Smurf is wearing glasses. The ceramic is flat like a magnet. He is solid white.
270/008	Papa Jumping			White	Flat Ceramic	2"		$1.50-2.00	Papa is jumping. His hands are out at his side. The ceramic is flat like a magnet. He is solid white.
270/009	Football Smurf			White	Flat Ceramic	1 1/2"		$1.50-2.00	The Smurf is running with a football tucked under his arm. He is wearing a football helmet. The ceramic is flat like a magnet. He is solid white.
270/010	Soccer Smurf			White	Flat Ceramic	1 1/2"		$1.50-2.00	The Smurf is kicking a soccer ball. The ceramic is flat like a magnet. He is solid white.
270/011	Golfer			White	Flat Ceramic	1 1/2"		$1.50-2.00	The Smurf is holding a golf club. The ceramic is flat like a magnet. He is solid white.
270/012	Smurf With Ruler			White	Flat Ceramic	1 1/2"		$1.50-2.00	The Smurf is wearing coveralls. He is running. He has a pencil behind one ear. He has a ruler in his hand. The ceramic is flat like a magnet. He is solid white.

Set 106: Chalkboards

106/001	Learning Smurf Chalkboard	H-G Industries	1/1/81	Blue	Chalkboard	11 1/2"L X 8 1/2" W	U.S.A.	$6.00-8.00	The chalkboard is blue. In the top left corner it has a Smurf face and in the top right corner it has Smurfette's face. The left side has the letters A through L going down, the right side has the numbers 1 through 12 going down. The bottom has a ruler in inches.
106/002	Smurfette Chalkboard	H-G Industries	1/1/81	Black	Chalkboard	18" X 12"	U.S.A.	$7.50-10.00	The chalkboard is black . On top is Smurfette's face with a red and yellow rainbow behind her. There is a red and yellow border going around the chalkboard under the picture of Smurfette. Smurfette has her hands on the border on top. Up in the corner it says Smurf in blue and white letters.
106/003	Smurf Chalkboard	H-G Industries	1/1/82	Black	Chalkboard	18" X 12"	U.S.A.	$7.50-10.00	The chalkboard is black . On top is a Smurf face with a red and yellow rainbow behind him. There is a red and yellow border going around the chalkboard under the picture of the Smurf. The Smurf has his hands on the border on top. Up in the corner it says Smurf in blue and white letters.
106/004	Papa Smurf Chalkboard	H-G Industries	1/1/82	Black	Chalkboard	18" X 12"	U.S.A.	$7.50-10.00	The chalkboard is black . On top is Papa Smurf's face with a red and yellow rainbow behind him. There is a red and yellow border going around the chalkboard under the picture of Papa. Papa has his hands on the border on top. Up in the corner it says Smurf in blue and white letters.
106/005	Mini Yellow Chalkboard	Peyo/I.M.P.S.	1/1/94	Yellow	Chalkboard	6" x 8"	Brussels	$4.50-6.00	The chalkboard has a yellow plastic frame. The top of the frame has the letters A-R, the bottom has S-Z and 1-0. In the corner of the chalkboard is a Smurf writing his ABC's. The chalkboard comes with a box of chalk and a pink sponge.

Set 215: Charms

215/001	Smurf Waving Silver Charm	Peyo		Silver	Charm		Germany	$7.50-10.00	The charm is made of silver. They have a Smurf waving. The Smurf isn't fully painted blue. His face is part silver colored. The Smurf has white pants and a white hat on. The earrings are post.
215/002	Smurf Face Silver Charm	Peyo		Silver	Charm		Germany	$7.50-10.00	The charm is silver. The face is painted half blue and the rest is silver. The Smurf has a white hat on. The face is made of silver.
215/003	Papa Pointing	M.B.M	1/1/82		Charm	1"	Italy	$4.50-6.00	Papa is standing and pointing a finger. He has a red hat and red pants on. The charm is on a display card.
215/004	Smurf Holding Pink Daisy	M.B.M.	1/1/82		Charm	1"	Italy	$4.50-6.00	The Smurf is standing and holding a pink daisy. The Smurf has a shy grin on his face. The charm is on a display card.
215/005	Smurfette	M.B.M	1/1/82		Charm	1"	Italy	$4.50-6.00	Smurfette is standing with a finger by her mouth. Smurfette is wearing a red dress, white shoes, and a white hat. The charm is on a display card.
215/006	Nude Smurf	M.B.M.	1/1/82		Charm	1"	Italy	$4.50-6.00	The Smurf is naked, except for a white hat. The Smurf has a hand over his front but his butt is showing. He has a startled expression on his face. The charm is on a display card.
215/007	Tennis Smurf	M.B.M.	11/20/98		Charm	1"	Italy	$1.50-6.00	The Smurf is standing and holding a gold tennis racket in the air. The charm is on a display card.
215/008	Gargamel And Azreal	M.B.M.	1/1/82		Charm	1"	Italy	$4.50-6.00	Gargamel is standing and he looks like he's slouched over. Gargamel is wearing a black shirt and red shoes. Azreal is by his feet. The charm is on a display card.
215/009	Smurf Smelling A Daisy	M.B.M.	1/1/82		Charm	1"	Italy	$4.50-6.00	The Smurf is standing with a yellow daisy to his nose. The Smurf has his eyes closed. The charm is on a display card.
215/010	Smurf With Gift	M.B.M.	1/1/82		Charm	1"	Italy	$4.50-6.00	The Smurf is carrying a red gift. The gift has yellow ribbon around it. The Smurf looks like he's offering the gift. The charm is on a display card.
215/011	Smurf Sticking Out Tongue	M.B.M.	1/1/82		Charm	1"	Italy	$4.50-6.00	The Smurf is standing with his thumbs in his ears. He is sticking out his tongue. The charm is on a display card.
215/012	Smurf Laughing	M.B.M.	1/1/82		Charm	1"	Italy	$1.50-6.00	The Smurf is standing and laughing. He is pointing a finger at something. The charm is on a display card.
215/013	Smurf Playing Guitar	M.B.M.	1/1/82		Charm	1"	Italy	$4.50-6.00	The Smurf is standing and playing a red and orange guitar. The Smurf has his mouth open like he's singing. The charm is on a display card.
215/014	Smurf Crying	M.B.M.	1/1/82		Charm	1"	Italy	$4.50-6.00	The Smurf is standing and crying. The Smurf has a tear running down his face. He has his hands by his side. The charm is on a display card.
215/015	Papa Smurf				Charm	1"		$2.00-3.00	Papa is standing and pointing a finger. Papa has a white hat and white pants on. Papa has a blue unpainted beard.

Set 110: Chatter Chums

110/001	Smurfette	W. Berrie	1/1/83		Pull String	7" High	Hong Kong	$18.00-25.00	Smurfette is sitting on a purple base. She is made of plastic, when you pull the string on the back, her face moves up and down. Smurfette says 7 different things.
110/002	Smurf	W. Berrie	1/1/83		Pull String	7" High		$18.00-25.00	A Smurf is sitting on an orange base. He is made of plastic, when you pull the string on the back, his face moves up and down. The Smurf says 7 different things.

Set 87: Christmas Items

087/001	Musical Smurfette Angel	Talbot Toys	1/1/82		Musical Toy	7" High	Hong Kong	$37.50-50.00	Smurfette is standing and wearing a white angel dress with red down the front. There are 2 white and gold wings on her back. Smurfette is holding a white candle that lights up. Plays 8 pre-programmed Christmas carols continuously.
087/002	Brainy Christmas Stocking	Peyo	1/1/83		Cloth Stocking	16"Long		$7.50-10.00	The stocking is white with red trim. In the center, Brainy is holding a stocking with 2 books inside it. There is a green circle with 2 red hearts behind Brainy. Above Brainy is says Merry Christmas! in blue letters and below him it says SMURF in white and blue.
087/003	Musical Santa Papa	Talbot Toys	1/1/82		Musical Toy	7" High	Hong Kong	$37.50-50.00	Papa is dressed in a red Santa outfit. He is wearing a red jacket, pants and hat with white trim. He is wearing a black belt with a gold buckle and black shoes. Papa has a brown sack with packages in it, over his left shoulder. He is carrying a gold lantern. The lantern lights up. Papa plays 12 Christmas songs over, continuously. He has a gold cord attached for hanging.
087/004	Smurf Houses Christmas Lights	Alderbrook/Peyo			Christmas Lights		Canada	$40.00-55.00	The set of lights are 10 plastic houses. The houses are red, yellow and white. The houses have different pictures on the front.
087/005	Smurf Heads Christmas Lights	Alderbrook			Christmas Lights		Taiwan	$30.00-40.00	The lights are of 10 Smurf heads. The Smurfs have white hats with red tassels on the ends.
087/006	Smurf Christmas Tags	Papercraft	1/1/82		Gift Tags		U.S.A.	$7.50-10.00	The package has 18 tags. There are 3 different designs. Papa and a Smurf are decorating a tree. A Smurf is carrying a sack of presents over his shoulder. A Smurf is carrying a green present with red ribbon. The tags are in a green package with Papa and a Smurf face on top.
087/007	Smurfette Stocking	Peyo	1/1/83		Cloth Stocking	16" Long		$7.50-10.00	The stocking is white with red trim. In the center, Smurfette is ice skating. Smurfette is wearing a white and green dress, red gloves, a white hat, and white skates. Above Smurfette it says "Have A Smurfy Christmas!" in red letters.
087/008	Smurf Christmas Bows	Papercraft	1/1/83		Bows		U.S.A.	$7.50-10.00	6 giant Christmas bows in the package. The package is blue and clear. The bows are all different.

Set 64: Clay- Models

064/001	Das Smurf Deluxe Modeling Clay Play set	Adica Pongo	1/1/82	Yellow Box	Modeling Clay		Italy	$15.00-20.00	Deluxe play set includes: 4 squish paints, clear varnish, plastic storage tray, 2 sets of accessories, 18 oz. clay, plastic Smurf and Smurfette mold, 1 brush, file, and a spatula and cutter.
064/002	Das Modeling Clay	DAS	1/1/82	Yellow Box	Modeling Clay		U.S.A.	$11.00-15.00	Includes: 1 brush, 1 Smurf mold, 2 sets of accessories, 3 paint colors, varnish, and 18 oz. of clay.
064/003	Smurf Het Gezellige Smurfendorp	Fila	1/1/95		Modeling Clay		Brussels	$11.00-15.00	Included in the box is 1 plastic mold to make a regular Smurf, paint, paintbrush, sandpaper, clay, and a sponge and plastic. The outside of the box has a Smurf making a statue of Smurfette. There are 3 Smurfs by the mold. A Smurf is getting painted.
064/004	Smurfette Het Gezellige Smurfendorp	Fila	1/1/95		Modeling Clay		Brussels	$11.00-15.00	Included in the box is 1 plastic mold to make a Smurfette, paint, paintbrush, sandpaper, clay, and a sponge and plastic. The outside of the box has a Smurf making a statue of Smurfette. There are 3 Smurfs by the mold. A Smurf is getting painted.

Set 246: Clips

246/001	Eitlerschlumpf	Ajena	1/1/83		Plastic Clip	1"	France	$11.00-15.00	The Smurf is 1" high. He has a yellow flower in his hat. He clips on to a pink barrette.
246/002	Winterschlumpf	Ajena	1/1/83		Plastic Clip	1"	France	$11.00-15.00	The Smurf is 1" high. He is wearing an orange scarf around his face. The Smurf has his hands wrapped around him like he's shivering. The Smurf hangs from a blue clip holder.
246/003	Schlumpf Volant	Ajena	1/1/83		Plastic Clip	1"	France	$11.00-15.00	The Smurf is 1" high. The Smurf has pink wings attached to his arms. The Smurf comes with a green clip that clips to a cup.
246/004	Kuchenschlumph	Ajena	1/1/83		Plastic Clip	1"	France	$11.00-15.00	The Smurf is 1" high. The Smurf is eating a piece of cake. The Smurf comes with a yellow circular clip that attaches to straws.
246/005	Schlafschlumpf	Ajena	1/1/83		Plastic Clip	1"	France	$11.00-15.00	The Smurf is 1" high. The Smurf is sleeping. The Smurf comes with a blue clip to attach to glasses.
246/006	Kochschlumpf	Ajena	1/1/83		Plastic Clip	1"	France	$11.00-15.00	The Smurf is 1" high. The Smurf is wearing a chefs hat. The Smurf is carrying a tray with a slice of cake on it over his head. He has a towel folded over his other arm. The Smurf comes with a green clip with a suction cup on it.
246/007	Guitar Smurf And Smurfette	Ajena	1/1/83		Plastic Clip	1"	France	$11.00-15.00	The Smurfs are 1" high. Smurfette is sitting on top of a round picture frame. A Smurf is next to the frame. The Smurf is playing a red and white guitar. The frame is yellow.
246/008	Dichter Schlumpf	Ajena	1/1/83		Plastic Clip	1"	France	$11.00-15.00	The Smurf is 1" high. The Smurf is standing and holding a white script and a red feather pen. The Smurf hooks on to the end of a pink paper clip.

Set 58: Clocks

058/001	Have A Smurfy Day Talking Alarm Clock	Bradley Time	1/1/82		Talking Alarm Clock	8 1/2" x 8 1/2"	Hong Kong	$25.00-35.00	The base of the clock is white. The clock looks like a log and has 2 Smurfs standing on top sawing. Papa is looking on. The Smurf figures are soft and 3 dimensional. The alarm talks in a Smurf's voice.
058/002	Smurf And Smurfette On A Log Animated Clock	Bradley Time	1/1/83		Animated Action	6" High	China	$35.00-45.00	The clock is metal. The outside of the clock is blue with 2 yellow bells on top and a gold handle. There are 2 gold legs on the bottom of the clock. The picture is of a Smurf and Smurfette sitting on a log teeter- totter. The log moves when the clock is wound. There are 2 Smurfs in the background with 3 mushroom houses. The numbers and the hands of the clock are white.
058/003	Have A Smurfy Day! Smurf On A Cloud	Bradley Time	1/1/83		Animated Action	6" High	Hong Kong	$35.00-45.00	The clock is metal. The outside of the clock is blue with 2 yellow bells on top and a gold handle. There are 2 gold legs on the bottom of the clock. The picture is a Smurf sitting on a white cloud with a rainbow in back. The background is yellow. The cloud says "Have A Smurfy Day " in red letters inside of it. The numbers and hands of the clock are blue.
058/004	A Smurfette's Work Is Never Done! Smurfette At A Desk	Bradley Time			Desk Clock	4"W X 5"H		$18.00-25.00	A desk top digital clock with a clear plastic stand. The picture on the front is of Smurfette wearing a white dress and sitting on a brown chair at a brown desk. Smurfette is typing on a red typewriter and there is a stack of papers on the desk. The background for the picture is white. The clock is red and is in the top corner.
058/005	Smurf On A Moon Wall Watch Clock	Bradley Time	1/1/84		Wall Watch Clock		Hong Kong	$18.00-25.00	The clock looks like a giant watch. The face of the clock is 6" around. There are 2 black straps, 1 on each side of the clock. The outside of the clock is gold. There is a picture of a Smurf sleeping on the moon inside the clock. The background is light blue and the numbers are dark blue.
058/006	Smurfette Handmade Wood Clock	Handmade			Wood Carved	20"		$7.50-10.00	The clock is dark stain. The top has the face with the numbers in gold Roman numerals. Smurfette is carved on the bottom. Smurfette has a white dress on.
058/007	Wind Surfer Smurf	Jerger		Light Blue	Alarm Clock	7" Tall	Germany	$35.00-45.00	The clock is metal. The outside of the clock is light blue with 2 yellow bells on top and a gold handle. There are 2 gold legs on the bottom of the clock. The picture inside the clock is a Smurf on an orange surfboard with a yellow sail. The Smurf has pink swim trunks on. The numbers and hands are black.
058/008	Smurfette Shaped Clock	Handmade			Wood Clock	17" High	Hand made	$18.00-25.00	The clock is 1 1/2" thick and 17" high. The clock is shaped like Smurfette. She has a white dress and a white hat on. The hat has the clock in it. There are gold Roman numerals for the clock. The finish on the clock is glossy.
058/009	Smurfette Plaster Clock	Handmade			Plaster Clock	8"	Hand made	$11.00-15.00	The clock has a plaster base. The clock has Smurfette on the front. Smurfette is wearing a white dress and holding a yellow and orange daisy. The background is white. The border is yellow and copper. The outside of the clock has gold Roman numerals.
058/010	Square Wood Wall Clock	Homemade			Wall Clock	8" x *"	Hand made	$7.50-10.00	The clock is made of wood. In the center is a Smurf jumping. Below him it says "Smurf" in white letters with blue trim. The numbers are black. The hands are black and made of metal.
058/011	Smurfette And Baby Smurf Dancing Alarm Clock	Peyo		White	Alarm Clock	5 1/2"		$35.00-45.00	The alarm clock is white with 2 white bells on top. The inside is white with black numbers. In the center is Smurfette dancing with Baby Smurf. Baby is wearing a pink sleeper.
058/012	Smurf Holding Hand In Air Digital Clock	Bradley Time			Digital Clock	4" x 3"	U.S.A.	$11.00-15.00	The clock is flat. The front has a Smurf holding a red stop sign. The sign has the window for the time. The Smurf is holding a hand in the air.
058/013	Smurf Trenmark Clock	Trenmark			Quartz Clock	8 1/4" x 10 1/4"		$35.00-45.00	The clock is rectangular- shaped and made of plastic. It has a picture of an orange and yellow mushroom house on it. A Smurf is sitting in the grass. Smurfette is dreaming of the time. A Smurf is jumping and another Smurf is laughing. The background is the sky and clouds.
058/014	Smurfette On Telephone Pocket Watch Clock	Elgin/Peyo		Blue	Wall Clock	15 1/4"		$40.00-55.00	The clock looks like a big pocket watch. Inside it has Smurfette standing and talking on an old red phone. There is a Smurf sitting and talking on a regular red phone. The background is white.
058/015	Smurf Chasing A Butterfly Alarm Clock	Peyo		Red	Alarm Clock	5 1/2"		$37.50-50.00	The alarm clock is red with 2 red bells on top. The inside is white with black numbers. In the center is a Smurf chasing a butterfly with a net.
058/016	Hammer Smurf Alarm Clock	Equity/Peyo	1/1/83	White	Alarm Clock	5"	China	$37.50-50.00	The clock is white on a white plastic base. The picture inside of the clock is of mushroom houses. A Smurf is hitting Brainy Smurf over the head with a hammer. Papa is running towards them.
058/017	Red Smurfette Clock	TAD	1/1/96	Red	Alarm Clock	3 1/2" Dia	Brussels	$18.00-24.00	The clock is red on the outside and white in the center. It is round. The center has a picture of Smurfette's face.
058/018	Blue Smurf Alarm Clock	TAD	1/1/96	Blue	Alarm Clock	3 1/2" Dia	Brussels	$18.00-24.00	The clock is blue on the outside and white in the center. It is round. The center has a picture of a Smurf's face.
058/019	Mushroom- Shaped Smurfette Clock	TAD	1/1/96	Pink/White	Alarm Clock	3 1/2" x 4"	Brussels	$16.50-22.00	The clock is in the shape of a mushroom house. The roof is pink and the sides and center are white. In the center is an orange heart with Smurfette posing. The numbers are red.
058/020	Smurfs Shaking Hands	Demons & Merveilles	1/1/96		Wall Clock	9' X 8 1/2"	Brussels	$22.50-30.00	The clock is square and made of plastic. It is yellow and has a purple border. In the center are 2 Smurfs shaking hands.
058/021	Smurf Walking Carrying A Candle	Demons & Merveilles	1/1/96		Wall Clock	9" x 8 1/2"	Brussels	$22.50-30.00	The clock is square and made of plastic. It is a picture of night with mushroom houses. It has a Smurf wearing a pink night shirt carrying a candle on it. He is walking out in the night.

Set 196: Clothes

196/001	Let's Party Shirt	C & A	1/1/96	Yellow	Shirt		Brussels	$14.25-20.00	The shirt is yellow with light blue around the collar and the sleeves. There are 4 Smurfs and Smurfette dancing on the front. All the Smurfs are wearing red and blue clothes. The top is open 1/4 of the way down and it laces up. The lace is light blue.
196/002	Let's Have Fun Polo Shirt	C & A	1/1/96	Yellow	Shirt		Brussels	$20.00-30.00	The shirt is tie-dye with a dark blue collar and sleeves. There are 2 buttons on the front. There is a big brown puppy on the front of the shirt. There is a Smurf riding the puppy. One Smurf looks like he's jumping off the dogs nose. There are 6 Smurfs running in front of the dog. There are 6 Smurfs running on the back.
196/003	The Smurfs V-neck Shirt	C & A	1/1/96	Yellow	V-Neck Shirt		Brussels	$12.75-17.00	The shirt is yellow with dark blue trim around the collar and sleeves. The shirt has a v neck. There are 2 mushrooms with 3 Smurfs on top, 3 Smurfs below the mushroom, and a Smurf flying through the air holding a mushroom.
196/004	Jokey's Surprise T-shirt	C & A	1/1/96	Yellow	T-shirt		Brussels	$18.00-25.00	The t-shirt is yellow. In the center on the front is a Smurf spattered with paint, a Smurf opening a trick box with a Smurf standing above him. Another Smurf is in the corner laughing.
196/005	Papa And A Smurf Running	ShowToons/Hanes	1/1/83	White	Boy's T-shirt	Size 4	U.S.A.	$4.50-6.00	The t-shirt is white with a light blue collar. In the left corner is Papa and a Smurf running. The Smurf crossed the finish line first.
196/006	Smurfin White Water Canyon Kings Island	Peyo	1/1/86	White	T - Shirt	2 - 4	U.S.A.	$11.00-15.00	The shirt is white and has a picture of Sassette and Nat going down a waterfall in a barrel. At the top of the waterfall is a tunnel. The shirt says "Smurfin White Water Canyon Kings Island".
196/007	Keep On Smurfin	Peyo	1/1/82	White	T - Shirt	10 1/2 - 12 1/2		$11.00-15.00	The shirt is white and has red trim around the sleeves and neck. There is a picture of Smurfette marching with a finger in the air. 5 Smurfs are marching behind her.
196/008	Baby Bunting (Green)	Handmade			Baby Bunting			$18.00-25.00	The bunting is white and green. It has different village scenes on it. Smurfs are eating at a table. A Smurf is pulling the pedals off a daisy. A Smurf is carrying a cake. The inside has white material with different colored stars. The bunting has a light green satin trim.
196/009	Baby Smurf Sweatsuit			Light Blue	Sweatsuit	Newborn	U.S.A.	$7.50-10.00	The sweat suit is light blue. The front of the shirt is white with a picture of Baby Smurf crawling after a butterfly on it. Baby is wearing a white sleeper. The butterfly is yellow and green. The pants are all blue.
196/010	Smurf Cupid Shirt	Stanley Desantis	1/1/97	Blue	T-shirt	Large	U.S.A.	$12.75-17.00	The shirt is blue. A Smurf and Smurfette are holding hands and dancing. There are hearts above her head. There is a cupid Smurf above Smurfette's head. He is holding a yellow bow and arrow.
196/011	Smurf Winter Boots	Peyo		Blue	Boots	size 11	Korea	$7.50-10.00	The boots are blue and lace- up. The top and bottom say Smurf and have Smurf faces around. The side has a picture of a Smurf riding a sled with another Smurf on his back. The inside is insulated.
196/012	Smurf Face Shirt	Stanley Desantis	1/1/97	White	t-shirt	Large	Brussels	$15.00-20.00	The shirt is white and has a big Smurf face in the middle. The Smurf is smiling. The shirt has a black collar and black border around the sleeves.
196/013	Red Smurf Pants	JC Pennys	1/1/82	Red	Kids Pants	Size 5	U.S.A.	$5.00-7.00	The pants are red corduroy. They have a Smurf stretch belt attached to the top. The belt is blue with different Smurfs pictured.
196/014	Super Smurf Shirt	Sears		White/ Blue	t-shirt	Size 4-5	U.S.A.	$6.00-8.00	The shirt is white with a blue collar and blue sleeves. The front has a picture of a Smurf wearing a red cape and flying through the air. The Smurf has a yellow "S" on his chest. Under the Smurf the shirt says "Super Smurf" in multi colors.
196/015	Smurf Pajama's	AT & TV Mrch..	1/1/97	White	Pajama's	Size 6	Brussels	$11.00-15.00	The pajama's are white with a red collar and red arms and leg cuffs. The pajamas have little red boxes all over with Smurfs in them. One Smurf is wearing a red shirt and the other isn't wearing a shirt. The front has black buttons.
196/016	Pro Smurf Sweatshirt	Peyo	1/1/82	Red & White	Sweatshirt	6 Months	U.S.A.	$6.00-8.00	The sweatshirt is red with a white front and red hood. On the front is a Smurf playing hockey. The Smurf is wearing a red and black uniform, a white helmet and white skates. The sweatshirt says "Pro Smurf" in red letters.
196/017	Sport Smurf Sweatshirt	Peyo	1/1/82		Sweatshirt	3T	U.S.A.	$6.00-8.00	The sweatshirt is green and blue with a white front. On the front is a picture of a Smurf playing basketball. The Smurf is holding the ball. The Smurf is wearing a white tank top with a green and blue stripe and white shorts. The sweatshirt doesn't have a hood. It says "Sport Smurf" on the front in yellow letters.,
196/018	Go For It Sweatshirt	Peyo	1/1/81	red & Gray	Hooded Sweatshirt	12 Months	U.S.A.	$6.00-8.00	The sweatshirt is red with a gray front and a red hood. The front has a Smurf roller-skating. The Smurfs' skates are blue with red wheels. It says "Go For It" in red letters.
196/019	Smurf Soccer Sweatshirt	Peyo	1/1/81	Red & White	Hooded Sweatshirt			$6.00-8.00	The sweatshirt is red with a white front and a red hood. The front has a Smurf kicking a soccer ball. The Smurf is wearing a red shirt, black shorts, red socks, and white shoes. It says "Smurf Soccer" in red letters under the grass.
196/020	Go For It Sweatshirt	Peyo	1/1/81	Red & Gray	Sweatshirt	7	U.S.A.	$6.00-8.00	The sweatshirt is red with a gray front. The front has a Smurf roller-skating. The Smurfs' skates are yellow with red wheels. It says "Go For It" in red letters.
196/021	Go For It Sweatshirt	Peyo	1/1/81	Red & Gray	Hooded Sweatshirt	2T	U.S.A.	$6.00 8.00	The sweatshirt is red with a gray front and a red hood. The front has a Smurf roller-skating. The Smurfs' skates are yellow with red wheels. It says "Go For It" in red letters.
196/022	Have A Smurfy Day Sweatshirt	Peyo	1/1/81	Green & Gray	Hooded Sweatshirt	2T		$6.00-8.00	The sweatshirt is green with a gray front and green hood. The front has a Smurf sitting on a cloud. There is a rainbow behind the Smurf. The sweatshirt says "Have A Smurfy Day" in orange letters.
196/023	Have A Smurf Day Sweatshirt	Peyo	1/1/81	Blue & White	Hooded Sweatshirt	Large	U.S.A.	$6.00-8.00	The sweatshirt is blue with a white front and blue hood. The front has a Smurf sitting on a cloud. There is a rainbow behind the Smurf. The sweatshirt says "Have A Smurfy Day" in orange letters.
196/024	True Blue Friends	Peyo	1/1/82	Lt. Blue & White	Sweatsuit	6 Months	U.S.A.	$11.00-15.00	The sweatshirt is light blue with a white front and light blue hood. The front has a Smurf handing Smurfette daisies. Smurfette is wearing a pink dress and pink shoes. The sweatshirt says "True Blue Friends" in pink and blue letters. The sleeves have white lace around the top. There is a pair of light blue pants to go with the sweatshirt.
196/025	Baby With Rattle Sweatshirt	Pampers	1/1/85	Lt. Blue & White	Hooded Sweatshirt	4T	U.S.A.	$6.00-8.00	The sweatshirt is light blue with a white hood. The front has a pocket. In the top corner of the sweatshirt is a picture of Baby Smurf laying on his back playing with a red rattle.
196/026	Super Smurf Sleeper	Lullaby Land	1/1/81	Red & White	Sleeper	12 Months	U.S.A.	$7.50-10.00	The sleeper is red with a white front. The sleeper has red snap- on pants. The front has a picture of a Smurf wearing a red cape and flying through the air. The Smurf has a yellow "S" on his chest. Under the Smurf the shirt says "Super Smurf" in blue colors.
196/027	Papa Sleeper	T. Tippee	1/1/81	White & Blue	Sleeper	Newborn	U.S.A.	$7.50-10.00	The sleeper is white with blue sleeves. In the corner is a picture of Papa Smurf standing on a mushroom and playing a horn.
196/028	Smurf Laying On Pillows Sleeper	T. Tippee	1/1/82	White & Yellow	Sleeper	Newborn Med	U.S.A.	$7.50-10.00	The sleeper is white with yellow sleeves and a yellow collar. The front has a picture of a Smurf laying on 2 pillows. The Smurf is resting his head on his hands. There are stars and a moon above him.
196/029	Skating Smurfette Sleeper	Peyo		Lt. Blue & White	Sleeper			$7.50-10.00	The sleeper is light blue with white on one side of the front. It has 3 different poses of Smurfette on ice-skates. The front has white snaps down to the feet.
196/030	Smurfs Sleeper	Sylva	1/1/83	Yellow & White	Sleeper	4	U.S.A.	$7.50-10.00	The sleeper is yellow with white on front. One side has Smurfette jumping rope. The other side has a Smurf laying in leaves. The sleeper says Smurfs in blue letters.
196/031	Smurf Nightgown	Peyo	1/1/82	White	Nightgown			$5.00-7.00	The sleeper is white with blue and red striped sleeves. The front has a green mushroom house with a red and yellow roof on it. A Smurf is standing and facing Smurfette pointing a finger.
196/032	Blue And White Stripe Nightgown	Peyo	1/1/82	Blue & White	Nightgown			$5.00-7.00	The nightgown is blue and white striped. The front has a Smurf offering Smurfette a pink daisy.
196/033	Smurfette Jersey	Peyo	1/1/82	Red & White	Jersey			$6.00-8.00	The shirt is red and white striped with red sleeves. The front corner has a picture of Smurfette standing and posing.
196/034	Have A Smurfy Day Shirt	Peyo	1/1/81	White & Orange	Long Sleeve Shirt	18 Months	U.S.A.	$7.50-10.00	The shirt is white with orange sleeves. The front has a Smurf sitting on a cloud. There is a rainbow behind the Smurf. The [illegible] says "Have A Smurfy Day" in orange letters.
196/035	Smurf Painters Night Shirt	Lullaby Land		Lt. Blue & White	Long Sleeve Shirt	3T	U.S.A.	$7.50-10.00	The shirt is light blue with white on the front. It has a picture of a Smurf on a ladder painting the letters to spell "SMURF". Another Smurf is painting a different letter.
196/036	Football Smurf Night Shirt	Lullaby Land		Yellow & White	Long Sleeve Shirt	3T	U.S.A.	$7.50-10.00	The shirt is yellow with a white front. It has a picture of a Smurf running and carrying a football under his arm. The Smurf is wearing a yellow shirt, red shorts, white shoes, and a red helmet.
196/037	SMURFS Nightgown	Peyo	1/1/81	White & Blue	Nightgown			$7.50 10.00	The nightgown is white with blue shoulders and cuffs. The front has 6 Smurfs carrying the letters to spell "SMURFS". Papa is sitting on the F. Smurfette is standing on the side watching.
196/038	SMILE Nightgown	Peyo	1/1/81	White & Red	Nightgown			$7.50-10.00	The night gown is white with red shoulders and cuffs. The front has 2 Smurfs holding hands. The Smurfs look happy. The top says "SMILE" in red letters.
196/039	Super Smurf T-shirt	Velva Sheen	1/1/81	Light Blue	T-shirt	10-12 Youth	U.S.A.	$5.00-7.00	The shirt is light blue. The front has a picture of a Smurf wearing a red cape and flying through the air. The Smurf has a yellow "S" on his chest. Under the Smurf the shirt says "Super Smurf" in multi colors.
196/040	Follow Me To Worlds Fair 1982 T-shirt	Renoble	1/1/81	White	T-shirt	10-12	U.S.A.	$5.00-7.00	The shirt is white [illegible] a Smurf walking on it. There is a rainbow behind him. The shirt says "Follow Me To Worlds Fair 1982 Knoxville, TN" in black letters.
196/041	Smurf Turtleneck	Peyo		White	Turtleneck Shirt	6	U.S.A.	$5.00-7.00	The turtleneck shirt is white. It has pictures of Smurfette, Papa, and a Smurf all over it.
196/042	Smurf Giving Smurfette Flowers Turtleneck Shirt	Sears		White	Turtleneck Shirt	6	U.S.A.	$5.00-7.00	The shirt is white with a picture of a Smurf handing Smurfette daisies. There are red hearts floating above the Smurf's head.
196/043	Smurf Of My Heart Undershirt	Peyo	1/1/81	White	Girls Undershirt	4 -6 Girls	U.S.A.	$2.00-3.00	The undershirt is white. The front has a picture of a Smurf handing Smurfette daisies. There are red hearts floating above the Smurf's head and Smurfettes. The shirt says "Smurf Of My Heart" in pink letters.
196/044	Smurfette Thermal Shirt	Peyo		Pink	Thermal UnderShirt	4 - 6	U.S.A.	$3.00-4.00	The shirt is pink and has long sleeves. The front has a picture of Smurfette ice-skating. The top says "Smurfette" in red letters.
196/045	Sledding Smurfs Thermal Undershirt	Showtoons	1/1/83		Thermal Undershirt	10-12	U.S.A.	$3.00-4.00	The shirt is white with red sleeves and blue cuffs. The front has a picture of a Smurf sledding. The Smurf on the sled has another Smurf sitting on his back.
196/046	Super Smurf Shirt	Peyo	1/1/82	White & Red	Long Sleeve Shirt		U.S.A.	$6.00-8.00	The shirt is white with long red sleeves. The front has a picture of a Smurf wearing a red cape and flying through the air. The Smurf has a yellow "S" on his chest. The shirt says "Super Smurf" in blue colors.

196/047	Just Smurfin Thru T-shirt	Velva Sheen		Navy Blue	T-shirt	Adult Small	U.S.A.	$11.00-15.00	The t-shirt has a picture of a Smurf carrying a knapsack over his shoulder. The Smurf is whistling while he is walking. The top of the shirt says "Just Smurfin Thru" in red letters.
196/048	Sleepy Smurf Sleeper	Lullaby Land		White & Blue	Sleeper	18 Months	U.S.A.	$7.50-10.00	The sleeper is white with blue trim. It has various pictures of tired Smurfs. One is laying down and sleeping. A Smurf is carrying a candle while sleepwalking. A Smurf is standing up yawning. The sleepers don't have feet. It snaps up the whole front.
196/049	Sleepy Smurf Sleeper	Lullaby Land		White	Sleeper	12 Months	U.S.A.	$7.50-10.00	The sleeper is white with blue trim. It has various pictures of tired Smurfs. One is laying down and sleeping. A Smurf is carrying a candle while sleepwalking. A Smurf is standing up yawning. The sleepers don't have feet. It snaps up the whole front.
196/050	Smurfette Sliding Down Rainbow Bibs	Peyo		Red	Overall Bibs	6x	U.S.A.	$9.00-12.00	The bibs are red. The front pocket has an embroidered patch of Smurfette on it. Smurfette is sliding down a rainbow.
196/051	Underwear	Showtoons/Hanes	1/1/83		Boys Underwear	3	U.S.A.	$1.50-2.00	The underwear is white with a blue waistband. It has pictures of various Smurfs playing sports.
196/052	Smurf Country Shirt	Lullaby Land	1/1/82	Blue & White	Long Sleeve Shirt	2T	U.S.A.	$3.50-5.00	The shirt is blue with a white front. On the front is a picture of a Smurf and Smurfette wearing cowboy outfits. The 2 are standing and facing each other.
196/053	Jokey Smurf Slipper Socks	BBC Imports			Slipper Socks	9 - 10	Taiwan	$2.00-3.00	The slipper socks are white with a blue and red stripe up by the ankle. The top has a picture of a Smurf carrying a pink gift box. Above the Smurf it says " Smurf" in red letters.
196/054	Papa Pointing Slipper Socks	BBC Imports			Slipper Socks	9 - 10		$2.00-3.00	The slipper socks are white with a blue and red stripe up by the ankle. The top has a picture of Papa standing and pointing at something. Above Papa it says " Smurf" in white letters with blue borders.
196/055	I Smurf You! Smurfette Shirt	Velva Sheen	1/1/82	Red	T-shirt	10-12	U.S.A.	$6.00-8.00	The t-shirt has a picture of Smurfette holding 4 different colored daisies. Smurfette is wearing a red dress and red shoes. The shirt says "I Smurf You!"
196/056	Smurf Neon Mushroom Black T-shirt	US T's	1/1/98	Black	T-shirt	XL	Mexico	$13.50-18.00	The shirt is black. It has a pink, green, and yellow mushroom on the front. There is a Smurf sitting under the mushroom. The Smurf is holding himself and laughing.
196/057	Papa Smurf White T-shirt	Soffe Shirts	1/1/98	White	T-shirt	Large	U.S.A.	$13.50-18.00	The shirt is white with a black collar and black cuffs on the arms. The front has an iron-on. It is Papa standing in grass pointing. The shirt says "Papa Smurf" in orange letters.
196/058	Pac-Man Smurf White T-shirt	Newsouth	1/1/98	white/Gray	Jersey	Large	U.S.A.	$13.50-18.00	The shirt is white with gray on it. The shirt has a Smurf standing while playing a pac-man video game.
196/059	Basketball Smurf Purple T-shirt	Soffe Shirts	1/1/98	Purple	T-shirt	XL	U.S.A.	$13.50-18.00	The shirt is purple with a black collar and black cuffs. On the front is an iron- on. There is a Smurf dunking a basketball into a hoop. The Smurf has on a yellow tank top and white pants.
196/060	Huge Smurf Blue T-shirt	Stanley Desantis	1/1/98	Blue	T-shirt	X-Large	U.S.A.	$13.50-18.00	The shirt is blue. It has a large Smurf standing on the front. The Smurf has his hands on his hips and his legs crossed.
196/061	A GIANT SMURF For Mankind Sleeper & Bib	Peyo/SEPP	1/1/81	White & Pink	Sleeper & Bib	Newborn	U.S.A.	$11.00-15.00	The sleeper is white with pink sleeves and pink trim. It has a bib that has a space Smurf on the front. The bib says "A GIANT SMURF For Mankind".
196/062	Smurf Standing By A Heart T-shirt	Screen Stars		White	Airbrush T-shirt	Large	U.S.A.	$11.00-15.00	The shirt is white. It has a picture of a Smurf standing by a heart. He is holding a yellow daisy behind his back. The background is a pink, orange, and yellow sky. The shirt is airbrushed.
196/063	Azreal T-shirt	Sierra Teez		Off White	T-shirt	XL		$11.00-15.00	The shirt has a big picture of Azreal on the front. He looks like he's on the prowl. He is licking his chops.
196/064	Gargamel And Azreal T-shirt	Sierra Teez		Off White	T-shirt	XL		$11.00-15.00	Gargamel is standing with a hand over his mouth. Azreal is hiding behind his feet. The picture is large.
196/065	Smurf With Hands Out T-shirt	Sierra Teez		Off White	T-shirt	XL		$11.00-15.00	The shirt has a Smurf in the middle. It only has his upper body. There is a square background. The square is red, yellow, and green.
196/066	Pink Ice skating Smurfette Earmuffs	Peyo		Pink/White	Earmuffs		Taiwan	$3.50-5.00	The earmuffs are white with pink fur around the ends. In the center on each muff is Smurfette ice-skating. Smurfette is wearing a red dress and yellow ice skates.
196/067	White Basic Smurf Wear One Piece	C & A	1/1/98	White	One- piece Baby Outfit	Baby	Brussels	$4.50-6.00	The outfit is white. It has sleeves. It snaps in the crotch. The front has a blue square on it. In the center of the square is a Smurf holding his hand out. It says Basic Smurf wear.
196/068	Blue Striped Basic Smurf Wear One-piece	C & A	1/1/98		Baby One-piece	Baby	Brussels	$4.50-6.00	The outfit is white with blue, red, and black stripes. It has sleeves. It snaps in the crotch. The front has a small square patch on it. In the center of the square is a Smurf face. Around the edge of the patch it says Basic Smurf wear.
196/069	Yellow And Blue Techno Smurf One Piece	C & A	1/1/98		One-piece Baby Outfit	Baby	Brussels	$4.50-6.00	The outfit is white with yellow and blue. It has sleeves. It snaps in the crotch. The front has a patch on it. In the center of the patch is a Smurf face. The Smurf has red and yellow stripes on his hat. The patch says Techno skate.
196/070	Smurf Faces Basic Wear One-piece	C & A	1/1/98		One- Piece Outfit	Baby	Brussels	$4.50-6.00	The outfit is white. It has sleeves. It snaps in the crotch. The front and back have Smurfs and Smurf faces on it. It says Basic Smurf wear.
196/071	Smurfette White Shirt	C & A	1/1/98	White	Shirt	Girl	Brussels	$9.00-13.00	The shirt is white and ribbed. It has Smurfette in a heart. Underneath it has Smurfette in a purple square. The pictures are embroidered. It says pretty active Smurf.
196/072	Desole Jel' Ai Encore Oublie! Apron	I.M.P.S./Peyo	1/1/95	Beige	Apron		Brussels	$7.50-10.00	The apron is beige with black trim. It has a picture of a Smurf cooking over an open fire. The pot is overflowing. The Smurf is looking worried. The apron says "Desole Jel' Ai Encore Oublie!" in red letters.

Set 242: Club Figures- SCCI

242/001	Papa And His Prehistoric Pets	Peyo			Rubber Figure	2" High	West Germany	$75.00-100.00	Papa is medium blue. Papa is standing with his hands out at his sides. On one hand is a green dinosaur and on the other hand is a blue dinosaur.
242/002	Smurf Holding A Pink Pig	Peyo	1/1/80		Rubber Figure	2" High	West Germany	$55.00-75.00	The Smurf has a medium blue body. The Smurf is holding his hand stretched out in front of him. The Smurf is holding a soft, rubber, pink pig in his hands.
242/003	Smurf Holding A Gold Pig	Schleich/Peyo	1/1/80		Rubber Figure	2" High	West Germany	$67.50-90.00	The Smurf has a medium blue body. The Smurf is holding his hand stretched out in front of him. The Smurf is holding a gold pig in his hands.
242/004	Smurf With A Dove	Schleich/Peyo	1/1/93		Rubber Figure	2" High	Germany/China	$25.00-35.00	The Smurf has a medium bright blue body. The Smurf is sitting with his legs crossed. The Smurf is holding one hand in the air. There is a white dove sitting in his hand.
242/005	Aerobic Smurfette With Ladybug	Schleich/Peyo	1/1/83		Rubber Figure	2" High	West Germany	$25.00-35.00	Smurfette has a medium blue body. She is wearing a pink bodysuit, white and pink shoes, a white hat and hot pink leg warmers. Smurfette has her hands out at her sides. Smurfette has a red ladybug on her fingertips.

Set 89: Coasters

089/001	6 Milkana Coasters	Dupuis		White	Felt Coasters	Circle	Germany	$25.00-35.00	There are 6 felt Milkana coasters to the set. The pictures are: Papa laughing and pointing at something -Smurfette in a flirting pose -Smurf holding a mug of beer -Brainy wearing black glasses and holding a finger in the air -Smurf sticking out his tongue -Smurf laughing holding his hand out in front of him.
089/002	6 Glaseruntersetzer Coasters	Giovanni Trimboli			Hard Plastic Coaster	Circle	Italy	$22.50-30.00	The coasters are hard plastic and there are 6 in the set. The pictures are: A chef Smurf running and holding a rolling pin in the air, wearing an apron and chef's hat. -Smurfette wearing a waitress dress and holding a tray with a glass and a bottle on it. -A Smurf carrying a big cake. -A Smurf holding a brown glass of beer. -A Smurf taking a bite out of a sandwich. -A Smurf carrying a brown jug and he has a dazzled expression on his face.

Set 43: Colorforms

043/001	Smurf Land Colorform	Colorform	1/1/81		Colorform Play set	16" x 12 1/2"	U.S.A.	$11.00-15.00	The cover has 14 Smurfs on the front and a big mushroom house. Super Deluxe Play Set.
043/002	Smurf Mini Play set	Colorform	1/1/81		Colorform Play set	8"W x 12"L	U.S.A.	$6.00-8.00	The cover of the box has a Smurf giving Smurfette flowers in front of a rainbow. Papa is standing on a mushroom and holding a red book with a Smurf sitting below him, and another Smurf with a yellow flower in his hat.
043/003	Smurfette Dress-Up Set	Colorform	1/1/82	Pink Box	Colorform Play set	8"W x 12"L	U.S.A.	$6.00-8.00	The box is pink. The cover has Smurfette holding a flower, standing on a mushroom with hearts floating around her.

Set 210: Comic Albums- Die Schlumpfe

210/001	Olympiade Der Schlumpfe	Bastei	1/1/92	Orange	Comic Album	11 1/2" x 8 1/2"	Germany	$6.00-8.00	The cover has a Smurf running and carrying a torch. There is another Smurf running and Brainy is behind him holding a finger in the air. Their are on a track.
210/002	Der Fliegerschlumpf	Bastei	1/1/92	Blue	Comic Album	11 1/2" x 8 1/2"	Germany	$6.00-8.00	The cover has a Smurf and Smurfette flying in a red airplane over Gargamel's castle. Gargamel is running under them shaking his fist.
210/003	Der Finanz-Schlumpf	Bastei	1/1/94	Orange	Comic Album	11 1/2" x 8 1/2"	Germany	$6.00-8.00	The cover has a Smurf holding a letter outside a safe. The safe is open. Another Smurf is carrying a gold coin. There is a bag of coins laying on the ground.
210/004	Der Zauber-Lehrling	Bastei	1/1/94	Orange	Comic Album	11 1/2" x 8 1/2"	Germany	$6.00-8.00	The cover has a Smurf sitting, looking dazzled, and holding a blown up test tube. There are broken things laying all around.
210/005	Die Schlumpf-Suppe	Bastei	1/1/94	Yellow	Comic Album	11 1/2" x 8 1/2"	Germany	$6.00-8.00	The cover has Smurfs swimming in a large pot. The pot is on a log fire. A Smurf is standing and blowing on the flames.

Set 124: Comic Books

124/001	Comic #1	Marvel Comics	1/1/81		Comic Books	10"L X 6 1/2"W		$2.00-3.00	The cover has a Smurf flying over Gargamel and Azreal in a red airplane.
124/002	Comic #2	Marvel Comics	1/1/81		Comic Books	10"L X 6 1/2"W		$2.00-3.00	The cover has a black bird carrying a Smurf. Papa, Smurfette and another Smurf are standing below looking panicky.
124/003	Comic #3	Marvel Comics	1/1/81		Comic Books	10"L X 6 1/2"W		$2.00-3.00	The cover is orange and has a Smurf carrying a present while running from 3 Smurf monsters.
124/004	#1 Large Comic Book	Marvel Comics	1/1/82		Comic Books	13"L X 10"W	U.S.A.	$7.50-10.00	The cover has a Smurf jumping on it. Papa, Smurfette and 5 other Smurfs are on the cover. Gargamel is peaking out of the bushes.

Set 126: Comic Books- Mini

126/001	Sticky Smurf	Marvel Comics	1/1/82		Mini Comics	2 1/4"W X 4 1/4"H	U.S.A.	$1.50-2.00	The cover is yellow with an orange border. There is a picture of a Smurf stuck to a cake on it.
126/002	Monster Smurf	Marvel Comics	1/1/82		Mini Comics	2 1/4"W X 4 1/4"H	U.S.A.	$1.50-2.00	The cover is light blue with a dark blue border. There is a picture of 3 monster Smurfs chasing a Smurf that's carrying a present.
126/003	Smurf Ball	Marvel Comics	1/1/82		Mini Comics	2 1/4"W X 4 1/4"H	U.S.A.	$1.50-2.00	The cover is white with a green border. There is a picture of Gargamel tromping through the Smurf village. 4 Smurfs are running to get away from Gargamel.
126/004	The Smurf Plane	Marvel Comics	1/1/82		Mini Comics	2 1/4"W X 4 1/4"H	U.S.A.	$1.50-2.00	The cover is light blue with an orange border. There is a picture of a Smurf flying over top of Gargamel and Azreal. Gargamel's castle is in the back.
126/005	The Smurf And The Evil Bird	Marvel Comics	1/1/82		Mini Comics	2 1/4"W X 4 1/4"H	U.S.A.	$1.50-2.00	The cover is blue with a yellow border. The cover has a big black bird swooping down on a bunch of Smurfs.
126/006	The Smurf Of Youth	Marvel Comics	1/1/82		Mini Comics	2 1/4"W X 4 1/4"H	U.S.A.	$1.50-2.00	The cover is yellow with a blue border. The cover has a Smurf in a river trying to save Smurfette.

Set 46: Compact Discs

046/001	Die Schlumpfe Tekkno Ist Cool	EMI	1/1/95	Dark Blue	Compact Disc	Brussels	$18.00-25.00	The cover has 4 Smurfs listening to a boom box and dancing around a campfire.
046/002	Die Schlumpfe Mesaparty	EMI	1/1/95	Pink	Compact Disc	Brussels	$18.00-25.00	The cover has a Smurf putting a record on a record player. Smurfette and 2 other Smurfs are dancing.
046/003	De Smurfen Party House His	EMI	1/1/96	Light Blue	Compact Disc	Brussels	$15.00-20.00	The cover has Smurfette and another Smurf wearing headphones with little cassette players, dancing. The Smurf has on purple pants and a red shirt. Smurfette is wearing a blue shirt and a green with pink flowered skirt. The background is green and blue.
046/004	Irene Moors & De Smurfen Smurf The House	EMI	1/1/95	Red	Compact Disc	Brussels	$15.00-20.00	The cover is red. The cover is a big Smurf holding a lady.
046/005	De Smurfen House Party	EMI	1/1/95		Compact Disc	Brussels	$15.00-20.00	The cover has a Smurf putting a record on a record player. Smurfette and 2 other Smurfs are dancing.
046/006	De Smurfen Surprise	EMI	1/1/96	Purple	Compact Disc	Brussels	$15.00-20.00	The cover has a Smurf carrying a yellow package with red ribbon. There are presents all over the cover.
046/007	Irene Moors & De Smurfen	EMI	1/1/94	White	Compact Disc	Holland	$11.00-15.00	The cover is white. It has a lady's face on the front. There are 5 Smurfs on the cover.
046/008	Het Smurfenfeest	Buma Stemra	1/1/95		Compact Disc	Germany	$15.00-20.00	The cover has a Smurf playing a big brown guitar. There are 6 Smurfs behind him playing different instruments. The cover says " Het Smurfenfeest met de allernieuwste liedjes uit Smurfenland".
046/009	De Smurfen Feest!	EMI	1/1/98		Compact Disc	Brussels	$22.50-30.00	The cover is all yellow and orange with different colored stars. It has Smurfette and a Smurf dancing. Both Smurfette and the Smurf are wearing headphones and walkmans. Smurfette is in a pink dress. There are 2 cd's in the package.
046/010	De Smurfenbus	EMI	1/1/99		Compact Disc	Brussels	$15.00-20.00	The cover has Papa driving a yellow school bus. A Smurf and Smurfette are in the school bus.
046/011	Vader Abraham Im Land der Schlumpfe	Karussel	1/1/95		Compact Disc	Netherlands	$11.00-15.00	The cover has Father Abraham on it and he is holding 2 Smurfs and Papa.
046/012	Die Schlumpfe Bleirt Bei Mir!	EMI	1/1/98	Orange	Compact Disc	Holland	$7.50-10.00	The cover is orange. It has a Smurf and Smurfette dancing. Smurfette is wearing a pink shirt and pink skirt. The Smurf is in yellow.
046/013	Die Schlumpfe Weltraumschlumpfe	EMI	1/1/97		Compact Disc	Holland	$7.50-10.00	The cover is dark purple with a purple planet on it. A Smurf is wearing an astronaut suit. The Smurf is carrying a Smurf flag.
046/014	Die Schlumpfe Samba	EMI	1/1/97	Orange	Compact Disc	Holland	$7.50-10.00	The cover is orange. It has Smurfette dancing with another Smurf. Another Smurf is playing a trumpet. Smurfette is wearing a pink shirt and green skirt.
046/015	Die Schlumpfe Auf Ne Party Gehn	EMI	1/1/97		Compact Disc	Brussels	$7.50-10.00	The cover has Snappy Smurf and Smurfette dancing on it. Smurfette is wearing purple shorts and a purple shirt. Snappy Smurf is wearing an orange shirt and purple pants.
046/016	Die Schlumpfe You're My Heart, You're my soul	EMI	1/1/99	Light Blue	Compact Disc	Brussels	$7.50-10.00	The cover is light blue. It has a Smurf dressed as a magician. The Smurf is holding a wand.
046/017	Die Schlumpfe Alles Banane! Vol. 3	EMI	1/1/96		Compact Disc	Holland	$15.00-20.00	The cover has 2 Smurfs and Smurfette rollerblading in a banana on it. Another Smurf is on a skateboard.
046/018	Die Schlumpfe Vol. 1 Der Winter! Vol. 4	EMI	1/1/96		Compact Disc	Holland	$15.00-20.00	The cover has 2 Smurfs and Smurfette snowboarding on it. Smurfette is holding Baby Smurf.
046/019	Die Schlumpfe Irre Galaktisch! Vol. 6	EMI	1/1/97		Compact Disc	Holland	$15.00-20.00	The cover is of outer space. It has a Smurf in a green space ship on it. Another Smurf and Smurfette are holding hands floating through space. A third Smurf is floating through the air holding a Smurf flag.
046/020	Die Schlumpfe Oh Du Schlumpfige! Vol. 8	EMI	1/1/98	Light Blue	Compact Disc	Holland	$15.00-20.00	The cover has Papa Smurf dressed as Santa. He is falling from the sky holding an umbrella. Papa is also holding a bag of toys.
046/021	Hitparade Der Schlumpfe	Karussel	1/1/95		Compact Disc	Germany	$11.00-15.00	The cover has Smurfette in the window of a mushroom house. There are 5 Smurfs playing instruments outside of the house.
046/022	Hitparade Der Schlumpfe 2	Karussell	1/1/95		Compact Disc	Germany	$11.00-15.00	The cover has a Smurf playing a guitar on it. In the background there are 5 other Smurfs playing instruments and a Smurf holding a yellow sign.
046/023	Die Hits Vom Land Der Schlumpfe	Karussell	1/1/82		Compact Disc	Germany	$11.00-15.00	The cover has Papa leading a band. Smurfette is sitting while playing a harp. Three other Smurfs are playing instruments. Another Smurf is laying in the grass sleeping.

Set 256: Computer Programs

256/001	Comic Book Artist	Lascaux	1/1/95	Computer Program	Belgium	$25.00-35.00	The program comes with a CD-ROM to make cartoon Smurf strips. The box is yellow and has 3 Smurfs on the front.
256/002	Der Schlumpf-Screen-Saver	Boeder Software	1/1/95	Screen Saver	Germany	$30.00-40.00	The Smurf screen saver has different Smurf pictures on it. The box is white and has various Smurfs on it. In the center is a Smurf house. Included are 2 disks.

Set 102: Cups- Plastic Tumblers

102/001	See Smurfette In Smurf Forest At Canada's Wonderland	Styroware	1/1/84	White	Promotional Cups	5" High	Canada	$4.50-6.00	A promotional cup from McDonald's for Canada's Wonderland theme park. The cup is white with a wrap around picture. Smurfette is dancing and there are three Smurfs singing. The cup says Smurfette on top in pink letters. On the bottom of the cup in pink letters it says "See Smurfette In Smurf Forest At Canada's Wonderland"
102/002	See Papa Smurf In Smurf Forest At Canada's Wonderland	Styroware	1/1/84	White	Promotional Cups	5" High	Canada	$4.50-6.00	A promotional cup from McDonald's for Canada's Wonderland theme park. The cup is white with a wrap around picture. The picture is of a mountain with a waterfall and it has a banner in front of the waterfall that says Canada's Wonderland. There is a rainbow behind the mountain and Papa is standing off to the side of the mountain. On the bottom of the cup in red letters it says "See Papa Smurf In Smurf Forest At Canada's Wonderland".
102/003	Smurf Fun Begins At Canada's Wonderland (Red Letters)	Styroware	1/1/85	White	Promotional Cups	5" High	Canada	$4.50-6.00	A promotional cup from McDonald's for Canada's Wonderland theme park. The cup is white with a wrap around picture. The picture is of a Smurf, Baby, and Smurfette riding on snails. In the background is a roller coaster and some mountains. On the bottom of the cup in red letters it says "Smurf Fun Begins At Canada's Wonderland."
102/004	Canada's Wonderland The One For All The Fun!	Styroware	1/1/89	White	Promotional Cups	5" High	Canada	$4.50-6.00	A promotional cup from McDonald's for Canada's Wonderland theme park. The cup is white with a wrap around picture. The picture is Papa pulling a chain on a cannon and a Smurf flying out. Smurfette is painting a picture of a snail face and Baby is sitting, squeezing the paint out of the tubes. There is a Smurf in a airplane pulling a banner that says Canada's Wonderland The One For All The Fun! Another Smurf is jumping on a pogo stick.
102/005	Smurf Fun Begins At Canada's Wonderland	Styroware	1/1/85	White	Promotional Cups	5" High	Canada	$4.50-6.00	A promotional cup from McDonald's for Canada's Wonderland theme park. The cup is white with a wrap around picture. The picture is of Papa and Smurfette in an air balloon. Baby has 3 balloons tied to him and he's floating in the air. The roller coaster, mountain, and castle are in the background. On the bottom of the cup in orange letters it says "Smurf Fun Begins At Canada's Wonderland."
102/006	See Baby In Smurf Forest At Canada's Wonderland	Styroware	1/1/84	White	Promotional Cups	5" High	Canada	$4.50-6.00	A promotional cup from McDonald's for Canada's Wonderland theme park. The cup is white with a wrap around picture. The picture is of Baby Smurf crawling under a rainbow after a butterfly while another Smurf is standing under the rainbow laughing. 4 Smurfs are putting up a Smurf forest sign. On the bottom of the cup in blue letters it says "See Baby In Smurf Forest At Canada's Wonderland."
102/007	Smurf Fun Begins At Canada's Wonderland (Blue Letters)	Styroware	1/1/85	White	Promotional Cups	5" High	Canada	$4.50-6.00	A promotional cup from McDonald's for Canada's Wonderland theme park. The cup is white with a wrap around picture. The picture is of Gargamel and Azreal running under the Smurf forest sign chasing the Smurfs with a net. The roller coaster, castle, and mountain is in the background. On the bottom of the cup in blue letters it says "Smurf Fun Begins At Canada's Wonderland."
102/008	The Mystery Continues At Canada's Wonderland	Styroware	1/1/86	White	Promotional Cups	5" High	Canada	$4.50-6.00	A promotional cup from McDonald's for Canada's Wonderland theme park. The cup is white with a wrap around picture. The picture is of Papa Smurfette, Baby and a Smurf dressed as Sherlock Holmes standing in front of the Smurf forest sign. The roller coaster, castle, and mountain is in the background. On the bottom of the cup in blue letters it says "The Mystery Continues At Canada's Wonderland."
102/009	The Smurf Family At Canada's Wonderland	Styroware	1/1/87	White	Promotional Cups	5" High	Canada	$4.50-6.00	A promotional cup from McDonald's for Canada's Wonderland theme park. The cup is white with a wrap around picture. The picture is of Papa pulling Smurfette on a snail while Baby is tied behind in a wagon in front of the Smurf forest sign. There is a Smurf on a skateboard holding a mirror. The roller coaster, castle, and mountain is in the background. On the bottom of the cup in blue letters it says "The Smurf Family At Canada's Wonderland."
102/010	The Smurfs Help	Styroware	1/1/84		Promotional Cups	5" High	Canada	$2.00-3.00	The cup is white and has a picture of Papa telling 8 Smurfs a story. The back of the cup has part of the story of the Smurfs and the magic flute written on it.
102/011	Scene From The Movie "Another Magic Flute"	Styroware	1/1/83		Promotional Cups	5" High	Canada	$2.00-3.00	The cup is white and has a picture of the stork carrying 2 Smurfs, and Johan and Peewit standing in a door. The back of the cup has part of the story of the Smurfs and the magic flute written on it.
102/012	Kings Dominion Cup	Sterling Products	1/1/84	White	Promotional Cups	5" High	U.S.A.	$3.50-5.00	The cup has a picture of Papa, a Smurf and Smurfette riding in a roller coaster. The roller coaster car is red and yellow. The bottom of the cup says "Kings Dominion" in red letters.

Set 180: Cutting Boards

180/001	Picnic On Blanket Scene	Fackelmann	1/1/95		Cutting Board	7"L X 8 1/2"W	Brussels	$11.00-15.00	The cutting board is made of a hard plastic. The picture is of Smurfette kneeling on a blanket while unpacking a picnic basket. Papa is pushing Baby Smurf in a tree swing. 3 other Smurfs are sitting around the blanket.
180/002	Papa And Smurfette At A Table	Fackelmann	1/1/95		Cutting Board	7"L X 8 1/2"W		$11.00-15.00	The cutting board is made of a hard plastic. The picture is of Smurfette and Papa sitting at a table holding drinking glasses. A Smurf is dressed as a waiter and is holding a loaf of bread. Jokey is walking towards the table with a present. The scene is outside.
180/003	Smurfs In Kitchen	Fackelmann	1/1/95		Cutting Board	7"L X 8 1/2"W	Brussels	$11.00-15.00	The cutting board is made of a hard plastic. There are 5 Smurfs cooking. A Smurf is decorating a cake, one is rolling dough, and another Smurf is kneading the dough. A Smurf is sticking a bread board into the fireplace. A Smurf is carrying a bucket of milk. The walls are orange.

Set 148: Dart Supplies

148/001	Velcro Dart Board	Synergistics Rsrch			Cloth Dart Board	18" X 15"	U.S.A.	$11.00-15.00	The dart ring is yellow and orange. The rings for 2 and 4 are yellow. The rings for 3 and 5 are orange. At the top of the ring is a Smurf face. The cloth on the outside of the dart ring is light blue. In the top corners are Papa and Smurfette. In the lower corners sitting in the grass is a Smurf sleeping and in the other corner is a Smurf sitting and waving. The balls are yellow with red velcro strips on them.
148/002	Cloth Dart Flights			White	Cloth Dart Flights	1 1/2"		$1.00-2.00	The dart flights are white with a picture of a Smurf throwing a red dart on them. There are 3 flights in the package.
148/003	Cloth Dart Mitten	Peyo			Cloth Dart Mitten	8" X 6"		$6.00-8.00	The mitten is yellow and has a Smurf in the middle. The Smurf is holding an orange bulls eye circle with a #10 in the middle. The mitten has 2 orange balls with red velcro stripes on them.

Set 206: Desk Sets

206/001	Smurfy Memo Dispenser With Pencil	Schleich	1/1/83	Blue	Memo Dispenser	3" X 5" X 2 1/2"H	Germany	$15.00-20.00	The memo dispenser is blue plastic. In the corner is a PVC Smurf standing and holding his hands together. Next to the Smurf is a spot to hold the pencil. Included is yellow paper with Poet Smurf on it.
206/002	Smurfy Paper clip Tray With Magnetic Fishing Line	Schleich	1/1/83	Blue	Paper clip Tray	3" X 3 " X 2 1/2"H	Germany	$15.00-20.00	The paper clip dispenser is made of plastic. There is a PVC Smurf sitting on a brown stump holding a yellow fishing pole. The fishing pole has a magnet on the end.
206/003	The Smurfy Pencil Pot	Schleich	1/1/83	Blue	Pencil Holder	3" X 3" X 3 1/2"H	Germany	$15.00-20.00	The pencil holder is made of plastic. The pencil holder has 3 slots. There is a PVC Smurf standing in front of the pencil slots. The Smurf is standing and holding a book. He is holding a finger in the air.
206/004	Smurfy Tray With Memo Holder	Schleich	1/1/83	Blue	Plastic	3" x 6" X 3"H	Germany	$15.00-20.00	The memo holder is made of plastic. There is a slot in back to hold paper. There are 3 little squares in front to hold tacks and paper clips. There is a PVC Smurfette standing in the corner. Smurfette is wearing a pink dress. She is holding a pad of paper and a pencil.

Set 13: Dishes

013/001	Coffee Cup	Deka Inc.	1/1/80	Multi	Plastic Dishes	3 1/2" high	U.S.A.	$3.00-4.00	The cup has a wrap around picture with a Smurf scuba diving in a yellow kettle, with Smurfette and 2 other Smurfs on it.
013/002	Cup	Deka Inc.	1/1/80	Multi	Plastic Dishes	4 3/4" high	U.S.A.	$3.00-4.00	The cup has a wrap around picture with, Smurfette, Papa, and 6 Smurfs on it.
013/003	Juice Cup	Deka Inc.	1/1/80	Multi	Plastic Dishes	4" high	U.S.A.	$3.00-4.00	The cup has a wrap around picture with, Papa, Smurfette, and 6 Smurfs on it.
013/004	Cereal Bowl	Deka Inc.	1/1/80	White	Plastic Dishes	5 1/2" W x 2 1/2" H	U.S.A.	$4.50-6.00	The bowl has the same picture on 2 sides. A Smurf giving Smurfette flowers, and a Smurf with a rolling pin chasing a Smurf with a cake.
013/005	Flat Plate	Deka Inc.		White	Plastic Dishes	8" diameter	U.S.A.	$4.50-6.00	Papa's sprinkling pepper in a bowl of food and a Smurf has heat coming out of his mouth. The plate is a dull plastic.
013/006	Shallow Bowl	Deka Inc.		White	Plastic Dishes	8" W x 1 1/4" Deep	U.S.A.	$4.50-6.00	The bowl has Smurfs written in red and 8 sport Smurfs on 2 sides. The bowl has a 1" flat surface around the ridge of the bowl that has the picture on it.
013/007	3 Section Lip Plate	Deka Inc.	1/1/80	White	Plastic Dishes	9 1/2" W x 1" Deep		$7.50-10.00	The plate has Smurfs written in red and 8 sport Smurfs on 2 sides. The plate has a 1" flat surface around the ridge of the plate that has the picture on it.
013/008	Soup Bowl	Deka Inc.	1/1/82	White	Plastic Dishes	1 1/2"H x 5 3/4 W	U.S.A.	$4.50-6.00	The bowl has a Smurf fishing on the inside.
013/009	Melmac Plate	Debonaire		White	Plastic Dishes	7 3/4"		$4.50-6.00	Papa's sprinkling pepper in a bowl of food and a Smurf has heat coming out of his mouth. The plate is melmac (a very hard, glossy plastic) and it curves up on the sides.
013/010	Water Squeeze Bottle	Fackelmann	1/1/96	White	Plastic Dishes	8"L X 2 1/2" W	Germany	$7.50-10.00	The water bottle is white with a red drink spout and cap. There are 5 Smurfs around the outside of the bottle. There are 2 Smurfette's playing in flowers, 1 on each side of the bottle. There is a Smurf playing with leaves, and one standing and shivering with a red scarf around his neck. Another Smurf is wearing snorkeling gear and has a yellow inner tube around his tummy.
013/011	Smurf Plastic Straw	Helm	1/1/83	Blue	Plastic Straw	9 1/2"	Hong Kong	$1.13 - 1.50	The straw has a Smurf face in the center. It is blue plastic and the top is bendable.
013/012	Smurfette Plastic Straw	Helm	1/1/83	Blue	Plastic Straw	9 1/2"	Hong Kong	$1.13 - 1.50	The straw has Smurfette's face in the center. It is blue plastic and the top is bendable.
013/013	Smurf Cookie Container	Decor		White	Plastic Dishes	6" High	Australia	$18.00-25.00	The container is white. It has a wrap around picture. There are chocolate chip cookies in the grass. A Smurf is standing and eating a cookie. A baker Smurf is chasing a Smurf with a rolling pin. Papa is standing and pointing at something and Smurfette is posing seductively. The container has a red push button suction top.
013/014	Smurf Tin Pail	Peyo.	1/1/97	Yellow	Tin Pail	9" High	Asia	$22.50-30.00	The pail is from some kind of food. It has a cover and a yellow handle. The pail has a windmill and mushroom houses in the background. It has Nat Smurf holding a rake. A Smurf with bandages around both of his hands. 2 Smurfs playing in the grass. Smurfette is standing in front of the windmill holding Baby. Papa is standing and pointing and there is a Smurf sitting in the flowers.
013/015	Smurf-A-Getti	Irwin Toy	1/1/84	White	Plastic Dishes	3 1/2"		$6.75-9.00	The cup is white. It has a picture of a Smurf holding a yellow kettle. Papa Smurf and Smurfette are standing next to the Smurf.
013/016	10 Smurfs On A Soup Bowl	BP		white	Plastic Dishes	6 1/2"	Australia	$7.50-10.00	The soup bowl has a Smurf picture all the way around it. The picture has Smurfette standing and holding a bowl of soup for Cook Smurf. Papa and a Smurf are holding a cake. A Smurf is sitting on a log, mad. Other Smurfs are doing various things.
013/017	Thin Chocolate Tin Soccer FuBballer Schlumpf	Peyo/Dupuis		Gold	Plastic Dishes	1/4"H x 4 1/2" Diam	Germany	$4.50-6.00	The tin is round and flat with a removable cover. The front has a picture of a Smurf kicking a soccer ball. There is a net behind the Smurf.
013/018	Baby Smurf Hot Bowl	Helly	1/1/94	Blue ? White	Plastic Dishes	8 1/2" dia	Germany	$4.50 - 6.25	The bowl is blue. The inside is white. The bowl has a plug to put hot water in to keep the bowl warm. The picture inside the bowl is of 2 Baby Smurfs in pink sleepers. One is crawling and holding a rattle. The other is sitting while holding a bottle and a block.
013/019	Baby Smurf Sippee Cup	Helly	1/1/94		Plastic Dishes	4"	Brussels	$2.50 - 3.75	The cup is clear with a blue top and bottom. It has a picture of Baby Smurf leaning on a big red and white beach ball.

Set 157: Door Hangings

157/001	Smurf In A Yellow Outfit	Papo	1/1/84		Door Hanging	5"	Hong Kong	$6.00-8.00	The Smurf is made of flat plastic. The Smurf is wearing a yellow jacket with a big gray button on the front and a white rope belt. He has on white pants, yellow socks, and black shoes. The Smurf has a white hat with a red flower. The Smurf has his hands out at his sides. The Smurf is hanging on a yellow rope.
157/002	Mushroom House With Signs	Peyo		White	Door Sign	6"L X 6 1/2" W	Hong Kong	$7.50-10.00	The door sign is flat plastic and in the shape of a mushroom house. It has a red and white roof with a Smurf looking out the window. There is a gray chimney. There is a red string attached for hanging. The bottom is white and has 2 hooks to hang signs. 1 sign says Keep Out. Missing the 2 other signs.

Set 116: Dress N' Style

116/001	Smurfette Dress 'n Style Doll	Applause			Dress 'n Style Doll	6"	Hong Kong	$18.00-25.00	Smurfette has rooted silky blonde hair, a personalized comb, moveable head, arms and legs. Smurfette has a pink dress on with a white cap and white shoes. Smurfette is made of a dark blue vinyl.
116/002	Papa Bendee Doll	Brabo	1/1/79		Rubber Smurf	4"		$4.50-6.00	Papa is wearing a red hat and red pants. Papa is made of rubber. Papa's head turns, his arms bend and his legs bend.
116/003	Cloth Smurfette Doll				Cloth Doll	7 1/2"	China	$6.75-9.00	The Smurf is made of polyester material. Smurfette has a white dress on with red dots going around the bottom. Smurfette's hair is a bright yellow yarn. She has 2 beady eyes that are connected. One arm is facing up and the other is facing down.
116/004	Brainy Judge Doll	Furga	1/1/82		Hard Plastic Doll	11"	Italy	$55.00-75.00	The doll is made of hard plastic. The Smurf is wearing a cloth red robe and a white ruffle collar. The Smurf has white pants and a white hat on but it is the plastic painted. The Smurf has black glasses painted on. The joints move on the doll.
116/005	Clown Doll	Furga	1/1/82		Hard Plastic Doll	11"	Italy	$55.00-75.00	The doll is made of hard plastic. The Smurf is wearing a cloth red jacket with tails. A striped bow. The Smurf has white pants and a white hat on but it is the plastic painted. The Smurf has black shoes. The Smurf's face is painted around the mouth like a clown. The joints move on the doll.
116/006	Baby Smurf Musical	Furga	1/1/80		Doll	18"	Italy	$40.00-55.00	Baby is wearing a yellow outfit. He has a record player inside.

Set 214: Earrings

214/001	Smurf Waving Gold Earrings	Peyo		Gold	Gold Earrings		Germany	$7.50-10.00	The earrings are solid gold posts. The earrings are a Smurf waving.

214/002	Smurf Waving Silver Earrings	Peyo		Silver	Silver Earrings		Germany	$7.50-10.00	The earrings are made of silver. They have a Smurf waving. The Smurf isn't fully painted blue. His face is part silver- colored. The Smurf has white pants and a white hat on. The earrings are post.
214/003	Smurf Face Silver Earrings	Peyo		Silver	Silver Earrings		Germany	$7.50-10.00	The earrings are silver posts. The face is painted half blue and the rest is silver. The Smurf has a white hat on. The face is made of silver.
214/004	Trumpet Smurf Plastic Earrings				Plastic Earrings	1" Diam	Taiwan	$2.00-3.00	The earrings are round, flat plastic. They are clear. The picture is of a Smurf playing a yellow trumpet. The Smurf is wearing a red shirt and white pants. There is a rainbow behind him. The earrings are wire and have a little pearl that is on a link to connect the plastic and the wire earring.
214/005	Papa Pointing Metal Earrings	S.E.P.P/Peyo	1/1/80		Metal Earrings		Taiwan	$2.00-3.00	The earrings are Papa standing and pointing a finger. He has a red hat and red pants on. The earrings are flat metal.
214/006	Smurf On Skateboard Metal Earrings				Metal Earrings			$2.00-3.00	The earrings have a Smurf standing on his hands on an orange skateboard. The earrings are flat metal.
214/007	Roller-skating Smurf Plastic Earrings				Plastic Wire Earring	1"		$2.00-3.00	The earrings are round, flat plastic. They are clear. The picture is of a Smurf roller-skating. There is a rainbow behind him. The earrings are wire and have a little pearl that is on a link to connect the plastic and the wire earring.

Set 156: Easter Items

156/001	Smurf Papier Mâché Eggs	Easter Unlimited	1/1/84		Papier Mâché Egg Kit		Hong Kong	$6.75-9.00	The box is light blue and has a picture of a Smurf sitting in front of a basket of eggs. There is a big egg that has an Easter bunny Smurf and a rainbow on it. The big egg is next to the basket. The kit includes: papier mâché egg molds, 4 different Smurf scenes, inside liners,

Set 159: Fabric

159/001	Light Blue Squares With Papa Standing In Front of Houses	Tigress Pride	1/1/82	Light Blue	Fabric	3Ft L X 3Ft 7"W	U.S.A.	$6.00-8.00	The material is light blue squares with white border. There are 4 different scenes. Papa standing in front of 4 mushroom houses. Smurfette picking daisies and a Smurf standing next to her with a pink daisy. A Smurf is laying in the grass with a shovel stuck in the ground. 3 Smurfs are walking in the grass (one has on a yellow shirt, another is holding a red mirror and the other is just walking). There are little pink flowers all over the material.
159/002	Light Pink Squares With Papa Standing In Front of Houses	Tigress Pride	1/1/82	Light Pink	Fabric	3FT 5IN x 3FT 9IN	U.S.A.	$6.00-8.00	The material is light pink squares with white border. There are 4 different scenes. Papa standing in front of 4 mushroom houses. Smurfette picking daisies and a Smurf standing next to her with a pink daisy. A Smurf is laying in the grass with a shovel stuck in the ground. 3 Smurfs are walking in the grass, one has on a yellow shirt, another is holding a red mirror, and the other is just walking. There are little dark pink flowers all over the material.
159/003	Red With Smurf And Smurfette Roller-skating - SMURFUN	Tigress Pride	1/1/82	Red & White	Fabric	3Ft 9"L X 2Ft 11"W	U.S.A.	$6.00-8.00	The material is red with little white flower buds all over it. In the center it's white with red checkered lines and a design. The design is of a Smurf sitting and talking on a dark blue telephone to Smurfette. Smurfette is standing and holding a red telephone. The Smurf is thinking of him and Smurfette roller-skating. The picture of him and Smurfette roller-skating is in a cloud and inside the cloud it says "SMURFUN". in red letters.
159/004	White With Red Dots Sports Scene	Tigress Pride	1/1/82		Fabric	3Ft L X 3Ft 9 1/2"W	U.S.A.	$6.00-8.00	The material is white and has red dots all over it. There are 9 different sport Smurfs: football Smurf, soccer Smurf, 2 basketball Smurfs, baseball Smurf, 2 running, a referee, and a Smurf leaping.
159/005	Light Blue With Boats	Tigress Pride	1/1/82	Light Blue	Fabric	1FT 8IN x 3FT 8IN	U.S.A.	$6.00-8.00	The material is light blue. The picture is of a Smurf fishing out the back of a boat. Papa is in another boat looking out with a telescope. A Smurf is standing on a stool turning the wheel. Smurf and Smurfette are on a raft.
159/006	Home Smurf Home	Tigress Pride	1/1/82	White	Fabric	3Ft 8IN x 2FT 10IN	U.S.A.	$6.00-8.00	The fabric is white with different colored dots. In the picture is Papa sitting in a rocking chair reading a book to 3 Smurfs and Smurfette. The Smurfs have milk and cookies sitting next to them.
159/007	Follow Your Smurf Wherever They Lead	Tigress Pride	1/1/82		Fabric	3FT 9IN x 2FT 10IN	U.S.A.	$6.00-8.00	The material is white with red dots. In the center is a large blue square. Inside the blue square is a rainbow with Smurfette and a Smurf sliding down it. There is a white cloud above the rainbow that says "Follow Your Smurf Wherever They Lead" in multi- colored letters.
159/008	Queen Smurfette					14"		$6.00-8.00	The material is cut in the shape of Smurfette. Smurfette is standing and wearing a purple gown. The material makes a pillow.

Set 239: Figures- 1" Blue

239/001	Telephone	Peyo	1/1/83	Bright Blue	Plastic Figure	1"		$3.50-5.00	The Smurf is standing, talking on a telephone. The phone is sitting by his foot. The figure is solid blue. The Smurf says "OMO" on the back.
239/002	Captain	Peyo	1/1/83	Bright Blue	Plastic Figure	1"		$3.50-5.00	Papa is solid blue. He is wearing a sailor hat and jacket. Papa is holding a telescope to his eye. The figure says "OMO" on the back.
239/003	Papa	Peyo		Dark Blue	Plastic Figure	1"		$6.00-8.00	Papa is solid blue. He is standing with his hands out at his side.
239/004	Brainy	Peyo		Dark Blue	Plastic Figure	1"		$6.00-8.00	The Smurf is solid blue. He is standing with his hands out at his side. The Smurf has glasses on.
239/005	Drummer	Peyo		Dark Blue	Plastic Figure	1"		$6.00-8.00	The figure is solid blue. The Smurf is holding 2 drumsticks and has a drum between his legs.
239/006	Mechanic	Peyo		Dark Blue	Plastic Figure	1"		$6.00-8.00	The figure is solid blue. The Smurf is standing, holding a wrench in his hand. The Smurf has suspenders holding his pants up.
239/007	Laughing	Peyo		Dark Blue	Plastic Figure	1"		$6.00-8.00	The figure is solid blue. The Smurf has one hand covering his mouth. The Smurf has his other arm extended out and is pointing at something. The Smurf has a grin on his face.
239/008	Gymnast	Peyo		Dark Blue	Plastic Figure	1"		$6.00-8.00	The figure is solid blue. The Smurf is holding a dumbbell above his head. The Smurf is wearing a tank top and pants.
239/009	Rock And Roll	Peyo		Dark Blue	Plastic Figure	1"		$6.00-8.00	The figure is solid blue. The Smurf is playing a guitar.
239/010	Jungle	Peyo		dark Blue	Plastic Figure	1"		$6.00-8.00	The figure is solid blue. The Smurf is holding a spear. The Smurf has wild hair and strange eye's. The Smurf is wearing a leaf wrap around his waist.
239/011	Flying	Peyo		Dark Blue	Plastic Figure	1"		$6.00-8.00	The figure is solid blue. The Smurf has wings attached to his back. The Smurf has his hands out at his sides with the wings extended.
239/012	Trumpet Player	Peyo		Dark Blue	Plastic Figure	1"		$6.00-8.00	The figure is solid blue. The Smurf is walking playing a trumpet. The Smurf has his eye's open.
239/013	Sledgehammer	Peyo		Dark Blue	Plastic Figure	1"		$6.00-8.00	The figure is solid blue. The Smurf is walking and dragging a sledgehammer behind his back.
239/014	Policeman	Peyo		Dark Blue	Plastic Figure	1"		$6.00-8.00	The figure is solid blue. The Smurf is dressed in a police uniform. He has a whistle in his mouth. The Smurf is holding a stick in his hand.

Set 197: Figures- Brass

197/001	Normal				Brass Figure	2" Tall		$15.00-20.00	The Smurf is solid brass. His hands are out at his sides and he is smiling.
197/002	Laughing				Brass Figure	2" Tall		$15.00-20.00	The Smurf is solid brass. He is standing and pointing with his right hand and his left hand is covering his mouth.
197/003	Sleepwalker				Brass Figure	2" Tall		$15.00-20.00	The Smurf is solid brass. He has his eyes closed and both hands out in front of him. The Smurf is wearing a night shirt and a hat with a tassel on top.
197/004	Naughty				Brass Figure	2" Tall		$15.00-20.00	The Smurf is solid brass. He has his thumbs in his ears. He has his tongue sticking out.
197/005	Jolly				Brass Figure	2" Tall		$15.00-20.00	The Smurf is solid brass. He has his left hand on his stomach and his right hand is up by his face. His mouth is open and he's laughing. His eye's are closed.
[illegible]	[illegible]				Brass Figure	2" Tall		$15.00-20.00	The Smurf is solid brass. He has a finger on his right hand covering his mouth. He has his eye's closed and he has a smile on his face. His left hand is behind his back.

Set 244: Figures- Fakes

244/001	Tennis Smurfette				Fake Rubber Figure	2" High		$2.00-3.00	Smurfette has a medium bright blue body. She is wearing a white tennis dress and shoes. Smurfette's dress has red dots on the bottom. Her hair is yellow. Smurfette is holding a red rubber tennis racket in her left hand.
244/002	Smurfette Carrying A Picnic Basket				Fake Rubber Figure	2" High		$2.00-3.00	Smurfette has a medium bright blue body. Smurfette is carrying a yellow picnic basket on her right arm. Smurfette is wearing a plain white dress, white hat and white shoes. Smurfette has short yellow hair.
244/003	Smurf Pulling A Rabbit From Hat				Fake Rubber Figure	2" High		$2.00-3.00	The Smurf has a medium bright blue body. The Smurf is wearing a red jacket, white shirt, white hat, red pants, black shoes and a black bow tie. The Smurf is holding a black hat in one hand and a gray rabbit in the other hand.

Set 50: Figures- Ferrero

050/001	Smurf With A Flower	El Ferrero	1/1/83		Hard Plastic Figures	1 1/2" High	Germany	$11.00-15.00	The Smurf is made of hard plastic. He has a red flower that he is holding with his mouth. The Smurf has his right hand in the air and his left hand is behind his back.
050/002	Blindfold Smurf	El Ferrero	1/1/83		Hard Plastic Figures	1 1/2" High	Germany	$7.50-10.00	The Smurf is made of hard plastic. The Smurf has a red blindfold over his eyes, both of his hands are out in front of him.
050/003	Smurf On A Red Jumping Ball	El Ferrero	1/1/83		Hard Plastic Figures	1 1/2" High	Germany	$7.50-10.00	The Smurf is made of hard plastic. The Smurf is sitting on a red ball.
050/004	Smurf In A Potato Sack	El Ferrero	1/1/83		Hard Plastic Figures	1 1/2" High	Germany	$7.50-10.00	The Smurf is made of hard plastic. The Smurf is inside of a brown potato sack.
050/005	Papa Referee	El Ferrero	1/1/83		Hard Plastic Figures	1 1/2" High	Germany	$7.50-10.00	The Smurf is made from hard plastic. Papa is wearing black pants and a black shirt, white shoes, red hat and has a red whistle in his mouth.
050/006	Smurf With Ice Cream Cone	El Ferrero	1/1/83		Hard Plastic Figures	1 1/2" High	Germany	$7.50-10.00	The Smurf is made of hard plastic. The Smurf is holding an ice cream cone in his right hand.

Set 274: Figures- France

274/001	Rock N Roll	Comic No Toxico			PVC	2"	Spain	$3.50-5.00	The Smurf is holding a white guitar with brown accents. The Smurf is wearing a white hat, white pants and black shoes.
274/002	Clown	Comic No Toxico			PVC	2"	Spain	$3.50-5.00	The Smurf is wearing yellow pants with a black pocket on the back, black suspenders, a red bow tie, white shoes with black toes, a white shirt, white gloves and a white hat. He has a thin line of red with white around for his mouth.
274/003	Mallet	Comic No Toxico			PVC	2"	Spain	$3.50-5.00	The Smurf is holding a mallet over his head. The mallet is a light brown.
274/004	Lover	Comic No Toxico			PVC	2"	Spain	$3.50-5.00	The Smurf is holding a bouquet of bright red daisies. The Smurf has a heart carved on his chest.
274/005	Artist	Comic No Toxico			PVC	2"	Spain	$3.50-5.00	The Smurf is holding a white paint brush with yellow paint on the end, in his right hand. In his left hand he's holding a white painter's palate, the colors are raised. The colors are brown, green, blue and yellow. The Smurf is smiling with his tongue sticking out. The Smurf has small black eye's.
274/006	Emperor	Comic No Toxico			PVC	2"	Spain	$3.50-5.00	The Smurf is wearing a yellow outfit, yellow crown, white hat, white shoes, gold jewelry and a red robe. He is holding his index finger on his right hand up in the air.
274/007	Bowler	Comic No Toxico			PVC	2"	Spain	$3.50-5.00	The Smurf is holding a dark red bowling ball.
274/008	Cricket	Comic No Toxico			PVC	2"	Spain	$3.50-5.00	The Smurf is wearing a white uniform with red trim. He has a brown bat. He is holding the bat on his left side.
274/009	Hammer	Comic No Toxico			PVC	2"	Spain	$3.50-5.00	The Smurf has his left hand wiping sweat from his forehead. His right hand is holding a hammer at his side. The hammer has a red handle and a black hammer. There is a bead of sweat running down the right side of the Smurf's face.
274/010	Oboist	Comic No Toxico			PVC	2"	Spain	$3.50-5.00	The Smurf is standing playing an orange oboe. The Smurf has his eye's open.
274/011	Angler	Comic No Toxico			PVC	2"	Spain	$3.50-5.00	The Smurf is holding a yellow fishing pole. The pole is part of the mold. The Smurf is wearing white pants and black shoes.
274/012	Pirate	Comic No Toxico			PVC	2"	Spain	$3.50-5.00	The Smurf is holding a silver sword in his right hand. He has a red belt with a black buckle around his waist. The Smurf has a black patch covering his left eye.
274/013	Scot	Comic No Toxico			PVC	2"	Spain	$3.50-5.00	The Smurf is wearing a red hat and red cape. The Smurf is playing a yellow bagpipe. The Smurf has green trim around his shoes.
274/014	Hairdresser	Comic No Toxico			PVC	2"	Spain	$3.50-5.00	The Smurf is holding a red comb and a silver scissors.
274/015	Conjuror	Comic No Toxico			PVC	2"	Spain	$3.50-5.00	The Smurf has a red cape around his neck. The Smurf is holding a yellow cloth in his left hand. In his right hand he has a black hat with a dark green cloth hanging out.
274/016	Policeman	Comic No Toxico			PVC	2"	Spain	$3.50-5.00	The Smurf is wearing a white hat with a black strap, a white jacket with black buttons, a black belt with a black buckle and white pants. He is holding a white stick in his left hand. He is holding a red whistle to his mouth with his right hand.
274/017	Rollerskater Smurfette	Comic No Toxico			PVC	2"	Spain	$3.50-5.00	Smurfette is wearing a white dress, a white hat, red bows in her hair and white and red roller-skates.
274/018	Graduation	Comic No Toxico			PVC	2"	Spain	$3.50-5.00	The Smurf is wearing a purplish maroon graduation cap with a yellow tassel and a purplish maroon gown. He is holding a white rolled diploma in his left hand and a black book bag in his right hand.
274/019	Judo	Comic No Toxico			PVC	2"	Spain	$3.50-5.00	The Smurf is wearing a white coat, white pants and a white hat. He has a black belt wrapped around his waist. The Smurf is standing with his hands at his side and his eye's closed.
274/020	Tennis Smurfette	Comic No Toxico			PVC	2"	Spain	$3.50-5.00	Smurfette is wearing a white dress, white shoes and a white hat. Smurfette is holding a red tennis racket.
274/021	Captain	Comic No Toxico			PVC	2"	Spain	$3.50-5.00	Papa is wearing a white coat with silver buttons, red pants, and a white hat with a red border and black visor. Papa is holding a silver telescope to his right eye.
274/022	Bricklayer	Comic No Toxico			PVC	2"	Spain	$3.50-5.00	The Smurf is wearing blue bib overalls and a white shirt. He has a red and gray brick in his left hand and under his left foot. He is holding a silver trowel in his right hand.
274/023	Telephone	Comic No Toxico			PVC	2"	Spain	$3.50-5.00	The phone is red with gold. The cradle is wide and gold. The hand piece is gold and he is holding it to his mouth with his right hand.
274/024	Blackboard	Comic No Toxico			PVC	2"	Spain	$3.50-5.00	Papa is standing, holding a yellow stick. He is wearing a red hat and red pants. The chalkboard isn't with the figure.

Set 202: Figures- Jubilee

202/001	Telephone Smurf	Schleich/Peyo/W.B.	1/1/80		Rubber Smurfs	2"	Hong Kong	$11.00-15.00	The Smurf has a dark blue body. The phone is red with gold. The cradle is small and copper. The hand piece is copper and he is holding it to his mouth with his right hand. The Smurf has 1980, Peyo with leaves around it stamped on the back of his hat.
202/002	Papa	Peyo	1/1/69		Rubber Figure	2"	West Germany	$11.00-15.00	Papa is standing with both hands out at his side. His body is medium blue. He has a red hat and pants on. The Smurf has 1969, Peyo with leaves around it stamped on the back of his hat.
202/003	Sailor	Schleich/Peyo	1/1/85		Rubber Figure		West Germany	$11.00-15.00	The Smurf has a medium blue body. He is wearing a white sailor suit with dark blue trim and a white sailor hat. The Smurf is carrying a brown sea bag over his right shoulder. The Smurf has 1985, Peyo with leaves around it stamped on the back of his hat. The Smurf has an orange dot on his foot.
202/004	40th Anniversary Jubilee Set	Schleich	1/1/98		Play set		Brussels	$60.00-80.00	The set comes in a nice 40th anniversary box with 5 Smurfs and band equipment. There is a gray platform to set the band on. There is a 40th banner that goes above the Smurfs. The 5 figures are drummer Smurf in a silver and green outfit. Keyboard player in a black, green and silver outfit. Singer in a black, yellow and silver outfit. Lead guitar in silver overalls. Bass guitarist in a green and silver outfit. Same molds as the 1998 individual figures..

Set 31: Figures- Poseables

031/001	Baby Smurf	Irwin Toys	3/1/96		Plastic Figures	3"	China	$6.00-8.00	Baby Smurf is wearing a pink outfit and is holding an ice cream cone, with a brown high chair.
031/002	Handy Smurf	Irwin Toys	3/1/96		Plastic Figures	3"	China	$6.00-8.00	Handy Smurf is wearing a white outfit and is holding a hammer, and has 2 tool boxes.
031/003	Artist Smurf	Irwin Toys	3/1/96		Plastic Figures	3"	China	$6.00-8.00	Artist Smurf is wearing a white outfit and is holding a paint brush in one hand and a can of green paint in the other. There is a brown easel with a picture of Smurfette painted on it.
031/004	Papa Smurf	Irwin Toys	3/1/96		Plastic Figures	3"	China	$6.00-8.00	Papa has a brown table and a silver pot of popcorn and there is a plate of fish on the table.
031/005	Smurfette	Irwin Toys	3/1/96		Plastic Figures	3"	China	$6.00-8.00	Smurfette is wearing a white dress and is holding a purple flower and she has a brown basket of flowers.

Set 163: Figures- Rubber 6"

163/001	Shy Smurf				Rubber Toy	6"High		$7.50-10.00	The Smurf is standing with a finger by his mouth and his other hand behind his back. He has his eyes closed. He has a light blue body, a white hat and white pants. He does not squeak.
163/002	Normal Smurf				Rubber Toy	6"High		$7.50-10.00	The Smurf is standing with his hands out at his sides. He is light blue with a white hat and white pants. The Smurf has a big smile on his face. He doesn't squeak.

Set 264: Finger Puppets

264/001	Black Smurf	Peyo		Black	Finger Puppet	2"	Italy	$11.00-15.00	The Smurf has a black face. He has red teeth. The Smurf has a white hat. The bottom is black and it fits on your finger.
264/002	Smurf Sticking Out Tongue Purple	Peyo		Purple	Finger Puppet	2"	Italy	$11.00-15.00	The Smurf has a blue face. He has his eye's crossed and he is sticking out his tongue. The bottom is purple and it fits on your finger.

Set 71: Folders

071/001	Time To Get Smurfin!	Mead	1/1/82	Orange	Cardboard Folders	12 1/2L x 9 1/2W	U.S.A.	$3.00-4.00	A Smurf is carrying books looking at a clock.
071/002	To Smurf Or Not To Smurf, That Is The Question	Mead	1/1/82	Yellow	Cardboard Folders	12 1/2L x 9 1/2W	U.S.A.	$3.00-4.00	Papa is holding 1 finger up in the air on his left hand.
071/003	Gargamel And Azreal	Mead	1/1/82	Light Blue	Cardboard Folders	12 1/2L x 9 1/2W	U.S.A.	$3.00-4.00	The cover has Gargamel holding his hands together and Azreal is walking in front of him. Gargamel's gray castle is in the background.
071/004	I Love Work... I Can Dream About It For Hours	Mead	1/1/82	Blue/Green	Cardboard Folders	12 1/2L x 9 1/2W	U.S.A.	$3.00-4.00	The cover has a Smurf leaning on a shovel daydreaming.
071/005	Smurf Sports	Mead	1/1/82	Green	Cardboard Folders	12 1/2L x 9 1/2W	U.S.A.	$3.00-4.00	The cover has 4 sport Smurfs. It has a basketball Smurf, soccer Smurf, football Smurf and baseball Smurf
071/006	Smurfy Friends Stay In Touch	Mead	1/1/82	Red	Cardboard Folders	12 1/2L x 9 1/2W	U.S.A.	$3.00-4.00	Smurfette is standing talking on the phone to a Smurf.
071/007	Smurf On A Carousel Horse	Mead	1/1/82	Light Blue	Cardboard Folders	12 1/2L x 9 1/2W	U.S.A.	$3.00-4.00	There is a Smurf on a yellow and orange carousel reaching for some rings. The folder has a scribble pad inside.
071/008	Smurf And Smurfette Roller-skating	Mead	1/1/82	Blue	Expandable Carryall	10"W X 15"L	U.S.A.	$3.00-4.00	The Smurf carryall has a velcro close tab. The picture is of Smurfette and another Smurf roller-skating. The Smurf looks like he's chasing Smurfette. The folder is light blue with a darker blue border.
071/009	Jokey Giving A Smurf A Present	Mead	1/1/82	Red	Expandable Carryall	10"W x 15"L	U.S.A.	$3.00-4.00	The Smurf carryall has a velcro close tab. The picture is Jokey carrying a yellow box with a red ribbon and another Smurf looking at the present with a worried expression on his face. The folder is light red with a darker red border.
071/010	Smurf 'Em Cowboy!	W. Berrie	1/1/82	Orange	Cardboard Folders	12 1/2L x 9 1/2W	U.S.A.	$3.00-4.00	The folder is orange. A Smurf is riding a horse. Smurfette is sitting on a cowboy. Smurfette is dressed in a cowgirl outfit. The Smurf is wearing a white cowboy hat and a red kerchief around his neck.
071/011	Dreamy Smurf	W. Berrie	1/1/82	Black	Cardboard Folders	12 1/2L x 9 1/2W	U.S.A.	$3.00-4.00	The folder is black. A Smurf is laying under a tree with a bubble above his head. The bubble has a pond scene. Smurfs are swimming, fishing, playing ball on the shore and standing around.

Set 19: Food Items

019/001	Smurf Cookies	L. Parein International	1/1/84		Cookies	5 OZ	U.S.A.	$7.50-10.00	The front of the box of cookies has a Smurf and Smurfette making cookies. The box comes with a sticker inside.
019/002	Smurf Chewable Vitamins	Mead Johnson	1/1/85		Chewable Vitamins	60 Tablets	U.S.A.	$6.00-8.00	There are several color vitamins with different Smurfs on them. They are in a glass bottle.
019/003	Smurf Gummy Goodies	General Foods			Gummy Goodies	.70 OZ	U.S.A.	$2.00-3.00	Raspberry, Orange and Lemon Flavors.
019/005	Orange Juice Box	Minute Maid	1/1/94	Orange	Cardboard Juice Box	2 1/2" x 3 1/4"	Brussels	$1.50-2.00	The box is orange and has Smurfette on all 4 sides, waving.
019/006	Multifruit Juice Box	Minute Maid	1/1/94	Yellow	Cardboard Juice Box	2 1/2" x 3 1/4"	Brussels	$1.50-2.00	The box is yellow and has a Smurf with his hands out at his side, he's happy. The picture is on all 4 sides.
019/007	Gum Ball Machine	Galoob	1/1/83		Gum Ball Machine		Hong Kong	$11.00-15.00	The gum ball machine is a mushroom. The top of the mushroom is orange with yellow spots, there is a rubber Smurf sitting on top of the gum ball machine. The bottom of the mushroom is clear and that's where you put the gum balls. The base is green. There is a bag of gum balls in the packages.
019/008	Smurfenmacaroni	Anco	1/1/87		Smurf Macaroni		Brussels	$3.50-5.00	The pasta noodles are shaped like all different Smurfs. The package has Smurfette holding a plate of the macaroni and a Smurf jumping in the air holding a fork.
019/009	Indian Head Ice Cream Container				Ice Cream Container	3" x 3"	France	$3.50-5.00	The container is in the shape of a Smurf head. The Smurfs face is blue. He is wearing a white hat with a yellow and red stripe around it and a red feather in his hat. The hat comes off and inside the head is were the ice cream is.
019/010	Smurfen Opzetbiskwie	Bolletje	1/6/97		Cookies	2 1/2"High	Holland	$6.00-8.00	The biscuits are in a red and blue package. On the front of the package is a Smurfette biscuit and a Smurf with a present biscuit.
019/011	Die Schlumpfe Bubble Gum	Look-o-Look			Bubble Gum		Holland	$5.00-7.00	There are 12 pieces in the package. The wrappers are light blue and have Papa and a Smurf face on them, or Smurfette and Gargamel's face on them. The gum comes with stickers
019/012	De Smurfen Tover Chocolade	Rademaker	1/1/97		Witte Chocolade		Amsterdam	$3.00-4.00	The bow is green and light blue. The front has a Smurf wearing a dark blue hat with yellow stars, blue pants with yellow stars and a yellow cape with blue stars. The Smurf is holding a hat and a candy stick.
019/013	Smurf-Berry Crunch Cereal	Post	1/1/86		Cereal	11oz.	U.S.A.	$18.00-25.00	The cover of the box has Papa Smurf eating a bowl of cereal. In the background 2 Smurfs and Smurfette are picking Smurf berries. The back of the box is advertising for 4 different posters inside. One side of the box is advertising for magnets.
019/014	Smurf-Berry Crunch Cereal	Post	1/1/85		Cereal Box	11oz.	U.S.A.	$18.00-25.00	The cover of the box has Papa Smurf eating a bowl of cereal. In the background 2 Smurfs and Smurfette are picking Smurf berries. The back of the box is advertising for Smurf rub on inside the box. One of the sides has a game on it.
019/015	Smurfette Chewable Vitamins	Mead Johnson	1/1/85		Vitamins	60 Tablets	U.S.A.	$6.00-8.00	There are several color vitamins with different Smurfs on them. They are in a glass bottle. The cap is Smurfette's head. Her head is made of a hard plastic.
019/016	Tover Chocolade	Rademaker	1/1/97		White Chocolate		Holland	$3.00-4.00	The box is green and light blue. The front has a Smurf wearing a dark blue hat with yellow stars, blue pants with yellow stars and a yellow cape with blue stars. The Smurf is holding a hat and a candy stick. There are 3 boxes per package. There are 3 packages each with a different puzzle in it.
019/017	Smolferne Samle-Kiks	Bolletje	1/1/97		Cookies		Brussels	$3.00-4.00	The box is red. It has various Smurfs on all sides of the box. In the center on the front is a Smurf biscuit and a Smurfette biscuit. The center of the box is blue. Inside are 2 round Smurf cardboard disk- like pogs.
019/018	Tover Chocolade (Red Box)	Rademaker	1/1/97	Red	White Chocolate		Holland	$3.00-4.00	The box is red and light blue. The front has a Smurf wearing a yellow hat with red stars, yellow pants with red stars and a red cape with yellow stars. The Smurf is holding a hat and a candy stick. There are 3 boxes per package. The box has a Smurf beach scene mini puzzle enclosed.
019/019	De Smurfen Gum	Dunkin		Pink	Gum	1" x 1 3/4"	Netherlands		The packages are pink. It has a Smurf running or a Smurf dancing, or Smurfette with head phones on, the outside wrapper. Inside is a transfer.
019/020	Smurfen Biscuits	Delacre	1/1/98		Biscuits	12 per pack	Brussels	$2.00-3.00	The box is yellow. It has a Smurf carrying a Smurf face biscuit. The biscuits are Smurf faces with chocolate hats.
019/021	Smurfen Aardbei - Fraise	Delacre			Cookies		Brussels	$2.00-3.00	The box has a picture of 4 Smurfs and Smurfette sitting on a blanket eating these cookies. Nat Smurf is carrying a basket of strawberries. The cookies are shown in the corner of the box. They are round and have Smurf faces in the middle.
019/022	Nectar Multifruit Juice Boxes	Coca Cola Company	1/1/98	Yellow	Juice Boxes	3 Boxes	Brussels	$3.00 - 3.50	The boxes are yellow. Each one has a different picture. Smurfette is disco-ing. Snappy Smurf is putting records on a record player. A Smurf is on a skateboard.

Set 42: Games

042/001	The Smurf River Ride Game	MB	1/1/88	Yellow	Board Games		U.S.A.	$11.00-15.00	The outside off the box has Gargamel and Azreal peeking out from a bush at the Smurfs. Papa, Smurfette and 2 Smurfs are floating on logs down a river.
042/002	Baby Smurf Game	MB	1/1/84		Board Games		U.S.A.	$9.00-12.00	The outside of the box has Baby Smurf in the middle with Papa, Smurfette and 4 other Smurfs fussing over Baby Smurf.
042/003	Smurf Puzzle Game	Playskool	1/1/82		Puzzle Games	11" x11"	U.S.A.	$7.50-10.00	There is a red plastic storage tray and 4 Smurf puzzles with a spinner. The cover shows a boy and girl playing the game.
042/004	Smurf Ahoy Game	MB	1/1/82		Ship Games		U.S.A.	$11.00-15.00	The cover of the box has Smurfs falling off a ship. The game has a blue plastic ship and cardboard Smurfs that you try to keep in the boat.
042/005	Smurfen Domino	Ravensburger	1/1/83		Domino Game		West Germany	$11.00-15.00	The box is orange and yellow. The front and back of the box has a Smurf with a red daisy in his hat holding 2 dominos.
042/006	Pin The Tail On Azreal	Unique	1/1/82		Wall Game	18" X 24"	Hong Kong	$3.50-5.00	The game has Azreal and you pin the tail on him. There is a blindfold, instructions and 16 tails included in the game.
042/007	Die Schlumpfe	Schmid			Card Games	4" x 2 1/2"	Germany	$7.50-10.00	The cards have Papa standing waving to a Smurf carrying a red and white knapsack. The Smurf is walking in the grass. There is a purple outline of a Smurf house in the background. The cards are in a yellow case with a clear cover.
042/008	A-Maze-ing Action Maze	Colorforms		Yellow	Maze Games		U.S.A.	$9.00-12.00	The game is on a card. The game is a circle with 3 halls. There are 3 balls and you try to get to center. The game has a red board.
042/009	Number Match-Ups	Playskool	1/1/82		Puzzle Game		U.S.A.	$4.50-6.00	24 2-piece puzzles with frame.
042/010	Rhyming Match-Ups	Playskool	1/1/82		Puzzle Games		U.S.A.	$4.50-6.00	24 2-piece puzzles with frame.
042/011	Les Schtroumpfs De Smurfen Maxi - Mini Cards	ASS			Mini Cards	2 1/2" X 1 3 3/4"	Germany	$7.50-10.00	The maxi mini cards are a promotional for ASS. Each card has a picture of a Smurf. The back of the cards are red and white. The top card of the pack has Papa standing with his hands out.
042/012	Rubic's Cube	Wonderful Puzzler			Board Games	2 1/4" X 2 1/4"	Taiwan	$6.00-8.00	The cube has 6 different pictures: There is a Smurf roller-skating playing a trumpet, a Smurf and Smurfette in an orange car, a Smurf snow skiing, a Smurf and Smurfette on a motorcycle, a Smurf playing baseball, and Smurfette with a Smurf playing in snow.
042/013	The Smurf Card Game	MB	1/1/82		Card Games		U.S.A.	$7.50-10.00	The cards are large with pictures of different Smurfs on them. The game is played like war. The back of the cards are white with blue in the center and they say SMURFS. There are flowers around the border of the blue area.
042/014	Zwarte Piet Cards	King	1/1/94		Card Game	4" x 2 1/2"	Netherlands	$7.50-10.00	The cards have a black Smurf on the front with 3 Smurf faces around him. The top and bottom of the cards are yellow and the center is purple. The cards are in a red case with a clear cover.
042/015	The Smurfs Games (Spanish)	Comansi, S.A.			Board Games	15 x 15	Spain	$37.50-50.00	The game board has a wood edge and the board has little squares that have Smurfs in them. The board comes in a green box with a picture of a Smurf giving Smurfette flowers. The pieces come in a separate box. There are little rubber Smurf figures for pieces. The board is 2 sided. El Parchis, De Los Pitufos.
042/016	Die Schlumpfe Quartett	Schmid	1/1/94		Card Game	4" x 2 1/2"	Brussels	$7.50-10.00	The cards have Papa standing waving to a Smurf carrying a red & white knapsack. There is another Smurf standing next to Papa looking troubled. The Smurf is walking down a dirt road into the woods. There is a purple outline of a Smurf house in the background. The cards have an orange border on the top and bottom. The cards are in an orange case with a clear cover.
042/017	Schwarzer Peter	Schmid	1/1/94		Card Games	4" x 2 1/2"	Brussels	$7.50-10.00	The cards have a black Smurf on the front with 3 Smurf f[illegible] him. The top and bottom of the cards are yellow and the center is purple. The cards are in a red case with a clear cover.
042/018	Die Schlumpfe Sport Domino	Schmid	1/1/94		Card Game	4" x 2 1/2"	Brussels	$11.00-15.00	The cards have 2 Smurfs running after an orange and white soccer ball. The top and bottom of the cards are blue and the center is yellow and purple. The cards are in a red case with a clear cover.
042/019	Schwarzer Peter Schlumpfe	ASS			Card Game	2 1/2L x 1 3/4W	Germany	$7.50-10.00	The cards are small. The front of the cards have different pictures. The front card has a black Smurf jumping in the air. The cards are in a red case with a clear top.
042/020	De TeleSmurfer	Infogrames	1/1/95		Computer Game		Brussels	$30.00-40.00	The cover of the case is blue. It has Smurfette standing and posing. Papa and a Smurf are standing in front of Smurfette. There are pipes in the background. The object of the game is to rescue Smurfette from Gargamel's castle.
042/021	Smurfette A Maze ing Action Maze	Colorforms			Board Games		U.S.A.	$8.25-12.00	The game is on a card. The game is a heart- shaped board with 3 balls. There are 3 balls and you try to get to the center where there is a picture of Smurfette. The heart board is yellow. The background has lovesick Smurfs on it.
042/022	Schlumpf Skat Mau-Mau und 66	ASS			Card Games	2 1/2" x 1 3/4"		$15.00-65.00	The top card has a Smurf dressed as a knight. The Smurf is holding a gray and red spear. The cards say "Schlumpf Skat Mau-Mau und 66" under the picture. The cards come in a case with a promo ASS Smurf figure. The Smurf is holding a heart card that says "[illegible]."
042/023	Smurf Ladder Game	Helm		Red/ White	Ladder Game	12" High	Hong Kong	$10.50-14.00	The game is a ladder. You tip the ladder upside down and a Smurf and Smurfette race to the bottom. The ladder is white and has a red stand.
042/024	Smurf Target Game	Galoob	1/1/83		Board Games		Hong Kong	$11.00-15.00	Gargamel is a big target cut- out figure. He holds a chain and lock. The Smurfs attached to the lock. When you hit the lock with the dart it releases the Smurfs and they slide to the ground.
042/025	Smurf Cube Puzzle	Peyo	1/1/85		Puzzle Cube	4" x 4"	Brussels	$9.00-13.00	The cube is large and has 6 different pictures on it. A different picture on each side. The cube comes apart. One side has a Smurf with white wings flying. Smurfette with a palate and paintbrush. Papa with a camera on a tripod. A Smurf holding a book with a butterfly flying above him. A Smurf carrying a yellow gift box. A Smurf playing a trumpet.
042/026	Smurfen Ringen Werpspel	Kortekaas Merch	1/1/83		Board Games	14" x 11"		$11.00-15.00	The board is cardboard. It has 3 Smurf faces, Papa and Smurfette's face. They all have plastic noses. Each Smurf has a point value. There are 5 plastic rings included.
042/027	Die Schlumpfe Und Die Zauberflote	ASS			Deck Of Cards	2 1/2" x 1 3/4"	Germany	$7.50-10.00	The front has Peewit playing a flute. Johan is falling backwards. There are 8 Smurfs and Papa laughing.
042/028	Traveler Smurf Cards Die Schlumpfe	ASS			Deck Of Cards	2 1/2" x 1 3/4"	Germany	$7.50-10.00	The front of the cards has a traveler Smurf sitting in the grass dreaming of the ocean and mountains.
042/030	Smurf Spin-A-Round Game	MB	1/1/83		Board Games		U.S.A.	$11.00-15.00	There is a big plastic Smurf in the center that holds the cards. The board is round.

042/031	The Smurf Game	MB	1/1/81		Board Games		U.S.A.	$11.00-15.00	The game board is 3 dimensional. The cover of the box has 12 different Smurfs pictured on it.
042/032	Schtroumpf Loto	Ravensburger	1/1/97	Yellow	Board Games	10 3/4" x 7 1/2"	Germany	$13.50-18.00	The box is yellow. The cover has a Smurf carrying a card that has a picture of a gift. There are cards with different pictures on the cover. The game has large cards with several pictures and then individual picture cards. The game is like bingo.
042/033	Le Grand Jeu Des Schtroumpfs	Ravensburger	1/1/97	Yellow	2 Board Games	9" x 13"	Germany	$17.25-23.00	The box is yellow. The cover has a Smurf dressed as an Indian. It shows 2 different board games.
042/034	Le Cauchemar Des Schtroumpfs Game Boy Cartridge	Infogrames	1/1/98	Purple	Game Cartridge		Japan	$30.00-40.00	The cover of the box is purple. It has a Smurf being spun on it. The Smurf has a troubled expression on his face. The cartridge is in color.

Set 233: Games- Atari & Coleco

233/001	Rescue In Gargamel's Castle	Coleco	1/1/82		Atari Game Cartridge		U.S.A.	$18.00-25.00	The object of the game is to rescue Smurfette from Gargamel's castle.
233/002	Smurf Paint 'n' Play Workshop	Coleco	1/1/84		Atari Game Cartridge		U.S.A.	$22.50-30.00	Paint 'n' Play Workshop.
233/003	Smurf's Save The Day	Coleco	1/1/83		Atari Game Cartridge		U.S.A.	$22.50-30.00	The game cartridge comes with cassettes. # 1 cassette is Harmony Smurf. # 3 Greedy Smurf.

Set 199: Games- Hand Held

199/001	Smurf Ball	Tomy	1/1/78	Red	Hand Held Game	2 1/2"L X 4 1/2"W	Hong Kong	$13.50-18.00	The game has a golf background. There are 4 Smurfs swinging golf clubs. The object is to get all the balls to the top circle. There are 4 flippers.
199/002	Smurf Look Alike	Tomy	1/1/75	White	Hand Held Game	2 1/2"L X 4 1/2"W	Hong Kong	$13.50-18.00	The game has a spinner and a button to stop the spinner. You try to match Smurfette and the Smurfs face. The outside of the front of the game is a flowered wall with Smurfette's and a Smurfs body.
199/003	Hat Trick	Tomy	1/1/77	Blue	Hand Held Game	2 1/2"L X 4 1/2"W	Hong Kong	$13.50-18.00	The game looks like a tick-tack-toe board. Each hole has a hat painted behind it. A Smurf is standing at the bottom holding a hat. The object is to fill as many holes as possible by shooting them with a flipper.
199/004	Schtroumpf	Tiger/Peyo	1/1/84		Hand Held Game	2 1/2"L X 4 1/2"W	Hong Kong	$25.00-35.00	The game is white and the top is in the shape of a Smurf's head. The Smurf has his hands on the side of the game window. The game window has a few mushroom houses and grass background. There is a clock in the game window. The Smurf chases butterflies.

Set 125: Gift Wrap

125/001	Christmas Carolers	Papercraft	1/1/82	Light Blue	Christmas Paper	2 20 X 30 IN.	U.S.A.	$6.00-8.00	The paper is light blue and has music written on it (Jingle Bells). Papa is standing on a mushroom leading a band. There are various Smurfs all over the paper playing different instruments.
125/002	Wreath Paper	Papercraft	1/1/82	White	Christmas Paper	2 20 X 30 IN.	U.S.A.	$6.00-8.00	The paper is white and has X-mas wreaths with Smurf face inside the wreaths. Papa is holding a candy cane, Smurfette is singing, a Smurf is putting a ribbon on his wreath and another Smurf is holding a Christmas list. All the Smurfs are in wreaths.
125/003	Christmas Tree Paper	Papercraft	1/1/82	Green	Christmas Paper	2 20 X 30 IN.	U.S.A.	$6.00-8.00	The paper is green. There is a Smurf carrying 4 presents to a tree, on the other side of the tree a Smurf is decorating it and Papa is standing writing in the Smurf Christmas list book. Smurfette is wearing ice skates and pulling a Smurf on a sled. There is another Smurf on a sled behind Smurfette.
125/004	Ice-skating Pond Scene	Papercraft	1/1/82		Christmas Paper	2 20 X 30 IN.	U.S.A.	$6.00-8.00	The paper has ponds with the Smurfs ice-skating. 2 Smurfs are making a snow man, Papa's skiing down a mountain, another Smurf is sledding down the mountain.
125/005	Sledding Scene	Papercraft	1/1/82	White	Christmas Paper	2 20 X 30 IN.	U.S.A.	$6.00-8.00	The paper is all snow and pine trees. There are Smurfs sledding. Smurfette is ice-skating with another Smurf.
125/006	Sports Scene	Papercraft	1/1/82		Any Occasion Paper	2 20 X 30 IN.	U.S.A.	$6.00-8.00	The paper is green and blue. There are different sport Smurfs. A baseball player, a soccer player, a basketball player, a football player, a Smurf on roller-skates and Smurfette dressed as a cheerleader.
125/007	Happy Smurf-Day	Papercraft	1/1/82		Birthday Paper	2 20 X 30 IN.	U.S.A.	$6.00-8.00	The paper is purple and blue. There is a clown Smurf carrying balloons, Smurfette holding ribbon, a Smurf carrying a present, a Smurf carrying a cake and a Smurf blowing a trumpet.
125/008	Sport Scene	Papercraft	1/1/82		Christmas Paper	25 SQ. FT.	U.S.A.	$6.00-8.00	The paper is green and blue. There are different sport Smurfs. A baseball player, a soccer player, a basketball player, a football player, a Smurf on roller-skates and Smurfette dressed as a cheerleader.
125/009	Christmas Wrap Roll With Carolers	Papercraft	1/1/82		Christmas Paper	3 Rolls	U.S.A.	$11.00-15.00	The package has 3 rolls of paper. 1 roll is light blue and has Christmas music with notes, Smurfette is holding a candy cane. Papa is on a mushroom holding a baton. Roll 2 is white and has a village scene out in the snow and Papa is dressed as Santa. Roll 3 is red and has Smurfs hanging holly around the Merry Christmas (Merry Christmas is in green letters).
125/010	Christmas Wrap With Village Scene	Papercraft	1/1/82		Christmas Paper	3 Rolls	U.S.A.	$11.00-15.00	The package has 3 rolls of paper. 1 roll is white with a Smurf driving a log car pulling 3 Smurfs on a trailer past the tree. The 3 Smurfs are playing X-mas carols, Smurfette is standing posing. Roll 2 is red and has green ornaments with different Smurfs inside. Roll 3 is white and has ponds with the Smurfs ice-skating. 2 Smurfs are making a snow man, Papa's skiing down a mountain, another Smurf is sledding down the mountain.
125/011	Blue Baby Smurf	Papercraft	1/1/82	Light Blue	Everyday Paper	2 20 X 30 IN.	U.S.A.	$6.00-8.00	The paper is light blue and has Baby Smurf all over. Baby is laying on a blanket drinking a bottle, Baby is on his knees holding a ball, Baby is in a carriage holding a rattle and teddy bear. Baby is crawling. The paper says Baby Smurf with flowers around it.
125/012	Pink Baby Paper	Papercraft	1/1/82	Pink	Everyday Paper	1 20 X 30 IN.	U.S.A.	$6.00-8.00	The paper is pink and has Baby Smurf all over. Baby is laying on a blanket drinking a bottle, Baby is on his knees holding a ball, Baby is in a carriage holding a rattle and teddy bear. Baby is crawling. The paper says Baby Smurf with flowers around it.
125/013	Smurfette Flower Paper	Papercraft	1/1/82		Christmas Paper	2 20 X 30 IN.	U.S.A.	$6.00-8.00	The paper has flowers all over. There are pictures of a Smurf giving Smurfette flowers. A Smurf holding the flowers behind his back. Smurfette standing in a flirty pose holding flowers.

Set 257: Glasses- Australian

257/001	Smurf Hitch Hiking	Peyo			High Ball Glass	3 1/2" High	Australia	$7.50-10.00	The glass is 3 1/2" high. The glass is on a little pedestal. It has a Smurf running and panting. He is hitchhiking. There is a 15km sign behind him. The Smurf is wearing a red shirt, clear pants, clear shoes and a clear hat.
257/002	Smurf On A Tricycle	Peyo			High Ball Glass	3 1/2" High		$7.50-10.00	The glass is 3 1/2" high. The glass is on a little pedestal. The Smurf is on a red tricycle. He is wearing a clear hat and clear pants. He has black goggles over his head. Australia
257/003	Drummer	Peyo			High Ball Glass	3 1/2" High	Australia	$7.50-10.00	The glass is 3 1/2" high. The glass is on a little pedestal. The Smurf is wearing a drum around his neck. The drum is red. He is holding a red drumstick in each hand. The Smurf is wearing a clear hat and clear pants.
257/004	Runner	Peyo			High Ball Glass	3 1/2" High	Australia	$7.50-10.00	The glass is 3 1/2" high. The glass is on a little pedestal. The Smurf is running through a finish line. He is wearing a clear shirt, a clear hat, red pants, and red shoes.

Set 173: Glasses- Canadian Pedestal

173/001	Sport Smurf	Peyo			Drinking Glass	4 3/8" High	Canada	$8.25-12.00	The glass has a small base, in the middle is a ridge and the top of the glass is bigger. The Smurf is on the top of the glass and only on one side. The Smurf is wearing a brown and yellow outfit and a white hat with a yellow star on the side. The Smurf is holding a yellow bat, white soccer ball, yellow tennis racket and a brown golf bag.
173/002	King Smurf	Peyo			Drinking Glass	4 3/8" High	Canada	$8.25-12.00	The glass has a small base, in the middle is a ridge and the top of the glass is bigger. The Smurf is on the top of the glass and only on one side. The Smurf is wearing a yellow hat, yellow crown, yellow pants and yellow shoes. The Smurf has a red robe with white trim. He is holding a red scepter with a white handle.
173/003	Fishing Smurf	Peyo			Drinking Glass	4 3/8" High	Canada	$8.25-12.00	The glass has a small base, in the middle is a ridge and the top of the glass is bigger. The Smurf is on the top of the glass and only on one side. The Smurf is sitting on a black stump. He is holding a yellow fishing rod with a red fish hanging on the end.
173/004	Drummer	Peyo			Drinking Glass	4 3/8" High	Canada	$8.25-12.00	The glass has a small base, in the middle is a ridge and the top of the glass is bigger. The Smurf is on the top of the glass and only on one side. The Smurf is standing, holding a white drum with red trim and 2 red drumsticks.
173/005	Lute	Peyo			Drinking Glass	4 3/8" High	Canada	$8.25-12.00	The glass has a small base, in the middle is a ridge and the top of the glass is bigger. The Smurf is on the top of the glass and only on one side. The Smurf is standing playing a large yellow lute.
173/006	Flying Smurf	Peyo			Drinking Glass	4 3/8" High	Canada	$8.25-12.00	The glass has a small base, in the middle is a ridge and the top of the glass is bigger. The Smurf is on the top of the glass and only on one side. The Smurf is wearing a red cape, gloves, shorts and shoes. There is a red patch with a yellow "S" in the middle, on his tummy.
173/007	Trumpet	Peyo			Drinking Glass	4 3/8" High	Canada	$8.25-12.00	The glass has a small base, in the middle is a ridge and the top of the glass is bigger. The Smurf is on the top of the glass and only on one side. The Smurf looks like he's walking. He is playing a yellow trumpet.
173/008	Guard With A Lance	Peyo			Drinking Glass	4 3/8" High	Canada	$8.25-12.00	The glass has a small base, in the middle is a ridge and the top of the glass is bigger. The Smurf is on the top of the glass and only on one side. The Smurf is holding a spear. The blade is silver and the handle is yellow.
173/009	Prisoner Smurf	Peyo			Drinking Glass	4 3/8" High	Canada	$8.25-12.00	The glass has a small base, in the middle is a ridge and the top of the glass is bigger. The Smurf is on the top of the glass and only on one side. The prisoner is wearing white pants with black stripes and a white hat with black stripes. There is a maroon chain and ball attached to his ankle.
173/010	Papa Bandleader	Peyo			Drinking Glass	4 3/8" High	Canada	$8.25-12.00	The glass has a small base, in the middle is a ridge and the top of the glass is bigger. The Smurf is on the top of the glass and only on one side. Papa is standing with his hands in the air and one is holding a yellow conductors stick. In front of Papa is a yellow music stand.
173/011	Hockey Smurf	Peyo			Drinking Glass	4 3/8" High	Canada	$8.25-12.00	The glass has a small base, in the middle is a ridge and the top of the glass is bigger. The Smurf is on the top of the glass and only on one side. The Smurf is wearing a yellow and black hockey uniform and red gloves. There is a white hockey net behind him.
173/012	Untitled	Peyo			Drinking Glass	4 3/8" High			

Set 213: Glasses- Canadian Water

213/001	Sport Smurf	TM		Water Glass	5 1/2" High	Canada	$8.25-12.00	The glass has indents going around the bottom. The glass has a picture in the middle and only on one side. The Smurf is wearing a brown and yellow outfit and a white hat with a yellow star on the side. The Smurf is holding a yellow bat, white soccer ball, yellow tennis racket and a brown golf bag.
213/002	King Smurf	TM		Water Glass	5 1/2" High	Canada	$8.25-12.00	The glass has indents going around the bottom. The glass has a picture in the middle and only on one side. The Smurf is wearing a yellow hat, yellow crown, yellow pants and yellow shoes. The Smurf has a red robe with white trim. He is holding a red scepter with a white handle.
213/003	Fishing Smurf	TM		Water Glass	5 1/2" High	Canada	$8.25 12.00	The glass has indents going around the bottom. The glass has a picture in the middle and only on one side. The Smurf is sitting on a black stump. He is holding a yellow fishing rod with a red fish hanging on the end.
213/004	Drummer	TM		Water Glass	5 1/2" High	Canada	$8.25 12.00	The glass has indents going around the bottom. The glass has a picture in the middle and only on one side. The Smurf is standing, holding a white drum with red trim and 2 red drumsticks.
213/005	Lute	TM		Water Glass	5 1/2" High	Canada	$8.25-12.00	The glass has indents going around the bottom. The glass has a picture in the middle and only on one side. The Smurf is standing playing a large yellow lute.
213/006	Flying Smurf	TM		Water Glass	5 1/2" High	Canada	$8.25-12.00	The glass has indents going around the bottom. The glass has a picture in the middle and only on one side. The Smurf is wearing a red cape, gloves, shorts and shoes. There is a red patch with a yellow "S" in the middle, on his tummy.
213/007	Trumpet	TM		Water Glass	5 1/2" High	Canada	$8.25-12.00	The glass has indents going around the bottom. The glass has a picture in the middle and only on one side. The Smurf looks like he's walking. He is playing a yellow trumpet.
213/008	Guard With A Lance	TM		Water Glass	5 1/2" High	Canada	$8.25-12.00	The glass has indents going around the bottom. The glass has a picture in the middle and only on one side. The Smurf is holding a spear. The blade is silver and the handle is yellow.
213/009	Prisoner Smurf	TM		Water Glass	5 1/2" High	Canada	$8.25-12.00	The glass has indents going around the bottom. The glass has a picture in the middle and only on one side. The prisoner is wearing white pants with black stripes and a white hat with black stripes. There is a maroon chain and ball attached to his ankle.
213/010	Papa Band leader	TM		Water Glass	5 1/2" High	Canada	$8.25-12.00	The glass has indents going around the bottom. The glass has a picture in the middle and only on one side. Papa is standing with his hands in the air and one is holding a yellow conductors stick. In front of Papa is a yellow music stand.
213/011	Hockey Smurf	TM		Water Glass	5 1/2" High	Canada		The glass has indents going around the bottom. The glass has a picture in the middle and only on one side. The Smurf is wearing a yellow and black hockey uniform and red gloves. There is a white hockey net behind him.
213/012	Spy Smurf	TM		Water Glass	5 1/2" High	Canada	$8.25-12.00	The glass has indents going around the bottom. The glass has a picture in the middle and only on one side. The Smurf is wearing a red cape, white pants, white hat and a black mask. He has a finger in front of his mouth.

Set 235: Glasses- France

235/001	Smurf Running A Marathon	Benediction	1/1/84	Small Mustard Glass	4" High	France	$9.00-13.00	A Smurf in a yellow shirt and white shorts is panting. He is standing by a red and white mile marker. The Smurf looks like running. The other side has a Smurf holding a record. The Smurf is wearing a red tank and white shorts. There is a bench on one side with a towel on it. There is also a dumbbell laying by the bench.
235/002	Smurf Shaking a Referee	Benediction	1/1/84	Small Mustard Glass	4" High	France	$9.00-13.00	The glass has a Smurf shaking another Smurf. The Smurf that is getting shook is wearing a black referee uniform. There is a mushroom house on one side. There is only 1 picture on the glass.
235/003	Smurfs On A Roller coaster	Benediction	1/1/89	Small Mustard Glass	4" High	France	$9.00-13.00	The Smurf glass has 2 Smurfs and Smurfette in a rollercoaster. There is only 1 rollercoaster car in the picture. The other side of the glass has 6 Smurfs with fireworks behind them.
235/004	Smurfs Carrying A Cob Of Corn	Benediction	1/1/87	Small Mustard Glass	4" High	France	$9.00-13.00	The glass has 2 Smurfs carrying a cob of corn. Smurfette is pushing a wheelbarrow. Another Smurf is rolling a radish. The other side of the glass has Smurfette standing in the rain. A Smurf is standing under a mushroom and another Smurf is carrying a leaf over his head. There is a bolt of lightening in the sky.
235/005	Smurf Looking at Self In Water	Benediction	1/1/83	Small Mustard Glass	4" High	France	$9.00-13.00	The glass has a Smurf laying on a bank looking in the water admiring himself. The other side of the glass has a Smurf sitting in the grass fishing. He has a yellow fishing pole. There is a red bobber floating in the water.
235/006	Smurfette Giving Papa A Present	Benediction	1/1/87	Small Mustard Glass	4" High	France	$9.00-13.00	The glass has Smurfette giving Papa a present. There is another Smurf walking up with a box of flowers. They are outside. The background is yellow. The other side has Smurfette dancing . She has a blowout in her mouth. Another Smurf wearing a party hat is also dancing. There is a Smurf playing a violin. He has a lovesick expression on his face and 2 hearts are floating above his head.
235/007	5 Smurfs Dancing Around A Christmas Tree	Benediction	1/1/86	Small Mustard Glass	4" High	France	$9.00-13.00	The glass has 5 Smurfs dancing around a Christmas tree. Baby is sitting on the side of the tree with a rattle. Baby is wearing a pink sleeper. There are a few presents by the tree. The other side of the glass has a Smurf standing on the side of a snowman. There is another Smurf carrying a rolling pin chasing a Smurf with a cake through the snow.
235/008	2 Smurfs Rolling A Snowball Off A Snowdrift	Benediction	1/1/90	Small Mustard Glass	4" High	France	$9.00-13.00	The glass has 2 Smurfs pushing a snowball off a snowdrift. There is a Smurf below the drift wiping snowballs. The other side has a Smurf singing with a blown up present in front of him. A Smurf is carrying another present to Brainy. The picture has half of the Christmas tree in it.
235/009	Smurf Blowing Fire From His Mouth	Benediction	1/1/85	Small Mustard Glass	4" High	France	$9.00-13.00	The glass has a Smurf leaping in the air with a flame coming out his mouth. Papa is standing in front of him with a pepper shaker. Papa was putting pepper on the food. The other side of the glass has Smurfette walking carrying a pink umbrella. There is a mushroom house behind her. It's not raining.
235/010	Smurf Flying A Kite	Benediction	1/1/90	Small Mustard Glass	4" High	France	$9.00-13.00	The glass has 2 Smurfs flying a kite. Baby Smurf is sitting in the grass watching. The kite is red and yellow. The other side of the glass has Smurfette painting a picture. There is a Smurf raking leaves with a paintbrush. There is a Smurf hiding behind an easel. They are outside in the grass.
235/011	Smurfette Standing On Top Of A Birthday Cake (30)	Benediction	1/1/88	Small Mustard Glass	4" High	France	$9.00-13.00	The glass is for the 30th anniversary. The one side has 2 Smurfs walking through the grass with a big cake. Smurfette is standing on top of the cake. The other side has 17 Smurfs piled on top of each other. Both sides have the number 30 on it with green leaves around the numbers.
235/012	Smurf Giving Smurfette Flowers	Dupuis	1/1/76	Small Mustard Glass	4" High	France	$9.00-13.00	The glass has a Smurf giving Smurfette a bouquet of flowers. The Smurf has a shy expression on his face.
235/013	3 Smurfs At The Beach	Peyo	1/1/90	Small Mustard Glass	4" High	France	$9.00-13.00	Smurfette is sitting in the sand at the edge of the water. A Smurf in the water is trying to splash Smurfette. There is another Smurf laying on an orange beach towel sunbathing. Smurfette is wearing an orange suit.
235/014	Smurf Carrying A Wood Beam	Benediction		Small Mustard Glass	4" High	France	$9.00-13.00	The one side of the glass has a Smurf walking carrying a yellow wood beam and a red saw. The Smurf is wearing a white tank top and white shorts. The other side of the glass has Papa standing outside a mushroom house pointing.
235/015	Smurflings Splashing In A Puddle	Benediction	1/1/91	Small Mustard Glass	4" High	France	$9.00-13.00	The glass has Snappy splashing in a puddle trying to get Slouchy wet. Slouchy is standing on the side with a grim expression on his face. The other side of the glass has 2 Smurfs playing marbles.
235/016	Papa, Smurfette And A Smurf Ice Skating	Benediction	1/1/86	Small Mustard Glass	4" High	France	$9.00-13.00	The one side of the glass has Papa and Smurfette holding hands ice-skating. There is another Smurf in front of them laying sprawled out on the ice. The other side of the glass has Papa dressed as Santa walking across a bridge. There are 2 Smurfs waiting at the end of the bridge for Papa.
235/017	Le Schtroumpf A Lanettes (Brainy)	Amora	1/1/96	Small Mustard Glass	4" High	France	$9.00-13.00	The front of the glass has Brainy Smurf holding up a finger. There is a book in front of him, [illegible] There is a yellow background behind Brainy. The bottom of the glass says "Le Schtroumpf A Lanettes". The other side has a Smurf hitting Brainy over the head with a mallet. A bunny is sitting watching.
235/018	Le Schtriumpf Gourmand (Greedy)	Amora	1/1/96	Small Mustard Glass	4" High	France	$9.00-13.00	The front of the glass has a Smurf holding a cake. A butterfly is flying above the cake. There is a yellow background behind Greedy. The bottom of the glass says "Le Schtroumpf Goumand". The back of the glass has a Smurf sitting on a blanket outside eating cake.
235/019	Le Schtroumpf Coquet (Vanity)	Benediction	1/1/96	Small Mustard Glass	4" High	France	$9.00-13.00	The front of the glass has a Smurf posing with a cracked mirror laying at his side. There is a pink background behind Vanity. The bottom of the glass says "Le Schtroumpf Coquet". The back of the glass has Vanity standing in front of a full length mirror admiring himself. The mirror is outside.
235/020	Le Schtroumpf Farceur (Jokey)	Benediction	1/1/96	Small Mustard Glass	4" High	France	$9.00-13.00	The front of the glass has a Smurf holding a present. There is a light blue background behind Jokey. [illegible] Farceur". The back of the glass has a Smurf squirting another Smurf in the face with water from a squirt gun.
235/021	Le Schtroumpfette (Smurfette)	Benediction	1/1/96	Small Mustard Glass	4" High	France	$9.00-13.00	The front of the glass has Smurfette posing. There is a pink background behind Smurfette. The bottom of the glass says "Le Schtroumpfette". The back of the glass has a Smurf giving Smurfette a bouquet of flowers. There are mushroom houses in the background. There is a bird, bee, chipmunk and butterfly in the picture.
235/022	Le Schtroumpf Musicien (Trumpet Smurf)	Benediction	1/1/96	Small Mustard Glass	4" High	France	$9.00-13.00	[illegible] There is a blue background behind trumpet Smurf. The bottom of the glass says "Le Schtroumpf Musicien". The back of the glass has a Smurf playing a trumpet by a tree. On the other side of the tree a Smurf is standing with his fingers in his ears.
235/023	Smurf And Dragon Eating A Book	Benediction	1/1/91	Small Mustard Glass	4" High	France	$9.00-13.00	The glass has a green dragon eating a book. Brainy Smurf is trying to get the book away. The other side of the glass has 3 Smurfs dressed in blue firemen uniforms. 2 of the Smurfs are on a water kettle and another Smurf is pulling the big kettle.
235/024	Smurf And Smurfette Sitting On A Log	Benediction	1/1/86	Small Mustard Glass	4" High	France	$9.00-13.00	The one side of the glass has a Smurf and Smurfette sitting on a log. There is a broken heart floating above the Smurfs head. Brainy Smurf is approaching them. The other side of the glass has 2 Smurfs eating ice cream. One is sneezing and his ice cream is flying towards the other Smurf. There is a concession stand behind the Smurfs. Papa is standing looking with a worried expression on his face.
235/025	Papa Is Putting Glasses On A Mole	Benediction	1/1/91	Small Mustard Glass	4" High	France	$9.00-13.00	The glasses have Papa putting glasses on a black mole. The other side of the glass has Smurfette and Papa looking real old. There is a Smurf in his pajama's standing looking shocked.
235/026	Smurfs Shaking Acorns From A Tree	Benediction	1/1/92	Small Mustard Glass	4" High	France	$9.00-13.00	The glass has a Smurf kicking on to a tree and squirrel falling from a tree. Another Smurf is running so the squirrel doesn't fall on him. The other side of the glass has the Smurf shaking acorns from the tree. Another Smurf is standing on the side licking his lips.
235/027	Smurf Raking Leaves	Benediction	1/1/91	Small Mustard Glass	4" High	France	$9.00-13.00	The glass has a Smurf raking leaves. Another Smurf is sneezing and the leaves are flying all over. The other side of the glass has a Smurf and Smurfette carrying an umbrella. The umbrella is red and yellow.

235/028	3 Smurfling Make Sand Castle	Benediction	1/1/91	Small Mustard Glass	4" High	France	$9.00-13.00	The glass has Nat, Sassette and another Smurf making a sand castle. They are all wearing bathing suits. The other side of the glass has Nat opening a cage and letting butterfly's out. Another Smurf is carrying a net trying to catch the butterflies.
235/029	Puppy Chasing A Stick	Benediction	1/1/93	Small Mustard Glass	4" High	France	$9.00-13.00	The glass has a Smurf throwing a stick and Puppy is running after it. The other side of the glass has Puppy running next to the Smurf. The Smurf is walking back towards another Smurf with a mad expression on his face and he's carrying the stick.
235/030	Slouchy Laying Against Puppy	Benediction	1/1/90	Small Mustard Glass	4" High	France	$9.00-13.00	The glass has a brown dog laying down trying to sleep. A Smurf is laying against the dogs stomach. The Smurf is wearing a white floppy hat, red shirt and white pants. The other side of the glass has Nat and Snappy trying to give Sassette a flower. Nat is pushing Snappy away.
235/031	Smurf Lighting Firecrackers In A Mole Hole	Benediction	1/1/94	Small Mustard Glass	4" High	France	$9.00-13.00	The glass has a Smurf lighting a bunch of firecrackers in a mole hole. The Smurf is thinking of the mole while he's lighting the fuses. There is a Smurf laying on the ground with his ears covered. The other side of the glass a Smurf crawling out of the mole hole with soot all over his face while 2 other Smurfs look at him confusingly.
235/032	Smurf Walking Towards A Door (Toc Toc)	Benediction	1/1/94	Small Mustard Glass	4" High	France	$9.00-13.00	The glass has a Smurf walking towards a door. On the other side of the glass the Smurf has the door open and he's flying backwards to avoid getting hit with a tennis ball. There is a Smurf outside the door holding a tennis racket.
235/033	Smurfs Are Carrying Apples	Benediction	1/1/91	Small Mustard Glass	4" High	France	$9.00-13.00	The glass has 2 Smurfs carrying a green apple. Another Smurf is rolling an apple and Papa is standing on the side watching. The other side of the glass has a Smurf rolling an apple that has bites out of it. The Smurf is chewing. Papa is looking shocked.
235/034	Smurf Kicking A Goal (Soccer)	Benediction	1/1/90	Small Mustard Glass	4" High	France	$9.00-13.00	The glass has 2 Smurfs running after a soccer ball. One Smurf is wearing yellow and the other is in a red uniform. Papa is wearing a red referee outfit. The other side of the glass has the Smurf in the red laying on the ground kicking the ball backwards into the net. The Smurf in the yellow is by the net trying to catch the ball. There is another Smurf in red on the side cheering.
235/035	Papa Standing On A Mushroom (30)	Benediction	1/1/88	Small Mustard Glass	4" High	France	$9.00-13.00	The glass has Papa standing on a mushroom with his hand in the air. There are various Smurfs standing around in the grass. The corner says (30) with leaves around it. The other side of the glass has the Smurflings singing and dancing around a campfire. The # 30 is on the top.
235/036	Smurf Fishing	Benediction	1/1/93	Small Mustard Glass	4" High	France	$9.00-13.00	The glass has a Smurf sitting on a rock fishing. The Smurf looks happy. Another Smurf is swimming. The other side of the glass has the Smurf yanking the fishing line back with a troubled look on his face. The Smurf stopped swimming and has a worried look on his face.
235/037	2 Smurfs Downhill Skiing	Benediction	1/1/89	Small Mustard Glass	4" High	France	$9.00-13.00	The glass has 2 Smurfs downhill skiing. The other side of the glass has the Smurflings playing by 2 ducks in a pond. Nat is playing a lute. Snappy is carrying a flower over to Sassette. Slouchy is resting on a tree stump. The 2 ducks are in love.
235/038	1 Smurf Downhill Skiing (Texaco Glass)	Applause/Peyo	1/1/88	Small Mustard Glass	4" High	France	$9.00-13.00	The glass has a Smurf zooming down a ski hill. The Smurf is wearing a red scarf. The picture goes around the whole glass. There is a matching place mat listed under place mats. Promotion for Texaco.
235/039	Gargamel And Azreal (Texaco Glass)	Applause/Peyo	1/1/88	Small Mustard Glass	4" High	France	$9.00-13.00	The glass has Gargamel holding a net standing by a tree. Azreal is by his feet. The picture goes around the whole glass. There is a matching place mat listed under place mats. Promotion for Texaco.
235/040	Smurfette Ringing School Bell (Texaco Glass)	Applause/Peyo	1/1/88	Small Mustard Glass	4" High	France	$9.00-13.00	The glass has Smurfette standing in the grass ringing a bell. Smurfette is carrying a purse in her other hand. The glass has a pencil, ruler, chalkboard and leaves floating around it. The picture goes around the whole glass. There is a matching place mat listed under place mats. Promotion for Texaco.
235/041	Smurf With Bucket Of Fish (Texaco Glass)	Applause/Peyo	1/1/88	Small Mustard Glass	4" High	France	$9.00-13.00	The glass has a Smurf standing in the grass holding a fishing pole. There is a bucket of fish sitting in front of him. The Smurf is wearing an orange fishing vest. There is a picnic basket and an umbrella sitting behind him. The picture goes around the whole glass. There is a matching place mat listed under place mats. Promotion for Texaco.
235/042	Le Grande Schtroumpf (Papa)	Amora	1/1/96	Small Mustard Glass	4" High	Brussels	$9.00-13.00	The front of the glass has a portrait of Papa. In front of him are test-tubes. The bottom of the glass says "Le Grande Schtroumpf". The picture on the back of the glass has Papa working at a lab table. Papa is holding test-tubes. There is a chalkboard and a book in back of him.
235/043	Smurflings In A Ship	Benediction	1/1/84	Small Mustard Glass	4" High	France	$9.00-13.00	The glass has Nat, Snappy, Slouchy and Sassette in a ship. They look like their in a storm. The other side of the glass has Sassette sitting on a bank fishing. Nat and Snappy are playing ball. Slouchy is sleeping under a big red and white umbrella.
235/044	Smurfette Mermaid	BEELI	1/1/83	Small Mustard Glass	4" High	France	$9.00-13.00	The glass has Smurfette as a mermaid. Smurfette has a red fish tail. She is laying in the water. The other side of the glass has a Smurf laying on the beach sunbathing. The Smurf has a white with red flower umbrella over him.
235/045	2 Smurfs Carolers And Papa	Peyo	1/1/88	Small Mustard Glass	4" High	Brussels	$9.00-13.00	Papa is holding a conductor stick. A Smurf is singing. Another Smurf is holding a music sheet and singing. They are all standing in the snow.
235/046	Tennis Smurfette/Soccer Smurf	BEELI	1/1/83	Small Mustard Glass	4" High	France	$9.00-13.00	One side of the glass has Smurfette hitting a tennis ball. She is on a red tennis court. The other side has a Smurf kicking a soccer ball in the grass. The Smurf is wearing yellow and black shorts.
235/047	Smurfs Carrying A Huge Birthday Cake	Peyo	1/1/88	Small Mustard Glass	4" High	France	$9.00-13.00	2 Smurfs are carrying a huge birthday cake. Baby Smurf is sitting on the ground blowing out a candle on a small cake.
235/048	Smurf Village	Dupuis	1/1/76	Small Mustard Glass	4" High	France	$9.00-13.00	A bunch of Smurfs (12) are playing outside. There are several mushroom houses in the background.
235/049	Smurfette Roller-skating/Smurf Running	BEELI	1/1/83	Small Mustard Glass	4" High	France	$9.00-13.00	On one side of the glass Smurfette is roller-skating. The other side has a Smurf running through a finish line. The Smurf is wearing a yellow shirt and white shorts.
235/050	Smurfette Ballerina/Smurf Surfing	BEELI	1/1/83	Small Mustard Glass	4" High	France	$9.00-13.00	One side of the glass has Smurfette dressed as a ballerina. She is posing. The other side of the glass has a Smurf riding a red surfboard.
235/051	Smurf Pulling A Wood Sheep	Belokapi/Maille	1/1/87	Small Mustard Glass	4" High	France	$9.00-13.00	A Smurf is pulling a wood sheep. The sheep has white fur. Smurfette is standing, holding Baby Smurf. They are watching the Smurf.
235/052	Jokey Carrying Present To Brainy	Belokapi/Maille	1/1/87	Small Mustard Glass	4" High	France	$9.00-13.00	Jokey Smurf is carrying a present towards Brainy. Grouchy Smurf is in the back watching. Papa is also watching. There are 2 mushroom houses in the background.
235/053	Jokey Carrying Present To Brainy	Benediction	1/1/84	Small Mustard Glass	4" High	France	$9.00-13.00	Jokey Smurf is carrying a present towards Brainy. Grouchy Smurf is in the back watching. Papa is also watching. There are 2 mushroom houses in the background.
235/054	Smurf Writing A Script	Benediction	1/1/86	Small Mustard Glass	4" High	France	$9.00-13.00	One side of the glass has a Smurf laying in the grass writing a script. He is using a white feather pen and is writing on yellow paper. The other side has Papa holding a yellow stick pointing at a chalkboard.
235/055	Soccer Smurf Red Shirt	Publiart		Small Mustard Glass	4" High	France	$9.00-13.00	The glass has a Smurf kicking a soccer ball. The Smurf is wearing a red shirt and black shorts. He is kicking a white and black soccer ball.
235/056	Soccer Smurf Orange Shirt	Publiart		Small Mustard Glass	4" High	France	$9.00-13.00	The glass has a Smurf kicking a soccer ball. The Smurf is wearing an orange shirt and black shorts. He is kicking a white and black soccer ball.
235/057	Soccer Smurf Yellow Shirt	Publiart		Small Mustard Glass	4" High	France	$9.00-13.00	The glass has a Smurf kicking a soccer ball. The Smurf is wearing a yellow shirt and black shorts. He is kicking a white and black soccer ball.
235/058	Smurfette Doing Laundry	Benediction	1/1/92	Small Mustard Glass	4" High	France	$9.00-13.00	The one side of the glass has Smurfette washing laundry in a washtub and a Smurf is hanging it up. The other side of the glass has a Smurf glaring at Handy Smurf after Handy just wiped his hands on the clean laundry.
235/059	Smurf Pole Vaulting	Benediction	1/1/84	Small Mustard Glass	4" High	France	$9.00-13.00	The glass has a Smurf jumping over a beam. The Smurf is landing on a red pillow. The Smurf is wearing a yellow tank top and white shorts. There is another Smurf standing by a bush looking troubled. The picture goes around the glass.
235/060	Smurflings Under An Umbrella	Benediction	1/1/89	Small Mustard Glass	4" High	France	$9.00-13.00	One side of the glass has Sassette, Nat and Slouchy hiding under a big blue umbrella. Snappy is walking towards them carrying a mushroom for an umbrella. Snappy has a mad expression on his face. The other side of the glass has a Smurf carrying 2 cherry's. 2 Smurfs are carrying a big green apple. Another Smurf is carrying a mushroom.
235/061	Papa Teaching School	Benediction	1/1/90	Small Mustard Glass	4" High	France	$9.00-13.00	One side of the glass has Papa standing at a chalkboard teaching math. Sassette is sitting at a desk. Snappy Smurf is looking at Sassette with a mad look on his face. The other side has Smurfette standing in the rain with 2 Smurfs rushing towards her carrying umbrella's.
235/062	3 Smurfs Mountain Climbing	Benediction	1/1/91	Small Mustard Glass	4" High	France	$9.00-13.00	One side of the glass has 3 Smurfs mountain climbing. The other side of the glass has a Smurf and Smurfette on a bicycle for 2. The Smurf is sweating as Smurfette sits on the back enjoying the ride. They are at a 3 marker.
235/063	Smurfs Rafting	Benediction	1/1/83	Small Mustard Glass	4" High	France	$9.00-13.00	The glass has a Smurf in a yellow rubber raft and another Smurf in a red rubber raft.
235/064	3 Smurfs In A Houseboat	Benediction	1/1/85	Small Mustard Glass	4" High	France	$9.00-13.00	One side of the glass has 3 Smurfs in a brown wood houseboat. There is a Smurf at the front of the boat looking through a telescope. The other side of the glass has Smurfette giving Baby a bath. Baby is in a washtub throwing water at a Smurf.
235/065		Benediction	1/1/90	Small Mustard Glass	4" High		$0.00-0.00	
235/066	Nat Riding A Turtle	Benediction	1/1/90	Small Mustard Glass	4" High	France	$9.00-13.00	One side of the glass has Nat riding a turtle. He is hanging grass over the turtle. The other side has a Smurf watering a daisy. The Smurf is pumping water through the hose. Another Smurf is holding a Smurf up to smell the flower.
235/067	Smurf Carrying Cake	Benediction		Small Mustard Glass	4" High	France	$9.00-13.00	The one side of glass has a Smurf carrying a big cake. The other side has Smurfette carrying a purse. There is a butterfly flying above her head.
235/068	Smurf Holding Daisies	Benediction		Small Mustard Glass	4" High	France	$9.00-13.00	One side of the glass has a Smurf holding a bunch of daisies. There are heart's floating above his head. The other side has Smurfette standing in the grass posing.
235/069	Baby Smurf Painting	Benediction	1/1/85	Small Mustard Glass	4" High	France	$9.00-13.00	The one side of the glass has Baby Smurf painting. Baby has paint all over him. There is a Smurf in a painter jacket standing behind watching. The other side of the glass has Papa sitting on a log reading Baby a book. Baby is sitting in the grass.
235/070	Smurfs Roasting An Apple Over A Fire	Benediction	1/1/87	Small Mustard Glass	4" High	France	$9.00-13.00	One side of the glass has a Smurf poking an apple with a fork. The apple is roasting over a fire. There is another Smurf laying on the ground blowing at the fire. There is a Smurf standing licking his lips and watching. The other side has Baby Smurf sitting in a highchair with Smurfette standing next to him. A Smurf is carrying a big present to Baby.
235/071	Smurf Dreaming About A Devil And Angel	Benediction	1/1/94	Small Mustard Glass	4" High	France	$9.00-13.00	One side of the glass has a Smurf leaning on a shovel. The Smurf is dreaming of a devil Smurf leaning on a shovel resting and a angel Smurf prepping a flower. The other side of the glass has a Smurf standing by a broken pot. He is dreaming of a devil Smurf laughing and a angel Smurf looking worried.

Set 258: Glasses- Hardee's

258/001	Hardee's Manager Glass Stein	Anchor Hocking	8/9/82	Promotional Glasses	8"		$25.00-35.00	The stein is 8" high. The front has a picture of Papa, Smurfette and 3 Smurfs marching in a line. These steins were given by Anchor Hocking to Hardee's employees at headquarters who were instrumental in the 1982 Smurf Promotion. Has paper production label on bottom with the date 8/9/82, the requisition number 4455 and the item 1153 along with the initials of the person who approved it. Stein is 8" tall and weighs 3 pounds!
258/002	Papa Smurf Party Glass	W. Berrie/Peyo	1/1/83	Promotional Glasses	6"	U.S.A.	$3.50-5.00	The glass has a party scene. Papa is in front of a punch bowl getting a glass of punch. There are 2 other Smurfs running towards the punch bowl holding cups. There is another Smurf in front of a mushroom house holding a mug. There are streamers and balloons around. All the Smurfs are wearing party hats.

258/003	Harmony Smurf Party Glass	W. Berrie/Peyo	1/1/83		Promotional Glasses	6"	U.S.A.	$3.50-5.00	The Smurf is playing a yellow guitar. There are 3 Smurfs looking troubled and running away from Harmony. There are streamers and balloons around. All the Smurfs are wearing party hats.
258/004	Baker Smurf Party Glass	W. Berrie/Peyo	1/1/83		Promotional Glasses	6"	U.S.A.	$3.50-5.00	The front has a Smurf carrying a huge birthday cake. He is wearing a chefs hat and a bib. There are 2 Smurfs walking behind Baker carrying plates. A Smurf is licking his lips and another carrying a fork. There are streamers and balloons around. All the Smurfs are wearing party hats.
258/005	Smurfette Party Glass	W. Berrie/Peyo	1/1/83		Promotional Glasses	6"	U.S.A	$3.50-5.00	The front of the glass has Smurfette unwrapping a present. There is a Smurf carrying another present to Smurfette. 2 Smurfs are by an unwrapped present. There is a Smurf standing watching Smurfette. There are streamers around. All the Smurfs are wearing party hats.
258/006	Handy Smurf Party Glass	W. Berrie/Peyo	1/1/83		Promotional Glasses	6"	U.S.A.	$3.50-5.00	Handy Smurf is nailing a poster of Azreal to a tree. There are 4 Smurfs waiting to play pin the tail on Azreal. There are streamers around. All the Smurfs are wearing party hats.
258/007	Clumsy Smurf Party Glass	W. Berrie/Peyo	1/1/83		Promotional Glasses	6"	U.S.A.	$3.50-5.00	Clumsy is dressed as a clown. He tripped over something and the bucket of punch is flying through the air. There are 3 Smurfs that stopped playing ball and are watching. There are streamers around. All the Smurfs are wearing party hats.
258/008	Lazy	W. Berrie/Peyo	1/1/82		Promotional Glasses	6"	U.S.A.	$3.50-5.00	Lazy Smurf is laying sleeping under a mushroom. There is a Smurf pushing a wheelbarrow behind him. He has a mad expression on his face.
258/009	Grouchy Smurf	W. Berrie/Peyo	1/1/82		Promotional Glasses	6"	U.S.A	$3.50-5.00	Grouchy Smurf is carrying an I HATE MUSIC sign. Grouchy is walking away from 3 Smurfs playing instruments.
258/010	Jokey Smurf	W. Berrie/Peyo	1/1/82		Promotional Glasses	6"	U.S.A.	$3.50-5.00	Jokey Smurf is standing laughing at a Smurf that opened a exploding box.
258/011	Smurfette	W. Berrie/Peyo	1/1/82		Promotional Glasses	6"	U.S.A.	$3.50-5.00	Smurfette is standing posing. There are 4 lovesick Smurfs standing admiring her.
258/012	Papa	W. Berrie/Peyo	1/1/82		Promotional Glasses	6"	U.S.A.	$3.50-5.00	Papa is sitting on the ground holding a exploding test-tube. Papa has a worried expression on his face.
258/013	Gargamel And Azreal	W. Berrie/Peyo	1/1/82		Promotional Glasses	6"	U.S.A.	$3.50-5.00	Gargamel and Azreal are chasing a Smurf. The Smurf looks troubled and dropped his basket.
258/014	Brainy	W. Berrie/Peyo	1/1/82		Promotional Glasses	6"	U.S.A.	$3.50-5.00	Brainy is holding a red book at his side. He has a finger in the air. There is a Smurf walking up behind Brainy with a mallet. Another Smurf has his ears plugged and the third Smurf is looking sleepy.
258/015	Hefty	W. Berrie/Peyo	1/1/82		Promotional Glasses	6"	U.S.A.	$3.50-5.00	Hefty is standing by a wheelbarrow full of tools and wood. The wheelbarrow lost its wheel. There is a Smurf rolling the broken wheel up a hill towards the wheelbarrow. Hefty has a heart tattoo on his arm.

Set 263: Glasses- Miscellaneous

263/001	Albino Smurf				Drinking Glass	4 3/4"		$11.00-15.00	The Smurf has a white body. He is wearing a blue hat and blue pants.
263/002	Schtroumpfissime	Belokapi/Maille	1/1/83		Drinking Glass	4 3/4"	France	$9.00-13.00	The front of the glass has a King Smurf holding a scepter. The Smurf is wearing a yellow hat, yellow crown, yellow pants and a red with white trim robe.
263/003	Smurf Jumping				Large Glass	7"H x 4"W		$7.50-10.00	The glass is a pedestal type glass, the bottom is narrower than the top. The glass has a blue outline of a Smurf jumping. The Smurf has a solid blue shirt. The picture is on both sides.
263/004	Smurfette And Smurf				Large Glass	7"H x 4" W		$7.50-10.00	The glass is a pedestal type glass, the bottom is narrower than the top. The glass has a blue outline of Smurfette flirting and a Smurf standing behind her with a flower.

Set 224: Greeting Cards

224/001	A Birthday Wish Just For You	Ambassador Cards	1/1/86		Greeting Card	7 3/4"L x 5 1/4"W	U.S.A.	$1.50-2.00	The front of the card has Smurfette standing, holding a pink present. Smurfette looks like she is flirting. There are trees and flowers behind her.
224/002	Heel Veel Liets......	Interstat	1/1/95	Pink	Greeting Card	7"L x 4 3/4"W	Brussels	$3.00-4.00	The card is pink and has Smurfette holding white and orange daisies. Smurfette looks like she is flirting. The background is pink. A button is included with the card. A Smurf is drawing a picture of Smurfette. The background of the button is red. Included is a blue envelope.
224/003	Joyeux Anniversaire	Les Editions	1/1/89		Greeting Card	7"L x 5"W	France	$5.00-7.00	The card has a hard flat Smurf pasted on the front. The Smurf has a hand on his chest and one hand out at his side. The background is a table and stove with a chimney in a yellow room. The card comes with a white envelope.
224/004	Smurfs Celebrate Their 40th Birthday Post-o-gram	I.M.P.S. & Puppy	10/23/98		Greeting Card	6" x 8 1/4"	Brussels	$3.50-5.00	The front has the Smurf and Papa with a birthday cake. Inside is a typed paper that says" IMPS and Puppy are pleased to announce that today, 23 October 1998, the Smurfs celebrate their 40th birthday. .... The post-o-gram came in a white envelope from IMPS.

Set 265: Greeting Stands- German

265/001	Unser Goldstuck! Smurfette	Schleich		Light Blue	Greeting Stand	4"	Germany	$11.00-15.00	Smurfette is standing on a light blue rectangular stand. On the front of the stand it says "Unser Goldstruck!" in red letters. Smurfette is standing with one hand out at her side and the other on her chest. Behind Smurfette is a cardboard background. The background is yellow with pink hearts. In the big pink heart it says "Herzliche Geburtstagsgrube.
265/002	Bleib So Frisch Und Gesund! Apple Smurf	Schleich		White	Greeting Stand	4"	Germany	$11.00-15.00	The Smurf is standing on a white rectangular stand. On the front of the stand it says "Bleib So Frisch Und Gesund!" in green letters. The Smurf is standing, holding a green apple. Behind the Smurf is a cardboard background. The background is yellow with green clovers. In one of the clovers it says "Alles Gute Zum Geburtstag".
265/003	Verzeihung! Devil Smurf	Schleich		White	Greeting Stand	4"	Germany	$11.00-15.00	The Smurf is standing on a white rectangular stand. On the front of the stand it says "Verzeihung!" in blue letters. The Smurf has red skin and has a devil suit on. Behind the Smurf is a cardboard background. The background is blue sky and white clouds. In the large cloud on top it says "Was Ist In Mich Gefahren?" in purple letters.
265/004	Kopf Hoch! Baby With Bowl And Spoon	Schleich		Pink	Greeting Stand	4"	Germany	$11.00-15.00	Baby Smurf is sitting on a pink rectangular stand. On the front of the stand it says "Kopf hoch!" in blue letters. Baby Smurf is sitting, holding a red spoon in one hand and a pink bowl in the other. He has yellow food all over. Behind the Baby is a cardboard background. The background is blue sky and white clouds. In the large cloud on top it says "Wenn's auch manchmal drunter und druber geht..." in purple letters.
265/005	Happi Borsdai Tu Ju! Cowboy	Schleich		Yellow	Greeting Stand	4"	Germany	$11.00-15.00	The cowboy is standing on a yellow rectangular stand. The stand says Happi borsdai tu ju!" in red letters. The Smurf figure is wearing a brown hat, brown vest and red pants. He is holding a rope over his shoulder. There is a cardboard background behind him. The cardboard is pink with blue and yellow dots. The card says "Hallo" in red.
265/006	Du Kleiner Wilder! Jungle Smurf	Schleich		Green	Greeting Stand	4"	Germany	$11.00-15.00	Jungle Smurf is standing on a light green rectangular stand. The Smurf has a orange body with black hair. He is holding a spear. The stand says "Du Kleiner Wilder in green. There is a cardboard background behind the Smurf. It is purple with different colored balloons. One balloon says "Ich Nag Dich.." in green letters.
265/007	Wir Werden Alle Nicht Junger! Papa	Schleich		Light Blue	Greeting Stand	4"	Germany	$11.00-15.00	Papa is standing on a light blue rectangular stand. The stand says "Wir Werde Alle Nicht Junger!" in dark blue letters. Papa is standing with his hands out at his side. There is a cardboard background. The background is dark blue with green clovers. There is a white bubble on top that says "Herzliche Geburtstags Gribe" in blue letters.
265/008	Grober Meister! Bandleader	Schleich		Light Blue	Greeting Stand	4"	Germany	$11.00-15.00	The Smurf is standing on a light blue rectangular stand. The Smurf has his hands in the air. He is holding a black plastic wand in his right hand. The Smurf has a red bow tie around his neck. The Smurf is standing on a green grass like base. The stand says "Grober Meister!" in dark blue letters. There is a pink cardboard background. There are balloons in the background. One balloon says "[illegible]" in green letters.
265/009	Herzlichen Gluckwunsch! Mailman	Schleich		Pink	Greeting Stand	4"	Germany	$11.00-15.00	Mailman Smurf is standing on a pink stand. The stand says "Herzlichen Gluckwunsch! " in red letters. The Smurf is blowing a yellow horn. He is carrying a white envelope. There is a green and pink cardboard background. The Background says "Zur Venlobung" in red.
265/010	Du List Reirend Und Charmant! Ballerina	Schleich			Greeting Stand	4"	Germany	$11.00-15.00	Smurfette is dressed as a ballerina. She is standing on a yellow rectangular stand. The stand says "Du List Reirend Und Charmant!" in green letters There is a purple cardboard background. The background has balloons. A green balloon says "Mein Geburtstag- Kompliment..." in green letters.
265/011	Wende Necht Bald Wieder Flugge! Smurferman	Schleich			Greeting Stand	4"	Germany	$11.00 15.00	The Smurf is dressed as superman. He is wearing a red cape, red shoes, white wrist bands, white underpants, white hat and has a white emblem with an orange "S" in the middle of his chest. The Smurf is standing on a white base. He has his left hand in the air like he's ready to fly. The Smurf is standing on a white rectangular stand. The stand says "Wende Necht Bald Wieder Flugge! " in blue letters. The background has a white bubble that says ("Zur Genesung") in blue letters.
265/012	Der Dokter Wird's Schon Richten! Doctor	Schleich			Greeting Stand	4"	Germany	$11.00-15.00	The Smurf is wearing a white doctor's coat with silver buttons. He has a thermometer in his left hand. The thermometer has red and black lines. He has a silver and red stethoscope in his right hand. The Smurf is standing on a light blue rectangular stand. The stand says "Der Dokter Wird's Schon Richten!" in dark blue letters. There is a purple cardboard background that has balloons decorating it. A green balloon says "Gute Bessering!" in green letters.

Set 182: Greeting Stands- German Soccer Team

182/001	Ha- Ess- Vau!!	Schleich/Peyo	1/1/80		Rubber PVC's	3" High	West Germany	$45.00-65.00	The Smurf has a medium blue body. He has his arms out at his sides. His right foot has a white soccer ball on the tip. He is wearing a white shirt, blue socks, black shoes and red shorts. The Smurf is on a blue triangle stand. The stand says "Ha- Ess- Vau!!" in black letters.
182/002	Vau Eff Ell!!	Schleich/Peyo	1/1/80		Rubber PVC's	3" High	West Germany	$45.00-65.00	The Smurf has a medium blue body. He has his arms out at his sides. His right foot has a white soccer ball on the tip. He is wearing a white shirt, white socks, black shoes and white shorts. The shirt and pants have a green stripe on them. The Smurf is on a green triangle stand. The stand says "Vau Eff Ell!!" in black letters.
182/003	Heja, Heja, FCK! 1.Fc Koln	Schleich/Peyo	1/1/80		Rubber PVC's	3" High	West Germany	$45.00-65.00	The Smurf has a medium blue body. He has his arms out at his sides. His right foot has a white soccer ball on the tip. He is wearing a white shirt, red socks, black shoes and white shorts. The Smurf is on a yellow triangle stand. The stand says "Heja, Heja, FCK!! 1.FC Koln" in black letters.
182/004	Ole - Ole VfB	Schleich/Peyo	1/1/80		Rubber PVC's	3" High	West Germany	$45.00-65.00	The Smurf has a medium blue body. He has his arms out at his sides. His right foot has a white soccer ball on the tip. He is wearing a white shirt, white socks, black shoes and white shorts. The shirt and pants have a red stripe on them. The Smurf is on a red triangle stand. The stand says "Ole - Ole VfB" in white letters.

182/005	I bin a Bayern - Fan	Schleich/Peyo	1/1/80		Rubber PVC's	3" High	West Germany	$45.00-65.00	The Smurf has a medium blue body. He has his arms out at his sides. His right foot has a white soccer ball on the tip. He is wearing a red shirt, red socks, black shoes and white shorts. The Smurf is on a red triangle stand. The stand says "I bin a Bayern - Fan" in white letters.
182/006	EINTRACHT...! EINTRACHT...!	Schleich/Peyo	1/1/80		Rubber PVC's	3" High	West Germany	$45.00-65.00	The Smurf has a medium blue body. He has his arms out at his sides. His right foot has a white soccer ball on the tip. He is wearing a red shirt, black socks, black shoes and black shorts. The shirt has black stripes. The Smurf is on a red triangle stand. The stand says "EINTRACHT...! EINTRACHT...!" in white letters.

Set 238: Greeting Stands- Promotional Figures

238/001	Schleich Wunscht Viel Gluck Im Neven Jahr!!	Schleich/Peyo	1/1/79		Promotional Figure	3" High	West Germany	$75.00-100.00	The Smurf is standing on a pink triangle stand. The Smurf has a medium blue body. The Smurf is holding a pink pig in his hands. The stand says "Schleich Wunscht Viel Gluck Im Neven Jahr!!" in black letters.
238/002	Jo, Cule - Field Hockey	Schleich/Peyo	1/1/80	White	Promotional Figure	3" High	West Germany	$18.00-25.00	The Smurf has a dark blue body. He is wearing a blue and red stripe shirt, red socks, blue and red stripe shorts, a white hat and black shoes. He is holding a tan hockey stick with a white ball on the end of the stick. The Smurf is standing on a white stand that says "Jo, Cule " in black letters.
238/003	Bonne Fette! - Secretary	Schleich/Peyo	1/1/80	Blue	Promotional Figure	3" High	West Germany	$11.00-15.00	Smurfette has a medium blue body. She is wearing a pink dress, white hat and white shoes. Smurfette is holding a brown pencil in her right hand. Smurfette is holding a white pad with no writing and a silver binder edge in her left hand. Smurfette is standing on a blue stand that says "Bonne Fete!" in black letters.
238/004	Recuerdo de ANDALUCIA Sign- bearer Smurf	Schleich/Peyo	1/1/80	Green	Promotional Figure	3" High	West Germany	$22.50-30.00	The Smurf is standing on a green triangle stand. He is holding a gray sign with a green and white flag on top. The stand says (Recuerdo de ANDALUCIA) in black letters.
238/005	D'una estada a Catulunya - Sign bearer Smurf	Schleich/Peyo	1/1/80	Yellow	Promotional Figure	3" High	West Germany	$22.50-30.00	The Smurf is standing on a yellow triangle stand. He is holding a gray sign with a red and yellow striped flag. The stand says (Duna estada a Catulunya) in black letters.
238/006	EUSKADI Zuretzat - Sign bearer	Schleich/Peyo	1/1/80	Red	Promotional Figure	3" High	West Germany	$22.50-30.00	The Smurf is standing on a red triangle stand. He is holding a gray sign with a red flag with green and white lines. The stand says (EUSKADI Zuretzat) in black letters.

Set 219: Greeting Stands- Triangle

219/001	Dad You're Out Of This World - Astro	Schleich/Peyo/W.B.		Blue	Smurf On Stand	3" High	Hong Kong	$9.00-12.00	The Smurfs body is medium blue. He is wearing a white astronaut outfit with a white hat and a clear plastic helmet. The Smurf has his hands extended at his side, the left hand has his finger pointing up. He has small feet and the right foot is lifted a little. He is standing on a blue stand that says "Dad You're Out Of This World" in white letters.
219/002	I'm Blue Without You - Crying	Schleich/Peyo/W.B.		Yellow	Smurf On Stand	3" High	Hong Kong	$9.00-12.00	The Smurf has a medium blue body. A tear is rolling down his face and he is wiping his face with his right hand. In his left hand he is holding a yellow handkerchief. He has on white pants and a white hat. The Smurf is standing on a yellow stand that says "I'm Blue Without You" in black letter.
219/003	Mom You're Too Nice For Words - Author	Schleich/Peyo/W.B.		Blue	Smurf On Stand	3" High	Hong Kong	$9.00-12.00	The Smurf has a medium blue body. He is standing, holding a white feather pen with black feathers in his right hand. In his left hand he is holding a white scroll with no writing on it. He is wearing white pants and a white hat. He is looking up with his eyes. The Smurf is standing on a blue stand that says "Mom You're Too Nice For Words" in white letter.
219/004	Love Me Tender - Rock N Roll	Schleich/Peyo/W.B.		Yellow	Smurf On Stand	3" High	Hong Kong	$9.00-12.00	The Smurf has a medium blue body. The Smurf is holding a light orange guitar with yellow strings. He has his eyes shut and his mouth open. The Smurf is standing on a yellow stand that says "Love Me Tender" in black letters.
219/005	You're First Class - Postman	Schleich/Peyo/W.B.		Yellow	Smurf On Stand	3" High	Hong Kong	$9.00-12.00	The Smurf has a light blue body. He is holding a yellow horn in his mouth with his right hand. He has a dark brown mailbag. He is holding a white envelope with a red heart on the flap of the envelope in his left hand. The Smurf is standing on a yellow stand that says "You're First Class" in black letters.
219/006	This Place Is Like A 3-Ring Circus - Clown	Schleich/Peyo/W.B.		Red	Smurf On Stand	3" High	Hong Kong	$9.00-12.00	The Smurf has a dark blue body. He is wearing yellow with red striped pants, black suspenders, a red bow tie, white shoes with black toes, a white shirt, white gloves and a white hat. He has a red mouth with white around it. The Smurf is standing on a red stand that says "This Place Is Like A 3-Ring Circus" in orange letters.
219/007	World's Greatest Soccer Player - Soccer Football	Schleich/Peyo/W.B.	1/1/78	Red	Smurf On Stand	3" High	Hong Kong	$9.00-12.00	The Smurf has a dark blue body. He has his arms out at his sides. His right foot has a white soccer ball on the tip. He is wearing a red shirt, red socks, yellow shorts and black shoes. The Smurf is standing on a red stand that says "World's Greatest Soccer Player" in orange letters.
219/008	Doctor's Orders... Get Well Soon!! - Doctor	Schleich/Peyo/W.B.		Blue	Smurf On Stand	3" High	Hong Kong	$9.00-12.00	The Smurf has a medium blue body. The Smurf is wearing a white doctor's coat with silver buttons. He has a thermometer in his left hand. The thermometer has blue lines. He has a silver and red stethoscope in his right hand. The Smurf is standing on a blue stand that says "Doctor's Orders... Get Well Soon!!" in white letters.
219/009	Happy Smurfday To You - Singer	Schleich/Peyo/W.B.		Blue	Smurf On Stand	3" High	Hong Kong	$9.00-12.00	The Smurf has a medium blue body. He is holding a yellow music sheet with both hands. The music sheet has notes on both sides. The Smurf is standing on a blue stand that says "Happy Smurfday To You" in white letters.
219/010	Happy Birthday - Gift	Schleich/Peyo/W.B.		Red	Smurf On Stand	3" High	Hong Kong	$9.00-12.00	The Smurf has a dark blue body. He is holding a yellow gift with a red bow in his left hand. He has a yellow and two red flowers in the other. The Smurf is standing on a red stand that says "Happy Birthday" in orange letters.
219/011	Happy Mother's Day - Gift	Schleich/Peyo/W.B.		Blue	Smurf On Stand	3" High	Hong Kong	$9.00-12.00	The Smurf has a dark blue body. He is holding a yellow gift with a red bow in his left hand. He has a yellow and two red flowers in the other. The Smurf is standing on a blue stand that says "Happy Mother's Day" in white letters.
219/012	Wanna Smurf Around? - Lover	Schleich/Peyo/W.B.	1/1/80	Blue	Smurf On Stand	3" High	Hong Kong	$9.00-12.00	The Smurf has a dark blue body. The Smurf is holding 3 red flowers with yellow centers and a bright green stem. The Smurf is holding the flowers in his left hand. He has his head turned to the right like he's posing. The Smurf is standing on a blue stand that says "Wanna Smurf Around?" in white letters.
219/013	I Love You - Lover	Schleich/Peyo/W.B.	1/1/78	Red	Smurf On Stand	3" High	Hong Kong	$9.00-12.00	The Smurf has a medium blue body. The Smurf is holding 3 reddish/orange flowers with yellow centers and a light green stem. The Smurf is holding the flowers in his left hand. He has his head turned to the right like he's posing. The Smurf is standing on a red stand that says "I Love You" in orange letters.
219/014	I Love You - Lover	Schleich/Peyo/W.B.		Red	Smurf On Stand	3" High	Hong Kong	$9.00-12.00	The Smurf has a dark blue body. The Smurf is holding 3 dark red flowers with yellow centers and a dark green stem. The Smurf is holding the flowers in his left hand. He has his head turned to the right like he's posing. The Smurf is standing on a red stand that says "I Love You" in orange letters.
219/015	World's Greatest Tennis Player - Tennis Star	Schleich/Peyo/W.B.		Red	Smurf On Stand	3" High	Hong Kong	$9.00-12.00	The Smurf has a dark blue body. He is wearing a white tennis shirt and white shorts. He has a yellow wrist band on his right wrist. He is holding a red plastic tennis racket in his right hand. The tennis racket has holes in it. The Smurf is standing on a red stand that says "World's Greatest Tennis Player" in orange letters.
219/016	There's A Word For A Person Like You - Terrific! - Pointing	Schleich/Peyo/W.B.	1/1/78	Blue	Smurf On Stand	3" High	Hong Kong	$9.00-12.00	The Smurf has a dark blue shiny body. He is standing, holding his left hand out pointing at something. The Smurf is standing on a blue stand that says "There's A Word For A Person Like You - Terrific!" in white letters.
219/017	World's Greatest Bowler - Bowler	Schleich/Peyo/W.B.		Red	Smurf On Stand	3" High	Hong Kong	$9.00-12.00	The Smurf has a dark blue body. He is holding a red bowling ball in his right hand. He is standing like he's going to throw the ball. The Smurf is standing on a red stand that says "World's Greatest Bowler" in orange letters.
219/018	You're My Favorite Flavor - Ice Lolly	Schleich/Peyo/W.B.		Yellow	Smurf On Stand	3" High	Hong Kong	$9.00-12.00	The Smurf has a dark blue body. He is eating a popsicle. The popsicle is red on the bottom and white on top. The popsicle is on a yellow stick. The Smurf is holding the popsicle in his right hand. He has his tongue sticking out of his mouth. The Smurf is standing on a yellow stand that says "You're My Favorite Flavor" in black letters.
219/019	Let's Patch Things Up - First Aid	Schleich/Peyo/W.B.	1/1/78	Blue	Smurf On Stand	3" High	Hong Kong	$9.00-12.00	The Smurf has a dark blue body. He is wearing a white uniform. The Smurf has a white bandage wrapped around a finger on his right hand. He is carrying a white medical bag with a red cross on it. The Smurf is standing on a blue stand that says "Let's Patch Things Up" in white letters.
219/020	World's Greatest Golfer - Golfer	Schleich/Peyo/W.B.		Yellow	Smurf On Stand	3" High	Hong Kong	$9.00-12.00	The Smurf has a dark blue body. He is wearing yellow pants, red socks, black shoes and no shirt. The Smurf is holding a gray golf club. The Smurf is standing on a yellow stand that says "World's Greatest Golfer" in black letters.
219/021	We Make A Great Pair - Card Player	Schleich/Peyo/W.B.	1/1/79	Blue	Smurf On Stand	3" High	Hong Kong	$9.00-12.00	The Smurf has a dark blue body. He has his left hand over his mouth like he's laughing. His left eye is closed and his right eye is open. He is holding 3 cards in his right hand, the face card is a heart. He has a club card stuffed in the front of his pants. The Smurf is standing on a blue stand that says "We Make A Great Pair" in white letters.
219/022	You're #1 In My Book - Teacher	Schleich/Peyo/W.B.		Yellow	Smurf On Stand	3" High	Hong Kong	$9.00-12.00	The Smurf has a dark blue body. He is holding a red book with white pages, black letters and a yellow bookmark in his right hand. He is holding a finger up in the air on his left hand. His eye's are shut and his mouth is open like he's declaring something. The Smurf is standing on a yellow stand that says "You're #1 In My Book" in black letters.
219/023	You Light Up My Nights - Candle	Schleich/Peyo/W.B.		Blue	Smurf On Stand	3" High	Hong Kong	$9.00-12.00	The Smurf has a dark blue body. He is wearing a white night shirt, brown shoes and a white hat with a red tassel. He is holding a candle in his right hand. The candle is white with a yellow fire The candle is sitting on a brown plate. The Smurf has his eye's open. The Smurf is standing on a blue stand that says "You Light Up My Nights" in white letters.
219/024	Reach Out And Touch Someone... Me! - Telephone	Schleich/Peyo/W.B.		Yellow	Smurf On Stand	3" High	Hong Kong	$9.00-12.00	The Smurf has a dark blue body. The phone is red with gold. The cradle is small and copper. The hand piece is copper and he is holding it to his mouth with his right hand. The Smurf is standing on a yellow stand that says "Reach Out And Touch Someone... Me! " in black letters.
219/025	Don't Hurry, Don't Rush... Do It Right When You Brush!	Schleich/Peyo/W.B.		Yellow	Smurf On Stand	3" High	Hong Kong	$9.00-12.00	The Smurf has a dark blue body. He is standing brushing his teeth with a yellow toothbrush. In his left hand he is holding a red tube of toothpaste. The toothpaste container has a white line on the front. The Smurf is standing on a yellow stand that says "Don't Hurry, Don't Rush.. Do It Right When You Brush!" in black letters.
219/026	Congratulations! I'll Drink To That - Beer	Schleich/Peyo/W.B.		Blue	Smurf On Stand	3" High	Hong Kong	$9.00-12.00	The Smurf has a dark blue body. He is holding a big yellow/orange mug of beer in his right hand. He has a big red mouth. The Smurf is standing on a blue stand that says "Congratulations! I'll Drink To That" in white letters.
219/027	I've Got A Surprise For... You Happy Birthday! - Present	Schleich/Peyo/W.B.		Red	Smurf On Stand	3" High	Hong Kong	$9.00-12.00	The Smurf has a dark blue body. He is holding a white present with red ribbon. The Smurf is holding the present with both hands off to his right side. The Smurf is standing on a red stand that says "I've Got A Surprise For... You Happy Birthday!" in orange letters.

219/028	Have A Yummy Birthday - Cake	Schleich/Peyo/W.B.		Red	Smurf On Stand	3" High	Hong Kong	$9.00-12.00	The Smurf has a dark blue body. He is holding a dark red cake with white frosting. The cake is on a yellow plate. The Smurf is licking his lips. His left foot is raised a little. The Smurf is standing on a red stand that says "Have A Yummy Birthday" in orange letters.
219/029	Haven't Heard From You... Drop Me A Line - Angler	Schleich/Peyo/W.B.		Yellow	Smurf On Stand	3" High	Hong Kong	$9.00-12.00	The Smurf has a dark blue body. He is standing, holding a tan rubber fishing pole. The fishing pole has a thick yellow line with a black hook that's attached to his right foot. The fishing pole is molded to him. The Smurf is standing on a yellow stand that says "Haven't Heard From You... Drop Me A Line" in black letters.
219/030	Sorry You're Under The Weather... - Umbrella	Schleich/Peyo/W.B.		Yellow	Smurf On Stand	3" High	Hong Kong	$9.00-12.00	The Smurf has a medium blue body. He is standing under an orange/red mushroom with white spots and an orange stem. He has a worried expression on his face. The Smurf is standing on a yellow stand that says "Sorry You're Under The Weather Get Well Soon" in black letters.
219/031	It's So Much Fun...-Roller-skate Girl	Schleich/Peyo/W.B.	1/1/80	Yellow	Smurf On Stand	3" High	Hong Kong	$9.00-12.00	Smurfette has a dark blue body. She is wearing a white dress and white and red roller-skates. Smurfette's hair is bright yellow and has red ribbons in it. Smurfette is standing on a yellow stand that says "It's So Much Fun To Smurf Around With You" in black letters.
219/032	I Love You Very Much! - Amour	Schleich/Peyo/W.B.		Blue	Smurf On Stand	3" High	Hong Kong	$9.00-12.00	The Smurf has a medium blue body. In his right hand he is holding a red heart with a yellow arrow through it. In his left hand he is holding a yellow bow. The Smurf has white wings and a white cloth on the front of his body. The Smurf is standing on a blue stand that says "I Love You Very Much!" in white letters.
219/033	I'm Aiming For You're Heart - Amour	Schleich/Peyo/W.B.		Red	Smurf On Stand	3" High	Hong Kong	$9.00-12.00	The Smurf has a medium blue body. In his right hand he is holding a red heart with an tanish/orange arrow through it. In his left hand he is holding and tanish/orange bow. The Smurf has white wings and a white cloth on the front of his body. The Smurf is standing on a red stand that says "I'm Aiming For You're Heart" in white letters.
219/034	World's Greatest Baseball Player - Baseball Batter	Schleich/Peyo/W.B.		Blue	Smurf On Stand	3" High	Hong Kong	$9.00-12.00	The Smurf has a dark blue body. He is wearing a white uniform with red sleeves, a red belt, red socks, a white hat with a red visor and black shoes. The Smurf is holding a light cream bat. The Smurf is standing on a blue stand that says "World's Greatest Baseball Player" in white letters.
219/035	Telephone - Moi! - Telephone	Schleich	1/1/80	White	Smurf On Stand	3" High	West Germany	$13.50-18.00	The Smurf has a dark blue body. The phone is red with gold. The cradle is small and copper. The hand piece is copper and he is holding it to his mouth with his right hand. The Smurf is standing on a white stand that says "Telephone - Moi! " in black letters.
219/036	Being With You Is An Upper!	Schleich	1/1/80	Red	Smurf On Stand	3" High	Hong Kong	$9.00-12.00	The Smurf has a medium blue body. The Smurf has a red and yellow hand-glider with dark brown straps. The Smurf is standing on a red stand that says "Being With You is An Upper! " in orange letters.
219/037	World's Best Secretary - Secretary	Schleich/Peyo/W.B.		Yellow	Smurf On Stand	3" High	Hong Kong	$9.00-12.00	Smurfette has a dark blue body. She is wearing a white dress, white hat and white shoes. Smurfette is holding a brown pencil in her right hand. Smurfette is holding a white pad with black writing and a silver binder edge in her left hand. Smurfette is standing on a yellow stand that says "World's Greatest Secretary" in black letters.
219/038	Happy Birthday - First Aid	Schleich	1/1/80	Yellow	Smurf On Stand	3" High	Hong Kong	$9.00-12.00	The Smurf has a medium blue body. He is wearing a white doctors uniform. The Smurf is carrying a white doctors bag. The Smurf has a white bandage around his finger. The Smurf is standing on a yellow stand. The stand says "Happy Birthday" in black letters.
219/039	I Love Soccer - Soccer Smurfette	Schleich/Peyo/W.B.	1/1/83	Red	Smurf On Stand	3" High	Hong Kong	$9.00-12.00	Smurfette has a medium blue body. She is wearing a white/light orange/ red shirt, shorts, shoes and red socks. Smurfette has a white soccer ball attached to her left foot. Smurfette is standing on a red stand that says "I Love Soccer" in orange letters.
219/040	You're A Knockout! - Boxer	Schleich/Peyo/W.B.		Blue	Smurf On Stand	3" High	Hong Kong	$9.00-12.00	The Smurf has a dark blue body. He is wearing a red tank top, dark green shorts, black shoes, white socks and dark brown and white boxing gloves. The Smurf is standing on a blue stand that says "You're A Knockout!" in white letters.
219/041	Fisherman's Wharf Monterey, California - Fisherman	Schleich/Peyo/W.B.	1/1/79	Red	Smurf On Stand	3" High	Hong Kong	$11.00-15.00	The Smurf has a medium blue body. He is standing, holding a tan rubber fishing pole. The fishing pole has a thick yellow line with a black hook that's attached to his right foot. The fishing pole is molded to him. The Smurf is standing on a red stand that says "Fisherman's Wharf Monterey, California" in orange letters.
219/042	You're Number One In My Book! - Teacher	Schleich	1/1/80	Red	Smurf On Stand	3" High	West Germany	$9.00-12.00	The Smurf has a medium blue body. He is holding a dark red book with white pages, black letters and a yellow bookmark in his right hand. He is holding a finger up in the air on his left hand. His eye's are shut and his mouth like he's declaring something. The Smurf is standing on a red stand that

Set 84: Hair Accessories

084/001	Roller-skating Smurfette Brush	W. Berrie	1/1/83	Pink	Plastic Brush	5" Long	Hong Kong	$2.00-3.00	The hair brush is pink with purple hearts all over. The end of the brush where the bristles are is in the shape of a heart and has a purple heart in the middle with Smurfette on roller-skates. The brush says Smurfette in white letters with purple outline. The bristles are clear plastic.
084/002	Roller-skating Smurfette Comb	W. Berrie	1/1/83	Pink	Plastic Comb	3"	Hong Kong	$3.00-4.00	The comb is pink with purple hearts all over. The comb is the shape of a heart and has a purple heart in the middle with Smurfette on roller-skates. The comb says Smurfette in white letters with purple outline.
084/003	Ballerina Smurfette Ponytail Holders	Applause	1/1/84	Pink	Ponytail Holders	1"	Taiwan	$3.00-4.00	The ponytail holders are pink rubber bands with a plastic heart attached. The hearts are pink and have Smurfette in a ballerina pose. There are 2 in the package.
084/004	Ballerina Smurfette Barrettes	Applause	1/1/84	Purple/Pink	Barrettes	2 1/2" Long	Taiwan	$3.00-4.00	The barrettes are purple and say Smurfette in white lettering. On the end of the barrettes are pink, plastic hearts. The hearts have Smurfette in a ballerina pose. There are 2 barrettes in the package.
084/005	Horn Smurf Barrettes	Howard Eldon, Ltd.			Barrettes	1 1/2" High	Taiwan	$3.00-4.00	The barrettes have a Smurf blowing a yellow horn. The Smurf looks like he is marching. The barrettes are plastic and have metal clips on the back. There are 2 barrettes in the package.
084/006	Smurf With His Hand In The Air Barrettes	Howard Eldon, Ltd.			Barrettes	1 1/2" High	Taiwan	$3.00-4.00	The barrettes have a Smurf that looks like he's walking and he has his left hand out in front of him. The barrettes are plastic with metal clips on the back. There are 2 barrettes in the package.
084/007	Smurfette & Smurf On Mushroom Barrette	Howard Eldon, Ltd.	1/1/82		Barrettes	1"	Taiwan	$3.00-4.00	The barrettes have Smurfette dressed in a red dress standing with her back to a mushroom. The mushroom is yellow with red spots and has a Smurf sitting on top. The Smurf has a red lip mark on his hat. The barrettes are plastic with metal clips on the back. There are 2 barrettes in the package.
084/008	Go Smurf Go! Football Player Brush	Peyo		Yellow	Plastic Brush	2 1/2"W X 4 1/2"L	U.S.A.	$1.50-2.00	The brush is square and has clear plastic bristles on the back. The brush is yellow with an blue outline of a Smurf with a football running on the front.
084/009	Howdy Smurf Mirror	A.R.C.		White	Hand Mirror & Comb	7 1/4"	U.S.A.	$3.00-4.00	The mirror is white, round and has a thin handle. There is a picture of a Smurf dressed as a cowboy and Smurfette as a cowgirl, looking at each other. Above the picture it says "Howdy Smurf" in black letters. There is a white comb with a thin handle.
084/010	It's Not Easy Being A Smurfette Mirror	A.R.C.	1/1/82	White	Hand Mirror & Comb	7 1/4"	U.S.A.	$4.50-6.00	The mirror is white, round and has a thin handle. The picture is of Smurfette standing with a daisy on both sides of her.
084/011	Just Smurfin' Along Mirror	A.R.C.		Yellow	Hand Mirror	7 1/4"	U.S.A.	$2.00-3.00	The mirror is yellow, round and has a thin handle. The picture is of Smurfette carrying a suitcase. Above the picture it says "Just Smurfin' Along" in red letters.
084/012	Have A Smurfy Day Mirror	A.R.C.	1/1/82	Yellow	Hand Mirror	7 1/4"	U.S.A.	$4.50-6.00	The mirror is yellow, round and has a thin handle. The picture is of a Smurf sitting on a cloud with a rainbow behind. Inside the cloud it says "Have A Smurfy Day" in red letters.
084/013	Boy's Baseball Play 'N Groom Set	A.R.C.		Yellow	Grooming Set		U.S.A.	$11.00-15.00	The groom set comes with a yellow, glove- shaped sponge with a picture of a Smurf carrying a baseball bat. The hairbrush is yellow and is the shape of a bat. The soap is cream color and the shape of a baseball. There is a yellow comb. The box is blue and has a picture of the Smurfs playing baseball.
084/014	Just Smurfin' Along Travel Twins	A.R.C.	1/1/82	White	Mirror & Comb	5"	Hong Kong	$9.00-12.00	The mirror is round and has Smurfette carrying a blue suitcase with the words "Just Smurfin' Along" in red letters. The handle on the mirror is round, the comb is white and looks like a banana comb. The box is blue and has a Smurf and Smurfette traveling in the corner.
084/015	Smurf Face Ponytail Holder				Ponytail Holder	1"		$1.00-2.00	The rubber band is red and on 2 ends there is a plastic Smurf face.
084/016	Smurfette Standing With Flowers Headband	Applause	1/1/84	Pink	Headband		Taiwan	$2.00-3.00	The headband is pink and has 2 purple hearts on the top. Each purple heart has a picture of Smurfette standing, holding a yellow daisy.
084/017	Cheerleader Ponytail Holders	Howard Eldon, Ltd.	1/1/82		Ponytail Holders	1"	Taiwan	$3.00-4.00	The ponytail holders are 2 yellow plastic hearts. The hearts each have Smurfette dressed in a red and white cheerleader outfit. The rubber bands are black.
084/018	Tennis Smurfette Barrette			Blue	Barrette Clip	3"		$1.50-2.00	The barrette is blue. It has a Smurfette rubber figure glued on. Smurfette is wearing a white tennis dress and white shoes. She is holding a red tennis racket.
084/019	Smurfette Flirting Ponytail Holders	Howard Eldon, Ltd	1/1/82		Ponytail Holders	1 1/2"	Taiwan	$3.00-4.00	The rubber bands are black with plastic Smurfette's on the end. Smurfette looks like she's flirting. Smurfette has her hands at her sides. Smurfette is wearing a red dress and white hat.
084/020	Smurfette Deluxe Dresser Set (Cheerleader)	W. Berrie/Peyo	1/1/82	White	Dresser Set		U.S.A.	$7.50-10.00	The set includes a brush, mirror and comb. The mirror is white with Smurfette in a pink cheerleader outfit on the back. Smurfette is holding 2 yellow pompoms. The brush has Smurfette written in yellow letters with a blue megaphone sitting next to her name.
084/021	Kings Dominion Smurfette Roller-skating Comb	W. Berrie/Peyo	1/24/84	Clear	Comb	5 3/4"	U.S.A.	$2.00-3.00	The comb is clear. On the end is a picture of Smurfette. Smurfette is wearing white roller-skates. There is a purple heart behind her. Above the heart it says "Kings Dominion" in black letters.
084/022	Smurf Walking Ponytail Holders	Howard Eldon, Ltd.	1/1/82	Blue	Ponytail Holder	1"	Taiwan	$3.00-4.00	The ponytail holder is blue. They are heart- shaped. They have a Smurf walking with his hands out in front of him.
084/023	Smurf's Up! Comb And Mirror	Applause/Peyo		White	Comb And Mirror	2 3/4" x 4 3/4"	Taiwan	$5.00-7.00	The comb is inside a compact. There is a mirror on one side and a pocket for the comb on the other. The cover has a Smurf on a yellow surfboard. The cover says "Smurf's Up!" in red letters.
084/024	Smurf 6-Piece Deluxe Hair Care Dresser Set	A.R.C.	1/1/82	White	Dresser Set		U.S.A.	$11.00-15.00	The set includes 2 white hair combs, a white brush that says Smurf on the back in blue letters, a white comb, a white hand mirror and a white trinket box. The mirror has a picture of Smurfette on the back . It says " It's Not Easy Being A Smurfette". The trinket box has Smurfette on a balcony with a Smurf below serenading her. The box is heart- shaped. It says "If You Love Someone ...Tell Them".
084/025	Papa Smurf Bobby Pins	Howard Eldon, Ltd.	1/1/80		Bobby Pins		Taiwan	$2.00-3.00	The end of the bobby pin has a metal charm of Papa Smurf standing and pointing. The bobby pin is gold. There is a set of 2.
084/026	Smurfette Bobby Pins	Howard Eldon Ltd	1/1/80		Bobby Pins		Taiwan	$2.00-3.00	The end of the bobby pin has a metal charm of Smurfette in a red dress. The bobby pin is gold. There is a set of 2.
084/027	Smurf Holding Daisy Bobby Pins	Howard Eldon Ltd	1/1/80		Bobby Pins		Taiwan	$2.00-3.00	The end of the bobby pin has a metal charm of a Smurf standing, holding a yellow flower behind his back. The Smurf has a red shirt and white pants. The bobby pin is gold. There is a set of 2.
084/028	Smurf Playing Horn White Barrettes	Howard Eldon Ltd	1/1/80		Barrettes		Taiwan	$2.00-3.00	The barrettes have a Smurf blowing a yellow horn. The Smurf looks like he is marching. The barrettes are white plastic. There are 2 barrettes in the package.
084/029	Smurf With Hands Behind Back Ponytail Holder	Peyo	1/1/82		Ponytail Holder	1"	Taiwan	$1.50-2.00	The ponytail holder is blue. They are heart- shaped. They have a Smurf walking with his hands behind his back.
084/030	Ceramic Smurf Face Ponytail Holder				Ponytail Holder	1"		$1.00-2.00	The ponytail holder is a Smurf face. It is ceramic. The Smurf has a dark blue face. He has a red tongue. The rubber band is red.

084/031	Cube Ponytail Holder				Ponytail Holder			$1.00-2.00	There is a cube on the end of a red rubber band. The cube is 6- sided. Each side has a different sport Smurf.
084/032	Smurfette In Pink Ponytail Holder	Peyo			Ponytail Holder	1 1/2"	Brussels	$1.00-2.00	The ponytail holder is blue elastic cloth. It has a plastic Smurfette glued to the front. Smurfette is wearing a pink shirt, red with pink star skirt, white shoes and a white hat with a red bow.

Set 61: Halloween Costumes & Masks

061/001	Smurfette Costume & Mask	Ben Cooper	1/1/82		Vinyl		Taiwan	$11.00-15.00	A white costume dress with blue sleeves. The costume has Smurfette sliding down a rainbow. Smurfette mask.
061/002	Smurf Costume & Mask	Ben Cooper	1/1/82		Vinyl		Taiwan	$11.00-15.00	The top of the costume has a big Smurf with 4 Smurfs around him. The top of the costume is red, white and yellow. The bottom of the costume is blue. Smurf mask.
061/003	Papa Smurf Costume	Ben Cooper			Vinyl	12 - 14	Taiwan	$7.50-10.00	The costume is red pants and a blue top. On the front of the shirt is Papa Smurf walking carrying a pumpkin basket. It says Papa Smurf in yellow letters above Papa.

Set 129: Happy Meal Boxes

129/001	Papa With Mushroom House	Benditt Marketing	1/1/88		Hardee's Box		U.S.A.	$3.00-4.00	The box is green and blue. The box has Papa standing outside a mushroom house on the front side. The back has a wishing well with 2 Smurfs, Smurfette and Baby Smurf around it. The sides have a Smurf chopping a tree and the other side has 6 Smurfs singing.
129/002	Baby Smurf Sitting Under A Tree	Benditt Marketing	1/1/88		Hardee's Box		U.S.A.	$3.00-4.00	The box is yellow and green. The box has a tree with Baby sitting underneath and 3 Smurfs piled up putting a bird in the tree. The back has a Smurf laying under a mushroom, Smurfette holding a apple and standing on ice skates, and Poet Smurf holding a piece of paper and a carrot pen. The side has a Smurf watering flowers and the other side has Papa and a Smurf crying.
129/003	Smurfs Singing	Benditt Marketing	1/1/88		Hardee's Box		U.S.A.	$3.00-4.00	The box is orange and pink. The front of the box has 2 Smurfs singing. The box has Handy, Greedy, Grouchy and Lazy on the back. The side has a Smurf in the bathtub. The other side has a Smurf carrying a green triangle and Brainy Smurf standing, holding a clip board.
129/004	2 Smurfs Swimming In A Pond	Benditt Marketing	1/1/88		Hardee's Box		U.S.A.	$3.00-4.00	The box is purple and blue. The front box has 2 Smurfs swimming in a swim hole. The back of the box has Painter Smurf painting a rainbow with 2 pots of gold. The side has Papa, Smurfette and a Smurf in a hot air balloon. The other side has Smurfette, Baby and 2 Smurfs holding balloons.
129/005	Smurf On An Orange Skateboard	Hardee's Food System	1/1/90		Hardee's Fun meal Box		U.S.A.	$3.00-4.00	The box is yellow and blue. The front of the box has a Smurf riding a wave on an orange surfboard. The back has Sassette on a green skateboard going off a ramp with Papa and Puppy looking on. One side has a Smurf sunbathing and the other side has the Smurfs in a fishing boat.
129/006	Quick Magic Box	Benditt Marketing	1/1/96		Quick Box		Germany	$3.00-4.00	The box looks like a mushroom house on all the sides. The front has a door and window. The back has a barrel with the water gutter draining into the barrel and a Smurf is standing next to Smurfette looking lovesick. One side has a Smurf carrying a present and the other side has Papa sitting in a green and brown chair.
129/007	Happy Meal House	McDonald's	1/1/98		Mc Donald's Box		Belgium	$3.00-4.00	The box makes a hexagon- shaped mushroom house. The house is yellow with a red and white roof. On top of the roof is flying Smurf. The box is for the promotional Smurfs for 1998.

Set 174: Hardee's Figures

174/001	Papa On A Red Skateboard	Applause/Peyo		Red	Rubber Figure	2 1/2" High		$7.50-10.00	Papa is standing on a red skateboard. He is wearing a white swim vest, red swim trunks and a red hat. Papa has a gold whistle around his neck. Papa is pointing a finger. There is a white sticker on the front of the skateboard that says "SMURF" in blue letters.
174/002	Smurf On Orange Skateboard	Applause/Peyo		Orange	Rubber Figure	2 1/2"		$7.50-10.00	The Smurf is standing on a purple lunch box on an orange skateboard. The Smurf is wearing orange and green swim trunks. The Smurf has his hands above his head. There is a green sticker on the front of the skateboard that says "Smurf" in blue letters.
174/003	Smurf On A Yellow Skateboard	Applause/Peyo		Yellow	Rubber Figure	2 1/2"		$7.50-10.00	A Smurf is standing on a yellow skateboard. The Smurf has a green horse inner-tube around his stomach. There is a green sticker on the front of the skateboard that says "Smurf" in blue letters.
174/004	Sassette On Green Skateboard	Applause/Peyo		Green	Rubber Figure	2 1/2"		$7.50-10.00	Sassette is standing on a green skateboard. Sassette is wearing a green swimsuit. Sassette has orange hair. There is an orange sticker on the front of the skateboard that says "Smurf" in blue letters.
174/005	Puppy On Blue Skateboard	Applause/Peyo		Blue	Rubber Figure	2 1/2"		$7.50-10.00	The dog is standing on a blue skateboard. The dog is brown with a cream colored spot, ear and snout. The dog has a pink tongue. The dog is wearing yellow swim trunks with pink designs. He is wearing blue sunglasses. There is a pink sticker on the front of the skateboard that says "Smurf" in blue letters.
174/006	Dog On Blue Skateboard	Applause/Peyo		Blue	Rubber Figure	2 1/2"		$7.50-10.00	The dog is standing on a blue skateboard. The dog is brown with a cream colored spot, ear and snout. The dog has a cream colored tongue. The dog is wearing yellow swim trunks with no designs. He is wearing blue sunglasses. There is a pink sticker on the front of the skateboard that says "Smurf" in blue letters.
174/007	Smurfette On A Purple Skateboard	Applause/Peyo		Purple	Rubber Figure	2 1/2"		$7.50-10.00	Smurfette is standing on a purple skateboard. Smurfette is wearing a 2 piece purple and white swimsuit. Smurfette has her hands out like she is surfing. There is a yellow sticker on the front of the skateboard that says "Smurf" in blue letters.

Set 161: Hats & Visors

161/001	Baseball Smurf Kings Island Cap	Monotex	1/1/84	Red	Baseball Cap		Hong Kong	$11.00-15.00	The baseball cap is red. In front its white and has a picture of a Smurf carrying a baseball bat and a glove. The Smurf has on orange shorts and a red shirt. It says Kings Island in red letters underneath the Smurf.
161/002	Smurfette Sun Visor	Carowinds	1/1/84		Sun Visor	Small	Taiwan	$7.50-10.00	The visor is blue plastic. The strap is white elastic. On the front is Smurfette's face. There is pink on each side of her face for the start of the straps.
161/003	Papa Smurf Baseball Hat			Navy Blue	Baseball Cap			$6.00-8.00	The cap is navy blue. The front has a patch with Papa Smurf on. The patch is yellow with blue trim. Papa is walking with one hand out in front of him.
161/004	Kings Island Red Visor	Monotex/Peyo	1/1/84	Red	Visor Sun		Hong Kong	$7.50-10.00	The visor is red with Kings Island written across in white. The front of the band is white and has a Smurf and Smurfette dancing. On each side of the Smurf and Smurfette are 5 Smurfs marching with their fingers in the air. The back of the band is red and white stretch elastic.
161/005	Kings Island Blue Visor	Monotex/Peyo	1/1/84	Blue	Sun Visor		Hong Kong	$7.50-10.00	The visor is light blue with Kings Island written across in white. The front of the band is white and has a Smurf and Smurfette dancing. On each side of the Smurf and Smurfette are 5 Smurfs marching with their fingers in the air. The back of the band is blue and white stretch elastic.
161/006	Kings Island White Visor	Monotex/Peyo	1/1/84	White	Sun Visor		Hong Kong	$7.50-10.00	The visor is white and made of cloth. On the top of the visor is Smurfette, a Smurf and Papa in a red and white rollercoaster. The band is white in front and says Kings Island in red letters.
161/007	Kings Island Soft Baseball Cap	Monotex/Peyo	1/1/84	White & Blue	Child's Cap		Hong Kong	$6.00-8.00	The cap is white with a light blue visor. The front of the cap has Papa jumping. On one side is a Smurf roller-skating and the other side has a Smurf riding a bike. The back has a Smurf on a pogo stick.
161/008	Les Jeux Des Schtroumpfs			White	Sailors Hat	53 cm.	France	$18.00-25.00	The hat is white. The hat has Smurfs doing various things. A Smurf is throwing darts at a tree. A Smurf is flying. A Smurf is fishing. A Smurf is flying a kite. A Smurf is walking. Smurfette is swinging and Papa is pushing her. A Smurf is jumping off another Smurfs back and a Smurf is standing, pointing, and laughing. Gargamel and Azreal are standing behind a tree, watching the Smurfs.
161/009	Kings Dominion Smurfette Visor	W. Berrie/Peyo	1/1/84	Red/Pink	Sun Visor	Children's	Taiwan	$7.50-10.00	The visor is red plastic. The strap is white elastic. On the front is Smurfette's face. The background is pink. On one side of Smurfette's face are 2 red hearts. The other side says Kings Dominion in red letters.
161/010	Kings Dominion Smurf Visor	W. Berrie/Peyo	1/1/84	Blue/Yellow	Sun Visor	Children's	Taiwan	$7.50-10.00	The visor is blue plastic. The strap is white elastic. On the front is a Smurf's face. The background is yellow. On one side of Smurf's face are 3 red stars. The other side says Kings Dominion in red letters.
161/011	King Dominion Smurf's Up! Painters Cap	Applause/Peyo		White/Blue	Painter Cap	Children's	Taiwan	$9.00-12.00	The cap has a picture of Smurfette sitting on a green beach towel. Smurfette is wearing a red bathing suit. A Smurf is laying next to her on a red and yellow striped towel. The Smurf is wearing white swim trunks and red glasses. The cap says "Smurf's Up!" in red letters on the top of the cap. On the front side the cap it says Kings Dominion in blue letters.
161/012	Jungle River	AT & TV Mrch..	1/1/98	Yellow	Baseball Hat	52 cm	Brussels	$9.00-12.00	The Smurf is embroidered on the front of a yellow cloth cap. The Smurf is wearing a tan and brown hiker outfit. The Smurf is carrying a butterfly net. The cap has an orange visor. The cap says "Jungle River".
161/013	Savannah Friends Wildlife Souvenirs	AT & TV Mrch..	1/1/98	Cream & Brown	Cloth Baseball Cap	54 CM	Brussels	$9.00-12.00	The cap is a cream- colored cloth with a brown visor. The front has a giraffe face with a Smurf face and Brainy's face embroidered on it. The giraffe is in the middle. The top of the hat says "Savannah Friends Wildlife Souvenirs" in red letters.
161/014	The Smurfs Classic	AT & TV Mrch..	1/1/98	Dk. Blue & Wht.	Cloth Baseball Cap	52 CM	Brussels	$9.00-12.00	The hat is dark blue and white. The front has a diamond shape with a Smurf face in the center. Above the Smurf's hat is a red rectangle with "The Smurfs" embroidered in it, in yellow letters. Below it says "Classic" in dark blue letters.
161/015	Dance Out	AT & TV Mrch..	1/1/98	Black & Red	Cloth Baseball Cap	54 CM	Brussels	$9.00-12.00	The cap is black with a red visor. It has a picture of a Smurf dancing on the front. The Smurf is wearing red pants, pink belt, white and pink shoes, a white hat and a yellow shirt with a white "S" on the front of the shirt. The Smurf's hands are out at his sides. On the bottom of the cap it says "Dance Out" in white letters>
161/016	Skate Smurf Xtreme Style	AT & TV Mrch..	1/1/98	Orange	Nylon Baseball Cap	54 CM	Brussels	$9.00 12.00	The cap is orange and made of nylon in front. The back is orange mesh. The front of the cap has a Smurf on a yellow skateboard. The Smurf is wearing a red shirt, yellow hat, green shorts and red/green/white shoes. The hat says "Skate, Smurf, Xtreme, Style" in white letters.
161/017	Let's Party Hat	Marini Silvano	1/1/98	Multi	Baseball Cap	Kids	Brussels	$11.00-15.00	The cap is different colors in each section. The front is red and has a Smurf with his finger in the air. It says "let's party" over top of the Smurf. The Smurf is wearing a yellow shirt and gray pants. The hat has 3 other Smurfs going around it.
161/018	Happy Smurf Red Hat	Marini Silvano	1/1/98	Red	Baseball Cap	Kids	Brussels	$11.00-15.00	The cap is red. On the front is a Smurf done in embroidery. The Smurf is holding his hands up in the air. He is standing on one foot.
161/019	Happy Smurf Black Hat	Marini Silvano	1/1/98	Black	Baseball Cap	Kids	Brussels	$11.00-15.00	The cap is black. On the front is a Smurf done in embroidery. The Smurf is holding his hands up in the air. He is standing on one foot.

Set 211: Hooks

211/001	Papa	Novocom			Figure Hook	2 1/2"	Italy	$11.00-15.00	The figure is flat, on the back, and has a hook on the bottom for hanging stuff on. Papa is holding his hands out at his sides. The hook has a glossy finish.
211/002	Judge	Novocom			Figure Hook	2 1/2"	Italy	$11.00-15.00	The figure is flat, on the back, and has a hook on the bottom for hanging stuff on. The Smurf is wearing a black robe, white hat and black glasses. The Smurf is pointing with his right hand.
211/003	Smurfette Flirting	Novocom			Figure Hook	2 1/2"	Italy	$11.00-15.00	The figure is flat, on the back, and has a hook on the bottom for hanging stuff on. Smurfette is wearing a white dress, white shoes and a white hat. Smurfette has one hand on her hair. She has long black eyelashes. Smurfette is in a flirting pose.
211/004	Naughty	Novocom			Figure Hook	2 1/2"	Italy	$11.00-15.00	The figure is flat, on the back, and has a hook on the bottom for hanging stuff on. The Smurf is standing with his thumbs in his ears and he is sticking out his tongue.
211/005	Present	Novocom			Figure Hook	2 1/2"	Italy	$11.00-15.00	The figure is flat, on the back, and has a hook on the bottom for hanging stuff on. The Smurf is holding a white package with red ribbon around it. The Smurf is holding it with both hands, off to the right.
211/006	Jolly	Novocom			Figure Hook	2 1/2"	Italy	$11.00-15.00	The figure is flat, on the back, and has a hook on the bottom for hanging stuff on. The Smurf is holding his hand on his chest while the other hand is in the air. The Smurf is laughing.

Set 158: Household Items

158/001	Thermometer	Fackelmann	1/1/96	White	Thermometer	7 1/2"L X 2 1/2"	Germany	$7.50-10.00	The thermometer is white. There is a picture of a Smurf wearing snorkeling gear and an inner-tube around his waist. The thermometer goes up to 55 C.
158/002	Lawn Sprinkler	HG	1/1/83		Lawn Sprinkler	9" Tall	Hong Kong	$15.00-20.00	The sprinkler is a Smurf standing on a yellow flower base with green petals around it. The Smurf is holding a red daisy over his head. The center of the daisy shoots the water. Hooks up to standard hoses.
158/003	Smurfette Broom	Peyo	1/1/82		Broom	22" Long	Hong Kong	$2.00-3.00	The broom is a small children's broom. It has Smurfette's face on the end by the bristles. The broom stick is blue. The bristles are pink.
158/004	Smurfette's Face Dustpan	Perma-Dust-Ette	1/1/81	Yellow	Dustpan	7 1/2" x 6"	U.S.A.	$2.00-3.00	The mini dustpan is yellow with Smurfette's face in the middle.
158/005	Smurfette Pushing A Vacuum Dustpan	Perma-Dust-Ette	1/1/81	Yellow	Dustpan		U.S.A.	$2.00-3.00	The mini dustpan is yellow. It has a picture of Smurfette pushing a pink carpet broom. Smurfette is wearing a white dress.
158/006	Jokey Smurf Planter			White	Planter	5 1/2" x 4 1/4"	Hand made	$4.50-6.00	The planter is oval in front with a planter in back. The front has a Smurf carrying an orange present. The planter says" This is for you." The border is blue.
158/007	Baby Wallpaper	Norwall	1/18/81		Wallpaper	53cm x 10m	Canada	$22.50-30.00	The wallpaper is white with pink flowers. It has a Smurf and Smurfette hugging. Smurfette pushing Baby in a stroller
158/008	Village Scene Wallpaper	Norwall	1/1/81		Wallpaper	53cm x 10m	Canada	$22.50-30.00	The wallpaper is white. It has a village scene on it. Smurfette in a balcony. Gargamel fishing in a pond.
158/009	Light Switch Cover	W. Berrie		White	Switch plate	6 1/4" x 4 1/4"		$15.00-20.00	The light switch plate is white and oval- shaped. It has 4 Smurfs on the front. One Smurf is laying in the grass. A Smurf is laughing. A Smurf is pointing, and a Smurf is playing the drums.
158/010	Smurfette Coat Rack			Pink	Coat Rack	22" x 20"	Hand made	$18.00-25.00	Smurfette is made of wood. She has a pink dress, white shoes and a white hat. In the center is a long yellow strip that has blue plastic hangers attached to it.
158/011	Smurf And Smurfette Wood Coat Rack				Coat Rack	14 1/2" x 7 1/4"	Hand Painted	$11.00-15.00	The back is wood. It is rectangle- shaped. A Smurf has his back facing Smurfette and he is holding 3 flowers. Smurfette has her back facing the Smurf. She has a happy expression on her face. There are 2 hooks for hanging.

Set 69: Houses- Smurf

069/001	Large Red Roof Mushroom House	Schleich	1/1/76		Plastic Houses	8" High	Hong Kong	$37.50-50.00	The house has a red roof with white spots. The bottom of the mushroom house is bone color. The door, chimney and 2 window rims are gray. The door and window shutters are brown.
069/002	Large Light Red Mushroom House	Schleich	1/1/76		Plastic Houses	8" High	West Germany	$37.50-50.00	The house has a light red roof with white spots. The bottom of the mushroom house is bone color. The door, chimney and 2 window rims are gray. The door and window shutters are brown.
069/003	Small Red Roof Cottage	Schleich	1/1/78		Plastic Houses	4" High	Hong Kong	$18.00-25.00	The house has a red roof with white spots. The bottom of the mushroom house is bone color. The door has gray trim. The door and 2 windows are brown.
069/004	Small Green Roof Cottage	Schleich	1/1/78		Plastic Houses	4" high	Hong Kong	$18.00-25.00	The house has a green roof with white spots. The bottom of the mushroom house is bone color. The door has gray trim. The door and 2 windows are brown.
069/005	Small Blue Roof Cottage	Schleich	1/1/78		Plastic Houses	4" high	West Germany	$22.50-30.00	The house has a blue roof with white spots. The bottom of the mushroom house is bone color. The door and windows are gray.
069/006	Small Smurfette's Cottage	Schleich	1/1/78		Plastic Houses	4" high	China	$18.00-25.00	A lavender roof with a light lavender on top and white spots that look faded. A light pink bottom with a lavender rim door and windows. The door is light pink. The door hinges are light gray.
069/007	Small Smurfette's Cottage	Schleich	1/1/78		Plastic Houses	4" high	Hong Kong	$18.00-25.00	A lavender roof with pink on top and white spots that are solid . A dark pink bottom with a lavender rim door and windows. The door is dark pink. The door hinges are pink.
069/008	Small Smurfette Cottage	Schleich	1/1/78		Plastic Houses	4" high	West Germany	$18.00-25.00	A lavender roof with a dull pink top and white spots that look faded. A light pink bottom with a lavender rim door and windows. The door is light pink. The door hinges are dark gray.
069/009	Farm House	Schleich			Plastic Houses		Germany	$75.00-100.00	The roof is light brown boards on 1 half and green, grassy on the other half. The outside walls are white with brown trim. A light gray chimney, with a stork, in a nest, on top.
069/010	Windmill	Schleich	1/1/81		Plastic Houses		West Germany	$55.00-75.00	The windmill is white with brown and blue sprayed on the walls. A dark green roof. The door and balcony are brown, 2 yellow windows and a red mushroom top canopy that connect to the side of the mushroom.
069/011	Gargamel's Castle	Schleich	1/1/81		Plastic Houses		West Germany	$82.50-110.00	The castle is 2 stories high on 1 half. The walls are blue with blue brick. The roof is yellow and brown. There is a brown door with black hinges and a brown balcony in back. The back of the castle is open to play inside. The castle is on a gray platform.
069/012	Small All Blue Cottage	Schleich	1/1/94	Blue	Plastic Houses	4" high	Germany	$11.00-15.00	The house has a blue roof with red on it. The bottom of the mushroom house is light blue. The door and windows are brown. The curtain on the door is yellow. The butterfly is blue.
069/013	Small Smurfette Cottage	Schleich	1/1/94		Plastic Houses	4" High	Germany/China	$11.00-15.00	The house has a dark purple roof with dark pink on it. The bottom of the mushroom house is light pink. The door and windows are brown. The curtain on the door is yellow. The butterfly is yellow.
069/014	Small Orange Roof Cottage	Schleich	1/1/94	Orange	Plastic Houses	4" High	Germany/China	$11.00-15.00	The house has a dark orange roof with light orange on top. The bottom of the mushroom house is light orange. The door and windows are brown. The curtain on the door is red. The butterfly is red.
069/015	Small Orange Roof Cottage With Snappy Smurf	Irwin	1/1/96	Orange	Plastic Houses	4" High	Germany/China	$15.00-20.00	The house has a dark orange roof with light orange on top. The bottom of the mushroom house is light orange. The door and windows are brown. The curtain on the door is red. The butterfly is red. Snappy Smurf comes with the cottage. He is wearing a yellow shirt with a thunder bolt on the front.
069/016	Small Purple Cottage With Baby Smurf	Schleich	1/1/96		Plastic Houses	4" High	Germany/China	$15.00-20.00	The house has a dark purple roof with dark pink on it. The bottom of the mushroom house is light pink. The door and windows are brown. The curtain on the door is yellow. The butterfly is yellow. Baby Smurf crawling, pushing a red car comes with the cottage.
069/017	Small All Blue Cottage With Smurf Playing A Flute	Schleich	1/1/96	Blue	Plastic Houses	4" high	Germany/China	$15.00-20.00	The house has a blue roof with red on it. The bottom of the mushroom house is light blue. The door and windows are brown. The curtain on the door is yellow. The butterfly is blue. A Smurf playing a flute is included with the cottage.
069/018	Large Bully House	Peyo/Bully			Plastic Houses	10" x 9"	Germany	$52.50-70.00	The bottom of the mushroom is yellow and the roof is red. The window trim is brown and they have green shutters. The door is green and has brown trim. The house has a red chimney.

Set 254: Iron- On

254/001	Smurfette Holding Baby	Zazehlovaci Obtisk			Iron - On		Spain	$5.00-7.00	The cloth has 2 different iron-on. Smurfette is holding Baby. Smurfette is standing in the grass. Baby Smurf is wearing a red sleeper. The other iron-on is Smurfette standing, with a flirty look on her face.
254/002	Papa Walking Under An Umbrella	Zazehlovaci Obtisk	1/1/89		Iron - On		Spain	$5.00-7.00	The cloth has 2 different iron-on. Papa is walking, carrying a yellow umbrella. The second one is of Smurfette. She is standing with a flirty look on her face.
254/003	Gargamel And Azrael	W.B./Peyo/ Introduct	1/1/83		Iron - On	1"		$1.50-2.00	Gargamel is standing, holding his hands together. Azreal is by his feet.
254/004	Smurf Crying	W.B./Peyo/ Introduct	1/1/83		Iron - On	3 3/4"		$1.50-2.00	The Smurf is standing, holding a yellow handkerchief. He is wiping his nose with his hand. He has tears running down his face.
254/005	Skier	W.B./Peyo/ Introduct	1/1/83		Iron - On	4"		$1.50-2.00	The Smurf is skiing. He has yellow ski's and brown poles. He is wearing a red scarf and yellow mittens.
254/006	Bicyclist	W.B./Peyo/ Introduct	1/1/83		Iron - On	4"		$1.50-2.00	The Smurf is riding a green bike with yellow tires. The Smurf is wearing a white hat, red shirt, yellow shorts, brown gloves, and red shoes.
254/007	Windsurfer	W.B./Peyo/ Introduct	1/1/83		Iron - On	4		$1.50-2.00	The Smurf is riding a green surfboard with a yellow sail. The Smurf is wearing white swim trunks.

Set 251: Jam Caps- De Smurfen

251/001	Schilder Smurfin	Andere Personogen	1/1/95		Jam Cap	1 1/2 Diam.	Brussels		Smurfette is standing, painting a picture on an easel. A Smurf is peeking around the side of the easel offering Smurfette a piece of toast. The background of the jam cap is blue, green, red, and yellow.

251/002	Smul Smurf	Andere Persongen	1/1/95		Jam Cap	1 1/2 Diam.	Brussels	The Smurf is standing, holding an ice cream cone. He is wearing a blue jacket with a hood. The jacket has white fuzzy trim. The background of the jam cap is pink, yellow, brown, and green circles.
251/003	Ruimte Smurf	Andere Persongen	1/1/95		Jam Cap	1 1/2 Diam.	Brussels	The Smurf is wearing a white space uniform. He is wearing an astronaut helmet. The background of the jam cap is blue and looks like space.
251/004	Grote Smurf	Andere Persongen	1/1/95		Jam Cap	1 1/2 Diam.	Brussels	Papa Smurf is standing at a lab table. He is mixing potions. The background is green.
251/005	Smurfin (Smurfette)	Andere Persongen	1/1/95		Jam Cap	1 1/2 Diam.	Brussels	Smurfette is standing, posing. She has a finger by her mouth. The background is yellow and red checker board.
251/006	Verliefde Smurfin	Andere Persongen	1/1/95		Jam Cap	1 1/2 Diam.	Brussels	Smurfette is giving a lovesick Smurf a kiss on the cheek. There are hearts floating above their heads. The background is pink.
251/007	Bril Smurf	Andere Persongen	1/1/95		Jam Cap	1 1/2 Diam.	Brussels	A Smurf is hitting Brainy Smurf over the head with a mallet. There are red and yellow stars floating above Brainy's head. The background is yellow and orange bolts.
251/008	Gargamel	Andere Persongen	1/1/95		Jam Cap	1 1/2 Diam.	Brussels	Gargamel is running, chasing Smurfs with a net. He has 3 Smurfs in the net. The background is green.
251/009	Natuur Smurf	Andere Persongen	1/1/95		Jam Cap	1 1/2 Diam.	Brussels	2 Smurfs are pulling a hay cart. Another Smurf is pushing the cart from behind. There is a Smurf sitting on top of the hay ride. The background is grass and sky.
251/010	Kok Smurf	Andere Persongen	1/1/95		Jam Cap	1 1/2 Diam.	Brussels	A Smurf is cooking over a grill. His pan of food is bubbling over. The Smurf has a worried expression on his face. The Smurf is wearing a chefs hat and an apron. The background is red, yellow and green checker board pattern.
251/011	Golf Smurf	Andere Persongen	1/1/95		Jam Cap	1 1/2 Diam.	Brussels	The Smurf is standing, swinging a golf club. The Smurf is wearing red and green checkered pants. The background is green.
251/012	Potige Smurf	Andere Persongen	1/1/95		Jam Cap	1 1/2 Diam.	Brussels	A Smurf is standing, shooting arrows. He is holding a brown bow. The background is a red and white dart board with numbers.
251/013	Knutsel Smurf	Andere Persongen	1/1/95		Jam Cap	1 1/2 Diam.	Brussels	The Smurf looks like a mechanic. He is wearing dirty white coveralls. The Smurf is carrying a hammer and an oil can. He has a pencil behind his ear. The background is 3 shades of blue.
251/014	Muziek Smurf	Andere Persongen	1/1/95		Jam Cap	1 1/2 Diam.	Brussels	A Smurf has a big drum strapped on his back. Another Smurf is walking behind him playing the drum. The background is 2 shades of orange and 2 shades of yellow.
251/015	Opa Smurf	Andere Persongen	1/1/95		Jam Cap	1 1/2 Diam.	Brussels	Grandpa Smurf is sitting in a rocking chair reading a red book. Grandpa has on a yellow hat and yellow pants with suspenders. He has a long white beard. The background is green with red with yellow circles.
251/016	Vlieg Smurf	Andere Persongen	1/1/95		Jam Cap	1 1/2 Diam.	Brussels	A Smurf is flying. He has white feathered wings attached to his arms. The background is blue sky and clouds. There are 2 blue birds flying next to him.
251/017	Ruimte Smurf	Andere Persongen	1/1/95		Jam Cap	1 1/2 Diam.	Brussels	There is a wild Smurf pointing at the Smurf in the space outfit. The wild Smurf has an orange body and long black hair. The background is blue sky and yellow ground.
251/018	Verliefde Grote Smurf	Andere Persongen	1/1/95		Jam Cap	1 1/2 Diam.	Belgium	Papa Smurf is kissing Smurfette's hand. The background is the night sky
251/019	Luilak Smurf	Andere Persongen	1/1/95		Jam Cap	1 1/2 Diam.	Brussels	A Smurf is sleeping in a hammock. He has a tray with food and a glass sitting on the ground next to him. The background is sky and ground.
251/020	Smul Smurf	Andere Persongen	1/1/95		Jam Cap	1 1/2 Diam.	Brussels	The Smurf is walking, carrying a cake. The background is pink with blue, green, red, yellow, and purple circles.

Set 104: Key chains- Acrylic

104/001	Cowboy Smurf	W. Berrie	1/1/81	Clear	Acrylic Key chain	1 1/2" X 1 1/2"	Hong Kong	$3.00-4.00	The key chain is flat plastic. There is a picture of a cowboy Smurf sitting on a brown saddle on a fence. The Smurf is swinging a rope in the air.
104/002	Sunny Side Up	W. Berrie	1/1/81	Clear	AM Radio	1 1/2" X 1 1/2"	Hong Kong	$3.00-4.00	The key chain is flat plastic. The picture is of a Smurf laying on a green towel on the beach. The sun is coming up behind him. Across the top of the key chain it says "Sunny Side Up" in orange letters.
104/003	Have A Smurfy Day!	W. Berrie	1/1/81	Clear	Acrylic Key chain	1 1/2" X 1 1/2"	Hong Kong	$3.00-4.00	The key chain is flat plastic. The picture is of a Smurf sitting on a white cloud with a rainbow behind him. Inside the cloud it says "Have A Smurfy Day!" in blue letters
104/004	Papa With Street Signs	W. Berrie	1/1/81	Clear	Acrylic Key chain	1 1/2" X 1 1/2"	Hong Kong	$3.00-4.00	The key chain is flat plastic. The picture is of Papa standing in the grass with 5 different street signs pointing every which way. Papa looks confused.
104/005	Shy Smurf With Yellow Daisy	W. Berrie	1/1/81	Clear	Acrylic Key chain	1 1/2" X 1 1/2"	Hong Kong	$3.00-4.00	The key chain is flat plastic. The picture is of a Smurf standing, holding a yellow daisy behind his back. The Smurf has a red shirt on and white pants.
104/006	Kings Island Smurfette Key chain	W. Berrie	1/1/84		Acrylic Key chain	1 1/2" X 1 1/2"		$7.50-10.00	The key chain is flat plastic. The picture is of a pink heart with Smurfette laying at the bottom of the heart with her head in her hands. In the middle of the heart it says "Kings Island" in purple letters.
104/007	Harmony Smurf	HMP			Acrylic Key chain	2 1/2"L X 1"W		$3.00-4.00	The key chain is flat plastic. The picture is of a Smurf standing, holding a yellow sheet of music. The Smurf is singing. The background is red. In black letters it says "Harmony Smurf".
104/008	Trumpet Player	W. Berrie	1/1/81		Acrylic Key chain	1 1/2" X 1 1/2"		$3.00-4.00	The key chain is flat plastic. The picture is of a Smurf walking, playing a yellow trumpet. The Smurf has a red shirt, white pants and a white hat.
104/009	Skier	W. Berrie	1/1/81		Acrylic Key chain	1 1/2" X 1 1/2"		$3.00-4.00	The key chain is flat plastic. The picture is of a Smurf wearing a white hat with an orange tassel, a red scarf, a yellow sweater, brown pants, and black shoes. He has white ski's on and is holding green and pink ski poles.
104/010	Smurf Sitting By A Mushroom House	W. Berrie	1/1/81		Acrylic Key chain	1 1/2" X 1 1/2"		$3.00-4.00	The key chain is flat plastic. The picture is of a Smurf sitting on a green patch of grass. There is a mushroom house in the back. The house is yellow with a red roof.
104/011	Dad's Key's	W. Berrie	1/1/81		Acrylic Key chain	1 1/2" X 1 1/2"	Hong Kong	$3.00-4.00	The key chain is flat plastic. The picture is of a Smurf driving a red car. There is a white cloud above his head that says "Dad's Key's" in yellow letters.
104/012	I'd Rather Be Playing Soccer!	W. Berrie	1/1/81		Acrylic Key chain	1 1/2" X 1 1/2"	Hong Kong	$3.00-4.00	The key chain is flat plastic. It has a picture of a Smurf bunting a soccer ball with his head. The Smurf is playing in the grass. The Smurf is wearing a yellow shirt and red shorts. The bottom of the key chain says "I'd Rather Be Playing Soccer!" in black letters.
104/013	Basketball Smurf	W. Berrie	1/1/81		Acrylic Key chain	1 1/2" X 1 1/2"		$3.00-4.00	The key chain is flat plastic. It has a picture of a Smurf jumping in the air with a basketball in his left hand. The basketball is red and yellow. The Smurf is wearing a red tank shirt, green shorts, white shoes, and a white hat.
104/014	Motorcycle Smurf	W. Berrie	1/1/81		Acrylic Key chain	1 1/2" X 1 1/2"		$3.00-4.00	The key chain is flat plastic. It has a picture of a Smurf riding a motorcycle off a ramp. The Smurf is wearing a yellow suit, a red scarf, a white hat, and purple goggles. The motorcycle is green and pink.
104/015	I'd Rather Be Playing Tennis!	W. Berrie	1/1/81		Acrylic Key chain	1 1/2" X 1 1/2"	Hong Kong	$3.00-4.00	The key chain is flat plastic. It has a Smurf swinging a tennis racket. The Smurf is wearing a yellow shirt, orange shorts, white shoes, and a white hat. The tennis racket is orange. He is standing in the grass. The key chain says " I'd rather be playing tennis" in black letters.
104/016	Don't Lose These Keys! (Papa)	W. Berrie	1/1/81	Clear	Acrylic Key chain	1 1/2" X 1 1/2"	Hong Kong	$3.00-4.00	The key chain is flat plastic. It has Papa standing with his hands on his hips. Papa has a mad expression on his face. There is a yellow bubble behind Papa. The bubble says "Don't Lose These Keys!" in orange letters.
104/017	Keep On Smurfin			Clear	Acrylic Key chain	1 1/2" X 1 1/2"	Hong Kong	$3.00-4.00	The key chain is flat plastic. It has 5 Smurfs marching, holding a finger in the air. The bottom of the key chain says "Keep On Smurfin" in red letters.
104/018	Have You Hugged Your Smurf Today?	W. Berrie	1/1/82	Clear	Acrylic Key chain	1 1/2" X 1 1/2"	Hong Kong	$3.00-4.00	The key chain is flat plastic. It has a Smurf standing with his hands behind his back. The Smurf has a shy grin on his face. The Smurf is standing in the grass. There is a red circle behind him. The key chain says "Have You Hugged Your Smurf Today?"
104/019	Roller Smurfin'			Clear	Acrylic Key chain	1 1/2" X 1 1/2"		$3.00-4.00	The key chain is flat plastic. It has Smurfette and a Smurf roller-skating. The Smurf is chasing Smurfette. The background is red with white clouds. The key chain says "Roller Smurfin" in red letters.
104/020	Want To Smurf Around?	W. Berrie	1/1/81	Clear	Acrylic Key chain	1 1/2" X 1 1/2"	Hong Kong	$3.00-4.00	The key chain is flat plastic. It has a Smurf sitting in the grass looking lovesick. He is holding a pink daisy. The Smurf has lip marks on his hat and foot. There are red hearts floating above his head. The Smurf has his tongue hanging out of his mouth. The key chain says "Want To Smurf Around?" in black letters.
104/021	I'd Rather Be Skiing!	W. Berrie	1/1/81	Clear	Acrylic Key chain	1 1/2" X 1 1/2"	Hong Kong	$3.00-4.00	The key chain is flat plastic. It has a Smurf skiing down a snow hill. The Smurf is wearing a white hat with an orange tassel, red scarf, green sweater, orange pants, and red gloves. The Smurf's ski's and poles are green. The key chain says "I'd Rather Be Skiing!" in black letters.
104/022	Watch Out I'm A Genius!	W. Berrie	1/1/82	Clear	Acrylic Key chain	1 1/2" X 1 1/2"	Hong Kong	$3.00-4.00	The key chain is flat plastic. It has a Smurf holding a stop sign that says "Watch Out". The Smurf is holding the sign in front of him and he is holding one hand in the air. The Smurf is wearing black glasses. In the grass it says "I'm A Genius!"
104/023	Mom's Keys	W. Berrie	1/1/81	Clear	Acrylic Key chain	1 1/2" X 1 1/2"	Hong Kong	$3.00-4.00	The key chain is flat plastic. It has Smurfette driving a red car. Above her in a white bubble it says "Mom's Keys" in blue letters.
104/024	I've Been Smurfed!	W. Berrie	1/1/82	Clear	Acrylic Key chain	1 1/2" X 1 1/2"	Hong Kong	$3.00-4.00	The key chain is flat plastic. It has a Smurf sitting in the grass looking dazzled. There are stars spinning around his head. The key chain says "I've Been Smurfed!"
104/025	Don't Lose These Keys! (Smurfette)			Clear	Acrylic Key chain	1 1/2" X 1 1/2"		$3.00-4.00	The key chain is flat plastic. It has Smurfette standing on a surfboard. The surfboard is blue, yellow, and pink with peach stars. There is a red arch behind Smurfette's head. Inside the arch it says "Don't lose these keys!" in white letters.
104/026	Jokey Smurf				Acrylic Key chain	2 1/2"L X 1"W		$3.00-4.00	The key chain is flat plastic. It has a Smurf carrying an umbrella playing in the rain. The Smurf is wearing a yellow rain coat. The background is yellow. The bottom says "Jokey Smurf" in black letters.
104/027	Smurf Carrying A Cake			Clear	Acrylic Key chain	1 1/2" X 1 1/2"		$3.00-4.00	The key chain is flat plastic. It has a Smurf walking, carrying a big yellow cake. The cake has white frosting with cherries on top. The cake is on a pink platter. The Smurf is licking his lips.
104/028	Smurfette Looking Dreamy	W. Berrie	1/1/81	Clear	Acrylic Key chain	1 1/2" X 1 1/2"	Hong Kong	$3.00-4.00	The key chain is flat plastic. It has Smurfette standing at the end of the rainbow looking dreamy. Smurfette is holding an orange daisy behind her back. Smurfette has red hearts floating around her.
104/029	Smurf Driving A Race Car			Clear	Acrylic Key chain	1 1/2" X 1 1/2"		$3.00-4.00	The key chain is flat plastic. It has a Smurf driving a race car. The car is blue, green and pink. The Smurf has a mean expression on his face. The background is pink with a yellow stripe.
104/030	Football Smurf			Clear	Acrylic Key chain	1 1/2" X 1 1/2"		$3.00-4.00	The key chain is flat plastic. It has a Smurf running with a green and white football. The Smurf is wearing a white helmet with a yellow star, white shorts with a pink stripe and white and green shoes. The background is pink with a yellow stripe.
104/031	Smurfette Skiing			Clear	Acrylic Key chain	1 1/2" X 1 1/2"		$3.00-4.00	The key chain is flat plastic. It has Smurfette downhill skiing. Smurfette is wearing a white dress and hat. Her ski's are green. The snow is white but the background is pink.
104/032	Papa Golfing			Clear	Acrylic Key chain	1 1/2" X 1 1/2"		$3.00-4.00	The key chain is flat plastic. It has Papa Smurf golfing. Papa is wearing a pink hat, pink pants and white shoes. The background is yellow and green.
104/033	Superman			Clear	Acrylic Key chain	1 1/2" X 1 1/2"		$3.00-4.00	The key chain is flat plastic. It has a Smurf wearing a pink cape, white hat, white pants and pink shoes. The Smurf is flying in front of a big yellow star. The background is green.

104/034	Soccer Smurf			Clear	Acrylic Key chain	1 1/2" X 1 1/2"		$3.00-4.00	The key chain is flat plastic. It has a Smurf kicking a white and black soccer ball. The Smurf is jumping at the ball to kick it. The Smurf is wearing a white hat, white pants and orange soccer shoes. The background is 2 shades of blue, mountains and sky.
104/035	Smurfs In A Hot Air Balloon	Rodekoratie B.V.	1/1/94		Acrylic Key chain	2" x 1 1/2"	Brussels	$3.00 - 3.50	The key chain is flat plastic. It has a picture of a Smurf falling out of a hot air balloon. Another Smurf is trying to hold on to the Smurf falling. The hot air balloon is orange and yellow. The background is blue.
104/036	Smurf Waving	Rodekoratie B.V.	1/1/94		Acrylic Key chain	2" x 1 1/2"	Brussels	$3.00 - 3.50	The key chain is flat plastic. It has a picture of a Smurf waving. The background is white.
104/037	Smurf Pulling Pedals From A Flower	Rodekoratie B.V.	1/1/94		Acrylic Key chain	2" x 1 1/2"	Brussels	$3.00 - 3.50	The key chain is flat plastic. It has a picture of a Smurf pulling leaves from a flower. He has hearts floating around his head. The background is green.
104/038	Smurf Spraying Perfume On Himself	Rodekoratie B.V.	1/1/94		Acrylic Key chain	2" x 1 1/2"	Brussels	$3.00 - 3.50	The key chain is flat plastic. It has a picture of a Smurf spraying perfume on his face. The background is white and says "SMURF" in red underneath him.
104/039	Smurf Raking	Rodekoratie B.V.	1/1/94		Acrylic Key chains	2" x 1 1/2"	Brussels	$3.00 - 3.50	The key chain is flat plastic. It has a picture of a Smurf raking his garden. The background is orange. The Smurf is wiping his head.
104/040	Papa	Rodekoratie B.V.	1/1/94		Acrylic Key chain	2" x 1 1/2"	Brussels	$3.00 - 3.50	The key chain is flat plastic. It has a picture of Papa standing, holding his hands behind his back.
104/041	Smurf With A Trick Box	Rodekoratie B.V.	1/1/94		Acrylic Key chain	2" x 1 1/2"	Brussels	$3.00 - 3.50	The key chain is flat plastic. It has a picture of a Smurf carrying a trick box. The box has Gargamel coming out of it. The Smurf looks startled. The background is white.
104/042	Smurf Mixing Honey	Rodekoratie B.V.	1/1/94		Acrylic Key chain	2" x 1 1/2"	Brussels	$3.00 - 3.50	The key chain is flat plastic. It has a picture of a Smurf stirring honey in a bowl. There are 2 bee's flying above his head. The background is yellow with green branches.
104/043	Smurf Drawing Smurfette	Rodekoratie B.V.	1/1/94		Acrylic Key chain	2" X 1 1/2"	Brussels	$3.00 - 3.50	The key chain is flat plastic. It has a picture of a Smurf drawing a picture of Smurfette. There are pencils laying on the floor. The background is green.
104/044	Gargamel Stirring Potion	Rodekoratie B.V.	1/1/94		Acrylic Key chain	2" x 1 1/2"	Brussels	$3.00 - 3.50	The key chain is flat plastic. It has a picture of Gargamel standing at a stove mixing a kettle of something. Azreal is by his feet licking his mouth. There is a cage hanging from the ceiling with Smurfs in it.

Set 260: Key chains- Carded Figure

260/001	Clown	Schleich/Peyo	1/1/78		Carded PVC Key chain	2"	Germany/China	$3.50-5.00	The Smurf has a dark blue body. He is wearing yellow with red striped pants, black suspenders, a red bow tie, white shoes with black toes, a white shirt, white gloves, and a white hat. He has a red mouth with white around it. There is a silver key chain attached to his head. The Smurf is on a blue card.
260/002	Teacher	Schleich/Peyo	1/1/78		Carded PVC Key chain	2"	Germany/China	$3.50-5.00	The Smurf has a dark blue shiny body. He is holding a red book with white pages, black letters and a yellow bookmark in his right hand. He is holding a finger up in the air on his left hand. His eye's and mouth are shut like he's declaring something. There is a silver key chain attached to his head. The Smurf is on a blue card.
260/003	French Fries	Schleich/Peyo	1/1/81		Carded PVC Key chain	2"	Germany/China	$3.50-5.00	The Smurf has a dark blue body. He is holding a white bundle of french fries in his left hand and a single french fry in his right hand. The french fries are bright yellow. There is a silver key chain attached to his head. The Smurf is on a blue card.
260/004	Baby With Car	Applause/Peyo			Carded PVC Key chain	2"	China	$3.50-5.00	Baby has a dark blue body. He is crawling, pushing a red car. Baby is wearing a white sleeper and a white hat. There is a silver key chain attached to his head. The Smurf is on a blue card.
260/005	Smurfette Stewardess	Schleich/Peyo	1/1/85		Carded PVC Key chain	2"	Germany/China	$3.50-5.00	Smurfette has a dark blue body. She is wearing a white dress, white shoes and a white hat with a yellow flower on each side. Smurfette is holding a red tray with a brown cup. The cup has white liquid in it and a blue straw. There is a silver key chain attached to her head. Smurfette is on a blue card.
260/006	Architect	Schleich/Peyo	1/1/89		Carded PVC Key chain	2"	Germany/China	$3.50-5.00	The Smurf has a dark blue body. He is wearing a yellow jacket, orange hard hat, white pants and green boots. The Smurf is holding a yellow ruler in his left hand and a white scroll in his right hand. He has an orange and brown pencil in his jacket pocket. There is a silver key chain attached to his head. The Smurf is on a blue card.
260/007	Hula Smurfette	Schleich/Peyo	1/1/88		Carded PVC Key chain	2"	Germany/China	$3.50-5.00	Smurfette has a medium blue body. Smurfette is wearing a tan hula skirt, green belt, red lei, and orange hat trim. Smurfette has her hands out by her side like she's dancing. There is a silver key chain attached to her head. Smurfette is on a blue card.
260/008	Snappy Smurfling	Applause/Peyo	1/1/87		Carded PVC Key chain	2"	China	$3.50-5.00	The Smurf has a medium bright blue body. He is wearing a white hat, white pants and a yellow shirt with a lightening bolt on the front. His hands are out in fists. He has an angry expression on his face. There is a silver key chain attached to his head. The Smurf is on a blue card.

Set 253: Key chains- Figure Clips

253/001	Bookworm	Schleich/Peyo			Clip Key chain	2"	Germany/China	$3.50-5.00	The Smurf has a medium blue body. He is holding a red book with yellow pages down by his side. His right hand is in the air with a finger up. The Smurf is wearing black glasses. The Smurf has a dark blue clip and a metal key ring attached to his head. The Smurf has a black dot on his foot.
253/002	Policeman	Schleich/Peyo	1/1/81		Clip Key chain	2"	Portugal	$3.50-5.00	The Smurf has a medium blue body. He is wearing a white hat with a black visor and strap, a white jacket with black buttons, a black belt with a black buckle and black pants. He is holding a white stick in his left hand. He is holding a red whistle to his mouth with his right hand. The Smurf has a dark blue clip and a metal key ring attached to his head.
253/003	Baby With Blocks	Schleich/Peyo	1/1/84		Clip Key chain	2"	West Germany	$3.50-5.00	The Smurf has a medium blue body. He is sitting, wearing a white sleeper. Baby has 3 blocks. The blocks are yellow, red and green. Baby has a red mouth. The Smurf has a red clip and a metal key ring attached to his head.
253/004	Gargamel With Lab Glass	Schleich/Peyo	1/1/82		Clip Key chain	2"	West Germany	$3.50-5.00	Gargamel has a dark flesh colored skin. Gargamel is wearing a black robe and maroon boots. Gargamel is holding a pink glass in his right hand and a purple lab glass in left hand. He looks like he's pouring one lab glass into the other. Gargamel has a red clip and a metal key ring attached to his head.
253/005	Azreal	Schleich/Peyo	1/1/91	Orange	Clip Key chain	2"	West Germany	$3.50-5.00	Azreal is orange and white. Azreal is in a prowling position with his tail up. He has his head up and he's licking his chops. Azreal has a red clip and a metal key ring attached to his head.

Set 141: Key chains- Miscellaneous

141/001	Soccer Player				Plastic Key chain	1 1/2"	Belgium	$3.50-5.00	The key chain is shaped like an award metal. The outside is gold and black and has a picture of a Smurf wearing white, kicking a soccer ball. The background behind the picture is green. It says in black letters "Fupball Macht Spab" below the Smurf. The key chain is metal.
141/002	Soccer Smurf With Pen			Yellow	Key chain	2"L x 1 1/4"W		$1.00-2.00	The key chain is yellow, flat and square- shaped. It has a Smurf kicking a soccer ball. The Smurf almost looks fake. There is a mini one, slid in a slot, on the plastic. The pen is white.
141/003	Smurf Playing Trumpet			White	Flat Plastic Key ring	3" x 1/2"	Hand made	$1.50-2.00	The key ring is a piece of flat white plastic. It has a raised Smurf design on it. The Smurf is walking, playing a yellow trumpet. The Smurf has a red shirt and white pants on.
141/004	Smurf Wearing Apron			White	Flat Plastic Key ring	3" x 1 1/2"	Hand made	$1.50-2.00	The key ring is a piece of flat white plastic. It has a raised Smurf design on it. The Smurf is standing, wearing a red apron.
141/005	Smurf With Gift			White	Flat Plastic Key ring		Hand made	$1.50-2.00	The key ring is a piece of flat white plastic. It has a raised Smurf design on it. The Smurf is walking, carrying a yellow package with red ribbon.
141/006	Smurf Holding Flowers			White	Flat Plastic Key ring	3" x 1 1/2"	Hand made	$1.50-2.00	The key ring is a piece of flat white plastic. It has a raised Smurf design. The Smurf is standing, leaning on a heart. His head is on his hand. He is holding a red flower in his other hand behind his back.
141/007	Smurf Holding Flower			White	Flat Plastic Key ring	3" 1 1/2"	Hand made	$1.50-2.00	The key ring is a piece of flat white plastic. It has a raised Smurf design. The Smurf is standing, holding a red flower behind his back. The Smurf is wearing a red shirt and white pants.
141/008	Smurf Sitting By A Mushroom House			White	Flat Plastic Key ring	3" x 1 1/2"	Hand made	$1.50-2.00	The key ring is a piece of flat white plastic. It has a raised Smurf design on it. The Smurf is sitting by a mushroom house. The house is red.
141/009	Smurf Playing Golf				Flat Metal Key chain	1 3/4"		$1.00-2.00	The key chain is flat and metal. The front has a Smurf wearing a white hat, white pants and red shoes. The Smurf is holding a white golf club. A metal key chain is attached at the top.
141/010	Smurf Wearing A Green Apron				Flat Metal Key chain	2 1/4"		$1.00-2.00	The key chain is flat and metal. The front has a Smurf wearing a green apron, white pants and a white hat. A metal key chain is attached at the top.
141/011	Smurf Holding Flowers				Flat Metal Key chain	2 1/4"		$1.00-2.00	The key chain is flat and metal. The front has a Smurf wearing a white shirt, white pants, purple shorts, and a white hat. The Smurf is holding a red daisy behind his back. A metal key chain is attached at the top.
141/012	Tennis Smurf With Rainbow				Flat Metal Key chain	1 1/4"		$1.00-2.00	The key chain is flat and metal. The front has a Smurf wearing a white shirt, red shorts, white shoes, and white hat. The Smurf is carrying a yellow tennis racket. There is a rainbow behind the Smurf. A metal key chain is attached at the top.
141/013	Smurf Skiing				Flat Metal Key chain	1"		$1.00-2.00	The key chain is flat and metal. The Smurf is wearing ski clothes and ski's. The Smurf is wearing a white hat with a pink tassel, red scarf, yellow sweater, brown pants, purple boots, and white ski's. The Smurf is holding 2 ski poles. The poles are green with pink ends. A metal key chain is attached at the top.
141/014	Smurf On Motorcycle				Flat Metal Key chain	1 1/4"		$1.00-2.00	The key chain is flat and metal. The Smurf is riding a green motorcycle with pink pipes. He is riding up a brown jump ramp. The Smurf is wearing purple goggles around his hat, red scarf, yellow pants and a yellow shirt. A metal key chain is attached at the top.
141/015	Troubled Smurf				Flat Key chain	2"	Taiwan	$1.50-2.00	The Smurf is made of a flat plastic. He looks troubled. The Smurf has a hand on his head. A silver key chain is attached to his head.
141/016	Smurfette Key chain				Knockoff key chain	1 1/2"			Smurfette is a small figure. Smurfette is standing with her one hand on her stomach and the other out at her side. She has a dark blue body. There is a silver key chain attached.
141/017	Smurf Key chain				Knock-off Key chain	1 3/4"			The Smurf is a hard plastic figure. He is standing with his hands out at his sides. The Smurf is white with a blue hat and white pants. There is a silver key chain attached.
141/018	Papa Smurf Golf Ball Key chain	Peyo			Golf ball Key chain	1 1/2"		$3.00-4.00	The golf ball has a picture of Papa Smurf swinging a golf club. The golf ball has a silver key chain attached to it.
141/019	Smurfette Golf ball Key chain	Peyo			Golf ball Key chain	1 1/2"		$3.00-4.00	The golf ball has a picture of Smurfette swinging a golf club. The golf ball has a silver key chain attached.
141/020	Smurf Golf ball Key chain	Peyo			Golf ball Key chain	1 1/2"		$3.00-4.00	The golf ball has a picture of a Smurf swinging a golf club. The golf ball has a silver key chain attached.

Set 79: Key chains- Pencil Sharpener

079/001	Biker			Key chains	1 1/2" High	Taiwan	$2.00-3.00	The Smurf has a dark blue body. The Smurf is riding a white framed bike with a black seat, black wheels with yellow in the center of the wheels. The sharpener is white and on the back of the Smurf. The Smurf is flat plastic. There is a metal key chain attached to the top.
079/002	Smurf With Flowers And A Heart			Key chains	1 1/2" High	Taiwan	$2.00-3.00	The Smurf has a medium blue body. The Smurf is standing, leaning against a red heart. He is holding a red daisy in his right hand. The Smurf has a dreamy look on his face. The sharpener is white and on the back of the Smurf. The Smurf is flat plastic. There in a metal key chain attached to the top.
079/003	Traveler Smurf			Key chains	1 1/2" High	Taiwan	$2.00-3.00	The Smurf has a medium blue body. The Smurf is walking, carrying a yellow knapsack on a white stick over his left shoulder. The Smurf is whistling. The sharpener is white and on the back of the Smurf. The Smurf is flat plastic. There is a metal key chain attached to the top.
079/004	Love-struck Smurf With A Yellow Daisy			Key chains	1 1/2" High	Taiwan	$2.00-3.00	The Smurf has a medium blue body. The Smurf is sitting, holding a yellow daisy in his right hand. There are red lip marks on his hat and foot. The Smurf has his tongue hanging out. The sharpener is blue and on the back of the Smurf. The Smurf is flat plastic. A metal key chain is attached to the top.
079/005	Love-struck Smurf With A Red Daisy			Key chains	1 1/2" High	Taiwan	$2.00-3.00	The Smurf has a medium blue body. The Smurf is sitting, holding a red daisy in his right hand. There are red lip marks on his hat and foot. The Smurf has his tongue hanging out. The sharpener is pink and on the back of the Smurf. The Smurf is flat plastic. There is a metal key chain attached to the top.
079/006	Jokey Smurf			Key chains	1 1/2" High	Taiwan	$2.00-3.00	The Smurf has a dark blue body. He is carrying a white gift with yellow ribbon in both hands. The Smurf has a sly grin on his face. The sharpener is pink and on the back of the Smurf. The Smurf is flat plastic. There is a metal key chain attached to the top.
079/007	Drummer			Key chains	1 1/2" High		$2.00-3.00	The Smurf has a light blue body. He is holding 2 yellow drum sticks. The Smurf has a red and yellow drum in front of him. The sharpener is black and on the back of the Smurf. The Smurf is flat plastic. There is a metal key chain attached to the top.
079/008	Smurf Sitting By Mushroom House			Key chains	1 1/2" High		$2.00-3.00	The Smurf has a light blue body. He is sitting in the grass. There is a white mushroom house in the background with a red roof. The sharpener is white and on the back of the Smurf. The Smurf is flat plastic. There is a metal key chain attached to the top.
079/009	Biker			Key chains	1 1/2" High		$2.00-3.00	The Smurf has a light blue body. The Smurf is riding a white framed bike with a black seat, black wheels with yellow in the center of the wheels. The sharpener is white and on the back of the Smurf. The Smurf is flat plastic. There is a metal key chain attached to the top.

Set 249: Key chains- PVC Figure

249/001	Normal	Peyo		Rubber Figure	2" High	West Germany	$3.00-4.00	The Smurf's body is light blue. He has a white hat and pants on. His hands are out at his sides and he has a thin black line for a mouth, that is smiling. There is a silver key chain attached to his head.
249/002	Spy	Schleich/Peyo	1/1/66	Rubber Figure	2" High	Hong Kong	$3.00-4.00	The Smurf has a dark blue body. He is holding 1 finger in front of his mouth. He has a medium red cape, white hat and white pants on. He has a black mask over his face. The Smurf has a silver key chain attached to his head.
249/003	Clown	Schleich/Peyo	1/1/78	Rubber Figure	2" High	Hong Kong	$3.00-4.00	The Smurf has a dark blue body. He is wearing yellow with red striped pants, black suspenders, a red bow tie, white shoes with black toes, a white shirt, white gloves, and a white hat. He has a red mouth with white around it. The Smurf has a silver key chain attached to his head.
249/004	Soccer Footballer	Schleich/Peyo	1/1/78	Rubber Figure	2" High	Hong Kong	$3.00-4.00	The Smurf has a dark blue body. He has his arms out at his sides. His right foot has a white soccer ball on the tip. He is wearing a red shirt, red socks, yellow shorts and black shoes. The Smurf has a silver key chain attached to his head.
249/005	Hang-Glider	Schleich/Peyo	1/1/78	Rubber Figure	2" High	Hong Kong	$3.00-4.00	The Smurf has a medium blue body. The Smurf has a red and yellow hand-glider with brown straps. The Smurf has a silver key chain attached to his head.
249/006	Hiker	Schleich/Peyo	1/1/78	Rubber Figure	2" High	Hong Kong	$3.00-4.00	The Smurf has a dark blue body. He is wearing a yellow jacket and socks. He has brown shoes and a brown walking stick. He has a red flower in his hat. There is a green backpack on his back. The Smurf has a silver key chain attached to his head.
249/007	Chef	Schleich/Peyo	1/1/78	Rubber Figure	2" High	Hong Kong	$3.00-4.00	The Smurf has a medium blue body. He is wearing a white chef's hat, shirt and pants. He is holding a dark brown ladle to his mouth with his right hand. The Smurf has a silver key chain attached to his head.
249/008	Lover	Schleich/Peyo	1/1/78	Rubber Figure	2" High	Hong Kong	$3.00-4.00	The Smurf has a medium blue body. The Smurf is holding 3 red flowers with yellow centers and a bright green stem. The Smurf is holding the flowers in his left hand. He has his head turned to the right like he's posing. The Smurf has a silver key chain attached to his head.
249/009	Artist	Schleich/Peyo	1/1/78	Rubber Figure	2" High	Hong Kong	$3.00-4.00	The Smurf has a medium blue body. He is holding a white paint brush with yellow paint on the end, in his right hand. In his left hand he's holding a white painter's palate, the colors are raised. The Smurf is smiling with his tongue sticking out. The Smurf has a silver key chain attached to his head.
249/010	Emperor	Schleich/Peyo	1/1/78	Rubber Figure	2" High	Hong Kong	$3.00-4.00	The Smurf has a light blue body. He is wearing a bright yellow outfit, a bright yellow crown, gold hat, gold shoes, gold jewelry, and a red robe. He is holding his index finger on his right hand up in the air. The Smurf has a silver key chain attached on top of his head.
249/011	Tennis Star	Schleich/Peyo	1/1/78	Rubber Figure	2" High	Hong Kong	$3.00-4.00	The Smurf has a dark blue body. He is wearing a white tennis shirt and white shorts. He has a yellow wrist band on his right wrist. He is holding a red plastic tennis racket in his right hand. The tennis racket has holes in it. The Smurf has a silver key chain attached to his head.
249/012	Bowler	Schleich/Peyo	1/1/79	Rubber Figure	2" High	Hong Kong	$3.00-4.00	The Smurf has a dark blue body. He is holding a red bowling ball in his right hand. He is standing like he's going to throw the ball. The Smurf has a silver key chain attached to his head.
249/013	Ice Lolly	Schleich/Peyo	1/1/79	Rubber Figure	2" High	Hong Kong	$3.00-4.00	The Smurf has a medium blue body. He is eating a popsicle. The popsicle is red on the bottom and white on top. The popsicle is on a yellow stick. The Smurf is holding the popsicle in his right hand. He has his tongue sticking out of his mouth. The Smurf has a silver key chain attached to his head.
249/014	Golfer	Schleich/Peyo/W.B.	1/1/79	Rubber Figure	2" High	Hong Kong	$3.00-4.00	The Smurf has a dark blue body. He is wearing yellow pants, red socks, black shoes and no shirt. The Smurf is holding a gray golf club. The Smurf has a silver key chain attached to his head.
249/015	Naughty	Bully /Peyo		Rubber Figure	2" High	West Germany	$3.00-4.00	The Smurf has a medium blue body. He has his thumbs in his ears. He has his tongue sticking out. His tongue is bright red. The Smurf has a silver key chain attached to his head.
249/016	Present	Schleich/Peyo	1/1/76	Rubber Figure	2" High	Hong Kong	$3.00-4.00	The Smurf has a dark blue body. He is holding a white present with red ribbon. The Smurf is holding the present with both hands off to his right side. The Smurf has a silver key chain attached to his head.
249/017	Beer	Schleich/Peyo/W.B.	1/1/75	Rubber Figure	2" High	Hong Kong	$3.00-4.00	The Smurf has a dark blue body. He is holding a big yellow/orange mug of beer in his right hand. He has a big red mouth. The Smurf has a silver key chain attached to his head.
249/018	Ballerina	Schleich/Peyo	1/1/78	Rubber Figure	2" High	Hong Kong	$3.00-4.00	Smurfette has a dark blue body. She is wearing a white ballerina dress and shoes. She is in a pose. Smurfette is standing on a green base. Smurfette has a silver key chain attached to her head.
249/019	Roller-skating Smurfette	Schleich/Peyo	1/1/80	Rubber Figure	2" High	Hong Kong	$3.00-4.00	Smurfette has a dark blue body. She is wearing a white dress and white and red roller-skates. Smurfette's hair is dark yellow and has red ribbons in it. Smurfette has a silver key chain attached to her head.
249/020	Superman	Schleich/Peyo	1/1/80	Rubber Figure	2" High	Hong Kong	$3.00-4.00	The Smurf has a dark blue body. He is wearing a yellow cape, red shoes, red gloves, white hat, white shorts, and a black mask over his eyes. The Smurf has a silver key chain attached to his head.
249/021	Baseball Batter	Schleich/Peyo	1/1/80	Rubber Figure	2" High	Hong Kong	$3.00-4.00	The Smurf has a dark blue body. He is wearing a white uniform with red sleeves, a red belt, red socks, a white hat with a red visor, and black shoes. The Smurf is holding a tan and brown bat. The Smurf has a silver key chain attached to his head.
249/022	Football (American)	Schleich/Peyo/W.B.	1/1/80	Rubber Figure	2" High		$3.00-4.00	The Smurf has a medium blue body. The Smurf is wearing a yellow helmet, a red shirt with a yellow number 3, white pants and black shoes. He is holding a dark brown football in his right hand. His left hand is out to block. The Smurf has a silver key chain attached to his head.
249/023	Secretary	Schleich/Peyo/W.B.	1/1/81	Rubber Figure	2" High	Hong Kong	$3.00-4.00	Smurfette has a dark blue body. She is wearing a white dress, white hat and white shoes. Smurfette is holding a brown pencil in her right hand. Smurfette is holding a white pad with black writing and a silver binder edge in her left hand. Smurfette has a silver key chain attached to her head.
249/024	Cheerleader	Schleich/Peyo/W.B.	1/1/81	Rubber Figure	2" High	Hong Kong	$3.00-4.00	Smurfette has a medium blue body. Smurfette is wearing a white hat, white shoes and a white dress with a light blue collar and cuffs. She is holding red pompoms. Smurfette is standing on a dark green platform. Smurfette has a silver key chain attached to her head.
249/025	Grouchy	Schleich/Peyo/W.B.	1/1/83	Rubber Figure	2" High	Hong Kong	$3.00-4.00	The Smurf has a medium blue body. He has his fist clenched at his sides. He has a mean frown on his face. The Smurf has a silver key chain attached to his head.
249/026	Soccer Smurfette	Schleich/Peyo/W.B.	1/1/83	Rubber Figure	2" High	Hong Kong	$3.00-4.00	Smurfette has a medium blue body. She is wearing a white/orange/ red shirt, shorts, shoes and red socks. Smurfette has a white soccer ball attached to her left foot. Smurfette has a silver key chain attached to her head.

Set 269: Key chains- PVC Promotional Figure

269/001	40th Anniversary Key chain	Schleich/Peyo	1/1/98	Key chain	2"	West Germany	$15.00-20.00	The Smurf is wearing a gold wreath around his neck. He is holding a hand out to shake hands. The Smurf has a medium bright blue body.Attached to the key chain is a yellow, flat diamond- shaped piece of plastic. The plastic says 1958-1998, 40, Die Schlumpfe, The Smurfs, Les Schtroumpfs.
269/002	Holiday On Ice - Skater	Schleich/Peyo		Key chain	2"		$22.50-30.00	The Smurf has a medium blue body. The Smurf is wearing a red scarf and yellow gloves. The Smurf has silver blades on his feet. He is laying on his back. The back of his hat says Holiday On Ice. A silver key chain is attached to his head.
269/003	ASB Doctor	Schleich/Peyo	1/1/78	Key chain	2"	Germany/China	$11.00-15.00	The doctor is wearing a white uniform. He has a white bandage wrapped around his finger. He is carrying a yellow bag. The bag has an asb sticker on it. There is a silver key chain attached to his head.

Set 140: Key chains- Slide Card

140/001	Smurf Pond Scene			Slide Cards Key ring	2" X 1 1/2"		Canada	The key chain has 5 slide cards, each having a different picture on it. The pictures are of a pond scene, a Smurf with a shovel, a Smurf giving Smurfette flowers, 2 Smurfs holding hands and running, a Smurf with a stop sign, a Smurf holding a mug of beer, love struck Smurf, a Smurf on a carousel horse, a Smurf serenading Smurfette, and a Smurf working at a cluttered desk. Each side is different. Each one has a different saying and a different color background. The key chain is metal.$3.00-4.00

140/002	Smurf With A Horn				Slide Cards Key ring	2" X 1 1/2"	Canada		The key chain has 5 slide cards, each having a different picture on it. The pictures are of a Smurf with a horn, drummer Smurf, Papa pointing with a blue beard, a Smurf with a cello, Smurfette standing and wearing a red dress, a Smurf with a yellow knapsack over his shoulder, a Smurf on a bicycle, a Smurf on roller-skates, a Smurf carrying a white present, and a Smurf carrying a red present. Both sides have different pictures. The background on all of them is white. The key chain is meta$3.00-4.00
140/003	Papa And A Smurf In An Airplane				Slide Cards Key ring	1 1/2"W X 3"L			The metal key chain has 6 slide cards, each having a different picture on both sides. The pictures are of a Smurf and Papa in an airplane, a Smurf and Smurfette playing baseball, swimming, on a red motorcycle, in a red boat, in a red car, on a rocking horse, playing kickball, playing tennis, and skiing, a Smurf on roller-skates playing a trumpet with Papa watching, and Gargamel scaring a Smurf.

Set 139: Keychain Slide Puzzle

139/001	Jokey Smurf Key chain				Slide Puzzle Key ring	2"L X 1 3/4"W	Taiwan	$1.50-2.00	The tray is tan and has 15 mini slide pieces. The picture is of a Smurf carrying a yellow package with a red bow and a fuse on the end.
139/002	Sassette Fristi Key chain	Fristi	1/1/94	Pink	Slide Puzzle Key ring	2" X 2"	Brussels	$2.00-3.00	The tray is pink and has 8 slide pieces. The picture is of Sassette in a pink outfit holding an orange Fristi soda bottle. A promotion for De fruitigste drink yogurt.
139/003	Cake Smurf				Slide Puzzle Key ring	2"L X 1 3/4"W	Taiwan	$1.50-2.00	The tray is peach and has 15 mini slide pieces. The picture is of a Smurf carrying a yellow cake with white frosting and pink cherries
139/004	Smurf Jumping	Puppy	1/1/94	Green	Slide Puzzle Key ring	2" X 2"	Hong Kong	$1.50-2.00	The tray is green and has 8 mini slide pieces. The picture is of a Smurf jumping. He has his hands in the air. The background is white.

Set 78: Key chains- Soft Rubber

078/001	Papa Waving	Schleich	1/1/95		Soft Rubber Key chain	3"	China	$4.50-6.00	Papa is standing, waving with his right hand. Papa is wearing a red hat and red pants. He has a bright medium blue body. The key ring is a sturdy metal. The design is on one side, on the other side is black rubber.
078/002	Smurf On A Skateboard	Schleich	1/1/95		Soft Rubber Key chain	3"	China	$4.50-6.00	The Smurf is standing on a dark blue skateboard with bright yellow stars and bright yellow wheels. The Smurf is wearing a purple hat with yellow dots, a lime green tank top, and yellow and purple shorts. He has lime green and purple sneakers on. He has a bright medium blue body. The key ring is a sturdy metal. The design is on one side, on the other side is black rubber.
078/003	Baby With A Teddy Bear	Schleich	1/1/95		Soft Rubber Key chain	2 1/2"	China	$4.50-6.00	Baby is sitting, holding a light blue teddy bear. Baby is wearing a pink sleeper with a pink hat. He has a bright medium blue body. The key ring is a sturdy metal. The design is on one side, on the other side is black rubber.
078/004	Baby Sitting On The Moon	Schleich	1/1/95		Soft Rubber Key chain	3"	China	$4.50-6.00	Baby is sitting on a bright yellow half moon with his hands out at his sides. Baby is wearing a pink sleeper and a pink hat. He has a bright medium blue body. The key ring is a sturdy metal. The design is on one side, on the other side is black rubber.
078/005	Mermaid Smurfette	Schleich	1/1/95		Soft Rubber Key chain	3"	China	$4.50-6.00	Smurfette has a red fish tail, a red star in her hat, bright yellow hair, a white shell necklace around her neck, and no shirt. Her right arm is over her chest. She has a flirty look on her face. She has a bright medium blue body. The key ring is a sturdy metal. The design is on one side, on the other side is black rubber.
078/006	Sexy Smurfette	Schleich	1/1/95		Soft Rubber Key chain	3"	China	$4.50-6.00	Smurfette is wearing a bright orange shirt with green flowers on the skirt, a light pink shirt, white shoes, and a white hat with a dark pink bow in front. She has one hand in her hair and one on her hip. She has a flirty look on her face. She has a bright medium blue body. The key ring is a sturdy metal. The design is on one side, on the other side is black rubber.

Set 34: Kinder Egg Play Sets

034/001	Cook Smurf With Mushroom House	Kinder	1/1/95		Kinder Egg Play sets	4"	Italy	$22.50-30.00	Mushroom house is 4" high and opens up and has 2 floors. The top floor has a bedroom and the bottom floor is a kitchen. The outside of the mushroom has a white base and a red roof. The Smurf is 1 1/2" high and he is wearing a green apron. The Smurf is made of hard plastic.
034/002	Farmer Smurf With A Tractor	Kinder	1/1/95		Kinder Egg Play sets		Italy	$22.50-30.00	The tractor is red with green wheels and 3 1/2" high by 3 1/2" long, and is pulling a yellow wagon with green wheels. The wagon is 2 1/2" high by 3" long. Farmer Smurf is 1 1/2" high and made of hard plastic. He sits in the tractor. He is wearing brown pants and a yellow straw hat.
034/003	Miller Smurf With A Windmill	Kinder	1/1/95		Kinder Egg Play Sets		Italy	$22.50-30.00	The windmill is 6 1/2" high. The windmill is white with a blue roof and orange blades ,window, and door. There is a little mushroom on the side. The Smurf is 1 1/2" high and made of hard plastic. The Smurf has a yellow miller bag in front of him.
034/004	Bricklayer Smurf With A Lighthouse	Kinder	1/1/95		Kinder Egg Play Sets		Italy	$22.50-30.00	The lighthouse is 7" high, the lighthouse is white with a yellow roof, door and balcony with green sticker bushes on the side. The Smurf is 1 1/2" high and made of hard plastic. The Smurf is holding a trowel in one hand and a brick in the other hand.
034/005	Lazy Smurf With Hammock	Kinder	1/1/95		Kinder Egg Play Sets		Italy	$22.50-30.00	2 trees are 6" high with a white hammock hung between them. The Smurf is 1 1/2" high but is laying in the hammock holding a pop bottle. He is made of hard plastic.
034/006	Smurf In Log Car	Kinder	1/1/91		Kinder Egg Play Sets		Italy	$25.00-35.00	The car is 3" high by 4 " long. The car is red with yellow wheels and a yellow grill. The front has 2 yellow candles on each side. The back has a mushroom roof. The interior is yellow. The Smurf stands inside. The front of the car has a red crank handle.
034/007	Smurf Tower	Kinder	1/1/91		Kinder Egg Play Sets	5 1/2"	Italy	$25.00-35.00	The bottom half of the tower is white. It has a red and white mushroom roof. There is a yellow chimney. The front has a yellow porch. The Smurf is hard plastic. He has a troubled look on his face. His arms are out at his sides.

Set 276: Kinder Egg Puzzles

276/001	Smurf In Tree	Peyo	1/1/96		Mini Puzzle	2 3/4" x 3 3/4"	Brussels	$2.00-3.00	A Smurf is hanging from a branch trying to pick an acorn. There is a Smurf under him pushing a wheel barrow. This is 1 puzzle of 4 to make a larger puzzle.
276/002	Smurf And Squirrel	Peyo	1/1/96		Mini Puzzle	2 3/4" x 3 3/4"			A Smurf is holding 2 acorns. A squirrel is walking up to the Smurf. The squirrel has a mad look on his face. This is puzzle 2 of 4 to make a larger puzzle.
276/003	2 Smurfs Trying To Catch A Falling Acorn	Peyo	1/1/96		Mini Puzzle	2 3/4" x 3 3/4"	Brussels	$2.00-3.00	2 Smurfs are holding a round disc trying to catch a falling acorn. This is puzzle 3 of 4 to make a larger puzzle.
276/004	Smurf Rolling An Apple	Peyo	1/1/96		Mini Puzzle	2 3/4" x 3 3/4"	Brussels	$2.00-3.00	A Smurf looks like he's rolling an apple. Another Smurf is standing, facing Smurfette. There are 2 berries on the side of the Smurf. This is puzzle 4 of 4 to make a larger puzzle.
276/005	Nat Strummin Strings	Peyo	1/1/96		Mini Puzzle	2 3/4" x 3 3/4"	Brussels	$2.00-3.00	Nat has a piece of wood attached to an upside down fry pan. He is strumming the strings. Another Smurf is standing, looking startled. This is puzzle 1 of 4 to make a larger puzzle.
276/006	Snappy Smurfling Playing Drums	Peyo	1/1/96		Mini Puzzle	2 3/4" x 3 3/4"	Brussels	$2.00-3.00	Snappy is playing drums made from kettles. There are outlines of mushroom houses in the background. This is puzzle 2 of 4 to make a larger puzzle.
276/007	Slouchy Smurfling Playing Saxophone	Peyo	1/1/96		Mini Puzzle	2 3/4" x 3 3/4"	Brussels	$2.00-3.00	Slouchy Smurfling is playing a saxophone. Sassette is dancing. This is puzzle 3 of 4 to make a larger puzzle.
276/008	Grandpa, Baby And Smurfette Dancing	Peyo	1/1/96		Mini Puzzle	2 3/4" x 3 3/4"	Brussels	$2.00-3.00	Grandpa, Baby and Smurfette are dancing. This is puzzle 4 of 4 to make a larger puzzle.

Set 33: Kinder Egg Smurfs- To Assemble

033/001	Soccer Referee With Flag	Kinder	1/1/91		Kinder Egg Kits	2"	Italy	$11.00-15.00	2" figure is made of small, colorful plastic parts that snap together. Kit includes decals and a full color cartoon pamphlet & instructions. Smurf is wearing a black uniform and is holding a red flag
033/002	Smurf With a jack-in-the-box	Kinder	1/1/91		Kinder Egg Kits	2"	Italy	$11.00-15.00	2" figure is made of small, colorful plastic parts that snap together. Kit includes decals & a full color cartoon pamphlet & instructions. Smurf is holding a yellow jack-in-the-box with a blue clown.
033/003	Smurfette	Kinder	1/1/91		Kinder Egg Kits	2"	Italy	$11.00-15.00	2" figure is made of small, colorful plastic parts that snap together. Kit includes decals & a full color cartoon pamphlet & instructions. Smurfette is wearing a white dress.
033/004	Smurf Offering A Box Of Kinder Eggs	Kinder	1/1/91		Kinder Egg Kits	2"	Italy	$11.00-15.00	2" figure is made of small, colorful plastic parts that snap together. Kit includes decals & a full color cartoon pamphlet & instructions. Smurf is holding a blue box with Kinder Eggs inside.
033/005	Winter Smurf	Kinder	1/1/91		Kinder Egg Kits	2"	Italy	$11.00-15.00	2" figure is made of small, colorful plastic parts that snap together. Kit includes decals & a full color cartoon pamphlet & instructions. Smurf is wearing pink mittens and a scarf.
033/006	Papa Referee	Kinder	1/1/91		Kinder Egg Kits	2"	Italy	$15.00-20.00	2" figure is made of small, colorful plastic parts that snap together. Papa is standing, holding a blue whistle in his right hand and a yellow pad in his left hand. Papa is wearing a black shirt with a yellow sticker on the front , a red hat and red pants.
033/007	Standing Soccer Smurf	Kinder	1/1/91		Kinder Egg Kits	2"	Italy	$11.00-15.00	2" figure is made of small, colorful plastic parts that snap together. The Smurf is standing, wearing an orange shirt with a yellow sticker on the front, white shorts and black shoes. The Smurf has a white soccer ball. There is a number 10 on the back of his shirt.
033/008	Soccer Smurf Squatting	Kinder	1/1/91		Kinder Egg Kits	2"	Italy	$11.00-15.00	2" figure is made of small, colorful plastic parts that snap together. The Smurf is squatting. He is wearing yellow shorts, a yellow shirt, black shoes and a white hat. The Smurf has a white soccer ball.
033/009	Vanity Smurf	Kinder	1/1/91		Kinder Egg Kits	2"	Italy	$11.00-15.00	2" figure is made of small, colorful plastic parts that snap together. The Smurf is standing, holding a blue mirror in his right hand. The Smurf has a white flower behind his left ear.
033/010	Smurf With Wheelbarrow	Kinder	1/1/96		Kinder Egg Kits	2"	Italy	$3.50-5.00	2" figure is made of small, colorful plastic parts that snap together. The Smurf is standing, pushing an orange wheelbarrow. The Smurf's eye's and mouth are painted on.
033/011	Smurf Carrying A Sprinkling Can	Kinder	1/1/96		Kinder Egg Kits	2"	Italy	$3.50-5.00	2" figure is made of small, colorful plastic parts that snap together. The Smurf is standing, holding a green watering can in his right hand. The Smurf's eye's and mouth are painted on.

033/012	Smurf With A Rake	Kinder	1/1/96		Kinder Egg Kits	2"	Italy	$3.50-5.00	2" figure is made of small, colorful plastic parts that snap together. The Smurf is standing, holding an orange rake in his right hand. The Smurf's eye's and mouth are painted on.
033/013	Smurf Leaning On A Shovel	Kinder	1/1/96		Kinder Egg Kits	2"	Italy	$3.50-5.00	2" figure is made of small, colorful plastic parts that snap together. The Smurf is standing, leaning on a green shovel. His left foot is resting on top of the shovel. The Smurf's eye's and mouth are painted on. The Smurf's eyes are closed.
033/014	Smurf With A Backpack	Kinder	1/1/96		Kinder Egg Kits	2"	Italy	$3.50-5.00	2" figure is made of small, colorful plastic parts that snap together. The Smurf is standing, wearing a yellow backpack. The Smurf's eye's and mouth are painted on.
033/015	Papa Bandleader	Kinder	1/1/96		Kinder Egg Kits	2"	Italy	$3.50-5.00	2" figure is made of small, colorful plastic parts that snap together. Papa is standing, holding a blue baton in his right hand. Papa is leading music with his left hand. Papa's eye's and mouth are painted on.
033/016	Smurf Playing an Accordion	Kinder	1/1/96		Kinder Egg Kits	2"	Italy	$3.50-5.00	2" figure is made of small, colorful plastic parts that snap together. The Smurf is standing, playing a red accordion. The Smurf's eye's and mouth are painted on.
033/017	Smurf Playing A Violin	Kinder	1/1/96		Kinder Egg Kits	2"	Italy	$3.50-5.00	2" figure is made of small, colorful plastic parts that snap together. The Smurf is standing, playing a red violin. The Smurf's eye's and mouth are painted on. The Smurf has his eye's closed.
033/018	Smurfette Waving	Kinder	1/1/96		Kinder Egg Kits	2"	Italy	$3.50-5.00	2" figure is made of small, colorful plastic parts that snap together. Smurfette is standing, waving. Smurfette has a white dress, white hat and white shoes. There is a sticker by her feet that has flowers .
033/019	Smurf Playing A Trumpet	Kinder	1/1/96		Kinder Egg Kits	2"	Italy	$3.50-5.00	2" figure is made of small, colorful plastic parts that snap together. The Smurf is standing, holding a yellow trumpet to his mouth.
033/020	Black Smurf With Whip	Peyo		Black	Kinder Egg Kits	2"	Italy	$15.00-20.00	2" figure is made of small, colorful plastic parts that snap together. The black Smurf is holding a hand in the air. In his hand he is holding a black whip.
033/021	Smurf With Torch	Componi Puffi			Kinder Egg Kits	3"	Italy	$8.25-12.00	2" figure is made of small, colorful plastic parts that snap together. The Smurf is standing, holding a yellow plastic torch in the air. The Smurf's arms

Set 40: Latch Hook Kits

040/001	Smurf Face	Wonder Art		Yellow	Yarn Latch Hook	15" x 15"	Hand made	$6.75-9.00	This latch hook is yellow with a Smurf face in the middle.
040/002	Smurf With A Mushroom House	Wonder Art		Multi	Yarn Latch Hook	18" W x 24" L	Hand made	$6.75-9.00	A Smurf is sitting in the grass looking at a yellow mushroom house with a red and white roof in the background.
040/003	Smurf With His Head In A Wreath	Wonder Art			Yarn Latch Hook	18" W x 24" L	Hand made	$9.00-12.00	A rusty color latch hook with a Smurf standing in the middle with his head through a green X-mas wreath.
040/004	Smurfette Smelling Flowers	Wonder Art		Light Blue	Yarn Latch Hook	18" W x 24" L	Hand made	$6.75-9.00	The latch hook is light blue and it has Smurfette kneeling in the grass smelling a flower.
040/005	Smurf Leaning On A Heart	Wonder Art		Yellow	Yarn Latch Hook	18" W x 24" L	Hand made	$6.75-9.00	The latch hook is yellow and it has a Smurf leaning on a red heart with an orange daisy in his other hand.
040/006	King Smurf Face	Wonder Art		Red	Yarn Latch Hook	15" x 15"	Hand made	$11.00-15.00	The latch hook is red with King Smurf's face in the middle. He has a rusty color hat with a yellow crown.
040/007	Man In The Moon Floor Rug	Wonder Art		Yellow	Yarn Latch Hook		Hand made	$15.00-20.00	The latch hook is yellow and has a Smurf sleeping on the moon. The latch hook is a big floor rug.
040/008	Smurf Face Pillow				Latch Hook Pillow	14 1/2" x 14 1/2"	Hand made	$4.50-6.00	This latch hook is yellow with a Smurf face in the middle. The pillow has a red flowered back.
040/009	Papa Face Pillow				Latch Hook Pillow	14 1/2" x 14 1/2"	Hand made	$4.50-6.00	This latch hook is yellow with Papa Smurf's face in the middle. Papa has a red hat on. The back of the pillow is red velvet.

Set 53: License Plates

053/001	S M U R F	TM	1/1/83	Red	Metal	2"L x 5"W	U.S.A.	$3.00-4.00	There are 4 Smurfs dancing between the letters of S M U R F. The license plate is red.
053/002	Smurfs Ride Free	TM	1/1/83	Dark Yellow	Metal	2"L x 5"W	U.S.A.	$3.00-4.00	There is a Smurf riding a red bicycle. The Smurf is wearing an orange shirt and shorts. The license plate is dark yellow.
053/003	I Brake For Smurfs	TM	1/1/83	Yellow	Metal	2"L x 5"W	U.S.A.	$3.00-4.00	A Smurf is holding a stop sign and has his other hand in the air. The license plate is yellow.
053/004	Keep On Smurfin	TM	1/1/83	Green	Metal	2"L x 5"W	U.S.A.	$3.00-4.00	A Smurf is on an orange leaf skateboard. The Smurf is wearing a yellow shirt and white pants. The license plate is green.
053/005	Honk If You Like Smurfs	TM	1/1/83	Gray	Metal	2"L x 5"W	U.S.A.	$3.00-4.00	Papa is holding a horn behind his back. The license plate is gray.
053/006	I Love Smurfs	TM	1/1/83	Pink	Metal	2"L x 5"W	U.S.A.	$3.00-4.00	A Smurf is on the left side holding a bouquet of flowers with a dreamy look on his face. There are little hearts behind his head. Smurfette is on the other side holding a flower and looking dreamy. There are hearts behind her head to. In the middle of the license plate it says "I (there is a red heart) Smurfs" in orange letters. The license plate is pink.
053/007	Kings Island Tina	TM	1/1/83	White	Plastic	2"L X 4" W	U.S.A.	$5.00-7.00	The license plate is made of plastic and is white. The license plate has King Smurf sitting on a mushroom under a tree in one corner. There are mushroom houses on the bottom. At the top it says "Kings Island" and in the middle it says "Tina" both are in black letters.

Set 220: Light Fixtures

220/001	Smurf On Moon	W. Berrie/Peyo	1/1/84	White	Round Light Globe	8 1/2" Dia	U.S.A.	$37.50-50.00	The globe is round and made of glass. It has a Smurf laying on 1/4 of a moon. There are red, yellow and blue stars around the moon. The Smurf has a happy expression on his face. The picture is on 2 sides.
220/002	Smurfs On Carousel Horses	W. Berrie/Peyo	1/1/84	White	Oval Light Globe	8 1/2" Dia	U.S.A.	$37.50-50.00	The globe is almost oval shape and made of glass. There are 3 Smurfs, Papa, Baby and Smurfette on carousel horses. The horses are all different. The picture goes around the whole globe.
220/003	Smurf Mushroom Lamp	Homemade			Ceramic Lamp		Hand made	$55.00-75.00	The lamp is a scared Smurf standing under a mushroom. The mushroom is yellow and brown with blue spots. The Smurf is holding the stem of the mushroom. The Smurf is on a round ceramic base.

Set 164: Lite Brite

164/001	Picture Play Lite	Janex Corp.	1/1/82	White	Lite- Brite	8" x 8"	Hong Kong	$25.00-35.00	The lite- brite is white with a red button in the top corner to light the board. There are 6 Smurf pictures and 12 black and white sheets. Comes with pink, blue, green, and yellow pegs. 1 Smurf sheet is unused the other 5 are used but in good condition.
164/002	Picture & Pegs Accessory Pack	Hasbro	1/1/83		Lite-Brite Pictures		U.S.A.	$18.00-25.00	The cover has Smurfette sitting on top of a pink lite - brite. In the back is sky, mountains and mushroom houses. 4 Smurfs are putting a picture together on the lite - brite. Included in the pack: 12 pictures, 8 blank sheets and 8 colors of pegs.

Set 65: Loc Blocs

065/001	Dr. Smurf's Office	Entex			Lego's		U.S.A.	$25.00-35.00	180 piece lego set to make a doctors office. Included is a vinyl Smurf, adjustable over head lamp and an instrument cabinet.
065/002	Papa Smurf's Laboratory	Entex	1/1/82		Lego's		U.S.A./Japan	$25.00-35.00	200 piece lego set to make Papa's laboratory. Included is a vinyl Papa Smurf.

Set 166: Luggage

166/001	Roller-skating Smurfette Suitcase	W. Berrie	1/1/83	Pink	Suitcase	12"L X 18"W	Taiwan	$22.50-30.00	The suitcase is pink with purple hearts all over it. On the front is a big purple heart with Smurfette in the middle. Smurfette has roller-skates on. Above the big purple heart it says "Smurfette" in white puffy letters. The suitcase has a purple handle and a purple border.
166/002	Soccer Smurf Duffel Bag	BBC Imports		Light Blue/White	Cloth Duffel Bag	10 /12"L X 12"W	Taiwan	$2.00-3.00	The bag is light blue with white ends, 2 white carry straps and a white removable shoulder strap. The front has a Smurf kicking a soccer ball. The Smurf is wearing a white shirt, white shoes, white hat and black shorts. It says "SMURF" in white letters above his head.
166/003	Smurf On A Cloud Have A Smurfy Day Duffel Bag	BBC Imports		Dark Blue	Cloth Duffel Bag	7"L x 10"W	Taiwan	$2.00-3.00	The bag is dark blue with red carry straps. The front has a Smurf sitting on a light blue cloud. The cloud says "Have A Smurfy Day" inside. There is a rainbow behind the Smurf . It says "SMURF" in red letters above his head.
166/004	Orange Suitcase	Dupuis	1/1/76	Orange	Suitcase	7" x 11"		$25.00-35.00	The suitcase is orange. It has a Smurf in the center. The Smurf is holding up one hand. The suitcase has black corners. It has a metal lock and keys.
166/005	Smurf Travel Case	ARC	1/1/82	Blue	Plastic Travel Case	6" x 9"		$10.50-14.00	The case is blue and made of plastic. It has a Smurf carrying a knapsack and Smurfette carrying a suitcase on the front. The picture is just an outline. The case comes with a yellow comb and mirror. The mirror has Smurfette carrying a blue suitcase. It says "Just Smurfin' Along" in red letters.
166/006	Yellow Hard Suitcase	BP	1/1/81		Hard Suitcase	7" x 11"	Australia	$25.00-35.00	The suitcase is yellow. It has pictures on all sides of it. The front has a Smurf kicking a ball to another Smurf. A doctor and Smurfette nurse standing by a Smurf holding a wrench. A Smurf on parallel bars. A Smurf lifting a weight. A Smurf dressed as superman. The corners and handle are yellow plastic. It has a metal lock and keys.
166/007	Smurfs Standing In Line At Puffi Aeroporto	I Puffi	1/1/83	Light Blue	Carrying Case	10 1/2" x 15"	Italy	$35.00-45.00	

Set 25: Lunch Boxes & Thermos

025/001	Queen Smurfette Being Honored (Metal)	King-Seeley Thermos	1/1/82	Blue	Metal Lunch Box	8 1/2" x & 7 1/2"		$135.00-180.00	Queen Smurfette is standing on a mushroom with Jokey, Papa, a harp Smurf, beer Smurf, cake Smurf, and King Smurf, all honoring her. There is a mushroom house in the background.
025/002	Queen Smurfette Being Honored (Plastic)	King-Seeley Thermos	1/1/82	Light Blue	Plastic Lunch Box	8 1/2" x 7 1/2"	U.S.A.	$11.00-15.00	Queen Smurfette is standing on a mushroom with Jokey, Papa, a harp Smurf, beer Smurf, cake Smurf and King Smurf, all honoring her. There is a mushroom house in the background. I have 3 different kinds: 1 latch, 2 latches, and one is a lunch pail that opens in half.
025/003	Smurflings Surfing	King-Seeley Thermos	1/1/82	Dark Blue	Plastic Lunch Box	8 1/2" x & 7 1/2"	U.S.A.	$11.00-15.00	I have 2 of these lunch boxes. One is light blue and one is dark blue. Both come with a thermos that is white with a Smurf face in the middle. The picture is of 4 Smurflings surfing in a tide. Lunch box says SMURF'S UP.

025/004	Smurfette	King-Seeley Thermos	1/1/82	Pink	Plastic Lunch Box	8 1/2" x & 7 1/2"	U.S.A.	$11.00-15.00	The picture is of Smurfette inside of a heart with one hand in her hair and her other hand is holding a yellow flower. The background for the picture is red and it has butterflies and flower in the corners. The lunch box has a pink thermos with Smurfette sitting on a mushroom.
025/005	Smurfs And Smurflings Eating	King-Seeley Thermos	1/1/87	Dark Blue	Plastic Lunch Box	8 1/2" x & 7 1/2"	U.S.A.	$15.00-20.00	The picture is of 6 Smurfs eating. The background is red and white checkered. The lunch box comes with a dark blue thermos with a Smurf eating a piece of pie.
025/006	Smurf Pond Scene	King-Seeley Thermos	1/1/82	Light Blue	Plastic Lunch Box	8 1/2" x & 7 1/2"	U.S.A.	$15.00-20.00	Gargamel and Azreal are hiding behind a tree watching the Smurfs. A Smurf is floating in the pond, one is fishing, one is jumping off a board, and Papa is holding a ball.
025/007	It's Lunch time	Unique	1/1/83	Brown	Lunch Bags	8 1/2" x & 7 1/2"	U.S.A.	$6.00-8.00	There is a Smurf holding a cookie and a lunch bag on the front of the paper bags. On the bottom it says "This Lunch Belongs To:" There are 25 Flat Bottom Bags.
025/008	Smurfette Lunch Bag	Applause/Peyo	1/1/84	Pink	Lunch Bag	8 1/2" x & 7 1/2"	China	$4.50-6.00	The lunch bag is cloth. It has Smurfette on the front. Smurfette is wearing a pink dress. There is a big purple heart behind her. The bag is pink with purple hearts. It has a velcro close top. The bag has a purple strap.
025/009	Smurf Serenading Smurfette Thermos	King-Seeley Thermos	1/1/82	Yellow	Thermos	8 1/2" x & 7 1/2"	U.S.A.	$4.50-6.00	A Smurf is serenading Smurfette. Smurfette is laying on top of a mushroom. The top Says "SMURF" in blue letters. The thermos is yellow with a white cap.
025/010	Smurf Giving Smurfette Flowers Thermos	King-Seeley Thermos	1/1/82	Yellow	Thermos	8 1/2" x & 7 1/2"	U.S.A.	$4.50-6.00	The thermos is yellow with a white cap. The thermos has a Smurf walking, carrying a red daisy towards Smurfette. Smurfette is standing in a bunch of daisies.
025/011	Drummer Smurf Lunch Box	Peyo-SEPP-BP	1/1/79	Green	Plastic Lunch Box	8 1/2" x & 7 1/2"	Australia	$10.00-25.00	The lunch box is green. It has a decal of a drummer Smurf on it. The Smurf has a red and white drum on a strap that's attached around his neck. He is holding 2 red drum sticks.
025/012	Papa Leaning On A Stump Lunch Box	BP		Light Blue	Plastic Lunch Box	8 1/2" x & 7 1/2"		$18.00-25.00	Papa Smurf is sitting, leaning against a tree stump. He is writing in a book. There is a butterfly flying above. There are flowers next to the stump.

Set 130: Magazines- German Die Schlumpfe Large

130/001	Start Frei Fur Die Grobe Sause	Bastei	1/1/92		German Comic Book	12"L X 8 1/3"W	Germany	$3.50-5.00	The cover is yellow with a picture of a Smurf in a log car with green wheels and a blue balloon for the exhaust. A Smurf is holding a checkered flag and Smurfette is standing on the other side of the car.
130/002	Ein Gespenst Wird Gebremst	Bastei	1/1/93		German Comic Book	12"L X 8 1/3"W	Germany	$3.50-5.00	The cover is blue and it has a picture of a ghost standing on a stool. There is a Smurf peaking in the door at the ghost. The Smurf dropped a mouse on the floor.
130/003	Ein Garten Ladt Die Schlumpfe Ein	Bastei	1/1/93		German Comic Book	12"L X 8 1/3"W	Germany	$3.50-5.00	The cover is pink and has a picture of Smurfette standing in flowers, a Smurf throwing a leaf, a Smurf wearing a scarf standing in snow, and a Smurf on a beach wearing goggles, flippers and an inner-tube around his waist.
130/004	Ein Grober Schreck Fur All Die Schlumpfe...	Bastei	1/1/93		German Comic Book	12"L X 8 1/3"W	Germany	$3.50-5.00	The cover is light blue and has a picture of a Smurf tangled in ropes hanging from a cliff. Another Smurf is climbing up the side of the mountain.
130/005	Ein Dino Zum Schmusen	Bastei	1/1/94		German Comic Book	12"L X 8 1/3"W	Germany	$3.50-5.00	The cover is off- white and has a picture of a Smurf sitting on a big green dinosaur head. There are 3 Smurfs by 3 mushroom houses below the dinosaur.
130/006	Gut Geschlumpft Ist Halbe Arbeit	Bastei	1/1/94		German Comic Book	12"L X 8 1/3"W	Germany	$3.50-5.00	The cover is light pink and has a picture of a big tree and grass. On one side of the tree 3 Smurfs are working with shovels and picks digging in the dirt. In front of the tree a Smurf is leaning on a shovel, sleeping with a wheelbarrow, and holding Smurf signs behind him.
130/007	Viel Geheule Um Eine Kleine Eule	Bastei	1/1/94		German Comic Book	12"L X 8 1/3"W	Germany	$3.50-5.00	The cover is blue and has a picture of an owl sitting on a bird house in a tree. There is a Smurf walking to the tree. The Smurf is wrapped in a blanket. The village is silhouetted in the background.
130/008	Lautes Gergrummel Im Tunnel	Bastei	1/1/94		German Comic Book	12"L X 8 1/3"W	Germany	$3.50-5.00	The cover is blue and gray. The picture is of 2 Smurfs in a cave digging while Gargamel's foot is coming through the top of the cave.
130/009	Ein Wespenstich Mit Folgen	Bastei	1/1/94		German Comic Book	12"L X 8 1/3"W	Germany	$3.50-5.00	The cover is green and has a picture of a Smurf with a bandage and a ladybug on his nose. The Smurf is sitting in a flower bed.
130/010	Die Verzaubert Traume	Bastei	1/1/94		German Comic Book	12"L X 8 1/3"W	Germany	$3.50-5.00	The cover is blue and has a picture of a Smurf sleeping in bed with an open book laying on his stomach and a candle burning on the night table. The Smurf is dreaming of a Smurf reading a book.
130/011	Viel Rabatz Auf Dem Rummelplatz	Bastei	1/1/94		German Comic Book	12"L X 8 1/3"W	Germany	$3.50-5.00	The cover is blue and green. The picture is of a Smurf playing a bean- ball toss and Brainy is behind the game. Brainy's glasses are flying off because he was hit with a ball. There is a Smurf standing, running the game and laughing.
130/012	Blumenstraub Und Hochzeitsschmaus	Bastei	1/1/94		German Comic Book	12"L X 8 1/3"W	Germany	$3.50-5.00	The cover is green and has a picture of Gargamel carrying daisies. There are pots tied to Gargamel's shirt and 3 Smurfs are behind a bush laughing.
130/013	Uberraschung Im Schlumpfdorf	Bastei	1/1/91		German Comic Book	12"L X 8 1/3"W	Germany	$3.50-5.00	The cover is yellow. The picture on the cover is of a Smurf opening a trick book and Gargamel's face popping out. There is a Smurf and Papa standing behind and laughing. Brainy is off to the side and he looks like he's giving a lecture. Also have one from 1979.
130/014	Gluckstag Fur Den Uberraschungs - Schulumpf	Bastei	1/1/91		German Comic Book	12"L X 8 1/3"W	Germany	$3.50-5.00	The cover is blue. The picture is of 2 Smurfs holding another Smurf up to smell a sunflower next to the mushroom house. There is a Smurf on the side of the house standing on a barrel trying to open the barrel.
130/015	Grober Wirbel Im Schlumpf - Dorf	Bastei	1/1/91		German Comic Book	12"L X 8 1/3"W	Germany	$3.50-5.00	The cover is pink. The picture is of a windmill with a swing attached to the blades. There are 4 Smurfs in the 4 swings with Papa and another Smurf looking on. The bottom of the windmill is cream and the top is blue.
130/016	Buhne Frei-Fur Die Schlumpferei	Bastei	1/1/91		German Comic Book	12"L X 8 1/3"W	Germany	$3.50-5.00	The cover is of an outside, night scene. There are 4 Smurfs sitting on a bench watching 2 Smurfs put on a puppet show.
130/017	Wo Geht's Denn Hier Ins Dorf?	Bastei	1/1/91		German Comic Book	12"L X 8 1/3"W	Germany	$3.50-5.00	The cover is yellow and green. The picture is of 5 Smurfs running around a tree stump. Brainy is trying to read a map. There is a Smurf under a bush laughing and a sign laying next to him.
130/018	Das Tolle Spiel Der Schlumpfe	Bastei	1/1/91		German Comic Book	12"L X 8 1/3"W	Germany	$3.50-5.00	The cover is blue and has a picture of a Smurf standing on a grassy bank, fishing. There is a Smurf in a scuba outfit wrapped to the end of the fishing pole.
130/019	Wirbel Um Schlumpfinchen	Bastei	1/1/91		German Comic Book	12"L X 8 1/3"W	Germany	$3.50-5.00	The cover is orange and yellow. The picture is of 2 Smurfs in a rollercoaster car and 3 other Smurfs are climbing up the track to work on the track.
130/020	Ein Kaninchen Fur Schlumpfinchen	Bastei	1/1/91		German Comic Book	12"L X 8 1/3"W	Germany	$3.50-5.00	The cover is lime green and orange. Papa, Smurfette and 2 other Smurfs are looking at a blue rabbit. The Smurfs have troubled expressions on their faces.
130/021	Der Verschlumpfte Zauberbrunnen	Bastei	1/1/91		German Comic Book	12"L X 8 1/3"W	Germany	$3.50-5.00	The cover is green. The picture is of Smurfette standing by a well, holding Baby Smurf. Baby is crying because his rattle is falling into the well. Another Smurf is standing on the other side of the well. There is a black bird sitting on top of the well.
130/022	Wer Schon Sein Will - Muss Schlumpfen!	Bastei	1/1/91		German Comic Book	12"L X 8 1/3"W	Germany	$3.50-5.00	The cover is yellow. The picture is of a Smurf sitting on a tree stump looking at himself in a mirror and crying. Gargamel and Azreal are hiding in the bushes.
130/023	Ein Schlumpf Kommt Selten Allein	Bastei	1/1/92		German Comic Book	12"L X 8 1/3"W	Germany	$3.50-5.00	The cover is yellow. The picture is of 7 Smurfs that all look identical and are walking around. Papa is looking at the Smurfs with a puzzled look on his face.
130/024	Drei Schlumpfe Aus Papier	Bastei	1/1/92		German Comic Book	12"L X 8 1/3"W	Germany	$3.50-5.00	The cover is pea green. The picture is of a Smurf standing with a broken mirror, a Smurf with food over his head, a Smurf with a bandage over his head, Painter with goop on his palate, and an angry Smurf running after 3 paper dolls.
130/025	Schreck In Der Abendstunde	Bastei	1/1/92		German Comic Book	12"L X 8 1/3"W	Germany	$3.50-5.00	The cover is dark blue. The picture is of Gargamel standing outside his castle looking angry. Papa is standing behind Gargamel and laughing.
130/026	Eine Nervensage Wird Ausgetrickst	Bastei	1/1/93		German Comic Book	12"L X 8 1/3"W	Germany	$3.50-5.00	The cover is light blue. The picture is of Jokey carrying a present with a fuse on the end. 2 Smurfs are running in the background. There are mushroom houses in the background.
130/027	Die Grobe Kataschlumpfe	Bastei	1/1/93		German Comic Book	12"L X 8 1/3"W	Germany	$3.50-5.00	The cover has Papa Smurf walking out of his house. Papa is gray from smoke and smoke is coming out of his house. Outside the house a Smurf has turned gray and 3 other Smurfs are standing around looking troubled.
130/028	Eis Schatten Spielt Verruckt	Bastei	1/1/93		German Comic Book	12"L X 8 1/3"W	Germany	$3.50-5.00	The cover is blue and has a picture of a Smurf sitting, hugging a globe (Earth). The globe is blue and green
130/029	Schlumpfinchens Zauberspruch	Bastei	1/1/93		German Comic Book	12"L X 8 1/3"W	Germany	$3.50-5.00	The cover is yellow and has a picture of a Smurf standing on a log hugging a monkey's face.
130/030	Gefahrliche Schatzsuche	Bastei	1/1/93		German Comic Book	12"L X 8 1/3"W	Germany	$3.50-5.00	The cover is light blue. The picture is of an ocean with a dolphin jumping out of the water. There is a Smurf riding on the dolphin. There is a Smurf in the water wearing snorkeling gear, and another Smurf is floating on a yellow plastic dolphin.
130/031	Ein Verhexter Plan	Bastei	1/1/93		German Comic Book	12"L X 8 1/3"W	Germany	$3.50-5.00	The cover is pink. The picture is of Smurfette standing on a scale looking startled. The springs are flying out of the scale. There is a Smurf sitting on the back of the scale eating.
130/032	Schnell Weg Ins Versteck	Bastei	1/1/93		German Comic Book	12"L X 8 1/3"W	Germany	$3.50-5.00	The cover is light blue. The picture is of a Smurf trapped under a large gold bell working on it. There is another Smurf dressed in an Easter bunny outfit holding a red egg in one hand, and handing a wrench to the Smurf under the bell.
130/033	Ein Schlumpf Wachst Uber Sich Hinaus	Bastei	1/1/93		German Comic Book	12"L X 8 1/3"W	Germany	$3.50-5.00	The cover is yellow. The picture is of a large Smurf sitting and taking really tiny Smurfs out of a pink box.
130/034	Die Schlumpfschule	Bastei	1/1/93		German Comic Book	12"L X 8 1/3"W	Germany	$3.50-5.00	The cover is of a beach scene. A Smurf is laying by the beach sun bathing. Another Smurf has a backpack, blankets, a shovel, and hammers on his back. He is climbing up the hill in the hot sun. The Smurf is panting.
130/035	Ein Fest Fur Ritter Ysengrin	Bastei	1/1/93		German Comic Book	12"L X 8 1/3"W	Germany	$3.50-5.00	The cover is blue. The picture is of a night party scene. Smurfette with black hair is kissing a Smurf. Smurfette is wearing a pink sun dress and dancing. A Smurf is carrying a cake. A Smurf is wearing a green clown outfit. Papa is walking around. There is another Smurf climbing up the lamp pole.
130/036	Die Schlumpfwehr, Die Ist Da!	Bastei	1/1/91		German Comic Book	12"L X 8 1/3"W	Germany	$3.50-5.00	The cover has 8 Smurfs and Smurfette running around in the rain with umbrellas.
130/037	Gurgelhals Gibt Auf	Bastei	1/1/91		German Comic Book	12"L X 8 1/3"W	Germany	$3.50-5.00	The cover has a mining Smurf leading Gargamel into the village. Gargamel is carrying a white surrender flag. 3 other Smurfs are scattering away and Papa is looking at Gargamel angrily.
130/038	Rotschlumpfchen Und Der Bose Wolf	Bastei	1/1/95		German Comic Book	12"L X 8 1/3"W	Germany	$3.50-5.00	The cover has Gargamel and Smurfette in the woods. Gargamel is dressed as a wolf and Smurfette is dressed as little red riding hood.
130/039	Auf Der Suche Nach Dem Goldenen Pilz	Bastei	1/1/94		German Comic Book	12"L X 8 1/3"W	Germany	$3.50-5.00	The cover has a Smurf picking mushrooms on it. Gargamel is standing in the bushes holding a net. Another Smurf is coming out of the bushes on the other side.
130/040	King Smurf Cover	Bastei	1/1/93		German Comic Book	12"L X 8 1/3"W	Germany	$3.50-5.00	The Book has 3 books in #23, 24, 25. 23 is Start frei fur die grobe Sause. 24 is In einem Land - wo alles grsser ist. 25 is Eine Nervensage wird ausgetrickst. The main cover has King Smurf on it and a Smurf is giving him a present.
130/041	Smurfs Playing In Rain Cover	Bastei	1/1/93		German Comic Book	12"L X 8 1/3"W	Germany	$3.50-5.00	The book has 3 books inside # 28, 29, 31. The main cover has 8 Smurfs and Smurfette playing in the rain on it. 28 isEin Schatten spielt verruckt. 29 isSchlumpfinchens Zauberspruch. 31 isDie Spiel - Maschine.

130/042	Smurfs Looking At A Rabbit Cover	Bastei	1/1/93		German Comic Book	12"L X 8 1/3"W	Germany	$3.50-5.00	The book has 3 books inside # 33, 32, 34. The main cover has Papa, Smurfette, and 2 other Smurfs looking at a blue rabbit. The Smurfs have troubled expressions on their faces. 33 is Ein schoner Raum in einem Baum. 32 is Ein Dieb im Schlumpfdorf. 34 is Ein verhexter Plan.
130/043	Smurfette Kissing A Smurf Cover	Bastei	1/1/93		German Comic Book	12"L X 8 1/3"W	Germany	$3.50-5.00	The book has 3 books inside # 35, 36, 37. The main cover has Smurfette giving a Smurf a kiss. Book 35 Schnell weg ins Versteck. Book 36 is Ein Schlumpf wachst uber sich hinaus. Book 37 is Ferien mit Hindernissen.
130/044	Vier Hexen Und Der Magische Stein	Bastei	4/1/99		German Comic Book	12"L X 8 1/3"W	Brussels	$2.00-3.00	The cover has Sassette riding a broom. She is flying over the Smurf village.

Set 41: Magic Talk Houses

041/001	Magic Talk Smurfette's House	Mattel Preschool	1/1/83	Purple	Talking Houses		Japan	$22.50-30.00	The back of the house folds down and has a mini record player. A plastic figure goes on top of the record player. The figures have the records on the bottom. The front door of the house opens and there is some moveable furniture. The house is lavender.
041/002	Magic Talk Papa Lab	Mattel Preschool	1/1/83	Light Blue	Talking Houses		Japan	$22.50-30.00	The back of the lab folds down and has a mini record player. A plastic figure goes on top of the record player. The figures have the records on the bottom. The front door of the lab opens and there is some moveable furniture. The lab is light blue.
041/003	Magic Talk Schoolhouse	Mattel Preschool	1/1/83	Yellow	Talking Houses		Japan	$25.00-35.00	The back of the house folds down and has a mini record player. A plastic figure goes on top of the record player. The figures have the records on the bottom. The front door of the house opens and there is some moveable furniture. The schoolhouse is yellow with 3 rooms.

Set 250: Magic Tricks

250/001	Wizard With Magic Disc And Beads	Toy Island	1/1/96		Magic Set	2"	Brussels	$11.00-15.00	The wizard is wearing a dark blue pointed hat with gold stars, and dark blue pants with gold stars. The figure is jointed. He comes with red, green and blue beads. Red, yellow, blue and green disc. The Smurf is in a pink package with a picture of the wizard Smurf on it.
250/002	Wizard With Blue Guillotine	Toy Island	1/1/96		Magic Set	2"	Brussels	$11.00-15.00	The wizard is wearing a dark blue pointed hat with gold stars, and dark blue pants with gold stars. The Smurf comes with a blue and brown guillotine.

Set 191: Magnets- 3-D

191/001	Disco Smurf	Peyo	1/1/93		Rubber Magnet	2 1/2" High	Brussels	$4.50-6.00	The Smurf has a light blue body. He looks like he's dancing. He is wearing dark blue pants with a yellow belt, and gray and white shoes. He has an orange shirt with a white collar on.
191/002	Disco Smurfette	Peyo	1/1/93		Rubber Magnet	2 1/2" High	Brussels	$4.50-6.00	Smurfette has a light blue body. She looks like she's dancing. Smurfette has a large pink bow in her hair. Her blouse is light pink and her skirt is orange with light green stars. Her shoes are white.
191/003	Papa Sea Captain	Peyo	1/1/93		Rubber Magnet	2 1/2" High	Brussels	$4.50-6.00	Papa has a light blue body. Papa is holding a yellowish, green telescope to his eye. Papa is wearing a white sailor jacket with yellow buttons, red pants and a red hat with a white hat band and a black visor.
191/004	Smurfette On A Beach Towel	Peyo	1/1/93		Rubber Magnet	2 1/2" High	Brussels	$4.50-6.00	Smurfette has a light blue body. She is sitting on a lime green towel. Smurfette is wearing a dark pink bathing suit. She has a red star on her hat.
191/005	Baby With Teddy Bear	Peyo	1/1/93		Rubber Magnet	2 1/2" High	Brussels	$4.50-6.00	The Smurf has a light blue body. Baby is sitting with a teddy bear sitting next to him. Baby is wearing a pink sleeper and hat. The teddy bear is light brown with white around his mouth, belly and bottom of his paws. The teddy's bow tie is navy blue.
191/006	Saxophone Smurf	Peyo	1/1/93		Rubber Magnet	2 1/2" High	Brussels	$4.50-6.00	The Smurf has a light blue body. He is standing and playing a saxophone. The saxophone is a bright gold and yellow. The Smurf is wearing a bright yellow shirt, green pants, a white hat, and dark blue with gray shoes.

Set 100: Magnets- Figures

100/001	Nat	Robinson Design Group			PVC Magnet	2 1/4"	China	$2.00-3.00	The magnet is half of a rubber PVC. Nat is standing with his hands behind his back. He is wearing tan pants, tan sash, a yellow straw hat, and no shoes.
100/002	Snappy	Robinson Design Group			PVC Magnet	2 1/4"	China	$2.00-3.00	The magnet is half of a rubber PVC. Snappy is standing with a mad look on his face and his hands on his hips. He is wearing a yellow shirt with a lightening bolt on the front.
100/003	Sassette	Robinson Design Group			PVC Magnet	2 1/4"	China	$2.00-3.00	The magnet is half of a rubber PVC. Sassette is holding a finger in the air. She is wearing a pair of pink pants with pink suspenders. Her hair is bright orange.
100/004	Slouchy	Robinson Design Group			PVC Magnet	2 1/4"	China	$2.00-3.00	The magnet is half of a rubber PVC. Slouchy is standing with his hand in his pocket. He is wearing a red shirt, white pants, and a droopy white hat.
100/005	Smurf With A Telephone	W. Berrie	1/1/83		PVC Magnet	2 1/4"	Hong Kong	$2.00-3.00	The magnet is half of a rubber PVC. The Smurf is holding a copper receiver to his mouth . The telephone is red and copper and is sitting next to his feet.
100/006	Mailman Smurf	W. Berrie	1/1/83		PVC Magnet	2 1/4"	Hong Kong	$2.00-3.00	The magnet is half of a rubber PVC. The Smurf is holding a gold horn in his mouth with his right hand. He has a brown mailbag. He is holding a white envelope with a red heart, on the flap of the envelope, in his left hand.
100/007	Smurf With A Ribbon Wrapped Around His Finger	W. Berrie	1/1/83		PVC Magnet	2 1/4"	Hong Kong	$2.00-3.00	The magnet is half of a rubber PVC. The Smurf has his left hand on his hip and the right hand is straight out at his side with a yellow ribbon wrapped around it.
100/008	Papa With Memo Pad And Feather Pen	W. Berrie	1/1/83		PVC Magnet	2 1/4"	Hong Kong	$2.00-3.00	The magnet is half of a rubber PVC. Papa is standing, holding a white memo pad in one hand and a white and black feather pen in the other hand.
100/009	Chef Smurf With A Spoon In Mouth	W. Berrie	1/1/83		PVC Magnet	2 1/4"	Hong Kong	$2.00-3.00	The magnet is half of a rubber PVC. The Smurf is wearing a white chef's coat with gold buttons, white pants, and a white chef's hat. He is holding a brown spoon to his mouth.
100/010	Brainy Holding A Finger In The Air	Robinson Design Group	1/1/83		PVC's	2 1/4"	Hong Kong	$2.00-3.00	The magnet is half of a rubber PVC. Brainy is holding a finger on his right hand in the air. He is wearing black glasses.
100/011	Gargamel	Robinson Design Group	1/1/83		PVC's	2 1/4"	Hong Kong	$2.00-3.00	The magnet is half of a rubber PVC. Gargamel is wearing a black outfit and red shoes. He looks like he's walking.

Set 190: Magnets- Flat

190/001	Drummer	Peyo			Flat Magnet	1 1/2"		$1.00-2.00	Drummer has a green and yellow drum around his neck. He has 2 brown drum sticks behind his ears. He is standing, holding a white paper with an award hanging off the end.
190/002	Cupid	Peyo			Flat Magnet	1 1/2"		$1.00-2.00	The Smurf looks like he's flying. He has white wings. The Smurf is holding a brown bow with a yellow arrow.
190/003	Flying Smurf	Peyo			Flat Magnet	1 1/2"		$1.00-2.00	The Smurf has red and white wings and looks like he's flying.
190/004	Jokey	Peyo			Flat Magnet	1 1/2"		$1.00-2.00	The Smurf is walking, carrying a yellow present. The present has a red ribbon and bow around it.
190/005	Black Smurf	Peyo			Flat Magnet	1 1/2"		$1.00-2.00	The Smurf is black. He is jumping in the air.
190/006	Papa	Peyo			Flat Magnet	1 1/2"		$1.00-2.00	Papa is standing and has a finger in the air. Papa is wearing red pants and a red hat.
190/007	Gargamel And Azreal	Peyo			Flat Magnet	1 1/2"		$1.00-2.00	Gargamel is standing, slouched over, with his hands held together. Azreal is at his feet.
190/008	Astro	Peyo			Flat Magnet	1 1/2"		$1.00-2.00	A Smurf is wearing a white spacesuit and a clear helmet. He is standing with his hands behind his back.
190/009	Fishing Smurf	Peyo			Flat Magnet	1 1/2"		$1.00-2.00	The Smurf is walking, carrying a fishing pole over his shoulder. The fishing pole is yellow and there is a red fish hanging off the line.
190/010	Smurfette Waving	Peyo			Flat Magnet	1 1/2"		$1.00-2.00	Smurfette is standing, waving with her left hand.
190/011	Smurfette Posing	Peyo			Flat Magnet	1 1/2"		$1.00-2.00	Smurfette is standing with a flirty look on her face. One hand is in her hair and the other is at her hip.
190/012	Soccer Smurf	Peyo			Flat Magnet	1 1/2"		$1.00-2.00	The Smurf is kicking a black and white soccer ball. He is wearing a red shirt, yellow shorts, red socks and white shoes.
190/013	Smurf Playing With Puppets	Peyo			Flat Magnet	1 1/2"		$1.00-2.00	The Smurf is standing, wearing hand puppets on both hands. One puppet is wearing red and the other puppet is wearing yellow.
190/014	Gardener	Peyo			Flat Magnet	1 1/2"		$1.00-2.00	The Smurf is wearing white bibs, brown shoes, and a yellow straw hat. The Smurf is leaning on a rake. He is wiping sweat from his face.
190/015	Smurf Juggling Balls	Peyo			Flat Magnet	1 1/2"		$1.00-2.00	The Smurf is juggling red, yellow and green balls.
190/016	Carpenter	Peyo			Flat Magnet	1 1/2"		$1.00-2.00	The Smurf is wearing white coveralls and a white work hat. The Smurf is carrying a gray toolbox and a yellow ruler in one hand. In the other hand he is carrying a hammer. The Smurf has a nail stuck in his mouth and a pencil behind his ear.
190/017	Runner	Peyo			Flat Magnet	1 1/2"		$1.00-2.00	The Smurf looks like he is running. Sweat is dripping from his face.
190/018	Brainy	Peyo			Flat Magnet	1 1/2"		$1.00-2.00	Brainy is standing with a finger in the air. He is wearing black glasses.
190/019	Lazy	Peyo			Flat Magnet	1 1/2"		$1.00-2.00	The Smurf is laying down in the grass. He has his hands behind his head.
190/020	Smurf Waving	Peyo			Flat Magnet	1 1/2"		$1.00-2.00	The Smurf is standing, waving with his right hand.
190/021	Smurf Smelling A Flower	Peyo			Flat Magnet	1 1/2"		$1.00-2.00	The Smurf is leaning over smelling a flower. The flower is white.
190/022	Naughty Smurf	Peyo			Flat Magnet	1 1/2"		$1.00-2.00	The Smurf is standing with his fingers in his ears and he is sticking his tongue out.
190/023	Grouchy	Peyo			Flat Magnet	1 1/2"		$1.00-2.00	Grouchy is standing, facing the left. He has a mean look on his face. His hands are fisted at his side.
190/024	Grouchy	Peyo			Flat Magnet	1 1/2"		$1.00-2.00	Grouchy is standing, facing the right. He has a mean look on his face. His hands are fisted at his side and he is slouched over.
190/025	Smurf With A Finger In The Air	Peyo			Flat Magnet	1 1/2"		$1.00-2.00	The Smurf is standing, holding one finger in the air. He has a happy look on his face.
190/026	Happy Smurf	Peyo			Flat Magnet	1 1/2"		$1.00-2.00	The Smurf is slightly turned to the left. He has a happy look on his face. He looks like he is jumping in the air. His hands are out at his sides.
190/027	Happy Smurf	Peyo			Flat Magnet	1 1/2"		$1.00-2.00	The Smurf is slightly turned to the right. He has a happy look on his face. He looks like he is jumping in the air. His hands are out at his sides.
190/028	Seeking Smurf	Peyo			Flat Magnet	1 1/2"		$1.00-2.00	The Smurf is leaning over and has a hand over his eye's like he is trying to see something far away.
190/029	Smurfette Holding Daisy's	Peyo	1/1/95		Flat Magnet	1 1/2" Circle	Brussels	$1.50-2.00	The magnet is round and flat. It has Smurfette standing, holding a bunch of white daisies. Smurfette has her head turned to the side and her eye's closed. The background is white.

190/030	Fristi Sassette	I.M.P.S.	1/1/98		Flat Magnet	4"	Brussels	$2.00-3.00	Sassette is standing, drinking a bottle of Fristi. Sassette is giving the thumbs up sign. She is wearing a pink outfit. Sassette has orange hair.

Set 171: Magnets- Square Ceramic

171/001	Smurfette				Ceramic Magnet	2" x 2"	Hand made	$2.00-3.00	The magnet is white and has a picture of Smurfette on it. Smurfette is dressed as a queen. She is wearing a red gown, pink robe, and a peach crown. Smurfette is holding a white daisy in her hand.
171/002	Smurf Fishing				Ceramic Magnet	2" x 2"	Hand made	$2.00-3.00	The magnet is white and has a picture of a Smurf fishing. The Smurf is standing in front of a little fishing hole. He has a brown stick with a yellow rope and a red and white bobber tied to the end for a fishing pole. There are green bushes behind the Smurf.
171/003	Gargamel Walking				Ceramic Magnet	2" x 2"	Hand made	$2.00-3.00	The magnet is white and has a picture of Gargamel walking on it. Gargamel is wearing a black outfit and red shoes. There are beads of sweat coming off Gargamel's head. There are green bushes in the background.
171/004	Gargamel And Azreal				Ceramic Magnet	2" x 2"	Hand made	$2.00-3.00	The magnet is white and has a picture of Gargamel and Azreal on it. Gargamel and Azreal are standing, looking troubled. Gargamel has one hand under his chin. Azreal is laying by his feet.
171/005	Smurf Soccer Player				Ceramic Magnet	2" x 2"	Hand made	$2.00-3.00	The magnet is white and has a picture of a Smurf kicking a soccer ball on it. The Smurf is wearing a red shirt, white shorts and a white hat. The Smurf is a kicking yellow and black soccer ball.
171/006	Smurf Playing a Horn				Ceramic Magnet	2" x 2"	Hand made	$2.00-3.00	The magnet is white and has a picture of a Smurf playing a horn on it. The Smurf is walking, playing a yellow horn.
171/007	Papa Running				Ceramic Magnet	2" x 2"	Hand made	$2.00-3.00	The magnet is white and has a picture of a Papa running with his hands out at his sides. Papa has red pants and a red hat on.
171/008	Papa Pointing				Ceramic Magnet	2" x 2"	Hand made	$2.00-3.00	The magnet is white and has a picture of Papa pointing with his left hand on it. Papa's other hand is at his side. Papa has a red hat and red pants on.
171/009	Papa Thinking				Ceramic Magnet	2" x 2"	Hand made	$2.00-3.00	The magnet is white and has a picture of Papa standing with a finger by his mouth. Papa's other hand is out at his side. There are green bushes behind Papa. Papa has a red hat and red pants on.
171/010	Gargamel And Azreal Looking Shocked				Ceramic Magnet	2" x 2"		$2.00-3.00	The magnet is white and has a picture of Gargamel and Azreal on it. Gargamel and Azreal have a shocked expression on their face.

Set 179: McDonald's- Promotional PVC's

179/001	Rock N Roll	Peyo	1/1/96		Rubber Smurf	2" High	China	$7.50-10.00	The Smurf has a medium blue body. The Smurf is holding a bright orange guitar with yellow strings. He has his eyes shut and his mouth open. The Smurf has the "McDonald's" emblem on the back of his hat.
179/002	Cheerleader	Peyo	1/1/96		Rubber Smurf	2" High	China	$7.50-10.00	Smurfette has a medium blue body. Smurfette is wearing a white hat, gold shoes and a white dress with a light blue collar and cuffs. The dress has gold dots on it. She is holding bright orange pompoms. Smurfette is standing on a dark green platform. Smurfette has the "McDonald's" emblem on the back of her hat.
179/003	Majorette	Peyo	1/1/96		Rubber Smurf	2" High	China	$7.50-10.00	Smurfette has a medium blue body. She is wearing a red dress with gold trim, gold boots and a white hat with a gold feather stuck in the side and a bright orange visor. Smurfette is carrying a bright orange baton with a gold tip in her right hand. Her left hand is on her hip. Smurfette is standing on a dark green stand. Smurfette has the "McDonald's" emblem on the back of her hat.
179/004	Jester	Peyo	1/1/96		Rubber Smurf	2" High	China	$7.50-10.00	The Smurf has a medium blue body. He is wearing a bright orange outfit, white shoes, white gloves and a white hat with gold stars. There are white ruffles around his neck. The outfit has yellow buttons and the shoes have yellow pompoms. The Smurf has a big, bright orange nose. The Smurf is holding a bright orange and yellow stick in his left hand. The Smurf has the "McDonald's" emblem on the back of his hat.
179/005	Present	Peyo	1/1/96		Rubber Smurf	2" High	China	$7.50-10.00	The Smurf has a medium blue body. He is holding a yellow present with bright orange ribbon. The Smurf is holding the present with both hands off to his right side. The Smurf has the "McDonald's" emblem on the back of his hat.
179/006	Cake	Peyo	1/1/96		Rubber Smurf	2" High	China	$7.50-10.00	The Smurf has a medium blue body. He is holding a bright orange cake with white frosting. The cake has bright orange, green and yellow berries on top. The cake is on a yellow plate. The Smurf is licking his lips. His left foot is slightly raised. The Smurf has the "McDonald's" emblem on the back of his hat.
179/007	Waiter	Peyo	1/1/96		Rubber Smurf	2" High	China	$9.00-12.00	The Smurf has a medium blue body. He is carrying a bright orange tray with a white McDonald's cup on it. On the front of the cup is a yellow "M". The glass has a yellow straw in it. The Smurf is carrying a white cloth in his left hand. The Smurf has the "McDonald's" emblem on the back of his hat.
179/008	Baker	Peyo	1/1/96		Rubber Smurf	2" High	China	$9.00-12.00	The Smurf has a medium blue body. He is wearing a white hat and a white apron around his waist. He is holding a tan bun with white seeds on a brown bread paddle. The Smurf has the "McDonald's" emblem on the back of his hat.
179/009	Smurf With A Big Mac	Peyo	1/1/96		Rubber Smurf	2" High	China	$9.00-12.00	The Smurf has a medium blue body. He is wearing a white hat and white pants. He is holding a big mac in front of him. The big mac has 2 dark brown patties, lettuce, onions and yellow cheese. The bun is tan and has white seeds on top. The Smurf has the "McDonald's" emblem on the back of his hat.
179/010	Smurf With "25"	Peyo	1/1/96		Rubber Smurf	2" High	China	$9.00-12.00	The Smurf has a medium blue body. He is wearing white pants and a white hat. He is leaning against a big number "25". The number "25" is bright orange with a yellow letter "M" on top. The Smurf has his legs crossed and his right hand is by his mouth. His left hand is around the letter "M". The Smurf has the "McDonald's" emblem on the back of his hat.
179/011	Smurf With Happy Meal Box	Peyo	1/1/97		Rubber Smurf	2" High	China	$7.50-10.00	The Smurf has a medium blue body. He is wearing white pants and a white hat. The Smurf is carrying a yellow happy meal box. The box has a smiley face on the front and a black M. The Smurf looks like he's scratching his chin with his other hand. The Smurf has the "McDonald's" emblem on the back of his hat.
179/012	Smurf On A Skateboard	Peyo	1/1/97		Rubber Smurf	2" High	China	$6.00-8.00	The Smurf is standing on a yellow skateboard. He is wearing a white hat with a red visor, white shirt, green hand pads, dark blue shorts, yellow knee pads and red/white shoes. The Smurf has a black M on his shirt. The Smurf has the "McDonald's" raised emblem on the back of his hat.
179/013	Smurf On Rollerblades	Peyo	1/1/97		Rubber Smurf	2" High	China	$6.00-8.00	The Smurf is rollerblading. He has his hands out at his sides. The Smurf is wearing a white hat, a white/red/blue shirt with a black "M" on the front, yellow hand pads, red shorts, dark blue knee pads, and red/yellow rollerblades. The Smurf has the "McDonald's" raised emblem on the back of his hat.
179/014	Smurfette On Rollerblades	Peyo	1/1/97		Rubber Smurf	2" High	China	$6.00-8.00	Smurfette is rollerblading. She has her hands out at her sides. Smurfette is wearing a white hat, a white/red/green shirt with a black "M" on the front, yellow hand pads, red pants, green knee pads, and yellow/red rollerblades. The Smurf has the "McDonald's" raised emblem on the back of his hat.
179/015	Smurf Holding A Snowboard	Peyo	1/1/97		Rubber Smurf	2" High	China	$6.00-8.00	The Smurf is holding a yellow snowboard. The Smurf is wearing a white hat, yellow glasses, red/green jacket, green pants, and red/white shoes. The snowboard has a black "M" on top. The Smurf has the "McDonald's" raised emblem on the back of his hat.
179/016	Basketball Smurf	Peyo	1/1/97		Rubber Smurf	2" High	China	$6.00-8.00	The Smurf is wearing a white tank top with dark blue trim and yellow stars on the front, white shorts with blue trim, a white hat, and dark blue/white shoes. The Smurf has a bright orange ball in his right hand. The ball has a black "M" on the front. The Smurf has the "McDonald's" raised emblem on the back of his hat.
179/017	Basketball Smurfette	Peyo	1/1/97		Rubber Smurf	2" High	China	$6.00-8.00	Smurfette is holding a bright orange ball on her hip. Smurfette is wearing a white shirt with orange trim and a black "M" on the front, white shorts with orange trim, a white hat, and dark blue/white shoes. The Smurf has the "McDonald's" raised emblem on the back of his hat.
179/018	Baseball Batter	Peyo	1/1/97		Rubber Smurf	2" High	China	$6.00-8.00	The Smurf is swinging a tan bat with a black "M" on the front. He is wearing a white hat with a red visor, a yellow shirt with green sleeves, red pants, and dark blue shoes. The Smurf has the "McDonald's" raised emblem on the back of his hat.
179/019	Baseball Pitcher	Peyo	1/1/97		Rubber Smurf	2" High	China	$6.00-8.00	The Smurf is getting ready to pitch a white ball. He has a brown glove on his other hand. The Smurf is wearing a white shirt with a black "M" on the front, red pants, white hat, and white shoes. The Smurf has the "McDonald's" raised emblem on the back of his hat.

Set 101: Memo Sets- Mini

101/001	Mini Measure Set (Yellow)	S.E.P.P.	1/1/82	Yellow	Mini Measure Set	5 1/2"L X 4 1/2" H	Hong Kong	$7.50-10.00	The set comes in a yellow plastic purse with a clear front and a handle. It has a pencil, a red triangle ruler, a green mini protractor and a mini scribble pad. The scribble pad has a Smurf sitting at a desk holding a ruler, the background is orange.
101/002	Mini Memo With Phone Book	S.E.P.P.	1/1/82		Mini Memo Set	5 1/2"L X 4 1/2" H	Hong Kong	$7.50-10.00	The set comes in a red and white plastic purse with a clear front and a handle. It has a blue pencil, a mini mushroom phone book, 2 pink plastic paper clips and a blue mini memo pad with a Smurf and Smurfette talking on a telephone on the front.
101/003	Mini Memo With A Cork board	S.E.P.P.	1/1/83		Mini Memo Set	5 1/2"L X 4 1/2" H	Hong Kong	$7.50-10.00	The set comes in a red and white plastic purse with a clear front and a handle. It has a mini tan cork board with a yellow back that says Smurfy memos and shows Papa holding a feather pen. There is a mini red memo pad with a Smurf holding a piece of paper and a yellow feather pen.
101/004	Mini Sketch (Green)	S.E.P.P.	1/1/82	Green	Mini Sketch Set	5 1/2"L X 4 1/2" H	Hong Kong	$7.50-10.00	The set comes in a green plastic purse with a clear front and a handle. It has a blue pencil, a blue pencil sharpener, a blue Smurf face pencil top, a white eraser and a yellow mini sketch pad. The sketch pad has a Smurf holding a piece of paper and a pencil.

Set 216: Minimates

216/001	Baby With Rocking Chair	Toy Island	1/1/96		Poseable Figure	2 1/2"	China	$6.75-9.00	Baby Smurf is wearing a pink sleeper. His arms and legs move. He has a yellow rocking chair with a red seat. The chair has a blue tray. There is a white bottle with a pink cap.
216/002	Fireman With Ladder	Toy Island	1/1/96		Poseable Figure	2 1/2"	China	$6.75-9.00	The Smurf is wearing a red shirt with a yellow scarf around his neck, red pants, black boots, and a red hat with a yellow tassel. The Smurf's arms, legs and head move. Included is a tan ladder and a gray hose with a red nozzle.
216/003	Sassette On A Surfboard	Toy Island	1/1/96		Poseable Figure	2 1/2"	China	$6.75-9.00	Sassette is wearing a pink 2 piece swimsuit, pink shoes and a white hat. She comes with a yellow surfboard that has plastic blue waves on each side. Sassette's arms, legs and head move.
216/004	Clown Smurf With A Tuba	Toy Island	1/1/96		Poseable Figure	2 1/2"	China	$6.75-9.00	The Smurf is wearing a yellow clown suit, green bow tie, white gloves, a white hat with pink dots, and a pink tassel. The Smurf has a big red nose and a big pink mouth. He comes with a large yellow tuba. The Smurf's arms, legs and head move.

216/005	Smurf Downhill Skier	Toy Island	1/1/96		Poseable Figure	2 1/2"	China	$6.75-9.00	The Smurf is wearing a pink and green shirt, pink pants, white shoes, white gloves, a white hat and black goggles. The Smurf comes with 2 green ski's attached to a snow hill and 2 green ski poles. The Smurf's arms, legs and head move.
216/006	Smurfette With Removable Dress	Toy Island	1/1/96		Poseable Figure	2 1/2"		$6.75-9.00	Smurfette is wearing a white body suit with a pink belt. She comes with a big white removable skirt. The skirt has a pink flower on the front. Smurfette's arms, legs and head move.
216/007	Smurf With Video Camera	Toy Island	1/1/96		Poseable Figure	2 1/2"	China	$6.75-9.00	The Smurf is wearing a white shirt, purple pants, white shoes, a white hat, and red glasses. The camera is dark pink and silver. The camera is sitting on a tan tripod. The Smurf's arms, legs and head move.
216/008	Smurfette With Stove	Toy Island	1/1/96		Poseable Figure	2 1/2"	China	$6.75-9.00	Smurfette is wearing a pink dress, white panties and pink shoes. The stove is white , silver and pink. Smurfette's arms, legs and head move.

Set 1: Mugs- Ceramic

001/001	Grouchy	W. Berrie	1/1/82		Coffee Mug	3 1/2"	Japan	$7.50-10.00	Full wrap around picture of a party scene. Grouchy is standing with his arms crossed over his chest and looking mad.
001/002	Jokey	W. Berrie	1/1/82		Coffee Mug	3 1/2"	Japan	$7.50-10.00	Full wrap around picture of Smurfs by exploded box.
001/003	Greedy	W. Berrie	1/1/82		Coffee Mug	3 1/2"	Japan	$7.50-10.00	Full wrap around picture of Smurfs eating at a table.
001/004	Papa	W. Berrie	1/1/82		Coffee Mug	3 1/2"	Japan	$7.50-10.00	Full wrap around picture of Papa Smurf leading a band.
001/005	Smurfette	W. Berrie	1/1/82		Coffee Mug	3 1/2"	Japan	$7.50-10.00	Full wrap around picture of Smurfs lined up waiting for Smurfette.
001/006	Lazy	W. Berrie	1/1/82		Coffee Mug	3 1/2"	Japan	$7.50-10.00	Full wrap around picture of all the Smurfs working except for Lazy.
001/007	I Love NY	W. Berrie	1/1/82		Smurf Travel America	3 1/2"	Korea	$7.50-10.00	Smurf carrying red apple.
001/008	I'm A Rocky Mountain Smurf	W. Berrie	1/1/82		Smurf Travel America	3 1/2"	Korea	$7.50-10.00	Hiker Smurf walking by a mountain.
001/009	California Smurfin'	W. Berrie	1/1/82		Smurf Travel America	3 1/2"	Korea	$7.50-10.00	Smurf sitting on beach towel sun-bathing.
001/010	I'm A Florida Sunshine Smurf	W. Berrie	1/1/82		Smurf Travel America	3 1/2"	Korea	$7.50-10.00	Smurf laying on beach towel sun-bathing.
001/011	I Smurf America	W. Berrie	1/1/82		Smurf Travel America	3 1/2"	Korea	$7.50-10.00	Smurf holding two American flags.
001/012	Smurfin' Thru The U.S.A.	W. Berrie	1/1/82		Smurf Travel America	3 1/2"	Korea	$7.50-10.00	Smurfs walking in front of a outline of the United States.
001/013	Smurf Loves Dixie	W. Berrie	1/1/82		Smurf Travel America	3 1/2"	Korea	$7.50-10.00	Smurf holding a Confederate flag.
001/014	Guess Who Loves Ya?	W. Berrie	1/1/82		Coffee Mugs	3 1/2"	Korea	$7.50-10.00	Smurf sitting on grass looking love-struck with hearts floating around his head.
001/015	Have A Happy Day	W. Berrie	1/1/81		Coffee Mugs	3 1/2"	Korea	$3.50-5.00	3 Smurfs, Smurf smelling a flower, Smurf doing a handstand, and a Smurf with a knapsack over his shoulder.
001/016	Lover Smurf	W. Berrie	1/1/81		Coffee Mugs	3 1/2"	Korea	$3.50-5.00	A Smurf holding yellow flowers with a heart in the background.
001/017	A Smurfing We Will Go!	W. Berrie	1/1/81		Coffee Mugs	3 1/2"	Korea	$3.50-5.00	Full wrap around picture of 5 Smurfs singing.
001/018	#1 GRAD	W. Berrie	1/1/82		Coffee Mugs	3 1/2"	Korea	$7.50-10.00	Graduation Smurf in a blue gown with diploma
001/019	SMILE SMILE SMILE	W. Berrie	1/1/81		Coffee Mugs	3 1/2"	Korea	$3.50-5.00	A full wrap around picture of 3 Smurfs laughing.
001/020	Jogger Smurfs	W. Berrie	1/1/82		Coffee Mugs	3 1/2"	Korea	$3.50-5.00	A full wrap around picture of 3 Smurfs jogging.
001/021	Sporty Smurf	W. Berrie	1/1/82		Coffee Mugs	3 1/2"	Korea	$3.50-5.00	A Smurf holding a bunch of sports equipment.
001/022	Super Smurf	W. Berrie	1/1/81		Coffee Mugs	3 1/2"	Korea	$3.50-5.00	Superman Smurf
001/023	OOPS! Happy Birthday	W. Berrie	1/1/82		Coffee Mugs	3 1/2"	Korea	$7.50-10.00	Smurf laying in a cake holding balloons.
001/024	Wow! Another Birthday!	W. Berrie	1/1/82		Coffee Mugs	3 1/2"	Korea	$7.50-10.00	Smurf sitting, holding three paint brushes.
001/025	Have A Bang-Up Birthday	W. Berrie	1/1/82		Coffee Mugs	3 1/2"	Korea	$7.50-10.00	Smurf holding a pink gift box, laughing.
001/026	Happy Birthday	W. Berrie	1/1/82		Coffee Mugs	3 1/2"	Korea	$7.50-10.00	Papa Smurf with a magic book and potions.
001/027	For Your Birthday	W. Berrie	1/1/82		Coffee Mugs	3 1/2"	Korea	$7.50-10.00	Smurfette carrying a birthday cake.
001/028	Hooray For Birthdays	W. Berrie	1/1/82		Coffee Mugs	3 1/2"	Korea	$7.50-10.00	Smurfette cheerleader
001/029	Happy Birthday!	W. Berrie	1/1/81		Coffee Mugs	3 1/2"	Korea	$7.50-10.00	A full wrap around picture of a Smurf carrying a cake, a Smurf carrying balloons, Jokey Smurf and King Smurf.
001/030	Is It Break Time Yet?	W. Berrie	1/1/81		Coffee Mugs	3 1/2"	Korea	$7.50-10.00	Smurf carrying books looking at a clock.
001/031	Bless This Mess!	W. Berrie	1/1/81		Coffee Mugs	3 1/2"	Korea	$7.50-10.00	Smurf sitting at a paper stacked desk.
001/032	I Love You	W. Berrie	1/1/82		Coffee Mugs	3 1/2"	Korea	$7.50-10.00	Smurf painting "I Love You" in a red heart.
001/033	You're My Sweetheart!	W. Berrie	1/1/82		Coffee Mugs	3 1/2"	Korea	$7.50-10.00	Smurf & Smurfette's faces in hearts.
001/034	Gotcha!	W. Berrie	1/1/82		Coffee Mugs	3 1/2"	Korea	$7.50-10.00	Full wrap around picture of a cupid Smurf shooting arrows at a heart.
001/035	I Love You	W. Berrie	1/1/82		Coffee Mugs	3 1/2"	Korea	$7.50-10.00	Smurf & Smurfette flying heart balloons saying "I Love You".
001/036	Smurfs & NBC	W. Berrie	1/1/82		Coffee Mugs	3 1/2"	Korea	$15.00-20.00	Smurf standing next to NBC logo.
001/037	Smurf Face	W. Berrie			Coffee Mugs	3 1/2"	Hand made	$2.00-3.00	Hand made coffee mug with raised Smurf face.
001/038	Painter Smurf	W. Berrie	1/1/82		Coffee Mugs	5"	Hand made	$2.00-3.00	Raised Painter Smurf on front.
001/039	Smurf	W. Berrie	2/1/82		Coffee Mugs	3"	Hand made	$2.00-3.00	Hand made coffee mug of a Smurf smiling.
001/041	Merry Christmas 1981	W. Berrie	1/1/81		Coffee Mugs	3 1/2"		$7.50-10.00	Papa Smurf is holding a candy cane and an ornament to hang on the tree and a Smurf is putting a star on top of the X-mas tree.
001/042	Smurf Chasing Hearts	W. Berrie	1/1/81		Coffee Mugs	3 1/2"	Korea	$7.50-10.00	The mug is white. The Smurf is carrying a net and chasing red hearts. There is a rainbow behind him.
001/043	Smurfs Are Fun!	Peyo	1/1/81	White	Coffee Mug	3 1/2"	New Zealand	$7.50-10.00	The mug has Smurfette standing, smelling an orange daisy. There is a mushroom house behind her. There are 2 Smurfs walking towards Smurfette. The picture goes around the whole cup. In the center it says "Smurfs Are Fun!" in orange letters.
001/044	We Smurf Things Big In Texas	W. Berrie	1/1/82		Coffee Mugs	3 1/2"	Korea	$15.00-20.00	The mug has a cowboy Smurf sitting on a brown saddle. The Smurf is twirling a yellow rope. The Smurf is wearing a brown cowboy hat and red bandanna around his neck. The mug says "We Smurf things Big In Texas"
001/045	SMURF - White Cup	Grindley	1/1/82	White	Coffee Mugs	3 1/2"	England	$7.50-10.00	The cup is white and has a picture of a Smurf jumping. The Smurf is black. Below the Smurf it says "SMURF" in white letters with black outline.
001/046	Beach Life	IMA	1/1/96	Yellow	Coffee Mugs	3 1/2"	Brussels	$7.50-10.00	The mug is yellow. On one side of the mug is a Smurf surfing on a green surfboard. In the middle of the mug it says "Beach Life." The other side of the mug has a Smurf face on it. The Smurf has a flower on his hat and sunglasses resting on his head. There is a small portrait picture inside of the mug the same as the Smurf face.
001/047	4 Different Smurfette Poses	Peyo	1/1/82	White	Coffee Mugs	3 1/4"	New Zealand	$7.50-10.00	The mug has 4 different poses of Smurfette. Smurfette is smelling a flower. Smurfette is hitting a tennis ball. Smurfette is laying on a beach blanket. Smurfette looks like she's standing there flirting. Smurfette is blue but her hair is white. There is a yellow stripe going around the cup.
001/048	SMURF - Light Blue Cup	Grindley	1/1/82	Light Blue	Coffee Mugs	3 3/4"	England	$7.50-10.00	The cup is light blue and has a picture of a Smurf jumping. The Smurf is black. Below the Smurf it says "SMURF" in light blue letters with black outline.
001/049	SMURF - Pink Cup	Grindley		Pink	Coffee Mugs	3 1/2"	England	$7.50-10.00	The cup is pink and has a picture of a Smurf jumping. The Smurf is black. Below the Smurf it says "SMURF" in pink letters with black outline.
001/050	Smurfette Posing			White	Coffee Mugs	3 1/2"		$7.50-10.00	Smurfette is standing with her hands in front of her. Smurfette's body is made of blue dots. It says "Smurfette" in yellow letters with blue trim. The same picture is on 2 sides. The cup is white.
001/051	Smurf Pole Vaulter	Peyo	1/1/89	White	Coffee Mugs	3 1/2"	France	$11.00-15.00	The cup has a Smurf jumping over a beam. The Smurf is landing on a red pillow. The Smurf is wearing a yellow tank top and white shorts. There is another Smurf standing by a bush looking troubled. The picture goes around the mug. There is a blue plastic cover for the cup.
001/052	4 Sports Smurfs	W. Berrie		White	Coffee Mugs	3 1/4"	New Zealand	$7.50-10.00	The cup is white with a yellow strip going around the center. There is a Smurf on a surfboard. A Smurf leaping over a fence. A Smurf on a bike. A Smurf jogging with sweat running down his face. The Smurfs outfits are all white.
001/053	Swinging Smurfs	W. Berrie	1/1/82		Coffee Mugs	3 1/2"	New Zealand	$11.00-15.00	The mug is called swinging Smurfs. It has Smurfette dancing with a boy Smurf. There is another Smurf playing a trumpet with a Smurf listening to the music. There are music notes floating all over the mug.
001/054	Smurf Forever	W. Berrie	1/1/82		Coffee Mugs	3 1/2"	New Zealand	$9.00-13.00	Smurfette and Papa are painting 2 Smurfs on a yellow poster. There is a green brick wall behind them.
001/055	Papa, Smurfette and A Smurf Playing Basketball	Dixan	1/1/91	White	Coffee Mugs	3 1/2"	France	$4.50-6.00	The mug has Smurfette, Papa and a Smurf playing basketball on it. Smurfette is passing the ball to Papa. The Smurf is trying to block Papa from making a basket. There is a yellow striped background.
001/056	Papa And Smurfette Ice-skating	Benediction	1/1/86	White	Coffee Mugs	3 1/2"	France	$4.50-6.00	One side of the mug has Papa holding Smurfette's arm and ice-skating. There is a Smurf laying on the ice. The other side has Papa dressed as Santa walking across a bridge. 2 Smurfs are waiting at the end of the bridge.
001/057	Smurfs On Skateboards	Dixan	1/1/90	white	Coffee Mugs	3 1/2"	France	$4.50-6.00	The 2 Smurfs and Smurfette are riding skateboards. There is an orange background.
001/058	Smurfs Eating Ice cream	Benediction	1/1/86		Coffee Mugs	3 1/2"	France	$4.50-6.00	The one side of the mug has a Smurf sneezing. He blew his ice cream into the other Smurfs face. They are standing by a yellow ice cream cart. Papa is watching with a worried expression on his face. The other side of the mug has a Smurf and Smurfette sitting on a log. Brainy Smurf is walking toward them.
001/059	Smurf Chasing Another Smurf With A Rolling Pin	Benediction	1/1/83	White	Coffee Mugs	3 1/2"	France	$4.50-6.00	A Smurf dressed as a cook is chasing a Smurf carrying a cake with a rolling pin. The other side has Smurfette mixing a pot of stew. A Smurf with a flower is following the scent.
001/060	Smurf Snowboarding	Porcelaine De Lologne	1/1/98	White	Coffee Mugs	3 1/2"	France	$15.00-20.00	The Smurf is on a brown snowboard. The Smurf is wearing a white hat, a green snow suit, yellow boots, and yellow goggles. The back of the mug says Les Schtroumpfs in yellow letters.
001/061	Smurf Skateboarding Large Mug	IMA	1/1/96	Green	Large Coffee Mugs	4"	France	$11.00-15.00	The mug is green. On one side of the mug is a Smurf riding on a blue skateboard. In the middle of the mug it says "Move Move Yo!" The other side of the mug has a Smurf face. The Smurf is wearing an orange hat with red circles. There is a small portrait picture inside of the mug of a Smurf wearing a red hat with green circles. There is a green square with yellow and red stripes behind his head.
001/062	Beach Life! Large Mug	IMA	1/1/96	Yellow	Large Coffee Mug	4"	France	$11.00-15.00	The mug is yellow. On one side of the mug is a Smurf surfing on a green surfboard. In the middle of the mug it says "Beach Life." The other side of the mug has a Smurf face. The Smurf has a flower on his hat and sunglasses resting on his head. There is a small portrait picture inside of the mug the same as the Smurf face.

001/063	Tiger Power! Large Mug	IMA	1/1/96	Orange	Large Coffee Mug	4"	France	$11.00-15.00	The mug is orange. On one side of the mug is a Smurf dressed in a multi- colored baseball uniform swinging a pink bat. The Smurf is wearing a green and orange hat that says "Tigers." In the middle of the mug it says "Tiger Power" in red letters with a red tiger face above it. The other side of the mug has a Smurf face on it. The Smurf is wearing a green and orange hat that says "Tigers" in red. There is a small portrait picture inside of the mug the same as the Smurf face.

Set 67: Musical Instruments

067/001	Horn	Ohio Art	1/1/82	Yellow	Plastic Horn	7 " Long		$6.00-8.00	A yellow horn with stickers on each side of a Smurf playing a horn.
067/002	Metal Drum		1/1/82	Blue	Metal Drum	8"Tall X 9"W		$15.00-20.00	The drum has a wrap around picture of 5 Smurfs playing different instruments. The Smurfs are on the grass, one is playing a drum, one is playing a cello, another is playing a horn, one is carrying a band stick, and the other one is playing the cymbals. The drum is light blue, the top and bottom are white and made of paper. The drum says "Smurf" in white letters with blue trim.
067/003	Smurf Mini Organ	Galoob	1/1/82	Blue/White	Musical Organ	4 1/2" X 3 X 1 1/4"	Hong Kong	$15.00-20.00	The organ is white and has a Smurf face on the front. The Smurf's hat says SMURF TUNES in blue letters. The top opens and has a numbered keypad inside. The organ comes with 3 music sheets. The mini organ is a shape of a Smurf's face.
067/004	Strummin' Smurf Guitar	Ohio Art	1/1/82	Light Blue	String Guitar	20"	U.S.A.	$15.00-20.00	The guitar is light blue with 4 clear strings. There is a picture of a Smurf singing and playing a guitar for Smurfette. The background is white and has 3 mushroom houses. Above the Smurf and Smurfette, it says "STRUMMIN' SMURF" in black puffy letters.
067/005	Orange Musical Ge-Tar	W. Berrie	1/1/82	Orange	Wind-up Guitar	14"	U.S.A.	$11.00-15.00	The guitar is orange and plays music. On the side is a knob to wind the guitar. There is a picture of a Smurf playing a yellow gutar and singing on the front. There is another Smurf standing on a hill in the background playing a drum. On the bottom it says "SMURF" in black and white puffy letters.
067/006	Smurf Small Plastic Drum	Ohio Art	1/1/82		Drum Set	18" High	U.S.A.	$22.50-30.00	The middle drum is white and blue plastic, 10". There are 2 small drums, 1 attached on each side, 1 green and white , and 1 red and white. On top is a cymbal and a metal bell. There are 2 yellow drum sticks. In the middle of the big drum is a picture of a Smurf playing the drums. The drum says "Tot Traps."
067/007	Smurf Drum Set	Ohio Art	1/1/82		Drum Set		U.S.A.	$95.00-125.00	The drum set includes: 20" bass drum, 9" snare drum, 6" bongo drum, 9" hissing Cymbal, 9" cymbal, 10" pain of drumsticks, plastic foot pedal with beater, sound block and bell.

Set 56: Musical Toys

056/001	Ferris Wheel	Durham	1/1/82	Blue/Pink	Musical Toys		Hong Kong	$25.00-35.00	The ferris wheel is blue. Papa, Smurfette and 2 Smurfs are in pink chairs. The arch that connects the ferris wheel to a crib is pink.
056/002	Blue TV	Ohio Art		Light Blue	Wind Up TV	10 1/4"L x 7 1/2"W	U.S.A.	$11.00-15.00	The TV winds and has the Flying Smurf moving picture with music.
056/003	4 Smurfs By A Campfire	Illco	1/1/82		Musical Toy		Hong Kong	$11.00-15.00	The toy is round, the top half has 4 Smurfs inside sitting around a campfire roasting hot dogs. The toy jingles when rolled.
056/004	Musical Smurf In The Box	Galoob	1/1/82	Blue	Smurf In The Box	5" X 5"	Hong Kong	$35.00-45.00	The box has blue trim with a small yellow top and yellow plastic handle to wind. There are 4 pictures on the sides of the box: A Smurf giving Smurfette flowers, a clown Smurf, Smurfette roller-skating, and a Smurf carrying a cake with another one serenading in front of a mushroom house. There is a clown Smurf wearing a red with white poke-a-dot shirt, that pops out of the top. The opening is square that the Smurf pops out of.
056/005	Musical Smurf In The Box	Galoob	1/1/82		Smurf In The Box	6"H X 5"W	Hong Kong	$35.00-45.00	The box has blue trim with a big yellow top and metal handle with a blue knob to wind. There are 4 pictures on the sides of the box: A Smurf giving Smurfette flowers, a clown Smurf, Smurfette roller-skating, and a Smurf carrying a cake with another one serenading in front of a mushroom house. There is a clown Smurf wearing a red with white poke-a-dot shirt, that pops out of the top. The opening is round that the Smurf pops out of and the material from his shirt goes inside..
056/006	Musical Smurf Radio	Illco	1/1/82	White	wind-up Radio	5" High	Macau	$15.00-20.00	The radio is white. The front has a small Smurf face with a blue knob under the Smurf's face. There is a blue antenna and a blue carrying handle. When you wind the radio it plays rock-a-bye-Baby and the Smurfs eyes move back and forth.
056/007	Peek-A-Boo Smurf	Illco	1/1/83	White/Blue	Musical Toy	6" High	Macau	$18.00-25.00	The toy is white and on the front it has a Smurf. The Smurf's hands move over the Smurf's eyes and his feet move when you wind the button. The Smurf has ears that stick out on the side of the toy. The Smurf plays a lullaby.
056/008	Carousel	Illco	1/1/83		Musical Carousel	9"H X 8"L	Hong Kong	$25.00-35.00	The carousel is a mushroom with 3 swings. Papa, Smurfette and a Smurf are sitting in yellow swings. They go around when you wind the knob. The stem of the mushroom is yellow and the top is red with yellow spots. The base is light blue with white on the side. There are stickers of mushrooms on the sides. The carousel plays music.
056/009	Roll Along Smurf	Illco	1/1/82	White	Wind-up Smurf	6"High	Hong Kong	$15.00-20.00	The Smurf is on wheels and rolls along and plays music when you wind him up. The Smurf is made of plastic. He is standing with both hands at his side. The Smurf is wearing a white outfit and hat. The hat has a red tassel and the middle of the front of his outfit has a red button.
056/010	Smurf On A Mushroom Music Box	Talbot Toys	1/1/82		Music Box	6 1/2"High	Hong Kong	$11.00-15.00	A 4" rubber Smurf is sitting on top of a mushroom box. The mushroom has a white bottom and the top is red with orange spots. The mushroom opens and has a mirror inside. The music plays when you open the box and wind it.
056/011	Smurf Tin Can				Tin Can	3"	Germany	$22.50-30.00	The tin can has a red and white top with a metal handle. There is a wrap around scene. A Smurf is laughing. Brainy is holding a fishing pole with a fish on the end. A Smurf is holding a hammer, a Smurf is holding a spoon, 2 Smurfs are running, and the last Smurf is blowing a horn. There are orange mountains in the background.
056/012	Plastic Papa With A Mini Record Player	Peyo			Papa Record Player	20" High	Italy	$55.00-75.00	Papa stands 20" high. He is made of plastic. There is a mini record player in Papa's back. Papa has a red plastic hat and red plastic pants. His beard is white and made of thread. Papa's arms and legs move.
056/013	Smurfette Poussah Musical	Kortekaas Merch	1/1/83		Roly - Poly	8"High	France	$15.00-20.00	The Smurfette sits 8" high and when it's moved it rocks and makes a music sound. Smurfette is standing, wearing a white gown with an orange border. Smurfette is holding an orange flower to her chest. Her eyes are closed and she has a smile on her face.
056/014	Musical Tree-Go-Round	Irwin	1/1/96		Tree Play set		Brussels	$30.00-40.00	The set includes: a musical tree-go-round, 1 Harmony figure, 2 transforming car rides, 1 xylophone, 2 swing rides, 1 Smurfscope, 1 music and ride activator push-button drum. The tree is brown with a red mushroom top.
056/015	Musical Cottage With Animated Picture	Illco	1/1/82		Musical Cottage	4 1/2"x 41/2"	Macau	$18.00-25.00	The wind-up is in the shape of a half of a cottage. The roof is red and the walls are white. In the center is a clear window. Inside is Papa holding a finger up. When wound, Papa rocks back and forth. There are 2 Smurfs and Smurfette behind him.
056/016	Schlumpf Musikhaus With Bully Figures	Lorenz Bolz			Swiss Music Box	5" High	Germany	$37.50-50.00	The music box is of a mushroom cottage. The cottage is yellow with a red roof. A blue Smurf is sticking out the window of the cottage. The cottage turns when wound. The cottage is on a green plastic base. There are 3 Smurfs on the base. One Smurf is blue and has a yellow trumpet in his mouth.

Set 192: Necklaces

192/001	Papa Pointing				Flat Metal Necklace			$9.00-12.00	Papa is standing, pointing at something. He has a red hat and red pants on. The charm is on a gold chain. The charm has a glossy finish.
192/002	Papa Eating An Ice cream Cone				Flat Metal Necklace			$5.00-7.00	Papa is standing, holding an ice cream cone. The ice cream is white and the cone is blue. He has a red hat and red pants on. Papa's eyes are white. The charm is on a gold chain.
192/003	Smurf Playing A Trumpet (Large)				Flat Metal Necklace			$5.00-7.00	The Smurf is walking, playing a yellow trumpet. He is wearing a red shirt, white pants and a white hat. The charm is on a gold chain. The charm has a glossy finish.
192/004	Smurf Playing A Trumpet				Flat Metal Necklace			$5.00-7.00	The Smurf is walking, playing a yellow trumpet. He is wearing a red shirt, white pants and a white hat. The charm is on a gold chain.
192/005	Large Roller-skating Smurf (White)				Flat Metal Necklace			$5.00-7.00	A Smurf is roller-skating. The roller-skates have white straps and black wheels. The charm is on a gold chain. The charm has a glossy finish.
192/006	Roller-skating Smurf (Yellow)				Flat Metal Necklace			$5.00-7.00	A Smurf is roller-skating. The roller-skates have yellow straps and orange wheels. The charm is on a gold chain.
192/007	Football Smurf				Flat Metal Necklace			$5.00-7.00	The Smurf looks like he's running. He is carrying a brown football under one arm. He is wearing a yellow shirt with a gold #1 on the front, yellow pants, white shoes and an orange hat with a yellow star on it. The charm is on a gold chain.
192/008	Smurf On A Skateboard				Flat Metal Necklace			$5.00-7.00	The Smurf is standing on his hands on the skateboard. The skateboard is orange with a yellow stripe and yellow wheels. The charm is on a gold chain.
192/009	Smurf Holding A Torch				Flat Metal Necklace			$5.00-7.00	The Smurf looks like he's running. He is holding a gold torch with a red flame in one hand. The Smurf is wearing a white shirt, red shorts, white shoes and a white hat. The charm is on a gold chain.
192/010	Smurf Leaning On A Heart				Flat Metal Necklace	1 1/4"		$5.00-7.00	The Smurf is standing, leaning against a red heart. He has his head in his hand. The other hand is holding a yellow daisy behind his back. The Smurf has a love-struck expression on his face. The charm is on a gold chain.
192/011	Hiker Smurf				Flat Metal Necklace			$5.00-7.00	The Smurf is holding a black walking stick. He is wearing a yellow jacket, white pants and black hiker boots. He has a green backpack on his back. The charm is on a gold chain.
192/012	Smurfette Holding A Heart				Flat Metal Necklace			$5.00-7.00	Smurfette is standing, holding a red heart in her hands. Smurfette has a flirty look on her face. Smurfette is wearing a white dress. The charm is on a gold chain.
192/013	Smurf With A Daisy				Flat Metal Necklace			$5.00-7.00	The Smurf is standing, holding a daisy behind his back. The daisy is yellow. The Smurf is wearing a red shirt and white pants. He has his head turned to the side with a smile on his face. The charm is on a gold chain.
192/014	Smurfette On Ice-skates				Flat Metal Necklace			$5.00-7.00	Smurfette is wearing pink ice-skates. Smurfette looks like she is ice-skating. Smurfette is wearing a white dress. The charm is on a gold chain.
192/015	Diecast Soccer Smurf	W. Berrie/Peyo	1/1/81		Diecast Necklace		U.S.A.	$9.00-12.00	The Smurf necklace is in the shape of a Smurf. It is made of a diecast metal. The Smurf is kicking a white and black soccer ball. The Smurf is wearing a red shirt, red socks, white shoes, white shorts, and a white hat.
192/016	Diecast Papa Pointing	W. Berrie/Peyo	1/1/81		Diecast Necklace		U.S.A.	$9.00-12.00	The Smurf necklace is in the shape of a Smurf. It is made of a diecast metal. Papa is standing, pointing at something. He has a red hat and red pants on. The charm is on a gold chain.
192/017	Diecast Cowboy				Diecast Necklace			$5.00-7.00	The Smurf pin is in the shape of a Smurf. It is made of a diecast metal. The Smurf is wearing brown chaps, white pants, white shoes, a brown vest, a brown cowboy hat, and a red handkerchief around his neck. The Smurf is standing with his hands at his hips. The charm is on a gold chain.
192/018	Cake Smurf				Plastic Necklace			$5.00-7.00	The necklace is a flat heart- shaped plastic. The heart is light blue and has a picture of a Smurf carrying a cake in the middle. The cake is yellow with white and red frosting and is on a pink plate. The Smurf is licking his lips. The charm is on a gold chain.

192/019	Ballerina 2 sided Necklace	Applause	1/1/84	Pink/Purple	Plastic Necklace		Taiwan	$7.50-10.00	The necklace is the shape of a heart and is 2- sided. 1 side is pink and has Smurfette wearing a ballerina outfit and posing. The second side is purple and has Smurfette standing, holding a yellow daisy. The charm is on a gold chain.
192/020	Smurfette	Howard Eldon			Flat Metal Necklace		U.S.A.	$9.00-12.00	Smurfette is standing with her hands in front of her. She is wearing a red dress. Her eyes are white with black dots. The charm is on a gold chain.
192/021	Torchbearer Smurf				Flat Metal Necklace	1 1/4"		$5.00-7.00	The Smurf looks like he's running. He is holding a gold torch with a red flame in one hand. The Smurf is wearing a white shirt, red shorts, white shoes, and a white hat. The charm is on a gold chain. The charm has a glossy finish.
192/022	Hiker				Flat Metal Necklace	1 1/4"		$5.00-7.00	The Smurf is holding a black walking stick. He is wearing a yellow jacket, no pants and black hiker boots. He has a green backpack on his back. The charm is on a gold chain. The charm has a glossy finish.
192/023	Smurf Wearing A Green Apron				Flat Metal Necklace			$5.00-7.00	The Smurf is standing with his hands at his side. He is wearing a green apron, white pants and a white hat. The charm is on a gold chain. The charm has a glossy finish.
192/024	Smurf Carrying A Yellow Tennis Racket				Flat Metal Necklace	1 1/4"		$5.00-7.00	The Smurf is carrying a yellow tennis racket in front of his chest. He is wearing a white shirt and red shorts. The Smurf has white legs and white feet. The charm is on a gold chain. The charm has a glossy finish.
192/025	Travel Smurf				Flat Metal Necklace	1 1/4"		$5.00-7.00	The Smurf is walking, carrying a black knapsack over his shoulder. The knapsack has a money symbol on the front. The Smurf is wearing a white tank shirt, yellow shorts and white shoes. The charm is on a gold chain. The charm has a glossy finish.
192/026	Smurf Holding A Flower				Flat Metal Necklace	1 1/4"		$5.00-7.00	The Smurf is standing, holding a red daisy behind his back. He is wearing a white shirt, a white hat, and purple shorts. The Smurf has white legs and white shoes. The charm is on a gold chain. The charm has a glossy finish.
192/027	Diecast Smurf Roller-skating				Flat Metal Necklace	1 1/4"		$5.00-7.00	The Smurf pin is in the shape of a Smurf. It is made of a diecast metal. The Smurf is wearing yellow pants, a white hat and green/white/ black roller-skates.
192/028	Silver Smurf Face Necklace	Peyo		Silver	Silver Necklace	36 cm	Germany	$11.00-15.00	The necklace is silver with a Smurf face pendant. The face is painted half blue and the rest is silver. The Smurf has a white hat. The face is made of silver.
192/029	Smurf Waving Silver Necklace	Peyo		Silver	Silver Necklace	36 cm	Germany	$11.00-15.00	The necklace is made of silver. The pendant is of a Smurf waving. The Smurf isn't fully painted blue. His face is part silver- colored. The Smurf has white pants and a white hat on.
192/030	Diecast Smurf Carrying A Cake			Blue	Diecast Necklace	1 1/4"		$5.00-7.00	The Smurf is carrying a cake. The complete figure and cake is blue. There is a white patch on his hat, pants and shoes. The necklace is 3- dimensional and made of metal.
192/031	Diecast Smurf Carrying Cake				Diecast Necklace	1 1/4"		$5.00-7.00	The Smurf is carrying a cake. The Smurf is wearing white pants and a white hat. The cake is white. The necklace is 3- dimensional and made of metal.
192/032	Diecast Soccer Player				Diecast Necklace	1 1/4"		$5.00-7.00	The Smurf necklace is in the shape of a Smurf. It is made of a diecast metal. The Smurf is kicking a white soccer ball. The Smurf is wearing a white shirt, white socks, white shoes, white shorts, and a white hat.
192/033	Diecast Smurf Playing A Trumpet				Diecast Necklace	1 1/4"		$5.00-7.00	The necklace is made of diecast metal. The Smurf is 3-D. The Smurf is wearing a white hat, and white pants. The Smurf is walking, playing a yellow trumpet.
192/034	Diecast Superman	W. Berrie/Peyo	1/1/81		Diecast Necklace	1 1/4"	U.S.A.	$9.00-12.00	The necklace is made of diecast metal. The Smurf is 3-D. The Smurf is wearing a red cape, white pants and a white hat. He has a red and yellow "S" symbol on his chest.
192/035	Diecast Cowboy				Diecast Necklace	1 1/4"		$5.00-7.00	The necklace is 3-D. It is made of a diecast metal. The Smurf is wearing white chaps, white pants, white shoes, white vest, a white cowboy hat, and a white handkerchief around his neck. The Smurf is standing with his hands at his hips. The charm is on a gold chain.
192/036	Diecast Drummer				Diecast Necklace	1 1/4"		$5.00-7.00	The necklace is made of diecast metal. The Smurf is 3-D. The Smurf is holding white drumsticks with red tips. He has a red/yellow/white drum in front of him. The Smurf is wearing a white hat and white pants. He is on a gold necklace.
192/037	Smurfette Holding A Daisy				Flat Metal Necklace	1"		$5.00-7.00	Smurfette is holding an orange daisy. She is wearing a white hat and white dress. The charm is on a gold necklace.
192/038	Travel Smurf				Necklace	1"		$5.00-7.00	The Smurf is walking, carrying an orange knapsack over his shoulder. The knapsack is on a yellow stick. The Smurf is wearing a white hat and white pants. The charm is on a gold chain.

Set 132: Needle Point Stitchery

132/001	Santa Papa Smurf	Wonder Art		Stitchery Kit	7" Dia	U.S.A.	$5.00-7.00	The kit is not completed. The kit, when finished, has a yellow background and has Papa dressed as Santa holding a green bag and a candy cane. The hoop is red.
132/002	Tennis Smurf	Wonder Art		Stitchery Kit	7" Dia	U.S.A.	$5.00-7.00	The kit is not completed. The kit, when finished, has a yellow background and has a tennis player in a white outfit holding a big, red, white, and black tennis racket. There is a red hoop.
132/003	Baby Smurf With A Rattle	Royal Paris	1/1/86	Needle Point	11 1/2" X 11 1/2"	France	$11.00-15.00	The Baby Smurf is stitched on white canvas. The picture is of Baby wearing a pink outfit, holding a red rattle and laying on a white and yellow pillow. The background is blue on top and green on the bottom.
132/004	Roller-skating Smurf	Wonder Art		Needle Point	7" Dia	U.S.A.	$3.50-5.00	The kit is completed. The background is yellow and has a red hoop. The picture is of a Smurf wearing orange and gray roller-skates with red wheels.
132/005	Dazzled Smurf			Needle Point	7" Dia		$3.50-5.00	The Smurf is stitched on white canvas. The picture is of a Smurf sitting, looking dazzled with yellow stars circling his head. There is no outside hoop.
132/006	Love-struck Smurf	Wonder Art		Needle Point	7" Dia		$3.50-5.00	The Smurf is stitched on white canvas. The picture is of a Smurf sitting, holding a white daisy. The Smurf has red lip marks on his hat and foot. There is a red hoop around the finished picture.
132/007	Smurfette With A Pink Daisy			Needle Point	6" Dia		$3.50-5.00	Smurfette is stitched on white canvas. The picture is of Smurfette standing, holding a pink flower. She is pulling the petals off the flower. There is a blue hoop around the finished picture.
132/008	Baseball Smurf	Wonder Art		Needle Point	7" Dia		$3.50-5.00	The kit is completed. The background is yellow and has a red hoop. The picture is of a Smurf wearing white shorts, a white shirt with a red "B" on the front, white shoes, white socks, and a red hat. The Smurf is walking, carrying a tan bat over his shoulder. The end of the bat has a brown glove hanging on it.
132/009	Soccer Smurf	Wonder Art		Needle Point	7" Dia		$3.50-5.00	The kit is completed. The background is yellow and has a red hoop. The picture is of a Smurf wearing black shorts, a white shirt, a white hat, and white shoes with black spikes on the bottom. The Smurf is kicking a white and black soccer ball.
132/010	Christmas Smurf With A Sack Of Presents	Wonder Art		Needle Point	7" Dia		$3.50-5.00	The kit is completed. The background is yellow and has a red hoop. The picture is of a Smurf wearing white pants and a red Santa hat with a white fuzzy pompom on the end. The Smurf is carrying a big green bag over his shoulder. The bag has a yellow horn and a blue ball sticking out.
132/011	Smurfs Stacked On Each Other			Needle Point Picture	18"L X 6"W	Hand made	$7.50-10.00	The picture is of 5 Smurfs stacked on each others shoulders. Smurfette is on top and Papa is on the bottom of the stack. The background is tan. The picture is done in needle point and is in a gold metal frame.
132/012	Smurfette Smelling A Red Daisy			Needle Point Picture	7"L X 5"W	Hand made	$3.50-5.00	The picture is needle point. Smurfette has a dark blue body. She is leaning over smelling a red daisy. Smurfette has a white dress, shoes and hat on. She is standing on a patch of green grass. The picture is in a metal gold frame.
132/013	Tennis Smurf Pillow			Cross Stitch	11 1/2" x 9 1/2"	Hand made	$3.50-5.00	The pillow has a Smurf holding a tennis racket on the front. The tennis racket is red. The Smurf is wearing a white shirt with red trim, white shorts, and white and red shoes. The back of the pillow is red velvet.
132/014	Smurf Opening Present	DMC	1/1/95			Holland	$5.00-7.00	The pattern is a Smurf standing, opening a present. In the background are mushroom houses.
132/015	Smurf Giving Smurfette Flowers			Needle Point			$3.50-5.00	A Smurf is standing, handing Smurfette a bouquet of yellow daisies. Smurfette is standing with her hands under her chin. The word LOVE is between them in multi- colors. The background is blue.
132/016	Lovesick Smurf With Smurfette	Dupuis	1/1/83	Framed Picture	17 1/2" x 21 1/2"	Germany	$75.00-100.00	A Smurf is standing, pulling petals off of a white daisy. Smurfette is standing, watching. The Smurf has a lovesick expression on his face. There are yellow hearts floating above his head. The picture is only painted canvas. It is in a burgundy matting with a wood frame.
132/017	Smurf Face Crochet			Crochet Face	4" x 3 1/2"			The Smurf face is made of yarn. He has 2 rolley eye's and a red yarn mouth. The Smurf's face is dark blue. He is crochet.
132/018	Smurf Face Crochet			Crochet Face	3 1/4" x 3"			The Smurf face is made of yarn. He has 2 rolley eye's and a black yarn mouth. The Smurf's face is light blue. He has a fuzzy nose. He is crochet.
132/019	Dazzled Smurf Pattern	Wonder Art		Cross Stitch	7" Diam	U.S.A.	$5.00-7.00	Dazzled Smurf is drawn on canvas. The picture is of a Smurf sitting, looking dazzled with yellow stars circling his head. MIP

Set 37: Night Lights

037/001	Smurf With Water Buckets	Zambezi Trading Co.	Round Night Light	2"	U.S.A.	$4.50-6.00	2 " Dia 1/2 sphere, with a clear, white-ish, background and a picture of a Smurf carrying 2 yellow buckets of water.
037/002	Flying Smurf	Zambezi Trading Co.	Round Night Light	2"	U.S.A.	$4.50-6.00	2 " Dia 1/2 sphere, with a clear, white-ish, background and a picture of a Smurf with white wings, flying.
037/003	Smurf Carrying A Cake	Zambezi Trading Co.	Round Night Light	2"	U.S.A.	$4.50-6.00	2 " Dia 1/2 sphere, with a clear, white-ish, background and a picture of a Smurf carrying a cake.
037/004	Smurfette With Hands Behind Her Back	Zambezi Trading Co.	Round Night Light	2"	U.S.A.	$4.50-6.00	2 " Dia 1/2 sphere, with a clear, white-ish, background and a picture of Smurfette standing with her hands behind her back (looks like she's flirting).
037/005	Papa Walking	Zambezi Trading Co.	Round Night Light	2"	U.S.A.	$4.50-6.00	2 " Dia 1/2 sphere, with a clear, white-ish, background and a picture of Papa walking with his hands down by his side.
037/006	Papa Running	Zambezi Trading Co.	Round Night Light	2"	U.S.A.	$4.50-6.00	2 " Dia 1/2 sphere, with a clear, white-ish, background and a picture of Papa running with one hand out in front of him and the other at his side.
037/007	Smurf With A Flower	Zambezi Trading Co.	Round Night Light	2"	U.S.A.	$4.50-6.00	2 " Dia 1/2 sphere, with a clear, white-ish, background and a picture of a Smurf pulling petals off a white daisy with 3 hearts floating above his head.
037/008	Papa Smurf Stain glass Night Light		Night Light	6"		$4.50-6.00	Papa looks like he is walking and his hands are at his side. Papa's wearing a red hat and red pants. The stain glass is attached to a white plug.
037/009	Smurf Running	Zambezi Trading Co.	Round Night Light	2"	U.S.A.	$4.50-6.00	2 " Dia 1/2 sphere, with a clear, white-ish, background and a picture of a Smurf running.
037/010	Smurf On Springs	Zambezi Trading Co.	Round Night Light	2"	U.S.A.	$4.50-6.00	2 " Dia 1/2 sphere, with a clear, white-ish, background and a picture of a Smurf wearing springs on his feet. The Smurf is leaping in the air. The springs are yellow.
037/011	Troubled Smurf	Zambezi Trading Co.	Round Night Light	2"	U.S.A.	$4.50-6.00	2 " Dia 1/2 sphere, with a clear, white-ish, background and a picture of a Smurf walking, looking troubled. The Smurf is looking down at the ground. He has his hands behind his back.
037/012	Brainy Pointing	Zambezi Trading Co.	Round Night Light	2"	U.S.A.	$4.50-6.00	2 " Dia 1/2 sphere, with a clear, white-ish, background and a picture of Brainy standing, pointing with his right hand. Brainy is wearing black glasses.

037/013	Papa Jumping In The Air	Zambezi Trading Co.			Round Night Light	2"	U.S.A.	$4.50-6.00	2 " Dia 1/2 sphere, with a clear, white-ish, background and a picture of Papa jumping in the air. Papa has both hands out at his sides. He has a big grin on his face.
037/014	Papa Pointing	Zambezi Trading Co.			Round Night Light	2"	U.S.A.	$4.50-6.00	2 " Dia 1/2 sphere, with a clear, white-ish, background and a picture of Papa standing, pointing. Papa is pointing with his left hand. He has his other hand at his side.
037/015	Lute	Zambezi Trading Co.			Round Night Light	2"	U.S.A.	$4.50-6.00	2 " Dia 1/2 sphere, with a clear, white-ish, clear background and a picture of a Smurf holding a yellow lute.
037/016	King	Zambezi Trading Co.			Round Night Light	2"	U.S.A.	$4.50-6.00	2 " Dia 1/2 sphere, with a clear, white-ish, background and a picture of a Smurf dressed as a king. The Smurf is running. The Smurf has a yellow hat, yellow crown and yellow pants on. The Smurf is wearing a red robe with white trim. The Smurf is carrying a red scepter.
037/017	Traveler	Zambezi Trading Co.			Round Night Light	2"	U.S.A.	$4.50-6.00	2 " Dia 1/2 sphere, with a clear, white-ish, background and a picture of a Smurf walking, carrying a knapsack over his shoulder. The Smurf is laughing. The knapsack is red.
037/018	Smiling Smurf	Zambezi Trading Co.			Round Night Light	2"	U.S.A.	$4.50-6.00	2 " Dia 1/2 sphere, with a clear, white-ish, background and a picture of a Smurf standing with his hands behind his back. The Smurf has a big grin on his face.
037/019	Smurfette Yawning	Zambezi Trading Co.			Round Night Light	2"	U.S.A.	$4.50-6.00	2 " Dia 1/2 sphere, with a clear, white-ish, background and a picture of Smurfette standing with her hands together in front of her. Smurfette has her eyes closed and her mouth open .
037/020	Smurf Snoozing	Zambezi Trading Co.			Round Night Light	2"	U.S.A.	$4.50 6.00	2 " Dia 1/2 sphere, with a clear, white-ish, background and a picture of a Smurf standing up, sleeping. There are Z's floating above his head.
037/021	Smurf With A Red Daisy On His Hat	Zambezi Trading Co.			Round Night Light	2"	U.S.A.	$4.50-6.00	2 " Dia 1/2 sphere, with a clear, white-ish, background and a picture of a Smurf walking with a hand on his head. The Smurf has a red daisy in his hat. The Smurf has a dreamy look on his face.
037/022	Smurf In Love	Zambezi Trading Co.			Round Night Light	2"	U.S.A.	$4.50-6.00	2 " Dia 1/2 sphere, with a clear, white-ish, background and a picture of a Smurf standing with a dreamy look on his face. There are 3 hearts floating above his head. There is a little clock in a cloud in front of his chest. The cloud says TICK, TICK.

Set 49: Notebooks- Spiral

049/001	Gargamel And Azreal	Mead	1/1/82	Light Blue	Spiral Notebooks	10 1/2" x 8"	U.S.A.	$3.50-5.00	The cover has Gargamel holding his hands together and Azreal is walking in front of him. Gargamel's gray castle is in the background.
049/002	Sports Smurfs	Mead	1/1/82	Green	Spiral Notebooks	10 1/2" x 8"	U.S.A.	$3.50-5.00	The cover has 4 sport Smurfs. It has a basketball Smurf, soccer Smurf, football Smurf, and baseball Smurf. The background is green.
049/003	Smurfy Friends Stay In Touch	Mead	1/1/82	Red	Spiral Notebooks	10 1/2" x 8"	U.S.A.	$3.50-5.00	Smurfette is standing, talking on the phone to a Smurf. The background is red.
049/004	To Smurf Or Not To Smurf, That Is The Question	Mead	1/1/82	Yellow	Spiral Notebooks	10 1/2" x 8"	U.S.A.	$3.50-5.00	Papa Smurf is standing, holding a finger up in the air. The background is yellow.
049/005	I Love Work.. I Can Dream About It For Hours	Mead	1/1/82	Light Blue/Green	Spiral Notebooks	10 1/2" x 8"	U.S.A.	$3.50-5.00	A Smurf is leaning on a shovel and dreaming. The background is light blue.
049/006	Time To Get Smurfin!	Mead	1/1/82	Orange	Spiral Notebooks	10 1/2" x 8"	U.S.A.	$3.50-5.00	A Smurf is carrying books and looking at a clock. The background is orange.
049/007	Smurfette	Mead	1/1/82	Red	Spiral Notebooks	10 1/2" x 8"	U.S.A.	$3.50-5.00	The cover has Smurfette standing, holding a yellow daisy. The background is red.
049/008	Smurfette Small Notepad	Mead	1/1/82	Red	Spiral Notepad	3 x 5	U.S.A.	$1.50-2.00	The cover has Smurfette standing, holding a yellow daisy. The background is red.
049/009	Papa Small Notepad	Mead	1/1/82	Yellow	Spiral Notepad	3 x 5	U.S.A.	$1.50-2.00	Papa Smurf is standing, holding a finger up in the air. The background is yellow.
049/010	Have A Smurfy Day! Small Notepad	Mead	1/1/82	Light Blue	Spiral Notepad	3 x 5	U.S.A.	$1.50-2.00	A Smurf is sitting on a cloud. He has a rainbow behind him. The notepad is light blue.
049/011	Smurf Small Notepad	Mead	1/1/82	Green	Spiral Notepad	3 x 5	U.S.A.	$1.50-2.00	The notepad has a Smurf standing with his hands in the air. The background is green.

Set 222: Ornaments- Alderbrook Christmas Plastic Figure

222/001	Skier	SEPP; Alderbrook			Plastic Ornament	3" High	Hong Kong	$11.00-15.00	The Smurf is made of hard plastic. He is wearing a yellow outfit with red gloves, red boots, a white hat, and orange goggles. The Smurf is holding orange, plastic ski poles and wearing orange, plastic ski's. The Smurf has a cloth red scarf. There is a gold cord attached for hanging.
222/002	Smurf On Sleigh	SEPP; Alderbrook			Plastic Ornament	3" High	Hong Kong	$11.00-15.00	The Smurf is made of hard plastic. The Smurf is sitting on a yellow plastic sleigh. He is wearing a white shirt, red pants, a red hat, and a cloth red scarf. His hat has a yellow pompom on the tip. He has plastic holly on the side of his hat. He is waving with his right hand. There is a gold cord attached for hanging.
222/003	Smurfette Wearing Santa Suit	SEPP; Alderbrook			Plastic Ornament	3" High	Hong Kong	$11.00-15.00	The figure is made of hard plastic. Smurfette is wearing a red jacket with white trim, a black belt with a gold buckle, white boots with 3 black buttons, and a red hat with white trim and a felt pompom on the tip. Smurfette is holding up a red sign with yellow paper on the front. The sign says "Merry Christmas From The Smurfs 1982" in black letters. There is a gold cord for hanging.
222/004	Papa Wearing Santa Suit	SEPP; Alderbrook			Plastic Ornament	3" High	Hong Kong	$11.00-15.00	The figure is made of hard plastic. Papa is wearing a red jacket with white trim, a black belt with a gold buckle, black boots with white trim, red pants, and a red hat with white trim and a felt pompom on the tip. Papa has a white fuzzy beard. Papa is holding up a red sign with yellow paper on the front. The sign says "Smurf And Kisses At Christmas 1982" in black letters. There is a gold cord for hanging.
222/005	Skier	SEPP; Alderbrook			Plastic Ornament	3" High	Hong Kong	$11.00-15.00	The Smurf is made of hard plastic. He is wearing a red outfit with white gloves, red boots, a white hat, and red goggles. The Smurf is holding yellow, plastic ski poles and wearing yellow, plastic ski's. The Smurf has a cloth white scarf. There is a gold cord attached for hanging.
222/006	Bell Ringer	SEPP; Alderbrook			Plastic Ornament	3" High	Hong Kong	$11.00-15.00	The Smurf is made of hard plastic. He is wearing red pants, a yellow shirt, white shoes, a red felt bow tie, and a white hat with a red pompom and plastic holly. The Smurf is holding a brass bell.
222/007	Ice Skater	SEPP; Alderbrook			Plastic Ornament	3" High	Hong Kong	$11.00-15.00	The Smurf is made of hard plastic. He is wearing red pants, a green shirt with a white "7", silver skates, and a white hat with a red pompom.
222/008	Angel	SEPP; Alderbrook			Plastic Ornament	3" High	Hong Kong	$11.00-15.00	The Smurf is made of hard plastic. The Smurf has a white hat with gold garland around it. The Smurf is holding a gold wand with a star on the end. The Smurf has gold and white wings. He is wearing gold pants and a red cloth scarf. The Smurf is standing on a cotton ball cloud. He has a gold string attached to his head for hanging.
222/009	Papa Holding A Candy cane	SEPP; Alderbrook			Plastic Ornament	3" High	Hong Kong	$11.00-15.00	The figure is made of hard plastic. Papa is wearing a red jacket with white trim, a black belt with a gold buckle, black boots with white trim, red pants and a red hat with white trim, and a felt pompom on the tip. Papa has a white fuzzy beard. Papa is holding a red and white plastic candy cane. There is a gold cord for hanging attached to his head.

Set 252: Ornaments

252/001	Papa Carrying Sack	Handmade	1/1/83		Glass	3 1/2" Diam	Hand Painted	$7.50-10.00	Papa is dressed as Santa. He is carrying a green sack over his shoulder. Papa is walking through snow. The ornament is frosted glass.
252/002	Smurf Pulling A Sled	Handmade	1/1/92		Glass	3 1/2" Diam	Hand Painted	$7.50-10.00	A Smurf is walking in the snow pulling a green sled. It looks like it's snowing. The Smurf is wearing a white and blue scarf, white hat, white pants, and red mittens. The ornament is frosted glass.
252/003	Smurf Caroling	Handmade	1/1/92		Glass	3 1/2" Diam	Hand Painted	$7.50-10.00	A Smurf is standing in a snowfall singing. The Smurf is holding a white book. There are music notes floating above his head. The ornament is frosted glass.
252/004	Smurf Skiing	Handmade	1/1/94		Glass	3 1/2" Diam	Hand Painted	$7.50-10.00	The Smurf is standing on a black ski. He is holding a black pole. There is a green package with a red bow sitting on the back of his ski. The Smurf is wearing a white hat, white pants and a red scarf. The ornament is frosted glass.
252/005	Smurfette Standing Under Mistletoe	Handmade			Glass Bell	3 " High	Hand Painted	$7.50-10.00	The ornament is bell shape. It has Smurfette standing with her hand on her hip. There is mistletoe hanging above her. The ornament is frosted glass.
252/006	Smurfette With Snowman	Handmade	1/1/93		Glass Ornament	3 1/2" Diam	Hand Painted	$7.50-10.00	Smurfette is standing, facing a snowman. The snowman has a black hat, black pipe and a red scarf. The ornament is frosted glass.
252/007	Smurf With Snowman	Handmade	1/1/93		Glass	3 1/2" Diam	Hand Painted	$7.50-10.00	The Smurf is walking away from the snowman on the right side. The snowman has a black hat, black pipe and red scarf. The ornament is frosted glass.
252/008	Smurf with Snowman	Handmade	1/1/94		Glass	3 1/2" Diam	Hand Painted	$7.50-10.00	The Smurf is walking away from the snowman on the left side. The snowman has a black hat, black pipe and red scarf. The ornament is frosted glass. The Smurf has snow under his feet and it's snowing.
252/009	Smurfette Under Mistletoe	Handmade	1/1/83		Glass Ornament	3"	Hand Painted	$7.50-10.00	It has Smurfette standing with her hand on her hip. There is mistletoe hanging above her. The ornament is frosted glass.
252/010	Smurf With Stockings	Handmade			Glass	3" High	Hand Painted	$7.50-10.00	The ornament is bell- shaped. A Smurf is standing next to 2 stockings. The stockings are hanging from a yellow piece of wood. The stockings are red. The ornament is frosted glass.
252/011	Blown Glass Blue Smurf	Handmade			Glass	3 1/2" High	Hand made	$3.00-4.00	The Smurf is blown glass. He has a light blue face. He is wearing a red hat, white shirt, and red pants. The Smurf has silver trim around his outfit.
252/012	Blown Glass Pink Smurf	Handmade			Glass	3 1/2"	Hand made	$3.00-4.00	The Smurf is blown glass. He has a light pink face. He is wearing a red hat, white shirt, and red pants. The Smurf has silver trim around his outfit.
252/013	Smurf Playing A Trumpet	Handmade			Glass	4" High		$3.50-5.00	A Smurf is standing, playing a yellow trumpet. The ornament is figure- shaped.
252/014	Smurf With Gingerbread Man	Handmade			Glass	3 3/4" High		$6.00-8.00	A Smurf is standing, holding a gingerbread man. The ornament is figure- shaped.
252/015	Papa Holding A Banner	Handmade			Glass	3 3/4"		$6.00-8.00	Papa is standing, holding a white banner. One side says " Merry Christmas" the other side says "Happy New Year". The ornament is figure- shaped.
252/016	Papa With Sack			Silver	Glass Ball	2 3/4"	Germany	$5.00-7.00	Papa is walking, carrying a brown sack of toys over his shoulder. Papa is wearing red pants and a red hat with a red tassel on the end. There is 1 white, frosted, glitter tree on each side of Papa.
252/017	Smurf Decorating Tree			Silver	Glass Ball	2 3/4"	Germany	$5.00-7.00	A Smurf is standing, putting a yellow star on the top of a tree. There is a white glittery shooting star above his head.
252/018	Smurf On Sled	Puffi	1/1/83	White	Plastic Ball	2 1/4"	Italy	$5.00-7.00	The ornament has a wrap around picture. It has a Smurf on a sled. Papa putting presents under a tree. A Smurf looking excited. The picture is all outdoors.
252/019	Smurfette Skater Kings Dominion Ornament	Peyo			Stain Glass	3 3/4" x 2 1/2"		$4.50-6.00	The ornament is flat stain glass. It has Smurfette wearing a red dress with green trim, white gloves, white skates, and a white hat with holly on it. Below Smurfette it says Kings Dominion in black letters. There is a gold cord for hanging.
252/020	Smurf Carrying A Gift Kings Dominion Ornament	Peyo			Glass	3 " x 2 1/2"		$4.50-6.00	The ornament is flat stain glass. It has a Smurf walking, carrying 2 X-mas presents. The Smurf is wearing a red scarf, white pants and a white hat with holly on the side. Below the Smurf it says Kings Dominion in black letters. There is a gold cord for hanging.
252/021	Smurfette Glass Figure				Glass	4"	Italy	$25.00-35.00	The ornament is in the shape of a figure. It is Smurfette. She has yellow fuzzy hair. Smurfette is wearing a white hat and a white and red dress.
252/022	Papa Glass Figure				Glass	4"	Italy	$25.00-35.00	The ornament is in the shape of a figure. Papa has a red cloth hat and red painted-on pants. Papa has a fuzzy white beard.

Set 170: Ornaments -- Satin Christmas

170/001	Merry Christmas 1982	S.E.P.P.	1/1/82		Satin Ball Ornament	3" Circle	U.S.A.	$9.00-12.00	The ball is white satin. There is a wrap around picture on plastic. The picture on one side has Papa sticking his head through a wreath. A Smurf and Smurfette are holding the wreath. The other side says "Merry Christmas 1982" in red letters. The background is blue sky and snow with green trees.
170/002	Happy Holidays	S.E.P.P.	1/1/82		Satin Ball Ornament	3" Circle	U.S.A.	$9.00-12.00	The ball is white satin. There is a wrap around picture on plastic. The picture is Smurfette making a snowman. The snowman has a red hat, green scarf, orange mittens, and is holding a candy cane. Papa is pulling an X-mas tree and waving. The ornament says "Happy Holidays" in red letters. The background is a mushroom house and snow.
170/003	Baby's 1st. Christmas 1982	S.E.P.P.	1/1/82		Satin Ball Ornament	3" Circle	U.S.A.	$9.00-12.00	The ball is white satin. There is a wrap around picture on plastic. The picture has a Smurf popping out of a jack-in-the-box. A Smurf is on a wood rocking horse. A toy Smurf is sitting next to a tree with presents. The ornament says "Baby's 1st. Christmas 1982" in red letters. The background is light green.
170/004	No 1 Teacher/Merry Christmas	S.E.P.P.	1/1/82		Satin Ball Ornament	3" Circle	U.S.A.	$9.00-12.00	The ball is white satin. There is a wrap around picture on plastic. The one side of the ornament has Papa sitting at a desk with a chalkboard on his side. Brainy Smurf is carrying him a present. The other side has the Smurf carrying the present to the school house. The background is snow. The ornament says "No. 1 Teacher" in yellow and "Merry Christmas" in red letters.
170/005	Merry Christmas	S.E.P.P.	1/1/82		Satin Ball Ornament	3" Circle	U.S.A.	$9.00-12.00	The ball is white satin. There is a wrap around picture on plastic. The picture has a Smurf standing next to a tree eating a candy cane. The other side has 2 Smurfs standing by a fireplace. The background is light green. The ornament says "Merry Christmas" on one side and "Wishing you beautiful moments and warm memories today and always."
170/006	Hope This Holiday Brings Health	S.E.P.P.	1/1/82		Satin Ball Ornament	3" Circle	U.S.A.	$9.00-12.00	The ball is white satin. There is a wrap around picture on plastic. The picture has 2 Smurfs and Papa carrying presents through the snow. The other side has a Smurf and Smurfette decorating a tree. The background is sky and snow. The ornament says "Hope this holiday brings health, happiness, and all your favorite things" in red letters.
170/007	Merry Christmas From The Smurfs	Alderbrook Industries	1/1/82		Satin Ball Ornament	3" Circle	Canada	$13.50-18.00	The ball is white satin. There is a wrap around picture on plastic. One side of the ornament has 3 Smurfs playing music instruments around a Christmas tree. The top says "Merry Christmas From The Smurfs". The other side has Papa standing on a brown mushroom. Smurfette is carrying flowers towards Papa and another Smurf is carrying a cake. There is a Christmas tree behind the Smurfs. The top says "Schtroumpf Noel".
170/008	Joyeux Noel De La Famille Schroumpf	Alderbrook	1/1/82		Satin Ball Ornament	3" Circle		$13.50-18.00	The ball is white satin. There is a wrap around picture on plastic. One side of the ornament has 2 Smurfs in front of a fireplace with Papa reading the story "Twas the night before Christmas". The other side of the ball has Brainy wearing red pajamas, standing in front of a tree, and another Smurf in green pajamas carrying a present towards the tree.
170/009	Peace On Earth	S.E.P.P.	1/1/94	White	Satin Ball Ornament	2" Circle		$6.75-9.00	The satin ball has a Smurf sitting by a mail box. An earth is floating above his head. The picture is in blue outline. The satin ball is white. It says "Peace On Earth" above the Smurf. The bottom says "Smurf Collectors Club International 1994".

Set 207: Ornaments- The Night Before Christmas

207/001	Hefty	W. Berrie/Peyo	1/1/83		Porcelain Ornament	4"	Japan	$11.00-15.00	The ornament is in the shape of a stocking. The stocking is green with red and pink stripes and a white top. Hefty Smurf is in the stocking. Hefty has a heart tattoo on his arm. He is holding a candy cane. The top of the stocking says "Hefty" in red letters. There is a gold cord for hanging.
207/002	Papa	W. Berrie/Peyo	1/1/83		Porcelain Ornament	4"	Japan	$11.00-15.00	The ornament is in the shape of a stocking. The stocking is green with red and white stripes and a white top. The stocking has a red heart in the center. Papa Smurf is in the stocking. Papa is waving. There are 2 candy canes in the back of the stocking. The top of the stocking says "Christmas 1983" in red letters. There is a gold cord for hanging.
207/003	Grouchy	W. Berrie/Peyo	1/1/83		Porcelain Ornament	4"	Korea	$11.00-15.00	The ornament is in the shape of a stocking. The stocking is red with green and white stripes and a white top. The stocking has green holly leaves in the center. Grouchy Smurf is in the stocking. Grouchy has a mean expression on his face. The top of the stocking says "Grouchy" in red letters. There is a gold cord for hanging.
207/004	Jokey	W. Berrie/Peyo	1/1/83		Porcelain Ornament	4"	Korea	$11.00-15.00	The ornament is in the shape of a stocking. The stocking is green with a red tip and heel. The top is white with red stripes. The stocking has a red flower in the center. Jokey Smurf is in the stocking. Jokey is holding a pink present with red ribbon. The top of the stocking says "Jokey" in red letters. There is a gold cord for hanging.
207/005	Smurfette	W. Berrie/Peyo	1/1/83		Porcelain Ornament	4"	Korea	$11.00-15.00	The ornament is in the shape of a stocking. The stocking is pink with a red tip and heel. The top is white with red trim. In the center of the stocking is a red heart with white dots. Smurfette is in the stocking. Smurfette has her eyes closed and a happy look on her face. The top of the stocking says "Christmas 1983" in red letters. There is a gold cord for hanging.
207/006	Brainy	W. Berrie/Peyo	1/1/83		Porcelain Ornament	4"	Korea	$11.00-15.00	The ornament is in the shape of a stocking. The stocking is red with green and white trim. The front of the stocking has a Christmas tree on it. Brainy is in the stocking. Brainy is wearing black glasses. The top of the stocking says "Brainy" in red letters. There is a gold cord for hanging.

Set 165: Outdoor Riding Toys

165/001	Smurfs -A-Poppin Train	Coleco			Riding Train		Canada	$22.50-30.00	The train is for a toddler. The train is white with blue wheels, a blue seat, yellow handle bars, and a yellow seat back/handle. The train car has a Smurf face on the front. By the handle bars is a clear dome with 3 balls inside. The seat flips up and has a storage area underneath. There are Smurf stickers all over the train car.
165/002	Sit -N- Spin	Hasbro	1/1/83		Sit -N- Spin			$35.00-45.00	The sit and spin has a brown bottom. The handle is in the shape of a mushroom. The base is blue and the top is red. There is a flat Smurf shape attached to the top of the handle. Smurfette is on one side and a Smurf is on the other side of the flat shape.
165/003	Smurf Snow Disc	Coleco	1/1/83	Light Blue	Snow Disc	24" Diam.	U.S.A.	$22.50-30.00	The snow disc is round. It has 2 yellow handles. In the center is a picture of a Smurf riding on the snow disc. The disc is light blue.
165/004	Smurf Skies				Snow Skies	23" Long	Austria	$18.00-25.00	The ski's are white and red with blue straps. They have white outlines of Smurf and Papa Smurf's face.
165/005	Smurf Leaning On Shovel Lawn Ornament	Peyo/Dupuis			Lawn Ornament	12" High	West Germany	$18.00-25.00	The Smurf is made of a hard plastic. He is standing, leaning on a yellow shovel. The Smurf has his eye's closed.
165/006	Tyrolese Smurf Lawn Ornament	Peyo/Dupuis			Lawn Ornament	12" High	West Germany	$18.00-25.00	The Smurf is made of a hard plastic. He is standing, smoking a red pipe. The Smurf is wearing green shorts with green suspenders and white shoes. The Smurf has a red feather in his hat.
165/007	Smurf-A-Boggin	Coleco	1/1/83	Blue	Toboggan		U.S.A	$35.00-45.00	The toboggan is blue. It has 2 Smurf faces on the front. The inside back has a Smurf sticker on it. The sticker says "Smurf-A-Boggin". It has pictures of Smurfs playing in the snow.
165/008	Super Smurf Piggyback Rider	Coleco	1/1/83	Blue	Riding Toy	20 x 7 1/2 x 15 1/2"	U.S.A	$22.50-30.00	The toy looks like a motorcycle. It is a Smurf that is sitting, riding a bike. You sit on the Smurfs back. There are 4 red wheels.
165/009	Smurf Mobile Pedal Car	Coleco	1/1/83	Blue	Pedal Car		U.S.A.	$110.00-150.00	The car is blue with black wheels. It has Smurf stickers all over it. The car has a white steering wheel, red honking horn, and an ignition key that clicks..
165/010	Gargamel On Treasure Chest Lawn Ornament	Hazero	1/1/83		Rubber Ornament	7"	Germany	$15.00-20.00	Gargamel is made of a hard plastic. He is sitting on a black and brown treasure chest. Gargamel is holding a finger in the air. There is a hole in the bottom to put sand in.
165/011	Papa With Barrel And Coin	Hazero	1/1/83		Plastic Ornament	7"	Germany	$15.00-20.00	Papa is made of a hard plastic. Papa is standing next to a tan barrel. Papa is holding a silver 5 cent coin in the air. There is an opening in the bottom to put sand in.
165/012	Smurf WhirllyGig	Homemade			Whirllygig	1 Foot	Hand made	$11.00-15.00	The Smurf's legs spin. The Smurf is made of wood. He is painted light blue with a white hat and white pants. There is a steel rod for sticking in the ground.

Set 212: Paper Clips- Large Figures

212/001	Alles Halb So Wichtig!!			Pink	Paper Clip	6"	Germany	$6.00-8.00	The clip is pink and large. On the end it has a Smurf figure glued on it. The figure is of a Smurf pointing. The clip says "Alles Halb So Wiichtig!!" In the center, in red letters.
212/002	Erledigt Sich Vo Selbst!!	Laurel		Blue	Paper Clip	6"	Germany	$6.00-8.00	The clip is blue and large. It has a Smurf figure on the end. The figure is of a Smurf sitting on a reddish/orange cushion. The clip says "Erledigt Sich Vo Selbst!!" in dark blue letters in the center.

Set 193: Paper Items & Advertisements- Miscellaneous

193/001	Hanging Smurfette	Peyo	1/1/83		Cardboard Decoration	14 1/2" High	Hong Kong	$7.50-10.00	Smurfette is cardboard. There is a hole on top to hang her. She is holding a white daisy.
193/002	Happy Smurfday Newspaper Article		10/25/88		Newspaper Article		Florida	$1.50-2.00	The article was in the Stuart, Florida newspaper. It is called Happy Smurfday. The article is about the Smurfs celebrating their 30th anniversary.
193/003	Ice Capade Bulletin	Peyo/Applause	1/25/87	Light Blue	Bulletin		Pennsylvania	$7.50-10.00	The bulletin has an ad for the Ice Capades in Hershey Park. It gives the time and cost per show.
193/004	The Official Smurf Fun Club Application		1/25/83		Club Application	8"L X 5"W	U.S.A.	$7.50-10.00	The ad is to join the Smurf fun club. The border of the paper is trees. At the top is a Smurf in a blue and yellow circle.
193/005	On Tour Baby Smurf Sticker	Helly	1/1/94		Promotional Sticker	4 1/2"	Brussels	$2.00-3.00	The sticker is triangle shape. The top has Baby Smurf sitting. Baby is wearing a pink sleeper. He is holding a bottle and his other hand is resting on a block. The bottom says " .... On Tour".
193/006	McDonald's 25th Anniversary Bag Blue	McDonald's Corp.	1/1/96	Blue	Happy meal Bag		France	$3.00-4.00	The bag is blue and has a Smurf face on the front. The top says "Ik Heb Lekker Alles OpgeSmurft!" The back has a game. It has little cutouts of the 25th anniversary Smurf figures.
193/007	McDonald's 25th Anniversary Bag	McDonald's Corp.	1/1/96	White	Happy meal Bag		France	$3.00-4.00	The front of the bag has a Smurf leaning against the number 25. The back has a story on it.
193/008	Smurf Magic Berries Ad	General Mills	1/1/88		Ad & Coupon	11" x 8 1/2"	U.S.A.	$6.00-8.00	The ad is for Smurf magic berries cereal. The ad has a blue box with Papa standing next to it, with a bowl of cereal in front of him. On the bottom is a coupon for the cereal.

193/009	Chef Boyardee Pasta Label	Chef Boyardee	1/1/85	Red	Label		U.S.A.	$1.50-2.00	The label is red. It has a Smurf dressed as a chef carrying a purple bowl of pasta. Smurfette is holding a spoon with sauce on it.
193/010	Family Footwear	BBC	1/1/89		Tag		China	$1.50-2.00	The tag has a house with a Smurf on the front. The Smurf is pointing. The front says "Smurf A Family Of Footwear". The inside opens and there is a story.
193/011	German Almanach 1994	La Poste	1/1/94		Map Book	8" X 10 1/2"	Germany	$13.50-18.00	The cover has Smurfs eating at a table. The months of January to June, listing events. The inside has maps. The back has Smurfs riding a swing and merry-go-round. The months of July to December, with events, is also on the back.
193/012	Skippy Peanut Butter Label	Designa Premium	1/1/95		Can Label			$4.50-6.00	The label on the front is white and has Papa Smurf dressed as a conductor. The front says Skippy Creamy Peanut Butter. The back has an advertisement for 4 Smurf under the cap liners.
193/013	Papa Smurf Shrinkum	W. Berrie/Peyo	1/1/85		Shrinkum	2 3/4"	U.S.A.	$3.00-4.00	The shrinkum is round and made from a light plastic. The front has Papa Smurf dressed as a conductor. There is a red border. Under Papa is a spot for a name. The shrinkum shrinks in the oven.
193/014	Smurfette Shrinkum	W. Berrie/Peyo	1/1/85	White	Shrinkum	2 3/4"	U.S.A.	$3.00-4.00	The shrinkum is round and made from a light plastic. The front has Smurfette wearing roller-skates and headphones. Under Smurfette is a spot for a name. The shrinkum shrinks in the oven.
193/015	Smurf Shrinkum	W. Berrie/Peyo	1/1/85	White	Shrinkum	2 3/4"	U.S.A.	$3.00-4.00	The shrinkum is round and made from a light plastic. The front has a Smurf in a red outfit. He has a helmet on and ice skates. Under the Smurf is a spot for a name. The shrinkum shrinks in the oven.
193/016	Baby Smurf Shrinkum	W. Berrie/Peyo	1/1/85	Yellow	Shrinkum	2 3/4"	U.S.A.	$3.00-4.00	The shrinkum is round and made from a light plastic. The front has Baby Smurf dressed in a white sleeper. Baby is crawling. Under Baby is a spot for a name. The shrinkum shrinks in the oven.
193/017	Scholler Ice Cream Ad Lid	Klebebild		Orange	Lid	2 1/2"		$6.75-9.00	The lid is a promotion for Eiskrem Ice Cream. The lid has a picture of a Smurf eating an ice cream bar. The Smurf is holding a blue flag in his other hand. The flag says "Scholler".
193/018	Smurf Picture Disc Ad	Peyo	1/1/82	White	Ad	7 1/2" x 4 1/2"		$1.50-2.00	Special ad for the Smurf's limited edition picture disc. The front of the ad has a picture of the record. The back of the ad tells about ordering and information about the record.
193/019	De Smurfen Sticker Package	Merlin Collections	1/1/95	Lime Green	Sticker Package	3 3/4" x 3 1/4"	England	$3.50-5.00	The package is lime green. The front has a big Smurf in the center. There are outlines of Smurfs behind him. The top says "De Smurfen" There are 2 Smurfs by the words. The back is pink and green. It has the licensing and date.
193/020	Die Schlumpfe Kommen In Die Stadt! Ticket	Unicef	1/1/96	Light Blue	Ticket	2 1/2" x 5 1/2"	Brussels	$1.50-2.00	The ticket is light blue. In the corner it has a Smurf on rollerblades. The Smurf is wearing a white hat with a visor, white tank top, purple shorts, and dark purple rollerblades. The ticket says : Die Schlumpfe Kommen In Die Stadt! in yellow letters. The rest says: Nurnberg Schlumpfenwelt Zeppelinfeld Familienvorstellung 12./13. Marz 97. 14.30 Uhr The ticket is worth 10DM's.
193/021	The Smurf Story Pamphlet	W. Berrie	1/1/79		Small Story Pamphlet	4" x 3"	U.S.A.	$1.50-2.00	The front cover has a Smurf village scene. The pamphlet is blue, green, yellow and white. Inside is a small story about the Smurfs and how they came to be. Inside, 3 Smurfs are working on the story.
193/022	The Untold Story Of History According to Smurfs	Applause/Peyo	1/1/84		Story Pamphlet	4" x 3"	U.S.A.	$3.50-5.00	The cover is red and rust- colored. The front says "The Untold Story Of History According To The Smurfs Vol." in gold letters. The pamphlet is a tri-fold. Inside it shows 6 history Smurfs and has stories next to them.
193/023	Pez Salesman Sheet	Pez	1/1/98	Yellow	AD Sheet	11 3/4" x 8 1/4"	Germany	$6.75-9.00	The paper is yellow. The front shows the new fluorescent Pez for 1998. The Pez are Papa, Brainy, regular Smurf, Gargamel and Smurfette. The sheet is in German. The front also shows the regular characters. The back shows a picture of the Smurf Pez box.
193/024	McDonald's Happy Meal Bag Smurfs 40th Ann.	McDonald's Corp.	1/1/98	Blue	Happy Meal Bag	11" x & 7"	Brussels	$3.00-4.00	The bag is blue. It has a 40th anniversary Smurf symbol. The back of the bag has Smurf cartoons. The front says Die Schlumpfe and has 2 Smurfs by the words. Ronald is on top.
193/025	Bubble Gum Fold Out	Dunkin	1/1/98		Fold Out	7 1/2" x 14"	Brussels	$1.50-2.00	The paper unfolds to different village scenes. You put the bubble gum stickers on to make pictures. The front cover has a Smurf on a skateboard.
193/026	1998 Schlumpfwelt Pamphlet	Schleich	1/1/98		Pamphlet	5 3/4" x 4 1/4"	Brussels	$1.50-2.00	The pamphlet shows the new 1998 figures and play sets. It also has colored pictures of various other Smurfs, houses, and Play Sets. 14 pages.
193/027	Smurf Award			Blue	Award	11" x 8 1/2"		$6.75-9.00	The paper is blue. It has a picture of a Smurf holding a feather and script on top. It says "Let it be known that: Has been officially SMURFED!!! For Now and Forever. This Day of . Signed The Smurfs".
193/028	1993 Smurf Pamphlet	Schleich	1/1/93		Pamphlet	5 3/4" x 4 1/4"	Brussels	$1.50-2.00	The pamphlet shows the new 1993 figures and Play Sets. It also has colored pictures of various other Smurfs, houses and Play Sets. The pamphlet unfolds.
193/029	1996 Smurf Pamphlet	Schleich	1/1/96		Pamphlet	5 3/4" x 4 1/4"	Brussels	$1.50-2.00	The pamphlet shows the new 1996 figures and Play Sets. It also has colored pictures of various other Smurfs, houses and Play Sets. The pamphlet has 13 pages.
193/030	1997 Smurf Pamphlet	Schleich	1/1/97		Pamphlet	5 3/4" x 4 1/4"	Brussels	$1.50-2.00	The pamphlet shows the new 1997 figures and Play Sets. It also has colored pictures of various other Smurfs, houses and Play Sets. The pamphlet has 13 pages.
193/031	1994 Smurf Pamphlet	Schleich	1/1/94	Green	Pamphlet	5 3/4" x 4 1/4"	Brussels	$1.50-2.00	The pamphlet shows the new 1994 figures and Play Sets. The cover has 2 Smurfs peeking out of the bushes. It also has colored pictures of various other Smurfs, houses and Play Sets. The pamphlet has 15 pages.
193/032	1995 Smurf Pamphlet	Schleich	1/1/95		Pamphlet		Brussels	$1.50-2.00	The pamphlet shows the new 1995 figures and Play Sets. It also has colored pictures of various other Smurfs, houses and Play Sets. The pamphlet has 15 pages.
193/033	Smurf Collectable Mini Catalog	W. Berrie/Peyo	1/1/82	Orange	Mini Catalog	3" x 5"	U.S.A.	$1.50-2.00	The catalog shows different Smurf figures and has a check spot to check them off if you have the figure. It shows super Smurfs, houses, Play Sets, and plush. The cover has a Smurf carrying a mushroom house to a treasure chest.
193/034	Smurf Collector's Catalogue	Schleich			Mini Catalogue	3 3/4" x 5 1/2"	Belgium	$1.50-2.00	The cover is yellow with a blue border. The top corner has a picture of a Smurf carrying a mushroom house to his treasure chest. The book has checklist. It shows 155 Smurf figures and 39 super Smurfs.
193/035	Catalogo Para Coleccionar Los Pitufos!!	Schleich			Mini Catalog	3 3/4" x 5 1/2"	Spain	$1.50-2.00	The cover is yellow with a blue border. It has a picture of a mushroom house with Smurfette, Papa and 4 Smurfs out front. The book has checklist. It shows 155 Smurf figures and 39 super Smurfs.
193/036	Marionnettes Schtroumpfs La Schtroumpfette	Hachette Jeunesse	1/1/84		Marionnette	4"	Germany	$22.50-30.00	The marionnette is of Smurfette. She is holding flowers in her hands. There are strings attached to her to move her body.
193/037	1994 Schleich Catalog	Schleich	1/1/84		Catalog	11 3/4" x 8 1/2"	Germany	$1.50-2.00	The catalog shows all different figures by Schleich. The catalog has 19 colored pages with Smurf items on them. The cover is red and has Papa Smurf, a girl, and 2 animal figures on it.
193/038	Montgomery Ward Christmas Values 1982	Montgomery Ward	1/1/82		Christmas Catalog	11" x 8"	U.S.A.	$18.00-25.00	The catalog has Papa Smurf sitting in front of a fire reading a book. The Smurfs are sitting on the floor listening. They have Christmas stockings hanging from the mantle. The catalog has 6 pages of Smurf things.
193/039	Smurf Advertising Pamphlet	Schleich	1/1/90		Advertisement Pamphlet	11 3/4' x 8 1/4"	Germany	$1.50-2.00	The pamphlet shows different rubber figures. There is only 4 pages. The cover has a Smurf face and says "Smurf" in 3 different languages. It has 2 animals on the bottom with grandma Smurf and a Smurf carrying a patriot flag.
193/040	Fun Club Newsletter # 1	Smurf Fun Club	1/1/83		Newsletters	11" x 8 1/2"	U.S.A.	$1.50-2.00	The headline on the front page is "Welcome To Our Village". The issue has an article about Peyo inside. There are games and activities in the newsletter.
193/041	McDonald's Hat	Karran Products	1/1/98		McDonald's Hat		Brussels	$2.00-3.00	The hat has Papa, Smurfette and a Smurf holding hands. It has a Smurf leaning on a rake. 3 Smurfs moving a yellow egg- shaped trampoline. There is a Smurf bouncing in the air above the yellow thing. The hat is paper and adjustable.
193/042	Jam caps Advertisement	H-B Prod, Inc.	1/1/95		Advertisement	8 1/4" x 11 3/4"	Germany	$7.50-10.00	The front shows number 81 through 100 of the De Smurfen jam caps in color. The back shows all different characters that you can buy in the jam caps.
193/043	Smurf Magic Berries Cereal Box	Post	1/1/88	Blue	Cereal Box	11" x 8"	U.S.A.	$18.00-25.00	The front of the box has Papa Smurf standing in front of a bowl of cereal. Papa is holding 3 berries in his hand. The box is blue. The back has a checker board with cardboard Smurf checkers.
193/044	Smurf Fun Club Newsletter #4	Smurf Fun Club	1/1/83		Newsletter	11" x 8 1/2"	U.S.A.	$1.50-2.00	The headline on the front page is" The Secret Of The Snow Smurf". The newsletter has games and activities inside. Inside is a list of Smurf figures.
193/045	Smurf Fun Club Newsletter #5	Smurf Fun Club	11/23/98		Newsletter	11" x 8 1/2"	U.S.A.	$1.50-2.00	The headline on the front page is "Smurf Sweetheart Of The Year." The newsletter has games and activities inside. Inside is pictures of new Smurfette merchandise.
193/046	Smurf's Fun Club Newsletter #6	Smurf Fun Club	1/1/83		Newsletter	11" x 8 1/2"	U.S.A.	$1.50-2.00	The headline on the front page is "The Smurf's Littlest Helper". The newsletter has games and activities inside. Inside it shows new Baby Smurf merchandise.
193/047	Smurf Fun Club Newsletter #7	Smurf Fun Club	1/1/83		Newsletter	11" x 8 1/2"	U.S.A.	$1.50-2.00	The headline on the front page is "Lazy's Nap time Nightmare." The newsletter has games and activities inside. The inside has an updated list of super Smurf figures. It also shows some Smurf merchandise.
193/048	Smurf Fun Club Newsletter #8	Smurf Fun Club	1/1/83		Newsletter	11" x 8 1/2"	U.S.A.	$1.50-2.00	The headline on the front page is "Create-A-Smurfs" or "Magnifique!" The newsletter has games and activities inside. The inside has the winners of a Smurf drawing contest.
193/049	Smurf Fun Club Newsletter #9	Smurf Fun Club	1/1/83		Newsletter	11" x 8 1/2"	U.S.A.	$1.50-2.00	The story is about the Smurfic Olympic Games. The newsletter has games and activities inside.
193/050	Smurf Fun Club Newsletter #10	Smurf Fun Club	1/1/83		Newsletter	11" x 8 1/2"	U.S.A.	$1.50-2.00	The story on the front page is about the Smurfs going to school. The newsletter has games and activities inside.
193/051	Jubilee Figure Package	Schleich/ Peyo			Figure Package	7 1/4" x 5 1/2"	Germany	$1.50-2.00	The package is from a jubilee Smurf figure. It advertises the Smurfs from 1965-1985. There are 2 stickers in the package. The package shows all 20 figures.
193/052	Smurf McDonald's Package From UK	McDonald's Corp.	1/1/98	White	Figure Package		London	$1.00-2.00	The package is white. It is from the Smurf Happy meal figures from McDonald's. The package shows 10 Smurf figures. It has the Smurf with the happy meal box on it.
193/053	Blue Smurf McDonald's Package	McDonald's Corp.	1/1/96		Figure Package	5" x 4 1/4"	Hong Kong	$1.00-2.00	The package is clear with a blue piece of Paper inside. The paper shows 2 series of 5 Smurfs for McDonald's. The 25th Smurf figure is shown.
193/054	Smurf Changing Picture	Applause/Peyo			Changing Picture	3" x 4"		$2.00-3.00	The picture is of 5 Smurfs and Smurfette outside. There are 2 mushroom houses behind. The picture is clear when in the frame. When pulled out it is colored.
193/055	Smurf Stamp	Philatelia	1/1/84		Postage Stamp	1" x 1 1/2"	Belgium	$7.50-10.00	The stamp is for 8 cents. The stamp has a Smurf carrying a mail sack. He is blowing a yellow horn.
193/056	6 Transfers	Peyo			Transfers		West Germany	$3.50-5.00	The transfers can go on plastic, wood or ceramic. There is a Smurf covering his mouth. A Smurf carrying a cake. A Smurf wearing red pants and he has his tongue hanging out. A Smurf holding a spear. A small Smurf jumping. A small Smurf with his hands hanging at his side.
193/057	Baby Smurf Birth Announcement	W. Berrie/Peyo	9/17/83	Light Blue	Paper	11 1/4" x 8 1/4"	U.S.A	$7.50-10.00	The page has a picture of a stork carrying Baby Smurf. The certificate has his height, parents, date of birth, sex, and place of Smurf.
193/058	#8 Newsletter From I.M.P.S.	I.M.P.S.	1/1/98		Newsletter	16" x 12"	Brussels	$2.00-3.00	Smurf newsletter #8. Headlines read 8 Celebrating 40 Years Of Smurfiness.
193/059	Smurf's With Trampoline McDonald's Flag	Peyo/I.M.P.S.	1/1/90		Flag	6" x 7 1/4"	Brussels	$1.00-2.00	The one side of the flag is green with 3 Smurfs holding a trampoline and another Smurf bouncing. The other side of the flag is red and yellow with the "M" symbol.

Set 15: Party Supplies

015/001	X-mas Table cover	Unique	1/1/82	Red	Paper Table Cloth	54" x 88"	U.S.A.	$3.50-5.00	The table cloth has a red border, with Papa and another Smurf decorating an X-mas tree.
015/002	X-mas Hot Or Cold Cups	Unique	1/1/82	Red	Paper cups	7 FL. OZ.	U.S.A.	$3.00-4.00	The cups have a red border, with Papa and another Smurf decorating an X-mas tree. (2 packages of 8 each).
015/003	X-mas Plates	Unique	1/1/82	Red	Plastic Coated Plate	7" Diam.	U.S.A.	$3.50-5.00	The plates have a red border, with Papa and another Smurf decorating an X-mas tree. (2 packages of 8 each).
015/004	X-mas Luncheon Napkins	Unique	1/1/82	Red	Napkins	13" x 13"	U.S.A.	$3.50-5.00	The napkins have a red border, with Papa and another Smurf decorating an X-mas tree. (1 package of 16).
015/005	X-mas Beverage Napkins	Unique	1/1/82	Red	Paper Napkins	10" x 10"	U.S.A.	$3.00-4.00	The napkins have a red border, with Papa and another Smurf decorating an X-mas tree. (1 package of 16).
015/006	White & Yellow Happy Smurfday Table cover	Unique	1/1/82	Yellow/White	Paper Table Cloth	54" x 88"	U.S.A.	$3.50-5.00	The table cover is white with a yellow border. A Smurf is carrying a birthday cake with a mushroom behind him.
015/007	White & Yellow Happy Smurfday Luncheon Napkins	Unique	1/1/82	Yellow/White	Paper Napkins	13 1/2" x 13 1/2"	U.S.A.	$3.50-5.00	The napkins are white with a yellow border. A Smurf is carrying a birthday cake with a mushroom behind him. (1 package of 16)
015/008	White & Yellow Happy Smurfday Paper Plates	Unique	1/1/82	Yellow/White	Plastic Coated Plate	7" Diam.	U.S.A.	$3.50-5.00	The plates are white with a yellow border. A Smurf is carrying a birthday cake with a mushroom behind him. (1 package of 8)
015/011	White & Yellow Happy Smurfday Cups	Unique	1/1/82	Yellow/White	Paper cups	7 FL. OZ.	U.S.A.	$3.50-5.00	Yellow cups with a Smurf carrying a cake. (1 package of 8)
015/012	Blue & Green Happy Smurfday Table Cover	Unique	1/1/83	Blue/Green	Paper Table Cloth	54" x 88"	U.S.A.	$5.00-7.00	Papa, Smurfette and 2 other Smurfs standing around a large birthday cake.
015/013	Blue & Green Happy Smurfday Beverage Napkins	Unique	1/1/82	Blue/Green	Paper Napkins	10" x 10"	U.S.A.	$3.00-4.00	Papa, Smurfette and 2 other Smurfs standing around a large birthday cake. (1 package of 16)
015/014	Bendee Flex Straws	Deka Inc.	1/1/82	White	Plastic Straws	7 1/2" Long	U.S.A.	$1.50-2.00	40 straws, in a clear package with a Smurf blowing a horn on the front.
015/015	Smurf Place Card Candy Cups	Unique	1/1/82	Red/Yellow	Candy Cups		U.S.A.	$3.50-5.00	The candy cups are red with a Smurf holding a yellow place card above the cup. (6 per pack).
015/016	Smurfette Place Card Candy Cups	Unique	1/1/82	Yellow/purple	Candy Cups		U.S.A.	$3.50-5.00	Smurfette is holding purple place cards above yellow cups.
015/017	You're Invited To A Smurfday Party (Clown Smurf)	Unique	1/1/82	Yellow/Green	Invitations	5 1/2" x 3 1/4"	Hong Kong	$3.50-5.00	The front of the invitations have a clown in a red outfit, dancing on the grass. The back has time, date, and place. (8 Per Pack)
015/018	You're Invited To A Smurfday Party (Mailman Smurf)	Unique	1/1/82	Yellow/Green	Invitations	5 1/2" x 3 1/4"	Hong Kong	$3.50-5.00	The front of the invitations have a mailman blowing a horn. The back of the invitations have a place for time, date, and place. (8 Per Pack)
015/019	I'm Having A Party (Smurfette)	Unique	1/1/82	Purple	Smurfette Invitation	5 1/2" x 3 1/4"	U.S.A.	$3.50-5.00	The front of the invitations have Smurfette talking on a telephone. The back has a place for time, date, and place. (8 Per Pack).
015/020	Smurf Blowouts	Unique	1/1/82		Blowouts		Taiwan	$4.50-6.00	(6 Per Pack)
015/021	Smurfette Beverage Napkins	Unique	1/1/82	Purple	Napkins	10" x 10"	U.S.A.	$3.00-4.00	The front of the napkins have Smurfette holding a flower. The napkins are purple and white.
015/022	Happy Smurfday Banner	Unique	1/1/82	White	Party Banner	64" x 9"	U.S.A.	$5.00-7.00	The banner is white with yellow and red lettering, and it has various Smurfs all over.
015/023	Happy Smurfday Hats	Unique	1/1/82	Yellow	Party Hats		Taiwan	$1.50-2.00	The hats are yellow.
015/024	Smurfette Birthday Balloons	Unique	1/1/82	Pink	Balloons	15"	U.S.A.	$2.00-3.00	(6 Big 15" Balloons Per Pack) Multi colors.
015/025	Smurf Birthday Balloons	Unique	1/1/82	Multi	Balloons	15"	U.S.A.	$2.00-3.00	Yellow package with different colored balloons inside. (6 Big 15" Balloons Per Pack)
015/026	Smurf Balloons	National Latex Co.	1/1/82	Multi	Balloons		U.S.A.	$2.00-3.00	The package is red with a Smurf giving Smurfette balloons. (5 Balloons Per Pack)
015/027	Orange Punch Ball Balloon	National Latex Co.	1/1/82	Orange	Punch Ball Balloon		U.S.A.	$1.50-2.00	The punch ball is orange and on one side a Smurf is blowing a horn, and the other side the Smurf is playing the drums.
015/028	Pink Punch Ball Balloon	National Latex Co.	1/1/02	Pink	Punch Ball Balloon		U.S.A.	$1.50-2.00	The punch ball is pink and on one side a Smurf is blowing a horn, and the other side the Smurf is playing the drums.
015/029	58 Piece Party Set	Randim Marketing	1/1/88		Party Set For 6		China	$15.00-20.00	The box has 6 hats, 6 blowouts (with Smurf face on), 6 loot bags, 10 balloons, 12 candles, 12 candle holders, 6 place mats. The picture on the hats, loot bags and place mats is of Smurfette sitting on a mushroom with Papa and a Smurf on each side of her.
015/030	Happy Smurfday Smurfette Table Center Piece	Unique	1/1/82		Paper Table Piece	13 1/2" High	Taiwan	$6.00-8.00	Smurfette is kneeling on a purple mushroom that opens up like an accordion. Smurfette is holding a yellow and purple happy Smurfday sign.
015/031	Smurf Party Favors Shakin' Smurf	Unique	1/1/82		Party Favors	5 In Pack	U.S.A.	$3.00-4.00	There are 5 red suction cups. There are little plastic squares with pictures on the front. The pictures are attached to the suction cups by a piece of clear plastic. You put these on a window and they shake. There are 2 pictures of a Grumpy Smurf, 2 shy Smurfs, and one Smurf laughing.
015/032	Happy Smurfday Horns	Unique	1/1/82		Horns	6" Long	Taiwan	$3.50-5.00	The horns are yellow. They say Happy Smurfday on the bottom in red letters. There are pictures of a Smurf walking, blowing a white horn on the horns.
015/033	Smurf Loot Bag (French)	Party time/Peyo/ Applause	1/1/88	White	Prize Bag	10 1/2"L X 7 3/4"W	China	$3.00-4.00	The loot bag is white and has a picture of Papa holding a present, Smurfette standing on a mushroom holding a teddy bear and pink lunch boy, and a Smurf standing, holding a gray thermos.
015/034	Smurfette Paper Plates	Unique	1/1/82	Purple	Paper Plates	9 In. Diam.	U.S.A.	$4.50-6.00	The paper plates are purple in the center and have Smurfette standing, holding a daisy. The border is purple with little daisies going around the plate.
015/035	Smurf Pill Puzzles	Unique	1/1/82		Party Favors	2 1/2" Diam.	Hong Kong	$4.50-6.00	There are 6 round plastic puzzle games. You try to get the balls into carved out holes. There are different pictures on each puzzle. The pictures are of a Smurf with a flower in his hat and he's jumping. Papa sweating. Superman Smurf. A Smurf sitting, looking dazzled. A Smurf jumping and laughing. A Smurf fishing.
015/036	Smurfette Party Pack	Unique	1/1/82	Purple/Pink	Paper Items		U.S.A.	$11.00-15.00	Includes: 8 plates, 8 cups, 8 napkins and & 1 large table cover. The napkins, plates and table cover are purple and have a picture of Smurfette standing in the middle with one hand on her hair. The cups are pink. The cups have a picture of Smurfette standing in the middle with her hand on her hair.
015/037	Smurfette Blowouts.	Unique	1/1/82	Purple	Blowouts		U.S.A.	$4.50-6.00	The blowouts are yellow and silver. The cardboard picture attached has Smurfette in a cheerleader outfit. The background is purple. There are 6 blowouts in the package.
015/038	Smurfette Plates.	Unique	1/1/82	Purple	Paper Plates	7" Diam.	U.S.A.	$3.50-5.00	The paper plates are purple in the center and have Smurfette standing, holding a daisy. The border is purple with little daisies going around the plate.
015/039	Smurfette Luncheon Napkins.	Unique	1/1/82	Purple	Napkins	13 1/2" X 13 1/2"	U.S.A.	$3.50-5.00	The front of the napkins have Smurfette holding a flower. The napkins are purple and white.
015/040	Smurfette Table cover.	Unique	1/1/82	Purple	Paper Tablecloth	54" x 88"	U.S.A.	$5.00-7.00	The table cover has Smurfette standing with one hand on her hair. There is a mushroom house behind her. The border is little daisies. The table cover is purple.
015/041	Smurfette Party Hats.	Unique	1/1/82	Purple	Party Hats		U.S.A.	$4.50-6.00	The hats look like crowns. The front has Smurfette wearing a long purple gown with white trim. Smurfette is wearing a white hat with a yellow crown. The crown has little daisies all over.
015/042	Happy Smurfday Smurf Centerpiece.	Unique	1/1/82		Paper Centerpiece	13 1/2"H x 8"W	Taiwan	$6.00-8.00	A Smurf is sitting on a yellow mushroom. The Smurf is holding a sign that says Happy Smurfday.
015/043	Smurf Loot Bag.	Unique	1/1/82	Yellow	Loot Bag		U.S.A.	$2.00-3.00	The Smurf loot bag is yellow. It has a Smurf carrying a white box with red ribbon. The bag says "Smurf Loot Bag" in red letters.
015/044	Smurf & Smurfette Mylar Balloon.	Unique	1/1/83	Pink	Balloon		U.S.A.	$5.00-7.00	The balloon is pink. A Smurf is leaning over kissing Smurfette's hand. There is a mushroom house behind the Smurf.
015/045	Smurf Piñata				Piñata	17" High	Mexico	$3.50-5.00	The Smurf is in sitting position. He has a blue body, white hat, and white pants. There is a wire for hanging.
015/046	Smurf Way Out Headset (Red)	R.O.C.		Red	Plastic Headband		Taiwan	$1.50-2.00	The headband is red. It has antennas with red stars on the end. In the center of the stars are pictures of a Smurf running, carrying a torch. The Smurf has on a red outfit.
015/047	Smurf Way Out Headset (Green)	R.O.C.		Green	Plastic Headband		Taiwan	$1.50-2.00	The headband is green. It has antennas with green stars on the end. In the center of the stars are pictures of a Smurf running, carrying a torch. The Smurf has on a red outfit.
015/048	Smurf Way Out Headset (Blue)	R.O.C.		Blue	Plastic Headband		Taiwan	$1.50-2.00	The headband is blue. It has antennas with blue stars on the end. In the center of the stars are pictures of a Smurf running, carrying a torch. The Smurf has on a red outfit.
015/049	Smurf Way Out Headset (Green)	R.O.C.		Green	Plastic Headband		Taiwan	$1.50-2.00	The headband is green. It has antennas with green circles on the end. In the center of the circles are pictures of a Smurf running, carrying a torch. The Smurf has on a red outfit.
015/050	Smurf Way Out Headset (Purple)	R.O.C.		Purple	Plastic Headband		Taiwan	$1.50-2.00	The headband is purple. It has antennas with purple circles on the end. In the center of the circles are pictures of a Smurf running, carrying a torch. The Smurf has on a red outfit.
015/051	Smurf Way Out Headset (Yellow)	R.O.C.		Yellow	Plastic Headband		Taiwan	$1.50-2.00	The headband is yellow. It has antennas with yellow stars on the end. In the center of the stars are pictures of a Smurf running, carrying a torch. The Smurf has on a red outfit.
015/052	Smurf Way Out Headset (Pink)	R.O.C.		Pink	Plastic Headband		Taiwan	$1.50-2.00	The headband is pink. It has antennas with pink stars on the end. In the center of the stars are pictures of a Smurf running, carrying a torch. The Smurf has on a red outfit.
015/053	Smurf Way Out Headset (Purple)	R.O.C.		Purple	Plastic Headband		Taiwan	$1.50-2.00	The headband is purple. It has antennas with purple stars on the end. In the center of the stars are pictures of a Smurf running, carrying a torch. The Smurf has on a red outfit.
015/054	Blue And Green Happy Smurfday Cups.	Unique	1/1/85		Paper Cups	7 Fl. Oz	U.S.A.	$3.50-5.00	Papa, Smurfette and 2 other Smurfs standing around a large birthday cake. The picture is on 2 sides of the cups. 1 package of 8 cups.
015/055	Party Game Punch Board.	Unique	1/1/82		Punch Board		U.S.A.	$2.00-3.00	The card has punch holes. You punch out the cardboard in a number and do what it says. Each number has a different Smurf picture on it. The board is yellow with a blue border.
015/056	Smurf Nut Cups.	Fotorama	1/1/83		Nut Cups	3" x 3"	Mexico	$3.50-5.00	The nut cups are shaped like mushroom houses. There are 8 nut cups with various pictures of the Smurfs on them. There are 7 different designs.
015/057	I Puffi Penne Magiche	Unique	1/1/82		Magic Box	6 1/4" x 8 1/4"	Brussels	$15.00-20.00	The box is blue with a red handle. It has a Smurf jumping on the front and back. Inside is packaged cards.
015/058	McDonald's Balloon	McDonald's	1/1/98	Blue	Balloon			$1.00-2.00	The balloon is blue. One side has the McDonald's symbol. The other side has a Smurf outlined in black. The Smurf is carrying a gift.
015/059	3 Smurfs Holding Hands Party Invitations	Party Time Products	1/1/82		Party Invitations	4" x 5"	Canada	$4.50-6.00	The invitations have 3 Smurfs walking, holding hands. The end Smurf is whistling. The other 2 Smurfs are smiling. Set of 6.
015/060	A Schtroumpfer La Fete! Invitation				Paper Items	5 1/2" x 3 1/4"	France	$5.00-7.00	The cover of the invitation has a Smurf party going on. They have a merry- go- round. There are Smurfs playing instruments. Smurfette and a Smurf eating a cake.

Set 272: Patches- Miscellaneous

272/001	Smurfette				Patch	5 1/2"		$2.00-3.00	The patch is Smurfette. She is standing, posing.
272/002	Smurf Face				Patch	3" x 3"		$1.50-2.00	The patch is of a Smurf face. It is embroidered. He has a bright blue face and a dark blue nose.

Set 142: Patches- Oval

142/001	Papa With Feather Pen			Pink	Cloth Patch	2"L X 1 1/2"W		$1.00-2.00	The patch is oval- shaped. The background is pink and it has a red border. Papa is standing in the middle of the patch, with his hand by his mouth.

Papa is holding a white feather pen with his right hand. The back of the patch is white.

142/002	Smurfette With A Daisy			Pink	Cloth Patch	2"L X 1 1/2"W		$1.00-2.00	The patch is oval- shaped. The background is pink and it has a blue border. In the middle of the patch Smurfette is standing, wearing a white dress. Smurfette is holding an orange daisy and is pulling the petals off. The back of the patch is white.
142/003	Smurf Sitting On Books			Pink	Cloth Patch	2"L X 1 1/2"W		$1.00-2.00	The patch is oval- shaped. The background is pink and it has a red border. In the middle of the patch a Smurf is sitting on three books. The books are green, pink and red. The back of the patch is white.
142/004	Smurf Drawing A Picture			Pink	Cloth Patch	2"L X 1 1/2"W		$1.00-2.00	The patch is oval- shaped. The background is pink and it has a blue border. In the middle of the patch a Smurf is laying on his stomach drawing a picture of Smurfette. The back of the patch is white.
142/005	Smurf With A Pad Of Paper And A Pencil			Pink	Cloth Patch	2"L X 1 1/2"W		$1.00-2.00	The patch is oval- shaped. The background is pink and it has a red border. In the middle of the patch a Smurf is standing, holding a red notepad and a green pencil. The back of the patch is white.
142/006	Smurf Carrying A Books On A Stick			Pink	Cloth Patch	2"L X 1 1/2"W		$1.00-2.00	The patch is oval- shaped. The background is pink and it has a blue border. In the middle of the patch a Smurf is carrying a stick with 2 books tied on it. The Smurf is whistling. The books are red and 1 is green. The back of the patch is white.
142/007	Brainy			Pink	Cloth Patch	2"L X 1 1/2"W		$1.00-2.00	The patch is oval- shaped. The background is pink and it has a red border. In the middle of the patch a Smurf wearing black glasses has a finger in the air. The Smurf looks like he is giving a speech. The back of the patch is white.
142/008	Smurfette Wearing A Swimsuit			Pink	Cloth Patch	2 1/2"L X 2"W		$1.00-2.00	The patch is oval- shaped. The background is pink and it has a blue border. In the middle of the patch Smurfette is sitting on a white blanket with red dots. Smurfette is wearing a red swimsuit. The back of the patch is white.
142/009	Baseball Smurf			White	Cloth Patch	2"L X 1 1/2"W		$1.00-2.00	The patch is oval- shaped. The background is white with red dots and it has a black border. In the middle of the patch a Smurf is walking, carrying a yellow bat with a brown glove hanging on the end. The Smurf is wearing red shorts and a white shirt with a brown "B" on the front. The back of the patch is white.

Set 143: Patches- Square

143/001	Smurf Laughing			Pink	Cloth Patch	2 1/2" X 2 1/2"		$1.00-2.00	The patch is square- shaped. The background is pink and it has a blue border. In the middle of the patch a Smurf is standing with his hand over his mouth laughing. The back of the patch is white.
143/002	A Smurf Looking Lost			Pink	Cloth Patch	2 1/2" X 2 1/2"		$1.00-2.00	The patch is square- shaped. The background is pink and it has a blue border. In the middle of the patch a Smurf is standing with his hands at his side. He has a puzzled look on his face. The back of the patch is white.
143/003	Smurf Sticking Out His Tongue			Pink	Cloth Patch	2 1/2" X 2 1/2"		$1.00-2.00	The patch is square- shaped. The background is pink and it has a blue border. In the middle of the patch a Smurf is standing with his fingers in his ears and he is sticking out his tongue. The back of the patch is white.
143/004	Smurf Carrying A Bunch Of Books			Pink	Cloth Patch	2 1/2" X 2 1/2"		$1.00-2.00	The patch is square- shaped. The background is pink and it has a blue border. In the middle of the patch a Smurf is carrying a bunch of books and pencils. The Smurf is looking at a clock on the wall. The Smurf has a worried look on his face. The back of the patch is white.
143/005	Mushroom House			Pink	Cloth Patch	3" X 3"		$1.00-2.00	The patch is square- shaped. The background is pink and it has a blue border. In the middle of the patch is an orange and dark pink mushroom house. The back of the patch is white.
143/006	Smurf Sitting On A Saddle			Pink	Cloth Patch	2 1/2" X 2 1/2"		$1.00-2.00	The patch is square- shaped. The background is pink and it has a blue border. In the middle of the patch is a green fence. There is a red and white blanket over the fence with a brown saddle over the blanket. A Smurf wearing a red bandanna around his neck and a white cowboy hat is sitting in the saddle, twirling a yellow rope. The back of the patch is white.
143/007	Cowboy Smurf			Pink	Cloth Patch	2 1/2" X 2 1/2"		$1.00-2.00	The patch is square- shaped. The background is pink and it has a blue border. In the middle of the patch a Smurf is standing, wearing a brown vest, a red bandanna around his neck , brown chaps and a white cowboy hat. He is standing in dirt and there is a yellow cactus behind him. The back of the patch is white.
143/008	Cowboy Smurf And Cowgirl Smurfette			Pink	Cloth Patch	3 1/4"W X 4 1/4"L		$1.00-2.00	The patch is square- shaped. The background is pink and it has a blue border. In the middle of the patch a Smurf is standing, wearing a brown vest, a red bandanna around his neck , brown chaps and a white cowboy hat. He is standing in dirt and there is a green cactus behind him. Smurfette is wearing a brown cowgirl dress with a red and white checkered vest, white cowgirl hat, and white boots. The two Smurf's are facing each other. The back of the patch is white.

Set 218: Patterns- Cloth Pillow Patterns Unfinished

218/001	Brainy With Book	Peyo	1/1/82		Pillow Pattern		U.S.A.	$4.50-6.00	The pattern is of Brainy Smurf walking, carrying a book. Brainy has a red shirt, white pants, black glasses and a white hat on. The book is orange and says Smurf Stories. The background is yellow.
218/002	Mushroom House	Tigress Pride	1/1/82		Pillow Pattern		U.S.A.	$4.50-6.00	The mushroom house is orange with a red, purple and white roof. Smurfette is in a window balcony on one side and a Smurf is in the balcony on the other side.
218/003	Papa Standing On A Mushroom	Tigress Pride	1/1/82		Pillow Pattern		U.S.A.	$4.50-6.00	Papa is standing on a mushroom. Papa has both hands straight out in front of him. The mushroom is orange, yellow and purple.
218/004	Papa Sailor	Peyo	1/1/82		Pillow Pattern		U.S.A.	$4.50-6.00	Papa is standing, looking through an orange telescope. Papa is wearing a white sailor hat, white sailor jacket and red pants.
218/005	Smurf Holding A Daisy	Peyo			Pillow Pattern	2 1/2 foot long		$4.50-6.00	The pattern is cut, it just needs stuffing. The Smurf is standing, holding a white daisy. It looks like he's pulling the petals off the daisy.
218/006	Smurfette	Peyo			Pillow Pattern	2 foot Long		$4.50-6.00	Smurfette is standing, posing. The pattern is cut, it just needs stuffing.
218/007	Smurf In Car				Pillow Pattern	2 Foot Long		$4.50-6.00	The pattern is cut, it just needs stuffing. A Smurf is driving a red car. The car has blue tires.

Set 146: Pen Set- Figures

146/001	Baseball Smurf	W. Berrie	1/1/82		Pen Set	6 1/2" High	Hong Kong	$11.00-15.00	The base is brown rubber with green on it to look like grass. There is a Smurf standing on the grass. The Smurf is wearing a red and white baseball uniform and holding a beige bat. There is a hole in the base to hold a pen that is shaped like a baseball bat. The pen is beige.
146/002	Cowboy Smurf	W. Berrie	1/1/82		Pen Set	6 1/2" High	Hong Kong	$11.00-15.00	The base is brown rubber with green on it to look like grass. There is a Smurf standing on the grass. The Smurf is wearing a brown cowboy outfit. There is a hole in the base to hold a pen. The end of the pen has a horse face on it.
146/003	Cupid	W. Berrie	1/1/82		Pen Set	6 1/2" High	Hong Kong	$11.00-15.00	The base is brown rubber with green on it to look like grass. There is a Smurf standing on the grass. The Smurf is dressed like cupid. He has white wings and a cloth over his front area. The Smurf is holding a red heart with an orange arrow through it. In his other hand he is holding a tan bow. There is a hole in the base to hold a pen. The end of the pen has a red heart that says "I Smurf You" in white letters.
146/004	Ballerina	W. Berrie	1/1/82		Pen Set	6 1/2" High	Hong Kong	$11.00-15.00	The base is brown rubber with green on it to look like grass. Smurfette is standing on the grass in a ballerina pose. Smurfette is wearing a white dress and white ballerina slippers. There is a hole in the base for a pen. The pen is green with a red flower on top.

Set 99: Pencil Huggers

099/001	Snappy With A Brown Guitar	Applause		Yellow	Rubber Pencil Hugger	7" Long	Taiwan	$3.50-5.00	The pencil is yellow and has a rubber PVC on the end. The PVC is Snappy wearing a yellow shirt and holding a brown guitar. He wraps around the top of the pencil.
099/002	Sassette With A Flower	Applause		Yellow	Rubber Pencil Hugger	7" Long	Taiwan	$3.50-5.00	The pencil is yellow and has a rubber PVC on the end. The PVC is Sassette holding a light pink daisy. Sassette has orange hair and is wearing dark pink bibs.
099/003	Nat	Applause		Yellow	Rubber Pencil Hugger	7" Long	Taiwan	$3.50-5.00	The pencil is yellow and has a rubber PVC on the end. The PVC is Nat. Nat is wearing a yellow straw hat, brown pants and a brown sash.
099/004	Slouchy	Applause		Yellow	Rubber Pencil Hugger	7" Long	Taiwan	$3.50-5.00	The pencil is yellow and has a rubber PVC on the end. The PVC is Slouchy. He is wearing a red shirt and white pants. His hat is droopy. Slouchy is looking up in the air.

Set 266: Pencil Sharpeners- PVC Figures On Tree Stump

266/001	Gift	Schleich	1/1/82		Pencil Sharpener	3" High		$11.00-15.00	The Smurf is a rubber figure. He is standing, holding a yellow present with red ribbon. In his other hand he is holding red and yellow daisies. The Smurf has a light blue body color. The Smurf is standing on a brown tree stump. The tree stump has a pencil sharpener in it.
266/002	Graduate With Briefcase	Schleich	1/1/82		Pencil Sharpener	3" High	West Germany	$11.00-15.00	The Smurf figure is standing on a brown tree stump. The tree stump has a pencil sharpener in it. He is wearing a glossy maroon graduation cap with a yellow tassel and a glossy maroon gown. He is holding a white rolled diploma in his left hand and a brown book bag in his right hand.

Set 234: Pencil Tops

234/001	Laughing Smurf Face			Blue	Rubber Pencil Top	1"		$2.00-3.00	The Smurf has his mouth open and he is laughing. The face goes on top of a pencil. The face is completely blue.
234/002	Qauck Smurf			Blue	Rubber Pencil Top	1"		$2.00-3.00	The Smurf is wearing a hat with a rim, big glasses and a bow tie. The face goes on top of a pencil. The face is completely blue.
234/003	Jungle Smurf			Blue	Rubber Pencil Top	1"		$2.00-3.00	The Smurf has long hair. His hair is different lengths. The Smurf doesn't have a hat on. The face goes on top of a pencil. The face is completely blue.
234/004	Cook Smurf			Blue	Rubber Pencil Top	1"		$2.00-3.00	The Smurf is wearing a chefs hat. He is licking his lips. The face goes on top of a pencil. The face is completely blue.
234/005	Baby Smurf With Butterfly				Rubber Pencil Top	2 1/2"		$3.50 5.00	The pencil top is a Baby Smurf rubber figure. Baby is sitting, holding a yellow butterfly. Baby is wearing a white sleeper.

234/006	Papa Smurf Holding A Finger In The Air				Rubber Pencil Top	2 1/2"		$1.50-2.00	Papa is standing, holding a finger in the air. He is made of a hard rubber. The bottom has a hole to put Papa on top of a pencil

Set 95: Pens & Pencils

095/001	Light Writer Pen	Larami	1/1/82		Lighted Pen	7" Long	Hong Kong	$11.00-15.00	The end of the pen has a Smurf on a brown stick. The Smurf is wrapped around the top. He is wearing a red shirt, white shorts and red an orange socks. The bottom of the pen is clear and it lights up. The whole pen is plastic.
095/002	Want To Smurf Around?	W. Berrie	1/1/83	Pink/Purple	Smurfette Pencils	7 1/2" Long		$1.00-2.00	The pencil is pink with purple hearts all over it. The pencil says "Want To Smurf Around?" in purple letters. There is a purple outline of Smurfette standing in a flirting pose with her hands at her side, in front of the saying. At the end it has the word Smurfette outlined in purple puffy letters.
095/003	Be Happy, Be Smurfy	W. Berrie	1/1/83	Pink/Purple	Smurfette Pencils	7 1/2" Long		$1.00-2.00	The pencil is pink with purple hearts all over it. The pencil says "Be Happy, Be Smurfy" in purple letters. There is a purple outline of Smurfette standing in a ballerina pose in front of the saying. At the end of the saying it has the word Smurfette outlined in purple puffy letters.
095/004	#1 Smurfette Fan	W. Berrie	1/1/83	Pink/Purple	Smurfette Pencils	7 1/2" Long		$1.00-2.00	The pencil is pink with purple hearts all over it. The pencil says "#1 Smurfette Fan" in purple letters. There is a purple outline of Smurfette standing, holding flowers in front of the saying. At the end of the saying it has the word Smurfette outlined in purple puffy letters.
095/005	I Love Smurfette	W. Berrie	1/1/83	Pink/Purple	Smurfette Pencils	7 1/2" Long		$1.00-2.00	The pencil is pink with purple hearts all over it. The pencil says "I Love Smurfette" in purple letters. There is a purple outline of Smurfette standing in a flirting pose with one hand in her hair, in front of the saying. At the end of the saying it has the word Smurfette outlined in purple puffy letters.
095/006	Keep On Smurfin!	W. Berrie	1/1/83	Pink/Purple	Smurfette Pencils	7 1/2" Long		$1.00-2.00	The pencil is pink with purple hearts all over it. The pencil says "Keep On Smurfin!" in purple letters. There is a purple outline of Smurfette on roller-skates, in front of the saying. At the end of the saying it has the word Smurfette outlined in purple puffy letters.
095/007	It's Not Easy Being A Smurfette	W. Berrie	1/1/83	Pink/Purple	Smurfette Pencils	7 1/2" Long		$1.00-2.00	The pencil is pink with purple hearts all over it. The pencil says "It's Not Easy Being A Smurfette" in purple letters. There is a purple outline of Smurfette laying on her stomach drawing a picture, in front of the saying. At the end of the saying it has the word Smurfette outlined in purple puffy letters.
095/008	Smurfette Pencil Holder	W. Berrie	1/1/83	Clear	Pencils	3 3/4"H X 2 3/4" W		$6.75-9.00	The pencil holder is square, made of a clear hard plastic. The front has a pink paper insert. The insert has a picture of a big purple heart. Inside the heart it says Smurfette in white letters and it has a picture of Smurfette on roller-skates.
095/009	Smurfy Rainbow Pencils	W. Berrie	1/1/83	White	Pencils	7" Long	Japan	$1.00-2.00	The pencil is white with rainbow stripes on each end. The lead is four colors in one (red, blue, green and purple). Across the middle of the pencil it says "My Smurfy Rainbow Pencils" in blue letters. The pencil is scented.
095/010	Smurfy Rainbow Pencil Holder	W. Berrie	1/1/83	Clear	Pencils	7 1/2" Long	Japan	$6.75-9.00	The pencil holder is round and made of soft, clear plastic. There is a white sticker on the front that says "Smurfy Rainbow Pencils". There are stripes on the bottom of the sticker in the color of the rainbow.
095/011	Smurfette Purple Pencil	Applause/Peyo		Purple	Pencil	7 1/2" Long	Japan	$1.00-2.00	The pencil is light purple. The pencil has pink hearts, white hearts and white daisies around it. The pencil says "Smurfette" in white letters at the top.
095/012	Smurfette Butterfly Pencil	Applause/Peyo		White	Pencil	7 1/2" Long	Japan	$1.00-2.00	The pencil is white. It has pink and yellow butterflies, red and yellow mushrooms around it. The top has a picture of Smurfette with a red butterfly on her hand. Smurfette has on a pink dress.
095/013	Smurfette Face Pencil	Applause/Peyo		Pink	Pencil	7 1/2" Long	Japan	$1.00-2.00	The pencil is pink. It has Smurfette's face all around it. There are little rainbows, yellow stars and blue stars around the pencil.
095/014	Rainbow Smurfette Pencil	Applause/Peyo		Rainbow	Pencil	7 1/2" Long	Japan	$1.00-2.00	The pencil is rainbow- colored. It has rainbow stripes running down it. Smurfette is on the pencil. She is wearing a pink dress, white roller-skates and headphones. The pictures of Smurfette go around the pencil.
095/015	School girl Pencil	Applause/Peyo		Light Blue	Pencils	7 1/2" Long	Japan	$1.00-2.00	The pencil is light blue. There are little pencils going around the pencil. Smurfette is carrying schoolbooks. The pencil says Smurfette in white all over.
095/016	Purple Heart Smurfette Pencil	Applause/Peyo		Silver	Pencil	7 1/2" Long	Japan	$1.00-2.00	The pencil is silver with purple hearts around it. It has a big purple heart on top with Smurfette's name inside in white. There is a picture of Smurfette in roller-skates and one of Smurfette holding a yellow daisy.
095/017	Pink Heart Smurfette Pencil	Applause/Peyo		Silver	Pencils	7 1/2" Long	Japan	$1.00-2.00	The pencil is silver with pink hearts around it. It has a big pink heart on top with Smurfette's name inside in white. There is a picture of Smurfette in roller-skates and one of Smurfette holding a yellow daisy.
095/018	Smurfs Alive! In Ice Capades Pencil	W. Berrie/Peyo		Blue & White	Promotional Pencil	9 3/4" Long		$9.00-12.00	The pencil is blue and white striped with a blue plastic hook on the end. The pencil has a flag on it. The flag is blue and says "Smurfs Alive! In Ice Capades" The flag has a picture of Papa Smurf coming out of a TV.
095/019	Red Schtroumpfs Marker	Editions	1/1/83	Red	Marker	6" Long	Holland	$1.00-2.00	The marker is thin. It is white and has a Smurf jumping, Smurfette dancing and a Smurf on a mushroom on the outside base. The marker colors red.
095/020	Yellow Schtroumpfs Marker	Editions	1/1/83	Yellow	Marker	6" Long	Holland	$1.00-2.00	The marker is thin. It is white and has a Smurf jumping, Smurfette dancing and a Smurf on a mushroom on the outside base. The marker colors yellow.
095/021	Green Schtroumpfs Marker	Editions	1/1/83	Green	Marker	6" Long	Holland	$1.00-2.00	The marker is thin. It is white and has a Smurf jumping, Smurfette dancing and a Smurf on a mushroom on the outside base. The marker colors green.
095/022	Red Smurf Pen			Red	Pen	5" Long		$1.50-2.00	The pen is red with a white end clip. It has a Smurf holding a flower on the side. The Smurf is holding the flower to his nose.
095/023	Blue And White Schlumpf Fountain Pen				Fountain Pen	5 1/4" Long		$18.00-25.00	The fountain pen is white with a blue cap. The pen has a picture of a Smurf standing, holding a finger to his mouth. This design is duplicated on the other side.
095/024	Fountain Pen With Poet Schtroumpfs	Michel Oks	1/1/83	Clear & Blue	Fountain Pen	5 1/4" Long	Germany	$25.00-35.00	The pen is see through. It has blue trim. The pen has several pictures of a Smurf holding a script and a yellow feather pen. The pen is metal and pointy on the tip. Included are cartridges of ink.

Set 236: Pewter Smurfs

236/001	Gold	Peyo	1/1/80	Silver	Pewter Figure	2" High		$30.00-40.00	The Smurf is standing with his hands out at his side. The Smurf is all silver except his hat. His hat is gold. The Smurf is standing on a flat silver base.
236/002	Drummer	Peyo	1/1/80	Silver	Pewter Figure	2" High		$30.00-40.00	The Smurf is holding drum sticks in front of him. There is a drum sitting between his legs. The Smurf is completely silver- colored.
236/003	Mechanic	Peyo	1/1/80	Silver	Pewter Figure	2" High		$30.00-40.00	The Smurf is holding a wrench in his left hand. He is wearing pants and suspenders. The figure is silver- colored.
236/004	Mirror	Peyo	1/1/80	Silver	Pewter Figure	2" High		$30.00-40.00	The Smurf is holding a mirror in his left hand. The Smurf is standing on a flat silver base. The Smurf is silver- colored.
236/005	Crying	Peyo	1/1/80	Silver	Pewter Figure	2" High		$30.00-40.00	The Smurf is wiping a tear from his eye. He is holding a handkerchief in his left hand. The Smurf is silver- colored. He is standing on a flat silver base.
236/006	Clown	Peyo	1/1/80	Silver	Pewter Figure	2" High		$30.00-40.00	The Smurf is wearing a clown outfit. He has a big bow-tie around his neck. The Smurf has a clown face painted on. He is pointing a finger with his left hand. The Smurf is silver- colored.
236/007	Smurfette	Peyo	1/1/80	Silver	Pewter Figure	2" High		$30.00-40.00	Smurfette is standing with her hand out at her side. Her other hand is on her chest. Smurfette is silver- colored. She is standing on a flat, silver base.
236/008	Singer	Peyo	1/1/80	Silver	Pewter	2" High		$30.00-40.00	The Smurf is standing, holding a music sheet. The music sheet is separate. The Smurf is silver- colored. He is standing on a flat, silver base.
236/009	Gift	Peyo	1/1/80	Silver	Pewter Figure	2" High		$30.00-40.00	The Smurf is standing, holding a present in one hand and flowers in the other hand. The Smurf is silver- colored. He is standing on a flat, silver base.
236/010	Trumpeter	Peyo	1/1/80	Silver	Pewter Figure	2" High		$30.00-40.00	The Smurf is walking, playing a trumpet. He has his eyes closed. The Smurf is silver- colored. He is standing on a flat, silver base.
236/011	Baseball Batter	Peyo	1/1/80	Silver	Pewter Figure	2" High		$30.00-40.00	The Smurf is standing, holding a baseball bat. He is wearing a baseball uniform. The Smurf is silver- colored. He is standing on a flat, silver base.
236/012	Kayak	Peyo	1/1/80	Silver	Pewter Figure	2" High		$30.00-40.00	The Smurf is in a kayak. He is holding a paddle. The figure is only half a Smurf body. The Smurf and the kayak are silver.
236/013	Cyclist	Peyo	1/1/80	Silver	Pewter Figure	2" High		$30.00-40.00	The Smurf is on a bicycle. He is wearing race gear. The Smurf and bike are silver- colored. The bike is on a flat, silver base.
236/014	Sunbather	Peyo	1/1/80	Silver	Pewter	2" High		$30.00-40.00	The figure is completely silver- colored. The Smurf is laying on his back with his hands behind his head. He has his eyes closed.
236/015	Painter	Peyo	1/1/80	Silver	Pewter	2" High		$30.00-40.00	The Smurf figure is silver. He is standing, holding a paintbrush in the air. The paint brush has a yellow tip. The Smurf is holding a palate in the other hand. The palate has red, green, white, blue and yellow paint on it.
236/016	Normal	Peyo	1/1/80		Pewter	2" High		$30.00-40.00	The Smurf is standing with his hands out at his side. The Smurf is all silver. The Smurf is standing on a flat, silver base.
236/017	Baby With Blocks	Peyo	1/1/80		Pewter	2" High			Baby is wearing a sleeper. Baby is sitting, playing with 3 blocks. The Smurf is all silver.
236/018	Smurf Carrying Lantern	Peyo	1/1/80		Pewter	2" High		$30.00-40.00	The Smurf is walking, carrying a lantern. He is completely silver.

Set 32: Pez

032/001	Papa Pez	Pez Candy	1/1/86	Red	Candy Dispenser	4" high	Austria	$7.50-10.00	Papa with red hat, base and feet.
032/002	Blue Smurf Pez	Pez Candy	1/1/86	Blue	Candy Dispenser	4" high	Austria	$7.50-10.00	Blue Smurf with feet and a white hat.
032/003	White Smurf Pez	Pez Candy	1/1/86	White	Candy Dispenser	4" high	Austria	$7.50-10.00	Smurf with white base and feet.
032/004	Yellow Smurfette Pez	Pez Candy	1/1/86	Yellow	Candy Dispenser	4" high		$7.50-10.00	Smurfette with yellow base and feet and a white hat . Painted eyes. Dark yellow hair. Smurfette's eyelashes are above her eyes and are painted black.
032/005	Red Smurf Pez	Pez Candy	1/1/86	Red	Candy Dispenser	4" high	Yugoslavia	$7.50-10.00	Smurf with white hat and a red base and feet.
032/006	White, No Eyes, Smurf Pez	Pez Candy	1/1/86	White	Candy Dispenser	4" high	Yugoslavia	$7.50-10.00	Smurf has no eyes and a white base and feet.
032/007	Smurfette Pez Bonbons	Hergestellt in Osterreich Von	1/1/86	Blue	Pez Bon Bons	4" high	Austria	$7.50-10.00	Smurfette has a white hat, blue base and blue feet. Painted eyes. Dark yellow hair. Smurfette's eyelashes are above her eyes and are painted black.
032/008	Papa Pez Bonbons	Hergestellt in Osterreich Von	1/1/86	Red	Pez Bon Bons	4" high	Austria	$7.50-10.00	Papa with red hat, base and feet.
032/009	Red Smurf Pez Bonbons	Hergestellt in Osterreich Von	1/1/86	Red	Pez Bon Bons	4" high	Austria	$7.50-10.00	Smurf with a white hat, red base and red feet.
032/011	Blue Smurfette Pez	Pez Candy	1/1/86	Blue	Candy Dispenser	4" high	Austria	$7.50-10.00	Smurfette has a white hat, blue base and blue feet. Painted eyes. Dark yellow hair. Smurfette's eyelashes are above her eyes and are painted black.
032/012	Blue Smurfette Pez	Hergestellt in Osterreich Von	1/1/97	Blue	Pez Bon Bons		Austria	$7.50-10.00	Smurfette has a white hat, blue base and blue feet. Painted eyes. Dark yellow hair. Smurfette's eyelashes are above her eyes and are painted black.
032/013	Red Papa Pez	Hergestellt in Osterreich Von	1/1/97	Red	Pez Bon Bons	4" High	Austria	$7.50-10.00	Papa with red hat, base and feet.
032/014	Red Smurf Pez	Hergestellt in	1/1/97	Red	Pez Bon Bons	4" High	Austria	$7.50-10.00	Smurf with white hat and a red base and feet.

032/015	White Smurf Pez	Osterreich Von Hergestellt in Osterreich Von	1/1/97	White	Pez Bon Bons	4" High	Austria	$7.50-10.00	Smurf has a white base and feet.
032/016	Blue Smurfette Pez	Peyo	1/1/87	Blue	Pez	4" High	Austria	$7.50-10.00	Smurfette has a white hat, blue base and blue feet. Smurfette has light yellow hair and is different from the regular Pez. Her eyes are small. Smurfette's eyelashes are not painted and are slanted down. Her lips have a little mouth cut out.
032/017	Chinese Papa Pez	Hergestellt in Osterreich Von	1/1/86	Red	Pez		China	$11.00-15.00	Papa has a red hat, red feet and a red base. Papa has big black eye's with white around them. Papa has a white beard and a blue mouth carved out.
032/018	Smurfette Pez Pink	Peyo	1/1/95	Pink	Pez	4" High	Brussels	$6.00-8.00	Smurfette has a painted on face. Her mouth is open and her tongue is showing. Smurfette has yellow medium length hair. Smurfette has a hot pink stem and feet. The Pez is on a green card.
032/019	Papa Pez	Peyo	1/1/95	Red	Pez	4" High	Brussels	$6.00-8.00	Papa has a painted on face. He has a white beard with a little red mouth in the center, showing. Papa has a light red stem and feet. The Pez is on a green card.
032/020	Smurf Pez Yellow	Peyo	1/1/95	Yellow	Pez	4" High	Brussels	$6.00-8.00	The Smurf has a painted on face. His mouth is open and his tongue is showing. The Smurf has a bright yellow stem and feet. The Pez is on a green card.
032/021	Gargamel	Peyo	1/1/95	Black	Pez	4" High	Brussels	$6.00-8.00	Gargamel has a painted on face. He has black hair and thick black eyebrows. Gargamel is on a black stem and he has black feet. The Pez is on a green card.
032/022	Brainy	Peyo	1/1/95	Orange	Pez	4" high	Brussels	$6.00-8.00	The Smurf has a painted on face. He has on black glasses. The Smurf has a bright orange stem and feet. The Pez is on a green card.

Set 117: Pez- Ball Games

117/001	Black Pez Ball Game	Pez		Black	Pez Ball Game	2 3/4" X 3 1/2"		$4.50-6.00	The game is flat and has a black back with a clear plastic front. There is a picture inside of Smurfette holding a Pez dispenser and Papa standing with his mouth open trying to catch it. There is a pink, blue and yellow Pez sign in back of Papa and Smurfette. There are 4 small holes in the picture and 4 different colored balls to try and get in the holes.
117/002	Blue Pez Ball Game	Pez		Blue	Pez Ball Game	2 3/4" X 3 1/2"		$4.50-6.00	The game is flat and has a blue back with a clear plastic front. There is a picture inside of Smurfette holding a Pez dispenser and Papa standing with his mouth open trying to catch it. There is a pink, blue and yellow Pez sign in back of Papa and Smurfette. There are 4 small holes in the picture and 4 different colored balls to try and get in the holes.
117/003	Lime Green Pez Ball Game	Pez		Lime Green	Pez Ball Game	2 3/4" X 3 1/2"		$4.50-6.00	The game is flat and has a lime green back with a clear plastic front. There is a picture inside of Smurfette holding a Pez dispenser and Papa standing with his mouth open trying to catch it. There is a pink, blue and yellow Pez sign in back of Papa and Smurfette. There are 4 small holes in the picture and 4 different colored balls to try and get in the holes.
117/004	Pink Pez Ball Game	Pez		Pink	Pez Ball Game	2 3/4" X 3 1/2"		$4.50-6.00	The game is flat and has a pink back with a clear plastic front. There is a picture inside of Smurfette holding a Pez dispenser and Papa standing with his mouth open trying to catch it. There is a pink, blue and yellow Pez sign in back of Papa and Smurfette. There are 4 small holes in the picture and 4 different colored balls to try and get in the holes.
117/005	Smurf Juggling Balls	Puppy	1/1/94		Ball Game	3" Diam		$3.50-5.00	The game is round with a yellow plastic border. In the center is a picture of a Smurf juggling balls. The background is red.

Set 66: Phonographs

066/001	A Smurf And Smurfette Dancing On A Phonograph	Vanity Fair	1/1/83	Red	Record Player	12 1/2" x10 1/2"	U.S.A.	$45.00-65.00	The cover has a Smurf and Smurfette holding hands and dancing on a phonograph. The inside of the cover is the same.
066/002	Smurf Band	Vanity Fair	1/1/82	Yellow	Record Player	12 1/2" x10 1/2"	U.S.A.	$45.00-65.00	Smurfette is standing on a mushroom with Papa below and 4 Smurfs are playing instruments. The top of the record player is yellow and the bottom is red.

Set 208: Pillows- Red Satin Heart

208/001	You've Got Me Head-Over-Heels!	W. Berrie/Peyo	1/1/83	Red	Satin Heart Pillow	5" x 5"	Taiwan	$3.50-5.00	The pillow is a red satin in the shape of a heart with a white lace border. On the front Smurfette is wearing a pink leotard doing a flip. On the bottom it says "You've Got Me Head-Over-Heels!" in white letters.
208/002	I'm Stuck On You Valentine!	W. Berrie/Peyo	1/1/83	Red	Satin Pillow	5" x 5"	Taiwan	$3.50-5.00	The pillow is a red satin in the shape of a heart with a white lace border. On the front is a Smurf with gum stuck on his foot. The Smurf is holding his head and laughing. On the bottom it says "I'm Stuck On You Valentine!" in white letters. The heart has a gold cord for hanging.
208/003	Valentine, You're The Smurfiest!	W. Berrie/Peyo	1/1/83	Red	Satin Pillow	5" x 5"	Taiwan	$3.50-5.00	The pillow is a red satin in the shape of a heart with a white lace border. On the front is Smurfette giving a Smurf a kiss. There are pink hearts floating around them. On the bottom it says "Valentine, You're The Smurfiest!" in white letters. The heart has a gold cord for hanging.
208/004	Smile... You've Just Been Smurfed!	W. Berrie/Peyo	1/1/83	Red	Satin Pillow	5" x 5"	Taiwan	$3.50-5.00	The pillow is a red satin in the shape of a heart with a white lace border. On the front is a Smurf carrying a yellow present. The present exploded and hearts are coming out. On the bottom it says "Smile.. You've Just Been Smurfed!" in white letters. There is a gold cord for hanging.

Set 88: Pillows- Shaped

088/001	Papa Standing On A Mushroom	Peyo			Cloth Pillow	18"		$6.00-8.00	Papa Smurf is standing on a mushroom. Papa has his hands out in front of him. The mushroom is purple and orange with a yellow stem.
088/002	Mushroom House	Peyo			Cloth Pillow	13" x 12"		$6.00-8.00	The mushroom house is orange with a red, purple and white roof. Smurfette is in a window balcony on one side and a Smurf is in the balcony on the other side.
088/003	Papa Sailor	Peyo			Cloth Pillow	15 1/2" x 11"		$7.50-10.00	Papa is standing, looking through an orange telescope. Papa is wearing a white sailor hat, white sailor jacket and red pants.
088/004	Papa Walking	Peyo			Cloth Pillow	18" x 12"		$6.00-8.00	Papa is walking. He has his hands out in front of him. He is looking over his shoulder. The background is yellow.
088/005	Smurfette Holding A Flower	Peyo			Cloth Pillow	18" x 8"		$6.00-8.00	Smurfette is standing, holding a pink daisy. Smurfette looks like she is flirting. She has one hand by her hair. Smurfette is wearing a white dress.
088/006	Smurf Holding A Flower	Peyo			Cloth Pillow	17" x 7 1/2"		$6.00-8.00	The Smurf is standing, holding a flower behind his back. The flower is red. The Smurf has a shy grin on his face.
088/007	Cowboy Smurf	Peyo			Cloth Pillow	12"		$4.50-6.00	The pillow is in the shape of a Smurf. The Smurf is wearing a white hat, brown cowboy vest, brown pants, brown boots and a red handkerchief around his neck.
088/008	Smurfette Cowgirl	Peyo			Cloth Pillow	12"		$4.50-6.00	The pillow is shaped like Smurfette. Smurfette is wearing a white hat, red and white cowboy shirt, brown vest dress, and white cowgirl boots.
088/009	Smurf Carrying A Sack Of Presents	Peyo			Cloth Pillow	16"		$6.00-8.00	The pillow is a Smurf carrying a red and white checkered sack. The sack has candy canes on it. There are candy canes, a yellow trumpet and a green ball in the sack. The Smurf has a red pompom on the end of his hat.
088/010	Smurfette Cheerleader	Peyo	1/1/82		Cloth Pillow			$7.50-10.00	Smurfette is wearing a pink cheerleader dress, white hat and white shoes with pink trim. Smurfette is waving red pompoms in the air.
088/011	Square White Smurf Pillow			White	Square Pillow	13" x 13"		$3.50-5.00	The pillow has a regular Smurf with different expressions on his face. Papa is on the pillow pouting. The Smurf is laughing, crying, sleepy, mad etc.

Set 167: Pins

167/001	Smurfette With Sausage	Ter Berke	1/1/92		Hat Pin	2.5 X 2cm	Brussels	$3.50-5.00	Smurfette is standing, holding a brown sausage in both hands. She has a white dress on. The pin is very glossy and smooth. The pin is a promotion for a meat company.
167/002	Smurf Rolling Dough	Ter Berke	1/1/92		Hat Pin	2.5 X 2cm	Brussels	$3.50-5.00	The Smurf is standing, leaning over a big piece of dough and rolling it. The dough is pink. The Smurf has a white apron tied around his waist. The pin is glossy and smooth. The pin is a promotion for a meat company.
167/003	Papa Carrying Bread	Ter Berke	1/1/92		Hat Pin	2.5 X 2cm	Brussels	$3.50-5.00	Papa is standing, holding a loaf of bread. Papa has a red hat, red pants and a white apron around his waist. Papa is holding the bread in the air with one hand. The bread is on a brown bread board. The pin is glossy and smooth. The pin is a promotion for a meat company.
167/004	Smurf Carrying A Big Sausage	Ter Berke	1/1/92		Hat Pin	2.5 X 2cm	Brussels	$3.50-5.00	A Smurf is walking, carrying a big brown sausage on his back. The Smurf has a white apron on. He has a red pencil stuck behind one ear. The pin is glossy and smooth. The pin is a promotion for a meat company.
167/005	Smurf Sharpening A Butcher Knife	Ter Berke	1/1/92		Hat Pin	2.5 X 2cm	Brussels	$3.50-5.00	The Smurf is standing, sharpening a butcher knife with a sharpening rod. The Smurf has an apron on. The pin is smooth and glossy. The pin is a promotion for a meat company.
167/006	Gargamel With Silverware And Bread	Ter Berke	1/1/92		Hat Pin	2.5 X 2cm	Brussels	$3.50-5.00	Gargamel is holding a fork and knife. There is a loaf of bread in front of him. Gargamel is leaning over the bread. The pin is smooth and glossy. The pin is a promotion for a meat company.
167/007	Smurf With A Rainbow				Flat Metal Pin	1" x 3/4"		$2.00-3.00	A Smurf is sitting on a white cloud. There is a red, orange, yellow, and green rainbow behind him. The Smurf has a happy expression on his face. The Smurf is medium blue. The pin is shiny paint.
167/008	Smurf On A Bicycle				Flat Metal Pin	2.5 X 2cm		$2.00-3.00	A Smurf is riding a brown bike with red wheels. The Smurf is looking over his shoulder while riding.
167/009	Smurf Kicking A Soccer ball				Flat Metal Pin	2.5 X 2cm		$2.00-3.00	The Smurf is standing on a green patch of grass kicking a soccer ball. The soccer ball is white and black. The Smurf is wearing a white and red striped shirt and white shorts. The Smurf has on black soccer shoes.
167/010	King Smurf				Flat Metal Pin	2.5 X 2cm		$2.00-3.00	The Smurf is standing, holding a red scepter. He is wearing a yellow robe, yellow crown, white hat and white pants. He looks like he's yelling at something.
167/011	Smurfette On A Surfboard				Flat Metal Pin	2.5 X 2cm		$2.00-3.00	Smurfette is standing on a surfboard. The surfboard is white, blue, yellow and red. Smurfette has a clear bubble helmet on her head. Smurfette is wearing white shoes, a white dress and hat. There are 2 orange stars on the surfboard.

167/012	Papa Pointing	S.E.P.P.	1/1/80	Flat Metal Pin	2.5 X 2cm		$7.50-10.00	Papa is standing, pointing at something. He has a red hat and red pants.
167/013	Roller-skating Smurf (Orange)	W. Berrie	1/1/79	Flat Metal Pin	2.2 X 1.8cm		$2.00-3.00	A Smurf is roller-skating. The roller-skates have orange straps and red wheels.
167/014	Smurf Playing A Trumpet	S.E.P.P.		Flat Metal Pin	2.4 X 2.1cm		$7.50-10.00	The Smurf is walking, playing a yellow trumpet. He is wearing a red shirt, white pants and a white hat.
167/015	Smurf On A Motorcycle			Flat Metal Pin	1" x 1"		$2.00-3.00	The Smurf is riding a brown motorcycle. He is riding up a yellow jump ramp. The Smurf is wearing black goggles around his hat and a tan scarf.
167/016	Travel Smurf			Flat Metal Pin	1"		$2.00-3.00	The Smurf is walking, carrying a black knapsack over his shoulder. The knapsack has a money symbol on the front. The Smurf is wearing a white tank shirt, yellow shorts and white shoes. The Smurf has white legs. The pin has a flat finish.
167/017	Gardener Smurf			Hat Pin	3/4"		$2.00-3.00	The Smurf is standing with his hands out at his sides. He is wearing a green work apron. The Smurf has white legs. The pin has a flat finish.
167/018	Smurfette (Red Dress)			Hat Pin	2.5 X 2cm		$2.00-3.00	Smurfette is standing with her hands in front of her. She is wearing a red dress. Her eyes are totally white.
167/019	Black Smurfette			Hat Pin	2.5 X 2cm		$2.00-3.00	Smurfette is standing with her hands in front of her. She is wearing a red dress. Her eyes are totally white. Her body is black and her hair is orange.
167/020	Peach Smurfette			Hat Pin	2.5 X 2cm		$2.00-3.00	Smurfette is standing with her hands in front of her. She is wearing a red dress. Her eyes are totally white. Her body is peach and her hair is red.
167/021	Roller-skating Smurf (White)		1/1/79	Hat Pin	2.5 X 2		$2.00-3.00	A Smurf is roller-skating. The roller-skates have white straps and red wheels. The pin is smooth and glossy.
167/022	Smurf Sitting Next To A Mushroom House			Flat Metal Pin	2.5 X 2cm		$2.00-3.00	The Smurf is sitting on a patch of green grass. In the back is a yellow and red mushroom house with a brown chimney.
167/023	Smurf Jumping			Hat Pin	2.5 X 2cm		$2.00-3.00	The Smurf is jumping in the air. He has his hands out at his sides. He has a yellow flower in his hat.
167/024	Football Smurf			Hat Pin	1"		$2.00-3.00	The Smurf looks like he's running. He is carrying a purple and red football under one arm. He is wearing an orange shirt with a #1, yellow pants, white shoes and a white hat with a yellow star on it. The Smurfs body is a dark blue. The pin is smooth and glossy.
167/025	Tennis Smurf			Hat Pin	3/4"		$2.00-3.00	The Smurf is carrying a yellow tennis racket. He is holding the tennis racket up in the air. The Smurf is wearing a green shirt and pink shorts.
167/026	Baseball Smurf			Hat Pin	2.5 X 2cm		$2.00-3.00	The Smurf is carrying a black bat over his shoulder. There is a red glove hanging off the end of the bat. The Smurf is wearing a yellow baseball shirt with a red "B" on the front, white pants and a white hat.
167/027	Tennis Smurf			Hat Pin	1"		$2.00-3.00	The Smurf is carrying a yellow tennis racket in front of his chest. He is wearing a white shirt and red shorts. The Smurf has white legs and white feet. The pin has a glossy finish.
167/028	Smurfette			Flat Metal Pin	2.5 X 2cm		$2.00-3.00	Smurfette is standing with her hands in front of her. She is wearing a red dress. Her eyes are white and black.
167/029	Smurfette			Flat Metal Pin	2.7 X 2cm		$2.00-3.00	Smurfette is standing with her hands in front of her. She is wearing a red dress. Her eyes are totally white. The pin is glossy and smooth.
167/030	Smurf With A Daisy	S.E.P.P.		Flat Metal Pin	2.5 X 2cm		$7.50-10.00	The Smurf is standing, holding a daisy behind his back. The daisy is yellow. The Smurf is wearing a red shirt and white pants. He has his head turned to the side with a smile on his face.
167/031	Ram Zodiac Pin	Peyo	1/1/92	Hat Pin	1" x 3/4"		$11.00-15.00	The Smurf is running, holding a battering ram. The Smurf has a ram head and Smurf body.
167/032	Jokey			Flat Plastic Pin	2.5 X 2		$2.00-3.00	The Smurf pin is a flat plastic. The Smurf is carrying a white box with red ribbon in both hands.
167/033	Diecast Jokey			Diecast Metal Pin	2.5 X 2		$3.50-5.00	The Smurf pin is in the shape of a Smurf. It is made of a diecast metal. The Smurf is carrying a yellow box with a red ribbon. The Smurf is holding the box in front of him. The Smurf has a blue body. He is wearing a white hat and pants.
167/034	Diecast Roller-skating Smurf			Diecast Metal Pin	2.5 X 2		$3.50-5.00	The Smurf pin is in the shape of a Smurf. It is made of a diecast metal. The Smurf has yellow roller-skates. He is wearing white pants and a white hat.
167/035	Diecast Smurf Holding Flowers	S.E.P.P.		Diecast Metal Pin	2.5 X 2		$7.50-10.00	The Smurf pin is in the shape of a Smurf. It is made of a diecast metal. The Smurf is holding 4 red daisies. The Smurf has a big smile on his face and his head is turned away from the flowers. The Smurf has a blue body. He is wearing white pants and a white hat.
167/036	Diecast Cowboy			Diecast Metal Pin	2.5 X 2	U.S.A.	$3.50-5.00	The Smurf pin is in the shape of a Smurf. It is made of a diecast metal. The Smurf is wearing brown chaps, white pants, white shoes, brown vest, a brown cowboy hat and a red handkerchief around his neck. The Smurf is standing with his hands at his hips. The charm is on a gold chain.
167/037	Diecast Painter			Diecast Metal Pin	2.5 X 2		$3.50-5.00	The Smurf pin is in the shape of a Smurf. It is made of a diecast metal. The Smurf is holding a palate in one hand and a paint brush in the other. The Smurf is holding a paintbrush with a yellow tip up in the air. The palate is white and has 6 colors on it, he is holding that at his side. The Smurf is wearing white pants and a white hat.
167/038	Diecast Trumpet Smurf			Diecast Metal Pin	2.5 X 2		$3.50-5.00	The Smurf pin is in the shape of a Smurf. It is made of a diecast metal. The Smurf is walking with his eyes shut, playing a yellow trumpet. The Smurf is wearing a white hat and white pants.
167/039	Diecast Soccer Player	Peyo/SEPP	1/1/81	Diecast Metal Pin	1 1/2" X 1"		$7.50-10.00	The Smurf pin is in the shape of a Smurf. It is made of a diecast metal. The Smurf is kicking a white and black soccer ball. The Smurf is wearing a red shirt, yellow shorts and red socks.
167/040	Diecast Drummer	Peyo/SEPP		Diecast Metal Pin	1 1/2" X 1"		$2.00-3.00	The Smurf pin is in the shape of a Smurf. It is made of a diecast metal. The Smurf has a yellow drum strap around his neck. He is playing a red, yellow, white and black drum. The drum sticks are white with red tips. The Smurf has a white hat and white pants on.
167/041	Smurf Sitting On A Cloud			Flat Metal Pin	1" X 3/4"		$2.00-3.00	A Smurf is sitting on a white cloud. There is a red, orange, yellow, and green rainbow behind him. The Smurf has a happy expression on his face. The Smurf is dark blue. The pin is flat paint.
167/042	Smurf Sitting On A Cloud	Peyo/SEPP	1/1/80	Flat Metal Pin	1" x 3/4"		$2.00-3.00	A Smurf is sitting on a pink cloud. There is a red, orange, yellow and green rainbow behind him. The Smurf has a happy expression on his face. The Smurf is light blue. The pin is flat paint.
167/043	Smurf Sitting On A Cloud			Hat Pin	1/2" x 3/4"		$2.00-3.00	A Smurf is sitting on a pink cloud. There is a red, orange, yellow and green rainbow behind him. The Smurf has a happy expression on his face. The Smurf is dark blue. The pin is glossy.
167/044	Smurfette			Hat Pin	1" High		$2.00-3.00	Smurfette is standing with her hands in front of her. She is wearing a red dress. Her eyes are white and black. Smurfette has a dark body. The paint is glossy.
167/045	Torchbearer			Flat Metal Pin	1 1/2"		$2.00-3.00	The Smurf looks like he's running. He is holding a gold torch with a red flame in one hand. The Smurf is wearing a white shirt, red shorts, white shoes and a white hat. The Smurfs legs are white. The pin has a glossy finish.
167/046	Torchbearer			Flat Metal Pin	1 1/2"		$2.00-3.00	The Smurf looks like he's running. He is holding a gold torch with a red flame in one hand. The Smurf is wearing a white shirt, red shorts, white shoes and a white hat. The Smurfs legs are white. The pin has a flat finish.
167/047	Torchbearer			Flat Metal Pin	1 1/2"		$2.00-3.00	The Smurf looks like he's running. He is holding a gold torch with a red flame in one hand. The Smurf is wearing a white shirt, red shorts, white shoes and a white hat. The Smurf's legs are blue. The pin has a flat finish.
167/048	Weight-lifter			Flat Metal Pin	1"		$2.00-3.00	The Smurf is lifting a gray weight bar over his head. The Smurf is wearing red pants, a red tank shirt and a white hat. The Smurf has a troubled expression on his face. The pin has a glossy finish.
167/049	Hiker			Flat Metal Pin	1"		$2.00-3.00	The Smurf is holding a black walking stick. He is wearing a yellow jacket, white pants and black hiker boots. He has a green backpack on his back. The Smurf has blue legs. The pin is flat paint.
167/050	Hiker			Flat Metal Pin	1"		$2.00-3.00	The Smurf is holding a black walking stick. He is wearing a yellow jacket, white pants and black hiker boots. He has a green backpack on his back. The Smurf has white legs. The pin is flat paint.
167/051	Hiker			Flat Metal Pin	1"		$2.00-3.00	The Smurf is holding a black walking stick. He is wearing a yellow jacket, white pants and black hiker boots. He has a green backpack on his back. The Smurf has white legs. The pin is has a glossy finish..
167/052	Baseball Smurf			Flat Metal Pin	1 1/4"		$2.00-3.00	The Smurf is carrying a black bat over his shoulder. There is a red glove hanging off the end of the bat. The Smurf is wearing a yellow baseball shirt with a red "B" on the front, white pants and a white hat. The pin has a flat finish.
167/053	Baseball Smurf			Flat Metal Pin	1 1/4"		$2.00-3.00	The Smurf is carrying a black bat over his shoulder. There is a red glove hanging off the end of the bat. The Smurf is wearing a yellow baseball shirt with a red "B" on the front, white pants and a white hat. The pin has a glossy finish.
167/054	Travel Smurf			Flat Metal Pin	1"		$2.00-3.00	The Smurf is walking, carrying a black knapsack over his shoulder. The knapsack has a money symbol on the front. The Smurf is wearing a white tank shirt, yellow shorts and white shoes. The Smurf has a white legs. The pin has a glossy finish.
167/055	Smurf Wearing Green Apron			Flat Metal Pin	1 1/4"		$2.00-3.00	The Smurf is standing with his hands out at his sides. He is wearing a green work apron. The Smurf has white legs. The pin has a flat finish.
167/056	Smurf Wearing A Green Apron			Flat Metal Pin	1 1/4"		$2.00-3.00	The Smurf is standing with his hands out at his sides. He is wearing a green work apron. The Smurf has white legs. The pin has a glossy finish.
167/057	Jokey Smurf			Flat Metal Pin	1 1/2" x 1"		$2.00-3.00	The Smurf is walking, carrying a present. The box is yellow and has a red ribbon around it. The Smurf has white pants and a white hat on. The pin has a glossy finish.
167/058	Smurf On A Motorcycle			Flat Metal Pin	1 1/4" x 1"		$2.00-3.00	The Smurf is riding a green motorcycle with pink pipes. He is riding up a brown jump ramp. The Smurf is wearing purple goggles around his hat, red scarf, yellow pants and a yellow shirt. The pin has a flat finish.
167/059	Smurf On A Motorcycle			Flat Metal Pin	1 1/4" x 1"		$2.00-3.00	The Smurf is riding a green motorcycle with pink pipes. He is riding up a brown jump ramp. The Smurf is wearing purple goggles around his hat, red scarf, yellow pants and a yellow shirt. The pin has a glossy finish.
167/060	Football Smurf			Flat Metal Pin	1"		$2.00-3.00	The Smurf looks like he's running. He is carrying a purple and red football under one arm. He is wearing a red shirt with a white #1 on the front, red pants, white shoes and a white hat with a yellow star on it. The Smurfs body is a medium blue. The pin is smooth and glossy.
167/061	Football Smurf			Flat Metal Pin	1"		$2.00 3.00	The Smurf looks like he's running. He is carrying a purple and red football under one arm. He is wearing a red shirt with a white #1 on the front, red pants, white shoes and a white hat with a yellow star on it. The Smurfs body is a medium blue. The pin has a flat finish.
167/062	Football Smurf			Flat Metal Pin	3/4"		$2.00-3.00	The Smurf looks like he's running. He is carrying a purple and red football under one arm. He is wearing a red shirt with a white #1 on the front, red pants, white shoes and a white hat with a yellow star on it. The Smurfs body is a medium blue. The pin is smooth and glossy.
167/063	Tennis Smurf			Flat Metal Pin	1 1/2"		$2.00-3.00	The Smurf is carrying a yellow tennis racket. He is holding the tennis racket up in the air. The Smurf is wearing a green shirt and pink shorts. The pin has a glossy finish.
167/064	Tennis Smurf			Flat Metal Pin	1 1/2"		$2.00-3.00	The Smurf is carrying a yellow tennis racket. He is holding the tennis racket up in the air. The Smurf is wearing a green shirt and pink shorts. The pin has a flat finish.
167/065	Tennis Smurf			Flat Metal Pin	1 1/4"		$2.00-3.00	The Smurf is carrying a yellow tennis racket in front of his chest. He is wearing a white shirt and red shorts. The Smurf has white legs and white feet. The pin has a flat finish.
167/066	Smurf Holding Hands In Air			Hat Pin	1"		$2.00-3.00	The Smurf is holding his hands in the air. He is wearing a red tank shirt, white hat and white pants. The pin has a flat finish.

167/067	Skiing Smurf			Flat Metal Pin	1"		$2.00-3.00	The Smurf is wearing ski clothes and ski's. The Smurf is wearing a white hat with a pink tassel, red scarf, yellow sweater, brown pants, purple boots and white ski's. The Smurf is holding 2 ski poles. The poles are green with pink ends. The pin has a glossy finish.
167/068	Skiing Smurf			Flat Metal Pin	1"		$2.00-3.00	The Smurf is wearing ski clothes and ski's. The Smurf is wearing a white hat with a pink tassel, red scarf, yellow sweater, brown pants, purple boots and white ski's. The Smurf is holding 2 ski poles. The poles are green with pink ends. The pin has a flat finish.
167/069	Skiing Smurf			Flat Metal Pin	3/4"		$2.00-3.00	The Smurf is wearing ski clothes and ski's. The Smurf is wearing a white hat with a pink tassel, red scarf, yellow sweater, brown pants, purple boots and white ski's. The Smurf is holding 2 ski poles. The poles are green with pink ends. The pin has a glossy finish.
167/070	Smurf With Red Flower Behind Back			Flat Metal Pin	1"		$2.00-3.00	The Smurf is standing, holding a red flower behind his back. The Smurf is wearing purple shorts, a white shirt, a white hat, and white shoes. The Smurf has a dark blue body. The pin has a glossy finish.
167/071	Smurf With Red Flower Behind Back			Flat Metal Pin	1"		$2.00-3.00	The Smurf is standing, holding a red flower behind his back. The Smurf is wearing purple shorts, white shirt, white hat, and white shoes. The Smurf has a dark blue body. The pin has a flat finish.
167/072	Smurf With Red Flower Behind Back			Flat Metal Pin	3/4"		$2.00-3.00	The Smurf is standing, holding a red flower behind his back. The Smurf is wearing blue shorts, white shirt, white hat, and white shoes. The Smurf has a dark blue body. The pin has a glossy finish.
167/073	Albino Smurf			Hat Pin	1 1/4"		$2.00-3.00	The Smurf is standing, holding a yellow flower behind his back. The Smurf is wearing purple shorts, white shirt, white hat, and white shoes. The Smurf has a white body. The pin has a glossy finish.
167/074	Basketball Smurf			Hat Pin	1"		$2.00-3.00	The Smurf is jumping in the air. He has his hand on a basketball above his head. The ball is yellow. The Smurf is wearing a red tank top, green shorts, white shoes, and a white hat. The pin has a glossy finish.
167/075	Basketball Smurf			Flat Metal Pin	1 1/4"		$2.00-3.00	The Smurf is jumping in the air. He has his hand on a basketball above his head. The ball is orange and yellow. The Smurf is wearing a red tank top, green shorts, white shoes, and a white hat. The pin has a glossy finish.
167/076	Basketball Smurf			Flat Metal Pin	1 1/2"		$2.00-3.00	The Smurf is jumping in the air. He has his hand on a basketball above his head. The ball is orange and yellow. The Smurf is wearing a red tank top, green shorts, white shoes, and a white hat. The pin has a glossy finish.
167/077	Basketball Smurf			Flat Metal Pin	1 1/2"		$2.00-3.00	The Smurf is jumping in the air. He has his hand on a basketball above his head. The ball is orange and yellow. The Smurf is wearing a red tank top, green shorts, white shoes, and a white hat. The pin has a flat finish.
167/078	J Pin			Flat Metal Pin	1"		$7.50-10.00	The pin is the letter J. The J is green. There is a Smurf standing inside the letter. The Smurf is holding a yellow lute. The pin has a glossy finish.
167/079	E Pin			Flat Metal Pin	1"		$7.50-10.00	The pin is the letter E. The E is red. There is a Smurf standing outside of the letter. The Smurf is walking, blowing a yellow horn. The pin has a glossy finish.
167/080	N Pin			Flat Metal Pin	1"		$7.50-10.00	The pin is the letter N. The N is cream color. There is a Smurf standing in the middle of the letter. The Smurf is walking, carrying a brown baseball bat. The Smurf is wearing a yellow shirt and white pants. The pin has a glossy finish.
167/081	Smurf Playing A Horn			Flat Plastic Pin	1 1/2"		$2.00-3.00	The Smurf pin is a flat, raised plastic. The Smurf is walking, playing a yellow horn. The Smurf is wearing a red shirt, white hat and white pants.
167/082	Smurf Wearing A Red Apron			Flat Plastic Pin	1 1/4"		$2.00-3.00	The Smurf pin is a flat, raised plastic. The Smurf is standing, wearing a red apron and white pants.
167/083	Smurf Walking In Sunset			Flat Metal Pin	1 1/4"		$2.00-3.00	A Smurf is walking, carrying a knapsack. There is a mushroom and sun behind him. The mushroom is red and pink. The sun is yellow with red rays. The Smurf has a blue face and blue pants. A white hat and white tummy.
167/084	Papa Plugging In A Cord	Legrand	1/1/92	Hat Pin	2.5 X 2	Brussels	$7.50-10.00	Papa is standing, holding 2 wire sockets he is going to plug together. The wires are gray. Papa is wearing a red hat and red pants. The Legrand symbol is on Papa's hat. The pin is glossy. From the Meilleurs Schtroumpfs collection.
167/085	Smurf Hitting Thumb With Hammer	Legrand	1/1/92	Hat Pin	2.5 X 2	Brussels	$7.50-10.00	The Smurf hit his thumb with a hammer. The hammer is silver on the end and has a brown handle. The Smurf has a painful expression on his face. He is wearing a white hat and white pants. The Legrand symbol is on his hat. The pin is glossy. From the Meilleurs Schtroumpfs collection.
167/086	Smurf Drilling	Legrand	1/1/92	Hat Pin	2.5 X 2	Brussels	$7.50-10.00	The Smurf is holding a yellow drill. He is drilling into a brown piece of wood. The Smurf is wearing white coveralls and a white hat with a visor. The Legrand symbol is on his hat. The pin is glossy. From the Meilleurs Schtroumpfs collection.
167/087	Handy	Legrand	1/1/92	Hat Pin	2.5 X 2	Brussels	$7.50-10.00	The Smurf is carrying a hammer in one hand and a gray tool box in the other. He has a yellow ruler tucked under his arm. He has a nail in his mouth. He is wearing gray coveralls and a white hat. The coveralls have the Legrand symbol on the front. The pin is glossy. From the Meilleurs Schtroumpfs collection.
167/088	Architect Smurf	Legrand	1/1/92	Hat Pin	2.5 X 2	Brussels	$7.50-10.00	The Smurf is holding a ruler in one hand and a ruler in the other. He is wearing a yellow jacket, white pants, white boots, and a white hat. The Smurf has on silver wire glasses. His hat has the Legrand symbol on it. The pin is glossy. From the Meilleurs Schtroumpfs collection.
167/089	Smurfette Turning A Light Bulb Into A Socket	Legrand	1/1/92	Hat Pin	2.5 X 2	Brussels	$7.50-10.00	Smurfette is turning a light bulb into a light socket. Smurfette is wearing a white dress and white hat. Her hat has the Legrand symbol on it. The pin is glossy. From the Meilleurs Schtroumpfs collection.
167/090	Smurf Sunbathing	W. Berrie/Peyo	1/1/79	Flat Metal Pin	2.5 X 2	Taiwan	$3.50-5.00	The Smurf is laying on a red towel. The sun is behind him. The Smurf is sleeping. He is wearing a white hat and green swim trunks.
167/091	Smurf On Skateboard	W. Berrie/Peyo	1/1/79	Flat Metal Pin	2.5 X 2	Taiwan	$3.50-5.00	The Smurf is standing on his hands riding a skateboard. The skateboard is red/yellow/orange. The Smurf is wearing white pants and a white hat. There is a white cloud behind the Smurf.
167/092	Smurf Leaning On A Heart Holding Flower	W. Berrie/Peyo	1/1/79	Flat Metal Pin	2.5 X 2	Taiwan	$3.50-5.00	The Smurf is leaning on a red heart. He has a yellow daisy behind his back. The Smurf looks like he's day dreaming.
167/093	Smurf Holding A Flower			Flat Plastic Pin	2.5 X 2		$2.00-3.00	The Smurf is standing, holding a red flower behind his back. The Smurf is wearing white pants, a red shirt and a white hat. The Smurf is grinning.
167/094	The Smurfs Irwin	Irwin		Flat Plastic Pin	1/2" x 1"		$3.00-4.00	The pin says "The Smurfs" in yellow letters. It has a Smurf laying on top of the letters and a Smurf leaning against the S. It says "Irwin" underneath in black letters. The background is red.
167/095	Papa Standing With Hands Behind Back	Mr. Pin	1/1/96	Hat Pin	1 1/4"	Brussels	$4.50-6.00	Papa Smurf is standing with his hands behind his back. He has a smug look on his face.
167/096	Soccer Smurf With Hands In The Air	Mr. Pin	1/1/96	Hat Pin	1 1/4"	Brussels	$4.50-6.00	The Smurf is standing in the grass. His foot is resting on a white and black soccer ball. The Smurf has his hands in the air. He is wearing a red shirt, white shorts, red socks, brown soccer shoes, and a white hat.
167/097	Love-struck Smurf	W. Berrie/Peyo	1/1/79	Flat Metal Pin	1"	Taiwan	$3.50-5.00	The Smurf is sitting, holding a yellow daisy. He has red lip marks on his foot and hat. The Smurf has his tongue hanging out of his mouth.
167/098	Smurfette Looking At A Daisy	W. Berrie	1/1/79	Flat Metal Pin	2.5 X 2	Taiwan	$3.50-5.00	Smurfette is standing in the grass looking at a red daisy. Smurfette has her hands folded under her chin. She is wearing a white dress and white hat.
167/099	Smurf JC's Pin	Anarb		Pin	1 2/1" x 1"	Taiwan	$9.00-12.00	The Smurf is standing, holding 3 balloons. Each balloon has a letter inside. The balloons spell (JC'S) underneath is a NH. The Smurf has a shy expression on his face.
167/100	N.H. Jaycees Kids	Anarb		Metal Pin	1"	Taiwan	$7.50-10.00	The pin is Baby Smurf. Baby is sitting, holding a red and yellow ball. Baby Smurf is wearing a white sleeper. His hat says "N.H. Jaycees". Baby's leg says "Kids".
167/101	Boxer	I.M.P.S.	1/1/93	Hat Pin	2.5 X 2	Brussels	$3.50-5.00	The Smurf is standing, wearing tan boxer gloves and looks like he is punching at something. The Smurf is wearing pink shorts and no shirt.
167/102	Smurf Minute Maid	Peyo/I.M.P.S.	1/1/92	Wood Pin	2.5 X 2	Brussels	$6.75-9.00	The pin has a Smurf sitting by a globe of the earth. Underneath it says "Minute Maid".
167/103	Astro Smurf With Rainbow	W. Berrie/Peyo	1/1/79	Pin	1"	Taiwan	$4.50-6.00	A Smurf is wearing an astronaut suit and helmet. He has a rainbow and a star behind him.
167/104	Smurf Playing Guitar	W. Berrie/Peyo	1/1/79	Pin	1"	Taiwan	$4.50-6.00	A Smurf is playing a brown guitar. The Smurf is singing.
167/105	LOVE Smurf With Flowers	W. Berrie/Peyo	1/1/79	Pin	2.5 X 2	Taiwan	$4.50-6.00	The Smurf is standing, holding red, yellow and green daisies. He has a shy expression on his face. The word "LOVE" is spelled in red, yellow and green letters below the flowers.
167/106	Johan And Peewit	Peyo/I.M.P.S.	1/1/91	Pin	1"	Brussels	$4.50-6.00	The pin has Johan's upper body and Peewit are peeking over his shoulder. Johan has a mad expression on his face.
167/107	Gargamel With Hands In Air	Peyo/I.M.P.S.	1/1/91	Pin	1"	Brussels	$4.50-6.00	Gargamel is walking with his hands in the air. He has a vicious expression on his face.
167/108	Smurf Carrying A European Flag	Peyo/I.M.P.S.	1/1/92	Pin	1"	Brussels	$4.50-6.00	A Smurf is walking, carrying a European flag. The flag is blue with gold stars. The stars make a circle.
167/109	Smurf Jumping	Peyo/I.M.P.S.	1/1/93	Pin	1"	Brussels	$4.50-6.00	The pin looks like the Smurf is getting ready to jump something. He has his arms at his side and his legs in front of him. He is wearing a white hat, white tank shirt and orange shorts.
167/110	Smurf Sweeping	Peyo/ I.M.P.S.	1/1/92	Pin	1"	Brussels	$4.50-6.00	The pin is a Smurf leaning on a broom. The Smurf has his eye's closed and a big grin on his face. The broom is yellow with a brown stick.
167/111	Smurf Falling	Peyo/I M PS	1/1/92	Pin	1"	Brussels	$4.50-6.00	The pin is a Smurf that is laying on his back. He looks like he is falling. He has his hands out at his side. He has a troubled expression on his face
167/112	Black Smurf Yelling	Le Grand	1/1/92	Pin	1"	Brussels	$4.50-6.00	The pin is glossy. It is a black Smurf. He is yelling. He has his hands fisted out at his sides
167/113	Teisseire Smurf	Peyo/I.M.P.S.	1/1/93	Pin	1"	Brussels	$7.50-10.00	The pin says Teisseire on the bottom. It has a Smurf holding an orange with a straw in the orange. The Smurf is licking his lips. The pin is a promotional for an orange juice company.
167/114	Smurf Bent Over Looking Up	Peyo/I.M.P.S.	1/1/92	Pin	1"	Brussels	$4.50-6.00	The pin is the back of a Smurf. He is bent over looking up. His hands are swung to one side.
167/115	Smurf Weight-lifter	Peyo/I.M.P.S.	1/1/93	Pin	2.5 X 2	Brussels	$4.50-6.00	The Smurf is standing, lifting a dumbbell over his head with one hand. The dumbbell is gray.
167/116	Dribbler basketball Smurf	Peyo/I.M.P.S.	1/1/93	Pin	1"	Brussels	$4.50-6.00	The Smurf is standing, dribbling a ball. He is holding one hand out to block. The Smurf is wearing a white hat, white shirt with a red stripe, green shorts, red socks, and green shoes.

Set 275: Pixi- Minis

275/001	Black Smurf And Angry Smurf	I.M.P.S.	1/1/96	Metal Pixi	3/4"	Paris	$22.50-30.00	The pixi's are metal and hand-painted. The black Smurf and the blue Smurf both have their hand's fisted and their arms are extended behind them. They are on a green base. The Smurfs are number 478. It comes with a certificate.
275/002	Chess Set Pixi-Mini Jeu D' Echecs	I.M.P.S.	1/1/96	Metal Pixi	3/4"	Paris	$187.50-250.00	The set is number 255. The chess set comes with a certificate. One side has a black king, a queen, 2 black Smurfs riding on horses, 2 black jesters, 2 Gargamel castles and 8 red mushrooms. The other side has the same Smurfs but they are blue. It has 2 mushroom houses and 8 cake pawns. The figures are mini and made of metal.
275/003	7 Piece Smurf Pixi Set	I.M.P.S.	1/1/96	Metal Pixi		Paris	$55.00-75.00	The pixi's are metal. The Smurfs are about 1/4" high. There are 5 Smurfs and Papa. Gargamel is 1 1/2" high. Azreal is 1" high. The pixi's come in a foam box. They are a numbered set. #626. They come with a certificate.

275/004	Brainy	I.M.P.S.	1/1/96		Metal Pixi	3/4"	Paris	$15.00-20.00	The pixi is metal and hand-painted. The Smurf is wearing black glasses. He is holding a finger in the air. There is a red book tucked under his arm. He is on a green base. The Smurf is number 350. It comes with a certificate.
275/005	Jokey	I.M.P.S.	1/1/96		Metal Pixi	3/4"	Paris	$15.00-20.00	The pixi is made of metal and hand-painted. It is a Smurf carrying a yellow package with red ribbon wrapped around it. He is on a green base. The Smurf is number 641. It comes with a certificate.
275/006	Smurf With Hands Behind Back	I.M.P.S.	1/1/96		Metal Pixi	3/4"	Paris	$15.00-20.00	The pixi is made of metal and hand-painted. The Smurf is walking with his hands behind his back. He is on a green base. They Smurf is number 1084. It comes with a certificate.
275/007	Smurfette With Flowers	I.M.P.S.	1/1/96		Metal Pixi	3/4"	France	$15.00-20.00	The pixi is made of metal and hand-painted. Smurfette is standing with some red daisies in her hands. She is on a green base. Smurfette is number 321. It comes with a certificate.

Set 105: Place mats

105/001	Smurfette Skating (Winter Scene)	W. Berrie	1/1/88		Texaco Place mat			$15.00-20.00	The place mat is a promotion for Texaco Gas. The place mat has Smurfette skating, she has a purple dress on. 4 Smurfs are downhill skiing, Papa and 2 Smurfs are on a toboggan. There are 3 houses in the background. (Matching glass is under glasses from France).
105/002	Gargamel And Azreal (Fall Scene)	W. Berrie	1/1/88		Texaco Place mat			$15.00-20.00	The place mat is a promotion for Texaco Gas. The place mat has Gargamel and Azreal hiding behind a tree watching the Smurfs. Papa and 5 Smurfs are taking sap from the trees in the village, 2 of the Smurfs are pulling a sled with a barrel of sap. There are three houses in the background. (Matching glass is under glasses from France).
105/003	Swim Hole (Summer Scene)	W. Berrie	1/1/88		Texaco Place mat			$15.00-20.00	The place mat is a promotion for Texaco Gas. The place mat has Smurfette and 3 Smurfs swimming, 1 Smurf fishing and 1 Smurf sleeping on the side of the pond. There is a picnic set up on the side with a house in the background. (Matching glass is under glasses from France).
105/004	Smurf Delight Place mat	W. Berrie		Cream	Place mat		U.S.A.	$7.50-10.00	The place mat has a Smurf holding a yellow bowl, a Smurf wearing an apron and a chef's hat is chasing another Smurf with a rolling pin. The Smurf being chased is carrying a cake. There is a table in the background with 2 cakes on it.
105/005	Smurftime	W. Berrie		Red	Place mat		U.S.A.	$7.50-10.00	The place mat has Smurfette standing, stirring food. There is a Smurf following the smell, holding a spoon, and another Smurf with a fork and knife.
105/006	Smurf Supper Scene	Peyo	1/1/84	Pink	2 Sided Place mat		U.S.A.	$7.50-10.00	The place mat is 2- sided. On one side, it's pink and has a fireplace with a table set up for dinner. Papa is holding up a glass, Smurfette is standing by a yellow gift box, Brainy is reading from a paper and a baker Smurf is carrying a cake. The other side has 3 games and there written in French.
105/007	Kitchen Scene	W. Berrie	1/1/88	Red	2 Sided Place mat		U.S.A.	$7.50-10.00	The place mat is 2- sided. On one side, it's red and has a silver stove and 2 brown tables in the room. One Smurf is reaching for a sack hanging from the ceiling. There is a Smurf standing next to a stove sampling what's in the kettle, and a Smurf wearing an apron and chef's hat is running through the door with a rolling pin. The tables have kettles and bowls on them. The other side has 3 games and there written in French.
105/008	Eat Drink And Be Smurfy	W. Berrie	1/1/88	Yellow	Place mat		U.S.A.	$7.50-10.00	The place mat is yellow and has a picture of 2 Smurfs and Papa sitting at a red and white checkered table, eating. There is another Smurf standing in front holding a mug of beer and the other is eating a sandwich. Across the front it says "Eat Drink And Be Smurfy."
105/009	Smurfette Ringing A Bell (Fall Scene)	W. Berrie	1/1/88		Texaco Place mat			$15.00-20.00	The place mat is a promotion for Texaco Gas. The place mat has Smurfette standing outside the schoolhouse ringing a bell. 3 Smurfs are hiding in the leaves and trees. Azreal is looking around the corner of the schoolhouse. (Matching glass is under glasses from France.)
105/010	Smurf And Smurfette Roller-skating	Party time	1/1/88		Paper Place mat	8 1/2"L x 13"W	U.S.A.	$7.50-10.00	The place mat has Smurfette roller-skating with a Smurf behind her. The Smurf looks like he's chasing Smurfette. They are out in the grass roller-skating. Smurfette is wearing a white dress, white hat and pink roller-skates. The Smurf is wearing a white hat, white pants and purple roller-skates. It says "Smurf" on top in white letters with a blue border.
105/011	Smurfette In A Tower French Place mat	Peyo/I.M.P.S.	1/1/88		French Place mat	11 1/2" x 17	Brussels	$4.50-6.00	The place mat has 2 games on it. It is a picture of a swim hole. Nat and Slouchy are fishing. Nat is standing on a boat dock looking at Smurfette who is in a tower. A Smurf is sleep walking towards the water.

Set 112: Plaques- German Wood

112/001	Viele Koche Verderben Die Kochin.				Wood Wall Plaque	2 3/4"H X 4 3/4"W	Germany	$11.00-15.00	The Smurf is half of a rubber PVC. The Smurf is wearing a white chef's coat with gold buttons, white pants and a white chef's hat. He is holding a brown spoon to his mouth. The Smurf is on a light- colored wood plaque with the saying "Viele Koche Verderben Die Kochin" in black letters on the left side. The saying means "too many cooks spoil the waitress."
112/002	Mud Und Satt, Wie Schon Ist Datt!!				Wood Wall Plaque	2 3/4"H X 4 3/4"W	Germany	$11.00-15.00	The Smurf is half of a rubber PVC. The Smurf is wearing a white chef's coat with gold buttons, white pants and a white chef's hat. He is holding a brown spoon to his mouth. The Smurf is on a light- colored wood plaque with the saying "Mud Und Satt, Wie Schon Ist Datt!!" in black letters on the left side. The saying means "tired and stuffed, what away to be."
112/003	Schweig, Wenn Du Mit Mir Sprichst!!				Wood Wall Plaque	2 3/4"H X 4 3/4"W	Germany	$11.00-15.00	The Smurf is half of a rubber PVC. The Smurf is holding a copper receiver to his mouth . The telephone is red and copper and is sitting next to his feet. The Smurf is on a light- colored wood plaque with the saying "Schweig, Wenn Du Mit Mir Sprichst!!" in black letters on the left side. The saying means "quiet, when you're talking to me."
112/004	Ich Bin Leider Nicht Da, …				Wood Wall Plaque	2 3/4"H X 4 3/4"W	Germany	$11.00-15.00	The Smurf is half of a rubber PVC. The Smurf is holding a copper receiver to his mouth . The telephone is red and copper and is sitting next to his feet. The Smurf is on a light- colored wood plaque with the saying "Ich Bin Leider Nicht Da, Und Weiz Auch Nicht, Wo Ich Gerade Bin!" in black letters on the left side. The saying means "I'm not home at this time, and I don't know where I'm at."
112/005	Ich Bin Das Ruckkgrat Der Firma: Lauter Wirbel!!				Wood Wall Plaque	2 3/4"H X 4 3/4"W	Germany	$11.00-15.00	The Smurf is half of a rubber PVC. The Smurf has his left hand out and he's pointing at something. The Smurf is on a light- colored wood plaque with the saying "Ich Bin Das Ruckkgrat Der Firma: Lauter Wirbel!!" in black letters on the left side. The saying means "I'm the backbone of my firm."
112/006	Es Gibt Viel Zu Tun, Fangt Schon Mal An…!				Wood Wall Plaque	2 3/4"H X 4 3/4"W	Germany	$11.00-15.00	The Smurf is half of a rubber PVC. The Smurf has his left hand out and he's pointing at something. The Smurf is on a light- colored wood plaque with the saying "Es Gibt Viel Zu Tun, Fangt Schon Mal An…!" in black letters on the left side. The saying means "there's lot's to do, go ahead and start."
112/007	Ich Bin Ganz Meiner Meinung!				Wood Wall Plaque	2 3/4"H X 4 3/4"W	Germany	$11.00-15.00	The Smurf is half of a rubber PVC. Brainy is holding a finger on his right hand in the air. He is wearing black glasses. The Smurf is on a light- colored wood plaque with the saying "Ich Bin Ganz Meiner Meinung!" in black letters on the left side. The saying means "I agree with myself, completely."
112/008	Die Wurde Des Menschen Ist Unfalzbar.				Wood Wall Plaque	2 3/4"H X 4 3/4"W	Germany	$11.00-15.00	The Smurf is half of a rubber PVC. Brainy is holding a finger on his right hand in the air. He is wearing black glasses. The Smurf is on a light- colored wood plaque with the saying "Die Wurde Des Menschen Ist Unfakbar." in black letters on the left side. The saying means "a person's dignity is unbelievable."
112/009	Unmogliches Wird Sofort Erledigt,…				Wood Wall Plaque	2 3/4"H X 4 3/4"W	Germany	$11.00-15.00	The Smurf is half of a rubber PVC. Papa is standing, holding a white memo pad in one hand and a white and black feather pen in the other hand. The Smurf is on a light- colored wood plaque with the saying "Unmogliches Wird Sofort Erledigt, Wunder Dauern Etwas Langer!" in black letters on the left side. The saying means "the impossible will be done right away, miracles take a bit longer."
112/010	Wir Haben Alles, Nur Keine Zeit!				Wood Wall Plaque	2 3/4"H X 4 3/4"W	Germany	$11.00-15.00	The Smurf is half of a rubber PVC. Papa is standing, holding a white memo pad in one hand and a white and black feather pen in the other hand. The Smurf is on a light- colored wood plaque with the saying "Wir Haben Alles, Nur Keine Zeit!" in black letters on the left side. The saying means "we have everything, except time."
112/011	Lazt Uns Mit Frischer Kraft Und Neuem …				Wood Wall Plaque	2 3/4"H X 4 3/4"W	Germany	$11.00-15.00	Gargamel is half of a rubber PVC. Gargamel is wearing a black outfit and red shoes. He looks like he's walking. Gargamel is on a light- colored wood plaque with the saying "Lazt Uns Mit Frischer Kraft Und Neuem Mut Der Arbeit Aus Dem Wege Gehen!" in black letters on the left side. The saying means "the main thing is that we're heading forwards, the direction doesn't matter."
112/012	Hauptsache Esgeht Vorwarts, Die Richtung Ist Egal!				Wood Wall Plaque	2 3/4"H X 4 3/4"W	Germany	$11.00-15.00	Gargamel is half of a rubber PVC. Gargamel is wearing a black outfit and red shoes. He looks like he's walking. Gargamel is on a light- colored wood plaque with the saying "Hauptsache Esgeht Vorwarts, Die Richtung Ist Egal!" in black letters on the left side. The saying means "fresh strength and new courage to avoid work."

Set 135: Plaques- Wood Wall

135/001	Brewer Smurf	Denise Oechsner	5/3/83		Wood Painting	7 1/2" X 9"	Hand made	$7.50-10.00	The plaque is wood. There is a picture of a Smurf wearing a white with blue stripe baseball uniform. On the front of the uniform shirt in black letters it says Brewer Smurf. The Smurf is wearing a white hat with a blue visor and the brewer symbol on the front of the hat. The Smurf has brown shoes. He is carrying a yellow bat with a brown glove over his shoulder. The background is blue and yellow.
135/002	Papa Holding A Finger In The Air				Wood Wall Plaque	12" X 9 1/2"	Hand made	$7.50-10.00	The picture is of Papa is standing, holding a finger in the air. Papa is wearing a red hat and pants. The picture is on stained wood and has a clear glossy finish.
135/003	Smurfette Ballerina		1/1/84		Wood Plague	7 1/2" X 5"	Hand Painted	$6.00-8.00	The plaque is oblong. Smurfette is painted on the front. She is wearing a white ballerina dress and shoes. Smurfette's hair is painted orange. She is standing in a ballerina pose. The plaque has a dark stain.
135/004	Smurf Holding Flowers		1/1/84		Wood Plaque	8" x 6"	Hand Painted	$6.00-8.00	The plaque is a rectangle. A Smurf with a shy grin holding 5 flowers is painted on the front. The plaque has a dark stain.
135/005	Tina's Room				Wood Wall Plaque	3 1/2"L X 6 1/2"W	Hand made	$2.00-3.00	The plaque is made of a light color wood. It has a clay Smurf glued in one corner. It says "Tina's" in black marker. It says room underneath in blue clay.

Set 147: Plastic & Rubber Smurfs

147/001	Smurfette Holding A Book	Peyo	1/1/84		Plastic Figure	4 1/2" High		$4.50-6.00	Smurfette is standing, wearing a white dress, white shoes and a white hat. Her body is bright blue. Smurfette is holding a red music book in her hands.
147/002	Smurf With A Guitar	Peyo	1/1/84		Plastic Figure	4 1/2" High		$4.50-6.00	The Smurf is standing, playing a yellow guitar. He has his eyes closed and his mouth open like he is singing. The Smurf has a bright blue body.
147/003	Papa Bandleader	Peyo	1/1/84		Plastic Figure	4 1/2" High		$4.50-6.00	Papa is standing with his hands out at his sides. He is wearing a white jacket with a yellow bow tie, red pants and a red hat. Papa has his eyes closed and his mouth open like he's singing. Papa has a bright blue body.

147/004	Smurf With Hands Over Head	Golde	1/1/84		Hard Rubber Figure	4 1/2" High	Brussels	$3.00-4.00	The Smurf is made of a hard rubber. He is holding both hands over his head. He has a happy expression on his face. The Smurf is on a blue base.
147/005	Smurfette	Golde/Peyo	1/1/84		Hard Rubber Figure	4 1/2" High	Brussels	$3.00-4.00	Smurfette is made from a hard rubber. She is standing, holding her hands together behind her back. Smurfette is wearing a white dress, white shoes and a white hat. She is standing on a blue base.
147/006	Smurf Wearing A Scarf	Golde/Peyo	1/1/84		Hard Plastic Figure	4 1/2" High	Brussels	$3.00-4.00	The Smurf is made of a hard plastic. The Smurf is wearing a white hat, yellow scarf and white pants. The Smurf is holding a hand over his mouth. He has a troubled expression on his face. The Smurf is standing on a blue base.

Set 172: Plates- Ceramic & Accessories Sets

172/001	Hot Air Balloon Plate	W. Berrie	1/1/82		Ceramic Plate	7 1/2" Circle	Japan	$15.00-20.00	The plate is white and has a picture of a Smurf and Smurfette in a hot air balloon. The balloon is pink, orange, yellow, and blue. The carrier part of the balloon is yellow and orange. Smurfette is leaning out of the balloon pointing and the Smurf is holding a telescope. The plate has blue behind the balloon with clouds. The plate says "Have A Smurfy Day!" on the bottom in blue letters. There is a blue border around the plate.
172/002	Hot Air Balloon Mug	W. Berrie	1/1/82		Ceramic Mug	3" High	Japan	$7.50-10.00	The cup is white and has 2 hot air balloons and a blimp. All the balloons and the blimp are pink, orange, yellow, and blue. The carrier parts are yellow and orange. One balloon has Gargamel and Azreal in it. The blimp has Brainy, pointing to Gargamel and Papa ,waving to the 2 Smurfs in the other balloon. The third balloon has Smurfette and another Smurf. The cup has blue background for the sky and white clouds. The cup has a blue border.
172/003	Hot Air Balloon Bell	W. Berrie	1/1/82		Ceramic Bell	6" High	Japan		The bell is white and has 2 hot air balloons and a blimp. All the balloons and the blimp are pink, orange, yellow, and blue. The carrier parts are yellow and orange. One balloon has Gargamel and Azreal in it. The blimp has Brainy, pointing to Gargamel and Papa, waving to the 2 Smurfs in the other balloon. The third balloon has Smurfette and another Smurf in it. The bell has a blue background for the sky and white clouds around each balloon. The bell is white with a blue border.$18.00-25.00
172/004	Hot Air Balloon Trinket Box	W. Berrie	1/1/82		Ceramic Jewelry Box	3" X 2 1/2"	Japan	$7.50-10.00	The box is made of ceramic and is rectangle in shape. The picture is of a Smurf and Smurfette in a hot air balloon on the cover. The balloon is pink, orange, yellow, and blue. The carrier part of the balloon is yellow and orange. Smurfette is leaning out of the balloon pointing and the Smurf is holding a telescope. The jewelry box has blue behind the balloon with clouds.
172/005	Smurf With Ice cream Cone Trinket Box	W. Berrie	1/1/82		Ceramic Jewelry Box	3" X 2 1/2"	Korea	$7.50-10.00	The box is rectangle and made from ceramic. The cover has a picture of a Smurf holding an ice cream cone. There are 5 scoops of ice cream in the cone. A scoop laying on the Smurfs head and 2 scoops on the ground. The background is blue. The box is white.
172/006	Have A Smurfy Day! Smurfette Trinket Box	W. Berrie	1/1/82		Ceramic Jewelry Box	3" X 2 1/2"	Japan	$7.50-10.00	The box is rectangle and made from ceramic. The cover has a picture of Smurfette standing, holding a bouquet of flowers. There is a rainbow behind her. Above the rainbow it says "Have A Smurfy Day!" in blue letters. The box is white.
172/007	Im Land Der Schlumpfe Plate		1/1/79		Ceramic Plate	7 1/2" Circle	Germany	$22.50-30.00	The plate has a picture of Father Abraham in the middle. There are 2 Smurfs and Papa Smurf on the plate. There is a mushroom house in the background. The bottom of the plate says "Wfr. Langenbach '79". The top of the plate says "Im Land Der Schlumpfe".
172/008	Smurf Plate With Various Smurfs	W. Berrie	1/1/82		Ceramic Plate	7 1/2" Circle	Japan	$15.00-20.00	The plate is white. It has a blue 1 " dotted border. Along the border are various Smurfs doing different things: Papa holding a strawberry. Smurfette holding an ice cream cone. Smurf slipping on a banana peal. Smurf holding a plate of chicken. Smurf blowing a bubble. Smurf holding ice cream that fell on the ground. Papa lifting a bucket of apples. Smurf juggling oranges. Smurfette carrying a pie. Smurf drinking a fountain soda. Smurf carrying a cake.
172/009	Smurf Doing Various Things Bowl	W. Berrie	1/1/82		Ceramic Bowl	4 1/2" Circle	Japan	$11.00-15.00	The bowl is white. It has a blue 1 " dotted border around the outside. Along the border are various Smurfs doing different things: Papa holding a strawberry. Smurfette holding an ice cream cone. Smurf slipping on a banana peal. Smurf holding a plate of chicken. Smurf blowing a bubble. Smurf holding ice cream that fell on the ground. Smurf juggling oranges. Smurfette carrying a pie. Smurf drinking a fountain soda. Smurf carrying a cake.
172/010	Smurfs Doing Various Things Cup	W. Berrie	1/1/82		Ceramic Cup	3" x 3"	Japan	$7.50-10.00	The bowl is white. It has a blue 1 " dotted border. Along the border are various Smurfs doing different things: Papa holding a strawberry. Smurfette holding an ice cream cone. Smurf blowing a bubble. Smurf holding ice cream that fell on the ground. Smurfette carrying a pie.
172/011	Merry Christmas 1982 Christmas Cup The Smurf Carolers	W. Berrie/Peyo	1/1/82		Limited Edition Mug	3 1/2"	Japan	$7.50-10.00	The front of the plate has a Smurf standing by a light post singing. Smurfette is standing next to them playing a cello. A Smurf is behind Smurfette playing a drum. And Papa is playing a trumpet chorus. The cup says "Merry Christmas 1982" in red letters. The cup is outlined in gold trim. The cup is the first of a limited edition.
172/011	Merry Christmas 1982 Christmas Plate The Smurf Carolers	W. Berrie/Peyo	1/1/82		Limited Plate	7"	Japan	$22.50-30.00	The front of the plate has 3 Smurfs standing by a light post singing. Smurfette is standing next to them playing a cello. Papa is in front leading the chorus. The front of the plate says "Merry Christmas 1982" in red letters. The plate is outlined in gold trim. The plate is the first of a limited edition.
172/012	Merry Christmas 1982 Smurf Music Box Smurf Carolers	W. Berrie/Peyo	1/1/82		Wind Up Music Box	7"H x 4" W	Japan	$63.75-85.00	The music box has a Smurf standing by a lamp post singing. Smurfette is standing next to the lamp post playing the cello. The front says "Merry Christmas" in white letter. The music box plays various different Christmas songs.
172/013	Merry Christmas 1983 Plate "The Night Before Christmas"	W. Berrie/Peyo	1/1/83		Limited Plate	7" Diam	Japan	$22.50-30.00	The plate has 3 Smurfs and Papa Smurf decorating a Christmas tree. Smurfette is decorating a rocking horse. The plate says "Merry Christmas 1983" in red letters. The plate is a limited edition. It is numbered 3178 of 18,000.
172/014	Merry Christmas 1983 Cup "The Night Before Christmas"	W. Berrie/Peyo	1/1/83		Limited Edition Mug	3 1/2"	Japan	$7.50-10.00	The cup has a Smurf and Papa decorating a Christmas tree. Smurfette is decorating a wood rocking horse. A Smurf is carrying a box of decorations. The top says "Merry Christmas 1983" in red letters. The rim of the cup and handle are outlined in gold trim. The cup is the second of a limited edition.
172/015	Merry Christmas 1983 Music Box "The Night Before…"	W. Berrie/Peyo	1/1/83		Wind Up Music Box	5 1/2"	Japan	$63.75-85.00	The music box is Papa sitting in a rocking chair reading the Night Before Christmas book. Smurfette is sitting on a stool listening to Papa read. There is a cup of milk and a plate of cookies in front of them on the ground. The front of the music box says "The Night Before Christmas" in white letters. The box plays various Christmas songs.
172/016	Baby With Bear - Bowl	Porcelaine De Lologne	1/1/98	White	Ceramic Bowl	5 1/4" Diam	France	$15.00-20.00	The bowl is white. The front has a picture of Baby Smurf sitting next to a teddy bear. Baby is wearing a pink sleeper. He has his arm around a brown bear. Inside is a small picture of a train. The back of the bowl says "Les Schtroumpfs" in yellow letter.
172/017	Baby With Bear - Cup	Porcelaine De Lologne	1/1/98	White	Ceramic Cup	3"	France	$15.00-20.00	The cup is white. The front has a picture of Baby Smurf sitting next to a teddy bear. Baby is wearing a pink sleeper. He has his arm around a brown bear. Inside is a small picture of a ball and block. The back of the cups says "Les Schtroumpfs" in yellow letter.
172/018	Baby With Bear - Soup Cup And Saucer	Porcelaine De Lologne	1/1/98	White	Cup And Saucer		France	$22.50-30.00	The cup is white and round. The cup is 3" high and 4" Diam. The front of the cup has a picture of Baby Smurf sitting next to a teddy bear. Baby is wearing a pink sleeper. He has his arm around a brown bear. Inside is a small picture of a train. The back of the cups says "Les Schtroumpfs" in yellow letters. The cup sits on a white plate. The plate says "Les Schtroumpfs" in yellow letters on top. It has a picture of a train, pacifier, block and ball.
172/019	Baby With Bear - Cereal Bowl	Porcelaine De Lologne	1/1/98	White	Ceramic Cereal Bowl	6 1/2" Diam	France	$15.00-20.00	The bowl is white. In the center of the bowl is a picture of Baby Smurf sitting next to a teddy bear. Baby is wearing a pink sleeper. He has his arm around a brown bear. The bowl inside says "Les Schtroumpfs" in yellow letters. It has a picture of a train, pacifier, block and ball. The bowl has a ridge on top.
172/020	Smurf Bumping Soccer ball With Head - Cup	Porcelaine De Lologne	1/1/98	White	Ceramic Coffee Cup	3"	France	$15.00-20.00	The cup is white. The front has a Smurf bumping the soccer ball with his head. The Smurf is wearing a yellow shirt and red shorts. Inside the cup is a small picture of a soccer ball. The back of the cup says "Les Schtroumpfs" in yellow letters.
172/021	Smurf Bumping Soccer ball With Head - Bowl	Porcelaine De Lologne	1/1/98	White	Ceramic Bowl	5 1/2" Diam	France	$15.00-20.00	The bowl is white. The front has a Smurf bumping the soccer ball with his head. The Smurf is wearing a yellow shirt and red shorts. In side the bowl is a small picture of a soccer ball. The back of the bowl says "Les Schtroumpfs" in yellow letters.
172/022	Smurf Bumping Soccer ball Soup Cup And Saucer	Porcelaine De Lologne	1/1/98	White	Cup And Saucer		France	$22.50-30.00	The cup is white and round. The cup is 3" high and 4" Diam. The front of the cup has a Smurf bumping a soccer ball with his head. The Smurf is wearing a yellow shirt and red shorts. Inside is a small picture of a soccer ball. The back of the cups says "Les Schtroumpfs" in yellow letters. The cup sits on a white plate. The plate says "Les Schtroumpfs" in yellow letters on top. It has a picture of a trophy, music notes and soccer ball.
172/023	Smurfette Kissing A Smurf - Cup	Porcelaine De Lologne	1/1/98	white	Ceramic Mug	3"	France	$15.00-20.00	The cup is white. The front has Smurfette giving a Smurf a kiss. The Smurf looks dazzled. He has hearts floating above his head. Smurfette is wearing a pink dress. In side the cup is a small picture of two lady bugs. The back of the cup says "Les Schtroumpfs" in yellow letters.
172/024	Smurfette Giving A Smurf A Kiss - Bowl	Porcelaine De Lologne	1/1/98	White	Ceramic Bowl	5 1/4"	France	$15.00-20.00	The cup is white. The front has Smurfette giving a Smurf a kiss. The Smurf looks dazzled. He has hearts floating above his head. Smurfette is wearing a pink dress. In side the cup is a small picture of the back of a squirrel. The back of the bowl says "Les Schtroumpfs" in yellow letters.
172/025	Smurfette Giving A Smurf A Kiss - Soup Cup And Saucer	Porcelaine De Lologne	1/1/98	White	Ceramic Cup & Saucer		France	$22.50-30.00	The cup is white and round. The cup is 3" high and 4" Diam. The front of the cup has Smurfette giving a Smurf a kiss. The Smurf looks dazzled. He has hearts floating above his head. Smurfette is wearing a pink dress. In side the cup is a small picture of the back of a squirrel. The back of the cup says "Les Schtroumpfs" in yellow letters. The cup sits on a white plate. The plate says "Les Schtroumpfs" in yellow letters on top. It has a picture of a squirrel and ladybugs on it.
172/026	Smurfette And Baby On A Snowboard - Cup	Porcelaine De Lologne	1/1/98	White	Ceramic Cup	3"	France	$15.00-20.00	The cup is white. On the front it has Smurfette and Baby sitting on a wood sled that is on a snowboard. Smurfette is wearing a green dress, pink scarf and pink boots. Baby is in a pink outfit. Smurfette and Baby are holding a snowball. Inside the cup is a small picture of a squirrel. The cup says "Les Schtroumpfs" in yellow letters on top.
172/027	Smurfette And Baby On A Snowboard - Bowl	Porcelaine De Lologne	1/1/98	White	Ceramic Bowl	5 1/4"	France	$15.00-20.00	The bowl is white. On the front it has Smurfette and Baby sitting on a wood sled that is on a snowboard. Smurfette is wearing a green dress, pink scarf and pink boots. Baby is in a pink outfit. Smurfette and Baby are holding a snowball. Inside the bowl is a small picture of a 2 ladybugs. The bowl says "Les Schtroumpfs" in yellow letters on top.
172/028	Smurf Skateboarding - Cup	Porcelaine De Lologne	1/1/98	white	Ceramic Mug	3 "	France	$15.00-20.00	The cup is white. On the front it has a Smurf on a skateboard. The Smurf is wearing a yellow and red hat, brown shirt, blue gloves, green shorts, and red and blue shoes. He has a brown and yellow skateboard. Inside the cup is a small picture of a Smurf racing on the skateboard. The cup says "Les Schtroumpfs" in yellow letters on top.
172/029	Smurf Skateboarding - Bowl	Porcelaine De Lologne	1/1/98	White	Ceramic Bowl	5 1/4"	France	$15.00-20.00	The bowl is white. On the front it has a Smurf on a skateboard. The Smurf is wearing a yellow and red hat, brown shirt, blue gloves, green shorts, and red and blue shoes. He has a brown and yellow skateboard. Inside the bowl is a small picture of a Smurf racing on the skateboard. The cup says "Les Schtroumpfs" in yellow letters on top.
172/030	Smurfette Skateboarding - Cup	Porcelaine De Lologne	1/1/98	white	Ceramic Cup	3"	France	$15.00-20.00	The cup is white. It has a picture of Smurfette on a skateboard. Smurfette is wearing a pink shirt, green shorts, blue gloves, blue and red shoes. The skateboard is yellow with red wheels. Inside the bowl is a small picture of a Smurf racing on the skateboard. The cup says "Les Schtroumpfs" in yellow letters on top.

172/031	Smurfette Skateboarding - Bowl	Porcelaine De Lologne	1/1/98	white	Ceramic Bowl	5 1/4"	France	$12.38 - 16.50	The bowl is white. It has a picture of Smurfette on a skateboard. Smurfette is wearing a pink shirt, green shorts, blue gloves, and blue and red shoes. The skateboard is yellow with red wheels. Inside the bowl is a small picture of a Smurf racing on the skateboard. The bowl says "Les Schtroumpfs" in yellow letters on top.
172/032	Smurfette Skateboarding - Soup Cup And Saucer	Porcelaine De Lologne	1/1/98	white	Soup Bowl And Saucer		France		The cup is white and round. The cup is 3" high and 4" Diam. It has a picture of Smurfette on a skateboard. Smurfette is wearing a pink shirt, green shorts, blue gloves, blue and red shoes. The skateboard is yellow with red wheels. Inside the bowl is a small picture of a music note. The back of the cup says "Les Schtroumpfs" in yellow letters. The cup sits on a white plate. The plate says "Les Schtroumpfs" in yellow letters on top. It has a picture of a trophy, music notes and a soccer ball.$22.50-30.00
172/033	Smurf Playing Guitar And Smurfette Singing - Cup	Porcelaine De Lologne	1/1/98	white	Ceramic Cup	3 "	France	$15.00-20.00	The cup is white. The cup has a picture of Smurfette singing and a Smurf playing a guitar. Smurfette is wearing a red halter and red shorts. The Smurf is playing a brown and green guitar. The back of the cup says "Les Schtroumpfs" in yellow letters.
172/034	Smurf Playing Guitar/Smurfette Singing/ Soup Cup & Plate	Porcelaine De Lologne	1/1/98	white	Soup Cup And Plate		France	$22.50-30.00	The cup is white. The cup has a picture of Smurfette singing and a Smurf playing a guitar. Smurfette is wearing a red halter and red shorts. The Smurf is playing a brown and green guitar. Inside the cup are small music notes. The back of the cup says "Les Schtroumpfs" in yellow letters. The cup sits on a white plate. The plate says "Les Schtroumpfs" in yellow letters on top. It has a picture of a trophy, music notes and a soccer ball.
172/035	Smurf Playing A Record Player - Plate	IMA	1/1/96	White	Ceramic Plate	7 3/4"	Brussels	$11.00-15.00	The plate is white. The Smurf is playing a record player. The Smurf is snapping his fingers. The Smurf is wearing an orange and green shirt and shoes, pink headphones and blue pants. The plate has music notes all over it.
172/036	Disco Smurf - Bowl	Porcelaine De Lologne	1/1/96	White	Ceramic Bowl	6 1/2"	Brussels	$11.00-15.00	The bowl is white. It has different colored music notes all over it. Inside has a picture of a Smurf dancing. The Smurf is wearing a green and pink checkered shirt, red shorts , blue glasses and green shoes.

Set 44: Play Sets- Flocked Figures Jointed

044/001	The Workers	Bikin Express	1/1/88		Flocked Vinyl	3 1/2" high	China	$25.00-35.00	There are 3 Smurfs and Papa all wearing denim work clothes. The play set comes with an ax, a shovel, a sledge hammer and a saw.
044/002	Cleaning Up Crew	Bikin Express	1/1/88		Flocked Vinyl	3 1/2" high	China	$25.00-35.00	Papa, Smurfette and 2 other Smurfs are wearing white overalls. There is a ladder and 2 brooms in the play set.
044/003	The Circus	Bikin Express	1/1/88		Flocked Vinyl	3 1/2" high	China	$25.00-35.00	Baby Smurf is dressed in gold, shiny pants with white suspenders. There are 3 other Smurfs: 2 are dressed in silver shirts and shiny red pants, the other 1 is wearing a silver vest and shiny purple pants. The play set comes with a barbell and trapeze ring.
044/004	Seaside	Bikin Express	1/1/88		Flocked Vinyl	3 1/2" high	China	$25.00-35.00	The play set has a blue air mattress. Baby Smurf is in an inner tube. Smurfette and 2 other Smurfs are wearing swimsuits.
044/005	The Choir	Bikin Express	1/1/88		Flocked Vinyl	3 1/2" high	China	$25.00-35.00	Smurfette and 3 Smurfs are dressed in white, black and silver jazz clothes. The play set comes with microphone stand.
044/006	Baby, Smurfette And Papa	Bikin Express	1/1/88		Flocked Vinyl	3 1/2" high	China	$18.00-25.00	The play set comes with 3 Smurfs. Papa, Smurfette in a white dress and Baby Smurf in a white jumper.
044/007	2 Smurfs And Brainy	Bikin Express	1/1/88		Flocked Vinyl	3 1/2" high	China	$18.00-25.00	The play set comes with 3 figures. Brainy and 2 regular Smurfs.
044/008	Smurf	Bikin Express	1/1/88		Flocked Vinyl	3 1/2" high	China	$7.50-10.00	1 Regular Smurf packaged separate.
044/009	The Musicians	Bikin Express	1/1/88		Flocked Vinyl	3 1/2" high	China	$25.00-35.00	Smurfette is wearing a white dress with black spots. There are 2 Smurfs wearing silver outfits with a white and black coat. The third Smurf is wearing a shiny purple outfit with a white and black coat. The set comes with a drum, trumpet and piano.

Set 12: Play Sets- Pop-Up

012/001	Gargamel's Castle	W. Berrie	1/1/83	Blue	Pop-up Play set	14" x 9 1/3"	Singapore	$11.00-15.00	A board unfolds into Gargamel's castle. Made of thin cardboard.
012/002	Smurf Sports Village	W. Berrie	1/1/83		Pop-up Play set	14" x 9 1/3"	Singapore	$11.00-15.00	A board unfolds into a sports village. Made of thin cardboard.
012/003	Smurf Deluxe Play Village	Lakeside's	1/1/81		Pop-up Play set	14" x 9 1/3"	U.S.A.	$18.00-25.00	Like a paper doll set. There are 4 houses, a bridge, a well, a camp fire, trees, bushes, 36 Smurfs and a village plan. This is still on the sheets not punched out.
012/004	Smurf Play Village	W. Berrie	1/1/83		Pop-up Play set	14" x 9 1/3"	Singapore	$11.00-15.00	A board unfolds into 2 mushroom houses. There is a well and bridge. Made of thin cardboard.

Set 70: Play Sets- Regular

070/001	Garden	Schleich	1/1/80		Plastic Play set		Hong Kong	$11.00-15.00	A green table, bench and 2 stools. A brown ladder and wheel barrow.
070/002	Fences	Schleich	1/1/80		Plastic Play set		West Germany	$11.00-15.00	16 Green locking tree stumps and 8 large, 8 small garden fences.
070/003	Gate	Schleich	1/1/80		Plastic Play set		West Germany	$11.00-15.00	2 Tree trunks with a gate between them and a board across the top to hold a bell. A yellow flower is also included.
070/004	Mushroom With Flower	Schleich	1/1/80		Plastic Play set		West Germany	$11.00-15.00	A red mushroom with a bone color stem, a yellow flower and a gray rock.
070/005	Sailboat	Schleich	1/1/80		Plastic Play set		West Germany	$15.00-20.00	A yellow boat with a red floor and bench, and a white cloth sail.
070/006	Gas Station	Schleich	1/1/80		Plastic Play set		West Germany	$35.00-45.00	A gray gas pump with a white and gray canopy over the top. Petrol Co. stickers on the pump.
070/007	Wishing Well	Schleich	1/1/80		Plastic Play set		Hong Kong	$15.00-20.00	A gray well with a green grass base. A tan bucket and a dark brown winch.
070/008	Snail Carriage	Schleich	1/1/80		Plastic Play set		Hong Kong	$15.00-20.00	A red body snail with a yellow shell. A tan square wagon that hooks to the snail and a round yellow wagon.
070/009	Forklift	Schleich	1/1/80		Plastic Play set		West Germany	$35.00-45.00	A brown and gray forklift.
070/010	Conveyor Belt	Schleich	1/1/80		Plastic Play set		West Germany	$35.00-45.00	A brown conveyor with a green belt. The miller bags are not included in this one.
070/011	Gate/ Fence And Figure	Schleich	1/1/97		Plastic Play set		Brussels	$11.00-15.00	2 tree trunks with a gate between them and a board across the top to hold a bell. A yellow flower is also included. Included is a Smurf figure pushing a wheelbarrow. The Smurf has a bright blue body. 8 tree stumps and 9 fence boards.
070/012	Wishing Well And Figure	Schleich	1/1/97		Plastic Play set		Brussels	$11.00-15.00	A gray well with a green grass base. A tan bucket and a dark brown winch. A yellow flower. A brown table and 2 stools and bench. The figure is a Smurf pushing a lawnmower. The Smurf has a bright blue body.
070/013	Snail Cart And Figure	Schleich	1/1/97		Plastic Play set		Brussels	$11.00-15.00	A red body snail with a yellow shell. A tan square wagon that hooks to the snail and a round yellow wagon. A red mushroom with a bone color stem, green grass patch and a gray rock. Comes with a Smurf sitting ring style with his hands in the air. The Smurf has a bright blue body.

Set 187: Play Sets- Transforming

187/001	Daisy Wheel & Snappy	Irwin	1/1/96		Transforming Play set		Brussels	$15.00-20.00	The daisy wheel is pink, yellow and red. It has green seats and a green stem. The daisy wheel transforms from a ferris wheel to a merry-go-round. The play set comes with Snappy Smurf. Snappy is wearing a yellow t-shirt with a cloud and lightening bolt on the front. He is holding a brown flag.
187/002	Dining Out & Greedy	Irwin	1/1/96		Transforming Play set		Brussels	$15.00-20.00	The diner is a food stand. It has a blue base with a white counter. The canopy is the shape of a mushroom. The diner changes into a fishing hole. The fishing hole is blue and the Smurf stands on a brown plank. The play set comes with Greedy Smurf. Included is a fishing pole, fish, bridge, mushroom stool, tray seat and hot dog basket.
187/003	Picnic Wish & Sassette	Irwin	1/1/96		Transforming Play set		Brussels	$15.00-20.00	Sassette is dressed in a pink outfit. The wishing well is brown and gray with a red mushroom canopy. The wishing well changes to a picnic table with a canopy. Included is a water bucket, table cloth and picnic accessories.
187/004	Saturn Ride & astrosmurf	Irwin	1/1/96		Transforming Play set		Brussels	$18.00-25.00	Space ride flips into a swinging ship ride. The outside of the ship is blue with a red circular top. Inside the ship Astro's in a smaller gray ship and he can swing. When taken out and put on top of the round red base he spins around. Astro is wearing a white outfit.

Set 83: Plush- Kinder Eggs

083/001	Papa	Ferrero	1/1/95		plush	7" High	Brussels	$11.00-15.00	Papa has a light blue body. He is wearing a bright red hat and pants. He is made from a short velvet- like material. Papa's beard is short, white and fuzzy and he has a red mouth.
083/002	Farmer Smurf (Nat)	Ferrero	1/1/95		plush	7" High	Brussels	$11.00-15.00	Nat has a light blue body. He is made from a short velvet- like material. He has a yellow hat on. He is wearing tan pants with a tan sash connected to the pants and no shoes.
083/003	Normal Smurf	Ferrero	1/1/95		plush	7" High	Brussels	$11.00-15.00	Normal Smurf has a light blue body. He is made from a short velvet- like material. The Smurf is wearing a white hat and white pants.
083/004	Baby Smurf	Ferrero	1/1/95		plush	7" High	Brussels	$11.00-15.00	Baby has a light blue body. He is made from a short velvet- like material. Baby Smurf is in a sitting position and is wearing a pink sleeper and a pink hat.
083/005	Sassette	Ferrero	1/1/95		plush	7" High	Brussels	$11.00-15.00	Sassette has a light blue body. She is made from a short velvet- like material. Sassette has a pink outfit on. Sassette has orange braided hair.
083/006	Brainy	Ferrero	1/1/90		plush	7" High	Brussels	$37.50-50.00	Brainy has a light blue body. He is made from a short velvet- like material. He is wearing black cloth glasses.
083/007	Papa	Ferrero	1/1/90		plush	7" High	Brussels	$37.50-50.00	Papa has a light blue body. He is made from a short velvet- like material. He has a fuzzy white beard and no mouth. Papa has a red hat and red pants.
083/008	Papa Valentine	Ferrero	1/1/97		plush	7" High	Brussels	$18.00-25.00	The Papa is a soft fuzzy material. He has a red hat and red pants. There is a lot of candy in the package. The box has hearts on it. Papa is in a bag with red ribbon tied on top.
083/009	Smurf Valentine	Ferrero	1/1/97		plush	7" High	Brussels	$18.00-25.00	The Smurf is a soft fuzzy material. He has a white hat and white pants. There is a lot of candy in the package. The box has hearts on it. The Smurf is in a bag with blue ribbon tied on top.

Set 35: Plush Smurf & Smurfette Wardrobe

035/001	Smurf Jogging Suit	W. Berrie	1/1/83	Orange	Plush Smurf Clothes	Hong Kong	$9.00-12.00	Outfit fits a floppy plush Smurf #640. The jogging suit is orange, the shirt has a decal in the left corner of a Smurf jogging.
035/002	Smurf Blue Jeans And White Shirt	W. Berrie	1/1/83		Plush Smurf Clothes	Hong Kong	$9.00-12.00	Outfit fits a floppy plush Smurf #640. The jeans are dark blue with a rainbow belt. The shirt is white and has 5 Smurfs walking across the front, says keep on Smurfin'.
035/003	Smurf Tennis Outfit	W. Berrie	1/1/83		Plush Smurf Clothes	Hong Kong	$9.00-12.00	Outfit fits a floppy plush Smurf #640. The shirt is a white polo with a decal in the left front corner of a tennis player. The shorts are red. The sun visor is white with a red border.
035/004	Smurf Beach Outfit	W. Berrie	1/1/83		Plush Smurf Clothes	Hong Kong	$9.00-12.00	Outfit fits a floppy plush Smurf #640. The tank top is red with a decal of a Smurf surfing on the front. The shorts are white and there are red sunglasses.
035/005	Smurf Pajamas	W. Berrie	1/1/83		Plush Smurf Clothes	Hong Kong	$9.00-12.00	Outfit fits a floppy plush Smurf #640. The outfit is a white night shirt and cap with blue poke-a-dots. The front of the night shirt has a sleepwalker in the left corner.
035/006	Smurf Sweater	W. Berrie	1/1/83		Plush Smurf Clothes	Hong Kong	$9.00-12.00	Outfit fits a floppy plush Smurf #640. The outfit is a red and yellow knit sweater and hat. In the left corner there is a decal of a Smurf wearing a sweater and cap
035/007	Smurfette Cheerleader Outfit	W. Berrie	1/1/83		Plush Smurf Clothes	Hong Kong	$9.00-12.00	Outfit fits a floppy plush Smurfette #642. The dress is pink with a white skirt attached, with purple border on the neck and sleeves. White slippers with purple pompom's on the top. The pompoms are white, pink and purple.
035/008	Smurfette Sun Dress	W. Berrie	1/1/83		Plush Smurf Clothes	Hong Kong	$9.00-12.00	Outfit fits a floppy plush Smurfette #642. The sun dress and visor is white with pink border.
035/009	Smurfette Bib Jeans	W. Berrie	1/1/83		Plush Smurf Clothes	Hong Kong	$9.00-12.00	Outfit fits a floppy plush Smurfette #642. The shirt is yellow terry- cloth. The bibs are denim blue.
035/010	Smurfette Plaid Skirt And Shirt	W. Berrie	1/1/83		Plush Smurf Clothes	Hong Kong	$9.00-12.00	Outfit fits a floppy plush Smurfette #642. The shirt is white with a red bow. The skirt is red with black and green plaid.
035/011	Smurfette Sleeper	W. Berrie	1/1/83	Pink	Plush Smurf Clothes	Hong Kong	$9.00-12.00	Outfit fits a floppy plush Smurfette #642. The sleeper is pink and the top of it is flowered, there is a zipper going down the front.
035/012	Hug Your Smurf Shirt	TM	1/1/83	White	Plush Smurf Clothes		$9.00-12.00	The shirt is white with blue trim. The shirt says "Hug Your Smurf" in blue letters. The shirt fits a 10" plush.
035/013	Smurfette Pink Sun dress	W. Berrie/Peyo	1/1/83	Pink	Plush Smurf Clothes		$9.00-12.00	Outfit fits a floppy plush Smurfette #642. The dress is pink with 2 pink bows and little dark pink berries all over. The sleeves have white lace. There is a matching pair of panties and a purse. The purse has a dark pink "S" on the front.
035/014	Smurfs Out Of This World T-shirt	W. Berrie	1/1/81	White	Plush Smurf Clothes	Taiwan	$9.00-12.00	The t-shirt fits 22" floppy. The shirt is white with Astro Smurf on the front. Astro is wearing a white spacesuit.. The Smurf is walking on an orange surface with orange rocks in the background. There are stars around the Smurf. The shirt says "Smurfs Out Of This World" in orange and black letters.
035/015	Let's Be Friends	W. Berrie	1/1/81	White	Plush Smurf Clothes	Taiwan	$9.00-12.00	The t-shirt fits 22" floppy. The shirt is white with a Smurfette and 2 Smurfs walking in the grass. The shirt says "Let's Be Friends" in yellow letters.
035/016	Let's Play!	W. Berrie	1/1/81		Plush Smurf Clothes	Taiwan	$9.00-12.00	The t-shirt fits 14" floppy. The shirt is white with a Smurf and 3 sport Smurfs on it. One Smurf is holding a basketball. A Smurf is carrying a baseball bat. The third is carrying a football.
035/017	Love Love Love T-shirt	W. Berrie	1/1/83	White	Plush Smurf Clothes		$9.00-12.00	The shirt is white. The front has a Smurf handing Smurfette a bunch of white daisies. There is a big red heart behind the Smurf and another big red heart behind Smurfette.

Set 29: Plush Smurfs

029/001	Devil Smurf	Applause	1/1/85	Red	Plush	10" High	Korea	$15.00-20.00	A red devil standing with black wings and a tail that looks like an arrow.
029/002	Papa Sea Captain	W. Berrie	1/1/83		Plush	10" High	Korea	$11.00-15.00	Papa Smurf in standing position wearing a white sea captains coat and hat and red pants.
029/003	Indian Brave Smurf	W. Berrie	1/1/83		Plush	10" High	Korea	$11.00-15.00	A Smurf in a standing position wearing tan pants, brown shoes and a tan head band with red feathers.
029/004	Clown Smurf	Applause	1/1/83		Plush	10" High	Korea	$15.00-20.00	A Smurf in a standing position, wearing a light green suit with yellow button, white ruffles around his neck, white shoes and he has a big red nose.
029/005	Cupid Smurf	W. Berrie	1/1/82		Plush	10" High	Korea	$11.00-15.00	A Smurf in a standing position wearing white wings, a white sash holding a red heart.
029/006	Ice Skater Smurf	Applause	1/1/80		Plush	7 3/4" High	Korea	$11.00-15.00	Blue Smurf with white hat and pants and silver metal ice blades on bottom of his feet.
029/007	Graduation Smurf	Applause	1/1/83		Plush	7" High	Korea	$9.00-12.00	A Smurf wearing a dark blue cap and shirt the shirt says "Class of '83". The Smurf is in a sitting position.
029/008	Vanity Smurf	Applause	1/1/83		Plush	6 1/2" High	Korea	$6.00-8.00	A Smurf wearing white pants and hat, holding a red mirror in his left hand. The Smurf is in a sitting position.
029/009	Leprechaun Smurf	Applause	1/1/84		Plush	6 1/2" High	Korea	$9.00-12.00	The Smurf is wearing a green hat, jacket and pants and is holding a four leaf clover in his right hand. The Smurf is in a sitting position.
029/010	Cowboy Smurf	Applause	1/1/83		Plush	7" High	Korea	$6.75-9.00	The Smurf is wearing a brown cowboy hat, a brown vest and boots, and red pants. The Smurf is in a sitting position.
029/011	Policeman Smurf	Applause	1/1/83		Plush	7" High	Korea	$6.75-9.00	The Smurf is wearing a black hat, coat and white pant. The Smurf is in a sitting position.
029/012	Easter Bunny Smurf	Applause	1/1/83		Plush	7" High	Korea	$6.00-8.00	The Smurf is dressed in a white Easter bunny suit. The Smurf is in a sitting position.
029/013	Ballerina Smurfette	Applause	1/1/82		Plush	7" High	Korea	$7.50-10.00	Smurfette is wearing a pink ballerina dress and shoes. She is in a sitting position.
029/014	Easter Bunny Smurfette	Applause	1/1/83		Plush	7" High	Korea	$6.00-8.00	Smurfette is wearing a pink Easter bunny outfit. She is in a sitting position.
029/015	Brainy Smurf	Applause	1/1/83		Plush	7" High	Korea	$6.00-8.00	The Smurf is wearing white pants, hat and is wearing black glasses. The Smurf is in a sitting position.
029/016	Gift Smurf	Applause	1/1/83		Plush	7" High	Korea	$6.00-8.00	A Smurf wearing white hat and pants, holding a pink gift box in both hands. The Smurf is in a sitting position.
029/017	Small Smurfette	Applause	1/1/81		Plush	6 1/2" High	Korea	$4.50-6.00	Smurfette is wearing a white dress that has pictures of Smurfette on the dress. Smurfette is in a sitting position.
029/018	Small Papa	Applause	1/1/81		Plush	6 1/2" High	Taiwan	$4.50-6.00	Papa is wearing a red hat and pants. He is in a sitting position.
029/019	Santa Smurf	Applause	1/1/81		Plush	9 1/2" High	Taiwan	$6.00-8.00	The Smurf is wearing red Santa outfit with white pants and is carrying a red sack on his back. The Smurf is in a sitting position.
029/020	Student Smurfette	Applause	1/1/83		Plush	9" High	Korea	$9.00-12.00	Smurfette is standing, wearing a white dress and panties, holding a pencil in one hand and a notepad in the other hand.
029/021	Elf Smurf	Applause	1/1/82		Plush	6" High	Korea	$3.50-5.00	The Smurf is wearing a red Santa hat and holding a candy cane in both hands. The Smurf is in a sitting position.
029/022	Baby Girl Smurf	Applause	1/1/84		Plush	!0 1/2" High	Korea	$11.00-15.00	The Smurf is wearing a pink dress and bonnet with a white flowery jumpsuit under the dress. The Baby Smurf is in a sitting position.
029/023	Baby Boy Smurf	Applause	1/1/84		Plush	!0 1/2" High	Korea	$11.00-15.00	The Baby Smurf is wearing a yellow jumpsuit with a white flowery shirt underneath the jumpsuit, he also has a yellow hat on. The Baby Smurf is in a sitting position.
029/024	Nat	Applause	1/1/88		Plush	8 1/2" High	China	$10.50-14.00	Nat is wearing a tan hat, pants and sash. He is in a sitting position.
029/025	Snappy	Applause	1/1/88		Plush	8 1/2" High	China	$10.50-14.00	Snappy is in a sitting position, wearing a yellow shirt with a cloud of lightening on it.
029/026	Slouchy	Applause	1/1/88		Plush	8 1/2" High	China	$10.50-14.00	Slouchy is wearing a white floppy hat and a red shirt. He is in a sitting position.
029/027	Medium Sassette	Applause	1/1/88		Plush	8 1/2" High	China	$10.50-14.00	Sassette is in a sitting position wearing a dark pink pants with suspenders.
029/028	Small Sassette	Applause	1/1/82		Plush	6 1/2" High		$6.00-8.00	Sassette is in a sitting position wearing a light pink pants with suspenders.
029/029	Gargamel	Applause	1/1/82		Plush	15 1/2" High	Korea	$15.00-20.00	Gargamel has a black outfit on and red pants.
029/030	Azreal	Virca	1/1/82	Orange	Plush	8" High	Korea	$11.00-15.00	Azreal is standing and he is orange with white under his neck.
029/031	Baseball Smurf	Applause	1/1/82		Plush	12" High	Korea	$9.00-12.00	Floppy Smurf is wearing a white with red trim shirt (Smurfs 3), white with red trim baseball cap, white pants and is holding a baseball in his right hand.
029/032	Soccer Smurf	Applause	1/1/82		Plush	12" High	Korea	$9.00-12.00	Floppy soccer Smurf is wearing a red with yellow trim shirt (Smurferoos), white hat, pants and has a white and black soccer ball on his right foot.
029/033	Football Smurf	Applause	1/1/81		Plush	12" High	Korea	$9.00-12.00	Floppy football Smurf is wearing a white jersey with red sleeves, white hat, pants and is holding a brown football in his right hand.
029/034	Johan	Applause	1/1/83		Plush	13" High	Korea	$15.00-20.00	Johan is standing with his hands at his side, wearing a blue cape, yellow shirt with white sleeves, red pants and brown shoes
029/035	Peewit	Applause	1/1/83		Plush	9" High	Korea	$13.50-18.00	Peewit is standing with his hands out at his sides and is wearing teal pants, teal shirt with white sleeves and orange shoes.
029/036	Lover Smurf & Smurfette	Applause	1/1/82		Plush	11" High	Korea	$15.00-20.00	A Smurf and Smurfette are standing with there arms around each other hugging.
029/037	Baby Boy Smurf Crawling	Applause	1/1/82	Yellow	Plush	5 1/2" High	Korea	$9.00-13.00	Baby Smurf is in a yellow sleeper with a yellow bonnet holding a pink and blue rattle in his right hand. Baby is crawling.
029/038	Baby Girl Smurf Crawling	Applause	1/1/82	Pink	Plush	5 1/2" High	Korea	$9.00-13.00	Baby Smurf is in a pink sleeper with a pink bonnet holding a pink and blue rattle in his right hand. Baby is crawling.
029/039	Teenager Smurf	Applause	1/1/93		Plush	10 1/2" High	Brussels	$13.50-10.00	Teenager Smurf is sitting, wearing a red shirt, lime green pants, purple and pink shoes and a white hat
029/040	Papa Smurf	Applause	1/1/82		Plush	11" High	China	$9.00-13.00	A regular Smurf in a sitting position but his eyes are close together looking up.
029/041	Classic Smurfette	Applause	1/1/85		Plush	10" High	Korea	$9.00-13.00	Tag says Classic Smurf. Smurfette has a satin looking white dress with ruffles. Smurfette's hair is loose, not braided, and her eyes are close together and embroidered on.
029/042	Mini Smurf	Applause	1/1/81		Plush	5 1/2" High	Korea	$3.00-4.00	A regular Smurf in a sitting position.
029/043	Jumbo Papa	Applause	1/1/82		Plush	3 Foot 8" High		$55.00-75.00	Jumbo Papa in a sitting position.
029/044	Large Clown	Applause	1/1/93		Plush	17 "High	Brussels	$37.50-50.00	Clown is sitting, wearing a yellow suit with blue buttons, white ruffles around his neck, white gloves, a dark blue hat with different color berries, an orange tassel on the tip of his hat and a big red nose.
029/045	Large Floppy Smurf	Applause	1/1/80		Plush	22" High	Korea	$15.00-20.00	Large floppy Smurf.
029/046	Jumbo Smurf	Applause	1/1/82		Plush	3 Foot 8" High		$45.00-65.00	A jumbo Smurf with a lighter blue body color.
029/047	XLarge Smurf	Applause	1/1/82		Plush	2 Foot		$25.00-35.00	A blue Smurf with round eyes and a cloth mouth.
029/048	Guam Smurf	Applause	1/1/82		Plush	19 " High		$30.00-40.00	The Smurf is sitting . He's got almond eyes that are slanted. His arms are thinner then a regular Smurf and his hat is narrower on top.
029/049	XLarge Papa	Applause	1/1/82		Plush	2 Foot		$15.00-20.00	A floppy Papa Smurf, his body is made from a nylon material, he has a furry hat and pants on.
029/050	XLarge Smurfette	Applause	1/1/82		Plush	2 Foot 1" High		$25.00-35.00	Smurfette has a dark blue body, dark yellow hair and is wearing a white with blue flower dress. Smurfette has almond eyes and a hard vinyl mouth. (possible handmade). Smurfette has floppy legs and is not 2 foot high when sitting.
029/051	XLarge Green Smurf	Applause	1/1/82		Plush	2 Foot 9" High		$11.00-15.00	A green Smurf with a red hat and pants, with yellow braided hair.
029/052	XLarge Yellow Smurf	Applause	1/1/82		Plush	2 Foot 4" High		$11.00-15.00	A yellow Smurf with white pants and hat.

029/053	Jumbo Smurfette	Ganz Bros.	1/1/82	Plush	2 Foot 5 Inches	Canada	$55.00-75.00	Smurfette is wearing a white dress that has pictures of Smurfette on the dress. Smurfette is in a sitting position.
029/054	Jumbo Spy Smurf	Ganz Bros.	1/1/82	Plush	3 Foot 5 Inches	Canada	$55.00-75.00	Spy Smurf is in a sitting position wearing a yellow cape, red mittens and boots, white pants and hat, and a black mask over his eyes.
029/055	Medium Smurfette	Applause	1/1/81	Plush	8" High	Korea	$4.50-6.00	Smurfette is wearing a white dress that has pictures of Smurfette on the dress. Smurfette is in a sitting position.
029/056	Large Smurfette	Applause	1/1/81	Plush	10 1/2" High	Korea	$6.00-8.00	Smurfette is wearing a white dress that has pictures of Smurfette on the dress. Smurfette is in a sitting position.
029/057	Small Smurf	Applause	1/1/80	Plush	7" High	Korea	$3.00-4.00	A regular Smurf in a sitting position.
029/058	Medium Smurf	Applause	1/1/80	Plush	8" High	Korea	$4.50-6.00	A regular Smurf in a sitting position.
029/059	Large Smurf	Applause	1/1/79	Plush	10 1/2" High		$6.00-8.00	A regular Smurf in a sitting position.
029/060	Mini Floppy Smurf	Applause	1/1/81	Plush	7 1/2" Long	Korea	$5.00-7.00	Mini floppy Smurf.
029/061	Medium Floppy Smurf	Schleich	1/1/81	Plush	13 1/2"	Korea	$7.50-10.00	Medium floppy Smurf.
029/062	Large Papa	Applause	1/1/81	Plush	10 1/2" High	Korea	$9.00-12.00	Papa is wearing a red hat and pants. He is in a sitting position.
029/063	Medium Papa	Applause	1/1/81	Plush	8" High	Korea	$7.50-10.00	Papa is wearing a red hat and pants. He is in a sitting position.
029/064	Mini Floppy Papa	Applause	1/1/81	Plush	7 1/2" Long	Korea	$6.00-8.00	Mini floppy Papa.
029/065	Smurf Holding A Baby	Applause	1/1/81	Plush	11" High	Korea	$10.50-14.00	A Smurf sitting, holding a mini Smurf in both hands.
029/066	Mini Spy Smurf	Ganz Bros.	1/1/83	Plush	6 1/2" High	Canada	$7.50-10.00	Spy Smurf is in a sitting position wearing a yellow cape, red mittens and boots, white pants and hat, and a black mask over his eyes.
029/067	Large Papa	I.M.P.S.	1/1/95	Plush	11" High	Brussels	$11.00-15.00	Papa is in a sitting position. He is wearing a red hat and red pants. His beard is a finer material then Papa's. He has small eyes with black dots in the middle and black eyebrows that are like an upside down V. Papa looks like he's frowning but he has a smile on his lips.
029/068	Large Smurf	I.M.P.S.	1/1/95	Plush	10" High	Brussels	$11.00-15.00	The Smurf is in a sitting position. He is wearing a white hat and white pants. He has small eyes with black dots in the middle and black eyebrows that are like an upside down V. The Smurf looks like he's frowning but he has a smile on his lips.
029/069	Large Smurfette	I.M.P.S.	1/1/95	Plush	11" High	Brussels	$11.00-15.00	Smurfette is in a sitting position. She is wearing a white dress and underpants with orange polka dots. Smurfette has yellow yarn hair with pink ribbons on each side. Smurfette is wearing a white hat and white shoes. She has small eyes with black dots in the middle and black eyebrows.
029/070	Mini Gargamel	W. Berrie	1/1/83	Plush	5 1/2"	Korea	$3.00-4.00	Gargamel has a black outfit on and red pants. Gargamel is mini size and floppy.
029/071	Baby Smurf	Hasbro	1/1/84	Plush	12" High	Hong Kong	$22.50-30.00	Baby has a blue soft-vinyl face, hands and tail. Baby has a cloth body. He is wearing a white sleeper that says Baby Smurf on the front in blue letters.
029/072	Tummy Huggin Papa Smurf	Irwin Toys	9/7/96	Plush	11" High	China	$18.00-25.00	Papa is wearing a red hat and pants. Papa has a soft hair- like white beard. He is in a sitting position. Push Papa's tummy and he plays a smurfy melody.
029/073	Hug-a-Smurf Boy	Irwin Toys	9/7/96	Plush	10" High	China	$11.00-15.00	The Smurf is a bright blue cloth- like material. He is standing, wearing a lime green hat, dark blue pants, white shoes with black laces and white shirt with a red "S" on the front.
029/074	Hug-a-Smurf Smurfette	Irwin Toys	9/7/96	Plush	10" High	China	$11.00-15.00	Smurfette is a bright blue cloth- like material. Smurfette is standing, wearing a bright pink shirt, dark blue bibs with suspenders, a white hat and red shoes with white laces. Smurfette's hair is yellow yarn. She has black eyelashes and eyebrows.
029/075	Hug-a-Smurf Papa	Irwin Toys	11/1/96	Plush	10" High	China	$11.00-15.00	Papa is a bright blue cloth- like material. He is standing, wearing a red hat, red pants and a light blue shirt with yellow and dark blue daisy on. Papa has a short fuzzy white beard. Papa has yellow sunglasses with black lenses. The sunglasses look like there are on upside down.
029/076	Berry Lovin' Baby Smurf	Applause	1/1/82	Plush		Brussels	$25.00-35.00	Baby Smurf is wearing a cloth pink outfit. He has a plastic face and a plastic pink hat. Baby says 3 different things when you squeeze his hand. Baby's face turns colors when you feed him. Baby comes with a bib, a bowl of blueberries, a spoon and a wash cloth.
029/077	White Smurf			Plush	2Ft. 4In	Germany	$7.50-10.00	The Smurf has a white face and white pants. His hat and tummy are navy blue. The Smurf has blue eyes and green eyebrows with long black eyelashes painted on. He is a hard plush.
029/078	Hard Large Papa			Plush	20"	Germany	$15.00-20.00	Papa is in a standing position. He is stuffed with a very hard stuffing. He has a red hat and red pants. His body is a light blue. He has a short soft white beard. He has white oval- shaped eyes with black circles for his pupils. He has thick black eyebrows.
029/079	Hard Large Smurf			Plush	20"	Germany	$15.00-20.00	The Smurf is in a standing position. He is stuffed with a very hard stuffing. He has a white hat and white pants. His body is a light blue. He has a black mouth with red material for a tongue. He has white oval- shaped eyes with black circles for his pupils. He has thick black eyebrows.
029/080	Angel Smurfette	W. Berrie/Peyo	1/1/83	Plush	6" High	Korea	$11.00-15.00	Smurfette is sitting, wearing a plain white sash dress. She is holding a brown plastic bow with a red arrow in her right hand. Smurfette has 2 white wings on her back.
029/081	Mini Smurfette	W. Berrie/Peyo	1/1/82	Plush	5" High		$3.00-4.00	Smurfette is wearing a white dress that has pictures of Smurfette on the dress. Smurfette is in a sitting position.
029/082	Large Papa			Plush	2Foot 2Inches		$35.00-45.00	Papa is in a standing position. He is stuffed with a very hard stuffing. He has a red hat and red pants. His body is a light blue. He has a short soft fuzzy white beard. He has white oval- shaped eyes with black circles for his pupils. He has a tail.
029/083	Small Teenager Smurf	Peyo	1/1/93	Plush	6" High	Brussels	$9.00-13.00	Teenager Smurf is sitting, wearing a red shirt, lime green pants, purple and pink shoes and a white hat.
029/084	Smurfette Cowgirl	Applause	1/1/84	Plush	6" High	Korea	$5.00-7.00	Smurfette is dressed as a cowgirl. She is wearing a white cowboy hat, white dress, red vest and a red bandanna around her neck. Smurfette has her hair in braids.
029/085	Mini Smurf	I.M.P.S.	1/1/97	Plush	6"	China	$3.50-5.00	The Smurf is in a sitting position. He is wearing a white hat and white pants. He has small eyes with black dots in the middle and black eyebrows that are like an upside down V. The Smurf looks like he's frowning but he has a smile on his lips.
029/086	Sleepy Smurf	Applause	1/1/84	Plush	6 1/2"	Korea	$11.00-15.00	The Smurf is in sitting position. The Smurf is wearing a white striped nightshirt, pink slippers, white underpants and a white hat with a pink pompom on the tip. The Smurf has sleepy looking eyes.
029/087	Smurf Plush Musical (De Singende Schlumpf)	Applause	1/1/95	Plush	8 1/2"	Brussels	$18.00-25.00	The Smurf is wearing a white hat and pants. The Smurf has beady eyes. He is in a sitting position. You push the Smurf's tummy and he plays the Smurf theme song.
029/088	Papa Smurf Musical (De Singende Schlumpf)	Peyo	1/1/95	Plush	8 1/2"	Brussels	$18.00-25.00	Papa is wearing a red hat and pants. Papa has a soft hair like white beard. He is in a sitting position. Push Papa's tummy and he plays a Smurf theme song.
029/089	Have A Smurfy Birthday! Mini Plush	Applause		Plush	5"	China	$3.00-4.00	The Smurf has a velour- like body. He has embroidered eye's. He is wearing a yellow shirt that says "Have A Smurfy Birthday".
029/090	I Smurf You! Mini Plush	Applause		Plush	5"	China	$3.00-4.00	The Smurf has a velour- like body. He has embroidered eye's. He is wearing a purple shirt that says "I Smurf You!"
029/091	I Smurf You Plush	W. Berrie	1/1/82	Plush	10"	Korea	$5.00-7.00	The Smurf is sitting, wearing a white shirt that has blue trim. The shirt says "I Smurf You" in blue letters.
029/092	Happy Birthday Medium Smurf	W. Berrie	1/1/82	Plush	10"	Korea	$5.00-7.00	The Smurf is sitting, wearing a white shirt with blue trim. The shirt says Happy Birthday in blue letters.
029/093	Smurf In Pink Outfit With Suction Cups	Puppy	1/1/88	Plush	10"	China	$16.00 - 21.50	The Smurf is wearing a pink terry- cloth hat and pink terry- cloth sleeper. He has suction cups on his feet and hands to stick him to the windows.

Set 247: Pogs- Les Schtroumpfs

247/001	Papa With Cornucopia	Avimage	1/1/95	Pog	1 1/2" Diam.	France		The pog has Papa with cornucopia. There is a green star behind Papa. The background is blue.
247/002	Smurfette Holding A Bouquet	Avimage	1/1/95	Pog	1 1/2" Diam.			Smurfette is holding a yellow bouquet of flowers. The background is light blue.
247/003	Smurf Plucking Petals From A Flower	Avimage	1/1/95	Pog	1 1/2" Diam.	France		A Smurf is standing, pulling petals off a white daisy. There are stars and hearts floating above his head. The background is blue.
247/004	Baby With Blocks	Avimage	1/1/95	Pog	1 1/2" Diam.	France		Baby Smurf is wearing a pink sleeper. He is sitting, playing with multi- colored blocks. The background is blue.
247/005	Smurf Cooking	Avimage	1/1/95	Pog	1 1/2" Diam.	France		The Smurf is mixing something in a brown bowl. He is using wire beater. The Smurf is wearing a chefs hat and a white apron. The Smurf is wearing gray springs on his feet. He has a troubled look on his face. The background is blue with pink stars.
247/006	Golfer	Avimage	1/1/95	Pog	1 1/2" Diam.	France		The Smurf is swinging a golf club. The Smurf has on yellow and red plaid pants. The background is blue with yellow triangles.
247/007	Smurf Playing Trumpet	Avimage	1/1/95	Pog	1 1/2" Diam.	France		A Smurf is playing a yellow trumpet. There is a green star behind the Smurf. The background is blue.
247/008	Poet	Avimage	1/1/95	Pog	1 1/2" Diam.	France		The Smurf is holding a yellow script and a white feather pen. The background is blue with pink squares.
247/009	Sassette Walking	Avimage	1/1/95	Pog	1 1/2" Diam.	France		Sassette is walking with her hands at her side. Sassette is wearing pink bib overalls. Sassette has orange hair. The background is blue with green circles.
247/010	Slouchy	Avimage	1/1/95	Pog	1 1/2" Diam.	France		Slouchy is walking. The Smurf has on a red shirt, droopy white hat and white pants. The Smurf is covering his mouth with his hand. He is yawning. The background is blue with teal triangles.
247/011	Snappy	Avimage	1/1/95	Pog	1 1/2" Diam.	France		The Smurf is wearing a yellow shirt with a lightening blot on the front. The Smurf looks like he's marching. The background is blue with purple triangles.
247/012	Nat With Caterpillar	Avimage	1/1/95	Pog	1 1/2" Diam.	France		Nat is wearing brown pants with a brown sash and a yellow straw hat. He is standing with a butterfly on his finger. A green caterpillar is crawling by his feet. The background is blue with red stars.
247/013	Smurf Throwing A Football	Avimage	1/1/95	Pog	1 1/2" Diam.	France		The Smurf is throwing a football. The Smurf is wearing white shorts, white shoes and a red and yellow striped shirt. The background is sparkly gold with a blue border.
247/014	Baseball Smurf	Avimage	1/1/95	Pog	1 1/2" Diam.	France		The Smurf is swinging a baseball bat at a brown ball. The Smurf is wearing a red and yellow shirt, white baseball pants with red stripes and a white baseball cap. The background is blue with yellow circles.
247/015	Rider Smurf	Avimage	1/1/95	Pog	1 1/2" Diam.	France		The Smurf is wearing a red jacket, white shirt, white pants, black boots and a black hat. The Smurf is carrying a brown saddle. The Smurf has a yellow star behind him. The background is blue.
247/016	Smurfette Rider	Avimage	1/1/95	Pog	1 1/2" Diam.	France		Smurfette is wearing a red jacket, white pants, black boots and a black hat. Smurfette is carrying a brown saddle. The background is blue with orange stars.
247/017	Indian Smurf	Avimage	1/1/95	Pog	1 1/2" Diam.	France		The Smurf is standing, wearing a head dress. The feathers on the head dress are white with black tips. The Smurf has red stripes painted on his face. The Smurf is standing with his arms crossed over his chest. The background is a sparkly silver with a blue border.
247/018	Smurf Carrying A Birthday Cake	Avimage	1/1/95	Pog	1 1/2" Diam.	France		The Smurf is carrying a 2 layer birthday cake with 7 candles on top. The Smurf is wearing a chefs hat and a white apron. The background is blue with yellow squares.

247/019	Smurf Mixing Something	Avimage	1/1/95		Pog	1 1/2" Diam.	France		The Smurf is mixing something in a brown pan. He is using wire beaters and wearing a chefs hat and a white apron. The background is blue with green stars.
247/020	Smurf Eating Cake	Avimage	1/1/95		Pog	1 1/2" Diam.	France		The Smurf is walking, eating a piece of cake. The background is a sparkly silver with a blue border.
247/021	Smurf Carrying Fishing Equipment	Avimage	1/1/95		Pog	1 1/2" Diam.	France		The Smurf is walking, carrying a yellow fishing bag and a fishing pole. There is a red star behind the Smurf. The background is blue.
247/022	Tennis Smurf	Avimage	1/1/95		Pog	1 1/2" Diam.	France		The Smurf is swinging a yellow tennis racket at a white ball. The Smurf is wearing a white shirt, white shorts, white hat and white shoes. The background is blue with green circles.
247/023	Tennis Smurfette	Avimage	1/1/95		Pog	1 1/2" Diam.	France		Smurfette is swinging a pink tennis racket at a white ball. Smurfette is wearing a white dress, white shoes and a white hat. The background is blue with red circles.
247/024	Serenade Smurf	Avimage	1/1/95		Pog	1 1/2" Diam.	France		The Smurf is on his knees singing. He has on hand on his chest and the other out at his side. There are hearts above his head. The background is sparkly silver with a blue border.
247/025	Snorkel Smurf	Avimage	1/1/95		Pog	1 1/2" Diam.	France		The Smurf is wearing red goggles, yellow flippers and a yellow inner-tube. The Smurf has a pink star behind him. The background is blue.
247/026	Butterfly Catcher	Avimage	1/1/95		Pog	1 1/2" Diam.	France		The Smurf is wearing a green shirt, green shorts and a yellow hat. He is carrying a butterfly net. There is a butterfly in front of him. The Smurf has a yellow stars behind him. The background is blue.
247/027	Smurf Writing On Chalkboard	Avimage	1/1/95		Pog	1 1/2" Diam.	France		The Smurf is standing and doing a math problem on a chalkboard. The Smurf is wearing black glasses. The background is blue with pink squares.
247/028	Papa Waving	Avimage	1/1/95		Pog	1 1/2" Diam.	France		Papa is waving. The background is sparkly gold with a blue border.
247/029	Smurf Wearing A Scarf	Avimage	1/1/95		Pog	1 1/2" Diam.	France		A Smurf is standing with his hands behind him. A snowball is flying towards his face. He is wearing red mittens and a yellow scarf. The background is blue with purple circles.
247/030	Smurf Getting Hit By Snowball	Avimage	1/1/95		Pog	1 1/2" Diam.	France		A Smurf is getting hit in the back of the head by a snowball. The Smurf is wearing a red scarf. He looks mad. The background is blue with pink circles.
247/031	Grandpa Smurf	Avimage	1/1/95		Pog	1 1/2" Diam.	France		The Smurf has a long white beard. He is wearing a yellow hat and yellow pants with suspenders. Grandpa is holding a brown stick cane. The background is blue with orange triangles.
247/032	Baby With Teddy Bear	Avimage	1/1/95		Pog	1 1/2" Diam.	France		Baby Smurf is sitting, holding a brown and white bear. Baby is wearing a pink sleeper. The background is sparkly silver with a blue border.
247/033	Smurf Thinking	Avimage	1/1/95		Pog	1 1/2" Diam.	France		The Smurf is walking with a finger in front of his mouth. He has a weird expression on his face. The background is blue with a big purple star behind the Smurf.
247/034	Smurf Crying	Avimage	1/1/95		Pog	1 1/2" Diam.	France		A Smurf is standing, crying. A tear is on his check. He is holding a white handkerchief. The background is blue with orange squares.
247/035	Jokey	Avimage	1/1/95		Pog	1 1/2" Diam.	France		The Smurf is carrying a big yellow box with a red bow around. The background is blue with a big purple star behind the Smurf.
247/036	Smurf With A Harp	Avimage	1/1/95		Pog	1 1/2" Diam.	France		The Smurf is kneeling. He is holding a brown harp. The Smurf has one hand out at his side. The background is sparkly silver with a blue border.
247/037	Fireman	Avimage	1/1/95		Pog	1 1/2" Diam.	France		The Smurf is wearing a red fireman outfit with white pants and purple shoes. The Smurf is running with a gray hose. He is spraying water. The background is blue with green stars.
247/038	Gargamel	Avimage	1/1/95		Pog	1 1/2" Diam.	France		Gargamel is sitting on a stool. He is resting his head in his hands. The background is blue with green triangles.
247/039	Azreal	Avimage	1/1/95		Pog	1 1/2" Diam.	France		Azreal is running like he's going to pounce. The background is blue with purple stars.
247/040	Smurf Shivering	Avimage	1/1/95		Pog	1 1/2" Diam.	France		The Smurf is standing, shivering. He has his arms wrapped around him. He is wearing a red scarf. Snowflakes are falling. The background is sparkly gold.
247/041	Papa With Hands Behind Back	Avimage	1/1/95		Pog	1 1/2" Diam.	France		Papa is standing with his hands behind his back. He has a smile on his face. The background is blue with green squares.
247/042	Brainy	Avimage	1/1/95		Pog	1 1/2" Diam.	France		Brainy is standing, pointing. Brainy is wearing black glasses. The background is blue with green squares.
247/043	Smurf Standing Under Leaf	Avimage	1/1/95		Pog	1 1/2" Diam.	France		The Smurf is standing under a green leaf. It is raining. The background is blue with pink triangles.
247/044	Gargamel Is Mad	Avimage	1/1/95		Pog	1 1/2" Diam.	France		Gargamel is standing with his fist clenched. He has a mad expression on face. The background is sparkly silver with a blue border.
247/045	Smurf Carrying A Tennis Racket And A Ball	Avimage	1/1/95		Pog	1 1/2" Diam.	France		The Smurf is carrying a yellow tennis racket. He is carrying a white and red ball. The Smurf is wearing a red floral shirt and green shorts. The background is blue with yellow circles.
247/046	Smurfing Playing With Leaves	Avimage	1/1/95		Pog	1 1/2" Diam.	France		The Smurf is playing with green leaves. The background is blue and pink with triangles.
247/047	Smurf Holding A Saw	Avimage	1/1/95		Pog	1 1/2" Diam.	France		The Smurf is standing, holding a saw. He has a pencil behind his ear. The background is blue with red squares.
247/048	Smurfette With Flowers	Avimage	1/1/95		Pog	1 1/2" Diam.	France		Smurfette is standing with little white daisies around her. Her hands are out at her sides. The background is sparkly silver.
247/049	Smurfette Looking Troubled	Avimage	1/1/95		Pog	1 1/2" Diam.	France		Smurfette is standing, looking troubled. Her hand is by her mouth. The background is blue with green triangles.
247/050	Puppy	Avimage	1/1/95		Pog	1 1/2" Diam.	France		Puppy is brown. He is running. The background is blue with a big orange star behind puppy.

Set 277: Pogs- Schtroumpf

277/001	Cupid	Caprice Des Dieux	1/1/96	Light Blue	Pog	1 1/2" Diam.	Brussels		The Smurf has white wings. He is shooting a bow and arrow. The background is light blue. Set of pogs cost $18.00-25.00.
277/002	Angry Smurf With Mallet	Caprice Des Dieux	1/1/96	Red	Pog	1 1/2" Diam.	Brussels		The Smurf is holding a mallet in the air. He has a mean expression on his face. The background is red.
277/003	Smurf Lifting Dumbbell	Caprice Des Dieux	1/1/96	Orange	Pog	1 1/2" Diam.	Brussels		The Smurf is lifting a dumbbell over his head. The Smurf has a red heart tattoo on his arm. The background is orange.
277/004	Trumpet Smurf	Caprice Des Dieux	1/1/96	Pink	Pog	1 1/2" Diam.	Brussels		The Smurf is standing, playing a yellow trumpet. The background is pink.
277/005	Azreal	Caprice Des Dieux	1/1/96	Light Blue	Pog	1 1/2" Diam.	Brussels		Azreal looks like he's on the prowl. He is licking his chops. The background is light blue.
277/006	Azreal	Caprice Des Dieux	1/1/96	Light Green	Pog	1 1/2" Diam.	Brussels		Azreal looks like he's scratching. The background is light green.
277/007	Gargamel With Clenched Fists	Caprice Des Dieux	1/1/96	Yellow	Pog	1 1/2" Diam.	Brussels		Gargamel is standing with his fists clenched. The pog has his whole body on. The background is yellow.
277/008	Gargamel	Caprice Des Dieux	1/1/96	Red	Pog	1 1/2" Diam.	Brussels		Gargamel has his fists clenched up by his face. The pog only shows his upper body. The background is red.
277/009	Vanity	Caprice Des Dieux	1/1/96	Pink	Pog	1 1/2" Diam.	Brussels		The Smurf is standing, admiring himself in a mirror. The Smurf has a white flower on his hat. He is holding a hand mirror. The background is pink.
277/010	Smurf Running With A Cake	Caprice Des Dieux	1/1/96	Orange	Pog	1 1/2" Diam.	Brussels		A Smurf is running with a cake. He is licking his lips. The background is orange.
277/011	Papa	Caprice Des Dieux	1/1/96	Peach	Pog	1 1/2" Diam.	Brussels		Papa is standing, holding a finger in the air. The background is peach.
277/012	Smurf Hiding His Face	Caprice Des Dieux	1/1/96	Pink	Pog	1 1/2" Diam.	Brussels		The Smurf is hiding his face with one hand. With his other he is pointing at something with his thumb. The background is pink.
277/013	Grumpy	Caprice Des Dieux	1/1/96	Light Blue	Pog	1 1/2" Diam.	Brussels		The Smurf is standing with his hands fisted at his sides. He has a grumpy look on his face. The background is light blue.
277/014	Jumping Black Smurf	Caprice Des Dieux	1/1/96	Yellow	Pog	1 1/2" Diam.	Brussels		The black Smurf is jumping. He has a mad expression on his face. The background is yellow.
277/015	Smurfette Holding Daisy's	Caprice Des Dieux	1/1/96	Light Green	Pog	1 1/2" Diam.	Brussels		Smurfette is standing, holding a bouquet of daisies. The flowers are yellow. Smurfette has little red hearts floating above her head. The background is light green.
277/016	Smurf Smelling Daisy	Caprice Des Dieux	1/1/96	Pink	Pog	1 1/2" Diam.	Brussels		A Smurf is smelling a white daisy. The background is pink.
277/017	Brainy Running	Caprice Des Dieux	1/1/96	Peach	Pog	1 1/2" Diam.	Brussels		Brainy Smurf is running with a finger out in front of him. The Smurf has on black glasses. The background is peach.
277/018	Smurf With Red Heart	Caprice Des Dieux	1/1/96	Light Green	Pog	1 1/2" Diam.	Brussels		The Smurf is standing, leaning on a big red heart. He has 3 smaller hearts floating above his head. The background is light green.
277/019	Baby With Rattle	Caprice Des Dieux	1/1/96	Light Blue	Pog	1 1/2" Diam.	Brussels		Baby Smurf is sitting, wearing a pink sleeper. He is holding a red rattle in one hand. Baby is pointing in the air with the other.
277/020	Smurf Pulling A Pillow	Caprice Des Dieux	1/1/96	Yellow	Pog	1 1/2" Diam.	Brussels		The Smurf is walking a long pulling a pillow behind him. The Smurf is yawning. The background is yellow.
277/021	Emperor	Caprice Des Dieux	1/1/96	Light Blue	Pog	1 1/2" Diam.	Brussels		The Smurf is wearing a red and white robe, orange hat, yellow crown and orange pants. The Smurf is holding a red and white scepter. The background is light blue.
277/022	Burnt Smurf	Caprice Des Dieux	1/1/96	Peach	Pog	1 1/2" Diam.	Brussels		The Smurf is carrying a box that blew up. He has black soot all over his face and body. There are stars floating above his head. The background is peach.
277/023	Happy Smurf	Caprice Des Dieux	1/1/96	Light Blue	Pog	1 1/2" Diam.	Brussels		The Smurf is standing with his hands in front of his chest. He is licking his lips. He has a smile on his face. The background is light blue.
277/024	Serenade Smurf	Caprice Des Dieux	1/1/96	Pink	Pog	1 1/2" Diam.	Brussels		The Smurf is down on his knee's serenading. He has one hand on his chest and the other out at his side. There are 3 hearts floating above his head. The background is pink.
277/025	Jumping Smurf	Caprice Des Dieux	1/1/96	Peach	Pog	1 1/2" Diam.	Brussels		The Smurf has his hands in the air. He is jumping. The Smurf has a happy expression on his face. The background is peach.
277/026	Smurf Carrying Present	Caprice Des Dieux	1/1/96	Teal	Pog	1 1/2" Diam.	Brussels		The Smurf is running, carrying a yellow box. The box has red ribbon around it. The background is teal.
277/027	Smurfette Posing	Caprice Des Dieux	1/1/96	Pink	Pog	1 1/2" Diam.	Brussels		Smurfette is standing with a hand on her hair. She has a flirty look on her face. The background is pink.
277/028	Handy Smurf	Caprice Des Dieux	1/1/96	Blue	Pog	1 1/2" Diam.	Brussels		The Smurf is wearing white overalls. He is carrying a hammer, tool box and a ruler. He has a nail in his mouth and a pencil behind his ear. The background is blue.
277/029	Masked Smurf	Caprice Des Dieux	1/1/96	Yellow	Pog	1 1/2" Diam.	Brussels		The Smurf is wearing a black mask, white hat, green shirt, red cap and white pants. He looks like he's flying. The background is yellow.
277/030	Papa Walking	Caprice Des Dieux	1/1/96	Teal	Pog	1 1/2" Diam.	Brussels		Papa Smurf is walking with his hands behind his back. The background is teal.
277/031	Astro Smurf	Caprice Des Dieux	1/1/96	Light Blue	Pog	1 1/2" Diam.	Brussels		A Smurf is wearing an astronaut uniform. He has a clear helmet on. The Smurf looks like he's floating. The background is light blue.

Set 184: Postcards

184/001	Smurf Fishing Postcard	Horn/Peyo			Postcards	4" x 5 3/4"	Germany	$2.00-3.00	A Smurf is standing on a bank fishing. His fishing pole is wrapped around another Smurfs foot. The other Smurf was swimming in the pond. He is hanging upside down from the fishing line.
184/002	Papa And Smurfette Looking In A Mirror Postcard	Horn/Peyo			Postcards	4" x 5 3/4"	Germany	$2.00-3.00	Smurfette is standing, looking in a mirror. On the other side of the mirror is a Smurf sticking out his tongue at Smurfette. Papa is standing on the side looking on.

184/003	Smurf Carrying A Present Postcard	Horn			Postcards	4" x 5 3/4"	Germany	$2.00-3.00	Jokey Smurf is carrying an orange box with a red ribbon. The box has a lit fuse on the top. There are mushroom houses in the background. There are 3 Smurfs looking frantic.
184/004	Boomerang Smurf Postcard	Horn			Postcards	4" x 5 3/4"	Germany	$2.00-3.00	A Smurf is walking a hill holding a boomerang and rubbing his head. There are 2 Smurfs on the hill looking around innocently. There are mushroom houses in the background.
184/005	Papa In His Lab Postcard	Horn			Postcards	4" x 5 3/4"	Germany	$2.00-3.00	Papa is standing at a table dumping the contents of a test tube into a pot. In his other hand he is holding a green test-tube that has smoke coming out. There is a Smurf standing behind Papa holding a blown up bag. There are test tubes laying all over.
184/006	Smurf Painting A Picture	Horn			Postcards	4" x 5 3/4"	Germany	$2.00-3.00	A Smurf is painting a frame around a picture of a mushroom house. The Smurf is wearing a white smock. He is holding a painters palate. There is another Smurf standing with his hands behind his back looking at the picture.
184/007	Smurf Dreaming Postcard	Horn			Postcards	4" x 5 3/4"	Germany	$2.00-3.00	A Smurf is sitting in the grass dreaming. Another Smurf is painting flowers in the dreamers Smurfs cloud. A third Smurf is standing looking inquizative.
184/008	Picnic Postcard	Horn			Postcards	4" x 5 3/4"	Germany	$2.00-3.00	A Smurf is sitting on the grass eating sweets. Brainy is standing behind talking and a third Smurf is sneaking up with a mallet.
184/009	Runner Smurf Postcard	Horn			Postcards	4" x 5 3/4"	Germany	$2.00-3.00	A Smurf is running. His butt is on fire. Another Smurf is holding a timer and the Smurf next to him his holding a record book. There is an Olympic flag on the side.
184/010	Matador Smurf Postcard	Horn			Postcards	4" x 5 3/4"	Germany	$2.00-3.00	A Smurf is in a bull ring with a snail. He is standing, holding a red cloth. There are a bunch of Smurfs watching over the fence.
184/011	Schlumpfe Die Geshichte Machten	Schleich/Peyo			Postcards	4" x 5 3/4"	Germany	$2.00-3.00	The postcard is dark blue with an American flag in one corner. The postcard has Ben Franklin Smurf, Abe Lincoln Smurf, Thomas Edison Smurf, George Washington Smurf, Paul Revere Smurf and Christopher Columbus Smurf on the front. On the back of the postcard is says when each one was famous.
184/012	Comicfiguren -Preiskatalog-		1/1/97	Light Blue	Postcards	4" x 5 3/4"	Germany	$2.00-3.00	The postcard is an advertisement for a Smurf catalog. The front has a picture of the Smurf with the wreath, baseball pitcher, baseball Smurfette and baseball catcher. The back has ordering information.
184/013	Merry Christmas From The Smurfs	Peyo	1/1/96		Postcards	4" x 5 3/4"	Brussels	$2.00-3.00	The front of the postcard has Smurfette leaning out her window listening to the Smurf singing X-mas songs. Papa is leading the chorus. There is snow falling and a decorated tree.
184/014	Ik Smurf Van Jou!	P.E.T.	1/1/95		Postcards	4" x 5 3/4"	Brussels	$2.00-3.00	The postcard has an angel Smurf on it. The Smurf is holding a red heart with 2 brown arrows through it. In his other hand he is holding a brown bow. The angel has yellow wings. The front of the postcard says "Ik Smurf Van Jou!"
184/015	Van Harte Beterschap!	P.E.T.	1/1/95		Postcards	4" x 5 3/4"	Brussels	$2.00-3.00	The postcard has Papa sitting in a rocking chair outside a mushroom house on it. Papa is wearing a nightgown. He has a cane next to his chair. Smurfette is standing in front of him holding flowers out. Smurfette is saying "Voor Jou. Met de allerliefste wensen!"
184/016	Voor Je Verjaadag Heb Ik Een Prachtig Cadeau Gekocht	P.E.T.	1/1/95		Postcards	4" x 5 3/4"	Brussels	$2.00-3.00	The postcard has a Smurf standing on a ladder behind a door. The door is open and he is pouring a bucket of water over the top. The Smurf is saying "Hartelijk gefeliciteerd!"
184/017	Ben Jij Jarig?	P.E.T.	1/1/95		Postcards	4" x 5 3/4"	Brussels	$2.00-3.00	The postcard has a Smurf laying in a hammock. The hammock is tied to 2 palm trees. Water is in the background. The Smurf is holding a beverage. The Smurf is saying "Ben Jij Jarig? Hartelijk gefeliciteerd!"
184/018	Hartelijk Gevelisiteert	P.E.T.	1/1/95		Postcards	4" x 5 3/4"	Brussels	$2.00-3.00	The postcard has a Smurf standing outside a mushroom house. The Smurf is holding a green pencil. On one side of him is a brown stool and the other side has gifts. The Smurf has a worried expression on his face. The Smurf is saying "Hoezo? Niet Goed?! Js Het Soms Vandaag niet?"
184/019	Hartelyk Geejeliciteerd!	P.E.T	1/1/95		Postcards	4" x 5 3/4"	Brussels	$2.00-3.00	The postcard has a Smurf painting the word Hartelyk Geejeliciteerd! on the canvas. Smurfette is standing on a log in front of the canvas. A Smurf is standing behind painter Smurf laughing. Smurfette is saying "Hoera, Jij Bent Jarig!"
184/020	Jk Voel Me Helemaal Niet Lekker... Als Jij Ziek Bent.	P.E.T	1/1/95		Postcards	4" x 5 3/4"	Brussels	$2.00-3.00	The postcard has a Smurf standing at the end of a bed. The Smurf has a tired expression on his face. The bed is pink. The Smurf is saying "Jk voel me helemaal niet lekker... als jij ziek bent."
184/021	Jk Wist Dat Ik Iets Vergeten Was...	P.E.T.			Postcards	4" x 5 3/4"	Brussels	$2.00-3.00	The postcard has a Smurf standing outside his door at night. The Smurf is in white pajama's. He is holding a candle. The Smurf is saying "Jk Wist Dat Ik Iets Vergeten Was... Alsnog Hartelijk Gefeliciteerd!"
184/022	En Natuurlijk Heeft Weer Niemand …	P.E.T.	1/1/95		Postcards	4" x 5 3/4"	Brussels	$2.00-3.00	The postcard has an angry Smurf standing behind a tree. On the other side of the tree Papa, Smurfette and other Smurfs are sneaking up behind the angry Smurf. Smurfette is carrying a birthday cake. A Smurf is carrying a gift. Angry Smurf is saying "En Natuurlijk Heeft Weer Niemand Aan Mijn Veraardag GeSmurft!"
184/023	Wij Wensen Jou... Een Knallende Verjaardag Toe!	P.E.T.	1/1/95		Postcards	4" x 5 3/4"	Brussels	$2.00-3.00	The postcard has 2 Smurfs working on a yellow package. The Smurfs are fixing the box to blow up. One Smurf says "Wij Wensen Jou.." The other Smurf said "...Een knallende verjaardag toe! Hartelijk Gefeliciteerd!"
184/024	Nu Je Groot Bent Moet Je Erg Je Best Doen...	P.E.T.	1/1/95		Postcards	4" x 5 3/4"	Brussels	$2.00-3.00	The postcard has Brainy Smurf standing outside holding a finger in the air. He is saying "Nu Je Groot Bent Moet Je Erg Je Best Doen Op School, Netjes Eten Aan Tafel, Altijd Beleefd Zijn En..." A Smurf is sneaking up behind Brainy with a mallet.
184/025	Attention! Voici Des Nouvelles... Toutes Fraiches!	Chromovogue	1/1/96		Postcards	4" x 5 3/4"	France	$2.00-3.00	The postcard has a picture of 2 Smurfs holding another Smurf up to smell the center of a flower. There is a Smurf behind a house that has a hose hooked to the flower he is getting ready to pump the water out of a barrel. The postcard says "Attention! Voici Des Nouvelles... Toutes Fraiches!"
184/026	Manger.. Boire.. Dormir..	Chromovogue	1/1/96		Postcards	4" x 5 3/4"	France	$2.00-3.00	The postcard has a picture of a Smurf waving. The Smurf is holding a spoon in his other hand. There are pots, bottles and test tubes sitting all over the floor. The postcard says "Manger.. Boire.. Dormir.. Il n'y a pas de petits plaisirs! & A bientot pour une petite bouffe!"
184/027	Pour Moi... C'est Toi Qui Comptes!	Chromovogue	1/1/96		Postcards	4" x 5 3/4"	France	$2.00-3.00	The postcard has a picture of 3 Smurfs standing on steps. The Smurfs are wearing winner metals. The 1st place Smurf is smiling. The 2nd place Smurf is scrawling and the 3rd place Smurf looks like he's panting. The postcard says "Pour Moi... C'est Toi Qui Comptes!"
184/028	Alors, Ca Boume?	Chromovogue	1/1/96		Postcards	4" x 5 3/4"	France	$1.00-2.00	The postcard has a picture of a Smurf carrying a large present. The box is ticking. The box has danger stickers on it. The postcard says "Alors, Ca Boume?"
184/029	Rigoler, C'est La Sante	Chromovogue	1/1/96		Postcards	4" x 5 3/4"	France	$1.00-2.00	The postcard has a picture of a Smurf doubled over from laughing. The Smurf is holding his stomach. The background has mushroom houses. The postcard says "Rigoler, C'est La Sante".
184/030	Oyez, Bons Schtroumpfs!	Chromovogue	1/1/96		Postcards	4" x 5 3/4"	France	$1.00-2.00	The postcard has a picture of a Smurf playing the drums. He has the drum strap around his neck. There are mushroom houses in the background. The postcard says "Oyez, Bons Schtroumpfs! Voici les toutes dernieres nouvelles... Rblam Rataataclic Bam Toc Pouet".
184/031	Carte Porte-Bonheur!	Chromovogue	1/1/96		Postcards	4" x 5 3/4"	France	$1.00-2.00	The postcard has a picture of a Smurf moving a wand over a hat. The Smurf is dressed as a magician. He is wearing a blue hat with yellow stars, yellow cape with blue stars and blue pants with yellow stars. The Smurf is on a stage with an orange curtain behind him. The postcard says "Carte Porte-Bonheur!".
184/032	Voice Des Nouvelles Schtroumpfees En Express!	Chromovogue	1/1/96		Postcards	4" x 5 3/4"	France	$1.00-2.00	The postcard has a picture of a Smurf in a red airplane. There are letters flying behind the plane. A bird is above it. The Smurf looks like he is on a mission. The postcard says "Par Avion & Voice Des Nouvelles Schtroumpfees En Express!".
184/033	C'est Super!	Chromovogue	1/1/96		Postcards	4" x 5 3/4"	France	$1.00-2.00	The postcard has a picture of a happy Smurf on it. The Smurf has his hands straight out at his sides and one foot lifted. There are mushroom houses in the background. The postcard says "C'est Super!"
184/034	JE DECLARE	Chromovogue	1/1/96		Postcards	4" x 5 3/4"	France	$1.00-2.00	The postcard has a picture of a Smurf in a sports coat and top hat reading a long piece of paper. The Smurf is on stage. He has one hand out at his side. The postcard says "JE DECLARE sans discours et sans detour, que tu es vraiment..."
184/035	Operation Anti-Stress... Attention Au Surmenage!	Chromovogue	1/1/96		Postcards	4" x 5 3/4"	France	$1.00-2.00	The postcard has a picture of a Smurf laying in a hammock sleeping. The Smurf has a plate with cookies and a drink sitting on the ground beside him. The postcard says "Operation Anti-Stress... Attention au Surmenage!".
184/036	Line Petite Carte Par Porteur Special!	Chromovogue	1/1/96		Postcards	4" x 5 3/4"	France	$1.00-2.00	The postcard has a picture of a Smurf delivering mail. Puppy is standing behind him holding an envelope. The Smurf is wearing dark blue pants and a dark blue mailman hat. He is holding letters in each hand. He has a bag over his shoulder. There are houses in the background. The postcard says "Pour Toi & Line Petite Carte Par Porteur Special!".
184/037	Bon Pour Un Plein De Super Bonheur …	Chromovogue	1/1/96		Postcards	4" x 5 3/4"	France	$1.00-2.00	The postcard has a picture of a Smurf driving a Smurfmobile into a gas station. Another Smurf is holding a gas hose waiting to fill the car. A mushroom is a garage. The postcard says "Bon Pour Un Plein De Super Bonheur Et Que Ca Schtroumpfe!".I
184/038	Salut L'artiste! T'es La Creme De La Creme...	Chromovogue	1/1/96		Postcards	4" x 5 3/4"	France	$1.00-2.00	The postcard has a picture of a Smurf painting a picture of a cake. The Smurf is wearing a painter smock and holding a palate. He has a picture hanging on the wall of a sunrise. The postcard says "Salut L'artiste! T'es La Creme De La Creme...".
184/039	Voici Quelques Notes De Gaiete Et De Bonheur	Chromovogue	1/1/96		Postcards	4" x 5 3/4"	France	$1.00-2.00	The postcard has a picture of a Smurf band. Papa is standing in front conducting the band. The postcard says "Voici Quelques Notes De Gaiete Et De Bonheur Avec Cette Schtroumphonie interprretee tout specialement en ton Honneur."
184/040	L'ALchimie Du Bonheur	Chromovogue	1/1/96		Postcards	4" x 5 3/4"	France	$1.00-2.00	The postcard has a picture of a Smurf at a table mixing potions. There are books, test-tubes and jars all over the room. The postcard says "L'ALchimie Du Bonheur c'est pas sorcier! ect..."
184/041	Ordonnance Du Grand Schtroumpf	Chromovogue	1/1/96		Postcards	4" x 5 3/4"	France	$1.00-2.00	The postcard is a picture of a Smurf circus. It has Smurfs on a merry-go-round. It has a Smurf dressed as a clown. It has a Smurf and Smurfette standing by a table with a cake. The postcard says "Ordonnance Du Grand Schtroumpf Pour Garder la Forme et vivre Heureux, Schtroumpfer une bonne dose de fete matin, midi et soir!"
184/042	Lang Zal-ie Smurfen In De Gloriaaaa!	P.E.T. Productions	1/1/95		Postcards	4" x 5 3/4"	Brussels	$1.00-2.00	The postcard has three Smurfs holding music sheets and singing. Papa is standing, pointing to the ear plug in his ear. There is a bubble above Papa's head that says "Gelukkig heeft een Smurf mij oordopjes gegeven". There are three presents sitting in front of the Smurfs. The top of the postcard says "Lang Zal-ie Smurfen In De Gloriaaaa!".
184/043	Gefelicteerd Met Je Verjaardag!	P.E.T. Productions	1/1/95		Postcards	4" x 5 3/4"	Brussels	$1.00-2.00	The postcard has a picture of 3 Smurfs reading books. There are piles of books all around them on the floor. The top of the postcard says "Gefelicteerd Met Je Verjaardag!"
184/044	Herzlichen Gluckwunsch	Helly	1/1/94		Postcards	4" x 5 3/4"	Germany	$1.00-2.00	The postcard is white with a blue border. In the center are 2 Baby Smurfs. One is holding a teddy bear. The other Baby is holding a bottle and block. Both Baby's are sitting, wearing pink sleepers. The postcard says "Herzlichen Gluckwunsch".
184/045	Herzlich Willkommen Im Baby-Club	Helly	1/1/94		Postcards	4" x 5 3/4"	Germany	$1.00-2.00	The postcard is white with a blue border. In the center is a picture of Baby Smurf in 4 poses. All 4 Baby's are wearing a pink sleeper. One Baby is crawling, holding a rattle. A Baby is sitting, holding a bottle and block. A Baby is sitting, holding a teddy bear. Another Baby is laying on his tummy with his thumb in his mouth. The postcard says "Herzlich Willkommen Im Baby-Club".
184/046	Walibi Schtroumpf 1	Walibi			Postcard Pack	4" x 3"	France	$2.00-3.00	The postcard pack has 10 postcards that show rides at the Smurf park in France.

Set 149: Posters

149/001	Geniuses Are Rarely Tidy	S.E.P.P.	1/1/81	Yellow	Wall Poster	19" X 13"	U.S.A.	$7.50-10.00	The poster is of a Smurf laying on the floor of his bedroom. The Smurf is drawing a picture on a tan piece of paper with an orange feather pen. There are books, papers, food, records, balls, clothes, pencils, drinking cups and sports equipment scattered all over. The bed post are brown with a blue mattress, the bed is unmade. On the top of the poster it says "Geniuses Are Rarely Tidy" in black letters. The background is yellow. The poster is in a black frame.
149/002	Smile... Friday's Coming	S.E.P.P.	1/1/81	Orange	Wall Poster	19" X 13"	U.S.A.	$7.50-10.00	The poster has a Smurf sitting at a brown wood desk. The Smurf is writing on a piece of paper, there is a stack of papers on the desk next to him. The trash can is over flowing with papers and there are papers all over the floor. The Smurf has sweat running down his face. On the top of the poster it says "Smile... Friday's Coming" in black letters. The background is orange. The poster is in a black frame.
149/003	T.G.I.F.	S.E.P.P.	1/1/81	Red	Wall Poster	19" X 13"	U.S.A.	$7.50-10.00	The poster is of a Smurf standing, holding a gray mug of beer. On top of the poster it says "T.G.I.F." in black letters with yellow outline. The background is red. The poster is in a black frame.
149/004	Girls Can Do Anything!	S.E.P.P.	1/1/81	Orange	Wall Poster	19" X 13"	U.S.A.	$7.50-10.00	The poster is of Smurfette wearing a white skirt, a white shirt, a purple jacket and a blue tie. Smurfette is outside a brown wood door that has a sign that says president. Smurfette is carrying a brown briefcase. On top of the poster it says "Girls Can Do Anything!" in black letters. The background is orange. The poster is in a black frame.
149/005	Smile! It's Good For You	S.E.P.P.	1/1/81	Yellow	Wall Poster	19" X 13"	U.S.A.	$7.50-10.00	The poster is of 2 Smurfs standing, holding hands and laughing. The Smurfs look like they are jumping in the grass. 1 Smurf is wearing an orange tank top and red shorts and the other is wearing a red tank top and white shorts. On the top of the poster it says "Smile! it's good for you" in black letters with white outline. The background is yellow. The poster is in a black frame.
149/006	If You Need A Friend... I'm Here	S.E.P.P.	1/1/81	Green	Wall Poster	19" X 13"	U.S.A.	$7.50-10.00	The poster is of a Smurf standing, holding a yellow daisy behind his back. The Smurf has a big smile on his face. On the top of the poster it says "If You Need A Friend... I'm Here" in black letters. The background is green. The poster is in a black frame.
149/007	Want To Smurf Around?	S.E.P.P.	1/1/81	Red	Wall Poster	19" X 13"	U.S.A.	$7.50-10.00	The poster is of a love-struck Smurf sitting, holding a yellow daisy. There are red lip marks on his hat and foot. The Smurf has his tongue hanging out of his mouth. On the top of the poster it says "Want To Smurf Around?" in black letters. The background is red. The poster is in a black frame.
149/008	Adults Not Admitted Unless Accompanied By A Child	S.E.P.P.	1/1/81	Green	Wall Poster	19" X 13"	U.S.A.	$7.50-10.00	The poster is of a Smurf standing, holding a big red STOP sign. He has a hand in the air. He has his mouth open like he's yelling STOP. On the top of the poster it says "Adults Not Admitted Unless Accompanied By A Child" in black letters. The background is green. The poster is in a black frame.
149/009	It Isn't Easy Staying On Top	S.E.P.P.	1/1/81	Green	Wall Poster	19" X 13"	U.S.A.	$7.50-10.00	The poster is of Papa and 22 other Smurfs piled on top of each other to make a pyramid. One Smurf is falling off the top. There are trees behind them. On the top of the poster it says "It Isn't Easy Staying On Top" in orange puffy letters with black outline. The background is green. The poster is in a black frame.
149/010	Have You Hugged Your Smurf Today?	S.E.P.P.	1/1/81	Green	Wall Poster	19" X 13"	U.S.A.	$7.50-10.00	The poster has a Smurf sitting on a log with his head in his hands. There is a big tree stump behind him and a bunch of green bushes. There is an orange daisy laying on the ground in front of him and a pink and orange daisy laying behind him. On the top it says "Have You Hugged Your Smurf Today?" in yellow letters. The poster is in a black frame.
149/011	Rock & Roll Lives	S.E.P.P.	1/1/82	Orange	Wall Poster	19" X 13"	U.S.A.	$7.50-10.00	The poster is of 3 Smurfs and Smurfette in a band. Smurfette is in front wearing a red dress and shoes holding a microphone singing. A Smurf on each side is playing a red and yellow guitar and singing. A Smurf in back is playing the drums and singing. The drums are red and pink and in the center is a big blue "S". The Smurfs are playing on a brown wood stage. The background is several shades of orange and yellow in the form of a rainbow. The poster is in a black frame.
149/012	The Smurfs	S.E.P.P.	1/1/82		Wall Poster	19" X 13"	U.S.A.	$7.50-10.00	he poster is of Papa, Smurfette, Handy, Vanity, Hefty, Jokey, Grumpy, Brainy and Cook Smurf sitting by a tree getting their picture taken. There is a wood sign hanging in the tree above the Smurfs that says "The Smurfs". Gargamel and Azreal are hiding behind the tree watching. A Smurf is under the cloth of the camera taking the picture. The poster is in a black frame.
149/013	Our Teacher Is The Greatest!	S.E.P.P.	1/1/82	Orange	Wall Poster	19" X 13"	U.S.A.	$7.50-10.00	The poster has Brainy, Smurfette, Hefty and 4 other Smurfs standing in front of a chalkboard. Hefty is carrying a red apple. There are papers all over the wood floor. The chalkboard is green with a wood trim. On the chalkboard it says "Our Teacher Is The Greatest!" in white chalk. The background is orange. The poster is in a black frame.
149/014	Smurfettes Do It Better!	S.E.P.P.	1/1/82	Yellow	Wall Poster	19" X 13"	U.S.A.	$7.50-10.00	The poster has Smurfette playing baseball. Smurfette is wearing a white and pink uniform. Smurfette is swinging a brown baseball bat at a white ball. There is a Smurf behind Smurfette wearing a green and yellow umpire uniform. There is a brown fence around the baseball diamond and there are Smurfs looking over the fence cheering. On the top it says "Smurfettes Do It Better!" in white letters with a black outline. The background is yellow. The poster is in a black frame.
149/015	Smurf Ahoy!	S.E.P.P.	1/1/82	Light Blue	Wall Poster	19" X 13"	me.	U.S.A.	The poster is of 8 Smurfs, Papa and Smurfette in a sail boat. On the side of the boat it says S.S. SMURF. Papa is in the back of the boat driving. 1 Smurf is fishing, 1 is swimming with another diving into the water, a Smurf is sleeping on the boat, Smurfette is in front waving and 1 Smurf is looking grumpy. Brainy and a Smurf with a telescope are on top of the sail pole with another Smurf climbing up. On top it says " Smurf Ahoy!" in red letters with black outline. The poster is in a black fra$7.50-10.00
149/016	Happy Birthday	S.E.P.P.	1/1/81	Orange	Wall Poster	19" X 13"	U.S.A.	$7.50-10.00	The poster has Papa Smurf standing on a chair decorating a birthday cake. There are 12 Smurfs gathered around watching. The background is orange and has balloons and streamers all over. It says "Happy Birthday" in white letters with pink border. The poster is in a black frame.
149/017	Homework Gives Me A Rash!!!	S.E.P.P.	1/1/81	Tan	Wall Poster	19" X 13"	U.S.A.	$7.50-10.00	The poster has a Smurf screaming. There are papers and books all over the floor. The background is tan. The letters "Homework Gives Me A Rash!!" are in black and red .
149/018	Even Geniuses Have Bad Days	S.E.P.P.	1/1/82	Brown	Wall Poster	19" X 13"	U.S.A.	$7.50-10.00	The poster is of Papa Smurf getting blown away from his work table by an explosion. There is a big yellow/orange cloud on the table and things are flying all over. The letters "Even Geniuses Have Bad Days" is in red. The background is brown.
149/019	Paint Your Own Rainbows	S.E.P.P.	1/1/82	Light Blue	Wall Poster	19" X 13"	U.S.A.	$7.50-10.00	The poster has a Smurf standing on a cloud painting a rainbow. The poster says "Paint Your Own Rainbows" in multi colors. The background is light blue.
149/020	Love Comes In Many Colors	S.E.P.P.	1/1/82	Yellow	Wall Poster	19" X 13"	U.S.A.	$7.50-10.00	The poster has a Smurf jumping in the air next to a red candy machine. The candy machine is spilling out all different color candy hearts. The background is yellow. The poster says "Love Comes In Many Colors" in red letters and "Colors" is in multi colors.
149/021	Die Schlumpfe Hier Erhaltlich!	Marburger	1/1/95		Wall Poster	19" X 13"	Brussels	$15.00-20.00	Promotional poster for German stamps. The poster has 2 mushroom houses with 3 Smurfs playing outside one house on it. Smurfette is kissing another Smurf in front of the other house. A Smurf wearing a pink night shirt carrying a candle is walking on the roof of the large house. Puppy is running between the Smurfs. At the top are 2 Smurfs by the word Die Schlumpfe. Die Schlumpfe Hier Erhaltlich! is in yellow letters. The poster also says Spiel Und Spab Mit Tollen Stempeln in black letters.
149/022	Once In A Blue Moon...	W. Berrie	1/1/84		Wall Poster	25 1/2"L X 17"W	U.S.A.	$15.00-20.00	The poster has purple, pink, orange and yellow stars. The Smurfs are standing on the stars with their catalog number under them. At the bottom is every catalog number and the Smurfs name and the date they were issued. There are 167 Smurf figures in the picture. The top says "Once in a blue moon... All the Smurfs get together for a portrait."
149/023	Ik Krijg Ze Allemaal Te Pakken	Infogrames			Wall Poster	23 1/2"L X 16 1/2"W	Germany	$11.00-15.00	The poster has Gargamel standing, holding a Game Boy box pack with his right hand. Gargamel has his other hand in the air and he is pointing. The poster has an orange background. The top of the poster says " Ik Krijg Ze Allemaal Te Pakken."
149/024	Snow Fall	W. Berrie	1/1/83		Wall Poster	11" X 17"	U.S.A.	$5.00-7.00	The poster has 15 Smurfs cleaning up the village after a snow fall. Papa is standing in the corner waving to a Smurf. The poster is out of a cereal box.
149/025	Jo Loves Aquests Colors			White	Wall Poster	13 1/2"L X 9 1/2"W		$6.00-8.00	The poster is white and has a Smurf unrolling yellow and red ribbon. The Smurf looks like he's marching. He is wearing a red hat, red and blue socks, red and blue shirt, blue shorts and brown shoes. The bottom says "Jo Loves Aquests Colors" in blue letters.
149/026	Go For It!	S.E.P.P.	1/1/81	Green	Wall Poster	19" X 13"	U.S.A.	$7.50-10.00	The poster is green with a Smurf riding on a brown carousel horse. He is reaching out for some rings on a stick. The poster says "Go For It" in black letters.
149/027	Love Is Something You Share...Not Own	S.E.P.P.	1/1/81	Pink	Wall Poster	19" X 13"	U.S.A.	$7.50-10.00	The poster is pink. It has a Smurf handing Smurfette red daisies. The poster says "Love Is Something You Share...Not Own" in black letters.
149/028	The World Of Smurfs	S.E.P.P.	1/1/80		Wall Poster	19" X 13"	U.S.A.	$11.00-15.00	The poster has a village scene. There are large Smurf houses and 2 small Smurf houses. There are Smurfs doing various things. There are 59 Smurfs in the picture. The top says "The World Of Smurfs" in white letters.
149/029	We All Grow Up To Be Cowboys	S.E.P.P.	1/1/81	Yellow	Wall Poster	19" X 13"	U.S.A.	$7.50-10.00	A Smurf is dressed as a cowboy. He is riding a mechanical bull. The Smurf is swinging a rope above his head. The top says "We All Grow Up To Be Cowboys" that looks like it's made from rope.
149/030	Do Not Disturb I'm Busy With Homework	S.E.P.P.	1/1/82	Orange	Wall Poster	19" X 13"	U.S.A.	$7.50-10.00	The poster has a Smurf sitting at a table. The Smurf is holding a soda and a remote for an Atari game. On top of the table is a TV. with a game on the screen. There are books and papers thrown all over. The radio is sitting on the floor. The top of the poster says "Do Not Disturb" in black. The bottom says "I'm Busy With My Homework" in yellow letters.
149/031	Fristi Poster	S.E.P.P.	1/1/81		Wall Poster	18" X 19"	Brussels	$7.50-10.00	The poster is pink and has a village scene in the center. It has the 40th anniversary symbols in the bottom corners. It says Fristi on top. There are 3 mushroom houses and it looks like the Smurfs are repairing the houses. There are 5 blue outlined Smurfs that aren't colored.
149/032	Festival De Los Pitufos!	Schleich/ Peyo	1/1/81		Wall Poster	24" x 37"	Spain	$15.00-20.00	The poster is a wood scene. It has a Smurf castle, the tree stump and the windmill. The picture has Papa and a bandleader going up in a hot air balloon. There are various Smurfs in the picture.
149/033	Creix La Familia Dels Autentics Barrufets Fuig De Les Imita	Schleich/Peyo	1/1/81		Wall Poster	24" x 37"	France	$15.00-20.00	The poster is a wood scene. It has a Smurf castle, the tree stump and the windmill. The picture has Papa and a band leader going up in a hot air balloon. There are various Smurfs in the picture.
149/034	Die Bunte Welt Der Schleich-Figuren!!!	Schleich/Peyo	1/1/81		Wall Poster	24" x 37"	Germany	$15.00-20.00	The poster is of a Smurf village. It has various Smurf figures pictured. It also has the Muppet's, Snorkels and other Schleich figures pictured.
149/035	Smurfland	S.E.P.P.	1/1/81		Wall Poster	24" x 37"		$15.00-20.00	The poster is of a Smurf village built into a mountain side. The picture has 200 different Smurf figures and 15 Play Sets and houses. Each Smurf is numbered in the picture and on top it has their names.

149/036	Smurf Campfire Poster	W. Berrie/Peyo	1/1/81		Wall Poster	11" x 17"	U.S.A.	$5.00-7.00	The poster has several Smurfs dancing around a campfire. There are Chinese lanterns hanging all over. The other Smurfs are partying. Papa is leading a band. The poster is out of a cereal box.
149/037	Smurfs Cleaning Up Snow	W. Berrie/Peyo	1/1/81		Wall Poster	11" x 17"	U.S.A.	$5.00-7.00	The Smurfs are all cleaning up snow in the village. The poster is out of a cereal box.
149/038	Smurfs Doing Yard Work	Verkerke Reproductions	1/1/98		Wall Poster	36" X 24"	Netherlands	$9.00-12.00	The poster has Smurfs doing yard work. 2 Smurfs are fixing the gate. A Smurf is watering some purple flowers. Smurfette is standing in the balcony of a mushroom house watching. There is a bird in the sky. The background is yellow. There are 15 Smurfs including Smurfette in the poster.
149/039	Smurfling Band	Verkerke Reproductions	1/1/98		Wall Poster	36" X 24"	Netherlands	$9.00-12.00	The poster has the Smurflings on a stage. Snappy is playing drums. Nat is playing a guitar. Slouchy is playing the trumpet. There are 11 Smurfs dancing in the grass below. Papa is in the corner of the poster.
149/040	Smurf With Dinosaurs	Sunshine Holland B.V.	1/1/81		Wall Poster	19 1/2 " X 15 1/2"	Brussels	$4.50-6.00	The picture is of 2 Smurfs, Papa, Grandpa and Smurfette riding a purple dinosaur. They are trying to get away from a big green dinosaur. The background is orange.
149/041	Smurfs On A Magic Carpet	Sunshine Holland B.V.	1/1/96		Wall Poster	19 1/2" X 15 1/2"	Brussels	$4.50-6.00	The poster has Smurfette, 2 Smurfs and Grandpa flying on a magic carpet above a castle. The sky is 2 shades of blue and yellow.
149/042	Gargamel Flying Above Smurf Village	Sunshine Holland B.V.	1/1/96		Wall Poster	19 1/2" X 15 1/2"	Brussels	$4.50-6.00	The poster is of Gargamel flying above the Smurf village. He is riding some kind of bike with propellers. The Smurfs are looking up at Gargamel with troubled expressions on there face.
149/043	Baby Smurf Flying	Sunshine Holland B.V.	1/1/96		Wall Poster	19 1/2" X 15 1/2"	Brussels	$4.50-6.00	The picture is of a Smurf flying in the air with a bird. The Smurf is wearing a white cloth diaper. Papa, Smurfette and 2 Smurfs are looking troubled. There are 2 mushroom houses and laundry hanging on the line.
149/044	Scrupples, Gargamel & Azreal Getting Picture Taken	Sunshine Holland B.V.	1/1/96		Wall Poster	19 1/2" X 15 1/2"	Brussels	$4.50-6.00	The picture is of a Smurf taking a picture of Gargamel, Azreal and Scrupples in the woods. Gargamel, Azreal and Scrupples have frazzled expression on there faces. Another Smurf is standing, laughing.
149/045	Smurfs With Blue Dinosaur	Sunshine Holland B.V.	1/1/96		Wall Poster	19 1/2" X 15 1/2"	Brussels	$4.50-6.00	The Smurfs are in a cave. Nat is sitting on the back of a dinosaur. The dinosaur is blue with wings. Grandpa, Sassette and Papa are standing on the side of the dinosaur.
149/046	Smurf Taking A Bath	Sunshine Holland B.V.	1/1/96	Yellow	Wall Poster	19 1/2" X 15 1/2"	Brussels	$4.50-6.00	A Smurf is sitting in a washtub scrubbing his back. The water is running out the sides of the tub. There is a Smurf peeking through the window laughing. The Smurf in the window is holding a drill. The background is yellow.
149/047	Smurf Serenading Smurfette	Sunshine Holland B.V.	1/1/96		Wall Poster	19 1/2" X 15 1/2"	Brussels	$4.50-6.00	Smurfette is in the balcony of a mushroom house. A Smurf is standing below serenading Smurfette. The Smurf has hearts floating above his head. The background is night sky.
149/048	Bigmouth And The Smurflings	Sunshine Holland B.V.	1/1/96		Wall Poster	19 1/2" X 15 1/2"	Brussels	$4.50-6.00	Nat, Sassette, Slouchy and Snappy are all sitting on a log. Bigmouth is in front of them holding grass. They are all in the woods.
149/049	Smurf Playing Trumpet	Sunshine Holland B.V.	1/1/96		Wall Poster	19 1/2" X 15 1/2"	Brussels	$4.50-6.00	A Smurf is playing a trumpet. 3 Smurfs are laying passed out by there instruments, on the ground. They are between the mushroom houses.
149/050	Smurfs On A Shoreline	Sunshine Holland B.V.	1/1/96		Wall Poster	19 1/2" X 15 1/2"	Brussels	$4.50-6.00	Papa, Smurfette and 3 Smurfs are standing on a shoreline with an Indian. All the Smurfs have troubled looks on their face. It is dark out and there is a big white moon.
149/051	Smurf Slumber Party	Sunshine Holland B.V.	1/1/96		Wall Poster	19 1/2" X 15 1/2"	Brussels	$4.50-6.00	Brainy Smurf and 5 other Smurfs are dressed in pajama's laying on brown mattresses. They have a candle burning in the window of the room.
149/052	Smurf Playing Harp	Sunshine Holland B.V.	1/1/96		Wall Poster	19 1/2" X 15 1/2"	Brussels	$4.50-6.00	A Smurf is kneeling in the grass playing a harp. There is a squirrel in the tree above him.
149/053	Baby Delivered From A Stork	Sunshine Holland B.V.	1/1/96		Wall Poster	19 1/2" X 15 1/2"	Brussels	$4.50-6.00	Baby Smurf is in a basket on a hill. A stork is flying in the sky. Papa, Smurfette and 7 Smurfs are running towards Baby. Gargamel and Azreal are peeking out through the bushes.
149/054	Smurfette And A Smurf On A Scale	Sunshine Holland B.V.	1/1/96		Wall Poster	19 1/2" X 15 1/2"	Brussels	$4.50-6.00	Smurfette is standing on a scale. The springs of the scale are flying out all over. A Smurf is sitting on the back of the scale eating sweats. Smurfette has a troubled look on her face.
149/055	Smurfette Kissing A Smurf	Sunshine Holland B.V.	1/1/96		Wall Poster	19 1/2" X 15 1/2"	Brussels	$4.50-6.00	Smurfette is kissing a Smurf on the nose. The Smurf is blushing and has hearts floating above his head. They are standing under a tree. The tree has 2 squirrels in love on the side of it.
149/056	Smurfs Drawing Hearts On A Tree	Sunshine Holland B.V.	1/1/96		Wall Poster	19 1/2" X 15 1/2"	Brussels	$4.50-6.00	7 Smurfs are drawing hearts on a tree. There are 2 squirrels and 2 butterflies with hearts above their heads. Smurfette and Baby Smurf are standing on the other side of the tree looking inquisitive.
149/057	Smurf In Bed Dreaming	Sunshine Holland B.V.	1/1/96		Wall Poster	19 1/2" X 15 1/2"	Brussels	$4.50-6.00	A Smurf is sleeping in bed. He is dreaming of a Smurf holding a book. The Smurf fell asleep with a book laying in his lap. There is a candle burning on the night stand. The background is blue.
149/058	De Smurfenbus Poster	EMI	4/1/99		Wall Poster	24" x 15 1/2"	Brussels	$7.50-10.00	The poster is an advertisement for a new CD and tape called De Smurfenbus. The picture is of Papa driving a school bus. Smurfette and a Smurf are riding the bus. The bottom of the poster has some of the new songs listed.
149/059	Walibi Schtroumpf Poster	Peyo			Wall Poster	21" x 31"	France	$11.00-15.00	The poster has Smurfs riding all different rides at a Smurf park. The top corner says Walibi Schtroumpf. It has a Smurf and a kangaroo waving.
149/060	Die Schlumpfe 40th Anniversary Poster	Peyo	1/1/98	white	Wall Poster	27 1/2" x 31"	Belgium	$15.00-20.00	The poster is white. It has a big 40th anniversary logo in the center. It says Die Schlumpfe 40 1958-1998. In the center is a Smurf that looks like he's jumping.

Set 77: Pull String Flapping Toys

077/001	Superman	Durham	1/1/82		Flapping Toys	2 1/2"	Macau	$6.00-8.00	Gently pull down and release, the Smurfs arms wiggle while the Smurf moves up the string. The Smurf makes a clickety-clack sound. The Smurf has both arms out at his sides. He is wearing a red shirt with a yellow s in the center, a red cape with a yellow collar, white pants and a white hat. The Smurf is painted right onto the plastic. There is a blue pull string.
077/002	Ice cream Smurf	Durham	1/1/82		Flapping Toys	2 1/2"	Macau	$6.00-8.00	Gently pull down and release, the Smurfs arms wiggle while the Smurf moves up the string. The Smurf makes a clickety-clack sound. The Smurf is holding an ice cream cone in his right hand and a yellow gift box in the left hand. He is wearing a red shirt with a black bow tie, white pants and a white hat. The Smurf is painted right onto the plastic. There is a blue pull string.
077/003	Boxer Smurf	Durham	1/1/82		Flapping Toys	2 1/2"	Macau	$3.50-5.00	Gently pull down and release, the Smurfs arms wiggle while the Smurf moves up the string. The Smurf makes a clickety-clack sound. The Smurf is wearing a yellow boxing glove on each hand, a yellow tank shirt, red shorts, white shoes and a white hat. The Smurf is painted right onto the plastic. There is a blue pull string.
077/004	Smurf With Silverware	Durham	1/1/82		Flapping Toys	2 1/2"	Macau	$3.50-5.00	Gently pull down and release, the Smurfs arms wiggle while the Smurf moves up the string. The Smurf makes a clickety-clack sound. The Smurf is holding a white knife in his right hand and a white fork in his left hand. He is wearing a red shirt with a yellow napkin tucked into the front, white pants, and a white hat. The Smurf is painted right onto the plastic. There is a blue pull string.
077/005	Tennis Smurf	Durham	1/1/82		Flapping Toys	2 1/2"	Taiwan	$3.50-5.00	Gently pull down and release, the Smurfs arms wiggle while the Smurf moves up the string. The Smurf makes a clickety-clack sound. The Smurf is a puffy sticker on the plastic and only has 1 flapping arm. He is holding a yellow tennis racket. The Smurf is wearing a yellow shirt, white pants and orange tennis shoes. The pull string is orange.

Set 134: Puppets- Hand

134/001	Handmade Papa Puppet				Cloth Hand Puppet	12"	Hand made	$3.50-5.00	Papa is made of a less fuzzy material. He has a long shaggy white beard. His body is light blue. Papa is wearing a red hat and the bottom part of his body is red. His eyes are white on top and black on the bottom.
134/002	Smurf Hand Puppet With Plastic Head	Bogi		Blue	Puppet	9" x 7 1/2"	Korea	$15.00-20.00	The Smurf has a plastic head. He has a white hat. The body is a soft cloth. The Smurf has two hands.
134/003	Smurf Wrap Around				Puppet	20"		$11.00-15.00	The Smurf is made of soft fuzzy material. The Smurf has long arms and legs that wrap around a person. The arms and legs velcro together. The Smurfs body is a puppet. He has a big cloth head.

Set 201: Purses

201/001	Smurf Giving Smurfette Flowers Square Purse	Peyo/SEPP		Pink	Cloth Purse	7 1/2"L X 10"W		$4.50-6.00	The purse is pink with 2 pink handles and 2 compartments. The purse is square. The front has a Smurf handing Smurfette 4 daisies. Smurfette is wearing a white dress with red dots. The Smurf has hearts floating above his head. The purse says "SMURF" on the front in blue letters.
201/002	Smurf Giving Smurfette Flowers Square Purse	Peyo - SEPP		Pink/Light Blue	Cloth Purse	7 1/2"L X 10"W		$4.50-6.00	The purse is pink and light blue with 2 light blue handles and 2 compartments. The purse is square. The front has a Smurf handing Smurfette 4 daisies. The background is pink. Smurfette is wearing a white dress with red dots. The Smurf has hearts floating above his head. The purse says "SMURF" on the front in multi colors.
201/003	Smurf At A School Desk Square Purse	Peyo/SEPP		Red	Cloth Purse	7 1/2"L X 10"W		$4.50-6.00	The purse is red with blue trim, 2 blue handles and 2 compartments. The purse is square. The front has a Smurf sitting at a school desk. The Smurf is holding a yellow pencil. Behind the desk is blue paper with a rainbow on it. The purse says "SMURF" on the front in blue letters.
201/004	Smurf Giving Smurfette Flowers Round Purse	BBC Imports	1/1/82	Red	Cloth Purse	7 1/2"L X 6"W	Hong Kong	$7.50-10.00	The purse is red with a blue shoulder strap. The purse is round on the bottom. The front has a fold over flap with a picture of a Smurf handling Smurfette 4 daisies. The background is red. Smurfette is wearing a white dress with red dots. The Smurf has hearts floating above his head. The purse says "SMURF" on the front in multi colors.
201/005	Papa Kissing Smurfette's Hand Heart Shape Purse	BBC Imports	1/1/82	Blue	Cloth Purse	6 1/2"L X 7 1/2"	Hong Kong	$2.00-3.00	The purse is blue with a red border and red shoulder strap. The strap is missing. The top zips shut. The front has Papa bent over kissing Smurfette's hand. Smurfette has her other hand in her hair. It says "SMURF" in white letters above the picture.
201/006	Smurf With A Pad And Pencil Square Purse	BBC Imports	1/1/82	Red	Cloth Purse	6"L X 7"W	Hong Kong	$4.50-6.00	The purse is red with blue trim and a blue shoulder strap. The purse has a fold over flap with a small see through window. On the flap is a picture of a Smurf standing, holding a red pad of paper and a yellow pencil. Above the Smurf it says "SMURF" in blue letters.
201/007	Smurf With A Pad And Pencil Square Purse	BBC Imports	1/1/82	Blue	Cloth Purse	6"L X 7"W	Hong Kong	$4.50-6.00	The purse is blue with red trim and a red shoulder strap. The purse has a fold over flap with a small see through window. On the flap is a picture of a Smurf standing, holding a red pad of paper and a yellow pencil. Above the Smurf it says "SMURF" in red letters.
201/008	Smurfette Purse	Applause/Peyo	1/1/84	Clear	Plastic Purse	5"L X 7"W	China	$7.50-10.00	The purse is rectangular. The purse is clear with pink hearts all over it. The purse has a clear shoulder strap and zips shut. Smurfette is in the corner of the front of the purse. She is standing, wearing a pink dress with a purple heart behind her. Her name is above her.

201/009	Kings Island Smurfette Purse	Peyo	1/1/84	Clear	Vinyl Purse	9" X 9"	Taiwan	$6.00-8.00	The purse is clear with little pink hearts all over it. The purse has a snap close and clear handles. The front has Smurfette walking, carrying a purple and pink umbrella. Smurfette is wearing a pink dress. There are yellow flowers by her feet. The purse says "Kings Island" in purple letters under Smurfette.
201/010	Smurf Clip On Pocket Purse	Peyo		Light Blue	Clip On Purse	4"L X 4 1/2"W		$5.00-7.00	The purse is light blue with dark blue trim. The purse zips with a flap over the zipper. The flap has a picture of a Smurf with his hands in the air. Behind the Smurf are rainbow colored lines. Above the Smurf is says "Smurf" in blue letters. There is a dark blue clip on the back to clip on to a belt.
201/011	Smurfette Clip On Pocket Purse	Peyo		Light Purple	Clip On Purse	4"L X 4 1/2"W		$5.00-7.00	The purse is light purple with dark purple trim. The purse zips with a flap over the zipper. The flap has a picture of Smurfette standing with a butterfly on her finger. Behind the Smurfette are lines that are rainbow colored. Above the Smurfette it says "Smurfette" in blue letters. There is a dark purple clip on the back to clip on to a belt.
201/012	Papa Clip On Pocket Purse	Peyo		Light Pink	Clip On Purse	4"L X 4 1/2"W		$5.00-7.00	The purse is light pink with dark pink trim. The purse zips with a flap over the zipper. The flap has a picture of Papa standing, holding a finger in the air. Behind the Papa are lines that are rainbow colored. Above the Papa it says "Papa Smurf" in blue letters. There is a dark pink clip on the back to clip on to a belt.
201/013	Smurf At A School Desk Square Purse	Peyo		Blue	Cloth Purse	7 1/2" x 10"		$3.00-4.00	The purse is blue with red trim, 2 red handles and 2 compartments. The purse is square. The front has a Smurf sitting at a school desk. The Smurf is holding a yellow pencil. Behind the desk is blue paper with a rainbow on it. The purse says "SMURF" on the front in red letters.
201/014	Smurf Giving Smurfette Flowers Square Purse	Peyo		Red	Cloth Purse	6" x 7 1/4"		$3.00-4.00	The purse is red with 2 pink flowers with blue trim and a blue strap. The front has a Smurf handing Smurfette 4 daisies. Smurfette is wearing a white dress with red dots. The Smurf has hearts floating above his head. The purse says "SMURF" on the front in blue letters. The purse has a fold over flap for closing.
201/015	Smurf Giving Smurfette Flowers Heart-Shaped Purse	Peyo		Red	Plastic Purse	6 1/4" x 7 1/2"		$3.00-4.00	The purse is red with 2 pink flowers with blue trim and a blue strap. The purse is heart- shaped. The front has a Smurf handing Smurfette 4 daisies. Smurfette is wearing a white dress with red dots. The Smurf has hearts floating above his head. The purse says "SMURF" on the front in blue letters. The purse has a fold over flap for closing.
201/016	Smurf Sitting In Cloud Purse	Peyo		Tan	Cloth Purse	6 1/4" x 7 1/4"		$3.00-4.00	The purse is tan with blue trim and a blue strap. A Smurf is sitting in a blue cloud. There is a rainbow behind him. The cloud says "Have A Smurfy Day". The purse has a flip open cover.
201/017	Smurf Sitting In Cloud Purse	Peyo		Blue	Cloth Purse	7 1/2"L X 10"W		$3.00-4.00	The purse is blue with red trim and a red strap. A Smurf is sitting in a blue cloud. There is a rainbow behind him. The cloud says "Have A Smurfy Day". The purse has a flip open cover.
201/018	Smurfette Pulling Pedals From Flower Purse	Peyo		Red	Cloth Purse	6" x 7"		$3.00-4.00	The purse is red with blue trim and a blue strap. Smurfette is standing, pulling the petals off from a daisy. The purse is square- shaped and has a flip over cover. It says "Smurfette" in red letters above Smurfette.
201/019	Smurfette Pulling Pedals From Flower Purse	Peyo		Tan	Cloth Purse	6" x 7"		$4.50-6.00	The purse is red with blue trim and a blue strap. Smurfette is standing, pulling the petals off from a daisy. The purse is square- shaped and has a flip over cover. It says "Smurfette" in red letters above Smurfette.

Set 229: Puzzle Blocks

229/001	Soft Puzzle Blocks	Talbot Toys	1/1/83		Soft Blocks		Taiwan	$15.00-20.00	The box has the letters A through I. All the sides have different pictures : Smurfette in a balcony with a Smurf below. Smurf riding a horse. Smurf giving Smurfette flowers. Astro Smurf. One side has the numbers 1 through 9.
229/002	Hard Puzzle Blocks	Kortekaas Merch	1/1/82		Puzzle Blocks	8 1/2 x 9 3/4"	Italy	$22.50-30.00	There are 20 small blocks in a red and clear carrying case. Each side of the block has a different picture on it. The blocks fit together to make 6 different pictures. There are picture cards enclosed with the blocks.

Set 62: Puzzles- 100 Pieces

062/001	Smurf School Scene	MB Puzzle		Pink Box	Cardboard Puzzles	16" x 11"	U.S.A.	$4.50-6.00	The puzzle has Brainy standing by a chalkboard, a Smurf sitting on a cactus, a Smurf sleeping with another Smurf tickling the Smurf sleeping with a weed, and another Smurf playing a horn. Total of seven Smurfs on the puzzle.
062/002	Baseball Scene	MB Puzzle		Green Box	Cardboard Puzzles	16" x 11"	U.S.A.	$4.50-6.00	The puzzle has Smurfette pitching a ball to a Smurf holding a bat, with an umpire behind and another Smurf out in the field by third base. The mountains are in the background. Total of 4 Smurfs on the puzzle.
062/003	Smurf Windmill And Snail Carriage Scene	Ravensburger	1/1/84		Cardboard Puzzles	16" x 11"	West Germany	$4.50-6.00	There are 8 Smurfs on the puzzle. A Smurf is sitting on potato sacks driving a snail carriage, 1 Smurf is standing in the door of the mushroom windmill. There is a stork in the air with a Smurf on it's back.
062/004	Smurfs Gathering Acorn Scene	Ravensburger	1/1/83		Cardboard Puzzles	16" x 11"	West Germany	$4.50-6.00	There are 13 Smurfs on the puzzle. The Smurfs are in the woods gathering acorns.
062/005	Rock Scene	Schmid	1/1/94		Cardboard Puzzles	42 x 28 cm	Germany	$4.50-6.00	The picture is of 3 Smurfs and Sassette in a rock band. There is a Smurf playing a keyboard, one playing the drum, a Smurf singing and Sassette is playing a pink guitar. There are Smurfs below the stage.
062/006	Cowboy & Indian Puzzle	Schmid	1/1/94		Cardboard Puzzles	16" x 11"	Germany	$4.50-6.00	The puzzle has 3 Smurfs dressed as Indians, a Smurf dressed as a cowboy and Smurfette dressed as a princess. There are tents and a chuck wagon in the background. All the Smurfs are by a campfire.
062/007	Smurfs Having A Picnic	MB Puzzle	1/1/88		Cardboard Puzzles	16" x 11"	U.S.A.	$2.00-3.00	Papa is standing, getting food out of a picnic basket. A Smurf is sleeping against a log. 2 Smurfs are playing catch with an acorn. Smurfette is sitting on a log eating a sandwich.
062/008	Grandpa And Sassette Water-skiing	MB Puzzle	1/1/87		Cardboard Puzzles	16" x 11"	U.S.A.	$2.00-3.00	Puppy is pulling Grandpa Smurf and Sassette on water ski's. Nat, Slouchy and Snappy are riding on Puppy's back.
062/009	Smurf's Camping	MB Puzzle	1/1/88		Cardboard Puzzles	16" x 11"	U.S.A.	$2.00-3.00	The Smurfs are camping. They have an all different colored tent set up by the river. Smurfette, Papa and another Smurf are fishing. There is a Smurf up in the tree.
062/010	Smurf's Sledding	MB Puzzle	1/1/83		Cardboard Puzzles	16" x 11"	U.S.A.	$2.00-3.00	There are 4 Smurf's, Papa and Baby Smurf sledding down a hill.
062/011	Smurfs River Rafting	MB Puzzle	1/1/83		Cardboard Puzzles	16" x 11"	U.S.A.	$2.00-3.00	3 Smurfs, Smurfette and Papa are river rafting. There is a Smurf sitting on the bank with Baby Smurf watching.
062/012	Smurf Rodeo	MB Puzzle			Cardboard Puzzles	16" x 11"	U.S.A.	$2.00-3.00	There is a Smurf riding a horse like a bull.. Papa, Smurfette and 2 other Smurfs are dressed in cowboy outfits watching. There are mountains and a tower in the background.
062/013	Smurf Wheelbarrow Races	MB Puzzle	1/1/88		Cardboard Puzzles	16" x 11"	U.S.A.	$2.00-3.00	Papa is holding a checkered flag. Smurfette is cheering 4 Smurfs on. There are 2 teams and the Smurfs are having wheelbarrow races.
062/014	5 Smurfs Marching Across A Bridge	MB Puzzle	1/1/82		Cardboard Puzzles	16" x 11"	U.S.A.	$2.00-3.00	Smurfette, Papa and 3 other Smurfs are marching across a bridge. The Smurfs are all holding a finger in the air. There is a Smurf in the river on a raft. There is another Smurf sitting on the bank fishing.
062/015	Smurf Castle	MB Puzzle	1/1/82		Cardboard Puzzles	16" x 11"	U.S.A.	$2.00-3.00	The castle is yellow and red. King Smurf is walking outside the castle. A Smurf is playing the drum and another is playing the trumpet. There are 3 Smurfs standing together waving and pointing. Another Smurf is standing on the side of the castle laughing.
062/016	Grandpa And Smurflings At Table	MB Puzzle	1/1/87		Cardboard Puzzles	16" x 11"	U.S.A.	$2.00-3.00	Slouchy, Snappy and Sassette are sitting at the table mixing bowls of food. Grandpa is drinking. Nat is carrying a bushel of tomatoes.
062/017	Smurflings Riding On Puppy	MB Puzzle	1/1/87		Cardboard Puzzles	16" x 11"	U.S.A.	$2.00-3.00	The Smurflings have a ladder backed up to Puppy. Sassette is sitting on Puppy's head. Nat is sitting on Puppy's back. Grandpa is climbing the ladder. Slouchy is leaning against the ladder and Snappy is standing in front of Puppy. Puppy is gray.
062/018	Smurf Orchestra	MB Puzzle	1/1/81		Cardboard Puzzles	16" x 11"	U.S.A.	$2.00-3.00	Papa's standing on a mushroom leading the band. There's a Smurf playing a guitar, a Smurf playing a saxophone, a Smurf playing a cello, a Smurf playing a piano, 1 holding a music sheet and another behind a log directing the band. 15 Smurfs total.
062/019	Smurf Birthday Party	MB Puzzle	1/1/88		Cardboard Puzzles	16" x 11"	U.S.A.	$2.00-3.00	Papa is standing by a cake. Nat and Sassette are sitting on a yellow couch. A Smurf is reaching for a balloon. Smurfette is standing by the table watching Papa.
062/020	Smurfs Mountain Climbing	MB Puzzle	1/1/83		Cardboard Puzzles	16" x 11"	U.S.A.	$2.00-3.00	Papa, Smurfette and 3 Smurfs are climbing up the side of a mountain. Papa is carrying Baby Smurf. There is a mountain in the background.
062/021	Smurf Olympic Scene	MB Puzzle	1/1/81		Cardboard Puzzles	16" x 11"	U.S.A.	$2.00-3.00	Smurfs are doing various things. Their pole vaulting, running, parallel bars, weight lifting ect. The 3 winners are standing on steps.
062/022	Smurf Village	MB Puzzle			Cardboard Puzzles	16" x 11"	U.S.A.	$2.00-3.00	Papa is standing on a mushroom. Jokey is carrying a present towards Brainy. A Smurf is reading a script. A Smurf is chasing a Smurf with a rolling pin. There are 12 Smurfs in the picture.
062/023	Baby In Dirt Crying	MB Puzzle	1/1/83		Cardboard Puzzles	16" x 11"	U.S.A.	$2.00-3.00	Baby is sitting in the dirt crying. His bottle is spilled next to him. Papa and 3 other Smurfs are standing around looking troubled.
062/024	Grandpa Hand gliding	MB Puzzle	1/1/87		Cardboard Puzzles	16" x 11"	U.S.A.	$2.00-3.00	Grandpa Smurf is hand gliding. The Smurflings and Puppy are on the ground watching. There are mountains in the background.
062/025	Smurfette Rafting	MB Puzzle			Cardboard Puzzles	16" x 11"	U.S.A.	$2.00-3.00	Smurfette is on a raft in the middle of nowhere. Papa is floating in a barrel. Superman Smurf is flying above. There is debris floating all around.
062/026	Smurf Carrousel	Ravensburger	1/1/98		Cardboard Puzzles	49 x 36cm	France	$7.50-10.00	The puzzle has a Smurf merry-go-round. There are lanterns hung on wire. There are a bunch of Smurfs partying by a campfire.
062/027	The Baby Smurf	Ravensburger	1/1/98		Cardboard Puzzles	16" x 11"	France	$7.50-10.00	The puzzle has Baby Smurf in a crib. There are Smurfs doing all different things in the room. Some are cooking, ironing, preparing bath water, hanging laundry, etc.

Set 154: Puzzles- 200 Piece

154/001	Smurf Weather Machine Puzzle	Ravensburger	1/1/83		Cardboard Puzzles	42 X 29,7 cm	Germany	$5.00-7.00	The puzzle is a picture of 4 seasons. In one corner 2 Smurfs are making a snowman, one Smurf is skiing and another is bundled up. Below that scene are 4 Smurfs on an ice pond. The other half of the puzzle is summer and has a Smurf in a boat, one fishing and a Smurf standing in the grass laughing. There is a Smurf windmill weather machine in the back. There is a total of 11 Smurfs on the puzzle.
154/002	Smurfs Dancing Around A Campfire	Ravensburger	1/1/83		Cardboard Puzzles	42 X 29,7 cm	Germany	$5.00-7.00	The picture on the puzzle is of a bunch of Smurfs dancing around a campfire and partying at night. There are 47 Smurfs in the picture.
154/003	Smurf Party Scene	Ravensburger	1/1/83		Cardboard Puzzles	42 X 29,7 cm	Germany	$5.00-7.00	The puzzle is in a blue box. It is a night scene with the shadow of 4 mushroom houses in the background. A Smurf is making a campfire, 4 are dancing, a Smurf is carrying a small cake and 2 other Smurfs are carrying a large cake. Smurfette is standing, talking to a baker, Jokey's carrying a present and Papa has a troubled look on his face. There are 17 Smurfs in the picture. 3 pieces are missing.

Set 63: Puzzles- 24 Pieces

063/001	Cowboy Scene	MB Puzzle	1/1/82	Purple Box	Cardboard Puzzles	15" x 12 1/2"	U.S.A.	$2.00-3.00	The puzzle has Smurfette dressed as a cowgirl talking to a Smurf that's dressed as a cowboy. There is another Smurf dressed as a cowboy sitting in a wagon. Total of 3 Smurfs on the puzzle.
063/002	Home Smurf Home	MB Puzzle	1/1/82		Cardboard Puzzles	15" x 12 1/2"	U.S.A.	$2.00-3.00	The puzzle has Papa Smurf sitting in a chair reading a story to a Smurf and Smurfette. There are milk and cookies sitting on the floor.
063/003	Smurf And Smurfette Giving Baby A Bath	MB Puzzle	1/1/83		Cardboard Puzzles	15" x 12 1/2"	U.S.A.	$2.00-3.00	A Smurf and Smurfette are giving Baby Smurf a bath. Baby is sitting in bubbles playing with a yellow rattle.
063/004	2 Smurfs Playing Chess	MB Puzzle	1/1/82		Cardboard Puzzles	15" x 12 1/2"	U.S.A.	$2.00-3.00	2 Smurfs are sitting at a table playing chess. There is a lantern hanging above their heads.
063/005	Smurf And Smurfette Roller-skating	MB Puzzle	1/1/82		Cardboard Puzzles	15" x 12 1/2"	U.S.A.	$2.00-3.00	A Smurf is chasing Smurfette. They are roller-skating through the grass.
063/006	Baby Smurf In Highchair	MB Puzzle	1/1/83		Cardboard Puzzles	15" x 12 1/2"	U.S.A.	$2.00-3.00	Baby Smurf is sitting in a highchair. A Smurf is crawling on the floor to pick up Baby's bottle. The Smurf has a bowl of food on his head. Smurfette is standing behind laughing.
063/007	Smurf Band	MB Puzzle	1/1/82		Cardboard Puzzles	15" x 12 1/2"	U.S.A.	$2.00-3.00	A Smurf is playing a cello. A Smurf is playing a trumpet and the 3rd Smurf is playing a drum. They are all standing in the grass.
063/008	Smurf At Beach	MB Puzzle	1/1/82		Cardboard Puzzles	15" x 12 1/2"	U.S.A.	$2.00-3.00	A Smurf is walking along the shore carrying a basket and an umbrella. A Smurf is wearing swim trunks. He is standing, holding a shovel. Brainy Smurf is sitting on a beach blanket under an umbrella.
063/009	Smurfs At Campfire	MB Puzzle	1/1/82		Cardboard Puzzles	15" x 12 1/2"	U.S.A.	$2.00-3.00	3 Smurfs are having a campfire. 2 Smurfs are roasting hot dogs and the 3rd Smurf is playing a guitar. There is Gargamel's castle in the background.
063/010	Smurfs In Garden	MB Puzzle	1/1/82		Cardboard Puzzles	15" x 12 1/2"	U.S.A.	$2.00-3.00	A Farmer Smurf is picking a tomato. Another Smurf is looking dreamy. The Farmer Smurf is thinking about when he planted the seeds.

Set 16: Puzzles- Miscellaneous

016/001	Baby Smurf With Rattle And A Block	MB	1/1/84	Yellow/orange	25 Piece Cardboard	11 1/2" x 14 1/2"	U.S.A.	$2.00-3.00	Baby Smurf is sitting, holding a circular rattle that has a bell in the middle, and there is an AB play block behind him.
016/002	Baby Smurf With A Ball	MB	1/1/84	Green	25 Piece Cardboard	11 1/2" x 14 1/2"	U.S.A.	$2.00-3.00	Baby Smurf is sitting, tossing a green and yellow ball in the air.
016/003	Picnic And Pond Scene	MB	1/1/84		450 Piece Puzzle	22" x22"	U.S.A.	$5.00-7.00	There are 17 Smurfs, including Papa and Smurfette, in the woods with the mountains behind them, swimming and having a picnic.
016/004	Olympic Scene And Work Out Scene	Ravensburger	1/1/83	Blue Box	2 X 20 Piece Puzzle	26,4 x 18,1 cm	West Germany	$3.50-5.00	The puzzle has 2 puzzles together. One is 2 Smurfs carrying flags with 5 runners between them and a bunch of Smurfs cheering over the fence. The second puzzle has 9 Smurfs working out.
016/005	250 Piece Band Scene	MB	1/1/82	Green Box	250 Piece Puzzle	19 7/8" x 13 7/8"	U.S.A.	$3.50-5.00	Papa's standing on a mushroom leading the band. There's a Smurf playing a guitar, a Smurf playing a saxophone, a Smurf playing a cello, a Smurf playing a piano, 1 is holding a music sheet and another behind a log directing the band. 7 Smurfs total.
016/006	320 Piece Band Scene	Ass-Puzzle	1/1/84		320 Piece Cardboard	40,0 x 50,0 cm	West Germany	$4.50-6.00	Papa's standing on a mushroom leading the band. There's a Smurf playing a drum, a Smurf playing a trumpet, a Smurf playing a cello, a Smurf playing a piano, 1 holding a music sheet with 2 others singing with him. There's a Smurf holding bells, 1 playing a flute. 15 Smurfs total.
016/007	130 Piece Mechanic Smurf Puzzle	MB	1/1/83		130 Piece Cardboard	19,5 x 24,5	West Germany	$3.50-5.00	The puzzle has a Smurf carrying a wrench. There are wood gears behind him and a yellow bell. In front of his feet are nuts and bolts.
016/008	De Ijverige Smurfen	Ravensburger	1/1/83		33 Piece Puzzle	11 1/2" x 14 1/2"	Spain	$7.50 10.00	There are 6 small puzzles in the box. The puzzles make different Smurfs. Nurse Smurfette, Painter, Postman, Gardner, Mechanic, and Sewing Smurf
016/009	De Smurfen Familie	MB	1/1/84		33 Piece Cardboard	11 1/2" x 14 1/2"	Spain	$7.50-10.00	There are 9 small puzzles in the box. The puzzles make different Smurfs: Papa looking mad, Smurfette, Astro, Brainy with a book, Football Smurf, Winter Smurf, Smurf playing the flute, Smurf standing, holding a flower and a Smurf laying, chewing a flower.
016/010	Baby Smurf Crawling With A Rattle	MB	1/1/84	Orange	25 Piece Cardboard	11 1/2" x 14 1/2"	U.S.A.	$2.00-3.00	Baby Smurf is crawling, holding a yellow circle rattle that has a gray bell in the middle. The puzzle is orange.
016/011	Village Figure Puzzle	ASS - Puzzle	1/1/84		140 Piece Cardboard	11 1/2" x 14 1/2"	West Germany	$3.50-5.00	The puzzle shows 27 different Smurfs that are the PVC figures. All the Smurfs are doing different things. Papa is standing in the door of a mushroom house.
016/012	Village Scene With Bell	Ravensburger	1/1/83		280 Piece Cardboard	49,3 x 36,2 cm	West Germany	$4.50-6.00	The puzzle is of a village scene. There are 22 Smurfs doing various things. There is a Smurf ringing a big bell. In the middle of the picture there is a well.
016/013	Olympic Scene	ASS - Puzzle	1/1/84		320 Piece Cardboard	11 1/2" x 14 1/2"	West Germany	$4.50-6.00	The scene is outside. There are 18 Smurfs doing various activities. Smurfette is wearing pink and jumping on a pink trampoline. There are 3 Smurfs standing on a winners step.
016/014	Orchestra Scene And Little Red Ridding Hood Scene	Ravensburger	1/1/83		2 X 20 Piece Puzzle	11 1/2" x 14 1/2"	West Germany	$3.50-5.00	The orchestra scene has Papa leading the group with 5 Smurfs playing instruments and 1 Smurf sleeping. They are set up on a tree stump. The second puzzle is a Smurf dressed as Grandma and another as little red riding hood. Grandma is sleeping in bed. There are 5 Smurfs around the stage.
016/015	Fest Der Schlumpfe	ASS	1/1/84		45 Piece Cardboard	19,5 X 21,5 cm	Germany	$3.50-5.00	The box has 3 puzzles, each with 45 pieces. There is a night scene of a Smurf standing on a lamp pole watching fire flies. The second puzzle has Papa leading a band. They are playing on a red table. The third puzzle is of 4 Smurfs playing with party favors at night.
016/016	Segelregatta Der Schlumpfe	ASS	1/1/84		63 Piece Cardboard		Spain	$3.50-5.00	The puzzle is of 2 sail boats with Smurfs riding in them. One boat has a red sail and the other has a yellow with green sail.
016/017	Smurf Decorating A Cake	Trefl	1/1/94		25 Piece Cardboard	130 X 200	Brussels	$2.00-3.00	The puzzle has a Smurf standing on a stool at a table decorating a cake. Smurfette is standing, watching. There is a red bowl with green batter sitting on the table.
016/018	Brainy Giving A Lecture	Trefl	1/1/94		54 Piece Cardboard	130 X 200	Brussels	$3.50-5.00	The puzzle has Brainy wearing a backpack. Brainy is standing in the woods and he has one hand in the air. Another Smurf is standing, looking angry with his fingers in his ears.
016/019	Dreamy Leaning On A Shovel Sleeping	Trefl	1/1/94		54 Piece Cardboard	130 X 200	Brussels	$3.50-5.00	Dreamy Smurf is standing, leaning against a shovel sleeping. He is standing under a red and yellow mushroom. The shovel is in the ground were he was digging a hole.
016/020	The Smurflings In A Boat	Golden	1/1/88		23 Frame Tray Puzzle	11 1/2" x 14 1/2"	U.S.A.	$2.00-3.00	The puzzle is of Sassette, Nat, Slouchy and Grandpa Smurf in a sail boat. Slouchy is holding a telescope, Nat is up on the sail pole looking out at the water, Sassette is standing in front pointing to something and Grandpa is steering the boat in back.
016/021	Smurfs Playing Chess	MB	1/1/83		450 Piece Cardboard	22" X 22"	U.S.A.	$5.00-7.00	The Smurfs made a chess board out of dirt. They are dressed as different chess pieces and are playing chess. The village is in the background. There are 33 Smurfs in the picture.
016/022	Smurf Band	Golden	1/1/88		25 Frame Tray Puzzle	11 1/2" x 14 1/2"	U.S.A.	$2.00-3.00	The puzzle has a Smurf playing a bubble organ. A Smurf is playing a guitar. Sassette is playing a tambourine. Slouchy Smurf is playing the drums. The background is pink.
016/023	Smurf Village Scene	MB	1/16/83		450 Piece Cardboard	11 1/2" x 14 1/2"	U.S.A.	$5.00-7.00	The picture is of several mushroom houses with Smurfs in the windows and outside on the paths. In the center of the house there is a well. There are 11 Smurfs, Smurfette and Papa in the picture.
016/024	Smurfs In Kitchen	MB	1/1/82		250 Piece Cardboard	19 7/8" x 13 7/8"	U.S.A.	$3.50-5.00	Smurfette, Papa and 6 Smurfs are in a kitchen. Papa is putting pepper on food. A Smurf has smoke coming out of his mouth. A Smurf is carrying a big green kettle. A Smurf is eating a waffle. Smurfette is standing, stirring a kettle of food. A Smurf has a spoon to his mouth and another Smurf is holding a beer mug.
016/025	Smurfs At The Beach	MB	1/1/82		250 Piece Cardboard	19 7/8" x 13 7/8"	U.S.A.	$3.50-5.00	Smurfette and a Smurf are surfing. 2 Smurfs are sitting on a beach blanket. There are 2 Smurfs playing ball and another Smurf holding an umbrella.
016/026	Smurflings Riding Puppy With Gargamel In Bushes	MB	1/1/87		60 Piece Cardboard	11" x 16"	U.S.A.	$2.00-3.00	The Smurflings are riding Puppy to get away from Gargamel. Gargamel is peeking through the bushes.

Set 131: Puzzles- Plastic

131/001	Smurfette			Pink	Plastic Tray Puzzle	13" X 11"		$4.50-6.00	The tray is a pink plastic. The puzzle is of Smurfette. Smurfette is in a flirty pose. There are 13 plastic pieces.
131/002	Smurf			Yellow	Plastic Tray Puzzle	13" X 11"		$3.50-5.00	The tray is a yellow plastic. The puzzle is of a Smurf. There are pieces missing from the puzzle.
131/003	Smurfette			Red	Plastic Tray Puzzle	13" X 11"		$3.50-5.00	The tray is a red plastic. The puzzle is of Smurfette. Smurfette is in a flirty pose. There are 13 plastic pieces.
131/004	Papa	Peyo	1/1/83	Light Blue	Plastic Tray Puzzle	13" X 11"	U.S.A.	$3.50-5.00	The tray is a light blue plastic. The puzzle is of Papa. Papa is in a standing position, pointing. There are 11 plastic pieces.

Set 111: Puzzles- Slide

111/001	Smurf's Playing Ball	Inter-Mundus	1/1/76		16 Piece Puzzle	7 3/4"H X 6 3/4"W	Netherlands	$9.00-13.00	The tray is white and has 16 slide pieces. The picture is 2 Smurfs playing in the grass with a red ball.
111/002	Smurfs Throwing A Present	Peyo/I.M.P.S.	1/1/95	Yellow	Slide Puzzle	5" x 6"	Hong Kong	$4.50-6.00	The plastic frame is yellow. The picture is of 3 Smurfs tossing a present to each other. The present has a fuse on it. The background is blue. The same picture is on the frame.

Set 20: Puzzles- Wood

020/001	Space Scene	MB Puzzle	1/1/82		Wood Puzzle	10 3/4 x 13 3/8	U.S.A.	$5.00-7.00	Smurfette and a Smurf are dressed in space suits standing in front of a space ship.
020/002	Smurf Village Scene	MB Puzzle	1/1/82		Wood Puzzle	10 3/4 x 13 3/8	U.S.A.	$5.00-7.00	Smurf village scene with a Smurf pushing a wheelbarrow in front of 3 mushroom houses.
020/003	Smurf Daydreaming Scene	MB Puzzle	1/1/82		Wood Puzzle	10 3/4 x 13 3/8	U.S.A.	$5.00-7.00	A Smurf is laying in front of a rainbow and mountains day dreaming.
020/004	Smurf Eating Scene	MB Puzzle	1/1/82		Wood Puzzle	10 3/4 x 13 3/8	U.S.A.	$5.00-7.00	The puzzle has a Smurf sitting at a table holding a yellow spoon in the air and he is licking his chops. Papa is standing, holding a green glass and eating a pink heart. A Smurf is eating a popsicle and another Smurf is carrying a huge cake. The Smurfs are in a yellow room with a window.

Set 4: Puzzles- Wood Tray

004/001	Papa Smurf	Playskool	1/1/82	Multi	12 Piece Wood Puzzle	9 1/4" x 11 1/2"	U.S.A.	$6.00-8.00	Papa with hands outstretched, jumping in the air with a meadow in background. Orange, yellow and lavender sky.
004/002	Milking Done	Playskool	1/1/82	Multi	10 Piece Wood Puzzle	9 1/4" x 11 1/2"	U.S.A.	$6.00-8.00	A Smurf carrying a bucket in each hand. Yellow and green background.
004/003	Skating Smurfette	Playskool	1/1/82	Multi	10 Piece Wood Puzzle	9 1/4" x 11 1/2"	U.S.A.	$6.00-8.00	Smurfette ice skating, wearing a red outfit, with an orange sky and snow in the background.
004/004	Hello Operator	Playskool	1/1/82	Purple	12 Piece Wood Puzzle	9 1/4" x 11 1/2"	U.S.A.	$6.00-8.00	Smurfette talking on a telephone, purple background.
004/005	Evening Serenade Smurf	Playskool	1/1/82	yellow/green	11 Piece Wood Puzzle	9 1/4" x 11 1/2"	U.S.A.	$6.00-8.00	Smurf standing, serenading in a meadow
004/006	Romeo, Romeo	Playskool	1/1/82	Light Blue	12 Piece Wood Puzzle	9 1/4" x 11 1/2"	U.S.A.	$6.00-8.00	Romeo Smurf declaring his love for Smurfette. Smurfette's in a balcony.
004/007	Spring Beauty	Playskool	1/1/82	Blue/Green	12 Piece Wood Puzzle	9 1/4" x 11 1/2"	U.S.A.	$6.00-8.00	Smurfette holding a bunch of flowers.
004/008	Soccer Star	Playskool	1/1/82	Yellow/Green	10 Piece Wood Puzzle	9 1/4" x 11 1/2"	U.S.A.	$6.00-8.00	Smurf wearing a red shirt and black shorts, kicking a gray soccer ball.
004/009	Adventure Bound	Playskool	1/1/82	Multi	13 Piece Wood Puzzle	9 1/4" x 11 1/2"	U.S.A.	$6.00-8.00	Smurf with a red knapsack walking through the village.
004/010	Clowning Around	Golden	1/1/82	Multi	9 Piece Wood Puzzle	9 1/4" x 11 1/2"	Sweden	$6.00-8.00	A Smurf and Smurfette dressed up like clowns
004/011	8 Smurfs And A Mushroom House	Simplex Toys	1/1/82	Wood Grain	10 Piece Wood Puzzle	12" x 8"	Holland	$6.00-8.00	10 Individual pieces each have red plastic knobs to lift. Little pieces include 6 Smurfs, Papa, Smurfette, and a mushroom house.
004/013	Good News	Playskool	1/1/82	Blue/Green	12 Piece Wood Puzzle	9 1/4 x 11 1/2	U.S.A.	$6.00-8.00	The puzzle has Papa Smurf walking next to an orange mushroom house. There is a Smurf on the side of the house jumping in the air.
004/015	Musical Smurfs	BP/Peyo	1/1/81	Wood	8 Piece Wood Puzzle	9 1/4 x 11 1/2	Australia	$7.50-10.00	8 Individual pieces each have red plastic knobs to lift. Little pieces include a Smurf playing a trumpet, a Smurf playing a flute, a Smurf playing guitar, a Smurf holding a music sheet, a Smurf playing the drums and Smurfette playing a harp. There is also a big mushroom house and an extra mushroom.

Set 259: PVC's- Carded Figures

259/001	Rock N' Roll	Schleich/Peyo	1/1/77	Carded Figures	2"	Hong Kong	$3.50-5.00	The Smurf has a dark blue body. The Smurf is holding a dark orange guitar with yellow strings. He has his eyes shut and his mouth open. The Smurf is on a green card with a picture of a mushroom.
259/002	Postman	Schleich/Peyo	1/1/78	Carded Figures	2"	Germany/China	$3.50-5.00	The Smurf has a bright blue body. He is holding a yellow horn in his mouth with his right hand. He has a dark brown mailbag. He is holding a white envelope with a red heart on the flap of the envelope in his left hand. The Smurf is on a blue card with a picture of 2 Smurfs and Smurfette on top.
259/003	Clown	Schleich/Peyo	1/1/78	Carded Figures	2"	Germany/China	$3.50-5.00	The Smurf has a dark blue body. He is wearing yellow with red striped pants, black suspenders, a red bow tie, white shoes with black toes, a white shirt, white gloves and a white hat. He has a red mouth with white around it. The Smurf is on a blue card with a picture of 2 Smurfs and Smurfette on top.
259/004	Hang-Glider	Schleich/Peyo		Carded Figures	2"	Germany/China	$3.50-5.00	The Smurf has a dark blue body. The Smurf has a red and yellow hand-glider with dark brown straps. The Smurf is on a blue card with a picture of 2 Smurfs and Smurfette on top.
259/005	Gift	Schleich	1/1/78	Carded Figures	2"	Hong Kong	$3.50-5.00	The Smurf has a dark blue body. He is holding a yellow gift with a red bow in his left hand. He has a yellow and two red flowers in the other hand. The Smurf is on a green card with a picture of a mushroom.
259/006	Lover	Schleich/Peyo	1/1/78	Carded Figures	2"	Hong Kong	$3.50-5.00	The Smurf has a dark blue body. The Smurf is holding 3 dark red flowers with yellow centers and a dark green stem. The Smurf is holding the flowers in his left hand. He has his head turned to the right like he's posing. The Smurf is on a green card with a picture of a mushroom.
259/007	Tennis Star	Schleich/Peyo	1/1/79	Carded Figures	2"	Germany/China	$3.50-5.00	The Smurf has a dark blue body. He is wearing a white tennis shirt and white shorts. He has a yellow wrist band on his right wrist. He is holding a red plastic tennis racket in his right hand. The tennis racket has holes in it. The Smurf is on a blue card with a picture of 2 Smurfs and Smurfette on top.
259/008	Telephone	Schleich/Peyo	1/1/80	Carded Figures	2"	Germany/China	$3.50-5.00	The Smurf has a dark blue body. The phone is dark red with gold. The cradle is thick, wide and copper colored. The hand piece is copper and he is holding it to his mouth with his right hand. The Smurf is on a blue card with a picture of 2 Smurfs and Smurfette on top.
259/009	Telephone	Schleich/Peyo	1/1/80	Carded Figures	2"	Hong Kong	$3.50-5.00	The Smurf has a dark blue body. The phone is bright red with gold. The cradle is small, thin and copper colored. The hand piece is copper and he is holding it to his mouth with his right hand. The Smurf is on a green card with a picture of a mushroom.
259/010	Umbrella	Schleich/ Peyo	1/1/79	Carded Figures	2"	Hong Kong	$3.50-5.00	The Smurf has a dark blue body. He is standing under an orange/red mushroom with white spots and an orange stem. He has a worried expression on his face. The Smurf is on a green card with a picture of a mushroom.
259/011	Roller-skating Smurfette	Schleich/ Peyo		Carded Figures	2"	Hong Kong	$3.50-5.00	Smurfette has a dark blue body. She is wearing a white dress and white and red roller-skates. Smurfette's hair is dark yellow and has red ribbons in it. Smurfette is on a green card with a picture of a mushroom.
259/012	Tennis Smurfette	Schleich/Peyo	1/1/81	Carded Figures	2"	Germany/China	$3.50-5.00	Smurfette has a dark blue body. She is wearing a white tennis dress and shoes. Smurfette's dress has red dots on the bottom. Her hair is dark yellow. Smurfette is holding a red plastic tennis racket in her left hand. Smurfette is on a blue card with a picture of 2 Smurfs and Smurfette on top.
259/013	Papa With Lab Glass	Applause/Peyo		Carded Figures	2"	China	$3.50-5.00	Papa has a medium blue body. Papa is holding a test tube in each hand. Papa is standing, pouring the green bottle into the tan bottle. Papa is wearing a dull colored red hat and pants. Papa is a red mold. Papa is on a blue card with a picture of 2 Smurfs and Smurfette on top.
259/014	Baby With Ice Cream	Schleich/Peyo		Carded Figures	2"	Germany/China	$3.50-5.00	Baby has a dark blue body. He is wearing a light blue sleeper and hat. Baby is sitting, eating an ice cream. The ice cream is pink/purple and on a brown stick. Baby is holding the ice cream in his right hand. The Smurf is on a blue card with a picture of 2 Smurfs and Smurfette on top.
259/015	Dentist	Schleich	1/1/84	Carded Figures	2"	China	$3.50-5.00	The Smurf has a dark blue body. He is wearing a white jacket with black buttons, white pants and a white hat. The Smurf is holding a red toothbrush with white bristles in his left hand. The Smurf is on a blue card with a picture of 2 Smurfs and Smurfette on top.
259/016	Angel	Schleich	1/1/85	Carded Figures	2"	China	$3.50-5.00	The Smurf has a dark blue body. He is wearing a long white robe. He has 2 white wings attached to his back. The Smurf is holding a finger up in the air on his right hand. The Smurf has a red mouth. The Smurf is on a blue card with a picture of 2 Smurfs and Smurfette on top.
259/017	Devil	Schleich/ Peyo	1/1/84	Carded Figures	2"	Germany/China	$3.50-5.00	The Smurf has a dark pink body. The Smurf has glossy red feet, pink ears, pink tail, black wings and white horns. The Smurf is on a blue card with a picture of 2 Smurfs and Smurfette on top.
259/018	Baby With Butterfly	Schleich/Peyo	1/1/84	Carded Figures	2"	Germany/China	$3.50-5.00	Baby has a dark blue body. Baby is wearing a white sleeper. He is sitting, holding a yellow butterfly in his right hand. The butterfly has red spots. The Smurf is on a blue card with a picture of 2 Smurfs and Smurfette on top.
259/019	Smurfette Stewardess	Schleich/Peyo	1/1/84	Carded Figures	2"		$3.50-5.00	Smurfette has a dark blue body. She is wearing a white dress, white shoes and a white hat with a yellow flower on each side. Smurfette is holding a red tray with a brown cup. The cup has white liquid in it and a blue straw. Smurfette is on a blue card with a picture of 2 Smurfs and Smurfette on top.
259/020	Baby With Bowl And Spoon	Schleich/Peyo		Carded Figures	2"	Germany/China	$3.50-5.00	Baby has a medium blue body. He is sitting, holding a pink bowl and a red spoon. He has yellow gobs spattered all over. Baby is wearing a white sleeper. The Smurf is on a blue card with a picture of 2 Smurfs and Smurfette on top.
259/021	Architect Smurf	Schleich	1/1/89	Carded Figures	2"	Germany/China	$3.50-5.00	The Smurf has a medium blue body. He is wearing a yellow jacket, orange hard hat, white pants and green boots. The Smurf is holding a yellow ruler in his left hand and a white scroll in his right hand. He has an orange and brown pencil in his jacket pocket. The Smurf is on a blue card with a picture of 2 Smurfs and Smurfette on top.
259/022	Hula Smurfette	Schleich/Peyo		Carded Figures	2"	Germany/China	$3.50-5.00	Smurfette has a medium blue body. Smurfette is wearing a tan hula skirt, green belt, red lei and orange hat trim. Smurfette has her hands out her side like she's dancing. Smurfette is on a blue card with a picture of 2 Smurfs and Smurfette on top.
259/023	Gargamel With Lab Glasses	Schleich/Peyo	1/1/82	Carded Figures	2"	Germany/China	$3.50-5.00	Gargamel has a dark flesh colored skin. Gargamel is wearing a black robe and maroon boots. Gargamel is holding a pink glass in his right hand and a purple lab glass in left hand. He looks like he's pouring one lab glass into the other. Gargamel is on a blue card with a picture of 2 Smurfs and Smurfette on top.
259/024	Slouchy Smurfling	Applause/Peyo	1/1/87	Carded Figures	2"	China	$3.50-5.00	The Smurf has a medium bright blue body. He is wearing a white droopy hat a white pants and red shirt. He has his right hand waving. The Smurf is on a blue card with a picture of 2 Smurfs and Smurfette on top.
259/025	Azreal The Cat	Schleich/Peyo	1/1/91	Carded Figures	2"	Germany/China	$3.50-5.00	Azreal is orange and white. Azreal is in a prowling position with his tail up. He has his head up and he's licking his chops. Azreal is on a blue card with a picture of 2 Smurfs and Smurfette on top.
259/026	Sitter	Schleich/Peyo	1/1/93	Carded Figures	2"	Germany/China	$3.50-5.00	The Smurf has a medium bright blue body. He is sitting with his legs crossed, looking up and gesturing towards the sky with his hands. The Smurf is wearing white pants and a white hat. The Smurf is on a blue card with a picture of 2 Smurfs and Smurfette on top.

Set 176: PVC's- Christmas Cord Smurf's

176/001	Smurf With Christmas Tree	Schleich/Peyo	1/1/81	Rubber Figures	2" High	Portugal	$4.50-6.00	The Smurf has a medium blue body. The Smurf is standing, hugging a green tree. The tree has red ornaments, a red stand and a yellow star on top. The Smurf is wearing white pants and a white hat. He has a gold cord attached for hanging.
176/002	Smurf With A Present	Schleich/Peyo	1/1/81	Rubber Figures	2" High	Portugal	$4.50-6.00	The Smurf has a medium blue body. The Smurf is wearing white pants, a white hat and a red scarf. The Smurf is carrying a big green box. The box has a red ribbon and a red bow around it. The Smurf has a gold cord attached for hanging.
176/003	Papa With A Sack Of Gifts	Schleich/Peyo	1/1/81	Rubber Figures	2" High	Portugal	$4.50-6.00	Papa has a medium blue body. He is wearing red pants and a red hat with a yellow tassel. In his right hand Papa is holding a red and white candy cane. In his left hand he is holding a tan sack with toys in it. There is a gold cord attached for hanging.
176/004	Smurf With A Sack Of Gifts	Schleich/Peyo	1/1/81	Rubber Figures	2" High	Portugal	$4.50-6.00	The Smurf has a medium blue body. He is wearing white pants and a white hat with a green tassel. The Smurf is carrying a brown sack over his left shoulder. The sack has 2 red and white candy canes, green package, a red package, an orange package and a yellow horn in it. The Smurf has a gold cord attached for hanging.

176/005	Smurf With Music Sheet And Candle	Schleich/Peyo	1/1/81		Rubber Figures	2" High	Portugal	$4.50-6.00	The Smurf has a medium blue body. He is wearing white pants, a white hat with a red tassel and a red scarf. The Smurf is holding a yellow sheet of music with black writing in front of him. In his right hand he is holding a candle out. The candle is on an orange plate. The candle is yellow with a red flame. The Smurf has a gold cord attached for hanging.
176/006	Smurf With Wreath	Schleich/Peyo/W.B.	1/1/81		Rubber Figures	2" High	Portugal	$55.00-75.00	The Smurf has a medium blue body. He is wearing white pants and a white hat. The Smurf has his head stuck through a big green wreath. The wreath has red berries and a red bow on top. The Smurf has a gold cord for hanging. There is an orange dot on the bottom of his foot.
176/007	Smurf Riding On A Candy cane	Schleich/Peyo	1/1/81		Rubber Figures	2" High	Portugal	$40.00-55.00	The Smurf has a medium blue body. He is wearing white pants and a white hat. The Smurf is sitting on a candy cane. The candy cane is red and white with a pink bow and green holly. The Smurf has his tongue hanging out of his mouth and he is waving with his left hand. There is a gold cord attached for hanging.
176/008	Smurf With Drum	Schleich/Peyo	1/1/81		Rubber Figures	2" High	Hong Kong	$25.00-35.00	The Smurf has a medium blue body. He is wearing white pants, a white hat, a green shirt with red cuffs and a red scarf. The Smurf has a white/red/yellow drum attached to his tummy. The Smurf is holding black drumsticks with red tops in both hands. There is a gold cord attached for hanging.
176/009	Smurfette With Music Sheet And Candle	Schleich/Peyo/W.B.	1/1/82		Rubber Figures	2" High	Portugal	$4.50-6.00	Smurfette has a medium blue body. She is wearing a green sweater with a red collar and red cuffs, a white skirt, a white hat, white shoes and red socks. Smurfette is holding a yellow sheet of music with black writing in her right hand. In her left hand Smurfette is holding a candle out at arms length. The candle is yellow with an orange flame. The candle is on a brown plate. Smurfette has her head turned towards the sheet of music. There is a gold cord attached for hanging.
176/010	Smurf Praying	Schleich/Peyo/W.B.	1/1/82		Rubber Figures	2" High	Portugal	$110.00-150.00	The Smurf has a medium blue body. The Smurf is wearing a red sleeper with red feet and a red hat with a white tassel. The Smurf is kneeling and has his hands folded, praying. There is a gold cord attached for hanging.
176/011	Smurfette Praying	Schleich/Peyo/W.B.	1/1/82		Rubber Figures	2" High	Portugal	$110.00-150.00	Smurfette has a medium blue body. Smurfette is wearing a pink sleeper with pink feet and a pink hat with a red tassel and green holly. Smurfette is kneeling and has her hands folded praying. There is a gold cord attached for hanging.
176/012	Smurfette With Candy cane	Schleich/Peyo/W.B.	1/1/82		Rubber Figures	2" High	Portugal	$35.00-45.00	Smurfette has a medium blue body. She is wearing a red coat with white trim, red pants, white shoes and a red hat with white trim and a white tassel. Smurfette is holding a big red and white candy cane. The candy cane has a green bow around it. There is a gold cord attached for hanging.
176/013	Smurfette With Candy cane	Schleich/Peyo/W.B.	1/1/82		Rubber Figures	2" High	Portugal	$35.00-45.00	Smurfette has a medium blue body. She is wearing a light red coat with white trim, light red pants, white shoes and a light red hat with white trim and a white tassel. Smurfette is holding a big light red and white candy cane. The candy cane has a green bow around it. There is a gold cord attached for hanging.

Set 203: PVC's- Color & Mold Variations

203/001	Tennis Star	Schleich/ Peyo	1/1/79		Rubber Figure	2"	Germany/China	$3.00-4.00	The Smurf has a medium blue body. He is wearing a white tennis shirt and white shorts. He has a yellow wrist band on his right wrist. He is holding a red-ish/orange plastic tennis racket in his right hand. The tennis racket has holes in it.
203/002	Angry Smurf				Rubber Figure	2"		$22.50-30.00	The Smurf has a black body. He has his arms extended, one is slanted down and the other hand is sideways. He has red eyes and his teeth are clenched. His teeth are red lines. He is a small mold. His feet are flat.
203/003	Mermaid	Schleich/Peyo	1/1/81		Rubber Figure	2"	West Germany	$15.00-20.00	Smurfette has a medium blue body and a blue and silver tail. Smurfette is sitting on a gray rock with dark blue and silver waves around the bottom of the rock. Smurfette has a red starfish on her hat. She is not wearing a necklace.
203/004	Captain	Schleich/Peyo	1/1/81		Rubber Figure	2"	West Germany	$75.00-100.00	Papa has a medium blue body. He is wearing a dark blue coat with yellow buttons, white pants, and a dark blue hat with a red border and black visor. Papa is holding a yellow telescope to his right eye.
203/005	First-Aid	Schleich/ Peyo	1/1/78		Rubber Figure	2"	West Germany	$11.00-15.00	The Smurf has a medium blue body. He is wearing a white uniform. The Smurf has a white bandage wrapped around a finger on his right hand. He is carrying a yellow medical bag, no cross on it.
203/006	Clown	Schleich/Peyo	1/1/78		Rubber Figure	2"	Germany/China	$4.50-6.00	The Smurf has a medium blue body. He is wearing yellow with red striped pants, black suspenders, a red bow tie, white shoes with black toes, a white shirt, white gloves and a white hat. He has a red mouth with white around it.
203/007	Hang-Glider	Schleich/ Peyo	1/1/78		Rubber Figure	2"	West Germany	$4.50-6.00	The Smurf has a medium blue body. The Smurf has a red and yellow hand-glider with dark brown straps.
203/008	Majorette	Schleich/Peyo	1/1/84		Rubber Figure	2"	West Germany	$11.00-15.00	Smurfette has a medium blue body. She is wearing a light pink dress with gold trim, hot pink boots and a white hat with yellow feather stuck in the side and a hot pink visor. Smurfette is carrying a hot pink baton with a gold tip in her right hand. Her left hand is on her hip. Smurfette is standing on a light green stand.
203/009	Hunter	Schleich/Peyo	1/1/78		Rubber Figure	2"	Hong Kong	$4.50-6.00	The Smurf has a dark blue body. He is holding a silver gun. On the end of the gun is a dark yellow bird. The Smurf has a green feather in his hat and a troubled look on his face.
203/010	Baby With Blocks	Peyo	1/1/84		Rubber Figure	2"	West Germany	$7.50-10.00	The Smurf has a medium blue body. He is sitting, wearing a yellow sleeper. Baby has 3 blocks. The blocks are dark yellow, red and dark blue. Baby has a red mouth.
203/011	Flirting Smurfette	Peyo/Bully			Rubber Figure	2"	West Germany	$18.00-25.00	Smurfette has a dark blue body. She is standing with her left hand on her right hip, her right hand is on her hair. She has light yellow hair and long black eyelashes.
203/012	Congratulations	Schleich/Peyo	1/1/80		Rubber Figure	2"	West Germany	$25.00-35.00	The Smurf has a medium blue body. He is holding his hands together in front of him. The Smurf has a smile on his face. The Smurfs hands are small and have a small space between them.
203/013	Shiver	Peyo			Rubber Figure	2"	West Germany	$7.50-10.00	The Smurf has a medium blue body. The Smurf is standing with his hands behind his back. He has a maroon scarf wrapped around his face.
203/014	Sitting	Peyo			Rubber Figure	2"	West Germany	$4.50-6.00	The Smurf has a medium blue body. He is sitting with his hands flat on both sides of him. He is wearing a white hat and white pants. He has a smile on his face. The Smurf is a small mold.
203/015	Doctor	Schleich/Peyo			Rubber Figure	2"	West Germany	$4.50-6.00	The Smurf has a medium blue body. The Smurf is wearing a white doctor's coat with silver buttons. He has a thermometer in his left hand. The thermometer has blue lines. He has a silver and red stethoscope in his right hand. The Smurf has a red dot on his foot.
203/016	Emperor	Schleich/Peyo			Rubber Figure	2"	West Germany	$11.00-15.00	The Smurf has a dark blue body. He is wearing a yellow outfit, yellow crown, gold hat, gold shoes, gold jewelry and a red robe. He is holding his index finger on his right hand up in the air. The Smurf is a yellow mold.
203/017	Candle	Schleich/Peyo	1/1/79		Rubber Figure	2"	West Germany	$7.50-10.00	The Smurf has a dark blue body. He is wearing a white night shirt, brown shoes and a white hat with a red tassel. He is holding a candle in his right hand. The candle is white with a yellow fire The candle is sitting on a brown plate. The Smurf has his eye's open.
203/018	Present	Peyo/Bully			Rubber Figure	2"	Hong Kong	$7.50-10.00	The Smurf has a dark blue body. He is holding a white present with reddish orange ribbon. The Smurf is holding the present with both hands off to his right side.
203/019	Pirate	Peyo/Bully			Rubber Figure	2"	West Germany	$3.00 - 4.25	The Smurf has a dark blue body. He is holding a silver sword in his right hand. He has a red belt with a black buckle around his waist. The Smurf has a black patch covering his left eye.
203/020	Cornucopia	Peyo/Bully			Rubber Figure	2"	West Germany	$3.00-4.00	The Smurf has a dark blue body. The Smurf is standing and leaning on a yellow horn with bright green leaves and red flowers in the end. The Smurf has his left foot in the air.
203/021	Baseball Batter	Schleich/Peyo/W.B.	1/1/81		Rubber Figure	2"	Hong Kong	$4.50-6.00	The Smurf has a dark blue body. He is wearing a white uniform with red sleeves, a red belt, red socks, a white hat with a red visor and black shoes. The Smurf is holding a flesh colored bat.
203/022	Tennis Smurfette	Schleich/Peyo	1/1/81		Rubber Figure	2"	West Germany	$4.50-6.00	Smurfette has a dark blue body. She is wearing a white tennis dress and shoes. Smurfette's dress has pink dots on the bottom. Her hair is dark yellow. Smurfette is holding a red plastic tennis racket in her left hand.
203/023	Smurfette Jump Roping	Schleich/Peyo	1/1/82		Rubber Figure	2"	West Germany	$4.50-6.00	Smurfette has a medium blue body. Smurfette is wearing a white dress, yellow panties, white hat and white shoes. Smurfette is holding a jump rope over her head. The jump rope is white and has red handles. The jump rope is a thick plastic.
203/024	Baby With Rattle	Schleich/Peyo		White	Rubber Figure	2"		$3.00 - 4.25	Baby has a dark blue body color. Baby is wearing a white sleeper and hat. He is crawling, carrying a dark red rattle with a yellow center in his right hand.
203/025	Baby With Teddy	Schleich	1/1/84		Rubber Figure	2"		$3.00 - 4.25	Baby has a medium blue body. Baby is wearing a white sleeper and white hat. Baby is sitting, holding a brown teddy bear. The teddy bear is brown and peach and he is wearing a red bow tie.
203/026	Baby With Butterfly	Schleich/Peyo	1/1/78		Rubber Figure	2"	West Germany	$3.00 - 4.25	Baby has a medium blue body. Baby is wearing a white sleeper. He is sitting, holding a yellow butterfly in his right hand. The butterfly has red spots.
203/027	Papa Pilot	Schleich/Peyo	1/1/85		Rubber Figure	2"	West Germany	$4.50-6.00	Papa has a medium blue body. He is wearing a brown pilot suit, yellow scarf, black shoes, a red hat and brown goggles. Papa has his thumb in the air on his right hand. He has his eye's closed. Papa has a red dot on his foot.
203/028	Rock -N- Roll	Schleich/Peyo			Rubber Figure	2"	West Germany	$9.00-12.00	The Smurf is has a medium blue body. He is playing a red and orange guitar with tan strings.
203/029	Emperor	Schleich/ Peyo			Rubber Figure	2"	West Germany	$11.00-15.00	The Smurf has a dark blue body. He is wearing a mustard colored outfit, gold crown, white hat, gold shoes, gold jewelry and a red robe. He is holding his index finger on his right hand up in the air.
203/030	Emperor	Schleich/Peyo	1/1/78		Rubber Figure	2"	Ger./Portugal	$3.00-4.00	The Smurf has a medium blue body. He is wearing a mustard colored outfit, yellow crown, gold hat, gold shoes, gold jewelry and a red robe. He is holding his index finger on his right hand up in the air. The Smurf is a yellow mold.
203/031	Gardener With Rake	Schleich/Peyo	1/1/82		Rubber Figure	2"	West Germany	$3.00-4.00	The Smurf has a medium, bright blue body. The Smurf is wearing a bright green apron, yellow straw hat, white pants and brown shoes. The Smurf is holding a brown plastic rake in front of him.
203/032	Tennis Player 2	Schleich/Peyo	1/1/77		Rubber Figure	2"		$3.00-4.00	The Smurf has a medium blue body. He is holding a red plastic tennis racket with big holes in his right hand. The Smurf is holding a yellow tennis ball in his left hand. He is standing like he's going to serve the ball. The Smurf is wearing a white tennis shirt and white shorts.

Set 245: PVC's- Dupuis Figures

245/001	Papa	Peyo & Ed Dupuis	1/1/57	First Rubber Smurf	2" High		$18.00-25.00	Papa is standing with both hands out at his side. His body is medium blue. He has a red hat and pants on. He is painted real sloppy. He is one of the first original Smurf figures.
245/002	Drummer	Peyo & Ed Dupuis	1/1/57	First Rubber Smurf	2" High		$18.00-25.00	The Smurf has a medium blue body. He has a black drum with red support strings and yellow edge rings. The drum sticks are yellow. He's holding one in each hand. The Smurf is in a squatting position with the drum in front of him. He is painted real sloppy. He is one of the first original Smurf figures.
245/003	Judge	Peyo & Ed Dupuis	1/1/57	First Rubber Smurf	2" High		$18.00-25.00	The Smurf has a medium blue body. The Smurf is wearing a black gown with a white bow tie. He is looking to his right, he has his mouth open and is wearing black glasses. His left hand is pointing. He is painted real sloppy. He is one of the first original Smurf figures.
245/004	Mirror	Peyo & Ed Dupuis	1/1/57	First Rubber Smurf	2" High		$18.00-25.00	The Smurf has a medium blue body. He is wearing white pants and a white hat. He is holding a red mirror in his left hand, his other hand is at his side. The inside of the mirror is white, the Smurf is facing the mirror and is smiling. He is painted real sloppy. He is one of the first original Smurf figures.
245/005	Smurfette	Peyo & Ed Dupuis	1/1/57	First Rubber Smurf	2" High		$18.00-25.00	Smurfette has a medium blue body. She is standing with her left hand on her chest and her right hand out at her side. Smurfette is wearing a white dress and shoes. She is painted real sloppy. She is one of the first original Smurf figures.
245/006	Naughty	Peyo & Ed Dupuis	1/1/57	First Rubber Smurf	2" High		$18.00-25.00	The Smurf has a medium blue body. He has his thumbs in his ears. He has his tongue sticking out. His tongue is maroon. He is painted real sloppy. He is one of the first original Smurf figures.

Set 181: PVC's- Easter Figures

181/001	Smurfette With An Easter Egg	Schleich/Peyo	1/1/84	Rubber Figure	2" High	Germany/China	$4.50-6.00	Smurfette has a medium blue body. She is standing with a light purple egg in front of her. There are pink flower spots on the egg. The egg is broken and there is a yellow chick, in the egg. Smurfette is wearing a white dress and has a purple flower in her hat.
181/002	Smurfette With An Easter Egg	Schleich/Peyo/Applause	1/1/84	Rubber Figure	2" High	Portugal	$4.50-6.00	Smurfette has a bright medium blue body. She is standing with a light purple egg in front of her. There are pink flower spots on the egg. The egg is broken and there is a yellow chick, in the egg. Smurfette is wearing a white dress and has a purple flower in her hat.
181/003	Smurf With A Easter Egg	Schleich/Peyo/ W.B/ Applause	1/1/84	Rubber Figure	2" High	Portugal	$4.50-6.00	The Smurf has a medium bright blue body. He is standing with an egg next to him. The egg is blue on the bottom, pink on top and has a yellowish/ orange bow around it. The Smurf is resting his right hand on the egg. The egg is resting on a bright green patch of grass.
181/004	Smurf With A Easter Egg	Schleich/Peyo	1/1/84	Rubber Figure	2" High	Germany/China	$4.50-6.00	The Smurf has a medium blue body. He is standing with an egg next to him. The egg is blue on the bottom, pink on top and has a yellowish/orange bow around it. The Smurf is resting his right hand on the egg. The egg is resting on a dark green patch of grass.
181/005	Smurf Painting An Easter Egg	Schleich/Peyo	1/1/93	Rubber Figure	2" High	Germany/China	$4.50-6.00	The Smurf has a medium bright blue body. He is wearing a rust jacket, with a black tie, a white hat and white pants. The Smurf is holding a brown paint brush with a green tip in his left hand. In his right hand he is holding a green Easter egg with yellow and rust detailing.
181/006	Smurf Painting An Easter Egg	Schleich/ Peyo	1/1/93	Rubber Figure	2" High	Germany/China	$4.50-6.00	The Smurf has a medium bright blue body. He is wearing a red jacket, with a black tie, a white hat and white pants. The Smurf is holding a brown paint brush with a yellow tip in his left hand. In his right hand he is holding a green Easter egg with yellow and red detailing.
181/007	Smurf In A Egg	Schleich/Peyo	1/1/93	Rubber Figure	2" High	Germany/China	$4.50-6.00	The Smurf has a medium blue body. He is wearing a pink bunny outfit. The Smurf is in an egg. The egg is white with orange detailing. The Smurf has some of the shell on his head.
181/008	Smurf In A White Bunny Costume	Schleich/Peyo	1/1/82	Rubber Figure	2" High	Portugal	$6.00-8.00	The Smurf has a dark blue body. He is wearing a white bunny costume. The inside of the bunny ears are dark pink. The Smurf is holding a green egg with a dark yellow ribbon and bow. The Smurfs right ear is attached to the bow on the egg.
181/009	Smurf In A White Bunny Costume	Schleich/Peyo	1/1/82	Rubber Figure	2"	Germany/China	$6.00-8.00	The Smurf has a dark blue body. He is wearing a white bunny costume. The inside of the bunny ears are light pink. The Smurf is holding a green egg with a yellow ribbon and bow. His ear is not attached to the egg.
181/010	Smurfette In A Pink Bunny Costume	Schleich/Peyo	1/1/82	Rubber Figure	2" High	Hong Kong	$6.00-8.00	Smurfette is wearing a dark pink bunny costume. The inside of her ears are rose color. Smurfette has dark yellow hair. Smurfette is holding a yellow egg in one hand. She is carrying a brown basket on her right arm. The basket has 1 green egg and 1 blue egg in it.
181/011	Smurfette In A Pink Bunny Costume	Schleich/Peyo	1/1/82	Rubber Figure	2"	Portugal	$6.00-8.00	Smurfette is wearing a pink bunny costume. The inside of her ears are dark pink color. Smurfette has yellow hair. Smurfette is holding a yellow egg in one hand. She is carrying a brown basket on her right arm. The basket has 1 green egg and 1 blue egg in it.
181/012	Smurfette In A Pink Bunny Costume	Schleich/ Peyo	1/1/82	Rubber Figure	2"	Germany/China	$6.00-8.00	Smurfette is wearing a light pink bunny costume. The inside of her ears are light pink color. Smurfette has pale yellow hair. Smurfette is holding a yellow egg in one hand. She is carrying a brown basket on her right arm. The basket has 1 green egg and 1 blue egg in it.
181/013	Smurfette With An Easter Egg	Schleich/Peyo	1/1/84	Rubber Figure	2"	Germany/China	$4.50-6.00	Smurfette has a medium blue body. She is standing with a dark purple egg in front of her. There are yellow flower spots on the egg. The egg is broken and there is a yellow chick, in the egg. Smurfette is wearing a white dress and has a red flower in her hat.
181/014	Smurf With A Easter Egg	Schleich/Peyo	1/1/84	Rubber Figure	2"	Germany/China	$4.50-6.00	The Smurf has a medium blue body. He is standing with an egg next to him. The egg is blue on the bottom, yellow on top and has a red bow around it. The Smurf is resting his right hand on the egg. The egg is resting on a dark green patch of grass.
181/015	Smurf In A Egg	Schleich/Peyo	1/1/93	Rubber Figure	2"	Germany/China	$4.50-6.00	The Smurf has a medium blue body. He is wearing a white bunny outfit. The Smurf is in an egg. The egg is yellow with a green stripe and red dots. The Smurf has some of the shell on his head.
181/016	Smurf Carrying An Egg On His Back	Schleich/Peyo	1/1/96	Rubber Figure	2"	China	$4.50-6.00	The Smurf has a medium blue body. He is carrying an egg on his back. The egg is dark pink, dark purple and yellow with red and yellow detailing.
181/017	Smurf Eating An Egg	Schleich/Peyo	1/1/96	Rubber Figure	2"	China	$4.50-6.00	The Smurf has a medium blue body. He is sitting with an egg between his legs. The Smurf is holding 2 pieces of the chocolate eggs in his hands. The outside of the eggs is dark purple and red. The egg has yellow and green detailing. The Smurf is eating the egg.

Set 177: PVC's- History Smurfs

177/001	Paul Revere	Schleich/Peyo	1/1/84	Rubber Smurf	2" High	Hong Kong	$11.00-15.00	The Smurf has a medium blue body. He is wearing a bright blue coat with tarnished gold trim, a bright blue hat with tarnished gold trim, white pants and a white tie. The Smurf is carrying a lantern in his left hand. The lantern is yellow/black and tarnished gold color. The Smurf has his other hand by his mouth like he's yelling.
177/002	Benjamin Franklin	Schleich/Peyo	1/1/84	Rubber Smurf	2" High	Macau	$18.00-25.00	The Smurf has a medium blue body. He is wearing a tan jacket, a red vest, white pants, brown glasses and a brown hat. The Smurf has white hair. He is holding a yellow kite with a tan string. The string has a gray key tied on the end.
177/003	Christopher Columbus	Schleich/Peyo	1/1/84	Rubber Smurf	2" High	Hong Kong	$13.50-18.00	The Smurf has a medium blue body. He is wearing a purple jacket with white trim, white pants, a white ruffle neck piece and a brown sailors hat with tan trim. The Smurf is holding a gray and black telescope in his left hand. He is pointing with his right hand. The Smurf has long tan hair.
177/004	Thomas Edison	Schleich/Peyo	1/1/84	Rubber Smurf	2" High	Macau	$18.00-25.00	The Smurf has a medium blue body. He is wearing a black coat, white pants, tan hat with a black stripe, a tan vest with red buttons and a red bow tie. The Smurf has white hair. He is wearing black glasses with white lenses. In his right hand he is holding a white light bulb with a silver end.
177/005	George Washington	Schleich/Peyo	1/1/84	Rubber Smurf	2" High	Hong Kong	$13.50-18.00	The Smurf has a medium blue body. He is wearing a dark blue coat with tarnished gold trim, a dark blue hat with tarnished gold trim, a white tie and white pants. The Smurf has white hair. He is holding an ax in his right hand. The ax is gray with a dark brown handle.
177/006	Abraham Lincoln	Schleich/Peyo	1/1/84	Rubber Smurf	2" High	Macau	$11.00-15.00	The Smurf has a medium blue body. He is wearing a black hat with a gray stripe, a black coat with gray trim, white pants and a red tie. The Smurf has a dark brown beard. He is holding a creme colored scroll with red writing in his left hand. He has a finger in the air on his right hand.

Set243. PVC's- Novelty Figures

243/001	Smurf Holding A Yellow Ladder	Schleich/Peyo	1/1/79	Rubber Figure	2" High	West Germany	$15.00-20.00	The Smurf has a medium blue body. The Smurf is the sign-bearing Smurf but he is holding a long yellow ladder in his hand. The ladder is a separate plastic piece. The Smurf has one hand covering his mouth.
243/002	Smurf Holding A Yellow Lightening Rod	Schleich/Peyo	1/1/79	Rubber Figure	2" High	West Germany	$15.00-20.00	The Smurf has a medium blue body. The Smurf is the sign-bearing Smurf but he is holding a long yellow lightening rod in his hand. The lightening rod is a separate plastic piece. The Smurf has one hand covering his mouth.
243/003	Smurf Holding A Red Lightening Rod	Schleich/Peyo	1/1/79	Rubber Figure	2" High	West Germany	$15.00-20.00	The Smurf has a medium blue body. The Smurf is the sign-bearing Smurf but he is holding a long red lightening rod in his hand. The lightening rod is a separate plastic piece. The Smurf has one hand covering his mouth.
243/004	Smurf Holding A Yellow Pacifier	Schleich/Peyo	1/1/79	Rubber Figure	2" High	West Germany	$15.00-20.00	The Smurf has a medium blue body. The Smurf is the sign-bearing Smurf but he is holding a yellow pacifier in his hand. The pacifier is a separate plastic piece. The Smurf has one hand covering his mouth
243/005	Smurf Holding A Red Pacifier	Schleich/Peyo	1/1/79	Rubber Figure	2" High	West Germany	$15.00-20.00	The Smurf has a medium blue body. The Smurf is the sign-bearing Smurf but he is holding a red pacifier in his hand. The pacifier is a separate plastic piece. The Smurf has one hand covering his mouth
243/006	Smurf Holding A Red Cup	Schleich/Peyo	1/1/79	Rubber Figure	2" High	West Germany	$15.00-20.00	The Smurf has a medium blue body. The Smurf is the sign-bearing Smurf but he is holding a red cup in his hand. The cup is a separate plastic piece. The Smurf has one hand covering his mouth.
243/007	Smurf Holding A Yellow Cup	Schleich/Peyo	1/1/79	Rubber Figure	2" High	West Germany	$15.00-20.00	The Smurf has a medium blue body. The Smurf is the sign-bearing Smurf but he is holding a yellow cup in his hand. The cup is a separate plastic piece. The Smurf has one hand covering his mouth.
243/008	Popsicle Smurf On A Skateboard	Schleich		Rubber Figure	2" High		$7.50-10.00	The Smurf has a medium blue body. He is eating a popsicle. The popsicle is red on the bottom and white on top. The popsicle is on a yellow stick. The Smurf is holding the popsicle in his right hand. He has his tongue sticking out of his mouth. The Smurf is standing on a blue plastic skateboard. The skateboard has red wheels.
243/009	Policeman On A Skateboard	Schleich		Rubber Figure	2" High		$7.50-10.00	The Smurf has a dark blue body. He is wearing a black hat, a black jacket with silver buttons, a brown belt with a silver buckle and white pants. He is holding a brown stick in his left hand. He is holding a silver whistle to his mouth with his right hand. The Smurf is standing on a blue plastic skateboard. The skateboard has red wheels.

243/010	CB Operator Smurf On A Skateboard	Schleich		Rubber Figure	2" High		$7.50-10.00	The Smurf has a medium blue body. He is holding a silver CB with a silver mike cord, dark brown mike and a black antenna. The Smurf has a dark brown strap across his chest to hold the CB. The Smurf is wearing a white hat with silver earphones and a silver and black headband. The Smurf is standing on a blue plastic skateboard. The skateboard has red wheels.
243/011	Sailor Smurf On A Skateboard	Schleich		Rubber Figure	2" High		$7.50-10.00	The Smurf has a medium blue body. He is wearing a white sailor suit with dark blue trim and a white sailor hat. The Smurf is carrying a brown sea bag over his right shoulder. The Smurf is standing on a blue plastic skateboard. The skateboard has red wheels.
243/012	Smurf Carrying A Round About Road Sign	Schleich/ Peyo	1/1/79	Rubber Figure	2" High	West Germany	$15.00-20.00	The Smurf has a medium blue body. The Smurf is the sign-bearing Smurf but he is holding a traffic sign. The traffic sign is yellow plastic. The top of it is circular. It has a blue sticker with white arrows going around in a circle. The Smurf is standing with his hand over his mouth.
243/013	Smurf Carrying A Hospital Sign	Schleich/Peyo	1/1/79	Rubber Figure	2" High	West Germany	$15.00-20.00	The Smurf has a medium blue body. The Smurf is the sign-bearing Smurf but he is holding a traffic sign. The traffic sign is yellow plastic. The top of it is square. It has a blue and white sticker with a red cross in the center. The Smurf is standing with his hand over his mouth.
243/014	Smurf Carrying A School Crossing Sign	W. Berrie/Peyo	1/1/79	Rubber Figure	2" High	West Germany	$15.00-20.00	The Smurf has a medium blue body. The Smurf is the sign-bearing Smurf but he is holding a traffic sign. The traffic sign is yellow plastic. The top of it a triangle. It has a white and red sticker with a blue picture of 2 kids running across a street. The Smurf is standing with his hand over his mouth.
243/015	Smurf Carrying A Pedestrian Road Sign	W. Berrie/Peyo	1/1/79	Rubber Figure	2" High	West Germany	$15.00-20.00	The Smurf has a medium blue body. The Smurf is the sign-bearing Smurf but he is holding a traffic sign. The traffic sign is yellow plastic. The top of it is a triangle. It has a white and red sticker with a picture of a blue pedestrian. The Smurf is standing with his hand over his mouth.
243/016	Smurf Holding A Stop Sign	W. Berrie/Peyo	1/1/79	Rubber Figure	2" High	West Germany	$15.00-20.00	The Smurf has a medium blue body. The Smurf is the sign-bearing Smurf but he is holding a traffic sign. The traffic sign is yellow plastic. The top of it is circular. It has a red sticker with white strip in the middle. The Smurf is standing with his hand over his mouth.
243/017	Smurf Carrying A 2 Way Traffic Sign	W. Berrie/Peyo	1/1/79	Rubber Figure	2" High	West Germany	$15.00-20.00	The Smurf has a medium blue body. The Smurf is the sign-bearing Smurf but he is holding a traffic sign. The traffic sign is yellow plastic. The top of it is circular. It has a red and white sticker with a blue arrow pointing up and a red arrow pointing down arrows. The Smurf is standing with his hand over his mouth.
243/018	Smurf Carrying A Train Crossing Sign	W. Berrie/Peyo	1/1/79	Rubber Figure	2" High	West Germany	$15.00-20.00	The Smurf has a medium blue body. The Smurf is the sign-bearing Smurf but he is holding a traffic sign. The traffic sign is yellow plastic. The top of it is a triangle. It has a white and red sticker with a blue triangle symbol in the center. The Smurf is standing with his hand over his mouth.
243/019	Smurf Carrying A Cedael Paso Sign	W. Berrie/Peyo	1/1/79	Rubber Figure	2" High	West Germany	$15.00-20.00	The Smurf has a medium blue body. The Smurf is the sign-bearing Smurf but he is holding a traffic sign. The traffic sign is yellow plastic. The top of it is a triangle upside down. It has a white and red sticker with the words "Cedael Paso" in blue. The Smurf is standing with his hand over his mouth.
243/020	Smurf With Thermometer	Endresss + Hauser	1/1/93	Rubber Figure	2" High	Germany	$55.00-75.00	The Smurf is made of polystone. The Smurf is standing, holding some kind of thermometer. The thermometer is on a long stick. The Smurf is wearing a gray jacket, white pants and gray boots. The Smurf has a white construction hat with the letters "EH" on the side. The Smurf is holding a white paper in his other hand.

Set 74: PVC's- Promotional Figures

074/001	Brainy - Carl Zeiss W. Germany	Schleich	1/1/69	Rubber PVC's	2" High	Portugal	$22.50-30.00	The Smurf has a bright medium blue body. He is standing with his hands out at his sides. The Smurf is wearing a red pair of glasses. He is a promotional for an eye glass company in Germany. Carl Zeiss is in blue letters on the back of his hat.
074/002	Torchbearer		1/1/78	Rubber PVC's	2" High		$18.00-25.00	The Smurf has a light blue body. He is carrying a silver torch with a red and yellow flame, he is carrying the torch in his right hand. The Smurf has a white shirt with a yellow and black horse emblem on the front. He has white shorts and shoes.
074/003	BP Cleaner	Schleich	1/1/79	Rubber PVC's	2" High	West Germany	$18.00-25.00	The Smurf has a light blue body. He is wearing white overalls with a green and yellow BP emblem on the front. He is holding a maroon rag in his left hand. There is an orange dot on the bottom of his foot.
074/004	ASB First-Aid	Schleich	1/1/78	Rubber PVC's	2" High	West Germany	$22.50-30.00	The Smurf has a medium blue body. He is wearing a white doctors uniform with a yellow dot on the front. A finger on his right hand is wrapped in a bandage. He is carrying a dark brown medical bag with an ASB emblem on the front. ASB means Arbeiter-Samanter-Bund a German paramedics.
074/005	Ass Card Player	Schleich	1/1/78	Rubber PVC's	2" High	West Germany	$45.00-65.00	The Smurf has a medium blue body. He has his left hand over his mouth like he's laughing, his left eye is closed and his right eye is open. He is holding 3 cards in his right hand, the face card is a heart with ASS on. He has a club card stuffed in the front of his pants. Ass is in red letters.
074/006	Thirsty - Staatl Fachingen	Schleich	1/1/78	Rubber PVC's	2" High	West Germany	$37.50-50.00	The Smurf has a medium blue body. He is holding a green bottle with Staatl Fachingen (stands for a mineral company) in blue on the front of the bottle.
074/007	Ice Skater - Holiday On Ice	Schleich	1/1/79	Rubber PVC's	2" High	Hong Kong	$25.00-35.00	The Smurf has a dark blue body. The Smurf is wearing a red scarf and yellow gloves. The Smurf has silver blades on his feet. He is laying on his back. The back of his hat says Holiday On Ice.
074/008	Bodybuilder - Sports & Fitness	Schleich	1/1/88	Rubber PVC's	2" High	West Germany	$25.00-35.00	The Smurf has a bright blue body. He is wearing gold posing trunks. He is flexing his muscles. The back of his hat says Sports & Fitness in black letters.
074/009	Scot - Silan	Schleich	1/1/79	Rubber PVC's	2" High	West Germany	$25.00-35.00	The Smurf has a medium blue body. He is wearing a red hat with a green tassel and red cape. He is standing, playing a yellow bagpipe. Silan is in blue letters on the bottom of his shoe. There is a green rope around the top of his shoes.
074/010	Baseball Batter - Silan	Schleich	1/1/81	Rubber PVC's	2" High	Portugal	$25.00-35.00	The Smurf has a dark blue body. He is wearing a white uniform with red sleeves, a red belt, red socks, a white hat with a red visor and black shoes. The Smurf is holding a tan bat. Silan is written on the bottom of one shoe in white letters.
074/011	Soccer Smurfette - Silan	Schleich	1/1/83	Rubber PVC's	2" High	West Germany	$25.00-35.00	Smurfette has a medium blue body. She is wearing a white, orange, and red shirt and shoes. She has on white shorts and a white hat. Smurfette is in a running position with a white soccer ball on her left foot. The back of her right foot has Silan written in black letters. Smurfette is standing on a bright green base.
074/012	Jogger - Silan	Schleich	1/1/82	Rubber PVC's	2" High	West Germany	$25.00-35.00	The Smurf has a medium blue body. He is wearing a red sweatsuit jacket, red shoes and white pants with a red stripe down the sides. The Smurf is in a running position and is standing on a bright green platform. Silan is in white letters on his right foot.
074/013	Papa With Book - Silan	Schleich	1/1/82	Rubber PVC's	2" High	West Germany	$30.00-40.00	Papa has a medium blue body. He is standing with a creme color book in his hands. The book is open and in the inside are numbers. Papa is wearing a red hat and red pants. Silan is in white letters on the bottom of his foot. There is a black dot on the bottom of Papa's foot.
074/014	Rugby - Omo	Schleich	1/1/80	Rubber PVC's	2" High	West Germany	$25.00-35.00	The Smurf has a medium blue body. He has his left hand out to block, his right hand is hugging a brown football to his side. The Smurf is wearing a bright yellow shirt, bright yellow socks, red shorts and black shoes. He has an exhausted look on his face. Omo is in black letters on the back of his shirt. There is a red dot on his foot.
074/015	Tennis Played 2 - Omo	Schleich	1/1/77	Rubber PVC's	2" High	West Germany	$25.00-35.00	The Smurf has a medium blue body. He is holding a plastic red tennis racket with big holes, in his right hand. He is holding a yellow tennis ball in his left hand. He is standing like he's going to serve the ball. He is wearing a white tennis shirt and shorts, with Omo in black letters on the front of his shirt. There is a red dot on the bottom of his foot.
074/016	Paracodin Smurf (red)	Schleich	1/1/79	Rubber PVC's	2 1/4" High	West Germany	$55.00-75.00	The Smurf has a medium blue body. He has a red nose and is breathing on his left hand. The Smurf is wearing a red scarf and white hat. He is standing on a white base that has Paracodin in red letters (the letters are worn off). Paracodin is a medicine company.
074/017	Paracodin Smurf (Yellow)	Schleich	1/1/79	Rubber PVC's	2 1/4" High	West Germany	$55.00-75.00	The Smurf has a medium blue body. He has a red nose and is breathing on his left hand. The Smurf is wearing a yellow scarf and yellow hat. He is standing on a white base that has Paracodin in red letters, the letters are worn off. Paracodin is a medicine company.
074/018	Ice Lolly With Scholler Flag	Schleich		Rubber PVC's	2" High	West Germany	$45.00-65.00	The Smurf has a medium blue body. He is holding a red with a white top popsicle in his right hand. In his left hand he's holding a white plastic flag pole with a blue flag on the end. The flag has Scholler in white letters across the middle.
074/019	GB Smurf	I.M.P.S.	1/1/80	Rubber PVC's	2" High	Brussels	$25.00-35.00	The Smurf has a bright, medium blue body. He is wearing a red hat and cape with a yellow star. He has a red ball with GB in white letters that he is pulling a yellow cloth off.
074/020	Money Smurf - Volks/Raiffeisen Bank	Schleich		Rubber PVC's	2" High	West Germany	$25.00-35.00	The Smurf has a medium blue body. He is standing, holding a 1 cent dark orange coin with Papa on the back of the coin. The Smurf is holding the coin in his right hand. He is wearing a white hat and white pants. There is a white x with a circle around in a black box and a white v in a black box on the back of the Smurf's hat. The symbols represent a German bank, Volks/Raiffeisen bank.
074/021	Miller Bag Smurf - Volks/Raiffeisen Bank	Schleich		Rubber PVC's	2" High	West Germany	$67.50-90.00	The Smurf has a medium blue body. He is holding a dull maroon miller bag in front of him. On the front of the miller bag is a maroon x and v in white boxes. The symbols represent a German bank, Volks/Raiffeisen bank. The Smurf has his eyes closed and a big grin on his face.
074/022	Smurf Holding A Piggy Bank - Volks/ Raiffeisen Bank	Schleich		Rubber PVC's	2" High	West Germany	$67.50-90.00	The Smurf has a medium blue body. He is standing, holding a dark orange bank (canister) with a x and v in black boxes on the front and white pants. He has his eyes closed and a big grin on his face. The symbols represent a German bank, Volks/Raiffeisen bank.
074/023	Phillips Light Bulb Smurf	Schleich	1/1/82	Rubber PVC's	2" High	West Germany	$95.00-125.00	The Smurf has a medium blue body. He is standing, holding a light bulb in both hands. The top of the light bulb is clear and the bottom is gray. Phillips is in silver letters on the gray part of the light bulb. There is a black dot on his foot.
074/024	75 Year Smurfs For BP	Schleich	1/1/79	Rubber PVC's	2" High	West Germany	$75.00-100.00	The two Smurfs have medium color bodies. One Smurf has his hands clasped around the other Smurfs hand shaking it. One Smurf has a green wreath around his neck and 75 painted on his stomach in white. The same Smurf has a green and yellow BP symbol on his hat. The Smurfs are standing on a yellow platform.
074/025	Smurf With OMO Soap box	Schleich/Peyo	1/1/82	Rubber PVC's	2" High	West Germany	$45.00-65.00	The Smurf has a medium blue body. He is standing, holding a white soap box in his hands. The soap box has OMO on the front in black letters. There is a red, blue, green and yellow circle under the OMO. The soap box is almost as big as the Smurf.
074/026	Ice Lolly With Schlumpfhausen Flag	Schleich/Peyo		Rubber PVC's	2" High	West Germany	$37.50-50.00	The Smurf has a medium blue body. He is holding a red with a white top popsicle in his right hand. In his left hand he's holding a yellow plastic flag pole with a yellow flag on the end. The flag has Schlumpf-hausen in blue letters across the middle.
074/027	Coca Cola Smurf	Peyo		Rubber PVC's	2" High	West Germany	$110.00-150.00	The Smurf has a medium blue body. He is holding a black bottle that looks like a coke bottle.
074/028	Rode Kruis Vlandereen First-Aid	Schleich/ Peyo	1/1/78	Rubber PVC's	2" High	Germany/China	$25.00-35.00	The Smurf has a medium blue body. He is wearing a white doctors uniform with a red cross on the front. A finger on his right hand is wrapped in a bandage. He is carrying a green medical bag with a white cross on the front. The Smurf is a key chain.
074/029	Heart - De Tout Mon Coeur	Schleich/Peyo	1/1/80	Rubber PVC's	2" High	West Germany	$25.00-35.00	The Smurf has a medium blue body. He is standing, holding a white and red plastic heart between his hands. The heart has black lettering with the phrase: DE TOUT MON COEUR.
074/030	Felis Navidad - Sign- bearer	Schleich/Peyo	1/1/79	Rubber PVC's	2" High	West Germany	$18.00-25.00	The Smurf has a dark blue body. He is standing, holding a gray plastic sign. There is a cardboard sign that says "Felis Navidad" in black letters. There is a picture of a Smurf singing next to the word. The card slides into the gray sign.

074/031	Bon Nadal - Sign- bearer	Schleich/Peyo	1/1/79	Rubber Figure	2" High	West Germany	$18.00-25.00	The Smurf has a dark blue body. He is standing, holding a gray plastic sign. There is a cardboard sign that says "Bon Nadal" in black letters. There is a picture of a Smurf singing next to the word. The card slides into the gray sign.
074/032	Feliz Aniversanrio - Sign- bearer	Schleich/Peyo	1/1/79	Rubber Figure	2" High	West Germany	$18.00-25.00	The Smurf has a dark blue body. He is standing, holding a gray plastic sign. There is a cardboard sign that says "Feliz Aniversanrio" in black letters. There is a picture of a Smurf carrying a cake next to the word. The card slides into the gray sign.
074/033	Schimmel Pianos	Schleich/Peyo	1/1/83	Rubber PVC's	2" High	West Germany	$37.50-50.00	The Smurf has a medium blue body. The piano is brown with white and black keys, a brown stool and a piece of white sheet music sitting on top of the piano. There is a sign on above the piano keys that says "Schimmel Pianos" in white letters. The sign is red.
074/034	Fluocaril - Toothbrush Smurf	Schleich/Peyo	1/1/79	Rubber PVC's	2" High	West Germany	$35.00-45.00	The Smurf has a dark blue body. He is standing, brushing his teeth with a yellow toothbrush. In his left hand he is holding a white tube of toothpaste. The toothpaste container says "Fluocaril bi-fluore 180" in green letters on the front.
074/035	Paracodin Smurf (Yellow)	Schleich/Peyo	1/1/79	Rubber PVC's	2" High	West Germany	$55.00-75.00	The Smurf has a medium blue body. He has a red nose and is breathing on his left hand. The Smurf is wearing a yellow scarf and white hat. He is standing on a white base that has Paracodin in red letters the letters that are written in marker. Paracodin is a medicine company.
074/036	Colgate	Schleich/Peyo/Puppy	1/1/92	Rubber PVC's	2" High	China	$18.00-25.00	The Smurf has a bright blue body. He is standing, brushing his teeth with a yellow toothbrush. In his left hand he is holding a red tube of toothpaste The toothpaste container says Colgate in white letters.
074/037	Philips White Light Bulb	Schleich/Peyo	1/1/80	Rubber PVC's	2" High		$100.00-133.00	The Smurf has a medium blue body. He is standing, holding a light bulb in both hands. The top of the light bulb is white with gold socket, Phillips is in gold letters on the gray part of the light bulb. The Smurf has his hands extended out in front of him.
074/038	Smurf In Wheelchair	Peyo/CE	1/1/94	Rubber PVC's	2" High	China	$35.00-45.00	The Smurf has a medium blue body. He sits in a tan wheelchair. His right leg has a white bandage around it. The Smurf is for a medical company in Europe.
074/039	Smurf With Crutch	Peyo/CE	1/1/94	Rubber PVC's	2" High	China	$35.00-45.00	The Smurf has a medium blue body. The Smurf is leaning on a tan crutch. The Smurfs right foot is in a white bandage. The Smurf is a promotional for a medical company in Europe.
074/040	30th Anniversary Smurf	Peyo/I.M.P.S.	1/1/88	Rubber PVC's	2 1/2" High	Brussels	$25.00-35.00	The Smurf is standing with his hands out at his side. He has a light blue body. He is standing on a gold trophy box. The front of the box says 30. He is a promotion for their 30th anniversary.

Set 73: PVC's- Regular Figures

073/001	Papa	Schleich/Peyo/W.B.	1/1/69	Rubber Figures	2" High	Hong Kong	$3.00-4.00	Papa is standing with both hands out at his side. His body is medium blue. He has a red hat and pants on.
073/002	Papa	Schleich/Peyo	1/1/69	Rubber Figures	2" High	Hong Kong	$3.00-4.00	Papa is standing with both hands out at his side. His body is dark blue. He has a red hat and pants on. I also have a 1980 Smurf with this same description.
073/003	Normal	Schleich/Peyo	1/1/65	Rubber Figures	2" High	Hong Kong	$3.00-4.00	The Smurfs body is light blue. He has a white hat and pants. His hands are out at his sides and he has a black thin line for a mouth, that is smiling.
073/004	Normal	Schleich/Peyo	1/1/65	Rubber Figures	2" High		$3.00-4.00	The Smurfs body is medium blue. He has a white hat and pants. His hands are out at his sides and he has a black thin line for a mouth, that is smiling.
073/005	Astro	Schleich/Peyo/W.B.	1/1/65	Rubber Figures	2" High	Hong Kong	$11.00-15.00	The Smurfs body is medium blue. He is wearing a white astronaut outfit with a white hat and a clear plastic helmet. The Smurf has his hands extended at his side, the left hand has his finger pointing up. He has small feet and the right foot is lifted a little.
073/006	Astro Bully	Bully /Peyo	1/1/65	Rubber Figures	2" High	West Germany	$18.00-25.00	The Smurfs body is medium blue. He is wearing a white astronaut outfit with a white hat and a clear plastic helmet and a red tie. The Smurf has his hands extended at his side, He has big feet and the left foot is in front of the right one a little. RARE!!!
073/007	Shiver	Schleich/Peyo	1/1/69	Rubber Figures	2" High	Hong Kong	$7.50-10.00	The Smurf has a medium blue body. The Smurf is standing with his hands behind his back. He has a red scarf wrapped around his face.
073/008	Shiver	Peyo	1/1/69	Rubber Figures	2" High	West Germany	$6.00-8.00	The Smurf has a light blue body. The Smurf is standing with his hands behind his back. He has a yellow scarf wrapped around his face.
073/009	Gold	Schleich/Peyo	1/1/65	Rubber Figures	2" High	Hong Kong	$15.00-20.00	The Smurf has a medium blue body. He is standing with his hands out at his side. He has a gold hat and gold pants. His mouth is a black line in the form of a smile.
073/010	Brainy	Schleich/Peyo	1/1/69	Rubber Figures	2" High	Hong Kong	$3.00-4.00	The Smurf has a dark blue body. He has black glasses and his mouth is in the form of an O. The Smurf is standing with his hands out at his side.
073/011	Brainy	Schleich/Peyo	1/1/69	Rubber Figures	2" High	Hong Kong	$4.50-6.00	The Smurf has a medium blue body. He has red glasses and his mouth is in the form of an O. The Smurf is standing with his hands out at his side.
073/012	Brainy	Schleich/Peyo	1/1/69	Rubber Figures	2" High	Hong Kong	$4.50-6.00	The Smurf has a dark blue body. He has red glasses and his mouth is in the form of an O. The Smurf is standing with his hands out at his side.
073/013	Brainy	Peyo	1/1/69	Rubber Figures	2" High	West Germany	$9.00-12.00	The Smurf has a light blue body. He has yellow glasses and his mouth is in the form of an O. The Smurf is standing with his hands out at his side.
073/014	Angry (Black)	Schleich/Peyo	1/1/66	Rubber Figures	2" High	West Germany	$22.50-30.00	The Smurf has a black body. He has his arms fully extended straight out at his sides. He has red eyes and his teeth are clenched. His teeth are black lines. He is a bigger mold. His left foot is lifted a little and in front of the right.
073/015	Angry (Black)	Peyo	1/1/66	Rubber Figures	2" High	West Germany	$22.50-30.00	The Smurf has a black body. He has his arms extended and slanted down a little at his sides. He has red eyes and his teeth are clenched. His teeth are black lines. He is a small mold. His left foot is slightly in front of the other.
073/016	Angry (Black)	Peyo	1/1/66	Rubber Figures	2" High	West Germany	$22.50-30.00	The Smurf has a black body. He has his arms extended and slanted down a little at his sides. He has black eyes and his teeth are clenched. His teeth are black lines. He is a small mold and shiny. His feet are round and close together.
073/017	Spy	Schleich/Peyo	1/1/66	Rubber Figures	2" High	Hong Kong	$4.50-6.00	The Smurf has a dark blue body. He is holding 1 finger in front of his mouth. He has a medium red cape, white hat and white pants on. He has a black mask over his face.
073/018	Spy	Peyo	1/1/66	Rubber Figures	2" High	West Germany	$4.50-6.00	The Smurf has a light blue body. He is holding 1 finger in front of his mouth. He has a maroon cape, white hat and white pants on. He has a black mask over his face. This Smurf has an orange dot on the bottom.
073/019	Drummer	Schleich/Peyo	1/1/66	Rubber Figures	2" High	Hong Kong	$4.50-6.00	The Smurf has a dark blue body. He has a red drum with yellow support strings and black edge rings. The drum sticks are white with red drum stick tops, he's holding one in each hand. The Smurf is in a squatting position with the drum in front of him.
073/020	Drummer	Schleich/Peyo	1/1/66	Rubber Figures	2" High	Hong Kong	$4.50-6.00	The Smurf has a light blue body. He has a red drum with yellow support strings and black edge rings. The drum sticks are white with dark red drum stick tops, he's holding one in each hand. The Smurf is in a squatting position with the drum in front of him.
073/021	Drummer	Schleich/Peyo	1/1/66	Rubber Figures	2" High	West Germany	$11.00-15.00	The Smurf has a light blue body. He has a white drum top, red support strings with black between them and yellow edge rings. The drum sticks are yellow, he's holding one in each hand. The Smurf is in a squatting position with the drum in front of him.
073/022	Prisoner	Peyo	1/1/65	Rubber Figures	2" High	West Germany	$7.50-10.00	The Smurf has a medium blue body. He has a white hat and pants on that have black stripes. He has a sad face and his hands are out at his sides. There is an orange dot on the bottom of his foot.
073/023	Prisoner	Schleich/Peyo	1/1/65	Rubber Figures	2" High	Hong Kong	$7.50-10.00	The Smurf has a dark blue body. He has a white hat and pants on that have black stripes. He has a sad face and his hands are out at his sides.
073/024	Laughing	Schleich/Peyo	1/1/70	Rubber Figures	2" High	Hong Kong	$7.50-10.00	The Smurf has a dark shiny blue body. He is standing and pointing with his right hand and his left hand is covering his mouth.
073/025	Mechanic	Schleich/Peyo	1/1/70	Rubber Figures	2" High	West Germany	$11.00-15.00	The Smurf has a medium blue body. He has no suspenders. He has white pants on and is holding a black wrench in his left hand. He is smiling.
073/026	Mechanic	Schleich/Peyo	1/1/70	Rubber Figures	2" High	Hong Kong	$7.50-10.00	The Smurf has a light blue body. He has dark green bib overalls on. The Smurf is holding a light gray wrench in his left hand. He has a smile on his face.
073/027	Mechanic	Schleich/Peyo	1/1/70	Rubber Figures	2" High	Hong Kong	$7.50-10.00	The Smurf has a medium blue body. He has dark green bib overalls on. The Smurf is holding a light gray wrench in his left hand. He has a smile on his face.
073/028	Mechanic	Schleich/ Peyo	1/1/70	Rubber Figures	2" High	Hong Kong	$7.50-10.00	The Smurf has a medium blue body. He has dark green bib overalls on. The Smurf is holding a dark gray wrench in his left hand. He has a smile on his face.
073/029	Mechanic	Schleich/Peyo	1/1/70	Rubber Figures	2" High	Portugal	$7.50-10.00	The Smurf has a medium blue body. He has lime green bib overalls on. The Smurf is holding a light gray wrench in his left hand. He has a smile on his face.
073/030	Lute	Schleich/Peyo	1/1/69	Rubber Figures	2" High	Hong Kong	$4.50-6.00	The Smurf has a medium blue body. He has white pants, a white hat and no shirt. He is holding a red lute on his right side. He has his eyes closed and his mouth in the shape of an O.
073/031	Lute	Peyo	1/1/69	Rubber Figures	2" High	West Germany	$6.00-8.00	The Smurf has a medium blue body. He has white pants, a white hat and no shirt. He is holding a maroon lute on his right side. He has his eyes closed and his mouth in the shape of an O. There is a orange dot on the bottom of his foot.
073/032	Lute	Peyo	1/1/69	Rubber Figures	2" High	Portugal	$7.50-10.00	The Smurf has a medium blue body. He has white pants, a white hat and no shirt. He is holding a tan lute on his right side. He has his eyes closed and his mouth in the shape of an O.
073/033	Sunbather	Peyo	1/1/70	Rubber Figures	2" High	Germany	$15.00-20.00	The Smurf has a medium blue body. The Smurf is laying on his back with his hands behind his head, his eyes are closed and he is smiling. The Smurf has on white bathing trunks with red stripes.
073/034	Sunbather	Peyo	1/1/70	Rubber Figures	2" High	Germany	$15.00-20.00	The Smurf has a medium blue body. The Smurf is laying on his back with his hands behind his head, his eyes are closed and he is smiling. The Smurf has on red bathing trunks with black stripes. There is an orange dot on the bottom of his left foot.
073/035	Sunbather	Peyo	1/1/70	Rubber Figures	2" High	Germany	$15.00-20.00	The Smurf has a medium blue body. The Smurf is laying on his back with his hands behind his head, his eyes are closed and he is smiling. The Smurf has on yellow bathing trunks with black stripes. There is an orange dot on his left foot.
073/036	Sunbather	Schleich/Peyo	1/1/70	Rubber Figures	2" High	Hong Kong	$11.00-15.00	The Smurf has a medium blue body. The Smurf is laying on his back with his hands behind his head, his eyes are closed and he is smiling. The Smurf has on light green bathing trunks with black stripes.
073/037	Sunbather	Schleich/Peyo	1/1/70	Rubber Figures	2" High	Hong Kong	$15.00-20.00	The Smurf has a medium blue body. The Smurf is laying on his back with his hands behind his head, his eyes are closed and he is smiling. The Smurf has on yellow bathing trunks with black stripes.
073/038	Sunbather	Schleich/Peyo	1/1/70	Rubber Figures	2" High	Portugal	$11.00-15.00	The Smurf has a medium blue body. The Smurf is laying on his back with his hands behind his head, his eyes are closed and he is smiling. The Smurf has on green bathing trunks with black stripes. There is a red dot on the left foot.
073/039	Earache	Schleich/Peyo	1/1/71	Rubber Figures	2" High	Hong Kong	$7.50-10.00	The Smurf has a medium blue body. He is standing with his fingers in his ears and his mouth is open forming an O.
073/040	Earache	Schleich/Peyo	1/1/71	Rubber Figures	2" High	Hong Kong	$7.50-10.00	The Smurf has a light blue body. He is standing with his fingers in his ears and his mouth is open forming an O.
073/042	Judge	Schleich/Peyo	1/1/71	Rubber Figures	2" High	Hong Kong	$9.00-12.00	The Smurf has a dark blue body. The Smurf is wearing a red gown with a white bow tie. He is looking to his right, he has his mouth open and is wearing black glasses. His left hand is pointing.

073/043	Judge	Peyo	1/1/71	Rubber Figures	2" High	Portugal	$9.00-12.00	The Smurf has a dark blue body. The Smurf is wearing a red/orange gown with a white bow tie. He is looking to his right, he has his mouth open and is wearing black glasses. His left hand is pointing. There is an orange dot on his foot.
073/044	Mirror	Schleich/Peyo	1/1/72	Rubber Figures	2" High	Hong Kong	$3.00-4.00	The Smurf has a dark blue body. He is wearing white pants and a white hat. He is holding a red mirror in his left hand, his other hand is at his side. The inside of the mirror is white, the Smurf is facing the mirror and is smiling.
073/045	Mirror	Schleich/Peyo/W.B.	1/1/72	Rubber Figures	2" High	Hong Kong	$3.00-4.00	The Smurf has a medium blue body. He is wearing white pants and a white hat. He is holding a red mirror in his left hand, his other hand is at his side. The inside of the mirror is white, the Smurf is facing the mirror and is smiling.
073/046	Crying	Peyo	1/1/72	Rubber Figures	2" High	West Germany	$9.00-12.00	The Smurf has a medium blue body. A tear is rolling down his face and he is wiping his face with his right hand. In his left hand he is holding a light yellow handkerchief. He has on white pants and a white hat.
073/047	Crying	Schleich/Peyo	1/1/72	Rubber Figures	2" High	Hong Kong	$9.00-12.00	The Smurf has a medium blue body. A tear is rolling down his face and he is wiping his face with his right hand. In his left hand he is holding a yellow handkerchief. He has on white pants and a white hat.
073/048	Flower	Schleich/Peyo/W.B.	1/1/72	Rubber Figures	2" High	Hong Kong	$15.00-20.00	The Smurf has a dark blue body. He is wearing white pants and a white hat. His left hand is behind his back and his right hand is out at his side. He has a big cloth red flower with a yellow center glued by his mouth. His eyes are closed and he's smiling.
073/049	Gymnast	Schleich/Peyo	1/1/72	Rubber Figures	2" High	Hong Kong	$11.00-15.00	The Smurf has a dark blue body. He is flexing his right arm and is holding a black dumb bell in his left hand. He has a red tank top and white pants on.
073/050	Gymnast	Schleich/Peyo	1/1/72	Rubber Figures	2" High	Hong Kong	$1l.00-15.00	The Smurf has a light blue body. He is flexing his right arm and is holding a black dumb bell in his left hand. He has a red/orange tank top and white pants on.
073/051	Gymnast	Peyo	1/1/72	Rubber Figures	2" High	West Germany	$18.00-25.00	The Smurf has a light blue body. He is flexing his right arm and holding a black dumb bell in his left hand. He has a yellow tank top and white pants on.
073/052	Sleepwalker	Schleich/Peyo	1/1/72	Rubber Figures	2" High	Hong Kong	$6.00-8.00	The Smurf has a dark blue body. He has his eyes closed and both hands out in front of him. The Smurf is wearing a white night shirt and a white hat with a yellow tassel on top.
073/053	Author	Schleich/Peyo	1/1/72	Rubber Figures	2" High	Hong Kong	$7.50-10.00	The Smurf has a dark blue body. He is standing, holding a white feather pen with black feathers in his right hand. In his left hand he is holding a white scroll with no writing on it. He is wearing white pants and a white hat. He is looking up with his eyes.
073/054	Author	Schleich/Peyo	1/1/72	Rubber Figures	2" High	Hong Kong	$7.50-10.00	The Smurf has a medium blue body. He is standing, holding a white feather pen with black feathers in his right hand. In his left hand he is holding a white scroll with no writing on it. He is wearing white pants and a white hat. He is looking up with his eyes.
073/055	Rock N Roll	Schleich/Peyo	1/1/77	Rubber Figures	2" High	Hong Kong	$4.50-6.00	The Smurf has a medium blue body. The Smurf is holding a light orange guitar with yellow strings. He has his eyes shut and his mouth open.
073/056	Rock N Roll	Schleich/Peyo	1/1/77	Rubber Figures	2" High	Hong Kong	$4.50-6.00	The Smurf has a dark blue body. The Smurf is holding a medium orange guitar with yellow strings. He has his eyes shut and his mouth open.
073/057	Rock N Roll	Schleich/Peyo	1/1/77	Rubber Figures	2" High	Hong Kong	$4.50-6.00	The Smurf has a dark blue body. The Smurf is holding a dark orange guitar with yellow strings. He has his eyes shut and his mouth open.
073/058	Rock N Roll	Schleich/Peyo	1/1/77	Rubber Figures	2" High	Hong Kong	$7.50-10.00	The Smurf has a dark blue body. The Smurf is holding a brownish/orange guitar with yellow strings. He has his eyes shut and his mouth open.
073/059	Watchman	Schleich/Peyo	1/1/77	Rubber Figures	2" High	Hong Kong	$4.50-6.00	The Smurf has a light blue body. The Smurf is holding a red lantern in his right hand. There is black fray on the lantern with yellow . His eyes are looking to the side and his mouth is in the form of an O.
073/060	Watchman	Schleich/Peyo	1/1/77	Rubber Figures	2" High	Hong Kong	$4.50-6.00	The Smurf has a dark blue body. The Smurf is holding a red lantern in his right hand. There is black fray on the lantern with yellow . His eyes are looking to the side and his mouth is in the form of an O.
073/061	Swimmer	Schleich/Peyo	1/1/80	Rubber Figures	2" High	Hong Kong	$15.00-20.00	The Smurf has a dark blue body. The Smurf has a yellow bathing trunk on with a red inner tube around him. His hands are together in front of his face.
073/062	Swimmer	Schleich/Peyo	1/1/77	Rubber Figures	2" High	West Germany	$4.50-6.00	The Smurf has a light blue body. The Smurf has a red bathing trunk on with a yellow inner tube around him. His hands are together in front of his face.
073/063	Sitting	Schleich/Peyo	1/1/79	Rubber Figures	2" High	Hong Kong	$4.50-6.00	The Smurf has a dark dull blue body. He is sitting with his hands flat on both sides of him. He is wearing a white hat and white pants. He has a smile on his face.
073/064	Sitting	Schleich/Peyo	1/1/78	Rubber Figures	2" High	Hong Kong	$4.50-6.00	The Smurf has a medium shiny blue body. He is sitting with his hands flat on both sides of him. He is wearing a white hat and white pants. He has a smile on his face.
073/065	Thinker	Schleich/Peyo	1/1/78	Rubber Figures	2" High	Germany	$11.00-15.00	The Smurf has a medium blue body. He is sitting with his left hand on his knee and he is resting his head on his right hand. He has a smile on his face. There is a red dot on the bottom of the Smurf. BIG MOLD!!!
073/066	Thinker	Schleich/Peyo	1/1/78	Rubber Figures	2" High	West Germany	$35.00-45.00	The Smurf has a light blue body. He is sitting with his left hand on his foot and he is resting his head on his right hand. He has a smile on his face. RARE!!! SMALL MOLD!!!
073/067	Gardner	Peyo	1/1/78	Rubber Figures	2" High	West Germany	$6.00-8.00	The Smurf has a medium blue body. He is wearing a green apron. His hands are out at his side and curled down.
073/068	Money	Schleich/Peyo	1/1/78	Rubber Figures	2" High	Hong Kong	$18.00-25.00	The Smurf has a dark blue body. He is standing, holding a 1 cent orange coin with Papa on the back of the coin in his right hand. He is wearing a white hat and white pants.
073/069	Money	Schleich/Peyo	1/1/78	Rubber Figures	2" High	Hong Kong	$18.00-25.00	The Smurf has a medium blue body. He is standing, holding a 1 cent light orange coin with Papa on the back of the coin in his right hand. He is wearing a white hat and white pants.
073/070	Torchbearer	Schleich/Peyo	1/1/78	Rubber Figures	2" High	Hong Kong	$25.00-35.00	The Smurf has a light blue body. He is holding a torch in his right hand. He is wearing a white shirt and shorts with a black waistband.
073/071	Torchbearer	Schleich/Peyo	1/1/78	Rubber Figures	2" High	West Germany	$3.00-4.00	The Smurf has a medium blue body. He is holding a torch in his right hand. He is wearing a white shirt and red shorts. There is a red dot on his foot.
073/072	Postman	Schleich/Peyo	1/1/78	Rubber Figures	2" High	Hong Kong	$3.00-4.00	The Smurf has a dark blue body. He is holding a yellow horn in his mouth with his right hand. He has a dark brown mailbag. He is holding a white envelope with a pink heart on the flap of the envelope in his left hand.
073/073	Postman	Peyo	1/1/78	Rubber Figures	2" High	West Germany	$3.00-4.00	The Smurf has a light blue body. He is holding a yellow horn in his mouth with his right hand. He has a reddish/brown mailbag. He is holding a white envelope with a red heart on the flap of the envelope in his left hand.
073/074	Postman	Peyo	1/1/78	Rubber Figures	2" High	West Germany	$45.00-65.00	The Smurf has a light blue body. He is holding a yellow horn in his mouth with his right hand. He has a brown mailbag. He is holding a white envelope with holly on the flap of the envelope in his left hand.
073/075	Ice Hockey	Schleich/Peyo	1/1/78	Rubber Figures	2" High	Hong Kong	$4.50-6.00	The Smurf has a medium blue body. The Smurf has a bright yellow hat, shirt and socks on. There is no number on his shirt. Black pants and skates, a brown hockey stick and red gloves.
073/076	Ice Hockey	Peyo	1/1/78	Rubber Figures	2" High	Hong Kong	$4.50-6.00	The Smurf has a medium blue body. The Smurf has a dull, dirty yellow hat, shirt and sock. There is a black # 6 on his shirt. Black pants and skates, a brown hockey stick and red gloves.
073/077	Clown	Schleich/Peyo	1/1/78	Rubber Figures	2" High	Hong Kong	$4.50-6.00	The Smurf has a dark blue body. He is wearing yellow with red striped pants, black suspenders, a red bow tie, white shoes with black toes, a white shirt, white gloves and a white hat. He has a red mouth with white around it.
073/078	Clown	Schleich/Peyo	1/1/78	Rubber Figures	2" High	Hong Kong	$4.50-6.00	The Smurf has a medium blue body. He is wearing yellow with red striped pants, black suspenders, a red bow tie, white shoes with black toes, a white shirt, white gloves and a white hat. He has a red mouth with white around it.
073/079	Smurfette	Schleich/Peyo	1/1/78	Rubber Figures	2" High	Hong Kong	$3.00-4.00	Smurfette has a dark blue body. She is standing with her left hand on her chest and her right hand out at her side. Smurfette is wearing a white dress and shoes.
073/080	Flirting Smurfette	Bully/Peyo	1/1/78	Rubber Figures	2" High	West Germany	$18.00-25.00	Smurfette has a light blue body. She is standing with her left hand on her right hip, her right hand is on her hair. She has light yellow hair and long black eyelashes.
073/081	Soccer Footballer	Schleich/Peyo	1/1/78	Rubber Figures	2" High	Hong Kong	$3.00-4.00	The Smurf has a dark blue body. He has his arms out at his sides. His right foot has a white soccer ball on the tip. He is wearing a red shirt, red socks, yellow shorts and black shoes.
073/082	Soccer Footballer	Schleich/Peyo	1/1/78	Rubber Figures	2" High	Hong Kong	$3.00-4.00	The Smurf has a medium blue body. He has his arms out at his sides. His right foot has a white soccer ball on the tip. He is wearing a red shirt, red socks, yellow shorts and black shoes.
073/083	Soccer Footballer - VFB Stuttgart	Schleich/Peyo	1/1/78	Rubber Figures	2" High	West Germany	$25.00-35.00	The Smurf has a medium blue body. He has his arms out at his sides. His right foot has a white soccer ball on the tip. He is wearing a white shirt, white socks and white shorts. The shirt and pants have a red stripe on them. He has on black shoes. The Smurf represents the VFB Stuttgart soccer team.
073/084	Soccer Footballer - Bayer Uerdingen	Schleich/Peyo	1/1/78	Rubber Figures	2" High	Germany	$25.00-35.00	The Smurf has a medium blue body. He has his arms out at his sides. His right foot has a white soccer ball on the tip. He is wearing a blue shirt, red socks and red shorts. He has on black shoes. The Smurf represents the Bayer Uerdingen soccer team.
073/085	Soccer Footballer - 1.FC Kaaiserslautern	Schleich/Peyo	1/1/78	Rubber Figures	2" High	Germany	$25.00-35.00	The Smurf has a medium blue body. He has his arms out at his sides. His right foot has a white soccer ball on the tip. He is wearing a red shirt, red socks and red shorts. The shirt and his shoe has an orange dot on it. He has on black shoes. The Smurf represents the 1.FC Kaaiserslautern soccer team.
073/086	Soccer Footballer	Schleich/Peyo	1/1/78	Rubber Figures	2" High	West Germany	$25.00-35.00	The Smurf has a light blue body. He has his arms out at his sides. His right foot has a white soccer ball on the tip. He is wearing a white shirt with a black dot in the right corner, white socks and black shorts. He has on black shoes. There is an orange dot on the bottom of his shoe.
073/087	Soccer Footballer - Eintracht Braunschweig	Schleich/Peyo	1/1/78	Rubber Figures	2" High	West Germany	$25.00-35.00	The Smurf has a light blue body. He has his arms out at his sides. His right foot has a white soccer ball on the tip. He is wearing a yellow shirt with a dark blue dot in the left corner, yellow socks and blue shorts. He has on black shoes, there is an orange dot on the bottom of the shoe.. The Smurf represents the Eintracht Braunschweig soccer team.
073/088	Soccer Footballer - Fc Bayern Munchen	Schleich/Peyo	1/1/78	Rubber Figures	2" High	West Germany	$25.00-35.00	The Smurf has a light blue body. He has his arms out at his sides. His right foot has a white soccer ball on the tip. He is wearing a red shirt with a white dot in the left corner, red socks and white shorts. He has on black shoes, there is an orange dot on the bottom of 1 shoe. The Smurf represents the Fc Bayern Munchen soccer team.
073/089	Soccer Footballer	Schleich/Peyo	1/1/78	Rubber Figures	2" High	West Germany	$25.00-35.00	The Smurf has a light blue body. He has his arms out at his sides. His right foot has a white soccer ball on the tip. He is wearing a red shirt with a white dot in the left corner, black socks and white shorts. He has on black shoes, there is an orange dot on the bottom of 1 shoe.
073/090	Soccer Footballer	Schleich/Peyo	1/1/78	Rubber Figures	2" High	West Germany	$3.00-4.00	The Smurf has a light blue body. He has his arms out at his sides. His right foot has a white soccer ball on the tip. He is wearing a yellowish/orange shirt with a black dot in the left corner, yellowish/orange socks and black shorts. He has on black shoes, there is an orange dot on the bottom of 1 shoe.

073/091	Soccer Footballer	Schleich/Peyo	1/1/78	Rubber Figures	2" High	West Germany	$3.00-4.00	The Smurf has a medium blue body. He has his arms out at his sides. His right foot has a white soccer ball on the tip. He is wearing a bright yellow shirt, bright yellow socks and orange shorts. He has on black shoes.
073/092	Soccer Footballer	Schleich/Peyo	1/1/78	Rubber Figures	2" High	West Germany	$3.00-4.00	The Smurf has a medium blue body. He has his arms out at his sides. His right foot has a white and black soccer ball on the tip. He is wearing a bright yellow shirt, bright yellow socks and red shorts. He has on black shoes.
073/093	Soccer Footballer - 1. FC Nurnberg	Schleich/Peyo	1/1/78	Rubber Figures	2" High	West Germany	$25.00-35.00	The Smurf has a light blue body. He has his arms out at his sides. His right foot has a white soccer ball on the tip. He is wearing a white with red stripe shirt, maroon socks and black shorts. He has on black shoes. The shirt has a black dot in the left corner and his shoe has an orange dot on it. The Smurf represents 1. FC Nurnberg soccer team.
073/094	Soccer Footballer	Schleich/Peyo	1/1/78	Rubber Figures	2" High	Hong Kong	$3.00-4.00	The Smurf has a medium blue body. He has his arms out at his sides. His right foot has a white soccer ball on the tip. He is wearing a red/orange shirt, red/orange socks, yellow shorts and black shoes.
073/095	Gymnast	Peyo	1/1/72	Rubber Figures	2" High	West Germany	$15.00-20.00	The Smurf has a light blue body. The Smurf is flexing his right arm and holding a dumb bell in the left. The Smurf has no shirt on.
073/096	Hand Glider	Schleich/Peyo/W.B.	1/1/78	Rubber Figures	2" High	Hong Kong	$4.50-6.00	The Smurf has a medium blue body. The Smurf has a red and yellow hand-glider with dark brown straps.
073/097	Doctor	Schleich/Peyo	1/1/78	Rubber Figures	2" High	Hong Kong	$4.50-6.00	The Smurf has a medium blue body. The Smurf is wearing a white doctor's coat with silver buttons. He has a thermometer in his left hand. The thermometer has red and black lines. He has a silver and red stethoscope in his right hand.
073/098	Doctor	Schleich/Peyo	1/1/78	Rubber Figures	2" High	Hong Kong	$4.50-6.00	The Smurf has a medium blue body. The Smurf is wearing a white doctor's coat with silver buttons. He has a thermometer in his left hand. The thermometer has blue lines on it. He has a silver and red stethoscope in his right hand.
073/099	Singer	Schleich/Peyo	1/1/78	Rubber Figures	2" High	Hong Kong	$4.50-6.00	The Smurf has a medium blue body. He is holding a yellow music sheet with both hands. The music sheet has notes on both sides.
073/100	Mallet	Schleich/Peyo	1/1/78	Rubber Figures	2" High	Hong Kong	$4.50-6.00	The Smurf has a medium blue body. The Smurf is holding a dark brown mallet over his left shoulder.
073/101	Mallet	Schleich/Peyo	1/1/78	Rubber Figures	2" High	West Germany	$6.00-8.00	The Smurf has a medium blue body. The Smurf is holding a light brown mallet over his left shoulder.
073/102	Gift	Schleich/Peyo	1/1/78	Rubber Figures	2" High	Hong Kong	$3.00-4.00	The Smurf has a dark blue body. He is holding a yellow gift with a red bow in his left hand. He has a yellow and two red flowers in the other hand.
073/103	Hiker	Schleich/Peyo	1/1/78	Rubber Figures	2" High	Hong Kong	$6.00-8.00	The Smurf has a dark blue body. He is wearing a yellow jacket and socks. He has brown shoes and a brown walking stick. He has a red flower in his hat. There is a green backpack on his back.
073/104	Chef	Schleich/Peyo	1/1/78	Rubber Figures	2" High	Hong Kong	$4.50-6.00	The Smurf has a medium blue body. He is wearing a white chef's hat, shirt and pants. He is holding a dark brown ladle to his mouth with his right hand.
073/105	Digger	Schleich/Peyo	1/1/78	Rubber Figures	2" High	Hong Kong	$4.50-6.00	The Smurf has a dark blue body. The Smurf is leaning on the shovel and the backs of his feet are off the ground. The shovel is big and is on his left side. The shovel has a red stick and a dark yellow shovel.
073/106	Digger	Schleich/Peyo	1/1/78	Rubber Figures	2" High	Hong Kong	$4.50-6.00	The Smurf has a medium blue body. The Smurf is leaning on the shovel and the backs of his feet are off the ground. The shovel is big and is on his left side. The shovel has a red stick and a light yellow shovel.
073/107	Digger	Bully/Peyo	1/1/78	Rubber Figures	2" High	West Germany	$40.00-55.00	The Smurf has a medium blue body. The Smurf is leaning on the shovel with his arms crossed. The shovel is small and right in front of him. The shovel has a red stick handle and a dark yellow shovel. The Smurf is leaning but his feet are close together and flat on the ground. SMALL MOLD!!!!!
073/108	Lover	Schleich/Peyo/W.B.	1/1/78	Rubber Figures	2" High	Hong Kong	$3.00-4.00	The Smurf has a medium blue body. The Smurf is holding 3 dark red flowers with yellow centers and a dark green stem. The Smurf is holding the flowers in his left hand. He has his head turned to the right like he's posing.
073/109	Artist	Schleich/Peyo	1/1/78	Rubber Figures	2" High	Hong Kong	$4.50-6.00	The Smurf has a medium blue body. He is holding a white paint brush with yellow paint on the end, in his right hand. In his left hand he's holding a white painter's palate, the colors are raised. The Smurf is smiling with his tongue is sticking out.
073/110	Emperor	Schleich/Peyo	1/1/78	Rubber Figures	2" High	Hong Kong	$6.00-8.00	The Smurf has a dark blue body. He is wearing a yellow outfit, yellow crown, gold hat, gold shoes, gold jewelry and a red robe. He is holding his index finger on his right hand up in the air.
073/111	Emperor	Schleich/Peyo	1/1/78	Rubber Figures	2" High	Hong Kong	$6.00-8.00	The Smurf has a light blue body. He is wearing a bright yellow outfit, a bright yellow crown, gold hat, gold shoes, gold jewelry and a red robe. He is holding his index finger on his right hand up in the air.
073/112	Trumpeter	Schleich/Peyo	1/1/80	Rubber Figures	2" High	Hong Kong	$4.50-6.00	The Smurf has a dark blue body. He is playing a tan trumpet. His left foot is lifted. His eyes are closed.
073/113	Trumpeter	Schleich/Peyo	1/1/74	Rubber Figures	2" High	Hong Kong	$4.50-6.00	The Smurf has a medium blue body. He is playing a tan trumpet. His left foot is lifted. His eyes are closed.
073/114	Trumpeter	Schleich/Peyo	1/1/74	Rubber Figures	2" High	Hong Kong	$4.50-6.00	The Smurf has a medium blue body. He is playing a mustard color trumpet. His left foot is lifted. His eyes are closed.
073/115	Trumpeter	Schleich/Peyo	1/1/74	Rubber Figures	2" High	West Germany	$4.50-6.00	The Smurf has a medium blue body. He is playing a bright yellow trumpet. His left foot is lifted. His eyes are closed.
073/116	Flautist	Schleich/Peyo	1/1/80	Rubber Figures	2" High	Hong Kong	$6.00-8.00	The Smurf has a dark blue body. He is playing a yellow flute. The Smurf is wearing white pants and a red shirt. He has his eyes closed.
073/117	Flautist	Schleich/Peyo	1/1/78	Rubber Figures	2" High	Hong Kong	$6.00-8.00	The Smurf has a medium blue body. He is playing a yellow flute. The Smurf is wearing white pants and a red/orange shirt. He has his eyes closed.
073/118	Flautist	Schleich/Peyo	1/1/80	Rubber Figures	2" High	West Germany	$6.00-8.00	The Smurf has a medium blue body. He is playing a yellow flute. The Smurf is wearing white pants and a maroon shirt. He has his eyes closed.
073/119	Tennis Star	Schleich/Peyo	1/1/78	Rubber Figures	2" High	Hong Kong	$3.00-4.00	The Smurf has a dark blue body. He is wearing a white tennis shirt and white shorts. He has a yellow wrist band on his right wrist. He is holding a red plastic tennis racket in his right hand. The tennis racket has holes in it.
073/120	Pointing	Schleich/Peyo/W.B.	1/1/79	Rubber Figures	2" High	Hong Kong	$11.00-15.00	The Smurf has a dark blue shiny body. He is standing, holding his left hand out pointing at something.
073/121	Pointing	Schleich/Peyo	1/1/79	Rubber Figures	2" High	Hong Kong	$11.00-15.00	The Smurf has a medium dull blue body. He is standing, holding his left hand out pointing at something.
073/122	Bowler	Schleich/Peyo	1/1/79	Rubber Figures	2" High	Hong Kong	$3.00-4.00	The Smurf has a dark blue body. He is holding a red bowling ball in his right hand. He is standing like he's going to throw the ball.
073/123	Bowler	Schleich/Peyo	1/1/79	Rubber Figures	2" High	Hong Kong	$3.00-4.00	The Smurf has a dark blue body. He is holding an orange bowling ball in his right hand. He is standing like he's going to throw the ball.
073/124	Cleaner	Schleich/Peyo	1/1/79	Rubber Figures	2" High	West Germany	$7.50-10.00	The Smurf has a medium blue shiny body. He is holding a dark red cloth in his left hand. He is not wearing a shirt, only white pants.
073/125	Cleaner	Schleich/Peyo	1/1/79	Rubber Figures	2" High	West Germany	$7.50-10.00	The Smurf has a medium blue dull body. He is holding a red cloth in his left hand. He is wearing white overalls. There is no emblem on the front. There is an orange dot on the bottom of his foot.
073/126	Ice Lolly	Schleich/Peyo	1/1/79	Rubber Figures	2" High	Hong Kong	$3.00-4.00	The Smurf has a medium blue body. He is eating a popsicle. The popsicle is red on the bottom and white on top. The popsicle is on a yellow stick. The Smurf is holding the popsicle in his right hand. He has his tongue sticking out of his mouth.
073/127	Ice Lolly	Schleich/Peyo	1/1/79	Rubber Figures	2" High	West Germany	$22.50-30.00	The Smurf has a medium blue body. He is eating a popsicle. The popsicle is half blue and half white. The popsicle is on a yellow stick. The Smurf is holding the popsicle in his right hand. He has his tongue sticking out of his mouth.
073/128	First-Aid	Schleich/Peyo	1/1/78	Rubber Figures	2" High	Hong Kong	$3.00 - 4.25	The Smurf has a dark blue body. He is wearing a white uniform. The Smurf has a white bandage wrapped around a finger on his right hand. He is carrying a white medical bag with a red cross on it.
073/129	First-Aid	Schleich/Peyo	1/1/78	Rubber Figures	2" High	West Germany	$11.00-15.00	The Smurf has a medium blue body. He is wearing a white uniform. The Smurf has a white bandage wrapped around a finger on his right hand. He is carrying a brown medical bag with a red cross on it. There is an orange dot on the bottom on his foot.
073/130	First-Aid	Schleich/Peyo	1/1/78	Rubber Figures	2" High	Germany	$11.00-15.00	The Smurf has a medium blue body. He is wearing a white uniform. The Smurf has a white bandage wrapped around a finger on his right hand. He is carrying a yellow medical bag with a red cross on it.
073/131	Golfer	Schleich/Peyo	1/1/79	Rubber Figures	2" High	Hong Kong	$3.00 - 4.25	The Smurf has a dark blue body. He is wearing yellow pants, red socks, black shoes and no shirt. The Smurf is holding a gray golf club.
073/132	Card Player	Schleich/Peyo/W.B.	1/1/78	Rubber Figures	2" High	Hong Kong	$3.00-4.00	The Smurf has a medium blue body. He has his left hand over his mouth like he's laughing. His left eye is closed and his right eye is open. He is holding 3 cards in his right hand, the face card is a heart. He has a club card stuffed in the front of his pants.
073/133	Thirsty	Peyo	1/1/79	Rubber Figures	2" High	West Germany	$30.00-40.00	The Smurf has a medium blue body. He is holding an orange bottle with an orange straw in front of him. The bottle doesn't have ribs. The Smurf has his eye's shut and a big grin on his face.
073/134	Champion	Schleich/Peyo	1/1/79	Rubber Figures	2" High	West Germany	$7.50-10.00	The Smurf has a medium blue body. He has a gold wreath around his neck. He is holding his right hand out to shake. His left hand is at his side. He has a big smile on his face.
073/135	Teacher	Schleich/Peyo	1/1/80	Rubber Figures	2" High	Hong Kong	$3.00 - 4.25	The Smurf has a dark blue shiny body. He is holding a red book with white pages, black letters and a yellow bookmark in his right hand. He is holding a finger up in the air on his left hand. His eye's are shut and his mouth like he's declaring something.
073/136	Teacher	Schleich/Peyo	1/1/80	Rubber Figures	2" High	Hong Kong	$3.00 - 4.25	The Smurf has a medium blue shiny body. He is holding a dark red book with white pages, black letters and a yellow bookmark in his right hand. He is holding a finger up in the air on his left hand. His eye's are shut and his mouth like he's declaring something.
073/137	Candle	Schleich/Peyo	1/1/79	Rubber Figures	2" High	Hong Kong	$7.50-10.00	The Smurf has a dark blue body. He is wearing a white night shirt, brown shoes and a white hat with a red tassel. He is holding a candle in his right hand. The candle is white with a yellow fire The candle is sitting on a brown plate. The Smurf has his eye's open.
073/138	Bandleader	Schleich/Peyo	1/1/80	Rubber Figures	2" High	West Germany	$9.00-12.00	The Smurf has a medium blue body. The Smurf has his hands in the air. He is holding a black plastic wand in his right hand. The Smurf has a red bow tie around his neck. The Smurf is standing on a green grass- like base. The Smurf has his eye's closed
073/139	Telephone	Schleich/Peyo	1/1/80	Rubber Figures	2" High	Hong Kong	$3.00-4.00	The Smurf has a dark blue body. The phone is red with gold. The cradle is small and copper. The hand piece is copper and he is holding it to his mouth with his right hand.
073/140	Telephone	Schleich/Peyo	1/1/80	Rubber Figures	2" High	West Germany	$4.50-6.00	The Smurf has a medium bright blue body. The phone is red with gold. The cradle is wide and gold. The hand piece is gold and he is holding it to his mouth with his right hand.
073/141	Tailor	Schleich/Peyo/W.B.	1/1/79	Rubber Figures	2" High	Hong Kong	$9.00-12.00	The Smurf has a medium blue body. He is sitting, mending a yellow shirt. He is holding a silver needle with red thread in his right hand. The shirt is laying in his lap. He has his eye's closed and a smile on his face.
073/142	Toothbrush	Schleich/Peyo/W.B.	1/1/79	Rubber Figures	2" High	Hong Kong	$3.00-4.00	The Smurf has a dark blue body. He is standing, brushing his teeth with a yellow toothbrush. In his left hand he is holding a red tube of toothpaste. The toothpaste container has a white line on the front.
073/143	Rugby	Schleich/Peyo	1/1/80	Rubber Figures	2" High	West Germany	$25.00-35.00	The Smurf has a medium bright blue body. He has his left hand out to block and his right hand is hugging a brown football to his side. The Smurf is wearing a yellow shirt, yellow socks, red shorts and black shoes. He has an exhausted look on his face.
073/144	Cricket	Schleich/Peyo	1/1/80	Rubber Figures	2" High	West Germany	$11.00-15.00	The Smurf has a medium blue body. He is wearing a white uniform. He has a brown bat with a red center that he's holding on his left side.
073/145	Congratulations	Schleich/Peyo	1/1/79	Rubber Figures	2" High	West Germany	$25.00-35.00	The Smurf has a medium blue body. He is holding his hands together in front of him. The Smurf has a smile on his face.

073/146	Football Player	Bully/Peyo	1/1/73	Rubber Figures	2" High	West Germany	$11.00-15.00	The Smurf has a medium blue body. The Smurf is wearing a white shirt, black pants and black shoes. The Smurf's right foot is slanted up and a white and black soccer ball is attached to his foot. His arms are out at his sides.
073/147	Jungle	Bully/Peyo	1/1/73	Rubber Figures	2" High	West Germany	$40.00-55.00	The Smurf has a burnt orange body. He is wearing a yellow cloth around his waist. He is holding a brown spear with a white end, in his right hand. The Smurf has black hair. He has white paint around his mouth. He has a green dot on the bottom of his foot.
073/148	Jungle	Schleich	1/1/73	Rubber Figures	2" High	West Germany	$25.00-35.00	The Smurf has a tan body. He is wearing a yellow cloth around his waist. He is holding a brown spear with a white end, in his right hand. The Smurf has black hair. He has white paint around his mouth.
073/149	Harp	Bully/Peyo	1/1/73	Rubber Figures	2" High	West Germany	$6.00-8.00	The Smurf has a medium blue body. He is holding a yellow harp on his right side. He has his eye's closed and his mouth in the form of an O.
073/150	Flying	Bully/Peyo	1/1/71	Rubber Figures	2" High	West Germany	$7.50-10.00	The Smurf has a medium blue body. He has white bird- like wings on his back. The wings extend out with his arms.
073/151	Trumpet Player	Bully/Peyo	1/1/72	Rubber Figures	2" High	West Germany	$7.50-10.00	The Smurf has a light blue body. He is standing, playing a yellow trumpet. His eye's are open. His feet are close together.
073/152	Cook	Bully/Peyo	1/1/73	Rubber Figures	2" High	West Germany	$9.00-13.00	The Smurf has a medium blue body. The Smurf is wearing a white apron around his waist, a white bow tie and a white chef's hat. He is holding an orange spoon in his right hand and a white pot with yellowish/orange stuff in his left hand. He has his tongue sticking out of his mouth.
073/153	Cook	Bully/Peyo	1/1/73	Rubber Figures	2" High	Hong Kong	$9.00-12.00	The Smurf has a dark blue body. The Smurf is wearing a white apron around his waist, a white bow tie and a white chef's hat. He is holding an orange spoon in his right hand and a white pot with yellowish/orange stuff in his left hand. He has his tongue sticking out of his mouth.
073/154	King	Schleich/Peyo	1/1/73	Rubber Figures	2" High	West Germany	$3.00 - 4.25	The Smurf has a medium blue body. The King is wearing a yellow hat, crown, pants and a red with white trim robe. He is holding a yellow with a red top scepter in his right hand.
073/155	King	Bully/Peyo	1/1/73	Rubber Figures	2" High	West Germany	$6.00-8.00	The Smurf has a medium blue body. The King is wearing a golden rod hat, crown, pants and a red with white trim robe. He is holding a yellow with a red top scepter in his right hand.
073/156	Quack	Schleich/Peyo	1/1/73	Rubber Figures	2" High	West Germany	$7.50-10.00	The Smurf has a medium blue body. He has a black robe, black glasses and a black hat on. He has a white bow tie around his neck. He is holding a white sringe in his right hand.
073/157	Quack	Bully/Peyo	1/1/73	Rubber Figures	2" High	West Germany	$11.00-15.00	The Smurf has a medium blue body. He has a black robe, black glasses and a black hat on. He has a white bow tie around his neck. He is holding a white sringe in his right hand.
073/158	Courting	Schleich/Peyo	1/1/73	Rubber Figures	2" High	West Germany	$7.50-10.00	The Smurf has a medium blue shiny body. He is holding a big white daisy with a green stem in his left hand. The daisy has a yellow center. His right hand is behind his back. He has a goofy grin on his face.
073/159	Courting	Bully/Peyo	1/1/73	Rubber Figures	2" High	Hong Kong	$11.00-15.00	The Smurf has a dark blue dull body. He is holding a big white daisy with a green stem in his left hand. The daisy has a yellow center. His right hand is behind his back. He has a goofy grin on his face.
073/160	Courting	Bully/Peyo	1/1/73	Rubber Figures	2" High	West Germany	$7.50-10.00	The Smurf has a medium blue shiny body. He is holding a big white daisy with a green stem in his left hand. The daisy has a yellow center. His right hand is behind his back. He has a goofy grin on his face.
073/161	Naughty	Schleich/Peyo/W.B.	1/1/73	Rubber Figures	2" High	Hong Kong	$6.00-8.00	The Smurf has a dark blue body. He has his thumbs in his ears. He has his tongue sticking out. His tongue is maroon.
073/162	Naughty	Bully/Peyo	1/1/73	Rubber Figures	2" High	West Germany	$6.00-8.00	The Smurf has a medium blue body. He has his thumbs in his ears. He has his tongue sticking out. His tongue is bright red.
073/163	Beer	Schleich/Peyo	1/1/75	Rubber Figures	2" High	Hong Kong	$3.00-4.00	The Smurf has a dark blue body. He is holding a big yellow/orange mug of beer in his right hand. He has a big red mouth.
073/164	Jolly	Bully/Peyo	1/1/74	Rubber Figures	2" High	West Germany	$11.00-15.00	The Smurf has a bright medium blue body. He has his left hand on his stomach and his right hand is up by his face. His mouth is open and he's laughing. His eye's are closed.
073/165	Biscuit	Bully/Peyo	1/1/74	Rubber Figures	2" High	West Germany	$7.50-10.00	The Smurf has a medium blue body. He is holding an orange cookie with a bite out. He is holding the cookie in his left hand. The Smurf is licking his lips. His right hand is on his stomach.
073/166	Tyrolese	Schleich/Peyo/W.B.	1/1/74	Rubber Figures	2" High	Hong Kong	$6.00-8.00	The Smurf has a shiny dark blue body. He is wearing green shorts and green suspenders. He has a red pipe in his mouth and a red feather in his hat. He is holding the pipe with his right hand.
073/167	Tyrolese	Bully/Peyo	1/1/74	Rubber Figures	2" High	West Germany	$7.50-10.00	The Smurf has a medium blue body. He is wearing bright green shorts and bright green suspenders. He has an orange pipe in his mouth and a red feather in his hat. He is holding the pipe with his right hand.
073/168	Shy	Bully/Peyo	1/1/74	Rubber Figures	2" High	West Germany	$12.00-16.00	The Smurf has a medium blue body. He has a finger on his right hand covering his mouth. He has his eye's closed and he has a smile on his face. His left hand is behind his back.
073/169	Hammer	Bully/Peyo	1/1/74	Rubber Figures	2" High	West Germany	$11.00-15.00	The Smurf has a medium blue body. He has his left hand wiping sweat from his forehead. His right hand is holding a hammer at his side. The hammer has a red handle and a black hammer. There is a bead of sweat running down the right side of the Smurf's face.
073/170	Handstand	Bully/Peyo	1/1/75	Rubber Figures	2" High	West Germany	$25.00-35.00	The Smurf has a medium blue body. He is standing on his hands. He is wearing red shorts and white shoes.
073/171	Handstand	Bully/Peyo	1/1/75	Rubber Figures	2" High	West Germany	$25.00-35.00	The Smurf has a medium blue body. He is standing on his hands. He is wearing yellow shorts and white shoes.
073/172	Cushion	Schleich/Peyo/W.B.	1/1/75	Rubber Figures	2" High	Hong Kong	$6.00-8.00	The Smurf has a dark blue body. He is sitting on a bright pink cushion. Both hands are flat on the cushion.
073/173	Cushion	Schleich/Peyo/W.B.	1/1/75	Rubber Figures	2" High	Hong Kong	$6.00-8.00	The Smurf has a medium blue body. He is sitting on a dark pink cushion. Both hands are flat on the cushion.
073/174	Present	Schleich/Peyo/W.B.	1/1/76	Rubber Figures	2" High	Hong Kong	$7.50-10.00	The Smurf has a dark blue body. He is holding a white present with red ribbon. The Smurf is holding the present with both hands off to his right side.
073/175	Woodcutter	Bully/Peyo	1/1/75	Rubber Figures	2" High	West Germany	$6.00-8.00	The Smurf has a medium blue body. He is standing, holding an ax. The Smurf is holding the ax with both hands off to his left side. The ax is silver and the handle is red.
073/176	Traveler	Bully/Peyo	1/1/75	Rubber Figures	2" High	West Germany	$11.00-15.00	The Smurf has a medium blue body. He is carrying a plastic orange stick with an orange knapsack over his right shoulder.
073/177	Painter	Schleich/Peyo/W.B.	1/1/76	Rubber Figures	2" High	West Germany	$11.00-15.00	The Smurf has a medium blue body. He is holding a white paint brush with red paint on the end, in his left hand. He is holding a painters palate and 2 brushes in his right hand. The paint on the painters plate is indented in the plate. The Smurf has his head turned to the left.
073/178	Jester	Schleich/Peyo	1/1/77	Rubber Figures	2" High	Hong Kong	$4.50-6.00	The Smurf has a dark blue body. He is wearing a green outfit, white shoes, white gloves and a white hat with gold stars. There are white ruffles around his neck. The outfit has yellow buttons and the shoes have yellow pompoms. The Smurf has a big red nose. The Smurf is holding a red and white stick in his left hand.
073/179	Jester	Bully/Peyo	1/1/77	Rubber Figures	2" High	West Germany	$9.00-12.00	The Smurf has a medium blue body. He is wearing a bright green outfit, white shoes, white gloves and a white hat with gold stars. There are white ruffles around his neck. The outfit has yellow buttons and the shoes have yellow pompoms. The Smurf has a big red nose. The Smurf is holding a red and white stick in his left hand.
073/180	Skier	Peyo	1/1/76	Rubber Figures	2" High	West Germany	$7.50-10.00	The Smurf has a medium blue body. He is wearing a yellow scarf. The Smurf has red plastic ski's on his feet and is holding white poles with red tips.
073/181	Conductor	Schleich/Peyo/W.B.	1/1/77	Rubber Figures	2" High	Hong Kong	$6.00-8.00	Papa has a medium blue body. He is wearing a black coat, red pants, red hat and a white bow tie. Papa is holding a yellow conductors stick in the air with his right hand. Papa also has his left hand in the air. He has his eyes closed. Papa is standing on a black base.
073/182	Tennis Player 2	Schleich/Peyo	1/1/77	Rubber Figures	2" High	West Germany	$11.00-15.00	The Smurf has a medium bright blue body. He is holding a bright, plastic yellow tennis racket with small holes in his right hand. The Smurf is holding a yellow tennis ball in his left hand. He is standing like he's going to serve the ball. The Smurf is wearing a white tennis shirt and white shorts.
073/183	Tennis Player 2	Bully/Peyo	1/1/77	Rubber Figures	2" High	West Germany	$11.00-15.00	The Smurf has a dull medium blue body. He is holding a clear, plastic yellow tennis racket with big holes in his right hand. The Smurf is holding a yellow tennis ball in his left hand. He is standing like he's going to serve the ball. The Smurf is wearing a white tennis shirt and white shorts. There is an orange dot on the bottom of his foot.
073/184	Bookworm	Schleich/Peyo/W.B.	1/1/83	Rubber Figures	2" High	Hong Kong	$3.00-4.00	The Smurf has a dark blue body. He is holding a red book with yellow pages down by his side. His right hand is in the air with a finger up. The Smurf is wearing black glasses.
073/185	Oboist	Bully/Peyo	1/1/78	Rubber Figures	2" High	West Germany	$7.50-10.00	The Smurf has a medium bright blue body. He is playing a long dark yellow obo. The Smurf has his eye's open and there both in the center like he's cross- eyed.
073/186	Sledgehammer	Schleich/Peyo	1/1/78	Rubber Figures	2" High	West Germany	$6.00-8.00	The Smurf has a shiny medium blue body. He is pulling a dark brown sledgehammer behind his back. His right foot is raised a little.
073/187	Sledgehammer	Bully/Peyo	1/1/78	Rubber Figures	2" High	West Germany	$6.00-8.00	The Smurf has a shiny medium blue body. He is pulling a tan brown sledgehammer behind his back. His right foot is raised a little.
073/188	Injured	Schleich/Peyo	1/1/78	Rubber Figures	2" High	Hong Kong	$4.50-6.00	The Smurf has a medium blue body. He has his left arm wrapped in a yellow bandage. He has a white bandage wrapped around his right foot. He is holding a brown cane in his right hand. There is a brown bandage on his hat. He has a troubled look on his face.
073/189	Injured	Bully/Peyo	1/1/78	Rubber Figures	2" High	Hong Kong	$6.00-8.00	The Smurf has a dark blue body. He has his left arm wrapped in a yellow bandage. He has a white bandage wrapped around his right foot. He is holding a tan cane in his right hand. There is a brown bandage on his hat. He has a troubled look on his face.
073/190	Ballerina	Schleich/Peyo	1/1/78	Rubber Figures	2" High	West Germany	$4.50-6.00	Smurfette has a medium blue body. She is wearing a white ballerina dress and shoes. She is in a pose. Smurfette is standing on a green base. There is a red dot on the bottom of the base.
073/191	Ballerina	Schleich/Peyo	1/1/78	Rubber Figures	2" High	Hong Kong	$3.00 - 4.25	Smurfette has a dark blue body. She is wearing a white ballerina dress and shoes. She is in a pose. Smurfette is standing on a green base. There is a red dot on the bottom of the base.
073/192	Head Cook	Schleich/Peyo	1/1/77	Rubber Figures	2" High	West Germany	$6.00-8.00	The Smurf has a medium blue body. He is wearing a white chef's hat and a white apron around his neck, a white apron around his waist, white pants and a yellow bow tie. He is holding a yellowish/orange rolling pin in the air with his left hand. He has a mad look on his face.
073/193	Head Cook	Bully/Peyo	1/1/77	Rubber Figures	2" High	West Germany	$7.50-10.00	The Smurf has a dark blue body. He is wearing a white chef's hat and a white apron around his neck, a white apron around his waist, white pants and a yellow bow tie. He is holding a yellowish/orange rolling pin in the air with his left hand. He has a mad look on his face.
073/194	Cake	Schleich/Peyo	1/1/78	Rubber Figures	2" High	Hong Kong	$3.00 - 4.25	The Smurf has a dark blue body. He is holding a dark red cake with white frosting. The cake is on a yellow plate. The Smurf is licking his lips. His left foot is raised a little.
073/195	Cake	Bully/Peyo	1/1/78	Rubber Figures	2" High	West Germany	$22.50-30.00	The Smurf has a dark blue body. He is holding a red cake with white frosting. The cake is on a yellow plate. The Smurf is licking his lips. His left foot is slightly raised.
073/196	Cake	Bully/Peyo	1/1/78	Rubber Figures	2" High	West Germany	$9.00-12.00	The Smurf has a medium blue body. He is holding an orange cake with white frosting. The cake is on a yellow plate. The Smurf is licking his lips. His left foot is slightly raised.

073/197	Angler	Schleich/Peyo	1/1/78	Rubber Figures	2" High	Hong Kong	$6.00-8.00	The Smurf has a dark blue body. He is standing, holding a tan rubber fishing pole. The fishing pole has a thick yellow line with a black hook that's attached to his right foot. The fishing pole is molded to him.
073/198	Angler	Bully/Peyo	1/1/78	Rubber Figures	2" High	West Germany	$6.00-8.00	The Smurf has a medium blue body. He is standing, holding a dark brown rubber fishing pole. The fishing pole has a thin dark yellow line with a black hook that's attached to a little piece of rubber off his right foot. The fishing pole is molded to him.
073/199	Angler	Bully/Peyo	1/1/78	Rubber Figures	2" High	West Germany	$6.00-8.00	The Smurf has a medium blue body. He is standing, holding a brown fishing pole. The line of the fishing pole is the same color as the pole. The pole is a separate piece of plastic.
073/200	Angler	Bully/Peyo	1/1/78	Rubber Figures	2" High	West Germany	$6.00-8.00	The Smurf has a medium blue body. He is standing, holding a plastic yellow fishing pole. The line of the fishing pole is the same color as the pole. The pole is a separate piece of plastic.
073/201	Archer	Schleich/Peyo	1/1/78	Rubber Figures	2" High	Hong Kong	$4.50-6.00	The Smurf has a dark blue body. He is holding 1 yellow and 1 red arrow in his left hand. He is holding a yellow bow with a white string in his right hand. The Smurf is wearing a white hat with a red feather stuck in the side.
073/202	Archer	Bully/Peyo	1/1/79	Rubber Figures	2" High	West Germany	$22.50-30.00	The Smurf has a medium blue body. He is holding 2 orange arrows in his left hand. He is holding a yellow bow in his right hand. The Smurf is wearing a white hat with a red feather stuck in the side. The bow and arrows are separate plastic pieces.
073/203	Scholar	Bully/Peyo	1/1/79	Rubber Figures	2" High	West Germany	$4.50-6.00	The Smurf has a medium blue body. The Smurf is wearing an orange backpack with black buckles on his back.
073/204	Pirate	Schleich/Peyo	1/1/78	Rubber Figures	2" High	Hong Kong	$3.00 - 4.25	The Smurf has a dark blue body. He is holding a silver sword in his right hand. He has a red belt with a black buckle around his waist. The Smurf has a black patch covering his left eye.
073/205	Scot	Schleich/Peyo	1/1/79	Rubber Figures	2" High	Hong Kong	$3.00-4.00	The Smurf has a dark blue body. He is wearing a red hat and red cape. The Smurf is playing a yellow bagpipe.
073/206	Scot	Bully/Peyo	1/1/79	Rubber Figures	2" High	West Germany	$7.50-10.00	The Smurf has a light blue body. He is wearing a red/orange hat and red/orange cape. The Smurf is playing a mustard color bagpipe.
073/207	Hunter	Schleich/Peyo/W.B.	1/1/78	Rubber Figures	2" High	Hong Kong	$4.50-6.00	The Smurf has a medium blue body. He is holding a silver gun. On the end of the gun is a yellow bird. The Smurf has a green feather in his hat and a troubled look on his face.
073/208	Carnival	Schleich/Peyo	1/1/79	Rubber Figures	2" High	Hong Kong	$7.50-10.00	The Smurf has a dark blue body. The Smurf is holding a red pole with a yellow Chinese lantern on the end of the pole in his left hand. In his right hand he is holding a yellow champagne glass. The Smurf is wearing a maroon mask over his eye's. He has a white tail.
073/209	Carnival	Schleich/Peyo/W.B.	1/1/79	Rubber Figures	2" High	Hong Kong	$7.50-10.00	The Smurf has a medium blue body. The Smurf is holding a red pole with a dark yellow Chinese lantern on the end of the pole in his left hand. In his right hand he is holding a dark yellow champagne glass. The Smurf is wearing a red mask over his eye's. He has a blue tail.
073/210	Sauna	Schleich/Peyo	1/1/79	Rubber Figures	2" High	Hong Kong	$3.00 - 4.25	The Smurf has a dark blue body. He is standing, wearing a red towel around his waist. In his left hand he is holding a dark yellow bar of soap. He is scrubbing his back with a tan scrub brush. He has a blue tail.
073/211	Sauna	Bully/Peyo	1/1/79	Rubber Figures	2" High	West Germany	$6.00-8.00	The Smurf has a medium blue body. He is standing, wearing a red towel around his waist. In his left hand he is holding a yellow bar of soap. He is scrubbing his back with an orange/tan scrub brush. He has a no tail painted, it's red.
073/212	Sauna	Bully/Peyo	1/1/79	Rubber Figures	2" High	West Germany	$6.00-8.00	The Smurf has a light blue body. He is standing, wearing a red towel around his waist. In his left hand he is holding a bright yellow bar of soap. He is scrubbing his back with a yellow scrub brush. He has a no tail painted, it's red.
073/213	Knight	Schleich/Peyo	1/1/78	Rubber Figures	2" High	Hong Kong	$4.50-6.00	The Smurf has a medium blue body. He is wearing a gold helmet with dark red hair. He has a gold badge around his neck. The Smurf is holding a silver sword in his right hand. His left hand is on his hip. The Smurf has a mean expression on his face.
073/214	Knight	Bully/Peyo	1/1/78	Rubber Figures	2" High	West Germany	$6.00-8.00	The Smurf has a dark blue body. He is wearing a copper helmet with red hair. He has a copper badge around his neck. The Smurf is holding a silver sword in his right hand. His left hand is on his hip. The Smurf has a mean expression on his face.
073/215	Hairdresser	Schleich/Peyo	1/1/79	Rubber Figures	2" High	Hong Kong	$3.00-4.00	The Smurf has a dark blue body. He is holding silver scissors in his left hand and a red comb in his right hand. He has a smile on his face and his eye's are closed.
073/216	Cupid	Bully/Peyo	1/1/80	Rubber Figures	2" High	West Germany	$11.00-15.00	The Smurf has a medium blue body. He is standing on a red heart platform. The Smurf has 2 white wings and is getting ready to shoot a dark yellow arrow. The bow is dark yellow. The bow and arrow is part of the figure.
073/217	Carpenter	Schleich/Peyo	1/1/79	Rubber Figures	2" High	Hong Kong	$3.00 - 4.25	The Smurf has a dark blue body. In his right hand he is holding a dark yellow board. In his left hand the Smurf is holding a silver saw with a red handle.
073/218	Carpenter	Schleich/Peyo	1/1/79	Rubber Figures	2" High	West Germany	$3.00 - 4.25	The Smurf has a medium blue body. In his right hand he is holding a dark yellow board. In his left hand the Smurf is holding a silver saw with a red handle.
073/219	Baker	Schleich/Peyo	1/1/79	Rubber Figures	2" High	Hong Kong	$4.50-6.00	The Smurf has a dark blue body. He is wearing a floppy white hat and a white apron around his waist. He is holding a tan, oval loaf of bread on a tan bread paddle.
073/220	Conjuror	Schleich/Peyo	1/1/79	Rubber Figures	2" High	Hong Kong	$4.50-6.00	The Smurf has a dark blue body and a red cape around his neck. The Smurf is holding a yellow cloth in his left hand. In his right hand he has a black hat with a dark green cloth hanging out.
073/221	Conjuror	Bully/Peyo	1/1/79	Rubber Figures	2" High	West Germany	$6.00-8.00	The Smurf has a medium blue body and has a red cape around his neck. The Smurf is holding a bright yellow cloth in his left hand. In his right hand he has a black hat with a bright green cloth hanging out.
073/222	Lion Tamer	Schleich/Peyo	1/1/79	Rubber Figures	2" High	Hong Kong	$4.50-6.00	The Smurf has a dark blue body. He has a tan whip in his right hand. In his left hand he has a green drum with a black top. The Smurf is wearing a tan cloth with black polka dots around his waist and red wrist bands. He has a troubled expression on his face.
073/223	Alchemist	Bully/Peyo	1/1/80	Rubber Figures	2" High	West Germany	$6.00-8.00	The Smurf has a dark blue body. He is wearing a long dark blue shirt with a backwards yellow "C". In his left hand he is holding a red genie lantern. In his right hand he is holding a white vase with a dark green snake coming out. The Smurf's hat has copper stars on it.
073/224	Alchemist	Bully/Peyo	1/1/80	Rubber Figures	2" High	West Germany	$6.00-8.00	The Smurf has a medium blue body. He is wearing a long dark blue shirt with a backwards yellow "C". In his left hand he is holding a red genie lantern. In his right hand he is holding a white vase with a light green snake coming out. The Smurf's hat has copper stars on it.
073/225	Cornucopia	Schleich/Peyo	1/1/79	Rubber Figures	2" High	Hong Kong	$3.00 - 4.25	The Smurf has a dark blue body. The Smurf is standing and leaning on a dark yellow horn with dark green leaves and red flowers in the end. The Smurf has his left foot in the air.
073/226	Umbrella	Bully/Peyo	1/1/80	Rubber Figures	2" High	West Germany	$7.50-10.00	The Smurf has a medium blue body. He is standing under a dark red mushroom with white spots and an orange stem. He has a worried expression on his face.
073/227	Umbrella	Applause/Peyo	1/1/80	Rubber Figures	2" High	China	$3.00-4.00	The Smurf has a medium blue body. He is standing under an orange/red mushroom with white spots and an orange stem. He has a worried expression on his face.
073/228	Smurferman	Schleich/Peyo	1/1/80	Rubber Figures	2" High	West Germany	$7.50-10.00	The Smurf has a dark blue body. He is wearing a red cape, red shoes, white wrist bands, white underpants, white hat and has a white emblem with an orange "S" in the middle of his chest. The Smurf is standing on a white base. He has his left hand in the air like he's ready to fly. There is an orange dot on the bottom of the base.
073/229	Smurferman	Schleich/Peyo	1/1/80	Rubber Figures	2" High	West Germany	$7.50-10.00	The Smurf has a medium blue body. He is wearing a dark red cape, dark red shoes, white wrist bands, white underpants, white hat and has a white emblem with an orange "S" in the middle of his chest. The Smurf is standing on a white base. He has his left hand in the air like he's ready to fly. There is a red dot on the bottom of the base.
073/230	Smurferman	Bully/Peyo	1/1/80	Rubber Figures	2" High	West Germany	$15.00-20.00	The Smurf has a medium blue body. He is wearing a red cape, red shoes, white wrist bands, white underpants, white hat and has a white emblem with an orange "S" in the middle of his chest. The Smurf is standing on a orange base. He has his left hand in the air like he's ready to fly.
073/231	Frogman	Schleich/Peyo	1/1/79	Rubber Figures	2" High	Hong Kong	$4.50-6.00	The Smurf has a medium blue body. He is wearing a dark orange scuba outfit. He has black flippers on his feet and a black face mask. The snorkel is dark red and the Smurf is holding a gray spear.
073/232	Frogman	Bully/Peyo	1/1/79	Rubber Figures	2" High	West Germany	$6.00-8.00	The Smurf has a medium blue body. He is wearing an orange scuba outfit. He has black flippers on his feet and a black face mask. The snorkel is bright red and the Smurf is holding a gray spear.
073/233	Ice Skater	Schleich/Peyo	1/1/80	Rubber Figures	2" High	Hong Kong	$3.00-4.00	The Smurf has a dark blue body. He is laying on his back with his feet in the air and his hands spread out. The Smurf is wearing a red scarf and dark yellow gloves. There are dark gray blades on his feet. He has a troubled look on his face.
073/234	Ice Skater	Bully/Peyo	1/1/80	Rubber Figures	2" High	West Germany	$6.00-8.00	The Smurf has a medium blue body. He is laying on his back with his feet in the air and his hands spread out. The Smurf is wearing a red scarf and yellow gloves. There are light gray blades on his feet. He has a troubled look on his face.
073/235	Cowboy	Schleich/Peyo	1/1/81	Rubber Figures	2" High	Hong Kong	$3.00 - 4.25	The Smurf has a dark blue body. He is wearing a brown hat, brown vest, brown shoes, red pants, white shirt with a red handkerchief around his neck and a black belt with a pocket for his brown gun. He has a white rope over his left shoulder and his right hand is out in the air.
073/236	Cowboy	Schleich/Peyo	1/1/81	Rubber Figures	2" High	Portugal	$11.00-15.00	The Smurf has a medium blue body. He is wearing a brown hat, brown vest, brown shoes, red pants, white shirt with a red handkerchief around his neck and a black belt with a pocket for his brown gun. He has a white rope over his left shoulder and his right hand is out in the air.
073/237	Policeman	Schleich/Peyo	1/1/81	Rubber Figures	2" High	Hong Kong	$4.50-6.00	The Smurf has a dark blue body. He is wearing a black hat, a black jacket with silver buttons, a brown belt with a silver buckle and white pants. He is holding a brown stick in his left hand. He is holding a silver whistle to his mouth with his right hand.
073/238	Policeman	Schleich/Peyo	1/1/81	Rubber Figures	2" High	West Germany	$6.00-8.00	The Smurf has a medium blue body. He is wearing a white hat with a black visor and strap, a white jacket with black buttons, a black belt with a black buckle and black pants. He is holding a white stick in his left hand. He is holding a red whistle to his mouth with his right hand.
073/239	Santa	Schleich/Peyo	1/1/81	Rubber Figures	2" High	Hong Kong	$7.50-10.00	Papa Smurf has a dark blue body. He is wearing a red Santa hat and coat. He has white pants and a yellow belt. He is carrying a mustard color bag over his shoulder. There is a doll with a light skin color face and bright orange hair in the bag.
073/240	Santa	Schleich/Peyo	1/1/81	Rubber Figures	2" High	Hong Kong	$7.50-10.00	Papa Smurf has a dark blue body. He is wearing a red Santa hat and coat. He has white pants and a yellow belt. He is carrying a mustard color bag over his shoulder. There is a doll with a dark skin color face and dull orange hair in the bag.
073/241	Heart	Schleich/Peyo	1/1/80	Rubber Figures	2" High	West Germany	$45.00-65.00	The Smurf has a medium blue body. He is standing, holding a white and red plastic heart between his hands. The heart has black lettering with the phrase: EIN HERZ FUR KINDER.
073/242	Roller-skating Smurfette	Schleich/Peyo	1/1/81	Rubber Figures	2" High	Hong Kong	$3.00-4.00	Smurfette has a dark blue body. She is wearing a white dress and white and red roller-skates. Smurfette's hair is dark yellow and has red ribbons in it.
073/243	Roller-skating Smurfette	Schleich/Peyo	1/1/81	Rubber Figures	2" High	Hong Kong	$3.00-4.00	Smurfette has a dark blue body. She is wearing a white dress and white and red roller-skates. Smurfette's hair is bright yellow and has red ribbons in it.

073/244	Superman	Schleich/Peyo	1/1/80	Rubber Figures	2" High	Hong Kong	$4.50-6.00	The Smurf has a dark blue body. He is wearing a yellow cape, red shoes, red gloves, white hat, white shorts and a black mask over his eyes. The Smurf is holding on to a yellow base with his right hand. He looks like he's flying.
073/245	Amour	Schleich/Peyo	1/1/80	Rubber Figures	2" High	Hong Kong	$4.50-6.00	The Smurf has a dark blue body. In his right hand he is holding a red heart with an orange arrow through it. In his left hand he is holding an orange bow. The Smurf has white wings and a white cloth on the front of his body.
073/246	Amour	Schleich/Peyo	1/1/80	Rubber Figures	2" High	Hong Kong	$4.50-6.00	The Smurf has a dark blue body. In his right hand he is holding a red heart with a tan arrow through it. In his left hand he is holding a tan bow. The Smurf has white wings and a white cloth on the front of his body.
073/247	Amour	Schleich/Peyo	1/1/81	Rubber Figures	2" High	Portugal	$4.50-6.00	The Smurf has a medium blue body. In his right hand he is holding a red heart with a yellow arrow through it. In his left hand he is holding a yellow bow. The Smurf has white wings and a white cloth on the front of his body.
073/248	Baseball Batter	Schleich/Peyo	1/1/80	Rubber Figures	2" High	Hong Kong	$4.50-6.00	The Smurf has a dark blue body. He is wearing a white uniform with red sleeves, a red belt, red socks, a white hat with a red visor and black shoes. The Smurf is holding a dark brown bat.
073/249	Baseball Batter	Schleich/Peyo	1/1/80	Rubber Figures	2" High	Hong Kong	$4.50-6.00	The Smurf has a dark blue body. He is wearing a white uniform with red sleeves, a red belt, red socks, a white hat with a red visor and black shoes. The Smurf is holding a tan brown bat.
073/250	Baseball Batter	Schleich/Peyo	1/1/80	Rubber Figures	2" High	Hong Kong	$4.50-6.00	The Smurf has a dark blue body. He is wearing a white uniform with red sleeves, a red belt, red socks, a white hat with a red visor and black shoes. The Smurf is holding a light cream bat.
073/251	Baseball Batter	Schleich/Peyo	1/1/80	Rubber Figures	2" High	Hong Kong	$4.50-6.00	The Smurf has a dark blue body. He is wearing a white uniform with red sleeves, a red belt, red socks, a white hat with a red visor and black shoes. The Smurf is holding a peach bat.
073/252	Graduation	Schleich/Peyo/W.B.	1/1/80	Rubber Figures	2" High	Hong Kong	$6.00-8.00	The Smurf has a bright medium blue body. He is wearing a light blue graduation cap with a yellow tassel and a light blue gown. He is holding a white rolled diploma in his left hand and a dark brown book bag in his right hand.
073/253	Graduation	Schleich/Peyo	1/1/81	Rubber Figures	2" High	Portugal	$9.00-12.00	The Smurf has a bright medium blue body. He is wearing a glossy maroon graduation cap with a yellow tassel and a glossy maroon gown. He is holding a white rolled diploma in his left hand and a brown book bag in his right hand.
073/254	French Fries	Schleich/Peyo/W.B.	1/1/80	Rubber Figures	2" High	Hong Kong	$4.50-6.00	The Smurf has a medium blue body. He is holding a white bundle of french fries in his left hand and a single french fry in his right hand. The french fries are yellowish/red.
073/255	French Fries	Schleich/Peyo	1/1/81	Rubber Figures	2" High	West Germany	$6.00-8.00	The Smurf has a medium blue body. He is holding a white bundle of french fries in his left hand and a single french fry in his right hand. The french fries are bright yellow.
073/256	American Football	Schleich/Peyo	1/1/80	Rubber Figures	2" High	Hong Kong	$9.00-12.00	The Smurf has a medium blue body. The Smurf is wearing a yellow helmet, a red shirt with a yellow number 3, white pants and black shoes. He is holding a dark brown football in his right hand. His left hand is out to block.
073/257	Field Hockey	Schleich/Peyo	1/1/80	Rubber Figures	2" High	Hong Kong	$3.00 - 4.25	The Smurf has a dark blue body. He is wearing a green shirt, green socks, dark brown shorts, a white hat and black shoes. He is holding a brown hockey stick with a white ball on the end of the stick.
073/258	Judo	Schleich/Peyo	1/1/81	Rubber Figures	2" High	Hong Kong	$3.00 - 4.25	The Smurf has a dark blue body. He is wearing a white judo outfit with a black belt and no shoes. The Smurf is standing, starting to bow with his eyes closed.
073/259	Judo	Schleich/Peyo	1/1/81	Rubber Figures	2" High	Hong Kong	$3.00 - 4.25	The Smurf has a medium blue body. He is wearing a white judo outfit with a black belt and no shoes. The Smurf is standing, starting to bow with his eyes closed.
073/260	Tennis Smurfette	Schleich/Peyo/W.B.	1/1/81	Rubber Figures	2" High	Hong Kong	$4.50-6.00	Smurfette has a dark blue body. She is wearing a white tennis dress and shoes. Smurfette's dress has red dots on the bottom. Her hair is dark yellow. Smurfette is holding a red plastic tennis racket in her left hand.
073/261	Halloween With Pumpkin	Schleich/Peyo	1/1/81	Rubber Figures	2" High	Hong Kong	$18.00-25.00	The Smurf has a dark blue body. He is holding an orange pumpkin with a black face and a green stem.
073/262	Halloween With Pumpkin	Schleich/Peyo/W.B.	1/1/81	Rubber Figures	2" High	Portugal	$18.00-25.00	The Smurf has a medium blue body. He is holding an orange pumpkin with a black face and a green stem.
073/263	Halloween With Pumpkin	Schleich/Peyo/W.B.	1/1/81	Rubber Figures	2" High	Portugal	$18.00-25.00	The Smurf has a dark blue body. He is holding an orange pumpkin with no face and a green stem. The Smurf has a brown dot on his foot.
073/264	Surfer	Schleich/Peyo	1/1/81	Rubber Figures	2" High	Hong Kong	$4.50-6.00	The Smurf has a dark blue body. The Smurf is wearing red swim trunks. He is carrying a plastic yellow surf board under his right arm.
073/265	Gardner With Rake	Schleich/Peyo/W.B.	1/1/82	Rubber Figures	2" High	Hong Kong	$4.50-6.00	The Smurf has a medium blue body. The Smurf is wearing a dark green apron, tan straw hat, white pants and dark brown shoes. The Smurf is holding a brown plastic rake in front of him.
073/266	Gardner With Rake	Schleich/Peyo/W.B.	1/1/82	Rubber Figures	2" High	Hong Kong	$4.50-6.00	The Smurf has a medium blue body. The Smurf is wearing a bright green apron, mustard color straw hat, white pants and dark brown shoes. The Smurf is holding a brown plastic rake in front of him.
073/267	Nurse	Schleich/Peyo	1/1/81	Rubber Figures	2" High	Hong Kong	$3.00-4.00	Smurfette has a dark blue body. She is wearing a white dress with a light blue apron. Smurfette is holding a silver/white/ red/ black needle in her right hand and a white bottle in her left. Smurfette has a black watch on her left wrist.
073/268	Nurse	Schleich/Peyo/W.B.	1/1/81	Rubber Figures	2" High	Hong Kong	$4.50-6.00	Smurfette has a medium blue body. She is wearing a white dress with a light blue apron. Smurfette is holding a silver/white/ red/ black needle in her right hand and a white bottle in her left. Smurfette has a black watch on her left wrist.
073/269	Nurse	Schleich/ Peyo	1/1/81	Rubber Figures	2" High	Germany/China	$7.50-10.00	Smurfette has a medium blue body. She is wearing a light blue dress with a white apron. Smurfette is holding a silver/white/ light blue needle in her right hand and a white bottle in her left. Smurfette has a silver watch on her left wrist.
073/270	Secretary	Schleich/Peyo	1/1/81	Rubber Figures	2" High	Hong Kong	$11.00-15.00	Smurfette has a dark blue body. She is wearing a pink dress, white hat and white shoes. Smurfette is holding a brown pencil in her right hand. Smurfette is holding a white pad with black writing and a silver binder edge in her left hand..
073/271	Secretary	Schleich/Peyo	1/1/81	Rubber Figures	2" High	Hong Kong	$9.00-12.00	Smurfette has a dark blue body. She is wearing a white dress, white hat and white shoes. Smurfette is holding a brown pencil in her right hand. Smurfette is holding a white pad with black writing and a silver binder edge in her left hand..
073/272	Captain	Schleich/Peyo	1/1/81	Rubber Figures	2" High	Hong Kong	$3.00-4.00	Papa has a dark blue body. He is wearing a white coat with gold buttons, red pants, and a white hat with a red border and black visor. Papa is holding a bronze telescope to his right eye.
073/273	Mermaid	Schleich/Peyo/W.B.	1/1/81	Rubber Figures	2" High	Hong Kong	$4.50-6.00	Smurfette has a dark blue body and a medium green tail. Smurfette is sitting on a dark gray rock with dark blue and white waves around the bottom of the rock. Smurfette has an orange starfish on her hat. She is wearing a white necklace.
073/274	Mermaid	Schleich/Peyo/W.B.	1/1/81	Rubber Figures	2" High	Hong Kong	$4.50-6.00	Smurfette has a medium blue body and a dark green tail. Smurfette is sitting on a gray rock with medium blue and white waves around the bottom of the rock. Smurfette has an orange starfish on her hat. She is wearing a white necklace.
073/275	Mermaid	Schleich/ Peyo	1/1/81	Rubber Figures	2" High	West Germany	$4.50-6.00	Smurfette has a medium blue body and a blue and silver tail. Smurfette is sitting on a gray rock with dark blue and silver waves around the bottom of the rock. Smurfette has a red starfish on her hat. She is wearing a white necklace.
073/276	CB Operator	Schleich/Peyo/W.B.	1/1/81	Rubber Figures	2" High	Hong Kong	$3.00 - 4.25	The Smurf has a medium blue body. He is holding a silver CB with a silver mike cord, dark brown mike and a black antenna. The Smurf has a dark brown strap across his chest to hold the CB. The Smurf is wearing a white hat with silver earphones and a silver and black headband.
073/277	Indian	Schleich/Peyo/W.B.	1/1/81	Rubber Figures	2" High	Hong Kong	$6.00-8.00	The Smurf has a medium blue body. The Smurf is wearing tan pants, brown and red moccasins, and a headdress with white and black feathers. The headdress has a tan border. The Smurf has 2 red lines painted on each side of his face. The Smurf has black eyebrows. He has his arms folded across his chest.
073/278	Indian	Schleich/Peyo	1/1/81	Rubber Figures	2" High	West Germany	$6.00-8.00	The Smurf has a medium blue body. The Smurf is wearing dark brown pants, red with white stripe moccasins, and a headdress with white/red/yellow/ green feathers. The headdress has a red and white border. The Smurf has 1 red line painted on each side of his face. The Smurf has no eyebrows. The Smurf has his arms folded across his chest.
073/279	Farmer With Sickle	Schleich/Peyo	1/1/81	Rubber Figures	2" High	Hong Kong	$3.00-4.00	The Smurf has a medium blue body. The Smurf is holding yellow and brown hay under his right arm. The Smurf is holding a dark brown and silver sickle in his left hand. The Smurf has a bead of sweat running down his face.
073/280	Baseball Catcher	Schleich/Peyo/W.B.	1/1/81	Rubber Figures	2" High	Hong Kong	$25.00-35.00	The Smurf has a medium blue body. The Smurf is wearing a white uniform, black vest, black shoes, black knee pads, black umpire mask, and red socks. The Smurf is wearing a dark brown mit on his left hand. The Smurf is kneeling.
073/281	Cowgirl	Schleich/Peyo	1/1/81	Rubber Figures	2" High	Hong Kong	$3.00-4.00	Smurfette has a dark blue body. She is wearing a white hat with a red hat tie, white dress, dark brown vest and dark brown boots. Smurfette is holding a tan lasso in her right hand.
073/282	Bricklayer	Schleich/Peyo	1/1/81	Rubber Figures	2" High	Hong Kong	$3.00 - 4.25	The Smurf has a medium blue body. The Smurf is wearing white pants and no shirt. He has a rust color brick in his left hand and under his left foot. He is holding a silver trowel in his right hand.
073/283	Bricklayer	Schleich/Peyo	1/1/82	Rubber Figures	2" High	West Germany	$11.00-15.00	The Smurf has a medium blue body. The Smurf is wearing dark blue bib overalls and a white shirt. He has a red and gray brick in his left hand and under his left foot. He is holding a silver trowel in his right hand.
073/284	Cheerleader	Schleich/Peyo/W.B.	1/1/81	Rubber Figures	2" High	Hong Kong	$6.00-8.00	Smurfette has a dark blue body. Smurfette is wearing a white hat, white shoes and a white dress with a light blue collar and cuffs. She is holding red pompoms. Smurfette is standing on a medium green platform.
073/285	Cheerleader	Schleich/Peyo/W.B.	1/1/81	Rubber Figures	2" High	Hong Kong	$6.00-8.00	Smurfette has a medium blue body. Smurfette is wearing a white hat, white shoes and a white dress with a light blue collar and cuffs. She is holding red pompoms. Smurfette is standing on a dark green platform.
073/286	Cheerleader	Schleich/Peyo/W.B.	1/1/81	Rubber Figures	2" High	Hong Kong	$6.00-8.00	Smurfette has a medium blue body. Smurfette is wearing a white hat, white shoes and a white dress with a light blue collar and cuffs. She is holding red pompoms. Smurfette is standing on a medium green platform.
073/287	Australian Football Player	Schleich/Peyo	1/1/80	Rubber Figures	2" High	Hong Kong	$25.00-35.00	The Smurf has a dark blue body. He is wearing a red shirt, white shorts, red socks, black shoes and a white hat. The Smurf is in a running position and is holding a dark brown football in his hands in front of him. The Smurf is standing on a bright green platform.
073/288	Graduate Smurfette	Schleich/Peyo/W.B.	1/1/81	Rubber Figures	2" High	Hong Kong	$4.50-6.00	Smurfette has a medium blue body. She is wearing a light blue gown, light blue cap with a yellow tassel, white hat and white shoes. Smurfette is holding a red book with white pages under her left arm. In her right hand she is holding a white scroll with a yellow ribbon tied around it.
073/289	Miller	Peyo	1/1/82	Rubber Figures	2" High	West Germany	$10.50-14.00	The Smurf has a medium blue body. He has a tan wheat bag with a picture of black wheat, in front of his feet. The Smurf has his eye's closed and a big smile on his face.

073/290	Santa Smurfette	Schleich/Peyo	1/1/81	Rubber Figures	2" High	Hong Kong	$12.00-16.00	Smurfette has a dark blue body. Smurfette is wearing a long red coat and hat with white trim. She is holding a green box with a red bow.
073/291	Patrol Crossing	Schleich/Peyo	1/1/83	Rubber Figures	2" High	West Germany	$12.00-16.00	The Smurf has a medium shiny blue body. The Smurf is holding a white round sign with a red center.
073/292	Traffic	Schleich/Peyo	1/1/83	Rubber Figures	2" High	West Germany	$12.00-16.00	The Smurf has a medium blue body. The Smurf is holding a pedestrian crossing guard sign. The sign is a triangle. The sign is white with a red border and black people in the middle.
073/293	Valentine Smurfette	Schleich/Peyo	1/1/82	Rubber Figures	2" High	Portugal	$4.50-6.00	Smurfette has a medium blue body. She is wearing a white dress, white hat with red hearts and no shoes. Smurfette has white wings. She is holding a brown bow with a red arrow attached.
073/294	Grouchy	Schleich/Peyo/W.B.	1/1/83	Rubber Figures	2" High	Hong Kong	$4.50-6.00	The Smurf has a medium blue body. He has his fist clenched at his sides. He has a mean frown on his face.
073/295	Hamburger	Schleich/Peyo/W.B.	1/1/83	Rubber Figures	2" High	Hong Kong	$4.50-6.00	The Smurf has a medium blue body. The Smurf is holding a hamburger in both hands in front of him. The hamburger has a tan bun, brown pattie, yellow cheese, green lettuce and red ketchup. The Smurf has his eye's shut and his mouth open.
073/296	Violin	Schleich/Peyo/W.B.	1/1/82	Rubber Figures	2" High	Hong Kong	$3.00 - 4.25	The Smurf has a medium blue body. The Smurf is wearing a light brown hat, red shirt, dark brown vest, dark brown shoes, blue pants and black belts. He is playing a light brown fiddle and an orange bow.
073/297	Apple #1 Teacher	Schleich/Peyo/W.B.	1/1/83	Rubber Figures	2" High	Hong Kong	$25.00-35.00	The Smurf has a medium blue body. He holding is a red apple with a green stem. The apple says #1 Teacher on the front in white letters.
073/298	Green Apple	Schleich/Peyo/W.B.	1/1/81	Rubber Figures	2" High	Hong Kong	$25.00-35.00	The Smurf has a medium blue body. He holding is a green apple with a brown stem.
073/299	Clumsy	Schleich/Peyo/W.B.	1/1/83	Rubber Figures	2" High	Hong Kong	$4.50-6.00	The Smurf has a medium blue body. He is laying on his stomach holding a tan basket. The basket is spilled with red and green apples rolling out.
073/300	Waiter	Schleich/Peyo	1/1/82	Rubber Figures	2" High	West Germany	$9.00-12.00	The Smurf has a medium blue body. He is carrying a silver tray with an orange and white glass. The glass has a yellow straw in it. The Smurf is carrying a white cloth in his left hand.
073/301	Waiter	Schleich/Peyo/W.B.	1/1/83	Rubber Figures	2" High	Hong Kong	$4.50-6.00	The Smurf has a dark blue body. He is carrying a brown tray with a white glass. The glass has a yellow straw in it. The Smurf is carrying a white cloth in his left hand.
073/302	Soccer Smurfette	Schleich/Peyo	1/1/82	Rubber Figures	2" High	Hong Kong	$6.00-8.00	Smurfette has a medium blue body. She is wearing a white/light orange/ red shirt, shorts, shoes and red socks. Smurfette has a white soccer ball attached to her left foot.
073/303	Soccer Smurfette	Schleich/Peyo	1/1/83	Rubber Figures	2" High	Hong Kong	$6.00-8.00	Smurfette has a medium blue body. She is wearing a white/orange/ red shirt, shorts, shoes and red socks. Smurfette has a white soccer ball attached to her left foot.
073/304	Papa With Lab Glasses	Schleich/Peyo/W.B.	1/1/82	Rubber Figures	2" High	Hong Kong	$3.00-4.00	Papa has a dark blue body. Papa is holding a test tube in each hand. Papa is standing, pouring the green bottle into the orange bottle. Papa is wearing a red hat and red pants.
073/305	Papa With Lab Glasses	Applause/Peyo		Rubber Figures	2" High	China	$3.00-4.00	Papa has a medium blue body. Papa is holding a test tube in each hand. Papa is standing, pouring the green bottle into the tan bottle. Papa is wearing a dull colored red hat and pants. Papa is a red mold.
073/306	Greedy	Schleich/Peyo/W.B.	1/1/83	Rubber Figures	2" High	Hong Kong	$3.00 - 4.25	The Smurf has a medium blue body. The Smurf is sitting on a red stool mixing light pink and white food with a gray spoon. The Smurf is licking his lips. He is wearing a white bib around his neck.
073/307	Greedy	Schleich/Peyo/W.B.	1/1/83	Rubber Figures	2" High	Hong Kong	$3.00 - 4.25	The Smurf has a medium blue body. The Smurf is sitting on a red stool mixing dark pink and white food with a brown spoon. The Smurf is licking his lips. He is wearing a white bib around his neck.
073/308	Baseball Pitcher	Schleich/Peyo/W.B.	1/1/83	Rubber Figures	2" High	Hong Kong	$86.25-115.00	The Smurf has a medium blue body. He is wearing a red shirt, white pants, red socks, black shoes, black belt and a white hat. The Smurf is holding a white baseball in his right hand. He is wearing a brown glove on the left hand. The Smurf is standing on a bright green plastic platform. The Smurf is standing like he's going to throw the ball.
073/309	Indian Smurfette	Schleich/Peyo/W.B.	1/1/83	Rubber Figures	2" High	Hong Kong	$11.00-15.00	Smurfette has a medium blue body. She is wearing a tan tunic, tan moccasins, tan ribbons in her hair and a white hat with a tan headband. The headband has a red feather stuck in it. Smurfette is holding a brown jug.
073/310	Indian Smurfette	Schleich/Peyo/W.B.	1/1/83	Rubber Figures	2" High	Hong Kong	$11.00-15.00	Smurfette has a medium blue body. She is wearing a goldenrod colored tunic, goldenrod moccasins, goldenrod ribbons in her hair and a white hat with a goldenrod headband. The headband has a red feather stuck in it. Smurfette is holding a darker brown jug.
073/311	Smurfette Jump Roping	Schleich/Peyo/W.B.	1/1/83	Rubber Figures	2" High	Hong Kong	$4.50-6.00	Smurfette has a medium blue body. Smurfette is wearing a white dress, yellow panties, white hat and white shoes. Smurfette is holding a jump rope over her head. The jump rope is white and has red handles.
073/312	Hot Dog	Schleich/Peyo/W.B.	1/1/83	Rubber Figures	2" High	Hong Kong	$6.00-8.00	The Smurf has a medium blue body. He is standing, holding a hot dog in his hands. The bun is brown with a red hot dog in the center and yellow mustard on top.
073/313	Quarterback	Schleich/Peyo/W.B.	1/1/83	Rubber Figures	2" High	Hong Kong	$40.00-55.00	The Smurf has a medium blue body. He is wearing an orange shirt with a red # 3 on the back and front, white pants, black spiked shoes and a white hat with an orange stripe and a red "S" on each side. He is holding a brown and white football. The Smurf is standing like he's going to throw the football.
073/314	Handy	Schleich/Peyo/W.B.	1/1/83	Rubber Figures	2" High	Hong Kong	$4.50-6.00	The Smurf has a medium blue body. He is wearing gray and white overalls, gray shoes and a white hat. The Smurf is holding a brown tool box with a white handle in his left hand and a yellow ruler under his left arm. He has a gray hammer with a brown handle in his right hand. There is an orange pencil behind his ear.
073/315	Jogger	Schleich/Peyo/W.B.	1/1/83	Rubber Figures	2" High	Hong Kong	$12.00-16.00	The Smurf has a dark blue body. He is wearing an orange and white jogging suit, white hat and light blue shoes. The Smurf is on a dark green plastic platform. The Smurf has a bead of sweat running down his face. His eyes are shut.
073/316	Jogger	Schleich/Peyo	1/1/82	Rubber Figures	2" High	Germany	$3.00 - 4.25	The Smurf has a medium blue body. He is wearing a red and white jogging suit, white hat and red shoes. The Smurf is on a bright green plastic platform. The Smurf has a bead of sweat running down his face. His eyes are shut. The Smurf has a black dot on his foot.
073/317	Schoolgirl	Schleich/Peyo/W.B.	1/1/83	Rubber Figures	2" High	Hong Kong	$6.00-8.00	Smurfette has a medium blue body. Smurfette is wearing a purple sweater, a pink with purple doted skirt, white shoes, white hat and pink ribbons in her hair. Smurfette is hugging brown book to her chest. She is standing on a green stand. Smurfette has no eyelashes.
073/318	Papa With Book	Schleich/Peyo/W.B.	1/1/82	Rubber Figures	2" High	Hong Kong	$13.50-18.00	Papa has a dark blue body. He is standing with a grayish/tan book in his hands. The book is open and in the inside are black lines. There is a red bookmark in the book. The outside of the book says "Magic" in black letters. Papa is wearing a red hat and red pants.
073/319	Clockwork	Schleich/Peyo/W.B.	1/1/83	Rubber Figures	2" High	Hong Kong	$7.50-10.00	The Smurf has a dark blue body. The Smurf has brown log ears, nose, feet and sticks in his hands. He has a wood tail but it is painted blue.
073/320	St. Patrick's Green	Schleich/Peyo	1/1/82	Rubber Figures	2" High	Portugal	$25.00-35.00	The Smurf has a medium blue body. He is wearing a green jacket, green pants, green shoes, green hat with an orange stripe, an orange bow tie and orange socks. The Smurf is holding a brown stick in his left hand and holding a light green shamrock with a green stem in his right hand.
073/321	St. Patrick's German	Schleich/Peyo	1/1/82	Rubber Figures	2" High	Portugal	$6.00-8.00	The Smurf has a very bright medium blue body. He is wearing a brown jacket, yellow shirt, red pants, black shoes, black hat with a yellow buckle on the front, a red bow tie and white socks. The Smurf is holding a dark brown stick in his left hand and holding a green shamrock with a brown stem in his right hand.
073/322	Thanksgiving Smurf	Schleich/Peyo	1/1/82	Rubber Figures	2" High	Portugal	$7.50-10.00	The Smurf has a medium blue body. He is wearing a gray jacket with white cuffs and a white collar, a yellow belt, white pants, a gray hat with a yellow buckle on the front and brown shoes with yellow buckles. The Smurf is carrying a red platter with a brown turkey.
073/323	Tracker	Schleich/Peyo/W.B.	1/1/83	Rubber Figures	2" High	Hong Kong	$6.00-8.00	The Smurf has a medium blue body. He is walking, carrying a dark brown stick in his right hand. The Smurf has a red feather in his hat.
073/324	Baby With Rattle White	Schleich/Peyo/W.B.	1/1/84	Rubber Figures	2" High	Hong Kong	$4.50-6.00	Baby has a bright, light blue body color. Baby is wearing a white sleeper and hat. He is crawling, carrying a red rattle with a yellow center in his right hand.
073/325	Papa With Pizza	Schleich/Peyo/W.B.	1/1/84	Rubber Figures	2" High	Hong Kong	$7.50-10.00	Papa has a medium blue body. Papa is wearing a white apron, red pizza hat and red pants. Papa is holding a brown platter with a tan/yellow pizza. The pizza has red dots on top.
073/326	Gargamel With Net	Schleich/Peyo	1/1/83	Rubber Figures	2" High	West Germany	$18.00-25.00	Gargamel has a dark flesh colored skin. He is wearing a dull black suit with a white patch and red boots. Gargamel is holding a dark yellow net. The net has a dark yellow handle. Gargamel is standing on a bright green stand.
073/327	Gargamel With Net	Schleich/Peyo/W.B.	1/1/83	Rubber Figures	2" High	Hong Kong	$15.00-20.00	Gargamel has a very light flesh colored skin. He is wearing a glossy black suit with a white patch and red boots. Gargamel is holding a dark yellow net. The net has a brown handle. Gargamel is standing on a dark green stand.
073/328	Smurfette With Comb And Mirror	Schleich/Peyo	1/1/83	Rubber Figures	2" High	West Germany	$4.50-6.00	Smurfette has a medium blue body. She is wearing a pink gown, white shoes and a white hat. Smurfette is holding a purple and white mirror in one hand while combing her hair with a purple comb.
073/329	Aerobic Smurfette	Schleich/Peyo/W.B.	1/1/84	Rubber Figures	2" High	Hong Kong	$6.00-8.00	Smurfette has a medium blue body. She is wearing a pink body suit, white and pink shoes, white hat and purple leg warmers. Smurfette has her hands out at her sides.
073/330	Aerobic Smurfette			Rubber Figures	2" High		$9.00-12.00	Smurfette has a medium blue body. She is wearing a pink body suit, white and pink shoes, white hat and hot pink leg warmers. Smurfette has her hands out at her sides.
073/331	Bullfighter	Schleich/Peyo/W.B.	1/1/84	Rubber Figures	2" High	Hong Kong	$11.00-15.00	The Smurf has a medium blue body. He is standing, holding a red cape.
073/332	Sailor	Schleich/Peyo	1/1/84	Rubber Figures	2" High	West Germany	$4.50-6.00	The Smurf has a medium blue body. He is wearing a white sailor suit with dark blue trim and a white sailor hat. The Smurf is carrying a brown sea bag over his right shoulder.
073/333	Baseball Smurfette	Schleich/Peyo/W.B.	1/1/84	Rubber Figures	2" High	Hong Kong	$75.00-100.00	Smurfette has a dark blue body. She is wearing a pink shirt with purple cuffs, white pants with pink stripes, purple socks, black shoes and a white hat with a purple visor. Smurfette is holding a creme colored bat.
073/334	Handy Plumber	Schleich/Peyo	1/1/83	Rubber Figures	2" High	Portugal	$7.50-10.00	The Smurf has a medium blue body. He is holding a silver wrench in his right hand. He is carrying a dull rust/ brown toolbox in his left hand.
073/335	Handy Plumber	Schleich/Peyo/W.B.	1/1/84	Rubber Figures	2" High	Hong Kong	$7.50-10.00	The Smurf has a glossy medium blue body. He is holding a silver wrench in his right hand. He is carrying a glossy rust/ brown toolbox in his left hand.
073/336	Majorette	Schleich/Peyo/W.B.	1/1/84	Rubber Figures	2" High	Hong Kong	$11.00-15.00	Smurfette has a medium blue body. She is wearing a pink dress with gold trim, purple boots and a white hat with yellow feather stuck in the side and a purple visor. Smurfette is carrying a purple baton with a gold tip in her right hand. Her left hand is on her hip. Smurfette is standing on a green stand.
073/337	Smurf With Dustpan	Schleich/Peyo	1/1/85	Rubber Figures	2" High	West Germany	$6.00-8.00	The Smurf has a medium blue body. He is holding a brown dustpan and a brown broom with yellow bristles. The Smurf is sweeping black dirt into the dustpan. He is bent over a little and has his eye's shut.

073/338	Smurf with Dustpan	Schleich/Peyo/Applause	1/1/84	Rubber Figures	2" High	Hong Kong	$6.00-8.00	The Smurf has a dark blue body. He is holding a dark brown dustpan and a dark brown broom with tan bristles. The Smurf is sweeping black dirt into the dustpan. He is bent over a little and has his eye's shut.
073/339	Smurfette With Ice Cream	Schleich/Peyo	1/1/84	Rubber Figures	2" High	Hong Kong	$6.00-8.00	Smurfette has a dark blue body. She is wearing a white dress with pink dots, a white hat and white shoes. Smurfette is holding an ice cream cone in front of her. The cone is brown with purple, pink and yellow ice cream.
073/340	Brainy Referee	Peyo	1/1/84	Rubber Figures	2" High	Hong Kong	$4.50-6.00	The Smurf has a medium blue body. He is wearing a white and black stripe shirt, white pants, black socks, white shoes, white hat, a black bow tie and a black belt with a silver buckle. The Smurf is holding a silver whistle to his mouth with one hand and pointing with the other hand. The Smurf is wearing black glasses.
073/341	Smurfette With Baby	Schleich/Peyo	1/1/84	Rubber Figures	2" High	Hong Kong	$4.50-6.00	Smurfette has a dark blue body. Baby has a light pale blue body. Smurfette is standing, holding Baby over her shoulder. Baby is sleeping. Smurfette is wearing a white dress, white shoes and a white hat. Baby is wearing a pink sleeper and pink hat. Baby is holding a red rattle with a yellow center.
073/342	Smurf With Mop And Pail	Schleich/Peyo	1/1/84	Rubber Figures	2" High	Hong Kong	$6.00-8.00	The Smurf has a medium blue body. The Smurf is holding a dark gray mop with a brown handle. There is a brown pail with a dark brown handle and edges by the Smurfs foot. The pail has dark gray water inside. The Smurf looks like he's whistling.
073/343	Smurfette Jogger	Schleich/Peyo	1/1/84	Rubber Figures	2" High	Hong Kong	$9.00-12.00	Smurfette has a medium blue body. Smurfette is wearing a pink jogging jacket with a purple collar, purple pants with pink stripes on the side, white hat and pink shoes with purple laces. Smurfette is standing on a green stand.
073/344	Graduation With Diploma	Schleich/Peyo/W.B.	1/1/84	Rubber Figures	2" High	Hong Kong	$75.00-100.00	The Smurf has a medium blue body. He is wearing a light blue gown, white shoes, a white hat with a light blue cap. The cap has a yellow tassel. The Smurf is holding a white diploma in his 2 hands. The diploma says "# 1 Grad" in red letters.
073/345	Smurfette With Thanksgiving pie	Schleich/Peyo/W.B.	1/1/82	Rubber Figures	2" High	Portugal	$7.50-10.00	Smurfette has a dark blue body. She is wearing a black pilgrim dress with a white collar, white cuffs and a white apron. Smurfette has a white pilgrim hat tied around her head. She is carrying a red and tan pie.
073/346	Thanksgiving With Ear Of Corn	Schleich/Peyo/ W.B/ Applause	1/1/84	Rubber Figures	2" High	Portugal	$11.00-15.00	The Smurf has a dark blue body. He is dressed as an Indian. He is wearing brown pants with red stitching, brown and yellow shoes and a yellow band around his left arm. The Smurf has a white hat with a red/brown/ yellow headband and a white feather with a red tip stuck in the headband. The Smurf is holding a yellow cob of corn in front of him. He has his eye's closed.
073/347	Smurfette Witch	Schleich/Peyo/W.B.	1/1/82	Rubber Figures	2" High	Portugal	$15.00-20.00	Smurfette has a dark blue body. She is wearing a purple witches dress, orange shoes with black stripes and a purple witches hat with an orange band and a yellow buckle. Smurfette is sitting on a tan broom stick with yellow bristles.
073/348	Smurf With Gargamel Mask	Schleich/Peyo/ W.B/ Applause	1/1/84	Rubber Figures	2" High	Portugal	$25.00-35.00	The Smurf has a dark blue body. The Smurf is wearing a Gargamel mask. The Gargamel mask has a light flesh color for skin. The Smurf is holding a yellow trick or treat bag in front of him. The bag is full of red/white/blue treats.
073/349	Smurf With Gargamel Mask	Schleich/Peyo/ W.B/ Applause	1/1/84	Rubber Figures	2" High	Portugal	$25.00-35.00	The Smurf has a dark blue body. The Smurf is wearing a Gargamel mask. The Gargamel mask has a dark flesh color for skin. The Smurf is holding a yellow trick or treat bag in front of him. The bag is full of red/white/blue treats.
073/350	Christmas Smurfette	Schleich/Peyo/ W.B/ Applause	1/1/84	Rubber Figures	2" High	Portugal	$15.00-20.00	Smurfette has a dark blue body. She is wearing a long green gown with white cuff, a white collar and a red bow. Smurfette is wearing a white hat and white shoes. She is holding a white package with red ribbon around it and a red bow.
073/351	Christmas Smurf With Lantern	Schleich/Peyo/ W.B/ Applause	1/1/84	Rubber Figures	2" High	Portugal	$11.00-15.00	The Smurf has a dark blue body. He is wearing a brown coat with a white collar, white cuffs and yellow buttons. His hat is brown with a white stripe and green leaves with red berries on one side. The Smurf has a red scarf and white pants. He is carrying a red lantern with a yellow flame in his left hand and a green package with a red ribbon in the right hand.
073/352	Baby With Rattle Pink	Schleich/Peyo	1/1/84	Rubber Figures	2" High	Hong Kong	$4.50-6.00	Baby has a medium body color. Baby is wearing a dark pink sleeper and hat. He is crawling, carrying a red rattle with a yellow center in his right hand.
073/353	Baby With Rattle Pink	Schleich/Peyo/W.B.	1/1/84	Rubber Figures	2" High	Hong Kong	$4.50-6.00	Baby has a bright light blue body color. Baby is wearing a light pink sleeper and hat. He is crawling, carrying a red rattle with a yellow center in his right hand.
073/354	Baby With Rattle Blue	Schleich/Peyo/W.B.	1/1/84	Rubber Figures	2" High	Hong Kong	$4.50-6.00	Baby has a bright light blue body color. Baby is wearing a light blue sleeper and hat. He is crawling, carrying a red rattle with a yellow center in his right hand.
073/355	Smurfette With Flute	Schleich/Peyo	1/1/85	Rubber Figures	2" High	Hong Kong	$6.00-8.00	Smurfette has a medium blue body. Smurfette is wearing a white dress with pink dots, white shoes and a white hat. She is holding a tan flute to her mouth.
073/356	Baby With Teddy	Schleich	1/1/85	Rubber Figures	2" High	Macau	$3.00 - 4.25	Baby has a medium blue body. Baby is wearing a white sleeper and white hat. Baby is sitting, holding a brown teddy bear. The teddy bear is brown/tan and he is wearing a red bow tie.
073/357	Baby With Ice Cream	Schleich/Peyo	1/1/84	Rubber Figures	2" High	West Germany	$4.50-6.00	Baby has a medium blue body. He is wearing a light blue sleeper and hat. Baby is sitting, eating an ice cream. The ice cream is pink/purple and on a brown stick. Baby is holding the ice cream in his right hand.
073/358	X-mas Smurf With A Gift	Schleich/Peyo/W.B.	1/1/82	Rubber Figures	2" High	Portugal	$11.00-15.00	The Smurf has a dark blue body. He is carrying a red and pink candy cane in his right hand. In his left hand he is carrying a green package with a red ribbon and red bow.
073/359	X-mas Smurfette	Schleich/Peyo/W.B.	1/1/82	Rubber Figures	2" High	Portugal	$11.00-15.00	Smurfette has a dark blue body. She is wearing a short red coat with white trim, a white collar and white buttons. She has white shoes, a red hat with white trim and holly on one side. Smurfette is carrying a green package with a red ribbon and red bow.
073/360	Dentist	Schleich/Peyo	1/1/85	Rubber Figures	2" High	Hong Kong	$4.50-6.00	The Smurf has a dark blue body. He is wearing a white jacket with black buttons, white pants and a white hat. The Smurf is holding a red toothbrush with white bristles in his left hand.
073/361	Smurfette Golfer	Schleich/Peyo	1/1/85	Rubber Figures	2" High	Hong Kong	$4.50-6.00	Smurfette has a medium blue body. Smurfette is wearing a white dress, white shoes and a white hat. Smurfette is swinging a gray golf club. There is a white ball on a gray tee in front of her. Smurfette is standing on a dark green stand.
073/362	Smurf Basketball	Schleich/Peyo	1/1/85	Rubber Figures	2" High	West Germany	$6.00-8.00	The Smurf has a medium blue body. He is wearing a red tank top with an orange #1 on the front, white shorts, a white hat, white shoes and orange socks. The Smurf has a brown ball in his right hand.
073/363	Angel	Schleich/Peyo	1/1/84	Rubber Figures	2" High	West Germany	$4.50-6.00	The Smurf has a medium blue body. He is wearing a long white robe. He has 2 white wings attached to his back. The Smurf is holding a finger up in the air on his right hand. The Smurf has a red mouth.
073/364	Angel	Schleich/Peyo	1/1/85	Rubber Figures	2" High	Hong Kong	$4.50-6.00	The Smurf has a light blue body. He is wearing a long white robe. He has 2 white wings attached to his back. The Smurf is holding a finger up in the air on his right hand. The Smurf has a black mouth.
073/365	Devil	Schleich/Peyo	1/1/84	Rubber Figures	2" High	West Germany	$4.50-6.00	The Smurf has a dark pink body. The Smurf has glossy red feet, pink ears, pink tail, black wings and white horns. There is a black dot on his foot.
073/366	Devil	Schleich/Peyo	1/1/84	Rubber Figures	2" High	West Germany	$4.50-6.00	The Smurf has a pink body. The Smurf has red feet, pink ears, pink tail, black wings and white horns. There is a black dot on his foot.
073/367	Baby With Blocks	Schleich/Peyo	1/1/84	Rubber Figures	2" High	West Germany	$3.00 - 4.25	The Smurf has a medium blue body. He is sitting, wearing a white sleeper. Baby has 3 blocks. The blocks are yellow, red and green. Baby has a red mouth.
073/368	Baby With Blocks	Schleich/Peyo	1/1/85	Rubber Figures	2" High	Hong Kong	$4.50-6.00	Baby has a light blue body. He is sitting, wearing a white sleeper. Baby has 3 blocks. The blocks are pink, yellow and light blue. Baby has a black mouth.
073/369	Baby With Car	Peyo/Applause	1/1/85	Rubber Figures	2" High	China	$4.50-6.00	Baby has a light blue body. He is crawling, pushing a red car. Baby is wearing a white sleeper and a white hat.
073/370	Smurf In Tuxedo	Schleich/Peyo	1/1/85	Rubber Figures	2" High	West Germany	$15.00-20.00	The Smurf has a medium blue body. He is wearing a white tux with a red bow tie and a red cumberbun.
073/371	Smurfette In Gown	Schleich/Peyo	1/1/84	Rubber Figures	2" High	West Germany	$15.00-20.00	Smurfette has a medium bright blue body. Smurfette is wearing a long white gown with a red belt and a red bow. Smurfette has a red flower in her hat.
073/372	Baby With Butterfly	Schleich/Peyo	1/1/84	Rubber Figures	2" High	West Germany	$3.00 - 4.25	Baby has a medium blue body. Baby is wearing a white sleeper. He is sitting, holding a yellow butterfly in his right hand. The butterfly has red spots.
073/373	Surprise Bag Green	Schleich/Peyo	1/1/82	Rubber Figures	2" High	West Germany	$37.50-50.00	The Smurf has a medium blue body. He is standing, holding a yellow and green wrapped cone.
073/374	Surprise Bag Pink	Schleich/Peyo	1/1/82	Rubber Figures	2" High	West Germany	$18.00-25.00	The Smurf has a medium blue body. He is standing, holding a pink and purple wrapped cone.
073/375	Surprise Bag Orange	Schleich/Peyo	1/1/82	Rubber Figures	2" High	West Germany	$37.50-50.00	The Smurf has a medium blue body. He is standing, holding an orange and gold wrapped cone.
073/376	Smurfette Stewardess	Schleich/Peyo	1/1/85	Rubber Figures	2" High	West Germany	$7.50-10.00	Smurfette has a medium blue body. She is wearing a white dress, white shoes and a white hat with a yellow flower on each side. Smurfette is holding a red tray with a brown cup. The cup has white liquid in it and a blue straw.
073/377	Papa Pilot	Schleich/Peyo	1/1/85	Rubber Figures	2" High	Germany	$4.50-6.00	Papa has a medium blue body. He is wearing a brown pilot suit, yellow scarf, black shoes, a red hat and brown goggles. Papa has his thumb in the air on his right hand. He has his eye's closed.
073/378	Baby With Bowl And Spoon	Schleich/Peyo	1/1/85	Rubber Figures	2" High	West Germany	$7.50-10.00	Baby has a medium blue body. He is sitting, holding a pink bowl and a red spoon. He has yellow gobs spattered all over. Baby is wearing a white sleeper.
073/379	Smurf With Accordion	Schleich/Peyo	1/1/87	Rubber Figures	2" High	West Germany	$4.50-6.00	The Smurf has a medium blue body. He is standing, holding a yellow and red accordion. The Smurf has his eye's shut.
073/380	Grandpa	Applause/Peyo	1/1/87	Rubber Figures	2" High	China	$4.50-6.00	The Smurf has a medium blue body. He is wearing yellow pants, a yellow hat and black glasses with silver chains. Grandpa has a white beard to the ground. He is holding a dark brown stick in his right hand.
073/381	Table Tennis Player	Schleich/Peyo	1/1/87	Rubber Figures	2" High	Germany	$6.00-8.00	The Smurf has a medium blue body. He is wearing a green shirt, black short, a white hat and white and black shoes. The Smurf is holding a red paddle in his right hand.
073/382	Fitness	Schleich/Peyo	1/1/87	Rubber Figures	2" High	West Germany	$7.50-10.00	The Smurf has a medium blue body. He is wearing red posing trunks. The Smurf is holding a silver and brown stretch exerciser. He has distinct muscles.
073/383	Architect Smurf	Schleich/Peyo	1/1/87	Rubber Figures	2" High	West Germany	$4.50-6.00	The Smurf has a medium blue body. He is wearing a yellow jacket, orange hard hat, white pants and green boots. The Smurf is holding a yellow ruler in his left hand and a white scroll in his right hand. He has an orange and brown pencil in his jacket pocket.
073/384	Wild	Schleich/Peyo	1/1/88	Rubber Figures	2" High	West Germany	$4.50-6.00	The Smurf has a medium bright blue body. The Smurf is hanging on a brown vine. He is wearing a green leaf hat and yellow shorts.
073/385	Hula Smurfette	Schleich/Peyo	1/1/88	Rubber Figures	2" High	West Germany	$4.50-6.00	Smurfette has a medium blue body. Smurfette is wearing a tan hula skirt, green belt, red lei and orange hat trim. Smurfette has her hands out to her side like she's dancing.
073/386	Gargamel With Lab Glasses	Schleich/Peyo	1/1/82	Rubber Figures	2" High	West Germany	$6.00-8.00	Gargamel has a dark flesh colored skin. Gargamel is wearing a black robe and maroon boots. Gargamel is holding a pink glass in his right hand and a purple lab glass in left hand. He looks like he's pouring one lab glass into the other.

073/387	Bully Gargamel	Peyo/Bully		Rubber Figures	2" High	West Germany	$37.50-50.00	Gargamel has a light flesh color skin. He is wearing a black robe with a white patch and red shoes. He has his right hand in the air and he's pointing. He is holding a green lab glass in left hand.
073/388	Snappy Smurfling	Applause/Peyo	1/1/87	Rubber Figures	2" High	China	$3.50-5.00	The Smurf has a medium bright blue body. He is wearing a white hat, a white pants and a yellow shirt with a lightening bolt on the front. His hands are out in fists. He has an angry expression.
073/389	Slouchy Smurfling	Applause/Peyo	1/1/87	Rubber Figures	2" High	China	$3.50-5.00	The Smurf has a medium bright blue body. He is wearing a white droopy hat a white pants and red shirt. He has his right hand waving.
073/390	Nat With Caterpillar	Applause/Peyo	1/1/87	Rubber Figures	2" High	China	$3.50-5.00	The Smurf has a medium bright blue body. He is wearing a brown pants with a brown sash. He has a yellow straw hat. There is a green caterpillar by his right foot
073/391	Sassette Smurfling	Schleich/Peyo	1/1/87	Rubber Figures	2" High	West Germany	$6.00-8.00	Sassette has a medium blue body. She has pink pants, pink feet, pink suspenders and a white hat. Sassette has red hair and freckles. She is standing with her hands on her hips.
073/392	Sassette Smurfling	Applause/Peyo	1/1/87	Rubber Figures	2" High	China	$3.50-5.00	Sassette has a medium blue body. She has pink pants, pink feet, pink suspenders and a white hat. Sassette has orange hair and no freckles. She is standing with her hands on her hips.
073/393	Puppy	Applause/Peyo	1/1/87	Rubber Figures	2" High	China	$3.00 - 4.25	The Puppy is brown with dark brown spots and both ears are dark brown. The dog is wearing a yellow collar. He has a tan colored muzzle, a dark brown nose and a red tongue.
073/394	Puppy	Schleich/Peyo	1/1/87	Rubber Figures	2" High	West Germany	$7.50-10.00	The Puppy is gray with brown spots and one brown ear. The dog is wearing a yellow collar. He has a cream colored muzzle, a dark brown nose and a red tongue.
073/395	Smoogle	Schleich/Peyo	1/1/88	Rubber Figures	2" High	West Germany	$9.00-12.00	Smoogle is a pink rabbit. He has pink floppy ears and a white face. Smoogle has his hands at his sides.
073/396	Chitter	Schleich/Peyo	1/1/88	Rubber Figures	2" High	West Germany	$9.00-12.00	Chitter is a gray squirrel. He has a white face, white eye's and a white patch on his tummy. He is holding a brown and green acorn.
073/397	Nanny	Applause/Peyo	1/1/89	Rubber Figures	2" High	China	$3.50-5.00	Nanny has a medium blue body. She is wearing a long pink dress, a dark pink shawl, white shoes and a white hat. Nanny has black glasses with white chains. Her hair is gray.
073/398	Patriot Smurf	Schleich/Peyo	1/1/89	Rubber Figures	2" High	West Germany	$12.00-16.00	The Smurf has a medium blue body. He is walking, carrying a dark blue flag. The flag has yellow stars forming a circle. The Smurf is standing on a bright green stand. The Smurf has a black dot on his foot.
073/399	Smurfette With Mouse	Schleich/Peyo	1/1/91	Rubber Figures	2" High	West Germany	$3.00 - 4.25	Smurfette has a medium blue body. She is wearing a white dress with pink dots, white shoes and a white hat. Smurfette is standing next to a gray and pink mouse. Smurfette has a black dot on her foot.
073/400	Azreal	Schleich/Peyo	1/1/91	Rubber Figures	2" High	Germany	$6.00-8.00	Azreal is orange and white. Azreal is in a prowling position with his tail up. He has his head up and he's licking his chops.
073/401	Bride	Schleich/Peyo	1/1/91	Rubber Figures	2" High	Germany	$3.00 - 4.25	Smurfette has a medium blue body. Smurfette is wearing a long white gown, white shoes and a white hat/veil with a pink flower hat band. Smurfette is holding a bouquet of pink flowers in her left hand. Her right hand is out to be held. There is a black dot on her foot.
073/402	Bridegroom	Schleich/Peyo	1/1/91	Rubber Figures	2" High	Germany	$3.00 - 4.25	The Smurf has a medium blue body. The Smurf is wearing a white tux, a white top hat, white shoes, a blue bow tie and a red carnation in his lapel. He is standing with his hand out to hold it. The Smurf has a black dot on his foot.
073/403	Video Smurf	Schleich/Peyo	1/1/91	Rubber Figures	2" High	Germany	$3.00 - 4.25	The Smurf has a medium blue body. He is standing, holding a black and gray camcorder to his right eye.
073/404	Handball	Schleich/Peyo	1/1/91	Rubber Figures	2" High	Germany	$4.50-6.00	The Smurf has a medium blue body. He is wearing a red shirt, white shorts, a white shirt and white and red shoes. The Smurf is holding a dark brown ball in his right hand. The Smurf has a black dot on his foot.
073/405	New Football Smurf	Schleich/Peyo	1/1/91	Rubber Figures	2" High	Germany	$3.00 - 4.25	The Smurf has a medium blue body. He has his arms out at his sides. His right foot has a white and black soccer ball on the tip. He is wearing a bright yellow shirt, bright yellow socks and red shorts. He has on black shoes.
073/406	Bodybuilder	Schleich/Peyo	1/1/93	Rubber Figures	2" High	Germany	$7.50-10.00	The Smurf has a medium bright blue body. He is wearing gold posing trunks. The Smurf has rippling muscles, his right arm is flexed and his left arm is straight out.
073/407	Gargamel With Raised Hands	Schleich/Peyo	1/1/93	Rubber Figures	2" High	Germany	$3.00 - 4.25	Gargamel has a light flesh color skin. He is wearing a black shirt with blue/yellow/green patches and red shoes. Gargamel has his hands up in the air like he's going to lunge at something.
073/408	Boxer	Schleich/Peyo	1/1/80	Rubber Figures	2" High	West Germany	$3.00 - 4.25	The Smurf has a medium bright blue body. He is wearing a yellow tank shirt, yellow socks, green shorts, a white hat and brown shoes. The Smurf has 2 red and white boxing gloves on. He is standing with one fist out and the other by his side. The Smurf has a black dot on his foot.
073/409	Scruple	Schleich/Peyo	1/1/93	Rubber Figures	2" High	Germany/China	$4.50-6.00	Scruple has a light flesh colored shin. He is wearing a light gray shirt, dark gray pants, blue shoes, a purple cape and a blue hat. Scruple's has orange hair and freckles. He is standing, pointing at something with his right hand.
073/410	Smurfette With Flower	Schleich/Peyo	1/1/93	Rubber Figures	2" High	Germany	$3.00 - 4.25	Smurfette has a medium bright blue body. She is wearing a white dress with pink dots, a white hat and white sandals with pink bows. Smurfette is holding a white daisy with a green stem in her left hand. Smurfette has a pink purse with white dots hanging on her right arm. There is a black dot on her foot.
073/411	Tramp	Schleich/Peyo	1/1/93	Rubber Figures	2" High	Germany	$3.00 - 4.25	The Smurf has a medium bright blue body. The Smurf is wearing a white cap with a gray patch and ripped white pants. The Smurfs hat is hanging over his right eye. His pants are torn and his right foot is sticking out. He is holding a yellow bag with red berries in his left arm. He is holding one red berry in his right hand. There is a black dot on the bottom of his foot.
073/412	Sitter	Schleich/Peyo	1/1/93	Rubber Figures	2" High	Germany/China	$7.50-10.00	The Smurf has a medium bright blue body. He is sitting with his legs crossed, looking up and gesturing towards the sky with his hands. The Smurf is wearing white pants and a white hat.
073/413	Papa With Hands Behind His Back	Schleich/Peyo	1/1/94	Rubber Figures	2" High	Germany	$3.00 - 4.25	Papa has a medium bright blue body. He is wearing a red hat and red pants. Papa is standing with his hands behind his back. He has his head turned to the right.
073/414	Gargamel With Hands On His Hips	Schleich/Peyo	1/1/94	Rubber Figures	2" High	Germany	$3.00 - 4.25	Gargamel has a light flesh colored skin. He is wearing a long black shirt with white patches and maroon boots. Gargamel is standing with his hands on his hips. He has an angry expression on his face. There is a black dot on his foot.
073/415	Schoolboy	Schleich/Peyo	1/1/94	Rubber Figures	2" High	Germany	$3.00 - 4.25	The Smurf has a medium bright blue body. He is wearing a yellow shirt, red belt, dark blue pants, a white hat and white and red shoes. The Smurf is carrying a green school bag in his right hand. The school bag has a yellow ruler sticking out the back. The Smurf has a black dot on his foot.
073/416	Stone Age Smurf	Schleich/Peyo	1/1/94	Rubber Figures	2" High	Germany	$3.00 - 4.25	The Smurf has a medium bright blue body. He is wearing a cream colored with brown blotched tunic and hat. He is carrying a dark brown bat over his right shoulder. There is a black dot on his foot.
073/417	Stone Age Smurfette	Schleich/Peyo	1/1/94	Rubber Figures	2" High	Germany	$3.00 - 4.25	Smurfette has a medium bright blue body. She is wearing a cream colored dress and hat with brown blotches. Smurfette has her right hand over her mouth. She looks like she has a troubled expression on her face. Smurfette has a black dot on the bottom of her foot.
073/418	Tattoo Smurf	Schleich/Peyo	1/1/94	Rubber Figures	2" High	Germany	$3.00 - 4.25	The Smurf has a medium bright blue body. The Smurf is standing with his hands on his hips. He has a red heart tattoo on his left shoulder. The Smurf has a sly look on his face. He is wearing white pants and a white hat. There is a black dot on the bottom of his foot.
073/419	Viking Smurf	Schleich/Peyo	1/1/94	Rubber Figures	2" High		$3.00 - 4.25	The Smurf has a medium bright blue body. He is wearing a tan and orange hat with white horns, tan tunic, white pants, a tan and orange wrist band and a dark brown belt with an orange buckle. The Smurf is holding a brown sword in his right hand. In his left hand he is holding a blue/orange/yellow shield. The Smurf has a black dot on his foot.
073/420	Monk	Schleich/Peyo	1/1/94	Rubber Figures	2" High	Germany	$3.00 - 4.25	The Smurf has a medium bright blue body. The Smurf is wearing a long brown robe with a hood, a white belt and tan sandals. The Smurf is holding his hands together like he's praying. There is a black dot on his foot.
073/421	Jockey Smurf	Schleich/Peyo	1/1/94	Rubber Figures	2" High	Germany	$3.00 - 4.25	The Smurf has a medium bright blue body. He is wearing white pants, a white shirt, a red jockey jacket, dark blue cap and dark blue shoes. The Smurf is carrying a tan whip in his left hand. In his right hand he is carrying a brown and gray saddle. There is a black dot on his foot.
073/422	Jockey Smurfette	Schleich/Peyo	1/1/94	Rubber Figures	2" High	Germany	$3.00 - 4.25	Smurfette has a medium bright blue body. Smurfette is wearing white pants, a white shirt, a red jockey jacket, dark blue cap and dark blue shoes. Smurfette is carrying a tan whip in her right hand. In her left hand she is carrying a brown and gray saddle. There is a black dot on her foot.
073/423	Slouchy With Cone	Schleich/Peyo	1/1/94	Rubber Figures	2" High	Germany	$3.00 - 4.25	The Smurf has a medium bright blue body. He is wearing a red shirt, white pants and a droopy white hat. The Smurf is holding a green cone with red and white dots on the cone. The top end of the cone has a yellow and blue wrapper.
073/424	Sassette With Cone	Schleich/Peyo	1/1/94	Rubber Figures	2" High	Germany	$3.00 - 4.25	Sassette has a medium bright blue body. She is wearing pink pants with pink suspenders and a white hat. Sassette has red hair. Sassette is holding a dark blue cone with yellow dots on the cone. The top end of the cone has a green and orange wrapper. Sassette has a black dot on her foot.
073/425	Saxophone	Schleich/Peyo	1/1/95	Rubber Figures	2" High	Germany	$3.00 - 4.25	The Smurf has a medium bright blue body. He is wearing a red shirt, purple pants, a white hat and white shoes with yellow accents. The Smurf has his eyes closed and is playing a gold saxophone. The Smurf has a black dot on his foot.
073/426	Techno Smurf	Schleich/Peyo	1/1/95	Rubber Figures	2" High	Germany	$3.00 - 4.25	The Smurf has a medium bright blue body. He is wearing a white tank top, light purple pants, white and blue shoes and a white floppy hat with a red patch. The red patch on his hat has a yellow "S" on it. The Smurf has hands out like he's waving. He has a big smile on his face.
073/427	Mobile Smurf	Schleich/Peyo	1/1/95	Rubber Figures	2" High	Germany	$3.00 - 4.25	The Smurf has a medium bright blue body. He is standing, holding a black cellular phone in his right hand. The phone has gray buttons. The Smurf is giving the thumbs up sign with his left hand. The Smurf is wearing a white hat and white pants. The Smurf has a black dot on the bottom of his foot.
073/428	Azreal Sitting	Schleich/Peyo	1/1/96	Rubber Figures	2" High	Germany/China	$3.00 - 4.25	Azreal is orange with white ears, a white face and a white spot on his tummy. Azreal is sitting and has a startled expression on his face.
073/429	Sport Swimmer	Schleich/Peyo	1/1/96	Rubber Figures	2" High	Germany/China	$3.00 - 4.25	The Smurf has a medium bright blue body. He is wearing red swim trunks, a white hat and yellow goggles with gray glasses. The Smurf is standing on a light blue stand and he looks like he's ready to dive off.
073/430	Sprinter	Schleich/Peyo	1/1/96	Rubber Figures	2" High	Germany/China	$3.00 - 4.25	The Smurf has a medium bright blue body. He is wearing a red tank shirt, black shorts with red stripes, a white hat and yellow shoes. The Smurf is crouched ready to start running.
073/431	In-line Skater	Schleich/Peyo	1/1/96	Rubber Figures	2" High	Germany	$3.00 - 4.25	The Smurf has a medium blue body. The Smurf is rollerblading. He is wearing a white hat with a visor, a dark blue shirt, green short, orange knee pads, orange and yellow palm guards, yellow and black rollerblades with silver wheels and blue sunglasses.
073/432	In-line Skater Smurfette	Schleich/Peyo	1/1/96	Rubber Figures	2" High	Germany	$3.00 - 4.25	Smurfette has a medium blue body. She is rollerblading. Smurfette is wearing an orange bandanna around her head, white hat, green shirt, orange shorts, dark blue knee pads, dark blue and orange palm guards, dark blue elbow pads and black/white rollerblades with silver wheels.

073/433	Disco Smurf	Schleich/Peyo	1/1/96		Rubber Figures	2" High	Germany/China	$3.00 - 4.25	The Smurf has a medium blue body. The Smurf looks like he's dancing. He is wearing purple pants, pink socks, white shoes, a white hat, a red shirt with a yellow "S" on the front and a yellow belt.
073/434	Disco Smurfette	Schleich/Peyo	1/1/96		Rubber Figures	2" High	Germany/China	$3.00 - 4.25	Smurfette has a medium blue body. She looks like she's dancing. Smurfette is wearing a white hat, pink shirt, a bright green skirt with red flowers and a white pair of high heeled shoes.
073/435	Girl With Baby	Schleich/Peyo	1/1/96		Rubber Figures	2" High	Germany	$3.00 - 4.25	The Smurf has a medium blue body. She has dark reddish brown hair and a yellow bow in back. She is wearing pink panties and no shirt. She is standing, holding a doll. The doll has yellow hair and flesh color skin. The doll is wearing an orange dress, white petticoat and brown shoes.
073/436	Boy With Truck	Schleich/Peyo	1/1/96		Rubber Figures	2" High	Germany	$3.00 - 4.25	The Smurf has a medium blue body. The Smurf is sitting on a toy truck. The Smurf is wearing a white diaper and white hat. The truck has a red cab, green hood, yellow bumper and black wheels.
073/437	Girl Bathing	Schleich/Peyo	1/1/95		Rubber Figures	2" High	Germany/China	$3.00 - 4.25	Smurfette has a medium bright blue body. She is wearing a pink hat and dark pink bows in her hair. She has no clothes on. She is kneeling. Smurfette is scrubbing her back with a tan scrub brush. In her right hand she has a hand full of white suds and a cream colored bar of soap.
073/438	Lead Guitar Player	Schleich/Peyo	1/1/97		Rubber Figures	2" High	Germany/China	$3.00 - 4.25	The Smurf has a medium blue body. The Smurf is wearing dark blue overall bibs, white hat, white shoes and green wrist bands. The Smurf is playing a red and tan guitar with black detail and silver strings. The Smurf has a dark brown guitar strap around his shoulder.
073/439	Bass Guitar Player	Schleich/Peyo	1/1/97		Rubber Figures	2" High	Germany/China	$3.00 - 4.25	The Smurf has a medium blue body. The Smurf is wearing a red shirt, black pants, a white belt, white hat, white shoes and green wrist bands. The Smurf is playing a tan and white guitar with black detail and silver strings. The Smurf has a dark brown guitar strap around his shoulder.
073/440	Singer	Schleich/Peyo	1/1/97		Rubber Figures	2" High	Germany/China	$3.00 - 4.25	The Smurf has a medium blue body. The Smurf is wearing an orange jacket, green shirt, brown pants, white shoes, white hat and black belt. The Smurf is holding a black and red microphone.
073/441	Snowboarder	Schleich/Peyo	1/1/97		Rubber Figures	2" High	Ger/Portugal	$3.00 - 4.25	The Smurf has a medium blue body. The Smurf is wearing an orange sweater, brownish gray pants, dark brown gloves, dark blue shoes and silver goggles. The Smurf is standing on a dark orange snowboard.
073/442	Snowboarder Smurfette	Schleich/Peyo	1/1/97		Rubber Figures	2" High	Ger/Portugal	$3.00 - 4.25	Smurfette has a medium blue body. The Smurf is wearing a green sweater, peach pants, dark blue gloves, dark blue shoes and whitish goggles. The Smurf is standing on a purple snowboard.
073/443	Soccer Player	Schleich/Peyo	1/1/97		Rubber Figures	2" High	Germany	$3.00 - 4.25	The Smurf is standing, holding a white soccer ball. He is wearing a red shirt with a black collar, black shorts, red socks and black shoes. The Smurf has a black 10 on the back.
073/444	Johan	Schleich/Peyo/W.B.	1/1/79		Rubber Figures	2" High	Hong Kong	$7.50-10.00	Johan is a boy. He has black hair. He is a tan and white shirt, a dark brown belt, dark brown shoes, red pants. Johan is holding a silver sword. His skin is painted peach.
073/445	Johan Bully	Bully/Peyo	1/1/79		Rubber Figures	2 1/2" High		$18.00-25.00	Johan is a boy. He has black hair. He is a tan and white shirt, a dark brown belt, dark brown shoes, red pants. Johan is holding a silver sword. Johan has flesh color skin. The mold is flesh colored. On the back of his shoe it says Johan.
073/446	Peewit	Schleich/Peyo/W.B.	1/1/78		Rubber Figures	2" High	Hong Kong	$7.50-10.00	Peewit has peach colored skin. He is a little boy with yellow hair. Peewit is wearing a blue shirt, blue shorts, orange stockings, orange shoes and an orange beanie hat. Peewit is playing a brownish/gray guitar.
073/447	Peewit	Schleich/Peyo/W.B.	1/1/78		Rubber Figures	2" High	Hong Kong	$7.50-10.00	Peewit has flesh skin. He is a little boy with yellow hair. Peewit is wearing a blue shirt, blue shorts, orange stockings, orange shoes and an orange beanie hat. Peewit is playing a brownish/gray guitar. The mold is flesh color.
073/448	Peewit	Schleich/Peyo/W.B.	1/1/78		Rubber Figures	2" High	Hong Kong	$7.50-10.00	Peewit has peach color skin. He is a little boy with yellow hair. Peewit is wearing a green shirt, green shorts, red stockings, red shoes and a red beanie hat. Peewit is playing a brownish/gray guitar.
073/449	Tourist	Schleich/ Peyo	1/1/98		Rubber Figures	2" High		$3.00 - 3.50	
073/450	Sportsman	Schleich/Peyo	1/1/98		Rubber Figures	2" High		$3.00 - 3.50	
073/451	Aerobic Smurfette	Schleich/ Peyo	1/1/98		Rubber Figures	2" High		$3.00 - 3.50	
073/452	Newsman	Schleich/Peyo	1/1/98		Rubber Figures	2" High		$3.00 - 3.50	The Smurf has a bright blue body. He is holding a newspaper in the air. He has a brown bag over his shoulders.
073/453	Smurf With Nameplate	Schleich/Peyo	1/1/98		Rubber Figures	2" High		$3.00 - 3.50	The Smurf has a bright blue body. He is wearing a yellow shirt, black pants and red shoes. He is holding a brown framed name plate with a white center.
073/454	Golf Smurf	Schleich/ Peyo	1/1/98		Rubber Figures	2" High		$3.00 - 3.50	The Smurf has a bright blue body. He is wearing a white hat, yellow shirt, orange pants, white sock and brown golf shoes. The Smurf is carrying a white golf ball in one hand. In his other hand he is carrying a green golf bag.

Set 204: PVC's- Solid Color Molds

204/001	Brainy	Peyo		Silver	Rubber Smurf	2"	West Germany	$25.00-35.00	The Smurf is silver. He has silver glasses and his mouth is in the form of an O. The Smurf is standing with his hands out at his side. The complete figure is silver. The mold is silver.
204/002	Judge	Peyo		Light Blue	Rubber Smurf	2"	West Germany	$25.00-35.00	The Smurf is light blue. The Smurf is wearing a light blue gown and bow tie. He is looking to his right, he has his mouth open and is wearing light blue glasses. His left hand is pointing. The complete Smurf is light blue. The mold is light blue.
204/003	Rock N Roll	Peyo		Pink	Rubber Smurf	2"	West Germany	$25.00-35.00	The Smurf is pink. The Smurf is holding a pink guitar. He has his eyes shut and his mouth open. The complete mold is pink.
204/004	Courting	Bully Peyo		Gopper	Rubber Smurf	2"	West Germany	$25.00-35.00	The Smurf is copper color. He is standing, holding a copper colored daisy. The complete mold is copper.
204/005	Beer	Peyo		Purple	Rubber Smurf	2"	West Germany	$25.00-35.00	The Smurf is purple. He is holding a mug of beer. The mug is purple. The complete Smurf is purple. The mold is white.
204/006	Present	Peyo		Pink	Rubber Smurf	2"	West Germany	$25.00-35.00	The Smurf is pink. He is holding a pink present. The Smurf is holding the present with both hands off to his right side. The mold is pink.

Set 178: Quick- Promotional PVC's

178/001	Rocking Chair	Schleich/Peyo	1/1/96		Rubber Smurf	2" High	Germany/China	$18.00-25.00	Papa has a medium blue body. Papa is smoking a tan and gray pipe. The rocking chair is brown and has a yellow cushion. The Quick emblem is on the back of Papa's hat.
178/002	Smurf In A Log Car	Schleich/Peyo	1/1/96		Rubber Smurf	2" High	Germany/China	$18.00-25.00	The Smurf has a medium blue body. The Smurf is wearing black goggles. The car is brown with dark brown wheels, yellow candles, a yellow steering wheel, and a yellow stem to the mushroom. The top of the mushroom is a dark red with white spots. The front of the car has a brown handle. The Quick emblem is on the back of the Smurf's hat.
178/003	Vanity Table	Schleich/Peyo	1/1/96		Rubber Smurf	2" High	Germany/China	$18.00-25.00	Smurfette has a medium blue body. Smurfette is sitting, wearing a white dress, holding a pink face puff. The table has a pink doily top, brown legs, a brown stool and a brown mirror frame with a silver mirror. The Quick emblem is on the back of Smurfette's hat.
178/004	Smurfette In Bath Tub	Schleich/Peyo	1/1/96		Rubber Smurf	2" High	Germany/China	$45.00-60.00	Smurfette has a medium bright blue body. She is wearing a pink hat and pink bows in her hair. She has no clothes on. Smurfette is scrubbing her back with a tan scrub brush. In her right hand she has a hand full of white suds. The tub is dark brown and the sprinkling can is yellow. The Quick emblem is on the back of her hat.
178/005	Smurf In Bed	Schleich/Peyo	1/1/96		Rubber Smurf	2" High	Germany/China	$18.00-25.00	The Smurf has a medium bright blue body. The Smurf is wearing a red night shirt and a white hat with a red tassel. The bed is dark brown with red and white mushrooms on the bed post. The mattress is yellow. The Smurf has the Quick emblem on the back of his hat.
178/006	Smurfette In Plug Car	Schleich/Peyo	1/1/96		Rubber Smurf	2" High	Germany/China	$18.00-25.00	Smurfette has a medium bright blue body. Smurfette is wearing a white dress and a pink ribbon in her hair. The car is dark pink with white/yellow daisy decals, yellow wheels and a tan heart- shaped steering wheel. The Quick emblem is on the side of Smurfette's hat.

Set 59: Radios

059/001	Square Smurf Radio With Headset	Power Tronic	1/1/82	Blue	AM Radio	2 1/4"W x 4 1/4"L	Hong Kong	$22.50-30.00	The radio is square and has a Smurf holding a white music sheet on the front. The headphones are blue and have a Smurf face on each end.
059/002	Smurf Face Radio	Power Tronic	1/1/82	Blue	AM Radio		Hong Kong	$15.00-20.00	The radio is AM. The radio is in the shape of a Smurf face with a white hat.
059/003	Smurf Radio And Speakers	Power Tronic	1/1/84	Blue	Plastic radio Set			$30.00-40.00	A plastic Smurf face AM/FM radio sits on a plastic base that has Smurfs written across in orange letters. Two speakers plug into the back of the Smurf. The speakers are square, blue with black on the front. One has Papa's face in front and the other has Smurfette's face in front.
059/004	Square Schlumpf Radio With Headset	Quelle	1/1/81	Blue	Plastic radio	2 1/4"W x 4 1/4"L	Hong Kong	$22.50-30.00	The radio is square and has a Smurf holding a white music sheet on the front of the radio. The headphones are blue and have a Smurf face on each end. The bottom says Schlumpf.
059/005	Smurfette Square Radio & Headphones	Power Tronic	1/1/82	Purple	AM radio	2 1/4"W x 4 1/4"L	Hong Kong	$22.50-30.00	The radio is square and has a Smurfette with a bird on her hand. The headphones are purple and have Smurfette's face on each end.

Set 232: Rain Gear

232/001	Umbrella	Peyo	1/1/89	Clear	Umbrella		Belgium	$15.00-20.00	The umbrella is clear. It has a picture of Snappy giving Sassette flowers. Nat playing a guitar. Slouchy is standing around on a river bank. The other side has Sassette fishing, a Smurf laying on a blanket under an umbrella and 2 other Smurfs playing ball.
232/002	Have A Smurfy Day Raincoat	Peyo		Blue	Raincoat	Size 3	Taiwan	$9.00-12.00	The coat is blue with a hood. The coat has a picture of a Smurf sitting on a yellow cloud. The Smurf has a rainbow behind him. The cloud says "Have A Smurfy Day!" in black letters. The front pocket has a Smurf sitting on a yellow cloud.
232/003	Yellow Boots Have A Smurfy Day!	W. Berrie/Peyo	1/1/84	Yellow	Rain Boots	13	U.S.A.	$3.50-5.00	The boots are yellow. The coat has a picture of a Smurf sitting on a white cloud. The Smurf has a rainbow behind him. The cloud says "Have A Smurfy Day!" in black letters.
232/004	Smurf Is Number 1 Vinyl Pink Raincoat	Peyo		Pink	Vinyl Raincoat	6X	Taiwan	$9.00-12.00	The jacket is pink and zips in front. The back has a picture of a Smurf sitting, holding a finger in the air. Above the Smurf it says "Smurf Is Number 1" The jacket has a hood and 2 pockets.

232/005	Put A Little Smurf In Your Life Rain Coat	Peyo/SEPP		Yellow	Vinyl Rain Coat	Size 5		$9.00-12.00	The coat is yellow. It has a Smurf that looks like he's jumping in the air on the back. The back says "Put A Little Smurf In Your Life". The front pocket has a picture of the same Smurf that's on the back.
232/006	Smurf Standing On A Cloud Red Boots	Peyo		Red	Rubber Boots	9 - 10	U.S.A.	$3.50-5.00	The boots are red. They have a Smurf standing on a cloud. The Smurf has his hands in the air. The cloud has a spot for a name.
232/007	Smurf Standing On A Cloud Blue Boots	Peyo		Blue	Rubber Boots	6	U.S.A.	$3.50-5.00	The boots are blue. They have a Smurf standing on a cloud. The Smurf has his hands in the air. The cloud has a spot for a name.

Set 5: Records- 33 & 45

005/001	Merry Christmas With The Smurfs	Starland Music	1/1/82	Blue	33 Record	33	Holland	$11.00-15.00	The cover has Smurfette leaning out her window listening to the Smurf singing X-mas songs.
005/002	Smurfing Sing Along Picture Disc	Starland Music	1/1/82		Picture Disc	33	Holland	$18.00-25.00	The record has a picture on each side. Side one has Smurfette in a window looking down at the Smurfs playing instruments. Side two has a Smurf in an airplane flying over a band.
005/003	Best Of Friends The Smurfs	Starland Music	1/1/82	White	33 Record	33	Holland	$9.00-12.00	The cover has a Smurf walking across it carrying a Smurf record.
005/004	Pitufos	Helix	1/1/82	White	33 Record, Spanish	33	Spain	$11.00-15.00	A Spanish record. The cover has a Smurf carrying a package across it
005/005	The Smurfs All Star Show	Starland Music	1/1/81	Yellow/ Red	33 Record	33	Holland	$9.00-12.00	The Smurf band is playing on a stage with red curtains on the side.
005/006	Smurfing Sing Song	Starland Music	1/1/80		33 Record	33	Holland	$9.00-12.00	Mushroom houses on the cover, with 3 Smurfs playing musical instruments.
005/007	Smurfs Party Time	Starland Music	1/1/83	Multi	33 Record	33	Holland	$9.00-12.00	The Smurfs are having a party on the cover.
005/008	Father Abraham in Smurfland	Starland Music	1/1/82		33 Record	33	Netherlands	$11.00-15.00	Smurf band on the cover with Father Abraham's face, red border.
005/009	Smurfenbier Vader Abraham	Dutch Record Co.	1/1/78		45 Record	33	Holland	$9.00-13.00	The 2 upper corners of the cover have a clown Smurf and a beer Smurf. The middle of the cover has Father Abraham holding a plush Papa and 2 other Smurfs.
005/010	La Valse Et La Marche Des	Starland Music	1/1/82		45 Record	33	France	$11.00-15.00	Has a big Smurf on the sleeve.
005/011	Smurfin! 10th Anniversary Commemorative Album	Starland Music	1/1/89		33 Record	33	U.S.A.	$18.00-25.00	The cover has a beach scene. Three Smurfs are dressed like Arabs. A Smurf and Smurfette are walking on a pier. A Smurf is surfing. Nat, Sassette and Snappy are jumping in the sand. A Smurf is playing the drums, another is playing a horn, and Brainy is singing.
005/012	The Smurfs And The Magic Flute	Starland Music	1/1/83		Video Disc	33	U.S.A.	$11.00-15.00	The cover has Smurfs outside a mushroom house doing various things. There is a film strip on one side of the picture.
005/013	Chick Corea Friends	Starland Music	1/1/82		33 Record	33	U.S.A.	$7.50-10.00	

Set 169: Rings

169/001	Smurfette	Howard Eldon, Ltd.			Metal Ring		U.S.A.	$3.50-5.00	Smurfette is standing with her hands in front of her. She is wearing a red dress. Her eyes are white with black dots. The ring has a gold band.
169/002	Papa Pointing	S.E.P.P.	1/1/80		Metal Ring		Taiwan	$3.50-5.00	Papa is standing, pointing at something. He has a red hat and red pants. The ring has a gold band.
169/003	Smurf Sitting On A Cloud				Metal Ring			$3.50-5.00	A Smurf is sitting on a white cloud. There is a pink, blue, red and yellow rainbow behind him. The Smurf has a happy expression on his face.
169/004	Smurf Playing A Trumpet	Peyo			Metal Ring			$5.00-7.00	The ring has a Smurf walking, playing a trumpet. The trumpet is yellow. The Smurf is wearing a red shirt, white hat and white pants.
169/005	Silver Smurf Face Ring	Peyo			Silver Ring	15,5mm	Germany	$7.50-10.00	The ring is made of silver. It has a Smurf face on the front. The face is only half painted blue the rest is silver. The Smurf has a white hat.
169/006	Smurf Waving Silver Ring	Peyo			Silver Ring	15,5mm	Germany	$5.00-7.00	The ring is made of silver. It has a Smurf waving on the front. The Smurf is painted half blue and the rest of his face is silver. He has a white hat and white pants.

Set 194: Roller-skates

194/001	Shoe Skates	Larami			Shoe Skates		Hong Kong	$11.00-15.00	The skates go over children's shoes. The front is red and has a Smurf face. The back of the skate is blue. The wheels are yellow and blue. The skates are adjustable and made of plastic. There are white adjustable straps to hold them on.
194/002	Blue Boot Skates	Peyo	1/1/82	Blue	Roller-skates	8		$18.00-25.00	The skates are blue boots with white accents. The side of the skates have a picture of a Smurf roller-skating.
194/003	White Boot Roller-skates	Peyo	1/1/82	White	Roller-skates	9		$18.00-25.00	The skates are white boots with blue accents. The side of the skates have a picture of a Smurf roller-skating.

Set 103: School & Office Supplies

103/001	Smurf Scotch Tape	Leadworks			Tape		U.S.A.	$3.50-5.00	The tape is clear and has all different Smurf designs on it. The tape is in a clear plastic holder.
103/002	Smurfette Heart-shaped Eraser (Smurfette Drawing)	W. Berrie	1/1/83	Pink	Eraser	2"	Taiwan	$2.00-3.00	The eraser is a pink heart with little purple hearts all over it. Smurfette is on the front of the eraser, she is laying on her stomach drawing a picture of a Smurf.
103/003	Blue Papa Bookmark	W. Berrie	1/1/83	Dark Blue	Plastic Bookmark	5 1/2"	Hong Kong	$2.00-3.00	The bookmark is blue plastic. On the top of the bookmark, Papa is standing, holding a finger in the air.
103/004	Red Smurf Bookmark	W. Berrie	1/1/83	Red	Plastic Bookmark	5 1/2"	Hong Kong	$2.00-3.00	The bookmark is red plastic. On the top of the bookmark a Smurf is standing with his hands at his side and he's winking.
103/005	Purple Smurfette Bookmark	W. Berrie	1/1/83	Purple	Plastic Bookmark	5 1/2"	Hong Kong	$2.00-3.00	The bookmark is purple plastic. On the top of the bookmark Smurfette is standing, holding a pink daisy.
103/006	Smurfette Clear Pencil Case	Peyo		Clear	Plastic Pencil Case	3 1/2"H X 9"L		$3.50-5.00	The pencil case is clear with little pink hearts all over it. Smurfette is in the bottom corner wearing a pink dress, standing in a purple heart. Above the big purple heart it says Smurfette in white letters with purple border.
103/007	Bicycle Smurf Eraser			White	Pencil Eraser	1 1/2"		$1.50-2.00	The eraser is white and has a picture of a Smurf riding a red bicycle on the front.
103/008	Smurfettes Do It Better Mini Notebook	W. Berrie	1/1/83		Mini Notebook	5 1/2"L X 3 1/2"W	U.S.A.	$3.00-4.00	The outside cover is pink with small purple hearts all over it. In the center is a big purple heart and it has Smurfette standing, holding a telephone. The paper inside is purple with lines. In the bottom corner it has a picture of Smurfette holding a pencil and paper.
103/009	Smurf Scribble Pad	Mead	1/1/82		Scribble Pad	12" X 14"	U.S.A.	$3.00-4.00	The cover has a Smurf with his hands in the air. The cover is orange with blue trim. The paper inside doesn't have lines.
103/010	16 Jumbo Smurf Crayons	Hasbro	1/1/83		Jumbo Crayons	16 Crayons	Israel	$6.75-9.00	The crayons are in a plastic box. There is a picture of a Smurf holding a crayon drawing the # 16.
103/011	8 Jumbo Crayons	Hasbro	1/1/83		Jumbo Crayons	8 Crayons	Israel	$4.50-6.00	The crayons are in a plastic box. There is a picture of a Smurf holding a crayon drawing the # 8.
103/012	Organizer With A Smurf Driving A Car	Lizen: El/Munchen		Red/Blue	Organizer	7" X 5 1/2"	Germany	$7.50-10.00	The organizer is red with blue trim and blue inside. The front of the cover has a Smurf driving a car and one standing on a curb by a traffic sign. Inside are colored pencils, rulers, a protractor, pens, a pencil sharpener, and an eraser.
103/013	Organizer With Smurfs At A Desk	Lizen: El/Munchen		Blue/Red	Organizer	7" X 5 1/2"	Germany	$7.50-10.00	The organizer is blue with red trim and blue inside. The front of the cover has 2 Smurfs sitting at a desk with a screen behind them with traffic signs on. The inside has elastic to hold pencils and rulers.
103/014	Mini Wipe Off Board	Peyo	1/1/84	White	Wipe Off Board	6"L X 4"W	Hong Kong	$10.50-14.00	The wipe off board is white and has a Smurf playing a trumpet walking across the bottom. At the top of the board it says "NOTES" in different colors. There is a red marker included.
103/015	Ruler Calculator - Clock	Bradley	1/1/83		Ruler & Calculator	12"L X 2"W	Hong Kong	$11.00-15.00	The ruler is white plastic. There is a digital clock on one end and a calculator on the other end. In the middle is a picture of Brainy Smurf teaching school. 3 Smurfs are sitting at desk fooling around and Papa is standing behind looking perturbed.
103/016	School Bus Pencil Case	W. Berrie	1/1/83		Pencil Case	5"L X 9"W	Hong Kong	$6.00-8.00	The pencil case is plastic. The outside looks like a yellow wood school bus. One side has Papa in the window, driving and 4 other Smurfs looking out the windows. Included is a blue eraser with a Smurf walking. The pencil case lip locks.
103/017	School Bus Pencil Case	W. Berrie	1/1/83		Pencil case	5"L X 9"W	Hong Kong	$6.00-8.00	The pencil case is plastic. The outside looks like a yellow wood school bus. One side has Papa in the window, driving and 4 other Smurfs looking out the windows. Included is a white eraser with Papa holding a finger up. A flicker action ruler with 4 Smurfs on it. The pencil case lip locks.
103/018	Pink Ruler	W. Berrie	1/1/83	Pink	Ruler	12" Long	Hong Kong	$4.50-6.00	The ruler is pink and see through. Down the center of the ruler is a sticker with a different Smurf by each number. The sticker has number 1 through 12 for each inch on the ruler.
103/019	Blue Ruler	W. Berrie	1/1/83	Blue	Ruler	12" Long	Hong Kong	$4.50-6.00	The ruler is blue and see through. Down the center of the ruler is a sticker with a different Smurf by each number. The sticker has number 1 through 12 for each inch on the ruler.
103/020	Papa Smurfy Glue	W. Berrie	1/1/83	Red	Glue	1.1 Fl. Oz	Taiwan	$3.00-4.00	The bottle is clear and has a red top. Papa is on the outside of the glue bottle. Papa is holding a finger in the air. There are 5 striped lines behind Papa. Above Papa's head it says "Papa Smurf" in blue letters
103/021	Roller-skating Smurfette Heart shaped Eraser	W. Berrie/Peyo	1/1/83	Pink	Eraser	2"	Taiwan	$2.00-3.00	The eraser is a pink heart with little purple hearts all over it. Smurfette is on the front wearing white roller-skates, a white dress and a white hat.
103/022	Jokey Smurfy Note Pad	Applause/Peyo	1/1/83	Orange	Note Pad	9 1/2"L X 7"W	U.S.A.	$7.50-10.00	The note pad is orange with dark orange lines. On the top is a picture of Jokey carrying a present. Every page has the same picture.
103/023	Brainy Smurfy Note Pad	Applause/Peyo	1/1/83	Yellow	Note Pad	9 1/2"L X 7"W		$7.50-10.00	The note pad is yellow with dark orange lines. On the top is a picture of Brainy Smurf holding a red book. Every page has the same picture. 50 sheets.
103/024	Papa Pitu-Notas	Peyo		White	Note Pad	9 1/2"L X 7"W	Mexico	$7.50-10.00	The note pad is white with blue lines. On the top is a picture of Papa holding a pencil and a script. The top of the paper says "Papa Pituto". Every page has the same picture.
103/025	Large Papa Eraser	W. Berrie/Peyo	1/1/83	Blue	Large Eraser	4"	Hong Kong	$2.00-3.00	The eraser is blue and flat. On the front is a picture of Papa Smurf holding a finger out in front of him. Papa has a red hat and red pants on.
103/026	Large Winking Smurf Eraser	W. Berrie/Peyo	1/1/83	Blue	Large Eraser	4"	Hong Kong	$2.00-3.00	The eraser is blue and flat. On the front is a picture of a Smurf standing with his hands at his sides. The Smurf is winking. The Smurf has a white hat and white pants on.
103/027	Smurfette Eraser In Lip Stick Tube	Applause/Peyo	1/1/84	Purple/Pink	Eraser	2 1/2" High	Japan	$3.00-4.00	The eraser is purple, long and in the shape of a heart. The case is a lip stick tube. The cap is clear with pink hearts all over. It has Smurfette on roller-skates with a big pink heart behind her. The eraser holder is purple.
103/028	Papa Smurf Stapler	Applause	1/1/84	Red/White	Stapler & Staples	2 1/4"	Hong Kong	$6.00-8.00	The stapler is white and red. The end of the stapler has a picture of Papa Smurf standing, holding a finger in the air. There are rainbow stripes in the center of the stapler and the opposite end from Papa is red. The stapler comes with a box of Smurfy staples. The stapler is in a clear case.

103/029	Smurfette Stapler	Applause	1/1/84	Purple/White	Stapler & Staples	2 1/4"	Hong Kong	$6.00-8.00	The stapler is white and purple. The end of the stapler has a picture of Smurfette standing with a butterfly on her finger. There are rainbow stripes in the center of the stapler and the opposite end from Smurfette is purple. The stapler comes with a box of Smurfy staples. The stapler is in a clear case.
103/030	Smurf Stapler	Applause	1/1/84	Blue/White	Stapler & Staples	2 1/4"	Hong Kong	$6.00-8.00	The stapler is white and blue. The end of the stapler has a picture of a Smurf jumping in the air. There are rainbow stripes in the center of the stapler and the opposite end from Smurf is blue. The stapler comes with a box of Smurfy staples. The stapler is in a clear case.
103/031	Papa Smurf Tape	Leadworks			Tape	10 Yards	U.S.A.	$3.50-5.00	The tape is clear and has all different Smurf designs on it. The tape is in a clear plastic holder. Papa Smurf is pictured on the card.
103/032	Smurfy Notes Folder	Peyo	1/1/83	Light Blue	Envelopes/Paper		U.S.A.	$6.00-8.00	There are 24 sheets of paper with a Smurf holding a pencil and script in the corner. The paper is blue but has 5 lines of rainbow colors at the top. There are 12 blue envelopes. The paper and envelopes are stored in a folder. The front of the folder has Papa on a stool, 3 Smurfs piled up holding a window. A Smurf carrying a ladder, Smurfette standing behind Papa, and Baby is crawling.
103/033	Purple Ruler	W. Berrie/Peyo	1/1/83	Purple	Ruler	12" Long	Hong Kong	$4.50-6.00	The ruler is purple and see through. Down the center of the ruler is a sticker with a different Smurf by each number. The sticker has number 1 through 12 for each inch on the ruler.
103/034	Smurfette Smurfy Glue	W. Berrie	1/1/83	Purple	Glue	1.1 Fl. Oz.	Taiwan	$4.50-6.00	The bottle is clear and has a purple top. Smurfette is on the outside of the glue bottle. Smurfette is holding a finger out in front of her with a butterfly on it. There are 5 striped lines behind Smurfette. Above Smurfettes head it says "Smurfy Smurf" in blue letters.
103/035	Smurf Glue	W. Berrie	1/1/83	Blue	Glue	1.1 Fl. Oz.	Taiwan	$4.50-6.00	The bottle is clear and has a blue top. A Smurf is on the outside of the glue bottle. The Smurf is jumping in the air. There are 5 striped lines behind the Smurf. Above the Smurfs head it says "Smurfy Smurf" in red letters.
103/036	Smurf Pencil Sharpener	Applause/Peyo		Blue	2 way Sharpener	2 1/2"L x 2 1/4"W	Japan	$6.00-8.00	The sharpener is a blue plastic base with a clear blue removable cover to empty. On the front is a picture of a Smurf. The Smurf is jumping in the air. There are 5 striped lines behind the Smurf and the rest of the background is silver.
103/037	Papa Pencil Sharpener	Applause/Peyo		Red	2 way Sharpener	2 1/2"L x 2 1/4"W	Japan	$6.00-8.00	The sharpener is a red plastic base with a clear red removable cover to empty. On the front is a picture of Papa. Papa is holding a finger out. There are 5 striped lines behind Papa and the rest of the background is silver.
103/038	Smurfette Pencil Sharpener	Applause/Peyo		Purple	2 way Sharpener	2 1/2"L x 2 1/4"W	Japan	$6.00-8.00	The sharpener is a purple plastic base with a clear purple removable cover to empty. On the front is a picture of Smurfette. Smurfette is holding a finger out with a butterfly on the end. There are 5 striped lines behind Smurfette and the rest of the background is silver.
103/039	Smurfette Tape Dispenser	Applause/Peyo		Purple	Tape Dispenser	2 1/2"	Japan	$6.00-8.00	The tape dispenser is a purple plastic. The dispenser is star- shaped. On the front is a picture of Smurfette. Smurfette is holding a finger out with a butterfly on the end. There are 5 striped lines behind Smurfette and the rest of the background is silver.
103/040	Papa Tape Dispenser	Applause/Peyo		Red	Tape Dispenser	2 1/2"	Japan	$6.00-8.00	The tape dispenser is a red plastic. The dispenser is star- shaped. On the front is a picture of Papa. Papa is holding a finger out. There are 5 striped lines behind Papa and the rest of the background is silver.
103/041	Smurf Tape Dispenser	Applause/Peyo		Blue	Tape Dispenser	2 1/2"	Japan	$6.00-8.00	The tape dispenser is a blue plastic. The dispenser is star- shaped. On the front is a picture of a Smurf. The Smurf is jumping in the air. There are 5 striped lines behind the Smurfette and the rest of the background is silver.
103/042	Papa Pencil Case	Applause/Peyo	1/1/84	Red	Pencil Case	5"	Japan	$6.00-8.00	The pencil case is red, see through and the shape of a fat pencil. Inside are 6 colored pencils. It has a picture of Papa on the front. Papa is holding a finger out. There are 5 rainbow striped lines under Papa's name. Under Papa's feet it says "Papa" in blue letters.
103/043	Smurf Pencil Case	Applause/Peyo	1/1/84	Blue	Pencil Case	5"	Japan	$6.00-8.00	The pencil case is blue, see through and the shape of a fat pencil. Inside are 6 colored pencils. It has a picture of a Smurf jumping in the air on the front. There are 5 rainbow striped lines under the Smurfs name. Under the Smurf's feet it says "Smurf" in blue letters.
103/044	Smurfette Pencil Case	Applause/Peyo	1/1/84	Purple	Pencil Case	5"	Japan	$6.00-8.00	The pencil case is purple, see through and the shape of a fat pencil. Inside are 6 colored pencils. It has a picture of Smurfette on the front. Smurfette is holding a finger out with a butterfly on the end. There are 5 rainbow striped lines under Smurfette's name. Under Smurfette's feet it says "Smurfette" in blue letters.
103/045	Smurfette Pitu-Notas	Peyo			Note paper		Mexico	$7.50-10.00	The paper has Papa Smurf in the corner holding a feather pen and paper. The box has Smurfette's face and says Pitu-Notas on top. The box is orange.
103/046	Postpapier	Peyo	1/1/95		Paper & Envelopes		Holland	$7.50-10.00	Included are 10 envelopes and paper. They are in a big envelope. The cover of the envelope has a Smurf standing on 2 other Smurfs trying to put an envelope in a mailbox. The mailbox is red. Papa is standing, helping one Smurf stay on another Smurf's back. The folder is light blue.
103/047	Smurfette On Telephone Heart Shape Eraser	Peyo	1/1/83	Purple	Eraser	2"	Taiwan	$1.50-2.00	The eraser is a purple heart with little pink hearts all over. Smurfette is on the front of the eraser. Smurfette is standing, holding a pink old fashion telephone.
103/048	Cheerleader Smurfette Heart Shape Eraser	W. Berrie/Peyo	1/1/83	Purple	Eraser	2"	Taiwan	$1.50-2.00	The eraser is a purple heart with little pink hearts all over. Smurfette is on the front of the eraser. Smurfette is in a cheerleader outfit. Smurfette is holding 2 pink Pompoms. Her cheerleader outfit is pink and white.
103/049	Ballerina Smurfette Heart Shape Eraser	W. Berrie/Peyo	1/1/83	Purple	Eraser	2"	Taiwan	$1.50-2.00	The eraser is a purple heart with little pink hearts all over. Smurfette is on the front of the eraser. Smurfette is standing in a ballerina pose. Smurfette is wearing a white and pink ballerina dress and pink slippers.
103/050	Roller-skating Smurfette Heart Shape Eraser	W. Berrie/Peyo	1/1/83	Pink	Eraser	2"	Taiwan	$1.50-2.00	The eraser is a pink heart with little purple hearts all over. Smurfette is on the front of the eraser. Smurfette is roller-skating. Smurfette is wearing a white dress and white roller-skates.
103/051	Posing Smurfette Heart Shape Eraser	W. Berrie/Peyo	1/1/83	Pink	Eraser	2"	Taiwan	$1.50-2.00	The eraser is a pink heart with little purple hearts all over. Smurfette is on the front of the eraser. Smurfette is wearing a white swimsuit with a pink banner across her chest. Smurfette has one hand on her hip and one in her hair.
103/052	Smurfette Glitter Glue	Applause/Peyo	1/1/84	Pink	Glitter Glue	.338 Fl Oz.	Japan	$3.50-5.00	The glue is in a tube. It is pink. The outside tube has a picture of Smurfette.
103/053	Smurf In Car Pencil Case			Red	Plastic Pencil Case	3 3/4"L x 7 1/2"W		$3.50-5.00	The pencil case is clear with red trim. It has a picture of a Smurf zooming by Smurfette. Smurfette looks like she's falling backwards. There is a startled bird in the air. The Smurf is driving a yellow car with blue wheels.
103/054	King Smurf Eraser	Peyo		Blue	Eraser	2" x 1 1/2"		$1.50-2.00	The eraser is blue. It has a Smurf dressed as a King carrying a scepter. The Smurf has a yellow hat, yellow crown, white shirt, yellow pants and a red with white trim robe.,
103/055	Smurfette Looking At Flowers Eraser			White	Eraser	1 1/2" x 1 1/2"		$1.50-2.00	The eraser is white. It has a picture of Smurfette standing, looking at a red daisy. Smurfette has her hands folded under her chin.
103/056	Smurf Blue Pencil Tube			Light Blue	Pencil Case	10" x 1 1/2"		$4.50-6.00	The Smurf pencil case is the shape of a pencil. The top unzips to put pencils in. It is light blue with a red tip and end. The tube has 8 different Smurfs and Papa.
103/057	Smurf Smurfy Note Pad	Applause/Peyo	1/1/83	Blue	Note Pad	9 1/2"L X 7"W	U.S.A.	$7.50-10.00	The note pad is blue with dark blue lines. The top has a picture of a Smurf jumping in the air. 50 sheets.
103/058	Smurf Calculator			Light Blue	Calculator	4 1/2" x 2"	Hong Kong	$11.00-15.00	Mini blue calculator. Slide the Smurf face up to display the window. The numbers are white and yellow.
103/059	Papa At Lab Table Silver Eraser Tin	Peyo/ Orange Story	1/1/97	Silver	Eraser Tin	2 1/2" x 2"		$8.25-12.00	The tin is silver and rectangular. The cover has Papa pouring liquid from a test tube into a bowl. The bottom of the tin has Papa standing by a chalkboard. There is a white and a pink eraser in the box.
103/060	Gargamel Orange Eraser Tin	Peyo/ Orange Story	1/1/97	Orange	Eraser Tin	2 1/2" x 2"		$10.50-14.00	The tin is orange and rectangle- shaped. The cover has Gargamel sitting on a stool and Azreal in front of him. The bottom of the tin has Azreal. There is a white and a green eraser in the box.
103/061	Bike Smurf And Happy Smurf Erasers			White	2 Erasers	1 1/2"	Taiwan	$1.50-2.00	There are 2 erasers in the package. One is a Smurf riding a red bike. The second eraser is a Smurf that is jumping in the air with a happy expression on his face.
103/062	Smurfette Smelling Flowers And LOVE Smurf 2 Erasers			White	2 Erasers	1 1/2"	Taiwan	$1.50-2.00	There are 2 erasers in the package. One is of Smurfette leaning over smelling a red daisy. The second eraser is of a Smurf holding a bouquet of daisies. The word LOVE is under the daisies.
103/063	Happy Smurf And Smurf Leaning On A Heart 2 Erasers			White	2 Erasers	1 1/2"	Taiwan	$1.50-2.00	There are 2 erasers in the package. One is a Smurf with his hands at his side and a happy expression on his face. The second eraser is of a Smurf leaning against a red heart. He is holding a yellow daisy behind his back.

Set 114: See 'N Say

114/001	Sport Scene	Mattel	1/1/83	Blue	Pull String	10" X 12"	Canada	$22.50-30.00	A big, round disc, with colored pictures of the Smurfs doing different sports activities. There is a red arrow in the middle with a Smurf pointing. Point the arrow at a Smurf and pull the string to here the Smurfs talk. 12 different sayings.
114/002	Talking Telephone	Mattel	1/1/78	Blue/White	Pull String		U.S.A.	$18.00-25.00	The telephone is blue and has a white receiver. There are 10 different Smurf pictures. In the middle of the dialer is a picture of a Smurf talking on the phone. When you pull the string the phone talks. Mine doesn't work.

Set 128: Shoe Laces

128/001	Smurf With Heart - White			White	Cloth Shoe Laces	40" Long		$1.50-2.00	The shoe laces are white and have Smurfs standing, leaning on a red heart. The design is repeated on the whole shoe lace.
128/002	Smurf Faces - White	Peyo		White/Blue	Cloth Shoe Laces	40" Long		$1.50-2.00	The shoe laces are white and have Papa's face, Smurfette's face and a Smurf face repeated on the whole lace. The laces also say SMURF in red letters repeatedly.
128/003	Smurf Faces - Dark Blue	Peyo		Dark Blue	Cloth Shoe Laces	40" Long		$1.50-2.00	The shoe laces are dark blue and have Papa's face, Smurfette's face and a Smurf face repeated on the whole lace. The laces also say SMURF in red letters repeatedly.
128/004	Smurf Faces - Gray	Peyo		Gray	Cloth Shoe Laces	40" Long		$1.50-2.00	The shoe laces are gray and have Papa's face, Smurfette's face and a Smurf face repeated on the whole lace. The laces also say SMURF in red letters repeatedly.
128/005	Smurf Laying With Head In Hands - Red	Peyo		Red	Cloth Shoe Laces	40" Long		$1.50-2.00	The shoe laces are red and have a Smurf laying on his stomach with his head in his hands. The Smurf design is repeated on the whole lace. The laces also say SMURF in white letters repeatedly.
128/006	Smurf Laying With Head In Hands - Light Blue	Peyo		Light Blue	Cloth Shoe Laces	40" Long		$1.50-2.00	The shoe laces are light blue and have a Smurf laying on his stomach with his head in his hands. The Smurf design is repeated on the whole lace. The laces also say SMURF in red letters repeatedly.
128/007	Smurf Laying With Head In Hands - White	Peyo		White	Cloth Shoe Laces	40" Long		$1.50-2.00	The shoe laces are white and have a Smurf laying on his stomach with his head in his hands. The Smurf design is repeated on the whole lace. The laces also say SMURF in red letters repeatedly. Some are in a yellow package with a Smurf wearing big red shoes on the front and some of the laces are loose.

128/008	Smurf Laying With Head In Hands - Yellow	Peyo		Yellow	Cloth Shoe Laces	40" Long		$1.50-2.00	The shoe laces are yellow and have a Smurf laying on his stomach with his head in his hands. The Smurf design is repeated on the whole lace. The laces also say SMURF in red letters repeatedly.
128/009	Smurf Jumping - White			White	Cloth Shoe Laces	40" Long			The shoe laces are white and have a Smurf that looks like he's jumping in the air. The Smurf has a red shirt on and yellow shorts. The Smurf design is repeated on the whole lace.

Set 137: Shoes

137/001	Blue Smurf Tennis Shoes	BBC Imports	1/1/82	Blue	Tennis Shoes	Size 8 1/2	Korea	$18.00-25.00	The tennis shoes are light blue and silver. The sides have a silver stripe with a Smurf face in. The velcro closer are blue and say Smurf sports. The box has several different sport Smurfs on the sides. The cover has Papa, Smurfette's face and a Smurf standing, pointing.
137/002	Blue Tennis Shoes	BBC Imports	1/1/82	Light Blue	Tennis Shoes	3		$11.00-15.00	The tennis shoes are light blue and white. The sides are white with a light blue stripe. The velcro closer are blue and have a Smurf face on it.
137/003	Smurfette Slippers		1/1/82	Blue & Yellow	Slippers		U.S.A.	$3.50-5.00	The slippers are blue and yellow. They are fuzzy. They have Smurfette's head on the front. Smurfette's head is made of plastic. There is a yellow cuff around the openings to the slippers.
137/004	Papa Smurf Slippers			Red & Blue	Slippers	7 - 8	U.S.A.	$3.50-5.00	The slippers are blue and red. They are fuzzy. They have Papa's head on the front. Papa's head is made of plastic. There is a red cuff around the openings to the slippers.
137/005	Smurf Slippers			Blue & White	Slippers	Medium	U.S.A.	$3.50-5.00	The slippers are blue and white. They are fuzzy. They have a Smurf head on the front. The Smurf's head is made of plastic. There is a blue cuff around the openings to the slippers.
137/006	Yellow Smurf Tongs	Peyo		Yellow	Tongs			$2.00-3.00	The tongs are yellow with yellow cloth straps. They say Smurf on the front tag.
137/007	Baseball Smurf Sandals	Peyo	1/1/82	Brown	Sandals	6		$7.50-10.00	The sandals are brown. They have a picture of a Smurf carrying a baseball bat on the front.
137/008	Baseball Smurf Sandals	Peyo		Brown	Sandals	3 & 7		$7.50-10.00	The sandals are brown. They have a picture of a Smurf carrying a baseball bat on the front.
137/009	Huge Papa Slippers	Peyo/I.M.P.S.	1/1/94		Slippers	Adult	France	$22.50-30.00	The slippers fit an adult. They are red and blue. They have Papa's head on the front. He has a fuzzy white beard and is wearing a red hat.
137/010	Huge Smurf Slippers	Peyo/ I.M.P.S.	1/1/94		Slippers	Adult	France	$22.50-30.00	The slippers fit an adult. They are white and blue. They have a big Smurf head on the front. The Smurf is wearing a white hat.

Set 217: Shopping Bags

217/001	Smurflings Riding Puppy	Peyo	1/1/88	White	Plastic Bag	19" x 18"	Brussels	$9.00-13.00	The bag has a brown dog with Sassette and Nat riding on his back. Slouchy is riding on the dogs head. Snappy is being pulled along by the dog's leash. The bottom of the bag has Smurf written in 7 different languages.
217/002	I Love The Smurfs Cupid Bag	Schleich/ Peyo		White	Plastic Bag	19" X 18"		$9.00-13.00	The bag has a cupid Smurf holding a red heart and a yellow bow. The bag Says "I Love The Smurfs". It also has Smurfs written in 7 different languages at the bottom. The bag has the same picture on both sides.

Set 21: Shrinky Dinks

021/001	Smurf Carrying Knapsack Collectible Figure Kit	MB	1/1/88	Light Blue	10 Shrinky Dinks		U.S.A.	$4.50-6.00	The box is blue and has a Smurf with a knapsack on the front.
021/002	Smurf Playhouse Deluxe Shrinky Dink Set	Colorforms	1/1/82	Yellow	24 Shrinky Dinks		U.S.A.	$11.00-15.00	The box is yellow and the front has a mushroom house with Smurfs around the house.
021/003	Superman Collector Set	Colorforms	1/1/81	Yellow	10 Shrinky Dinks		U.S.A.	$4.50-6.00	The box is yellow and has a picture of superman Smurf on the front. Included 5 designed sheets, 3 colored pencils, snap on bases and instruction book.
021/004	Papa Collector Set	Colorforms	1/1/83	Yellow	10 Shrinky Dinks		U.S.A.	$4.50-6.00	The box is yellow and has a picture of Papa Smurf on the front. Included 5 designed sheets, 3 colored pencils, snap on bases and instruction book.
021/005	Jumping Smurf Collector Set	Colorforms	1/1/81	Orange	10 Shrinky Dinks		U.S.A.	$4.50-6.00	The box is orange and has a picture of a Smurf jumping on the front. Included 5 designed sheets, 3 colored pencils, snap on bases and instruction book.
021/006	Smurfette Collector Set	Colorforms	1/1/83	Pink	8 Shrinky Dinks		U.S.A.	$4.50-6.00	The box is pink and has Smurfette with her hands out. Included are 8 stand up Smurfette shrinky dinks, 4 colored pencil and slide on bases.

Set 26: Silverware

026/001	Child's Smurf Fork	Danara		Silver	Stainless Steel	5 1/2" High		$3.00-4.00	The top of the fork has a Smurf engraved in it.
026/002	Toddler Smurfette Soup Spoon	Danara		Silver	Stainless Steel	4 1/2" high		$3.00-4.00	The top of the spoon has Smurfette engraved in it.
026/003	Toddler Smurf Fork	Danara		Silver	Stainless Steel	4 1/2" high		$3.00-4.00	The top of the fork has a Smurf engraved in it.
026/004	Child's Soup Spoon	Danara		Silver	Stainless Steel	5 1/2" High		$3.00-4.00	The top of the spoon has a Smurf engraved in it.
026/005	Child's Smurf Teaspoon	Danara		Silver	Stainless Steel	5 1/2" High		$3.00-4.00	The top of the spoon has a Smurf engraved in it.
026/006	Baby Smurf Spoon	Danara		Silver	Stainless Steel	5 1/2" High		$6.00-8.00	The top of the spoon has Smurf engraved in it. The spoon has white soft plastic covering the spoon part.
026/007	Child's Materne Teaspoon	I.M.P.S.	1/1/92	White	Stainless Steel	5 1/2" High		$7.50-10.00	The handle of the spoon is white plastic covering the stainless steel and it has a picture of a Smurf holding Smurf berries.
026/008	Smurf Cutlery Set	Peyo	1/1/96		Stainless Steel		Brussels	$18.00-25.00	The set includes a knife, fork, large spoon and small spoon. The handles all have different designs of Smurfs. The knife has a Smurf doing a handstand. The fork has a Smurf singing. The small spoon has a Smurf walking with his hands behind his back. The large spoon has a Smurf sitting. The set comes in a blue box with a Smurf on a sled on the cover.

Set 39: Sleeping Bags

039/001	Smurfs Carrying A Cake Preparing For A Party	Wenzel		Dark Blue	Single Sleeping Bag	Single Sleeping Bag		$30.00-40.00	The sleeping bag is dark blue with a yellow inside. The picture is of a village. A Smurf is driving a red car, 2 Smurfs are carrying a cake out of an orange mushroom house, Lazy sleeping on a hill, and other Smurfs are relaxing or playing. 12 Smurfs in the picture.
039/002	Papa Smurfs Birthday Party Scene	Wenzel		Light Blue	Single Sleeping Bag	Single Sleeping Bag		$25.00-35.00	The sleeping bag is yellow inside. The outside has 2 yellow mushroom houses with orange roof. Papa is standing in front of a table with a cake, Smurfette and a few other Smurfs are singing, Jokey's bringing a present. 11 Smurfs in the picture.
039/003	Smurfette America's Sweetheart!	Peyo	1/1/83	Pink/Purple	Single Sleeping Bag	Single Sleeping Bag		$25.00-35.00	The outside of the sleeping bag is pink and has small purple hearts all over. In the middle is a big purple heart with Smurfette's upper body and head inside the heart. Smurfette is wearing a white hat and dress. Above the heart it's says "Smurfette" in white letters with purple border. Below the heart it says "America's Sweetheart!" in purple letters. The inside of the sleeping bag is light purple.

Set 72: Soakies

072/001	Papa Smurf		1/1/91		Shampoo Containers	8 1/2" High	Brussels	$11.00-15.00	The container is a shampoo bottle. Papa is wearing red pants and a red hat. He is standing with his hands at his side.
072/002	Baker Smurf		1/1/91		Shampoo Containers	8 1/2" High	Brussels	$11.00-15.00	The container is a shampoo bottle. The Smurf is wearing a white chef's hat, is standing with his tongue out, and has his hands at his side.
072/003	Smurf		1/1/91		Shampoo Containers	8 1/2" High	Brussels	$11.00-15.00	The container is a shampoo bottle. The Smurf is standing with his mouth open and his hands at his side.

Set 228: Squeeze Bottles

228/001	Yellow Walk Pot	Peyo/Applause		Yellow	Plastic bottle	5 1/2" x 4"		$4.50-6.00	The bottle is square. In the corner has a Smurf face and Smurfette face. It says "Smurf's Up!" in red letters. The bottle has a white cap and clear straw. The bottle has a dark blue clip.
228/002	Blue Walk Pot	Applause/Peyo		Light Blue	Plastic bottle	5 1/2" x 4"		$4.50-6.00	The bottle is light blue. The bottle is square. In the corner has a Smurf face and Smurfette face. It says "Smurf's Up!" in red letters. The bottle has a white cap and clear straw. The bottle has a white clip

Set 76: Squeeze Toys

076/001	Papa Standing	Danara	1/1/84		Rubber Squeakies	4 1/2" High	Hong Kong	$11.00-15.00	The toy is a soft rubber and when squeezed it makes a squeaking noise. Papa is standing with his hand on his tummy. He is wearing a red hat and red pants. Papa has a dark blue body.
076/002	Smurf Carrying A Candle	Danara	1/1/84		Rubber Squeakies	4 1/2" High	Hong Kong	$6.00-8.00	The toy is a soft rubber and when squeezed it makes a squeaking noise. The Smurf is wearing a pink pair of pajamas, pink slippers and is carrying a white candle with an orange flame. The Smurf has his mouth open and is looking up with his eyes. The Smurf has a dark blue body.
076/003	Flirting Smurfette	Danara	1/1/84		Rubber Squeakies	4 1/2" High	Hong Kong	$6.00-8.00	The toy is a soft rubber and when squeezed it makes a squeaking noise. Smurfette is standing with her hands at her sides. She is wearing a white dress, white shoes and a white hat. Smurfette has yellow hair and a flirty look on her face. Smurfette has a dark blue body.
076/004	Cowboy Smurf	Danara			Rubber Squeakies	4 1/2" High	Hong Kong	$6.00-8.00	The toy is a soft rubber and when squeezed it makes a squeaking noise. The Smurf is wearing a white cowboy hat, white pants, a brown vest and a red bandanna around his neck. He is standing with his hand on his hips. The Smurf has a dark blue body.
076/005	Baby Smurf Sleeping	Danara			Rubber Squeakies	4 1/2" High	Taiwan	$6.00-8.00	The toy is a soft rubber and when squeezed it makes a squeaking noise. Baby is laying on his tummy with his hands folded in front of him sleeping. Baby has a light blue body. He is wearing a yellow sleeper and yellow hat.

076/006	Baby With A Rattle	Danara	1/1/84		Rubber Squeakies	4 1/2" High	Taiwan	$6.00-8.00	The toy is a soft rubber and when squeezed it makes a squeaking noise. Baby is laying on his tummy holding a white rattle with a yellow bell in the center of the rattle. Baby has a light blue body. He is wearing a white sleeper and white hat.
076/007	Baby Sitting Sucking His Thumb	Danara	1/1/84		Rubber Squeakies	4 1/2" High	Taiwan	$6.00-8.00	The toy is a soft rubber and when squeezed it makes a squeaking noise. Baby has a light blue body. He is sitting, sucking his thumb and he has sleepy eyes. Baby is wearing a bright green hat and sleeper.
076/008	Schlroumpfs Regular Smurf	Fiba	1/1/83		Rubber Squeakies	6" High	Italy	$11.00-15.00	The Smurf is made of rubber and when squeezed he makes a noise. The Smurf has a blue body, white hat and white pants. The Smurfs head turns. He is in a standing position with his hands at his side.
076/009	Smurf Holding A Cake	Peyo/SEPP			Rubber Squeakies	7" High		$9.00-12.00	The Smurf is standing, holding a pink cake in his right hand. The Smurf has white pants, a white chef's hat and a white apron tied around his waist. The Smurf is licking his lips.
076/010	Baby With Bottle	Danara	1/1/84	Pink	Rubber Squeakies	4 1/2" High	Taiwan	$6.00-8.00	The toy is a soft rubber and when squeezed it makes a squeaking noise. Baby Smurf is sitting, holding a white bottle in his right hand. Baby is wearing a pink sleeper and a pink hat. Baby has light blue skin.
076/011	Smurf Sitting With Kisses On Head	Lanco			Rubber Squeakies	3" High	Spain	$7.50-10.00	The toy is a soft rubber and when squeezed it makes a squeaking noise. The Smurf is sitting with kisses on his head. The Smurf is holding a picture of Smurfette. The Smurf has big bulgy eye's and his tongue is hanging out.
076/012	Smurf With Flower	Lanco			Rubber Squeakies	4 1/2" High	Spain	$7.50-10.00	The toy is a soft rubber and when squeezed it makes a squeaking noise. The Smurf is leaning over, he has his eye's closed. He is smelling a red flower.

Set 267: Stamp & Ink Pads- PVC Figures

267/001	Indian	Schleich	Stamp & Ink Pad	3"	Germany	$22.50-30.00	The Smurf figure is standing on a plastic book. The book opens to reveal an ink pad and a stamper. The Smurf is wearing tan pants, brown and red moccasins, and a headdress with white and black feathers. The headdress has a tan border. The Smurf has 2 red lines painted on each side of his face. The Smurf has black eyebrows. He has his arms folded across his chest.
267/002	Mailman	Schleich	Stamp & Ink Pad	3"	Germany	$22.50-30.00	The Smurf figure is standing on a plastic book. The book opens to reveal an ink pad and a stamper. The Smurf is holding a yellow horn in his mouth with his right hand. He has a dark brown mailbag. He is holding a white envelope with a pink heart on the flap of the envelope in his left hand.

Set 98: Stampers

098/001	Sleepy Smurf Carrying Candle	Marburger	1/1/95		Ink Stampers	2" X 1.5"	Brussels	$2.00-3.00	The stamp is rubber with a green plastic handle. The Smurf is sleep walking. He is carrying a candle. The stamp is in a colorful box.
098/002	Two Smurfs Smiling Shaking Hands	Marburger	1/1/95		Ink Stampers	2" X 1.5"	Brussels	$2.00-3.00	The stamp is rubber with a green plastic handle. Two Smurfs are standing, shaking hands. The stamp is in a colorful box.
098/003	Smurf Carrying Gift	Marburger	1/1/95		Ink Stampers	2" X 1.5"	Brussels	$2.00-3.00	The stamp is rubber with a blue plastic handle. A Smurf is walking, carrying a present. The stamp is in a colorful box.
098/004	Soccer Smurf With Foot On Ball, Holding A Trophy	Marburger	1/1/95		Ink Stampers	2" X 1.5"	Brussels	$2.00-3.00	The stamp is rubber with a blue plastic handle. The Smurf is holding a trophy and has a soccer ball attached to his right foot. The stamp is in a colorful box.
098/005	Shy Smurf Looking Behind Him, Fingers At Mouth	Marburger	1/1/95		Ink Stampers	2" X 1.5"	Brussels	$2.00-3.00	The stamp is rubber with a blue plastic handle. The Smurf is looking behind him and has his fingers at his mouth. The stamp is in a colorful box.
098/006	Smurf Smiling And Waving	Marburger	1/1/95		Ink Stampers	2" X 1.5"	Brussels	$2.00-3.00	The stamp is rubber with a green plastic handle. The Smurf is standing, waving. The stamp is in a colorful box.
098/007	Lovesick Smurf	Marburger	1/1/95		Ink Stampers	2" X 1.5"	Brussels	$2.00-3.00	The stamp is rubber with a green plastic handle. The Smurf is standing, pulling leaves of a daisy. He has hearts floating around his head. The stamp is in a colorful box.
098/008	Bad Smurf	Marburger	1/1/95		Ink Stampers	2" X 1.5"	Brussels	$2.00-3.00	The stamp is rubber with a green plastic handle. The Smurf is standing with his fingers in his ears. He is sticking out his tongue. The stamp is in a colorful box.
098/009	Smurfette	Marburger	1/1/95		Ink Stampers	2" X 1.5"	Brussels	$2.00-3.00	The stamp is rubber with a blue plastic handle. Smurfette is holding an umbrella. The stamp is in a colorful box.
098/010	Smurfette Kissing Dumb Struck Smurf	Marburger	1/1/95		Ink Stampers	2" X 1.5"	Brussels	$2.00-3.00	The stamp is rubber with a blue plastic handle. Smurfette is kissing a Smurf. The stamp is in a colorful box.
098/011	Tired Smurf	Marburger	1/1/95		Ink Stampers	2" X 1.5"	Brussels	$2.00-3.00	The stamp is rubber with a green plastic handle. The Smurf is standing, covering his mouth from a yawn. The stamp is in a colorful box.
098/012	Smurf Windsurfer	Marburger	1/1/95		Ink Stampers	2" X 1.5"	Brussels	$2.00-3.00	The stamp is rubber with a blue plastic handle. The Smurf is riding a surf board. The stamp is in a colorful box.
098/013	Smurf In A Hammock	Marburger	1/1/95		Ink Stampers	2" X 1.5"	Brussels	$2.00-3.00	The stamp is rubber with a blue plastic handle. The Smurf is sleeping in a hammock. The stamp is in a colorful box.
098/014	Smurf Weight-lifter	Marburger	1/1/95		Ink Stampers	2" X 1.5"	Brussels	$2.00-3.00	The stamp is rubber with a green plastic handle. The Smurf is holding a dumbbell. The stamp is in a colorful box.
098/015	Smurf Swimmer	Marburger	1/1/95		Ink Stampers	2" X 1.5"	Brussels	$2.00-3.00	The stamp is rubber with a green plastic handle. The Smurf is wearing a dinosaur swim ring. The stamp is in a colorful box.
098/016	Smurf Bike Rider	Marburger	1/1/95		Ink Stampers	2" X 1.5"	Brussels	$2.00-3.00	The stamp is rubber with a blue plastic handle. The Smurf is riding a bike. He looks like he's flying off the back. The stamp is in a colorful box.
098/017	Smurf Skateboarding	Marburger	1/1/95		Ink Stampers	2" X 1.5"	Brussels	$2.00-3.00	The stamp is rubber with a blue plastic handle. The Smurf is skateboarding. The stamp is in a colorful box.
098/018	Smurf Skier	Marburger	1/1/95		Ink Stampers	2" X 1.5"	Brussels	$2.00-3.00	The stamp is rubber with a blue plastic handle. The Smurf is wearing a ski outfit and ski's. He looks like he's jumping. The stamp is in a colorful box.
098/019	Naughty Smurf Tube Stamper	TNP		Blue	Ink Stampers	1 1/2"L X 1"W	U.S.A.	$2.00-3.00	The stamp is a tube. The end of the cap has ink. The stamper is of a Smurf sticking out his tongue with his thumbs in his ears.
098/020	Smurf Stamp-A-Smurf	W. Berrie		light Blue	Ink Stampers	1 1/2"L X 1"W	Hong Kong	$4.50-6.00	The stamp is a tube. The end of the cap has ink. The stamper is of a Smurf face. The outside of the tube is light blue.
098/021	Papa Stamp-A-Smurf	W. Berrie		Red	Ink Stampers	1 1/2"L X 1"W	Hong Kong	$4.50-6.00	The stamp is a tube. The end of the cap has ink. The stamper is of Papa Smurf's face. The outside of the tube is red.
098/022	Big Mouth Stamp-A-Smurf	W. Berrie		Green	Ink Stampers	1 1/2"L X 1"W	Hong Kong	$4.50-6.00	The stamp is a tube. The end of the cap has ink. The stamper is of Bigmouth's face. The outside of the tube is green.

Set 240: Stickers- Miscellaneous

240/001	Smurfling Sticker	Schleich/Peyo		Pink	Sticker	2 1/4" x 3 3/4"		$1.00-2.00	The sticker has Slouchy Smurf leaning against a tree. Slouchy is wearing a droopy white hat, tee shirt and white pants. The sticker says "Schlumpfe-Linge., Smurflings., P'tit Schtroumpfe." The sticker is pink.
240/002	Le Schtroumpf Magique GB	Peyo	1/1/90		GB Sticker	3 1/2" circle	Brussels	$1.00-2.00	The sticker is white with a picture of a Smurf magician. The Smurf is wearing a red cape and hat with yellow stars. He is pulling a yellow with red-stared towel from a red ball. The ball says "GB".
240/003	Here' What I Think....	Peyo	1/1/83		Sticker			$1.00-2.00	The sticker has a Smurf sticking out his tongue. He has his thumbs in his ears. The background is orange. The sticker says "Here' What I Think..." in yellow letters. 100 Stickers.
240/004	Make Each Day A.... BLAST!!	Peyo			Sticker	1 1/2" x 1 1/2"		$1.00-2.00	The sticker has a picture of a Smurf carrying a yellow package with a fuse connected. The sticker says "Make Each Day A... Blast!".
240/005	I Hate Smoking!	Peyo	1/1/83		Sticker	2" x 1 1/4"		$1.00-2.00	The sticker is of a Smurf blowing smoke out of his ears. The Smurf looks angry. There is a cloud above his head that says "I Hate Smoking!"
240/006	You're The Smurfiest	Peyo	1/1/83		Sticker	2" x 2"		$1.00-2.00	The sticker is of Papa standing on a tree stump. Papa has his hands in the air. The sticker says "You're The Smurfiest".
240/007	S.W.A.K. Smurfed With A Kiss	Peyo			Sticker	1 1/2" x 1 1/2"		$1.00-2.00	The sticker has a Smurf holding a yellow daisy. He has red lip marks on his hat and foot. The sticker says "S.W.A.K. Smurfed With A Kiss".
240/008	Have A Smurfy Day!	Peyo	1/1/83		Sticker			$1.00-2.00	The sticker has a Smurf sitting on a cloud. The sun and a rainbow are behind him. The Smurf has his hands out at his side. The cloud says "Have A Smurfy Day!"
240/009	Everything Is Coming Up Smurfy!	Peyo			Sticker	2" circle		$1.00-2.00	The sticker has Smurfette picking pink daisies. The daisies have Smurf faces in the center. The sticker says "Everything Is Coming Up Smurfy!"
240/010	Be Happy	Peyo	1/1/83		Sticker	2" x 1 1/2"		$1.00-2.00	The sticker is of Smurfette standing, holding a bouquet of balloons. Smurfette is waving with her free hand. The sticker says "Be Happy".
240/011	Have You Hugged Your SMURF Today?	Peyo	1/1/83		Sticker	2" x 2"		$1.00-2.00	The sticker has a Smurf sitting on a log. He is resting his head in his hands. There are bushes behind him. The sticker says "Have You Hugged Your Smurf Today?"
240/012	Good Work!	Peyo			Sticker	2" x 2"		$1.00-2.00	The sticker is of a Smurf sitting at a desk. The desk is yellow. There is a chalkboard behind him. The Smurf is holding a paper up with a star on it.
240/013	Smurfette Sliding Down A Rainbow	Peyo	1/1/83		Sticker	2" x 2"		$1.00-2.00	The sticker is Smurfette is sliding down a rainbow. Smurfette has her eye's closed and a happy expression on her face.
240/014	Smurfette Sliding Down A Rainbow Glitter Sticker	Peyo			Stickers	1 1/2" x 1 1/2"		$1.00-2.00	The sticker is Smurfette sliding down a rainbow. Smurfette has her eye's open and a happy expression on her face. The sticker is all glittery.
240/015	Smurfen Sticker Boekje	IJsselstein	1/1/87		Sticker Book	3 1/4" x 4 1/2"	Brussels	$3.50-5.00	The book has a white cover with a red border. It has various Smurf pictures. The inside has activities and stickers.
240/016	Smurf With Feather In Hat	Peyo -S.E.P.P.			Sticker	3 3/4" x 3"		$1.00-2.00	The sticker is of a Smurf waving. He has a yellow feather stuck in his white hat.
240/017	Smurfette Playing A Harp	Peyo -S.E.P.P.			Sticker	4" x 2 1/2"		$1.00-2.00	Smurfette is sitting, playing a red harp. Smurfette is wearing a white dress with red hearts.
240/018	Mailman	Peyo -S.E.P.P.			Sticker	3 1/2" x 3 1/2"		$1.00-2.00	The Smurf is holding a white envelope with a red heart seal. He has a brown mail bag over his shoulder. The Smurf is pointing a finger with his free hand.
240/019	Badminton	Peyo -S.E.P.P.			Sticker	3 1/2" x 3 1/2"		$1.00-2.00	The Smurf is holding a tennis racket. A birdie hit him in the nose. The Smurf is wearing a white tennis outfit, red and white shoes and black glasses.
240/020	Smurf Holding Daisies	Peyo -S.E.P.P.			Sticker			$1.00-2.00	The Smurf is holding 3 daisies. He is standing with one hand behind his back.
240/021	Smurf Playing Flute	Peyo -S.E.P.P.			Sticker	4" x 3"		$1.00-2.00	The Smurf is standing, playing a brown flute. The Smurf has his eye's closed.
240/022	Smurf Oboe	Peyo -S.E.P.P.			Sticker	4" x 2 3/4"		$1.00-2.00	The Smurf is sitting on a brown wood stool. The Smurf is playing an orange oboe.
240/023	Smurf With Shot Put	Peyo -S.E.P.P.			Sticker	3 3/4" x 2 3/4"		$1.00-2.00	The Smurf is standing, going to throw a shot put (a black ball) . The Smurf is wearing a yellow tank shirt, yellow with a red stripe shorts, red shoes and a white hat.
240/024	Smurf Playing Cricket	Peyo -S.E.P.P.			Sticker	4" x 3"		$1.00-2.00	The Smurf is holding a yellow cricket bat. He is hitting at a red ball. There are yellow stakes by his feet. The Smurf is wearing a white shirt, white pants, white knee pads, white shoes and a white hat.
240/025	Smurf Painter	Peyo - SEPP			Sticker	3 3/4" x 2 1/2"		$1.00-2.00	The Smurf is standing, holding a paint brush and palate. The Smurf is wearing a yellow painters coat and a red/black bow tie.
240/026	Die Schlumpfe Kommen In Die Stadt!	Unicef	1/1/97		Postcard Sticker	5 3/4" x 4"	Brussels	$1.50-2.00	The postcard sticker has a picture of a tent. It has Papa, Brainy, Smurfette and 2 other Smurfs playing instruments. The background is the sky and grass. The bottom of the sticker has dates on it.
240/027	The Smurfs	Price Group	1/1/97		Sticker Book	3 1/4"x 4"	Brussels	$4.50-6.00	The cover has 3 Smurfs on it. One Smurf is standing against the letter S. Another Smurf is laying, sleeping on the RFS. The third Smurf is pushing a ladybug through the grass.

240/028	The Smurfs Green Sticker Sheet	Introduct Holland B.V.	1/1/97	Green	23pc. Sticker Sheet	6 1/2" x 6 1/4"	Brussels	$2.00-3.00	The sticker sheet is green. It has 23 different Smurfs. In the top corner is a Smurf pushing a wheelbarrow with a carrot in it. The other top corner has a Smurf laying, sleeping.
240/029	The Smurf Pink Sticker Sheet	Introduct Holland B.V.	1/1/97	Pink	27pc. Sticker Sheet	6 1/2" x 6"	Brussels	$2.00-3.00	The sticker sheet is pink. It has 27 different Smurf stickers. In the top corner is a Smurf rollerblading. The other corner has a Smurf running, carrying a yellow gift box.
240/030	Various Smurf Stickers	Spec. Products	1/1/92	Red	20 pc. Sticker Sheet	9 1/4" x 5 1/4"	Brussels	$2.00-3.00	The sticker sheet is red. It has 20 different Smurf stickers. There is 1 Papa, 4 Smurfette's, The top row has a Smurf playing a trumpet. A Smurf carrying a fishing pole and tackle box. A Smurf singing and playing a guitar. Brainy Smurf holding a finger in the air.
240/031	Baby Smurf Riding Papa's Back Quick Sticker	Quick	1/1/96	Light Blue	Sticker	2 1/2" x 2"	Brussels	$1.00-2.00	The sticker has Papa Smurf crawling with Baby on his back. Baby is in a pink sleeper. The background is light blue.
240/032	Fishing Smurf Quick Sticker	Quick	1/1/96	Light Green	Sticker	2 1/2" x 2"	Brussels	$1.00-2.00	The Smurf is walking, carrying a fishing pole over one shoulder and a yellow tackle box over the other. The background is light green
240/033	Smurfette Playing In Flowers Quick Sticker	Quick	1/1/96	Pink	Sticker	2 1/2" x 2"	Brussels	$1.00-2.00	Smurfette is walking through a bunch of white daisies. Smurfette has her hands out at her side. The background is pink
240/034	Vanity Smurf Quick Sticker	Quick	1/1/96	Yellow	Sticker	2 1/2" x 2"	Brussels	$1.00-2.00	The Smurf is admiring himself in a hand mirror. The background is yellow.
240/035	Brainy At Chalkboard Quick Sticker	Quick	1/1/96	Light Blue	Sticker	2 1/2" x 2"	Brussels	$1.00-2.00	Brainy Smurf is wearing black glasses. He is kneeling in front of a chalkboard. He is doing a math problem.
240/036	Butterfly Catcher Quick Sticker	Quick	1/1/96	Light Green	Sticker	2 1/2" x 2"	Brussels	$1.00-2.00	The Smurf is carrying a butterfly net. The Smurf is chasing a butterfly. The Smurf is wearing a green hunter outfit. The Smurf is wearing a yellow hat.
240/037	Smurf Carrying A Cake Quick	Quick	1/1/96	Yellow	Sticker	2 1/2" x 2"	Brussels	$1.00-2.00	The Smurf is carrying a cake. The Smurf is running with the cake. The background is yellow.
240/038	Spy Sticker	S.E.P.P.	1/1/78		Sticker	4 1/2"		$1.50-2.00	The Smurf is wearing a red cape and a black mask. He is holding a finger in front of his mouth to be quit.
240/039	Shy Smurf With Daisy	S.E.P.P.	1/1/78		Sticker	4 3/4"		$1.50-2.00	The Smurf is holding a big yellow daisy. The Smurf has his head turned to the side and he has a shy grin on his face.
240/040	Soccer Smurf	S.E.P.P.	1/1/78		Sticker	5"		$1.50-2.00	The Smurf is kicking a white and black soccer ball. The Smurf is wearing a white shirt, red shorts, white shoes and a white hat.
240/041	Papa	S.E.P.P.	1/1/78		Sticker	4 1/4"		$1.50-2.00	Papa is standing, holding his hand out. Papa is wearing a red hat and red pants.
240/042	Jokey	S.E.P.P.	1/1/78		Sticker	4 1/2"		$1.50-2.00	The Smurf is walking, carrying a yellow gift box with red ribbon.
240/043	Clown	S.E.P.P.	1/1/78		Sticker	4 1/4"		$1.50-2.00	The Smurf is wearing a green hat with yellow stars, green pants, white shoes, white gloves, white ruffles and a big red nose. The Smurf is holding a red and white candy stick.
240/044	Smurf Playing Harp	S.E.P.P.	1/1/78		Sticker	4 1/2"		$1.50-2.00	The Smurf is standing, playing a yellow harp. He looks like he's singing.
240/045	King	S.E.P.P.	1/1/78		Sticker	4 1/2"		$1.50-2.00	The Smurf is wearing a yellow hat with a red crown, yellow pants, a red robe with white trim. He is holding a red and white mushroom scepter.
240/046	Large Smurfette				Sticker	6"		$3.50-5.00	Smurfette is standing, flirting. The sticker is large.
240/047	Puffi Parade Sticker Package	Peyo	1/1/85	Light Blue	Panini	3 1/4" x 4"		$2.00-3.00	The package has majorette Smurfette, a Smurf playing the drums, and a Smurf playing a horn on the front.
240/048	Smurfen Sticker Boekje #1	IJsselstein	1/1/87		Sticker Book	3 1/4" x 4 1/2"	Brussels	$3.50-5.00	The book is white with a blue border. It has various Smurfs pictured on the cover. The corner has a Smurf holding flowers.
240/049	35 Small Stickers On Postcard (Smurf Jumping)	Peyo/ I.M.P.S.	1/1/91	White	Postcard Stickers	6" x 4"	Brussels	$1.50-2.00	The postcard has 35 small squares with different Smurfs in each sticker. The top corner has a Smurf jumping in the air. The other top corner has a Smurf laughing.
240/050	9 Postcard Stickers (Papa)	Van Wavery Prod/Peyo	1/1/83	Yellow	Postcard Stickers	6" x 4"	Brussels	$1.50-2.00	The postcard has 9 square stickers. Papa is in the top left corner. Gargamel mixing potion is in the top right corner.
240/051	Postcard Stickers (Naughty Smurf Large Sticker)	Van Wavery Prod/Peyo	1/1/83	Pink	Postcard Sticker	6" x 4"	Brussels	$1.50-2.00	The postcard has 20 square stickers. The postcard has 19 small stickers and 1 large sticker of a Smurf sticking out his tongue.

Set 185: Stickers- National Gas Station

185/001	Smurf With Feather In Hat	Peyo - SEPP			Sticker	2"		$1.50-2.00	The Smurf is wearing a white hat with a yellow feather stuck in the side. He is waving.
185/002	Vanity Smurf	Peyo - SEPP			Sticker	2"		$1.50-2.00	The Smurf is standing, holding a red mirror. The Smurf has an orange daisy in his hat.
185/003	Doctor Smurf	Peyo - SEPP			Sticker	2"		$1.50-2.00	The Smurf is wearing a white doctors uniform. He has a red and gray stethoscope around his neck.
185/004	Cook Smurf	Peyo - SEPP			Sticker	2"		$1.50-2.00	The Smurf is wearing a white chef's hat, white pants, a white apron and a red bow tie. The Smurf is holding a yellow spoon with something red on it.
185/005	Laughing Smurf	Peyo - SEPP			Sticker	2"		$1.50-2.00	The Smurf is standing, holding his stomach. He is laughing. He has his eye's closed.
185/006	Soccer Smurf	Peyo - SEPP			Sticker	2"		$1.50-2.00	The Smurf is kicking a white and black soccer ball. The Smurf is wearing a yellow shirt, black shorts, red socks, black cleat shoes and a white hat.
185/007	Papa Smurf	Peyo - SEPP			Sticker	2"		$1.50-2.00	Papa is standing, holding a finger in the air. He is wearing a red hat and red pants.
185/008	Smurfette Playing A Harp	Peyo - SEPP			Sticker	2"		$1.50-2.00	Smurfette is sitting on a pink stool playing a pink harp. Smurfette is wearing a white dress with red hearts.
185/009	Spy Smurf	Peyo - SEPP			Sticker	2"		$1.50-2.00	The Smurf is wearing a red cape and black mask.
185/010	Sitting Smurf	Peyo - SEPP			Sticker	2"		$1.50-2.00	The Smurf is sitting with his head resting in his hand.
185/011	Badminton Smurf	Peyo - SEPP			Sticker	2"		$1.50-2.00	The Smurf is standing, holding a badminton racket. The birdie is on the end of his nose. The Smurf is wearing a white tennis outfit and black glasses.
185/012	Thinking Smurf	Peyo - SEPP			Sticker	2"		$1.50-2.00	The Smurf is standing with a finger by his mouth. He has his eye's looking up like he's thinking.
185/013	Painter Smurf	Peyo - SEPP			Sticker	2"		$1.50-2.00	The Smurf is holding a pink palate with different colored paints on it. In his other hand he is holding a paint brush with red paint on the tip. The Smurf is wearing a yellow smock, white pants, white hat and a black bow tie with red dots.
185/014	Oboe Smurf	Peyo - SEPP			Sticker	2"		$1.50-2.00	The Smurf is standing, playing an orange oboe. He has his eye's closed.
185/015	Drummer Smurf	Peyo - SEPP			Sticker	2"		$1.50-2.00	The Smurf is sitting behind a red and white kettle drum. He is holding red and white drum sticks in both hands.
185/016	Tuba Player	Peyo - SEPP			Sticker	2"		$1.50-2.00	The Smurf is sitting on a dark brown stool playing a yellow tuba.
185/017	Bicycle Smurf	Peyo - SEPP			Sticker	2"		$1.50-2.00	The Smurf is riding a red bicycle. The bike has yellow wheels. The Smurf has red goggles around his head.
185/018	Smurf With Flowers	Peyo - SEPP			Sticker	2"		$1.50-2.00	The Smurf is holding a three yellow daisies. His other hand is behind his back.
185/019	Nurse Smurfette	Peyo - SEPP			Sticker	2"		$1.50-2.00	Smurfette is wearing a white dress with a red cross on the front, white nylons, white shoes and a white hat. She is holding a hot water bottle in one hand and a needle in the other.
185/020	Track Jumper Smurf	Peyo - SEPP			Sticker	2"		$1.50-2.00	The Smurf looks like a long jumper. He is wearing a yellow shirt with red trim, white hat, white shorts and white shoes.
185/021	Shot-Put Smurf	Peyo - SEPP			Sticker	2"		$1.50-2.00	A Smurf is getting ready to throw a black ball. The Smurf is standing in a ring. He is wearing a yellow tank top, yellow shorts with a red braided belt, red shoes with yellow laces and a white hat.
185/022	Cricket Smurf	Peyo - SEPP			Sticker	2"		$1.50-2.00	The Smurf is wearing a white uniform, white shoes with cleats and yellow and red gloves. The Smurf is holding a yellow mallet. There are yellow stakes next to his feet. There is a red ball in the air.
185/023	Mailman	Peyo - SEPP			Sticker	2"		$1.50-2.00	The Smurf has a brown mail bag over his shoulder. He is holding a white envelope with a red heart on the front in his hand.
185/024	Shiver Smurf	Peyo - SEPP			Sticker	2"		$1.50-2.00	The Smurf is standing, wearing a red scarf and shivering.

Set 118: Stickers- Puffy

118/001	Papa Pointing	W. Berrie	1/1/80	Blue	Puffy Stickers	9 On A Sheet	Taiwan	$2.00-3.00	Assortment # 2. The stickers are on a blue background. Include are: Papa pointing, Smurfette standing in flowers, King Smurf walking with his eyes shut, Jokey carrying a pink present, a Smurf jumping, a Smurf playing a red/yellow/orange drum, flying Smurf, Astro Smurf and a Smurf carrying an orange knapsack.
118/002	Smurf Eating A Waffle	W. Berrie	1/1/80	Blue	Puffy Stickers	9 On A Sheet	Taiwan	$2.00-3.00	Assortment # 3. The stickers are on a blue background. Include are: A Smurf eating a waffle, Jokey carrying a yellow box with a fuse on the end, Smurfette holding flowers, a Smurf playing a trumpet, a serenade Smurf, a Smurf sitting, King Smurf yelling, Greedy carrying a cake and a Smurf on roller-skates.
118/003	Sunbather Smurf	W. Berrie	1/1/80	Blue	Puffy Stickers	9 On A Sheet	Taiwan	$2.00-3.00	Assortment # 4. The stickers are on a blue background. Include are: a Smurf laying on a beach blanket, Smurfette on a surfboard, a Smurf on a bicycle, Smurfette looking at a flower, Smurf with the word love, a Smurf sitting in front of a mushroom house, lovesick Smurf, Smurfette on roller-skates, and a Smurf leaning on a heart holding yellow flowers.
118/004	Gargamel	W. Berrie	1/1/81	Blue	Puffy Stickers	8 On A Sheet	Taiwan	$2.00-3.00	Assortment # 5. The stickers are on a blue background. Include are: Handy, Hefty, Grouchy, Vanity, Gargamel, Clumsy, Azreal and Brainy.
118/005	Bigmouth	W. Berrie	1/1/82	Blue	Puffy Stickers	8 On A Sheet	Taiwan	$2.00-3.00	Assortment # 7. The stickers are on a blue background. Include are: Bigmouth, Hogatha, Gargamel and Azreal, Spy, Grouchy sticking his tongue out, Smurfette in a flirting pose, mushroom house and Papa band leader.
118/006	Smurfette	W. Berrie	1/1/82	Blue	Puffy Stickers	9 On A Sheet	Taiwan	$2.00-3.00	Assortment # 8. The stickers are on a blue background. The whole sheet is all Smurfette. Include are: tennis Smurfette, cheerleader, singing Smurfette, nurse, Smurfette on the telephone, Smurfette twirling a baton, Smurfette picking petals off a daisy and Smurfette roller skating.
118/007	Sports Smurfs	W. Berrie	1/1/80		Puffy Stickers	9 On A Sheet	Taiwan	$2.00-3.00	Assortment # 1. The stickers are on a blue background. Included are: A football Smurf, a Smurf with a yellow tennis racket, a baseball player, a Smurf on a skateboard, a Smurf on a motorcycle, a skier, a weight-lifter, a soccer player and a basketball player.
118/008	Jokey	W. Berrie	1/1/81		Puffy Stickers	8 On A Sheet	Taiwan	$2.00-3.00	Assortment # 6. The stickers are on a blue background. Included are: Jokey, Brainy, Smurfette, Greedy, Lazy, Papa looking startled, Azreal and Gargamel.
118/009	Boxer Schlumpf Knibbel-Bilder	Mundi Paper	1/1/83		Puffy Stickers	12 On A Sheet	Germany	$5.00-7.00	The stickers are on a white background. Included are: Boxer, Smurf with a crutch and bandages, Mailman with a horn, Chimney Sweep Smurf, Smurf with daisies, Papa holding paper with a flower on, Smurf on a leaf skateboard, Smurf in a go-kart, Alchemist Smurf, Gas Attendant, Smurf in a car and a Smurf carrying a cake.
118/010	Smurf With A Scroll And Feather Pen	W. Berrie	1/1/83		Puffy Stickers	12 On A Sheet	Germany	$5.00-7.00	The stickers are on a white background. Included are: Smurf with a scroll and feather pen, Papa holding paper with a flower on, Smurf on a hobby horse, Roller-skating Smurfette, Smurf with flowers, Gargamel leaning over a pot, Smurf playing a guitar, mushroom house, Smurf on a surfboard, Smurf holding a radio, Smurf on a bicycle and a Smurf carrying a wrench.
118/011	Painter Schlumpf Knibbel-Bilder	W. Berrie	1/1/83		Puffy Stickers	12 On A Sheet	Germany	$5.00-7.00	The stickers are on a white background. Included are: Painter, Smurf with pencil, Jester, Smurf with rake, Smurf on a motorcycle, Clown with balloons, Smurfette secretary, Smurf with a present, Smurf laughing, Papa panting and sweating, a Smurf sitting on a stool looking sleepy and Superman Smurf.

118/012	Smurfette Stirring A Pot Of Food				Puffy Stickers	5 On A Sheet	Taiwan	$2.00-3.00	The stickers have Smurfette stirring a pot of food. A Smurf looks like he's choking. A Smurf is looking lovesick and is carrying a flower. A Smurf face. 3 Smurfs singing, two of the Smurfs are in red .
118/013	Smurf Playing A Flute	Bolletje	1/1/95		Puffy Stickers	1 Sticker		$1.00-2.00	The sticker is of a Smurf standing, playing a yellow flute. The sticker is individual.
118/014	Big Papa	Mundi Paper	1/1/83		Puffy Sticker	5"	Brussels	$7.50-10.00	The sticker is of a huge Papa. Papa is standing with his hands at his side. He has a big smile on his face.
118/015	Big Smurf	Mundi Paper	1/1/83		Puffy Stickers	5 1/2"	Brussels	$7.50-10.00	The sticker is of a big Smurf. He is standing in the grass with his hands at his side. The Smurf is smiling.
118/016	Big Smurfette	Mundi Paper	1/1/83		Puffy Stickers	5 1/2"	Brussels	$7.50-10.00	The sticker is a big Smurfette. Smurfette is standing in the grass. She is holding a pencil and a notepad. Smurfette has a flirty look on her face.

Set 119: Stickers- Scratch 'n Sniff

119/001	Cherry	W. Berrie	1/1/83		Scented Stickers	12 Per Sheet	U.S.A.	$2.00-3.00	12 Scented self-stick Smurfy stickers. There are 3 different stickers in the package: a Smurf carrying a cherry, Smurfette holding a cherry and Papa running with a piece of cherry pie.
119/002	Orange	W. Berrie	1/1/83		Scented Stickers	12 Per Sheet	U.S.A.	$2.00-3.00	12 Scented self-stick Smurfy stickers. There are 3 different stickers in the package: a Smurf juggling oranges, Smurfette holding a section of an orange and a Smurf using an orange as a ball.
119/003	Strawberry	W. Berrie	1/1/83		Scented Stickers	12 Per Sheet	U.S.A.	$2.00-3.00	12 Scented self-stick Smurfy stickers. There are 3 different stickers in the package: A Smurf that dropped his ice cream, Smurfette eating an ice cream cone and Papa holding a strawberry.
119/004	Banana	W. Berrie	1/1/83		Scented Stickers	12 Per Sheet	U.S.A.	$2.00-3.00	12 Scented self-stick Smurfy stickers. There are 3 different stickers in the package: a Smurf slipping on a banana peel, Smurfette eating a banana split and Papa carrying 3 banana's over his shoulder.
119/005	Peppermint	W. Berrie	1/1/83		Scented Stickers	12 Per Sheet	U.S.A.	$2.00-3.00	12 Scented self-stick Smurfy stickers. There are 3 different stickers in the package: a Smurf brushing his teeth, Papa running with a candy cane and Smurfette licking a peppermint.
119/006	Grape	W. Berrie	1/1/83		Scented Stickers	12 Per Sheet	U.S.A.	$2.00-3.00	12 Scented self-stick Smurfy stickers. There are 3 different stickers in the package: a Smurf eating grape toast, Smurfette eating a plum and a Smurf pouring grape juice.
119/007	Chocolate	W. Berrie	1/1/83		Scented Stickers	12 Per Sheet	U.S.A.	$2.00-3.00	12 Scented self-stick Smurfy stickers. There are 3 different stickers in the package: a Smurf carrying a piece of chocolate cake, a Smurf eating a popsicle and Smurfette stirring a kettle of chocolate.
119/008	Rootbeer	W. Berrie	1/1/83		Scented Stickers	12 Per Sheet	U.S.A.	$2.00-3.00	12 Scented self-stick Smurfy stickers. There are 3 different stickers in the package: a Smurf standing, holding a mug of rootbeer, a Smurf sitting on a blanket holding a mug of rootbeer and Papa holding a test-tube.
119/009	Peanut Butter	W. Berrie	1/1/83		Scented Stickers	12 Per Sheet	U.S.A.	$2.00-3.00	12 Scented self-stick Smurfy stickers. There are 3 different stickers in the package: a Smurf jumping over peanut, Smurfette holding peanut butter toast and a Smurf chopping a peanut with an ax.
119/010	Rose	W. Berrie	1/1/83		Scented Stickers	12 Per Sheet	U.S.A.	$2.00-3.00	12 Scented self-stick Smurfy stickers. There are 3 different stickers in the package: a Smurf looking in a mirror, Smurfette holding 4 roses and Papa smelling a rose.
119/011	Spearmint	W. Berrie	1/1/83		Scented Stickers	12 Per Sheet	U.S.A.	$2.00-3.00	12 Scented self-stick Smurfy stickers. There are 3 different stickers in the package: a Smurf drinking something green, Smurfette carrying leaves and a Smurf on a leaf skateboard.
119/012	Bubble Gum	W. Berrie	1/1/83		Scented Stickers	12 Per Sheet	U.S.A.	$2.00-3.00	12 Scented self-stick Smurfy stickers. There are 3 different stickers in the package: Papa blowing a bubble with his gum. A Smurf blowing a bubble with his gum, a Smurf with gum all over his face.

Set 189: Stickers- Vitamins

189/001	Smurfette & Brainy	Peyo	1/1/83	Red	Flicker Stickers	1 1/2" Diam.	U.S.A.	$3.00-4.00	There are 2 stickers on the sheet. Smurfette is posing and her eyes flicker. Brainy is holding a hand up and his eyes flicker. The backgrounds are red.
189/002	Gargamel & Hefty	Peyo	1/1/83	Yellow	Flicker Stickers	1 1/2" Diam.	U.S.A.	$3.00-4.00	There are 2 stickers on the sheet. Gargamel is mixing potion and the potion flickers to smoke. Hefty is lifting a dumbbell and the dumbbell flickers. The backgrounds are yellow.
189/003	Grouchy & Jokey	Peyo	1/1/83	Yellow	Flicker Stickers	1 1/2" Diam.	U.S.A.	$3.00-4.00	There are 2 stickers on the sheet. Grouchy is standing with his arms crossed over his chest and it flickers his arms to his side. Jokey is thinking of a present in a cloud and the present flickers to a star. The backgrounds are yellow.
189/004	Handy & Lazy Smurf	Peyo	1/1/83	Pink	Flicker Stickers	1 1/2" Diam.	U.S.A.	$3.00-4.00	There are 2 stickers on the sheet. Handy is hammering a nail. The hammer flickers. Lazy is digging with a shovel and the shovel flickers. The backgrounds are pink.
189/005	Greedy & Papa	Peyo	1/1/83	Light Blue	Flicker Stickers	1 1/2" Diam.	U.S.A.	$3.00-4.00	There are 2 stickers on the sheet. Greedy is holding a Smurf berry and the sticker flickers to him chewing. Papa has his hand by his mouth like he's thinking and his mouth and eye's flicker. The backgrounds are light blue.

Set 115: Store Displays & Display Cases

115/001	Huge Papa	Applause	8/28/94		Fiberglass Statue	3Foot 2In Tall	Germany	$275.00-375.00	Extra-ordinary huge over 3 foot tall Papa Smurf with a white fluffy beard. Papa holds a sign that says "The new Smurfs are here" in German. Made special only for Smurf dealers.
115/002	Papa Smurf Dealer Key Chain Display	Schleich	1/1/95		Heavy Cardboard	18"Tall & 16"Wide	Brussels	$35.00-45.00	Papa smiles holding & pointing to a yellowish/white sign that holds 6 different key rings. Full color glossy.
115/003	A Smurf Christmas PVC Box	Schleich	1/1/83		Display Box	8" X 5" X 3"	Portugal	$15.00-20.00	The sides of the box has all different color mushroom houses covered with snow. The flap in back that stands up has Smurfette in a red Santa suit sitting in a sleigh. In front if the sleigh a Smurf is carrying a gift and a candy cane. Above the Smurf and Smurfette it says A Smurf Christmas in red letters.
115/004	Smurf PVC Display Box	Schleich			Display Box	8 1/2" X 6" X 3 1/2"	Germany	$15.00-20.00	The sides of the boxes show all different Smurfs doing different things. The flap has a big Papa in a yellow star. The flap also has 3 mushroom houses with a bunch of Smurfs around.
115/005	Papa Smurf Clip-On Display Box	W. Berrie	1/1/82		Display Box	8 1/2" X 6 1/2" X 3 1/2"	Korea	$11.00-15.00	The box has Papa on 2 sides and all 3 sides say Papa Smurf Clip-On. The flap has a mushroom house with Papa outside it holding a yellow heart. The heart says Papa Smurf Clip-On inside it.
115/006	Die Schlumpfe Kinderstempel	Marburger	1/1/95		Display Box	11" X 4" X 2 1/2"	Germany	$15.00-20.00	The box is green and yellow and has different Smurfs on it. The box holds 36 stamps. Full color graphics of all the Smurf stamps.
115/007	Smurf Collectors Center	Peyo	1/1/80	Yellow	Store Display			$45.00-60.00	The display is made of plastic. The top has 45 square compartments for PVC's. The top has a piece of clear plexi glass to cover the compartments. The bottom is open to throw more figures in. The top says Smurf collectors center and has a Smurf on each side. One Smurf is holding a circle to put the price on. 1Foot 9"High X 1Foot 7 1/2" W.
115/008	Olympic Stadium	Schleich	11/28/96		Dealer Display	20"L X 11"W x 14"H	Germany	$22.50-30.00	The display is made of sturdy, glossy cardboard. The cardboard is in the shape of a playing field with 2 tier stands. The background shows various Smurfs in the grandstands. You can display PVC's in the fields and stands.
115/009	Smurf Display Case	W. Berrie	1/1/83	Yellow	Display Box	13 1/2" x 8 1/2"	U.S.A.	$25.00-35.00	The case is yellow and has a raised Smurf on the front. There is a yellow handle. The case open to hold 33 Smurfs. There is clear plastic in front of the cubicles.
115/010	Smurf Candy Container Box	Topps	1/1/82		Display Box	7 1/2"L X 9 1/2"W	Hong Kong	$7.50-10.00	The box is blue and green. The front flap has Papa standing on a mushroom pointing. Papa is outside a school house. On the other side of the house is Smurfette posing and a Smurf walking toward her. There is a wood sign that says "Containers Filled With Candy". The box holds 18 containers.
115/011	Super Smurf Spin Collectors Center	W. Berrie/Peyo	1/1/83	Yellow	PVC Display			$75.00-100.00	The display is made of plastic. The display has 6 sides. The display is 22" High. The display is yellow and on each side is compartments to hold Smurfs and a clear piece of plex glass to go over the compartments. The display has a Smurf on 3 sides pointing to the saying "Super Smurf". The display holds 66 Smurfs.
115/012	Welcome To Smurf Land	W. Berrie/Peyo	1/1/79		Store Display	20" High	U.S.A.	$55.00-75.00	The display is of a Smurf standing with his arms out. He is standing in grass with yellow daisies around his feet. The bubble says "Welcome To Smurf Land". The display stands 20" high.
115/013	Super Smurf Collector Center	W. Berrie	1/1/80	Yellow	Store Display		U.S.A.	$60.00-80.00	The display is made of plastic. The top has 3 compartments for super Smurf boxes. The top compartments are open from end to end.. The bottom 4 slots to put more boxes in.. The top says Super Smurf collectors center and has a Smurf on one end. One Smurf is holding a circle to put the price on.
115/014	Kinder Egg Box	Peyo	1/1/90		Display Box	2 1/2" x 5 1/2"	Germany	$1.50-2.00	The front of the box has Smurfette holding a daisy. There is another Smurf holding a jack-in-a-box. The box is white, blue and red. It says "Kinder Surprise".
115/015	Die Schumpfe Display For Candy In The Tubes			Multi	Display	12" x 12 1/2"	Holland	$22.50-30.00	The display is flat cardboard. The display is for the candy that comes in the tubes. The front has a picture of various Smurfs playing with M&M's in the grass. Papa is holding an M&M. 2 Smurfs are playing catch with an M&M. Gargamel and Azreal are walking after the M&M's. There is a Smurf in a red plane at the top and above his head it says "The Smurfs". The bottom half holds 20 candy tubes.
115/016	Limited Collector Series Display	Applause	1/1/95	Blue	Display	14"L X 11"W	U.S.A.	$18.00-25.00	The display is a flat piece of cardboard with a stand on the back to hold it up. The display has 16 Smurfs on it. It says the release dates of the Smurfs. The background is dark blue night sky. The display says 1985 Smurf Limited Collector Series, The Most Lovable Smurfs Ever! Look For These Upcoming Releases:
115/017	Mushroom House Store Display	Applause			Store Display			$40.00-55.00	The backboard of the display has 3 mushroom houses with compartments cut out for Smurfs. The mushroom houses have pink and white roof and off white walls. The back says "Smurf Collector figures by applause". The bottom has green tray toy Smurfs. 12 1/2" x 15 1/2" x 16"H
115/018	McDonald's Display	McDonald's	1/1/98		McDonald's Display		Europe	$125.00-175.00	The display is from McDonald's in the UK It is a promotion for the Smurfs 40th anniversary. The display is a mushroom house with steps off on one side and a oven on the other side. The display comes with 10 Smurf figures. Smurf with big mac, Smurf baker, Smurf waiter, majorette, cheerleader Smurfette, jester, guitar Smurf, Smurf with happy meal box, Smurf carrying cake and Smurf carrying gift. The background of the display is sky. The top says "Smurfs 40th Anniversary".

115/019	1998 Das Jubilaumsdisplay	Schleich	1/1/98	Yellow	Display Box	5" x 8" x 15"	Germany	$60.00-80.00	The box is yellow. The cover has a Smurf band pictured. A Smurf is on a keyboard. 2 Smurfs are playing guitar. A Smurf is playing drums. A Smurf is singing. The cover pops up.
115/020	Hardee's Glass Display	W. Berrie/Peyo	1/1/82	Blue/Yellow	Hardee's Display	18" x 18"	U.S.A	$75.00-100.00	The display holds 8 Smurf glasses. The glasses make a horseshoe. In front is Papa holding a band stick in the air. The back is blue cardboard. It says "SMURFS" in red letters. Below that it says "Kids Love Them, Collect all 8" in white letters.
115/021	Smurf Mobile Store Display	Peyo/I.M.P.S.	1/1/91		Hanging Display		Brussels	$45.00-60.00	There are 5 pieces to the set. They all hang individually. There is a Smurf smelling a flower. Smurfette posing. A Smurf making a funny face. 6 Smurfs running. A Smurf carrying a present in front of 2 mushroom houses with 2 other Smurfs scattering away.
115/022	Kinder Egg Card Display				Card Display	6 1/4" x 9"	Brussels	$9.00-12.00	The card says (Mit Toller Uberraschung). The card has an open egg with Papa Smurf plush inside. There is also a red airplane shown in the picture.
115/023	Puffi Charm Display Card	M.B.M.	1/1/82		Display Card	6" x 8"	Italy	$55.00-75.00	The card has Smurfette in the corner. Smurfette is holding a pink daisy. The card has 12 different Smurf charms on it. The charms are listed separately under the charm category
115/024	Smurf Tattoo Box	Topps	1/1/82		Display Box	8 1/2" x 5" x 2"		$7.50-10.00	The box is red. It has pictures of Smurf tattoos on the cover and sides.
115/025	Bobbing Head Smurf Display Box	Simex			Display Box	3"	West Germany	$15.00-20.00	The fold up flap has Papa Smurf with a flute, a Smurf playing a guitar and a Smurf playing a trumpet. The box shows Smurfs holding different things.
115/026	Jewelry Display	Howard Eldon, Ltd.			Display	14" x 12"	U.S.A.	$40.00-55.00	The display is white plastic. The front has clear plastic hooks. It has a picture of mushroom houses with 4 Smurfs playing instruments and a Smurf carrying a knapsack over his shoulder.
115/027	McDonald's 40th Anniversary Translite	McDonald's Corp.	1/1/98	Red	Translite	21" x 21 1/2"	Brussels	$40.00-55.00	The translite is large, flat, plastic. The front has a Smurf holding a big yellow happy meal box. The top has the 40th anniversary symbol for the Smurfs. The top also says Ronald McDonald HAPPY MEAL.
115/028	#1 Smurfy Teacher Figure Box	Schleich/Peyo/W.B.	1/1/83		Display Box	8 1/2" x 5" x 7"	Hong Kong	$25.00-35.00	The top of the box has a Smurf carrying an apple in front of a chalk board. The bottom sides of the box have a bunch of Smurfs in school with Papa standing in front of the chalk board.
115/029	Ask Here About Your Smurf Collectors Poster/Disp. Card	Applause/Peyo	1/1/84	Dark Blue	Display Card	15" x 11 1/2"	U.S.A.	$15.00-20.00	The display is a flat piece of cardboard with a stand on the back to hold it up. The display is dark blue. Up in the corner is Baby Smurf laying in the crest of the moon. There are all different colored stars on the board. It says "Ask here for your Smurf Collector Poster... available once in a blue moon" in white letters.
115/030	Smurf Open And Close Sign	Applause/Peyo	1/1/84		Display Card	12"	U.S.A.	$15.00-20.00	The sign is round. One side has a Smurf laying in the grass sleeping, holding a "closed Gone Smurfin'" sign. There is a rainbow above him. The other side has a Smurf sitting in the grass holding a "Open Smurf On In" sign. There is a rainbow above him.
115/031	Untitled								

Set 152: Sun Catchers

152/001	Smurf Sitting On A Log				Sun Catcher	10 1/2" X 7"		$11.00-15.00	The sun catcher has a Smurf sitting on a brown log. The Smurf has his head resting in his hands. The background is blue and green. There is a silver chain to hand the sun catcher.

Set 36: Super- Super Play Sets

036/001	Western Chuck Wagon	Schleich	1/1/84		Plastic Play set		Hong Kong/Macau	$150.00-200.00	The wagon is yellow with a gray cover. The wheels are a light brown. The cowboy Smurf sits in front of the wagon holding a yellow rope. The cowboy Smurf is medium blue colored. The wagon is pulled by a reddish/orange and yellow snail. There are barrels, a cactus and a kettle over a wood fire with the play set.
036/002	Moon Landing	Schleich	1/1/84		Plastic Play set		Hong Kong	$110.00-150.00	The rocket is red with a silver top and bottom. There is a Smurf driving a red cart, with a gray battery box on the back. The cart is pulling a gray wagon. There are 5 gray rocks and a telescope.
036/003	Gargamel's Lab	Schleich	1/1/83		Plastic Play set		West Germany	$15.00-20.00	There is a brown table with a stool, a white book stand with an orange book on it. A gray stove with a white back, there is a big black fork and spoon to hang next to the stove. A PVC Gargamel in a sitting position holding a white book. There are potion bottles to sit on the table.
036/004	Smurfette Bedroom	Schleich	1/1/83		Plastic Play set		West Germany	$15.00-20.00	A tan closet, bed side dresser and base boards on the bed, a white mattress, a little white and red lamp, a make-up table with a pink top and tan legs and around the rim of the mirror, and a green chest. A PVC Smurfette in a sitting position holding a powder sponge.
036/005	Smurfette Bedroom	Schleich	1/1/83		Plastic Play set		Hong Kong	$15.00-20.00	A light brown closet, bed side dresser and base boards on the bed, a pink mattress, a little white and pink lamp, a make-up table with a pink top and brown legs and around the rim of the mirror, and a light brown chest. A PVC Smurfette in a sitting position holding a powder sponge.
036/006	Tree-Stump	Schleich			Latex Play set		West Germany	$165.00-225.00	A latex tree stump with a hole through the bottom. The tree-stump is on a cardboard base with trees around. Limited # made.
036/007	Reissued Western Chuck Wagon	Schleich/Peyo	1/1/94		Play set		Germany/China	$15.00-20.00	The wagon is yellow with a gray cover. The wheels are dark brown. The cowboy Smurf sits in front of the wagon holding a white rope. The cowboy Smurf is a bright blue. The wagon is pulled by a red and yellow snail. There are barrels, a cactus and a kettle over a wood fire with the play set.
036/008	Drummer	Schleich/Peyo	1/1/97		Play set		China	$15.00-20.00	The Smurf has a medium blue body. The Smurf is wearing a yellow tank top, purple pants, white shoes, white hat and a brown wrist band. The Smurf has a red and silver 3 piece drum set with 2 gold cymbals. The Smurfs drum set is on a black platform.
036/009	Smurf Latex Castle	Mafi			Latex Castle	9" High	Germany	$150.00-200.00	The castle is made of a soft rubber. The castle is built into the side of a cave. The back of the castle has an entrance like a cave. The castle is red and yellow with 2 gray chimneys. It has 3 brown doors but only one door opens. The is a tree on each side. The castle is on a hard cardboard base.

Set 127: Super Smurfs

127/001	Sledder	Schleich	1/1/78		PVC's W/ Accessories	2"	West Germany	$4.50-6.00	The Smurf has a dark blue body. The Smurf is wearing a red-orange scarf. The Smurf is laying on a yellow sled.
127/002	Chimney Sweep	Schleich	1/1/78		PVC's W/ Accessories	2"	West Germany	$6.00-8.00	The Smurf has a medium blue body. The Smurf is wearing a black cap, black coat, black shoes, a red scarf and gray pants. The Smurf has a black ladder and brush. The Smurf has a red dot on the bottom of his foot.
127/003	Tricycle	Schleich	1/1/78		PVC's W/ Accessories	2"	Germany	$4.50 - 6.50	The Smurf has a medium blue body. The Smurf is riding a red tricycle with yellow wheels. The Smurf has black glasses on.
127/004	Skateboard	Schleich	1/1/78		PVC's W/ Accessories	2"	Hong Kong	$7.50-10.00	The Smurf has a dark blue body. The Smurf has a red shirt and white pants. The Smurf is standing on a green leaf skateboard. The skateboard has black wheels.
127/005	Skateboard	Schleich	1/1/78		PVC's W/ Accessories	2"	Hong Kong	$4.50 - 6.50	The Smurf has a light blue body. The Smurf has an orange, red shirt and white pants. The Smurf is standing on a green leaf skateboard. The skateboard has black wheels.
127/006	Skier	Schleich	1/1/78		PVC's W/ Accessories	2"	Hong Kong	$7.50-10.00	The Smurf has a dark blue body. He is wearing a bright yellow tassel on the end of a white cap, a bright yellow jacket, red orange pants, white socks and black boots. The Smurf has gray ski's and poles.
127/007	Skier	Schleich	1/1/78		PVC's W/ Accessories	2"	West Germany	$5.00-7.00	The Smurf has a medium blue body. He is wearing a dark yellow tassel on the end of a white cap, a dark yellow jacket, red pants, white socks and black boots. The Smurf has gray ski's and poles.
127/008	Gardener	Schleich	1/1/78		PVC's W/ Accessories	2"	Hong Kong	$11.00-15.00	The Smurf has a dull, dark blue body. The Smurf is wearing a green apron and white pants. The Smurf has a tan square wheelbarrow.
127/009	Gardener	Schleich	1/1/78		PVC's W/ Accessories	2"	West Germany	$11.00-15.00	The Smurf has a glossy, medium blue body. The Smurf is wearing a green apron and white pants. The Smurf has a brown square wheelbarrow.
127/010	Angler	Schleich	1/1/79		PVC's W/ Accessories	2"	West Germany	$35.00-45.00	The Smurf has dark blue body. The Smurf is sitting on a reddish brown stump. The fishing rod is dark yellow plastic. The fish is dark red plastic.
127/011	Angler	Schleich	1/1/79		PVC's W/ Accessories	2"	Hong Kong	$37.50-50.00	The Smurf has medium blue body. The Smurf is sitting on a dark brown stump. The fishing rod is yellow plastic. The fish is dull red plastic. The Smurf has a red tongue. There is an orange dot on the bottom of the stump.
127/012	Sign- bearer	Schleich	1/1/79		PVC's W/ Accessories	2"	Hong Kong	$30.00-40.00	The Smurf has a medium blue body. He is standing, holding a red plastic sign. The sign comes with 4 cards. The cards are blue and white. They say Happy Smurfday, I Smurf You, Let's Go Smurfing and one is white with no writing.
127/013	Butterfly Catcher	Schleich	1/1/79		PVC's W/ Accessories	2"	West Germany	$25.00-35.00	The Smurf has a medium blue body. The Smurf is carrying a yellow and brown net. Most of the net is brown and the end of the pole were the net connects is brown. The butterfly is yellow with red markings and has black eyes.
127/014	Butterfly Catcher	Schleich	1/1/79		PVC's W/ Accessories	2"	West Germany	$25.00-35.00	The Smurf has a medium blue body. The Smurf is carrying a yellow and brown net. Most of the net is yellow and the end of the pole were the net connects is yellow. The butterfly is yellow and has black eyes.
127/015	Car Driver	Schleich	1/1/78		PVC's W/ Accessories	2"	Hong Kong	$7.50-10.00	The Smurf has a dark blue body. He is wearing a bright yellow helmet with black dots. He has a red car with a brown open steering wheel and yellow wheels.
127/016	Car Driver	Applause	1/1/90		PVC's W/ Accessories	2"	China	$7.50-10.00	The Smurf has a medium blue body. He is wearing a dark yellow helmet with black dots. He has an orange car with a solid green steering wheel and gray wheels.
127/017	Car Driver	Schleich	1/1/78		PVC's W/ Accessories	2"	West Germany	$22.50-30.00	The Smurf has a medium blue body. He is wearing black goggles. He has a yellow car with a black open steering wheel and orange wheels.
127/018	Gargamel And Azreal	Schleich	1/1/78		PVC's W/ Accessories	2"	Hong Kong	$18.00-25.00	Gargamel's skin is peach color. Gargamel is wearing a black smock and red shoes. He is holding a blue bottle and a dark green bottle. Azreal is orange.
127/019	Gargamel And Azreal	Schleich	1/1/78		PVC's W/ Accessories	2"	Hong Kong	$18.00-25.00	Gargamel's skin is a pink-ish peach color. Gargamel is wearing a black smock and red shoes. He is holding a blue bottle and a dark green bottle. Azreal is orange.
127/020	Smurf In Cage	Schleich	1/1/80		PVC's W/ Accessories	2"	West Germany	$22.50-30.00	The Smurf has a medium blue body. He stands in a dark brown cage, with gray bars, with a dark brown door, a gray latch and a gray key.
127/021	Chain Gang	Schleich	1/1/80		PVC's W/ Accessories	2"	Hong Kong	$37.50-50.00	The Smurf has a dark, dark blue body. He is wearing white pants and yellow shoes. The Smurf is holding a silver pick with a brown handle. The Smurf has a black ball and chain attached to a leg iron. The Smurf comes with a gray pile of rock
127/022	Chain Gang	Schleich	1/1/80		PVC's W/ Accessories	2"	Hong Kong	$11.00-15.00	The Smurf has a dark blue body. He is wearing white pants and a darker yellow pair of shoes. The Smurf is holding a silver pick with a brown handle. The Smurf has a black ball and chain attached to a leg iron. The Smurf comes with a gray pile of rock.
127/023	Hobby Horse	Schleich	1/1/80		PVC's W/ Accessories	2"	Hong Kong	$7.50-10.00	The Smurf has a dark blue body. The Smurf is wearing a red jacket and a black hat. He is riding a tan and brown horse. The horse has a black harness.

127/024	Wind Surfer	Schleich	1/1/80	PVC's W/ Accessories	2"	Hong Kong	$11.00-15.00	The Smurf has a medium blue body. He is wearing red swim trunks. The surfboard is white, the sail is yellow and has red writing.
127/025	Fireman	Schleich	1/1/81	PVC's W/ Accessories	2"	Hong Kong	$11.00-15.00	The Smurf has a medium blue body. He is wearing a red helmet, a yellow jacket and black pants and shoes. The water drum and nozzle is a dull gray. The hose is black.
127/026	Fireman	Schleich	1/1/81	PVC's W/ Accessories	2"	West Germany	$11.00-15.00	The Smurf has a medium bright blue body. He is wearing a light gray helmet, a black jacket with a red collar, black pants and brown shoes. The water drum and nozzle is a silver. The hose is black.
127/027	Photographer	Schleich	1/1/81	PVC's W/ Accessories	2"	West Germany	$11.00-15.00	The Smurf has a bright medium blue body. He is wearing a black hat , black pants, black shoes, a yellow hat band, yellow bow tie and a red jacket. The camera stand is tan and the camera is black.
127/028	Go-Cart	Schleich	1/1/81	PVC's W/ Accessories	2"	Hong Kong	$7.50-10.00	The Smurf has a dark blue body. The Smurf is wearing a green helmet with black dots. The cart is red with black wheels, a black steering wheel, a dark gray engine and dark gray roll bar.
127/029	Go-Cart	Schleich	1/1/81	PVC's W/ Accessories	2"	West Germany	$11.00-15.00	The Smurf has a medium blue body. The Smurf is wearing a green helmet with black dots. The cart is red with black wheels, a black steering wheel, a light gray engine and light gray roll bar. The Smurf has a red dot on the bottom of his foot.
127/030	Row Boat	Schleich	1/1/81	PVC's W/ Accessories	2"	Hong Kong	$7.50-10.00	The Smurf has a dark blue body. The boat is a light brown and has tan seats and tan oars.
127/031	Row Boat	Schleich	1/1/81	PVC's W/ Accessories	2"	West Germany	$7.50-10.00	The Smurf has a medium blue body. The boat is a dark brown and has tan seats and tan oars. The Smurf has a red dot on his foot.
127/032	School Desk	Schleich	1/1/81	PVC's W/ Accessories	2"	Hong Kong	$7.50-10.00	The Smurf has a dark blue body. He is holding a red pencil. The desk is brown and has white paper on top. The chair is brown.
127/033	School Desk	Schleich	1/1/81	PVC's W/ Accessories	2"	Hong Kong	$7.50-10.00	The Smurf has a medium blue body. He is holding a red pencil. The desk is brown and has white paper on top. The chair is brown.
127/034	Rocking Horse	Schleich	1/1/81	PVC's W/ Accessories	2"	West Germany	$11.00-15.00	The Smurf has a medium blue body. The horse is brown with dark brown hair and hoofs. The harness and stirrups are yellow, the saddle is red and the rocker is red. The Smurf has a black dot on the bottom of his foot.
127/035	Airplane	Schleich	1/1/81	PVC's W/ Accessories	2"	Hong Kong	$11.00-15.00	The Smurf has a medium blue body. The Smurf is wearing a red scarf and black goggles. The airplane is gray with a yellow propeller, yellow wheels and red wings.
127/036	Volleyball Smurf	Schleich	1/1/81	PVC's W/ Accessories	2"	Hong Kong	$11.00-15.00	The Smurf has a dark blue body. He is wearing a white shirt, white socks, brown shoes and red with white striped shorts. The grass is green and the net and poles are yellow. The Smurf has a white ball attached to his hands.
127/037	Blackboard	Schleich	1/1/81	PVC's W/ Accessories	2"	Hong Kong	$11.00-15.00	Papa has a dark blue body. He is wearing a red hat, red pants and black glasses. Papa is holding a yellow pointer. The chalkboard is black with writing on. The chalkboard has teacher written on it with one math problem. The chalkboard stand is brown.
127/038	Lawnmower Smurf	Schleich	1/1/81	PVC's W/ Accessories	2"	Hong Kong	$11.00-15.00	The Smurf has a medium blue body. The lawnmower has a brown handle, yellow wheels and a yellow blade.
127/039	Hammock Smurf	Schleich	1/1/84	PVC's W/ Accessories	2"	West Germany	$11.00-15.00	The Smurf has a bright medium blue body. He is wearing red shorts and is holding a yellow glass with a white straw in one hand. The hammock is white. There are 2 brown stumps and a lime green base to hold the hammock up.
127/040	Smurfette With Shopping Cart	Schleich	1/1/81	PVC's W/ Accessories	2"	Hong Kong	$11.00-15.00	Smurfette has a dark blue body and dark yellow hair. Smurfette is wearing a dark pink dress with white dots. The cart is a dark gray and has black wheels.
127/041	Smurfette With Shopping Cart	Schleich	1/1/81	PVC's W/ Accessories	2"	Hong Kong	$11.00-15.00	Smurfette has a dark blue body and dark yellow hair. Smurfette is wearing a pale pink dress with white dots. The cart is a dark gray and has black wheels.
127/042	Papa In Rocking Chair	Schleich	1/1/81	PVC's W/ Accessories	2"	Hong Kong	$7.50-10.00	Papa has a dark blue body. Papa is smoking a tan and gray pipe. The rocking chair is brown and has a yellowish/orange cushion.
127/043	Piano Smurf	Schleich	1/1/83	PVC's W/ Accessories	2"	Hong Kong	$11.00-15.00	The Smurf has a medium blue body. The piano is brown with white and black keys, a brown stool and a piece of white sheet music sitting on top of the piano.
127/044	Scooter	Schleich	1/1/82	PVC's W/ Accessories	2"	Hong Kong	$9.00-12.00	The Smurf has a medium blue body. The scooter is red and has black tires and a black handlebar.
127/045	Motorcross	Schleich	1/1/82	PVC's W/ Accessories	2"	Hong Kong	$15.00-20.00	The Smurf has a medium blue body. The Smurf is wearing black goggles, a green shirt, red scarf, white pants and brown gloves. The motorcycle is red with black tires, a gray handlebar and a gray kick stand.
127/046	Motorcross	Schleich	1/1/82	PVC's W/ Accessories	2"	West Germany	$11.00-15.00	The Smurf has a bright medium blue body. The Smurf is wearing black goggles, a yellow suit, yellow hat and brown gloves. The motorcycle is red with black tires, a black handlebar and a black kick stand.
127/047	Smurf In A Log Car	Schleich	1/1/82	PVC's W/ Accessories	2"	West Germany	$11.00-15.00	The Smurf has a medium bright blue body. The Smurf is wearing black goggles. The car is brown with a brown cushion, dark brown wheels, dark yellow candles, a dark yellow steering wheel and a dark yellow stem to the mushroom. The top of the mushroom is a dark red with 5 white spots. The Smurf has a red dot on his foot.
127/048	Smurf In A Log Car	Schleich	1/1/83	PVC's W/ Accessories	2"	Hong Kong	$11.00-15.00	The Smurf has a dark blue body. The Smurf is wearing black goggles. The car is dark brown with a yellow cushion, brown wheels, white candles on a brown stand, a yellow steering wheel and a yellow stem to the mushroom. The top of the mushroom is a red with 13 white spots. The front of the car has a white and brown crank handle.
127/049	Helicopter	Schleich	1/1/82	PVC's W/ Accessories	2"	West Germany	$11.00-15.00	The Smurf has a medium bright blue body. He is wearing black goggles and a red and white scarf. The helicopter is a brown walnut shell with orange wheels, gray blades, a dark brown seat and a clear plastic window. The Smurf has a red dot on his foot.
127/050	Vanity Table Smurfette	Schleich	1/1/82	PVC's W/ Accessories	2"	Hong Kong	$11.00-15.00	Smurfette has a medium blue body. Smurfette is sitting, wearing a white dress holding a pink face puff. The table has a pink doily top, dark brown legs, a dark brown stool and a dark brown mirror frame with a silver mirror.
127/051	Bath Tub	Schleich	1/1/82	PVC's W/ Accessories	2"	West Germany	$11.00-15.00	The Smurf has a medium bright blue body. He is wearing a red pair of shorts. The Smurf has a brownish green scrub brush with white soap and soap suds. The tub is dark brown and the sprinkling can is yellow.
127/052	Bath Tub	Schleich	1/1/83	PVC's W/ Accessories	2"	Hong Kong	$11.00-15.00	The Smurf has a medium blue body. He is wearing a bright red pair of shorts. The Smurf has a tan scrub brush with white soap and soap suds. The tub is brown and the sprinkling can is orange.
127/053	Bicycle Smurfette	Schleich	1/1/83	PVC's W/ Accessories	2"	Hong Kong	$13.50-18.00	Smurfette has a dark blue body. Smurfette is wearing a white hat, white dress and a pink ribbon in her hair. The bike is red with black wheels, gray pedals and a gray handlebar.
127/054	Trapeze	Schleich	1/1/83	PVC's W/ Accessories	2"	West Germany	$11.00-15.00	The Smurf has a medium bright blue body. The tree is brown with green leaves, green grass, red and white mushrooms on the side, yellow rope on top with a tan swing.
127/055	Smurfette In Kitchen	Schleich	1/1/82	PVC's W/ Accessories	2"	West Germany	$7.50-10.00	Smurfette has a medium blue body. Smurfette is wearing a white dress with a red apron and red hair ribbons. Smurfette is holding a red salt shaker and a black fry pan with eggs. The stove is silver, black and gray.
127/056	Artist With Easel	Schleich	1/1/83	PVC's W/ Accessories	2"	West Germany	$11.00-15.00	The Smurf has a medium bright blue body. He is holding a white brush with red tip, white palette with different colors on it, a green brush tip and a brown brush tip. The canvas has a picture of Smurfette and the easel is dark brown.
127/057	Smurf In Bed	Schleich	1/1/82	PVC's W/ Accessories	2"	West Germany	$11.00-15.00	The Smurf has a medium bright blue body. The Smurf is wearing a red night shirt and a white hat with a red tassel. The bed is dark brown with red and white mushrooms on the bed post. The mattress is yellow.
127/058	Smurfette In Plug Car	Applause	1/1/84	PVC's W/ Accessories	2"	China	$11.00-15.00	Smurfette has a medium bright blue body. Smurfette is wearing a white dress and a pink ribbon in her hair. The car is bright pink with white/yellow daisy decals, gray wheels and a lavender steering wheel.
127/059	Lifeguard	Schleich	1/1/83	PVC's W/ Accessories	2"	Hong Kong	$11.00-15.00	The Smurf has a medium blue body. The Smurf is wearing red shorts and holding a dark yellow megaphone. The chair is white, the life buoy is white and blue and the flag is yellow with a black dot in the middle and a brown flag pole.
127/060	Lifeguard	Schleich	1/1/83	PVC's W/ Accessories	2"	West Germany	$11.00-15.00	The Smurf has a medium bright blue body. The Smurf is wearing red shorts and holding a bright yellow megaphone. The chair is white, the life buoy is white and the flag is red and white with a red dot in the middle and a red flag pole.
127/061	High Diver	Schleich	1/1/83	PVC's W/ Accessories	2"	Hong Kong	$11.00-15.00	The Smurf has a medium blue body. He is wearing red shorts. The pool is dark brown with silver stripes, a dark brown ladder and a yellow diving board.
127/062	National Gas Station	Schleich	1/1/85	PVC's W/ Accessories	2"	West Germany	$35.00-45.00	The Smurf has a light blue body. He is wearing white overalls with a blue, white and yellow national logo on the front. The Smurf is holding a red cloth in his left hand. The gas pump is white, gray, blue and yellow with a black hose and a gray nozzle. The pump is on a gray base.
127/063	Tea Set	Schleich	1/1/84	PVC's W/ Accessories	2"	West Germany	$9.00-12.00	Smurfette has a medium blue body. Smurfette is wearing a white dress with a dark pink apron. Smurfette is holding a purple teapot and cup. The table is pink with brown legs. There is a white dish with brown food and another purple coffee cup.
127/064	River Raft	Schleich	1/1/84	PVC's W/ Accessories	2"	Hong Kong	$11.00-15.00	The Smurf has a dull medium blue body. He is wearing white tattered pants. The raft is brown with yellowish tan ropes holding the logs together. The flag pole is reddish brown with a white flag.
127/065	River Raft	Schleich	1/1/84	PVC's W/ Accessories	2"	West Germany	$11.00-15.00	The Smurf has a shiny medium blue body. He is wearing white tattered pants. The raft is dark brown with tan ropes holding the logs together. The flag pole is dark brown with a white flag.
127/066	Jokey With Trick Box	Schleich	1/1/82	PVC's W/ Accessories	2"	West Germany	$18.00-25.00	The Smurf has a medium bright blue body. He is holding a light yellow box with a reddish orange ribbon and bow. Inside the box is Gargamel's head on a spring. Gargamel's head is a peach color.
127/067	Jokey With Trick Box	Schleich	1/1/83	PVC's W/ Accessories	2"	Hong Kong	$18.00-25.00	The Smurf has a medium blue body. He is holding a dark yellow box with a dark red ribbon and bow. Inside the box is Gargamel's head on a spring. Gargamel's head is pink.
127/068	Stork Delivering Baby	Schleich	1/1/85	PVC's W/ Accessories	2"	West Germany	$9.00-12.00	The stork is white with black wings, an orange beak and orange feet. The stork is carrying a Baby Smurf in a dark yellow blanket. The Baby Smurf has a medium blue body. The nest is tan.
127/069	Computer Smurf	Schleich	1/1/94	PVC's W/ Accessories	2"	Germany	$4.50 - 6.50	The Smurf has a medium bright blue body. He is sitting at a gray computer with a yellow screen. A picture of Papa Smurf reading a book is in the middle of the screen.
127/070	Cyclist	Schleich	1/1/79	PVC's W/ Accessories	2"	West Germany	$11.00-15.00	The Smurf has a light blue body color. He is wearing black shorts, black shoes, black gloves and a yellow shirt with a white tag on his back. The tag has a black number 6 on it. The bike has dark red bars, gray tires with white spokes, a silver handbar and silver pedals. The stand to hold the bike is dark brown.

127/071	Cyclist	Schleich	1/1/79		PVC's W/ Accessories	2"	Hong Kong	$11.00-15.00	The Smurf has a medium blue body color. He is wearing black shorts, black shoes, black gloves and a yellow shirt with a white tag on his back. The tag has a black number 6 on it. The bike has red bars, gray tires with white spokes, a silver handbar and silver petals. The stand to hold the bike is dark brown.
127/072	Kayak	Schleich	1/1/79		PVC's W/ Accessories	2"	West Germany	$7.50-10.00	The Smurf has a medium bright blue body. The Smurf is wearing a red and white helmet and a white shirt with yellow sleeves. The kayak is green and the oar is brown.
127/073	Discus Thrower	Schleich	1/1/79		PVC's W/ Accessories	2"	Hong Kong	$9.00-12.00	The Smurf has a medium blue body. He is wearing a yellow tank top, black shorts and white shoes. The Smurf is holding a red discus and is standing on a round gray platform.
127/074	Fencer	Schleich	1/1/80		PVC's W/ Accessories	2"	West Germany	$11.00-15.00	The Smurf has a bright light blue body. The Smurf is wearing a yellow plastic helmet and holding a gold rubber fencing stick.
127/075	Ice Hockey With Net	Schleich	1/1/79		PVC's W/ Accessories	2"	West Germany	$11.00-15.00	The Smurf has a medium blue body. He is wearing a bright yellow jersey, bright yellow socks, a bright yellow and white helmet with black buttons, black shorts, red gloves and black skates with silver blades. He has a brown hockey stick, a brown puck and the net is white.
127/076	Pole Vaulter	Schleich	1/1/79		PVC's W/ Accessories	2"	Hong Kong	$18.00-25.00	The Smurf has a medium blue body. He is wearing a red tank top, yellow shorts, a white hat and white shoes. The poles and sticks are red and the bases are bright green.
127/077	Weight-lifter	Schleich	1/1/80		PVC's W/ Accessories	2"	Hong Kong	$11.00-15.00	The Smurf has a dark blue body. The Smurf is wearing a dark brown body suit with a yellow belt, black shoes, white wrist sweatbands and white socks. The Smurf is holding a silver weight bar with silver weights on the ends.
127/078	Weight-lifter	Schleich	1/1/94		PVC's W/ Accessories	2"	West Germany	$4.50 - 6.50	The Smurf has a medium bright blue body. The Smurf is wearing a dark pink body suit with a tan belt, black shoes and tan wrist sweatbands. The Smurf is holding a light gray weight bar with light gray weights on the ends.
127/079	Boxer	Schleich	1/1/80		PVC's W/ Accessories	2"	Hong Kong	$9.00-13.00	The Smurf has a dark blue body. He is wearing a red tank top, green shorts, black shoes,white socks and dark brown and white boxing gloves. The base is medium brown and the punching sack is rust color.
127/080	Boxer	Schleich	1/1/80		PVC's W/ Accessories	2"	Hong Kong	$9.00-13.00	The Smurf has a medium blue body. He is wearing a red tank top, darker green shorts, black shoes, white socks and dark brown and white boxing gloves. The base is medium brown and the punching sack is rust color.
127/081	Bars Gymnast	Schleich	1/1/80		PVC's W/ Accessories	2"	Hong Kong	$11.00-15.00	The Smurf has a medium blue body. He is wearing a white tank top, white pants with white feet pads, a white hat and white wrist sweatbands. The bars are a medium brown and the platform is a brownish red.
127/082	Bars Gymnast	Schleich	1/1/80		PVC's W/ Accessories	2"	West Germany	$11.00-15.00	The Smurf has a medium blue body. He is wearing a white tank top, white pants with white feet bands, a white hat and white wrist sweatbands. The bars are a dark brown almost black color and the platform is a brownish red. The Smurf has an orange dot on his foot.
127/083	Rings Gymnast	Schleich	1/1/80		PVC's W/ Accessories	2"	Hong Kong	$15.00-20.00	The Smurf has a medium blue body. He is wearing a white tank top, white pants with white feet bands, a white hat and white wrist sweatbands. The plastic yellow rings are attached to a yellow bar by red ropes.
127/084	Hurdler	Schleich	1/1/80		PVC's W/ Accessories	2"	West Germany	$11.00-15.00	The Smurf has a medium bright blue body. He is wearing a green tank top, yellow shorts, white socks, black shoes and a white hat. The hurdle is brown with white stripes and the platform is red.
127/085	Basketball Player	Schleich	1/1/80		PVC's W/ Accessories	2"	Hong Kong	$9.00-13.00	The Smurf has a medium blue body. He is wearing a dark yellow tank top, dark yellow socks, red shorts, black shoes and a white hat. The backboard is brownish green with a yellow net and a green stand. The ball is brown.
127/086	Blackboard	Schleich/ Peyo	1/1/81		PVC's W/ Accessories	2"	West Germany	$7.50-10.00	Papa has a medium blue body. He is wearing a red hat, red pants and black glasses. Papa is holding a yellow pointer. The chalkboard is black with writing on it. The chalkboard has math problems on it in American numbers. The chalkboard stand is tan.
127/087	Keyboarder	Schleich	1/1/97		PVC's W/ Accessories	2"	China	$4.50 - 6.50	The Smurf has a medium blue body. The Smurf is wearing a long green jacket, white jacket, orange tie, brown pants, white shoes, white hat and orange an brown glasses. The Smurf is playing a gray keyboard.

Set 136: Tattoos

136/001	24 Smurf Tattoos	Topps	1/1/82		Rub On Tattoos	1 Sheet Of 24	U.S.A.	$2.00-3.00	The tattoos are in yellow packages that show different Smurfs on the front. In red writing it says 24 Tattoos, in blue it says SMURF across the front. There are 12 different packages. I have 8.
136/002	Space Traveling Smurf Tattoo	W. Berrie/Peyo			Rub On Tattoos	1 Individual Tattoo			The Smurf is in a spacesuit. He is holding a yellow flower in his hand.
136/003	Roller-skating Smurf Tattoo	W.B./Peyo			Rub On Tattoos	1 Individual Tattoo			The Smurf is wearing brown strap on skates with pink wheels.
136/004	Motorcyclist Smurf Tattoo	W.B./Peyo			Rub On Tattoos	1 Individual Tattoo			The Smurf is riding a brown motorcycle. The Smurf is wearing a green jacket and white pants.
136/005	Ice Skating Smurf	W.B./Peyo			Rub On Tattoos	1 Individual Tattoo			The Smurf is wearing orange ice skates, a yellow shirt, green pants, an orange scarf and a white hat.
136/006	Triple Scoop Smurf	W.B./Peyo			Rub On Tattoos	1 Individual Tattoo			The Smurf is carrying a triple scoop ice cream cone. The ice cream is yellow, pink and blue.
136/007	Lazy Smurf Tattoo	W.B./Peyo			Rub On Tattoos	1 Individual Tattoo			Lazy is laying in the grass sleeping.
136/008	Tennis Smurf Tattoo	W.B./Peyo			Rub On Tattoos	1 Individual Tattoo			The Smurf is holding a brown tennis racket. He is swinging at a yellow tennis ball. The Smurf is wearing a white tennis shirt and shorts.
136/009	Slam-Dunk Smurf Tattoo	W. Berrie/Peyo			Rub On Tattoos	1 Individual Tattoo			The Smurf is slam dunking a yellow ball. The Smurf is wearing a red tank top, white shorts, white shoes and a white hat.
136/010	Rescue Smurf Tattoo	W.B./Peyo			Rub On Tattoos	1 Individual Tattoo			The Smurf is wearing a yellow cape, and a red superman outfit. The Smurf has a black mask covering his eyes.
136/011	Clown Smurf Tattoo	W. Berrie/Peyo			Rub On Tattoos	1 Individual Tattoo			The Smurf is wearing a green clown outfit, a white hat with an orange tassel and a big red nose. The Smurf is holding an exploding test tube.
136/012	Traveler Smurf Tattoo				Rub On Tattoos	1 Individual Tattoo			The Smurf is carrying a stick over his shoulder. The stick has a red knapsack on the end.
136/013	Smurfette Singing Tattoo	Peyo			Rub On Tattoos	1 Individual Tattoo			Smurfette is standing, holding a music sheet. Smurfette is singing.
136/014	Nat With Butterfly	Peyo			Rub On Tattoos	1 Individual Tattoo			Nat is wearing brown pants with a brown sash and a yellow hat. There is a butterfly in the air.
136/015	Smurf With Harp	Peyo			Rub On Tattoos	1 Individual Tattoo			A Smurf is holding a yellow harp.
136/016	Papa Bandleader Tattoo	Peyo			Rub On Tattoos	1 Individual Tattoo			Papa is dressed as a conductor. He is wearing a black jacket, red pants, and a red hat. He is holding a yellow stick.
136/017	Smurf Carrying A Cake Tattoo	Peyo			Rub On Tattoos	1 Individual Tattoo			A Smurf is carrying a cake. He is wearing a white chef's hat, a white bow tie, a white apron and white pants.
136/018	Smurf Getting Hit By Snowball	Peyo			Rub On Tattoos	1 Individual Tattoo			The Smurf is getting hit in the back of the head by a snow ball. The Smurf is wearing a white hat, white pants and a green scarf.
136/019	Skier Smurf Tattoo	Peyo			Rub On Tattoos	1 Individual Tattoo			The Smurf is wearing brown ski's and holding tan poles. The Smurf is wearing a red hat, red gloves, a green scarf, green shirt and white pants.,
136/020	Proud Smurf Tattoo	Peyo			Rub On Tattoos	1 Individual Tattoo			The Smurf is holding his hands in together in the air. He looks happy. There are green leaves on his hat.
136/021	Smurfette Tattoo	Peyo			Rub On Tattoos	1 Individual Tattoo			Smurfette is standing with her hands out at her sides. She has a happy expression on her face.
136/022	Dreamy Smurf Tattoo	Peyo	1/1/82		Rub On Tattoos	1 Individual Tattoo			The Smurf is standing with a dreamy expression on his face. He is holding his hands together in front of him. The Smurf has a white and red daisy on his hat.

Set 226: Tea Sets

226/001	Smurf Cook N' Serve Set	Peyo	1/1/81		Plastic Dishes		Italy	$15.00-20.00	The dishes are white, yellow and red. There is a serving for 4. 4 plates with a Smurf carrying a cup and plate. 4 smaller plates with Smurfette pouring something into a cup. 4 coffee cups with Smurfette pouring something into a cup. 4 regular cups with yellow and blue stripes. A plastic burner. 2 kettles with yellow and blue stripes. A pitcher with both the Smurf and Smurfette on. 4 yellow spoons, forks and knifes.
226/002	Smurf Tea And Serve Set	Peyo	1/1/81		Plastic Dishes		Italy	$15.00-20.00	The dishes are white, yellow and red. There is a serving for 3. 3 plates with a Smurf carrying a cup and plate. 3 smaller plates with Smurfette pouring something into a cup. 3 coffee cups with Smurfette pouring something into a cup. A pitcher with both the Smurf and Smurfette on. 3 yellow spoons, forks and knifes.
226/003	Smurfette Cook N' Serve	Peyo	1/1/83		Plastic Dishes		Italy	$15.00-20.00	The dishes are white, yellow and red. There is a serving for 4. 4 plates with Smurfette sitting at a table eating cake. 4 smaller plates with a Smurf carrying a cake. 2 coffee cups with Smurfette sitting at a table eating cake and 2 cups with the Smurf carrying the cake. 4 regular cups with yellow and blue stripes. A plastic burner with Smurfette's face in the center. 2 kettles with yellow and blue stripes. A pitcher with Smurfette sitting at a table eating cake. 4 yellow spoons, forks and knifes.
226/004	Los Pitufos	Peyo/Romagosa		Orange	Plastic Dishes		Spain	$30.00-40.00	There is a metal platter. The platter has Papa sitting at a table. A Smurf is carrying a cake towards Papa. Papa has his tongue hanging out. There are 2 cups Smurfette and Smurfs face. They have hearts around there head. There are 2 little plates with Smurf faces around. In the center is Smurfette. There are 2 white spoons. There is a water bottle with Smurfette stirring a kettle of food and a Smurf carrying a cake. The backgrounds is orange.

Set 121: Telephones

121/001	6 1/2" Smurf Standing On A Yellow Base Telephone	H-G Industries	1/1/82	Yellow	Play Phone	9 1/2"High	U.S.A.	$15.00-20.00	A 6 1/2"High rubber Smurf is standing, holding a sign that says Smurf me. The Smurf is standing on a yellow base and when you turn the knob to dial the Smurf goes around in circles. The hand piece is red and the dial piece is red.
121/002	6 1/2" Smurf Standing On A Red Base Telephone	H-G Industries	1/1/82	Red	Play Phone	9 1/2" High	U.S.A.	$15.00-20.00	A 6 1/2"High rubber Smurf is standing, holding a sign that says Smurf me. The Smurf is standing on a red base and when you turn the knob to dial the Smurf goes around in circles. The hand piece is yellow and the dial piece is yellow.
121/003	Schtroumpf Phone	Eurostil	1/1/83		Play Phone	10"L X 8"W	France	$37.50-50.00	The phone sits 10" high. The receiver is blue and the base is white with blue. There is a blue door to open and a Smurf swings out. Smurfette is standing on a blue balcony and she pops up when a button is pushed up. There is a blue dial with Smurfette's picture in the center and a picture of a mushroom house with hands for a clock. There are 3 white buttons towards the bottom: 1 opens the change drawer and 2 make Smurfette pop-up.
121/004	Talking Flip Phone	Irwin Toys	1/1/96		Talking Phone	7" x 4 1/2"	China	$18.00-25.00	The phone is in the shape of a mushroom house. The top is red and white. The bottom is blue with a yellow door on the front. You open the door to display the number pad. The phone says 3 different sayings.
121/005	Smurf & Mushroom House Telephone	Enterprise			Real Telephone	13" High		$150.00-200.00	The hand piece is a hard plastic Smurf. The Smurfs sits inside an orange mushroom house. The Smurf house has a night light. The side of the house has a door that opens to reveal a number pad. Comes with a little red phone book.

Set 68: Toys- Miscellaneous

068/001	Red Rolling Wheel Baby Toy	W. Berrie	1/1/82	Red	Baby Toy			$11.00-15.00	There are 2 red wheels and in the middle is a 5" Smurf that rocks inside when the toy is rolled.
068/002	Blue Plastic Airplane	Illco	1/1/83	Light Blue	Plastic Airplane	11" Long	Thailand	$22.50-30.00	The airplane is light blue and has 2 doors on one side that open and a sliding track to slide the Smurfs through. There are 2 Smurfs, Papa and Smurfette that slide inside the plane. The airplane has 4 windows on each side. In front there is a sticker of 2 pilots.
068/003	Blue Plastic School Bus	Illco	1/1/82	Light Blue	Plastic School Bus	8" Long x 3" High	Thailand	$11.00-15.00	The bus is light blue and has 2 doors on one side that open. There are 3 Smurfs that slide inside the bus. There are 4 windows on each side. There is a sticker of Papa and Smurfette in the front window and a sticker of 2 Smurfs in the back.
068/004	Yellow Plastic School Bus	Illco	1/1/82	Yellow	Plastic School Bus	8" Long x 3" High	Thailand	$11.00-15.00	The bus is yellow and has 2 doors on one side that open. There are 3 Smurfs that slide inside the bus. There are 4 windows on each side. There is a sticker of Papa and Smurfette in the front window and a sticker of 2 Smurfs in the back.
068/005	Dark Blue Skier Earmuffs	S.E.P.P.	1/1/82	Dark Blue	Earmuffs		Taiwan	$3.50-5.00	The earmuffs are dark blue and fuzzy. There is a puffy sticker on each side of a Smurf downhill skiing.
068/006	Blow Up Smurf		1/7/83		Inflatable Smurf	16" High	Taiwan	$7.50-10.00	The Smurf is made of plastic and is inflatable. The Smurfs arms are out at his side. He has a white hat and white pants.
068/007	Space Traveler Rocket	Zima	1/1/83	Yellow	Plastic Rocket	6 1/2" Long	Taiwan	$37.50-50.00	The rocket is yellow with a blue tip and exhaust on the back. There is a Smurf sitting in the middle with a black steering wheel in front of him. There are 3 wheels on the bottom to roll across the floor with.
068/008	Walkie Talkies	Power Tronic	1/1/82	Blue	Walkie Talkies	4 3/4"H X 2 3/4"W	Hong Kong	$30.00-40.00	There are 2 blue, rectangular walkie talkies with a Smurf in a spacesuit on the front. Features are: solid state, telescopic antenna, push and hold talk button, on/off volume control, belt clip on, battery operated.
068/009	Smurf Inflatable Chair	Helm			Vinyl Chair		Taiwan	$11.00-15.00	The chain is inflatable. The seat is white with Smurf written on top in yellow. The back of the chair is a Smurf and the arms of the chair is the Smurfs arms. The Smurf is wearing a red, white and blue shirt.
068/010	Spinning Top	Ohio Art	1/1/82	Dull green	Metal Top	10" High	U.S.A.	$11.00-15.00	The top has a picture of 5 different Smurfs around. There is a Smurf giving Smurfette flowers, a Smurf carrying a knapsack, a Smurf sitting on the grass and a Smurf dressed in a baseball outfit. The top has a red wood handle, the bottom is all different color stripes. The colors on the top are dull.
068/011	Spinning Top	Ohio Art	1/1/82	Bright Green	Metal Top	10" High	U.S.A.	$18.00-25.00	The top has a picture of 5 different Smurfs around. There is a Smurf giving Smurfette flowers, a Smurf carrying a knapsack, a Smurf sitting on the grass and a Smurf dressed in a baseball outfit. The top has a red wood handle, the bottom of the toy is all different color stripes. The colors on the top are bright.
068/012	Smurf Head Watering Can	H-G Industries	1/1/82	Blue/White	Watering Can	6" High	U.S.A.	$7.50-10.00	The watering can is in the shape of a Smurfs head. He has a white hat that has 4 holes in the top to pour water out of. The handle is white. The watering can is plastic.
068/013	Smurf In A Red Fire Car	Zima	1/1/83		Rolling Toy		Macau	$40.00-55.00	The fire car is red with brown bumpers and tan wheels. There is a Smurf wearing a red fire hat sitting in the middle. The license plate on the car says Fire Chief.
068/014	Smurf Figure Air Freshener	Creation			Air Freshener	3 1/2"High	France	$15.00-20.00	Smurf figure on air freshener. Smurf holds a red helmet with a black & white "S" and signals "thumbs up". The Smurf is mounted on a light green stand.
068/016	Bully Smurf Top				Spinning Top			$35.00-45.00	There are 3 small blue rubber Smurfs that spin around a mushroom house. The house is yellow with a red top and brown trim. There is a Smurf in the window in the house. One Smurf is pushing a yellow wheelbarrow. One is carrying a green knapsack and the other one is carrying a red shovel. The top has a red base, clear top with the scene inside and a yellow knob on top.
068/017	Smurf On Red Bicycle	Helm		Red	Push Toy	4"H X 4"L	Hong Kong	$7.50-10.00	A plastic Smurf is riding a red bicycle. The tires are black. The Smurfs legs move back and forth when you push the bicycle.
068/018	Smurf Push Puppet (Yellow Base)	Helm			Push Puppet	4"High	Hong Kong	$6.00-8.00	There is a Smurf standing on a yellow base. When you push the bottom of the base the Smurf falls over. His neck, arms and legs flop.
068/019	Yellow Liquid Hourglass			Yellow	Liquid Hourglass	5"High		$7.50-10.00	The hourglass is square and has orange liquid inside. When you tip it upside down the liquid drips on to a wheel and turns the wheel. The top and bottom are plastic. The top has a picture of a Smurf giving another Smurf a present. The bottom has a Smurf kicking a soccer ball.
068/020	Smurf Figure Air Freshener	Creation			Air Freshener	3 1/2"High	France	$15.00-20.00	Smurf figure on air freshener. Smurf holds a red helmet with a black & white "S" and signals "thumbs up". The Smurf is mounted on a black stand.
068/021	Piggyback Stroller	Coleco	1/1/82		Stroller		U.S.A.	$45.00-60.00	The stroller is a plastic blue Smurf on his hands and knees. There are 4 white wheels with stickers in the middle of different Smurfs. There is a white vinyl seat with Smurfs all over. The stroller handle is white and made of metal. A plush fits in the seat.
068/022	Smurf Doll Carriage	Coleco	1/1/82		Doll Carriage		U.S.A.	$52.50-70.00	The carriage has a white frame with a white metal handle. There are 4 white plastic wheels with Smurf head caps in the center. The carriage basket is white vinyl with various Smurfs all over.
068/023	Smurf-Ups	Hasbro	1/1/83		Pop-Ups		U.S.A.	$18.00-25.00	The pop-ups is log shape. There are 4 mushrooms. Under each mushroom is a different Smurf. The yellow dial is broke off.
068/024	Smurf-A-Scope	Helm	1/1/82	Red	Picture Tube	7 1/4" Long	Hong Kong	$9.00-12.00	The scope is a round tube. The outside has a picture of 3 Smurfs walking and Smurfette holding flowers. You look inside and turn the bottom and the picture in side changes.
068/025	Smurf Toy Tooter	Coleco	1/1/82		Toy Box	Big	Canada	$67.50-90.00	The toy tooter looks like a cart. It has a metal handle to push it around. The front and back panel are blue plastic. The sides are white. The 2 side panels have a picture of Smurfette standing in the middle of Papa and another Smurf. They are standing in the grass.
068/026	Smurf Copper Coin	Peyo		Copper	Copper Coin	2 1/4" circle	Paris	$75.00-100.00	The coin has Peyo's face on one side. The other side has Papa, Smurfette and a Smurf holding hands. There is a mushroom house etched in the coin behind the Smurfs. It says on the bottom "Les Schtroumpfs". It comes with a plastic stand.
068/027	Smurf White Car	Peyo		White	Mini Toy Car	1"H x 3"L		$1.50-2.00	The car is white and made of plastic. It has blue wheels. The front has a Smurf face. The 4 fenders all have a red heart on.
068/029	Smurf Slide	Cave Of The Mounds			Slide	2" x 2"			The slide is from the cave of the mounds. It has the Smurf figures set out in a corner of the cave.
068/030	ETPOYMPO-ITAPEA Smurf Figures From Greece	Peyo/ Pyroplastic Toys	1/1/85		Figures With Parts		Greece	$40.00-55.00	There are 6 plastic figures in the box. 2 have red hats and red pants. 4 have white pants and white hats. The figures come with accessories. They are in a blue box with a pink present on the front.

Set 138: Trading Cards

138/001	Smurf Super Cards	Topps	1/1/82		36 Trading Cards		U.S.A.	$25.00-35.00	The trading cards are in a clear package. The top of the package is red and says Smurf in yellow. The cards have different pictures and sayings.

Set 55: Trains

055/001	Choo-Choo Train With Headlights	Durham	1/1/81		Battery Operated	8 Feet Of Track	Hong Kong	$30.00-40.00	The train is battery operated and has 8 feet of track. The engine is blue with a blue fuel car. There is a red passenger car and white passenger car.
055/002	Crib Train	Durham	1/1/82	Light Blue	Musical Caboose	7" x 7"	Hong Kong	$15.00-20.00	The caboose rolls along the rail of a baby crib. The caboose is light blue with yellow wheels. There is a Smurf face on the front and 2 small Smurfs in the windows on each side.
055/003	Mini Train Set	Durham	1/1/82		Battery Operated			$11.00-15.00	The train has 2 blue straight tracks, 1 blue straight track, 2 blue curve track, 1 blue locomotive, 1 red box car, 1 blue coal tender and 1 white caboose. There are little stickers of Smurfs on the cars. The train is a mini set.
055/004	Smurf Land wind-up Train	Durham	1/1/82		wind-up Train	Over 18" Long	Hong Kong	$15.00-20.00	wind-up loco with 3 detachable cars. The locomotive is white and blue, there is a picture of a Smurf engineer. The next car is a red circus car with Papa's picture on the side, and a plastic Papa face on top. The next car is a blue circus car with Smurfette's picture on the side, and a plastic Smurfette face on top. The next car is a white circus car with Smurf picture on the side, and a plastic Smurf face on top. Each car is 3 3/4" Long.
055/005	Smurf Express	Peyo	1/1/82		wind-up Train	3"L X 2"H	Hong Kong	$3.50-5.00	The Smurf cabin is dark blue . The bottom of the train is white. The smoke stack is red. The front boiler is yellow and the wheels are red. There is a plastic Smurf sitting on top in the back of the cabin.
055/006	Mini Yellow Engine	Galoob	1/1/82		wind-up Train	2 1/2"Long	U.S.A.	$4.50-6.00	The mini engine is yellow and has a Smurf face on the front. There is a picture of a Smurf waving in the window on both sides. The smoke stack and wheels are red. There is a 2 piece track that snaps together. There is a cardboard play mat with a picture of 2 mushroom houses.
055/007	Rubber Smurf Train	Peyo			Rubber Train	3 1/2" X 3 1/2"	Germany	$40.00-55.00	There are 4 rubber train cars and 4 rubber Smurfs that fit in the cars. The first car is a blue caboose with yellow wheels and has a Smurf standing with his hand by his mouth. There is a red car with teal wheels and a Smurf is sitting. The third train car is yellow with green wheels and has a Smurf standing, holding a finger in the air. The last car is green with red wheels and has a Smurf standing, blowing a yellow trumpet. The train is a total of 12" long. The train and Smurfs are flat rubber pieces.

Set 38: Trays- Lap

038/001	Three Smurfs Eating Scene	W. Berrie	1/1/82	Yellow	Metal TV Trays	17" Long 13" Width	U.S.A.	$9.00-13.00	The tray is made of metal and stands 6" high. The tray is yellow and has Papa sitting in the middle of a table eating with a Smurf on each side of him.
038/002	Smurf Giving Smurfette Flowers	W. Berrie	1/1/82	Blue	Metal TV Trays	17" Long 13" Width	U.S.A.	$7.50-10.00	The tray is made of metal and stands 6" high. The tray is blue and it has a Smurf giving Smurfette yellow flowers with Papa standing on the side watching. There is a mushroom house in the background.
038/003	Sport Smurf Serving Tray	Willow			Metal Serving Tray	11 1/2" x 16"	Australia	$9.00-13.00	The tray has a red ridge. In the center is a beach. There are 2 different pictures of Smurfette on the beach. A Smurf is swimming and another Smurf is wind surfing. Above the beach is grass. It has a Smurf blowing a whistle. A Smurf jumping a fence. A Smurf sleeping. A Smurf leaping out of a bush. A Smurf running through a finish line.

Set 162: Trinket Boxes

162/001	Let's Be Friends	W. Berrie			Heart Box	3"L X 3 1/2"W	U.S.A.	$6.00-8.00	The box is small, white and heart- shaped with a removable cover. The cover has a picture of a Smurf looking lovesick handing Smurfette 4 red daisies. Smurfette is wearing a white dress with yellow dots and red ribbons in her ponytails. Underneath the Smurf and Smurfette it says "Let's Be Friends" in black letters.

162/002	Love Is Something You Share ...Not Own	W. Berrie			Heart Box	5 1/2"L X 5"W	U.S.A.	$6.00-8.00	The box is large, white and heart- shaped with a removable cover. The cover has a picture of a Smurf looking lovesick handing Smurfette 4 red daisies. Smurfette is wearing a white dress with yellow dots and red ribbons in her ponytails. Underneath the Smurf and Smurfette it says "Love Is Something You Share ...Not Own " in black letters.
162/003	Gargamel And Azreal			White	Candle Trinket Box	5"		$7.50-10.00	The trinket box is ceramic. It is heart- shaped. It has a picture of Gargamel and Azreal. They both have a troubled look on his face. The inside has a white candle in it.
162/004	Papa Looking Troubled				Candle Trinket Box	4"		$7.50-10.00	The trinket box is ceramic. It is heart- shaped. It has a picture of Papa. He is holding a hand up by his beard. He has an inquisitive look on his face. There are 2 mushrooms. The inside has a white candle in it.
162/005	Schoolgirl				Candle Trinket Box	3"		$7.50-10.00	The trinket box is ceramic. It is heart- shaped. It has a picture of Smurfette carrying a brown school bag. Smurfette is wearing a brown jacket, white dress, red bow-tie, white shoes and a white hat. The inside has a white candle in it.

Set 200: Video Movies

200/001	The Smurfs And The Magic Flute	Peyo	1/1/83		VHS Movie		Brussels	$11.00-15.00	The movie is in a collector case. The cover has a film strip with different pictures on one side. Below the strip are 6 Smurfs and Papa in the Smurf village. The Smurfs are doing various things.
200/002	L' Isola Del Tesoro	Cinehollywood	1/1/92		VHS Movie		Italy	$11.00-15.00	The movie is in a plastic case. The cover has Sassette yelling at a pirate. Another Smurf is running up to Sassette. Papa is standing, watching. The pirate is standing in front of a treasure chest. The movie needs to be translated.
200/003	Gargamel's Giant/Smurfiplication	Worldvision	1/1/87	Orange	VHS Movie		U.S.A.	$11.00-15.00	The cover of the box is orange and has a Smurf's upper body. The Smurf is wearing a white vest.
200/004	Baby's First Christmas	Command Performance	1/1/90	Yellow	VHS Movie		U.S.A.	$11.00-15.00	The box cover has Papa Smurf dressed as Santa and Smurfette dressed as an elf. Baby Smurf is sitting in Santa's lap with a long white list.
200/005	The Secret Of Shadow Swamp	Kid Klassic	1/1/89	Pink	VHS Movie		U.S.A.	$11.00-15.00	The cover of the box has a Smurf and Smurfette standing in a swamp.
200/006	Village Tales	Kid Klassic	1/1/88		VHS Movie		U.S.A.	$11.00-15.00	The cover of the box has a Smurf standing with his hand on his hip. The cover is yellow and pink. There are 2 cartoons on the tape: Symbols Of Wisdom and Jokey's Shadow.
200/007	Pussywillow Pixies	Kid Klassic	1/1/90	Green	VHS Movie		U.S.A.	$11.00-15.00	The box cover is green with a Smurf standing with his hands on his hip.
200/008	Never Smurf Off Till Tomorrow	Kid Klassic	1/1/90	Red	VHS Movie		U.S.A.	$11.00-15.00	The cover of the box is red. It has a picture of a Smurf standing with his hands on his hip.
200/009	The Whole Smurf And Nothing But The Smurf	Kids Klassic	1/1/87	Pink	VHS Movie		U.S.A.	$11.00-15.00	The cover of the box is pink. It has a Smurf feeding Smurfette grapes. Another Smurf is fanning the Smurf with a leaf.
200/010	Cartoon All-Star To The Rescue	Buena Vista Home Video	1/1/90		VHS Movie		U.S.A.	$16.50-22.00	The cover of the box has all different cartoon characters on it. There are 4 Smurfs in the picture.
200/011	The Smurfs And The Magic Flute	Vestron Video	1/1/83		VHS Movie		U.S.A.	$15.00-20.00	The movie is in a cardboard box. The cover has a film strip with different pictures on one side. Below the strip are 6 Smurfs and Papa in the Smurf village. The Smurfs are doing various things.

Set 97: View- Master & Reels

097/001	Flying Smurf Gift Set	View Master Int.	1/1/82		View-Master		U.S.A.	$15.00-20.00	The box is yellow and has a Smurf flying in one corner and a Smurf on springs in the other corner. The set comes with a 3-D viewer and 3 reels.
097/002	Smurfette	View Master Int.	1/1/82		View-Master Reels		U.S.A.	$3.00-4.00	The package has Smurfette standing in a door with Papa and 6 love-struck Smurfs waiting outside. There are 3 reels with a total of 21 pictures.
097/003	Traveling Smurf	View Master Int.	1/1/82		View-Master Reels		U.S.A.	$2.00-3.00	The package has a Smurf carrying a knapsack leaving the village. There are 3 reels with a total of 21 pictures.
097/004	Show Beam Smurphony In C	View Master Int.	1/1/82	Red	Show Beam	11" Long		$15.00-20.00	The show beam is like a long tube. It has a cartridge with a show, Smurpony In C. The show beam lights and shows the movie on a wall.
097/005	The Smurf's Apprentice Show Beam Cartridge	View Master Int.	1/1/84		Show Beam Cartridge		U.S.A.	$7.50-10.00	The cartridge fits in a show beam. The card has Papa sitting at a lab table with test tubes around.
097/006	Smurf Theater	View Master Int.	1/1/82		3-D Smurf Theater		U.S.A.	$18.00-25.00	The view master is in a yellow canister. The canister has Gargamel and 5 Smurfs on the front. The view master is blue and sits on a table.

Set 150: Wall Hanging Pictures & Plaques

150/001	Baseball Smurf				Wood Framed	12 1/2" X 10 1/4"		$4.50-6.00	The picture is of a Smurf wearing a white baseball uniform. The Smurf has a dark blue body. The Smurf is holding a white baseball bat. The picture is in the glass. There is red foil behind the glass. The top of the picture says "SMURF" with the red foil. The Smurf is standing on a red base. The base is foil. The frame is a dark wood stain.
150/002	Jumping Smurf				Wood Framed	13" X 11 1/2"			The picture is of a Smurf jumping. The Smurf has a blue body. He is wearing a white hat and white pants. Below the Smurf it says "SMURF" in white and blue letters. The picture is on the glass. The frame is a medium wood stain. The blue parts are foil.
150/003	Smurfette				Wood Framed	11 1/2" X 10 1/2'			The glass is black and has an outline of Smurfette standing. Behind the glass is foil. Smurfette has a bright blue foil body. Yellow foil hair, silver foiled dress and shoes. The frame is a dark wood stain.
150/004	Mini Jumping Smurf				Wood Framed	9 1/2 X 6 3/4"			The picture is of a Smurf jumping. The Smurf has a blue body. He is wearing a white hat and white pants. Below the Smurf it says "SMURF" in white and blue letters. The picture is on the glass. The frame is a medium wood stain.
150/005	Smurfs At A Pond				Wood Framed	10"L X 8"W			The picture is of 3 Smurfs at a pond. One Smurf is sitting on a pink towel with an orange and red umbrella over him. There is a Smurf standing on the bank of the water pointing and holding a red shovel. Another Smurf is walking by the water carrying a tan picnic basket and pink and orange umbrella. The picture is in a gold metal frame.
150/006	Baby Smurf Sitting On A Mushroom	Hanna - Barbera	1/1/85	Black & White	Wood Framed	12"L X 10"W	U.S.A.		The picture is in black and white. Baby Smurf is sitting on a mushroom. Baby is holding a finger in the air. Papa and Smurfette are standing on one side of the mushroom. On the other side is a Smurf with a tape measure around his neck and a Smurf jumping in the air with a present behind him. All the Smurfs have an excited expression on their face. The picture is in a dark wood frame.
150/007	Smurfette				Ceramic Plaque	6" Circle			The plaque has Smurfette in the middle. Smurfette is standing with her hands at her side. She has a white dress. The background is white. This is hand painted.
150/008	Smurfette				Wood Framed	6"L x 5"W			Smurfette is standing with a cloud behind her and flowers. Smurfette has her hands at her side. The background of the picture is red. The picture has a foil behind it. The picture is in a medium brown wood frame.
150/009	Papa Smurf				Wood Framed	12" x 10"			Papa is walking with his hands behind his back. Papa is looking over his shoulder as he's walking. The top of the picture says "Papa Smurf" in white and blue letters. The background is black. The picture is in a brown wood frame.
150/010	Smurfball				Wood Framed	12" x 10 1/4"			The picture is of a Smurf wearing a white football jersey with a red #1 on the front and a white football helmet with a red star. The Smurf has a dark blue body. The Smurf is holding a white and red football. The picture is in the glass. There is red foil behind the glass. The bottom of the picture says "Smurfball" with the red foil. The frame is a dark wood stain. The red stuff on the picture is red foil.
150/011	Smurfette Foil				Wood Framed	13 3/4" x 11 3/4"			The glass is white and has an outline of Smurfette standing. Behind the glass is foil. Smurfette has a bright blue foil body. Yellow foil hair, silver foiled dress and shoes. The frame is a dark wood stain.
150/012	Smurfette Foil				Wood Framed	9 1/4" x 6 1/4"			The glass is white and has an outline of Smurfette standing. Behind the glass is foil. Smurfette has a bright blue foil body. Yellow foil hair, silver foiled dress and shoes. The frame is a dark wood stain.
150/013	Smurfette				Wood Framed	12" x 10"			The picture has Smurfette in the middle. Smurfette is standing with her hands at her side. She is wearing a white dress. The background is black. Below her it says "Smurfette" in white and blue letters.
150/014	Smurf Jumping				Wood Framed	6" x 5"			The picture is of a Smurf jumping. The Smurf has a blue body. He is wearing a white hat and white pants. Below the Smurf it says "SMURF" in white and blue letters. The frame is a medium wood stain.
150/015	Jumping Smurf				Sparkle Framed	5" x 4"			The picture is of a Smurf jumping. The Smurf has a blue body. He is wearing a white hat and white pants. Below the Smurf it says "SMURF" in white and blue letters. The picture is on the glass. The frame is gold and sparkly.
150/016	Jumping Smurf				Wood Framed	3" x 1"			The picture is of a Smurf jumping. The Smurf has a blue body. He is wearing a white hat and white pants. The Smurfs hat and pants are white foil. The background is dark blue. Below the Smurf it says "SMURF" in red foil letters. The picture is on the glass. The frame is a metal frame.
150/017	Want To Smurf Around? (Love-struck Smurf)				Metal Framed	10" X 8"			The picture is all foil. It is of a Smurf sitting, holding a yellow foil daisy. The Smurf has his tongue hanging out of his mouth. There are lip marks on his hat and shoe. The Smurfs body is blue foil. The background is red foil. The picture says "Want To Smurf Around?" in blue foil letters.
150/018	Surfin Smurf				Sparkly Frame	6" x 5"			The picture is all blue and white. It is of a Smurf on a surfboard riding a wave. The bottom says "Surfin' Smurf". The picture is in a green sparkly frame.
150/019	Cowboy Smurf				Sparkly Frame	6" x 5"			A Smurf is sitting, riding a wood horse. The picture is black, blue and white. The bottom says Cowboy Smurf. The picture is in a green sparkly frame.
150/020	Smurf Pointing				Metal Framed	10" x 8"			The picture is of a Smurf standing and pointing. The Smurf has a happy expression on his face. The background is silver like a mirror. The picture is in a metal frame.
150/021	Smurf Laying Under A Rainbow				Wood Framed	9 1/4" x 11 1/4"			The picture is of a Smurf laying sleeping under a rainbow. The Smurf is laying in the grass with flowers and trees around him. The background is silver. It has a white matting and a wood frame.
150/022	Papa Smurf				Wood Framed	11 1/4" x 9 1/4"			The picture is of Papa. He is standing with his hands out at his side like he's flying. Papa is blue but his clothes and beard are clear. Papa doesn't look like a normal Smurf. The top says "Smurf" the bottom says "Papa".
150/023	Papa Oil Painting				Oil Painting	21" x 14"	Mexico	$11.00-15.00	The painting is of Papa Smurf walking, holding a finger in the air. The painting is on black canvas. The picture is in a wood frame.
150/024	Smurfette Oil Painting				Oil Painting	21" x 14"	Mexico	$11.00-15.00	The painting is of Smurfette standing, posing. She has one hand on her hip and another hand behind her head. The picture is on black canvas. The canvas is in a wood frame.
150/025	Jokey Smurf Oil Painting				Oil Painting	21" x 14"	Mexico	$11.00-15.00	The painting is of Jokey Smurf. He is carrying a yellow present. The painting is on black canvas. It is in a wood frame.

150/026	Smurf And Smurfette In Woods Oil Painting				Oil Painting	14" x 21"	Mexico	$11.00-15.00	The painting is of Smurfette standing, facing a Smurf. The Smurf has his head turned slightly away from Smurfette with a shy expression on his face. The background is woods. The painting is on black canvas. The canvas is in a wood frame.
150/027	Smurfette And Smurf In Front Of House Oil Painting				Wood Framed	14" x 21"	Mexico	$11.00-15.00	The painting is of Smurfette standing, facing a Smurf. The Smurf has his head turned slightly away from Smurfette with a shy expression on his face. The bushes and a mushroom house. The painting is on black canvas. The canvas is in a wood frame.
150/028	Basketball Smurf Medallion				Ceramic Medallion	3" Diam		$3.00-4.00	The medallion is round and has a blue piece of ribbon on top for hanging. The medallion has a picture of a Smurf holding a basketball. The Smurf is wearing white shorts and a white shirt with a yellow an orange stripe.
150/029	Smurfette Medallion				Ceramic Medallion	3" Diam		$3.00-4.00	The medallion is round and has a blue piece of ribbon on top for hanging. The medallion has a picture of Smurfette holding a pink and an orange pompom. Smurfette is on a football field. She is wearing a red dress.
150/030	BFW Smurf Plaque			Red	Plastic Plaque	2 1/4" x 3 1/4"		$3.00-4.00	The plaque is flat and red. The front has an outline of a Smurf and half a sun. It says BFW ECKERT.
150/031	Shy Smurf Cloth Hoop Wall Hanging	Homemade			Cloth Picture	5 1/2"		$1.50-2.00	There is a piece of fabric material in a hoop. It has white and red lace around it. In the center of the material is a Smurf holding a pink daisy behind his back. He has a shy expression on his face.
150/032	Lazy Smurf Cloth Hoop Wall Hanging	Homemade			Cloth Picture	5 1/2"		$1.50-2.00	There is a piece of fabric material in a hoop. It has white and red lace around it. In the center of the material is a Smurf laying in the grass sleeping. There is a shovel stuck in the dirt next to him.
150/033	Papa Smurf Cloth Hoop Wall Hanging	Homemade			Wood Framed	5 1/2"		$1.50-2.00	There is a piece of fabric material in a hoop. It has white and red lace around it. In the center of the material is Papa Smurf standing with his hands behind his back. There are 2 mushroom houses behind Papa.

Set 151: Wall Mirrors

151/001	Want To Smurf Around?				Framed Wall Mirrors	12" X 12"		$7.50-10.00	The mirror has a picture of a Smurf sitting, holding a white daisy. The Smurf has lip marks on his hat and foot. The Smurf has a love-struck expression on his face. The background is red. Above the Smurf it says "Want to Smurf around?" in black letters. The mirror is in a wood frame.
151/002	T G I S Thank Goodness It's Smurf Day!				Framed Wall Mirrors	12" X 12"		$13.50-18.00	The picture on the mirror is of a Smurf holding a beer. The Smurf has white pants and a white hat. The beer mug is black. Above the Smurf it says " T G I S Thank goodness it's Smurf day!" in blue letters. The mirror is in a medium wood stained frame.
151/003	Love-struck Smurf				Wall Mirrors	6" X 6"		$7.50-10.00	The mirror has a white and blue Smurf painted on it. The Smurf is sitting, holding a flower. He has lip marks on his hat and foot. The mirror is not framed. The mirror says "Smurf" in blue letters.
151/004	Smurfette Sliding Down A Rainbow				Framed Wall Mirrors	12" X 10"		$7.50-10.00	Smurfette is sliding down a rainbow. Smurfette is wearing a white dress and shoes. The mirror is in a wood frame.
151/005	A Smurf A Day, Keeps The Drearies Away				Framed Wall Mirrors	6" X 6"		$7.50-10.00	The mirror has a Smurf walking, holding a finger in the air. The mirror says "A Smurf A Day Keeps The Drearies Away" in blue letters.
151/006	Torchbearer				Framed Wall Mirrors	13 1/2" x 13 1/2"		$7.50-10.00	The Smurf is wearing white shorts and a white hat. He is running, carrying a torch. The torch is a black outline on the mirror.
151/007	Love-struck Smurf				Framed Wall Mirrors	12" X 10"		$7.50-10.00	The mirror has a Smurf is sitting, holding a yellow flower. He has red lip marks on his hat and foot. The mirror is in a wood frame.
151/008	Smurf Posing				Metal Frame Mirror	6" x 4"		$7.50-10.00	The mirror has a Smurf posing. He has his head turned to one side and has a big grin on his face. His hands are behind his back. The Smurf has a white award on his chest that says "Smurf". The mirror is in a metal frame.
151/009	Jumping Smurf				Mirrors	6 1/2" x 5"		$7.50-10.00	The picture is of a Smurf jumping. The picture is a blue outline. The bottom says "Smurf" in blue outlined letters.
151/010	Smurfette				Framed Wall Mirrors	6" x 5"		$3.50-5.00	The mirror has an outline of Smurfette standing in the grass. The outline is blue. The bottom of the mirror says "Smurfette" in blue outlined letters.

Set 85: Wallets & Coin Purses

085/001	Roller-skating Smurfette Heart Shape Coin Purse	W. Berrie	1/1/83	Pink	Cloth Coin Purse	4"	China	$3.00-4.00	The coin purse is pink with little purple hearts all over. The coin purse is in the shape of a heart. It has a purple zipper and purple border. On the front is a purple heart with Smurfette on roller-skates in the middle. It says Smurfette in white letters with purple outline right above the picture.
085/002	Roller-skating Smurfette Wallet	W. Berrie	1/1/83	Pink	Cloth Wallet		China	$3.00-4.00	The wallet is pink with little purple hearts all over. The wallet is a tri-fold. On the front is a purple heart with Smurfette on roller-skates in the middle. It says Smurfette in white letters with purple outline right above the picture. There is a spot for coins, credit cards and for dollar bills inside the wallet. The wallet has a velcro close.
085/003	Love-struck Smurf Red Wallet	W. Berrie		Red	Tri-Fold Wallet			$3.00-4.00	The wallet is red with black trim. The front of the wallet has a Smurf sitting, holding a yellow daisy. The Smurf has a dumb expression on his face. There are red lip marks on his hat and foot. The wallet opens and has a coin spot, a picture holder, an extra pocket and a spot for dollar bills.
085/004	Make The Dream A Colorful Truth Coin Purse	W. Berrie		White	Cloth Coin Purse			$2.00-3.00	The coin purse is white with a blue zipper and a blue border. There is a picture of a Smurf sleeping on a cloud with a rainbow behind on both sides of the coin purse. In black letters it says "Make The Dream A Colorful Truth".
085/005	Have A Smurfy Day! (Blue)	BBC Imports	1/1/81	Light Blue	Tri-Fold Wallet		Taiwan	$7.50-10.00	The wallet is blue with blue trim. The front of the wallet has a Smurf sitting on a cloud. There is a rainbow behind the Smurf. The wallet opens and has a coin spot, a picture holder, an extra pocket and a spot for dollar bills. Have A Smurfy Day! is in black letters inside the cloud. The wallet flap Says SMURF in different colors.
085/006	Have A Smurfy Day! (Red/Yellow)	BBC Imports	1/1/81	Red/Yellow	Tri-Fold Wallet		Taiwan	$7.50-10.00	The wallet is yellow with a red front and dark blue trim. The front of the wallet has a Smurf sitting on a cloud. There is a rainbow behind the Smurf. The wallet opens and has a coin spot, a picture holder, an extra pocket and a spot for dollar bills. Have A Smurfy Day! is in black letters inside the cloud. The wallet flap is yellow Says SMURF in different colors.
085/007	Smurfette Sliding Down A Rainbow	W. Berrie	1/1/81	Red/Yellow	Tri-Fold Wallet		Taiwan	$7.50-10.00	The wallet is red with a yellow front and dark blue trim. The front of the wallet has Smurfette sliding down a rainbow. The wallet opens and has a coin spot, a picture holder, an extra pocket and a spot for dollar bills. The wallet flap is red says SMURFETTE in different colors.
085/008	Smurf Watering Plants	W. Berrie			Coin Purse	3 1/4"H X 4 3/4"W		$3.00-4.00	The coin purse is plastic with a slide zipper to close it. There is a picture of a Smurf watering 2 pink flowers. There are green hills in the back with 2 brown horses. The border is pink and green.
085/009	Smurfette Clear Plastic Purse	Applause	1/1/84	Clear	Plastic Purse	5" X 7"	China	$3.00-4.00	The purse is clear with little pink hearts all over. Smurfette is in the middle wearing a pink dress standing in a purple heart. Above the big purple heart it says Smurfette in white letters with purple border.
085/010	Papa Pointing With Other Hand At His Side	W. Berrie		Red	Tri-Fold Wallet			$6.00-8.00	The wallet is red with a picture of Papa standing, pointing on the front. The wallet opens and has a coin spot, a picture holder, an extra pocket and a spot for dollar bills.
085/011	Papa Smurf	W. Berrie	1/1/83	Pink	Tri-Fold Wallet		China	$6.00-8.00	The wallet is light pink with a dark pink border. Papa is on the outside of the wallet. Papa is holding a finger in the air. There are 5 striped lines behind Papa. Above Papa's head it says "Papa Smurf" in blue letters. The wallet opens and has a coin spot, a picture holder, an extra pocket and a spot for dollar bills.
085/012	Smurfette Posing Coin Purse	W. Berrie/Applause	1/1/84	Clear	Coin Purse	3 1/2"	China	$2.00-3.00	The coin purse is clear with little pink hearts all over. The coin purse is in the shape of a heart. It has a pink zipper and purple border. On the front is a purple heart with Smurfette standing in the middle with her hand out. Smurfette has a pink dress on. It says Smurfette in white letters with purple outline right above the picture.
085/013	Smurf Waving Coin Purse	W. Berrie			Coin Purse	4" X 4"		$3.00-4.00	The coin purse is in the shape of a Smurf. It looks like it's made from leather. The Smurf is walking and is waving with his left hand. The zipper is in his hat.
085/014	Biker Kings Island Wallet	W. Berrie	1/1/84	Red	Tri-Fold Wallet		Taiwan	$7.50-10.00	The wallet is red with blue trim. The front has a Smurf on a bike. There is a yellow, purple, blue and white stripe behind him. The Smurf has a red bike. He is wearing a yellow shirt and white shorts. The wallet says "Kings Island" in white letters.
085/015	Ice Skater Smurfette	W. Berrie		Red	Tri-Fold Wallet		Taiwan	$3.00-4.00	The wallet is red with black trim. The wallet has Smurfette on the back. Smurfette is wearing ice skates and a white dress.
085/016	Traveler Smurf Coin Purse	W. Berrie		Clear	Coin Purse	3"L x 4 1/2"W		$2.00-3.00	The purse has a clear center with a yellow top and bottom. The Smurf is walking with a brown backpack on his back. He is carrying pink flowers in his hands.
085/017	Schoolboy Smurf Coin Purse	W. Berrie		Clear	Coin Purse	3"L x 4 1/2"W		$2.00-3.00	The purse has a clear center with a yellow top and bottom. The Smurf is carrying an orange book bag. He is waving with his left hand.
085/018	Smurfette Wallet	W. Berrie	1/1/83	Pink	Vinyl Wallet		Hong Kong	$3.00-4.00	The wallet is pink and folds in half. A snap holds the wallet closed. On the front is Smurfette standing, pulling the petals off a daisy. Smurfette is wearing a white dress with pink dots. The back of the wallet says "Smurfette" in multi colors. There is a ring of flowers around Smurfette's name.
085/019	3 Smurf's Holding Hands Heart- shaped Purse	BBC Imports	1/1/82	Red	Purse	5" x 6 1/4"		$3.50-5.00	The purse is red with white trim and a white handle. It is heart- shaped. The front has 3 Smurfs holding hands. The front Smurf is smiling. The middle Smurf is laughing and the last Smurf is whistling a tune.
085/020	Smurfette Carrying Umbrella Kings Island Coin Purse	W. Berrie	1/1/83	Pink	Coin Purse	3 1/2" x 4 "	U.S.A.	$3.50-5.00	The coin purse is in the shape of a little purse. It has a pink handle and a flip up cover. The front has Smurfette sitting at a vanity. She is looking at her self in the mirror. The other side has Smurfette strolling, carrying an umbrella. Underneath it says Kings Island.
085/021	Smurf And Smurfette Yellow Fanny Pack	W. Berrie	1/1/94	Yellow	Fanny Pack		Brussels	$3.50-5.00	The purse goes around your waist. It is yellow with a blue zipper and blue trim. It has a yellow strap for around the waist. The front has a Smurf and Smurfette holding hands, waving. Smurfette is wearing a red skirt with yellow flowers, a white blouse, a white hat and a red bow. The Smurf is wearing a red shirt with a yellow "S" on the front, blue pants , a yellow belt and red an white shoes.
085/022	Smurf Carrying A Coin - Coin Purse	Peyo	1/1/94	Yellow	Coin Purse	3 1/2"	Brussels	$3.50-5.00	The coin purse is round. It is yellow with red trim. It has a picture of a Smurf walking, carrying a yellow coin.

Set 93: Watches

093/001	Smurf Time Smurfette Watch	Bradley Time	1/1/83		Wind Watch	1"Face	Hong Kong	$22.50-30.00	The band is blue plastic. The watch is silver metal with a picture of Smurfette on the front. Smurfette's arms move to tell the hour and minute. The background is yellow and says Smurf time in orange letters. The numbers are blue. The watch is a wind up wrist watch.

093/002	Smurf Time Smurf Watch	Bradley Time	1/1/83		Wind Watch	1"Face	Hong Kong	$22.50-30.00	The band is blue plastic. The watch is silver metal with a picture of a Smurf on the front. The Smurf's arms move to tell the hour and minute. The background is yellow and says Smurf time in orange letters. The numbers are blue. The watch is a wind up wrist watch.
093/003	Have A Smurfin' Day!	Bradley Time	1/1/83	Blue	Digital Watch	1"Face	Hong Kong	$22.50-30.00	The band is blue plastic. The back of the watch is silver metal. The front has a little window in the center where it says the time. On top of the window a Smurf is laying, smiling and below Papa is pointing. The background is light blue and says "have a Smurfin' day!" in black and red letters on the bottom of the watch.
093/004	Love A Smurf Today!	Bradley Time	1/1/83		Digital Watch	1"Face	Hong Kong	$15.00-20.00	The band is blue plastic. The back of the watch is silver metal. The front has a little window in the center where it says the time. On top of the window a Smurf is laying, smiling and below Smurfette is standing in a flirty pose. The background is light blue and says "Love a Smurf today!" in black and red letters on the bottom of the watch
093/005	Good Day Sunshine!		1/1/83	Yellow	Digital Watch	1"Face	Spain	$7.50-10.00	The band is yellow plastic. The back of the watch is silver metal. The front has a little window in the center where it says the time. On top of the window it says "Good day sunshine!" in red letters. The background is white. The picture on the watch is of a white and a yellow Smurf, 1 is in a red sailboat and the other is on a red surfboard. The Smurfs aren't blue.
093/006	Mountain Scene		1/1/83	Dark Blue	Digital Watch	1"Face	Spain	$7.50-10.00	The band is dark blue plastic. The back of the watch is silver metal. The front has a little window in the center where it says the time. The background is white and blue. The picture on the watch is of a mountain in the background with a Smurf sitting in front of a mushroom house and Papa standing, pointing.
093/007	Pitufos		1/1/83	Red	Digital Watch	1"Face	Spain	$7.50-10.00	The band is red plastic. The back of the watch is silver metal. The front has a little window in the center where it says the time. The background is white. The picture on the watch is of a Smurf carrying a yellow and white birthday cake. The Smurf is red and not blue.
093/008	Soccer Smurf	Royal	1/1/83	Red	Digital Watch	1"Face	Hong Kong	$15.00-20.00	The band is red plastic. The back of the watch is silver metal. The front has a little window in the center where it says the time. The background is white. The picture on the watch is of a Smurf wearing red shorts, yellow shoes and a yellow shirt kicking a soccer ball at a net.
093/009	Large Plastic Kid's Wrist Watch	Durham	1/1/81		Wind Watch	1 1/2"Face	Taiwan	$9.00-12.00	The band is a thin white plastic. The back of the watch is dark blue. The watch is 1/2" high and the face is 1 1/2". There is a picture of a Smurf walking in the center and the watch hands are black. The background is white and the numbers are black.
093/010	Large Plastic Kid's Wrist Watch	Durham	1/1/81		Wind Watch	1 1/2"Face	Hong Kong	$7.50-10.00	The band is a thin white plastic. The back of the watch is light blue. The watch is 1/2" high and the face is 1 1/2". There is a picture of a Smurf walking in the center and the watch hands are red. The background is white and the numbers are black.
093/011	Flying Smurf Watch	Frontier			Wind Watch	1"Face		$15.00-20.00	The band is blue fake leather. The watch is silver metal with a picture of a Smurf flying. The Smurf has red and yellow wings attached to his arms. The background behind the Smurf is white and the numbers are black.
093/012	Smurfling Glitter Watch	Peyo	1/1/83		Digital Watch	4"H X 5 1/2" L	Brussels	$11.00-15.00	The watch has a glitter band. The plastic has Nat, Slouchy, Sassette and Snappy walking on both sides. The picture is the same on both sides. The blue liquid with the glitter dried up. The watch has a velcro band.
093/013	Smurf Mushroom Watch	Watch It	1/1/96		Watch		China	$11.00-15.00	The watch has a green band. The band says "Smurfs" in yellow letters. The buckle is blue. The watch is a pink mushroom house with a purple roof. The watch opens up to display a digital clock. There is a mini removable Smurf that sits inside.
093/014	Good Day Sunshine (Smurfs Surfing)	Formost Trading Co.		Black	Digital Watch	1" Face	Hong Kong	$7.50-10.00	The band is black fake leather. The back of the watch is silver metal. The front has a little window in the center to display the time. The background is white. The picture on the watch is of a Smurf on a yellow surfboard and another Smurf in a red sail boat. The top of the watch above the window says "Good Day Sunshine" in red letters.
093/015	King Smurf	Quartz		Black	Digital Watch	1" Face	Hong Kong	$7.50-10.00	The watch has a fake black leather band. The front of the watch is blue with a window in the center for the time display. Above the window is the Smurfs head. Below the window is the rest of his body. He is dressed as a king and is carrying a red scepter.
093/016	Pink Cowboy And Cowgirl Watch			Pink	Digital Watch	1" Face	Hong Kong	$7.50-10.00	The watch has a fake pink leather band. The center is metal. The front of the watch has a window for the time to be displayed. The picture is of a pink/blue/ yellow mushroom house. The display window is in the center of the house. At the bottom of the watch is a Smurf and Smurfette dressed in cowboy outfits. The watch only shows the 2 Smurfs upper body.
093/017	Black Cowboy And Cowgirl Watch			Black	Digital Watch	1" Face	Hong Kong	$7.50-10.00	The watch has a black plastic band. The face of the watch is black plastic. The front of the watch has a window for the time to be displayed. The picture is of a pink/blue/ yellow mushroom house. The display window is in the center of the house. At the bottom of the watch is a Smurf and Smurfette dressed in cowboy outfits. The watch only shows the 2 Smurfs upper body.
093/018	Light Blue Cowboy And Cowgirl Watch			Light Blue	Digital Watch	1" Face	Hong Kong	$7.50-10.00	The watch has a light blue plastic band. The face of the watch is light blue plastic. The front of the watch has a window for the time to be displayed. The picture is of a pink/blue/ yellow mushroom house. The display window is in the center of the house. At the bottom of the watch is a Smurf and Smurfette dressed in cowboy outfits. The watch only shows the 2 Smurfs upper body.
093/019	Have A Smurfin Day			Light Blue	Digital Watch	1" Face	Hong Kong	$7.50-10.00	The band is light blue plastic. The back of the watch is silver metal. The front has a little window in the center where it says the time. On top of the window a Smurf is laying, smiling and below Papa is pointing. The background is light blue and says "Have a Smurfin' day!" in black and red letters on the bottom of the watch.
093/020	Smurf Pulling Leaves From Flower Dark Blue	TAD	1/1/95	Dark Blue	Watch		Brussels	$22.50-30.00	The watch has a dark blue band. The face is white. It has a lovesick Smurf pulling petals off from a flower. Hearts are floating above his head.
093/021	Papa Cornucopia Red Band	TAD	1/1/95	Red	Watch		Brussels	$22.50-30.00	The watch has a red band. The face is white. It has Papa pushing a cornucopia. The cornucopia has flowers coming out.
093/022	Smurf In Hammock White Band	TAD	1/1/95	White	Watch		Brussels	$22.50-30.00	The band and face are white. A Smurf is sleeping in a hammock. A plate of food is laying beside him.
093/023	Smurfs Hauling A Large Drum	TAD	1/1/95	Blue	Watch		Brussels	$22.50-30.00	The watch band is blue with light blue and white spots. The band has a Smurf face on each end closest to the face. The face is white. It has a Smurf hauling a huge drum on his back. Another Smurf is walking behind playing the drum.
093/024	Snappy Smurf	TAD	1/1/95	Red	Watch		Brussels	$22.50-30.00	The watch band is red with writing, stars and dots all over it. The face is white. It has a Snappy Smurf dressed in a light blue shirt, purple pants, light blue shoes and a droopy white hat. YO is written next to the Smurf.
093/025	Smurfette Face Blue Watch	TAD	1/1/95	Blue	Watch		Brussels	$22.50-30.00	The watch band is blue with light blue and white spots. The band has Smurfette's face on each end closest to the watch face. The face is white. It has Smurfette's face in the center between the numbers.
093/026	Smurfette Kissing A Smurf	TAD	1/1/95	Clear/Pink	Watch		Brussels	$22.50-30.00	The watch band is clear with pink and light blue hearts. The face is pink with a white center. The center has Smurfette giving a Smurf a kiss. There are hearts floating above there heads.
093/027	Pink Smurfette Watch	TAD	1/1/95	Pink	Watch		Brussels	$22.50-30.00	The watch is pink. The band has pink and blue hearts on it. In the center of the watch is an orange heart with Smurfette posing. She has a hand in her hair. The numbers are pink on a white background.
093/028	Smurfette And Smurf Green Band Watch	Corvair	1/1/95	Green/Red	Watch		Brussels	$22.50-30.00	The watch has a green band with Smurf across it. The Smurfs are all holding hands. The face is yellow inside and red outside. In the center Smurfette and a Smurf are holding hands.
093/029	Smurf Carrying A Cake	TAD	1/1/95		Watch		Brussels	$22.50-30.00	The watch band is dark blue. It has 3 Smurf faces on both sides of the face. The middle is pink with a white center. In the center is a Smurf carrying a cake.
093/030	Smurfette Face Open Up Watch	TAD	1/1/95	Pink	Watch		Brussels	$11.00-15.00	The strap is a pink nylon. In the center is plastic. The plastic is Smurfette's face. The watch opens to reveal the time.
093/031	Heart- shaped Smurfette Watch	TAD	1/1/95	Pink	Memo Kits		Brussels	$11.00-15.00	The strap is a pink nylon. In the center is plastic. The plastic is in the shape of a heart. It has Smurfette's upper body. Smurfette has her hand in her hair. The watch opens to reveal the time.
093/032	Astro Smurf Open Up Watch	TAD	1/1/95		Watch		Brussels	$11.00-15.00	The strap is a white nylon. In the center is plastic. The plastic is of sky and a Smurf floating in an astronaut uniform. The watch opens to reveal the time.
093/033	Cupid Open Up Watch	TAD	1/1/95		Watch		Brussels	$11.00-15.00	The strap is a light blue nylon. In the center is plastic. The plastic is a Smurf dressed as cupid. The Smurf is holding a bow and a heart. The watch opens to reveal the time.

Set 205: Weebles

205/001	Smurfette	Peyo	1/1/84		Weeble	2 1/4" High		$15.00-20.00	Smurfette is sitting on a white round base. Smurfette has her hands out at her side. Smurfette has a white dress, white hat and white shoes. Smurfette has slanted eye's.
205/002	Papa	Peyo	1/1/84		Weeble	2 1/4" High		$15.00-20.00	Papa is sitting on a white round base. Papa has his hands out at his side. Papa has red pants and a red hat. Papa's eye's are close together.
205/003	Brainy Smurf	Peyo			Weeble	2 1/4" High		$15.00-20.00	The Smurf is sitting on a white round base. The Smurf has his hands out at his side. The Smurf is wearing black glasses, a white hat and white pants.

Set 122: Whistles

122/001	Papa Smurf Whistle	Helm		24" Neck cord Whistle	2"	Hong Kong	$7.50-10.00	The whistle is a 2" Papa figure. Papa is wearing red pants and a red hat. There is a red neck cord attached.
122/002	Gargamel Whistle	Helm		24" Neck cord Whistle	2"	Hong Kong	$7.50-10.00	The whistle is a 2" Gargamel figure. Gargamel is wearing a black outfit and red shoes. There is a red neck cord attached.
122/003	Smurf Whistle	Helm		24" Neck cord Whistle	2"	Hong Kong	$7.50-10.00	The whistle is a 2" Smurf figure. The Smurf is wearing a white hat and white pants. There is a red neck cord attached.
122/004	Smurfette	Helm		24" Neck cord Whistle	2"	Hong Kong	$7.50-10.00	The whistle is a 2" Smurfette figure. Smurfette is wearing a white dress. There is a red neck cord attached.
122/005	Lime Green Smurf Whistle	Peyo	Lime Green	24" Neck cord Whistle	2"	Taiwan	$2.00-3.00	The whistle is lime green. It has a decal of a Smurfs face. The Smurf is whistling with his lips. The whistle is on a lime green cord.
122/006	Yellow Smurf Whistle	Peyo	Yellow	24" Neck cord Whistle	2"	Taiwan	$2.00-3.00	The whistle is yellow. It has a decal of a Smurfs face. The Smurf is whistling with his lips. The whistle is on a yellow cord.

Set 81: Wind-up Toys

No.	Name	Manufacturer	Variant	Type	Date	Size	Origin	Price	Description
08/001	Fun House	Galoob		Wind-up Toys	1/1/82	3"High	Hong Kong	$15.00-20.00	The mushroom house is lavender with a dark blue roof. The toy is a Smurf connected to the top of the house that flies around it when you wind it up. The Smurf is plastic and has orange wings on his back.
08/002	The Smurf Chase	Galoob		Wind-up Toys	1/1/83	8 1/2" Long	Hong Kong	$18.00-25.00	Gargamel is dressed in black laying on his tummy holding a brown box in front of him. When you wind him up a Smurf rolls out of the box and then back in. The toy is made of plastic.
08/003	Smurf With A Wheelbarrow	Galoob		Wind-up Toys	1/1/82	5"L X 3"H	Hong Kong	$11.00-15.00	Wind the Smurf up and he pushes a wheelbarrow. The wheelbarrow is brown and there is a Smurf holding on to it. The toy is plastic.
08/004	Smurf On A Train Cart	Galoob		Wind-up Toys	1/1/82	4 1/2"L X 4 1/2"H	Hong Kong	$11.00-15.00	A Smurf is standing on a brown train cart. Wind up the cart and it rolls, the Smurfs arms go up and down with the hand.
08/005	Smurf Runabout Red Car	Galoob	Red	wind-up Toys	1/1/82	4"L X 3"H	Hong Kong	$11.00-15.00	The car is red with gray lights, gray bumpers and black tires. There is a white and black #1 sticker on the side of the car. The Smurf is sitting in the middle holding a black steering wheel. Wind up and it rolls across the floor.
08/006	Airplane	Galoob		Wind-up Toys	1/1/82	3"H X 3 1/2"L	Hong Kong	$3.50-5.00	The airplane is red with yellow landing gear and red wheels. The propeller is blue. There is a rubber Smurf sitting in the middle of the plane. Don't work.
08/007	3 Smurfs Jumping Rope	Galoob		Wind-up Toys	1/1/82	3"H X 5"L	Hong Kong	$15.00-20.00	2 Smurfs are standing by daisies, holding a yellow rope. When you wind the rope turns and the Smurf in the middle jumps. The Smurfs are standing on a light green base.
08/008	Smurf On A Swing	Galoob		Wind-up Toys	1/1/82	5 1/2" High	Hong Kong	$15.00-20.00	A Smurf is sitting on a yellow swing. The swing is in a green tree. When you wind it. The tree is on a light green base.
08/009	Smurfs On A Teeter Totter	Ganz Bros.		Wind-up Toys	1/1/82	4 1/2"L	Hong Kong	$15.00-20.00	2 Smurfs are sitting on a yellow teeter totter. The teeter totter is over a brown log. Wind it up and the teeter totter goes up and down.
08/010	Smurf Swimmer	Puffi		Wind-up Toys	1/1/83	5"	Hong Kong	$15.00-20.00	The Smurfs wearing red swim trunks and a white hat. The Smurfs arms and legs move to swim.
08/011	Smurf With Pull Cart	Galoob		Wind-up Toys	1/1/82	5"	Hong Kong	$11.00-15.00	Wind the Smurf up and he pulls a cart. The cart is brown and there is a Smurf holding on to it. The toy is plastic.

Set 107: Wind- Up Walkers

No.	Name	Manufacturer	Type	Date	Size	Origin	Price	Description
107/001	Papa	Galoob	Wind-up Walker	1/1/82	3" High	Hong Kong	$4.50-6.00	Papa is standing, holding a yellow package under his left arm. Papa winds-up and walks across the floor.
107/002	Smurfette	Galoob	Wind-up Walker	1/1/82	3" High	Hong Kong	$4.50-6.00	Smurfette is standing, holding yellow and red flowers on her left side. The Smurf winds-up and walks across the floor.
107/003	Smurf	W. Berrie	Wind-up Walker	1/1/80	3" High	Hong Kong	$7.50-10.00	The Smurf is standing with both hands at his side. The Smurf winds-up and walks across the floor.
107/004	Smurf With A Present	Galoob	wind-uP Walker	1/1/82	3" High	Hong Kong	$4.50-6.00	The Smurf is standing, holding a red gift with a yellow bow in both hands in front of him. The Smurf is wearing a white hat and white pants. The Smurf winds-up and walks across the floor.
107/005	Gargamel	Galoob	Wind-up Walker	1/1/82	3 1/2" High	Hong Kong	$4.50-6.00	Gargamel is standing with both hands clasped together in front of him. Gargamel is wearing a purple gown and red shoes. Gargamel winds-up and walks across the floor.
107/006	Flying Smurf	Galoob	Wind-up Walker	1/1/82	3" High	Hong Kong	$4.50-6.00	The Smurf is standing with both arms out at his sides holding red and yellow wings. The Smurf winds-up and walks. The Smurf is wearing a white hat, white pants and has yellow belts around his chest holding the wings on. The Smurf winds-up and walks across the floor.
107/007	Drummer Smurf	Galoob	Wind-up Walker	1/1/82	3" High	Hong Kong	$4.50-6.00	The Smurf is standing, playing a red, white and blue drum. The drum is attached at his waist, with his 2 blue drum sticks in his hands. The Smurf is wearing a white hat, white pants and red bow tie. The Smurf winds-up and walks across the floor.
107/008	Guitar Smurf	Galoob	Wind-up Walker	1/1/82	3" High	Hong Kong	$7.50-10.00	The Smurf is standing, playing a yellow guitar. The Smurf is wearing a white hat and white pants. The Smurf winds-up and walks his upper body rocks. The Smurf winds-up and walks across the floor.
107/009	Smurf Playing A Trumpet	Galoob	Wind-up Walker	1/1/82	3" High	Hong Kong	$7.50-10.00	The Smurf is standing, playing a yellow trumpet. The Smurf is wearing a white hat, white pants and red bow tie. The Smurf is playing a yellow trumpet.

Set 255: Window Transfers

No.	Name	Manufacturer	Type	Date	Size	Origin	Price	Description
255/001	Smurfs Going To School	Patented Design	Window Transfers	1/1/86	5 1/4" x 6 1/2"	Brussels	$3.50-5.00	The picture is of 2 Smurfs walking down a path to a schoolhouse. One Smurf is wearing black glasses. He is carrying a brown backpack. The Smurf is waving. The Smurf behind is running. He is carrying a red book bag.
255/002	Smurf Carrying A Present To Baby	Patented Design	Window Transfers	1/1/86	5 1/4" x 6 1/2"	Brussels	$3.50-5.00	The transfer is of a Smurf carrying a big yellow present to Baby. Baby is sitting in the grass under a tree.
255/003	Smurf Carrying a Cake to Papa	Patented Design	Window Transfers	1/1/86	5 1/4" x 6 1/2"	Brussels	$3.50-5.00	The transfer is of a Smurf carrying a cake with lots of candles on to Papa. The Smurf is winking. Papa is standing in the grass with his hands behind him looking at the cake.
255/004	Smurf Playing A Trumpet For A Clown	Patented Design	Window Transfers	1/1/86	5 1/4" x 6 1/2"	Brussels	$3.50-5.00	A Smurf is standing on a dirt path playing a yellow trumpet. There is a blue bird sitting on the end of the trumpet. Another Smurf is dressed as a clown. The clown is dancing in the path. The clown is wearing an orange outfit.
255/005	Smurf And Smurfette Sitting At A Table	Patebrd Design	Window Transfers	1/1/86	5 1/4" x 6 1/2"	Brussels	$2.00-3.00	A Smurf and Smurfette are sitting at a table outside. The table has a yellow table cloth. There is a birthday cake in the center of the table. The Smurf is holding a wine glass.
255/006	5 Smurfs	Robbedoes	Window Transfers		6" x 4"	Brussels	$2.00-3.00	The card has 5 1 1/2" Smurfs on it. There is a green Smurf with a long brown beard. There is a Smurf with a long tail and he has hair all over. There is a green and purple striped Smurf. There is a brown thing that is wearing a Smurf hat. There is a Smurf holding a spoon looking at a large egg.

Set 91: Wood Carved Smurfs

No.	Name	Type	Date	Size	Origin	Price	Description
091/001	Papa Sitting In A Chair	Wood Carved	1/1/92	7" High	Hand made	$25.00-35.00	Papa is sitting in a chair holding a book. The book has writing in it. On Papa's right side is a barrel with two books sitting on top of it. The base is wood and painted brown. Papa is painted blue and is wearing a red hat and red pants. The whole statue is wood and is painted.
091/002	Papa Book End	Wood Book End		6 1/2" X 5"	Hand made	$2.00-3.00	Papa is carved out of wood. He is holding one finger in the air. Papa is light blue. He has a red hat and red pants.

Set 75: Zipper Pulls

No.	Name	Manufacturer	Variant	Type	Size	Origin	Price	Description
075/001	Brainy	Applause		Rubber 1/2 Figures	2" High	China	$2.00-3.00	The figure is a half figure with a yellow clip on the back. Brainy is wearing black glasses. He has his right hand in the air pointing down.
075/002	Nat With A Caterpillar	Applause		Rubber 1/2 Figures	2" High	China	$2.00-3.00	The figure is a half figure with a yellow clip on the back. Nat is wearing brown pants, brown sash, a yellow straw hat and no shoes. He is waving with his right hand and there is a green caterpillar by his right foot.
075/003	Snappy With A Ball	Applause		Rubber 1/2 Figures	2" High	China	$2.00-3.00	The figure is a half figure with a yellow clip on the back. Snappy is holding a red ball in his right hand. He has a sly look on his face and is wearing a yellow shirt with a lightning bolt on the front.
075/004	Slouchy On A Skateboard	Applause		Rubber 1/2 Figures	2" High	China	$2.00-3.00	The figure is a half figure with a yellow clip on the back. Slouchy is wearing a droopy white hat, white pants and a red shirt. He is riding a dark brown skateboard with black wheels.
075/005	Sassette With A Teddy Bear	Applause		Rubber 1/2 Figures	2" High	China	$2.00-3.00	The figure is a half figure with a white clip on the back. Sassette is wearing pink pants with pink suspenders. She has dark orange hair. Sassette is carrying a brown teddy bear.
075/006	Smurfette With Flowers	Applause		Rubber 1/2 Figures	2" High	China	$2.00-3.00	The figure is a half figure with a yellow clip on the back. Smurfette is standing, holding a red and an orange daisy behind her back. Smurfette is wearing a white dress and white shoes. They have a yellow clip. She has red ribbons in her hair.
075/007	Smurf Waving	Peyo	Silver	Silver Zipper Pull	2" High	Germany	$7.50-10.00	The zipper pull is made silver. The Smurf waving. His face is part silver- colored. The Smurf isn't fully painted blue. The Smurf has white pants and a white hat on. The hooks are silver.
075/008	Diecast Painter	Applause		Rubber 1/2 Figures	2" High	U.S.A.	$4.50-6.00	The zipper pull is made of diecast metal. The Smurf is 3-D. The Smurf is holding a paint brush in one hand and a palate in the other.

Index